AF740635

Ohio State University Mathematical Research Institute Publications 2

Editors: Gregory R. Baker, Walter D. Neumann, Karl Rubin

The Arithmetic of Function Fields

Proceedings of the Workshop at The Ohio State University
June 17–26, 1991

Editors

David Goss
David R. Hayes
Michael I. Rosen

Walter de Gruyter · Berlin · New York 1992

Editors:

David Goss
Department of Mathematics
The Ohio State University
Columbus, Ohio 43210-1174, USA

Michael I. Rosen
Department of Mathematics
Brown University
Providence, RI 02912, USA

David R. Hayes
Department of Mathematics and Statistics
University of Massachusetts
Amherst, MA 01003, USA

Series Editors:

Gregory R. Baker, Walter D. Neumann, Karl Rubin
Department of Mathematics, The Ohio State University, Columbus, Ohio 43210-1174, USA

1991 Mathematics Subject Classification: Primary: 11G09. Secondary: 11R58.

Printed on acid-free paper which falls within the guidelines of the ANSI to ensure permanence and durability.

Library of Congress Cataloging in Publication Data

The arithmetic of function fields : proceedings of the workshop at The Ohio State University, June 17–26, 1991 / editors, David Goss, David R. Hayes, Michael I. Rosen.
p. cm. – (Ohio State University Mathematical Research Institute publications, ISSN 0942–0363 ; 2)
Includes bibliographical references.
ISBN 3-11-013171-4 (alk. paper)
1. Drinfeld modules – Congresses. 2. Fields, Algebraic – Congresses. I. Goss, David, 1952– . II. Hayes, David (David R.) III. Rosen, Michael I. (Michael Ira), 1938– IV. Series.
QA247.3.A75 1992
512′.74–dc20 92-29651
CIP

Die Deutsche Bibliothek – Cataloging in Publication Data

The **arithmetic of function fields** : proceedings of the workshop at the Ohio State University, June 17–26, 1991 / ed. David Goss ... – Berlin ; New York : de Gruyter, 1992
(Ohio State University Mathematical Research Institute publications ; 2)
ISBN 3-11-013171-4
NE: Goss, David [Hrsg.]; Ohio State University <Columbus, Ohio>; International Mathematical Research Institute <Columbus, Ohio>: Ohio State University ...

© Copyright 1992 by Walter de Gruyter & Co., D-1000 Berlin 30.

All rights reserved, including those of translation into foreign languages. No part of this book may be reproduced in any form – by photoprint, microfilm, or any other means – nor transmitted nor translated into a machine language without written permission from the publisher.

Printing: Gerike GmbH, Berlin. Binding: Dieter Mikolai, Berlin. Cover design: Thomas Bonnie, Hamburg. Printed in Germany.

Preface

This volume consists of contributions from participants in a workshop on the Arithmetic of Function Fields which took place under the auspices of the International Mathematical Research Institute at Ohio State University from June 17-26, 1991. It was attended by over 90 mathematicians from more than 15 countries.

The workshop on the Arithmetic of Function Fields was generously funded by grants from the National Science Foundation, the National Security Agency, the International Research Institute at Ohio State, the Office of the Dean of Arts and Sciences at The Ohio State University, and the Office of the Vice-President for Research at The Ohio State University. We wish to express our sincere appreciation to these organizations and individuals.

A primary topic of the workshop was the arithmetic of Drinfeld modules which is still a new area of research. As such, many of the contributions are of an expository nature and serve as an introduction to non-experts. Therefore, this volume will be useful to researchers in the area — both new and old — and to those who are simply curious! Indeed, the last article in the volume is a "dictionary" to help the reader understand the remarkable similarities between Drinfeld modules and classical objects such as elliptic curves. This dictionary is only meant to serve as a guide; as the subject is evolving so rapidly it will, in fact, soon be out of date.

The Research Institute. The International Mathematical Research Institute at Ohio State University was founded in 1989 to support a program of visiting research scholars in mathematics at Ohio State and to run Workshops and Special Emphasis Programs on topics of particular importance and timeliness. A Research Semester on Low Dimensional Topology was the first major program of the Institute. Since then the Institute has supported workshops on, among others, Nearly Integrable Wave Phenomena in Nonlinear Optics, Quantized Geometry, the Arithmetic of Function Fields, L-Functions Associated to Automorphic Forms, and Geometric Group Theory. The Institute is currently supporting about 20–30 other research visitors (mostly short term) per year. The Institute publishes a preprint series as well as this book series, which is devoted to research monographs, lecture notes, proceedings, and other mathematical works arising from activities of the Research Institute.

Acknowledgements. First and foremost, the editors thank The Ohio State University for its support of this program through the Research Institute. We thank our participants for their enthusiasm and contributions to the workshop and this volume. We also thank the non-academic staff of the Mathematics Department for their help in the organization and running of the Research Semester, particularly Marilyn Howard (administration and visas), Marilyn Radcliff (expenses), and Terry England for the numerous papers she has typed into TeX. The TeX macros were written by Larry Siebenmann and edited by Walter Neumann, and we wish to heartily thank Walter Neumann for his wizardry with TeX.

Without his help this volume would not be possible. We also wish to thank Jeremy Teitelbaum who contributed some graphics work on a paper not his own. Finally we wish to thank the referees of the proceedings papers for their invaluable comments and service.

David Goss, for the editors, June 1992.

Contents

A Brief Introduction to Drinfeld Modules

David R. Hayes

Throughout, we work over a global base-field $\mathbf{k}$ of characteristic $p > 0$ and field of constants $\mathbb{F}_r$, where $r = p^m$. We distinguish a place ∞ of $\mathbf{k}$ called the *place at infinity*, and we let '$\mathbf{A}$' denote the ring of elements of $\mathbf{k}$ which have only ∞ as a pole. The local ring of the completion $\mathbf{K}$ of $\mathbf{k}$ at ∞ is isomorphic to the ring of formal power series $\mathbb{F}_{r^d}[[\pi]]$, where $d = d_\infty = \deg \infty$ and π is a uniformizer at ∞. Every non-zero element $x \in \mathbf{K}$ can be expanded in a Laurent series

$$x = \sum_{\nu \geq n} c_\nu \pi^\nu \tag{0.1}$$

with each $c_\nu \in \mathbb{F}_{r^d}$ and $n = \mathrm{v}_\infty(x)$. We define $\deg x = -d_\infty \cdot \mathrm{v}_\infty(x)$, and we put $N(x) = r^{\deg x}$. We further set $\mathrm{sgn}(x) = c_n$, the leading coefficient in the expansion of x. It is convenient also to define $\mathrm{sgn}(0) = 0$.

We regard sgn as a multiplicative map from $\mathbf{K}$ onto the copy of $\mathbb{F}_{r^d}$ which is contained in $\mathbf{K}$ itself. The map sgn depends upon the choice of uniformizer π, but the reader can easily prove that there are only finitely many such functions. We say that the element $x \in \mathbf{K}$ is *positive* if $\mathrm{sgn}(x) = 1$.

The ring $\mathbf{A}$ is a Dedekind domain with finite class number $h(\mathbf{A}) = \#\mathbf{Pic}(\mathbf{A}) = d_\infty h$, where h is the class number of the function field $\mathbf{k}$. Let $\Im(\mathbf{A})$ be the group of fractional ideals of $\mathbf{A}$. Since $\Im(\mathbf{A})$ is freely generated by the prime ideals $\mathfrak{p}$ in $\mathbf{A}$, we can define a group morphism $\deg\colon \Im(\mathbf{A}) \to \mathbb{Z}$ by specifying that each $\deg \mathfrak{p}$ is the $\mathbb{F}_r$-dimension of the residue class field at $\mathfrak{p}$. For $\mathfrak{a} \in \Im(\mathbf{A})$, we put $N(\mathfrak{a}) = r^{\deg \mathfrak{a}}$.

The algebraic closure $\overline{\mathbf{K}}$ of $\mathbf{K}$ is not complete, but the completion $\mathbf{C}$ of $\overline{\mathbf{K}}$ is algebraically closed. We view $\mathbf{C}$ as the function field analog of the complex numbers $\mathbb{C}$.

A Tiny Bit of History

The earliest papers in the arithmetic of function fields focused on the case $\mathbf{k} = \mathbb{F}_r(T)$ with $\mathbf{A} = \mathbb{F}_r[T]$ and with the place at infinity more or less taken for granted. Of these early efforts, the most important are Dedekind's **[D]** systematic development, Kornblum's **[K]** proof of the polynomial analog of Dirichlet's Theorem and Artin's **[A]** detailed exposition of the arithmetic of quadratic extensions of $\mathbb{F}_r(T)$. Artin's paper (his doctoral dissertation) makes explicit use of the completion at infinity, and, following Dedekind, introduces a sign-function sgn. The class numbers which Artin computes in this paper are class numbers of rings, and his L-functions are all missing their Euler factors over ∞.

The great flowering of the theory of global function fields associated with the names of Artin, Chevalley, Hasse, F. K. Schmidt, Weil and many others that flourished in the 1920's and 1930's emphasized the symmetry of the set of places of $\mathbf{k}$. Choosing a distinguished place at infinity was for them like puncturing a sphere, thereby losing both compactness and the aesthetics of the subject. The crowning achievement of this era was the Riemann Hypothesis which had been conjectured by Artin in **[A]**, proved by Hasse for elliptic curves, and proved in general by Weil in 1940 **[W]**, just as World War II was beginning in Europe.

This symmetric point of view was studiously ignored, however, by Carlitz in all his papers on function fields during the 1930's. His willingness to work with a distinguished infinite place removed an obstacle that might have prevented him from introducing the first Drinfeld module, the *Carlitz module*, in 1938 **[C4]**, and using it to give an explicit construction of the class field theory of $\mathbb{F}_r(T)$. Actually (see **[H1]**), Carlitz did not construct all of the maximal abelian extension of $\mathbb{F}_r(T)$ by this method. Naturally enough, he missed the part that comes from the place at infinity!

The moral of this bit of history is that you should always learn as many points of view as possible on any mathematical enterprise, and you should keep them all in mind when you think about a problem.

Part I: Basic Theory of Drinfeld Modules with Coefficients in a Field over A

Drinfeld's papers **[Dr1]** and **[Dr2]** provide the foundation for the theory. Other sources are the notes of Deligne and Husemoller **[D–H]**, the survey papers of Goss **[Go1]** and **[Go2]** and Gekeler's book **[Ge1]**. The presentation below more or less follows that of **[H2]**.

1. Motivation: The Carlitz module. We find our first motivation in the rational function field $\mathbf{k} = \mathbb{F}_r(T)$ with $\mathbf{A} = \mathbb{F}_r[T]$ and ∞ equal to the unique pole of T. We ask how might one compute a factorial in $\mathbb{F}_r[T]$? Carlitz answered this question in 1932 (see **[C2]**). For $j \geq 0$, put $[j] = T^{r^j} - T$ and define

$$L_k = \prod_{j=1}^{k} [j] \qquad \text{and} \qquad F_k = \prod_{j=0}^{k-1} [k-j]^{r^j}. \tag{1.1}$$

The product of the monic polynomials in $\mathbb{F}_r[T]$ of degree n is F_n; and the product of the non-zero polynomials of degree strictly less than n is $(-1)^n \cdot F_n/L_n$. Carlitz derived these results by a method that he used to prove much more. He showed that the product

$$\Psi_n(z) = \prod_{\substack{A \in \mathbf{A} \\ \deg A < n}} (z - A) \tag{1.2}$$

expands into the $\mathbb{F}_r$-linear polynomial

$$\Psi_n(z) = \sum_{i=0}^{n} (-1)^{n-i} \begin{bmatrix} n \\ i \end{bmatrix} z^{r^i} \tag{1.3}$$

in z where

$$\begin{bmatrix} n \\ i \end{bmatrix} = \frac{F_n}{F_i \cdot L_{n-i}^{r^i}}.$$

Dividing (1.2) and (1.3) by $(-1)^n \cdot F_n/L_n$, we obtain

$$z \cdot \prod_{\substack{A \neq 0 \\ \deg A < n}} \left(1 - \frac{z}{A}\right) = \sum_{i=0}^{n} (-1)^i \frac{L_n}{F_i \cdot L_{n-i}^{r^i}} z^{r^i}. \tag{1.4}$$

It is natural to try to compute the limit as $n \to \infty$ in (1.4). The left-hand side is easily seen to converge to an element of the completion $\mathbf{K}$, and so we try to compute the limit on the right in $\mathbf{K}$. Following Carlitz, we define $\mathbf{i}$ to be some fixed $(r-1)$-st root of $T^r - T$ in $\overline{\mathbf{K}}$, and for $i \geq 0$, we put

$$\boldsymbol{\pi}_i = \mathbf{i}^{r^i - 1}/L_i\,.$$

Then $\boldsymbol{\pi} = \lim_{i \to \infty} \boldsymbol{\pi}_i$ exists in $\mathbf{K}$ and has the infinite product representation

$$\boldsymbol{\pi} = \prod_{j=1}^{\infty} \left(1 - \frac{[j]}{[j+1]}\right).$$

We may now rewrite (1.4) as

$$z \cdot \prod_{\substack{A \neq 0 \\ \deg A < n}} \left(1 - \frac{z}{A}\right) = \frac{1}{\mathbf{i}\boldsymbol{\pi}_n} \sum_{i=0}^{n} (-1)^i \frac{(\mathbf{i}\boldsymbol{\pi}_{n-i})^{r^i}}{F_i} z^{r^i},$$

which encourages us to believe (as we may in fact prove!) that

$$e_{\mathbf{A}}(z) = z \cdot \prod_{\substack{A \neq 0 \\ A \in \mathbf{A}}} \left(1 - \frac{z}{A}\right) = \frac{1}{\xi} \sum_{i=0}^{\infty} (-1)^i \frac{(\xi z)^{r^i}}{F_i},$$

where $\xi = \mathbf{i}\boldsymbol{\pi}$. If we now replace z with z/ξ, we obtain

$$e_{\xi\mathbf{A}}(z) = z \cdot \prod_{\substack{A \neq 0 \\ A \in \mathbf{A}}} \left(1 - \frac{z}{\xi A}\right) = \sum_{i=0}^{\infty} (-1)^i \frac{z^{r^i}}{F_i}. \tag{1.5}$$

The function $e_{\xi\mathbf{A}}(z)$ has many remarkable properties. We note that it is an $\mathbb{F}_r$-linear function on $\mathbf{C}$, and that it admits $\xi\mathbf{A}$ as a group of periods. We leave it to the reader to reflect on the obvious analogies between the product expansion (1.5) and the familiar Weierstraß product expansion of $e^{2\pi i z} - 1$ as a function on $\mathbb{C}$. Carlitz showed that $e_{\xi\mathbf{A}}(z)$ satisfies "complex multiplications" under the action of the elements of $\mathbf{A}$. For example,

$$e_{\xi\mathbf{A}}(Tz) = T e_{\xi\mathbf{A}}(z) - e_{\xi\mathbf{A}}(z)^r. \tag{1.6}$$

This is how the first Drinfeld module was discovered.

The ring $\mathbf{A}$ is a discrete additive subgroup of $\mathbf{K}$, and $\xi\mathbf{A}$ is a discrete additive subgroup of $\mathbf{C}$. We are justified in referring to these subgroups as *lattices*, and we call the functions $e_{\mathbf{A}}(z)$ and $e_{\xi\mathbf{A}}(z)$ the *exponential functions* associated to $\mathbf{A}$ and $\xi\mathbf{A}$ respectively. The element ξ has the remarkable property that it multiplies the lattice $\mathbf{A}$ into a lattice whose exponential function has a power series expansion with rational coefficients.

2. Motivation: Stark's conjectures. Our second motivation is valid for any $\mathbf{k}$ and any choice of ∞. Let $\mathfrak{a}$ be any non-zero fractional ideal of $\mathbf{A}$. We think of $\mathfrak{a}$ as a lattice in $\mathbf{C}$. For every $z \in \mathbf{C} - \mathfrak{a}$ and every s in the complex half-plane $\mathrm{Re}(s) > 1$, the series

$$Z_{\mathfrak{a}}(s,z) = \sum_{a\in\mathfrak{a}} \frac{1}{N(a+z)^s}$$

and

$$V_{\mathfrak{a}}(s) = \sum_{a\in\mathfrak{a}-\{0\}} \frac{1}{N(a)^s}$$

are absolutely convergent and define holomorphic functions of s in that half-plane. As one can show easily from the Riemann–Roch Theorem, both these functions are holomorphic in the whole complex plane except for simple poles at the points $s = 1 + 2\pi i k/d\log r$, $k \in \mathbb{Z}$. (Both are in fact rational functions of r^{-ds}.) Further, $Z_{\mathfrak{a}}(0,z) = 0$ for every $z \in \mathbf{C} - \mathfrak{a}$, and $V_{\mathfrak{a}}(0) = -1$. We call $Z_{\mathfrak{a}}(s,z)$ the *partial zeta-function* of $\mathfrak{a}$. From its definition, $Z_{\mathfrak{a}}(s,z)$ is periodic in z with $\mathfrak{a}$ as a group of periods.

According to the conjectures of Stark, the derivatives of the functions $Z_{\mathfrak{a}}(s,z)$ at $s = 0$ encode important information about those class fields of $\mathbf{k}$ in which ∞ splits completely. For $\mathrm{Re}(s) > 1$, we can compute the derivative term by term:

$$(\log r)^{-1} \cdot \left[\frac{d}{ds} Z_{\mathfrak{a}}(s,z)\right] = -\sum_{a\in\mathfrak{a}} \deg(a+z) \cdot r^{-s\deg(a+z)}$$

$$= -\deg z \cdot r^{-s\deg z} - \sum_{a\in\mathfrak{a}-\{0\}} \deg(1+z/a) \cdot r^{-s\deg(a+z)} - H(s)$$

where

$$H(s) = \sum_{a\in\mathfrak{a}-\{0\}} \deg(a) \cdot r^{-s\deg(a+z)}.$$

Now $\deg(1+z/a) = 0$ for all but finitely many elements $a \in \mathfrak{a}$ and so

$$(\log r)^{-1} \cdot \left[\frac{d}{ds} Z_{\mathfrak{a}}(s,z)\right]_{s=0} = d\mathrm{v}_\infty(z) + d \cdot \sum_{a\in\mathfrak{a}-\{0\}} \mathrm{v}_\infty(1+z/a) - H(0)$$

$$= d\mathrm{v}_\infty(e_{\mathfrak{a}}(z)) - H(0)$$

where

$$e_{\mathfrak{a}}(z) = z \cdot \prod_{a\in\mathfrak{a}-\{0\}} \left(1 + \frac{z}{a}\right)$$

is the *exponential function* of $\mathfrak{a}$. To evaluate $H(0)$, we note that for $\mathrm{Re}(s) > 1$

$$\begin{aligned} H(s) &= \sum_{\substack{a \neq 0 \\ \deg a \leq \deg z}} \deg a \cdot r^{-s \deg(a+z)} + \sum_{\substack{a \\ \deg a > \deg z}} \deg a \cdot r^{-s \deg a} \\ &= G(s) - (\log r)^{-1} \cdot \frac{dV_{\mathfrak{a}}}{ds} \end{aligned}$$

where $G(s)$ is a polynomial in r^{-s} such that $G(0) = 0$. Thus $H(0)$ is independent of z, and we have derived the expression

$$(\log r)^{-1} \cdot \left[\frac{d}{ds} Z_{\mathfrak{a}}(s, z) \right]_{s=0} = dv_{\infty}(e_{\mathfrak{a}}(z)) + (\log r)^{-1} \cdot \left. \frac{dV_{\mathfrak{a}}}{ds} \right|_{s=0}.$$

Once again an exponential function $e_{\mathfrak{a}}(z)$ associated to a lattice in $\mathbf{C}$ has occurred in a very natural context. What about the $dV_{\mathfrak{a}}/ds$ term in the above formula? In fact, it equals $dv_{\infty}(\xi(\mathfrak{a}))$, where $\xi(\mathfrak{a})$ is an "invariant" that maps $\mathfrak{a}$ onto a "canonical lattice" $\xi(\mathfrak{a})\mathfrak{a}$ in $\mathbf{C}$. For more details, the reader can consult **[H5]**.

3. The category of Drinfeld modules in a field K over $\mathbf{A}$. A *field over* $\mathbf{A}$ is an $\mathbb{F}_r$-algebra morphism $\delta: \mathbf{A} \to K$, where K is an overfield of $\mathbb{F}_r$. In practice, δ is either an inclusion or else a reduction map modulo some non-zero prime ideal of $\mathbf{A}$.

Definition. A *Drinfeld* $\mathbf{A}$*-module over* K is a non-trivial representation of $\mathbf{A}$ as a ring of endomorphisms on the additive group scheme $\mathbf{G}_{a/K}$ such that the induced representation on the tangent space at zero is given by δ.

Just as the integers of an imaginary quadratic number field can act as a ring of endomorphisms on an elliptic curve, $\mathbf{A}$ can act as endomorphisms on $\mathbf{G}_{a/K}$. For this reason, Drinfeld called such representations *elliptic modules* in his famous paper **[Dr1]** in which Drinfeld modules were first defined in a general context.

In characteristic zero, the additive group scheme has very few endomorphisms, but this is not true in characteristic p. In fact, $\mathrm{End}(\mathbf{G}_{a/K})$ is isomorphic to the *left twisted polynomial ring* $K[\mathbf{F}]$ in the Frobenius endomorphism $\mathbf{F}$ on $\mathbf{G}_{a/K}$. The elements of $K[\mathbf{F}]$ are polynomials in $\mathbf{F}$ with coefficients written on the left; but multiplication in $K[\mathbf{F}]$ is *twisted* by the rule

$$\mathbf{F}w = w^p\mathbf{F} \tag{3.1}$$

for all $w \in K$. Let $D: K[\mathbf{F}] \to K$ be the map which assigns to every twisted polynomial τ the constant coefficient of τ. Then D is just the "derivative at zero"—i.e. the restriction of an endomorphism to the space of tangent vectors. We may therefore restate our definition as follows.

Definition. A *Drinfeld* $\mathbf{A}$*-module* over K is a ring morphism $\phi: \mathbf{A} \to K[\mathbf{F}]$ that satisfies $D \circ \phi = \delta$ and which is non-trivial in the sense that not all the twisted polynomials in the image of ϕ are constant polynomials.

A Drinfeld module is always a monomorphism and therefore defines a faithful representation of $\mathbf{A}$ on $\mathbf{G}_{a/K}$. Because δ is $\mathbb{F}_r$-linear, ϕ actually takes its values in the subring $K[\mathbf{F}^m]$ of $K[\mathbf{F}]$.

We write ϕ_x for the twisted polynomial associated to $x \in \mathbf{A}$ by ϕ. If R is any K-algebra, then (taking points) ϕ_x acts as an $\mathbb{F}_r$-linear map on the additive group of K. We write $\phi_x(t)$ for the action of ϕ_x on $t \in R$. Of course, under the taking of points, multiplication in $K[\mathbf{F}]$ corresponds to composition of linear maps.

Example: The Carlitz module. Take $\mathbf{A} = \mathbb{F}_r[T]$, and let δ be the inclusion map from $\mathbf{A}$ into $\mathbf{C}$. If τ is any non-constant polynomial in $\mathbf{C}[\mathbf{F}^m]$ with $D(\tau) = T$, then there is a unique Drinfeld module $\phi\colon \mathbf{A} \to \mathbf{C}$ such that $\phi_T = \tau$. If we take $\tau = T - \mathbf{F}^m$, we obtain the Carlitz module mentioned in §1 above. The action of ϕ_T on an element t in some K-algebra is just $\phi_T(t) = Tt - t^r$ (cf. (1.6)).

In **[H1]**, the Carlitz module was defined by $T \mapsto T + \mathbf{F}^m$. This choice is better for many purposes because it produces a *normalized* module (see §12). The two Carlitz modules are isomorphic in a sense that we will now explain.

Definition. Let ϕ and ϕ' be Drinfeld $\mathbf{A}$-modules over K. An *isogeny* from ϕ to ϕ' is a twisted polynomial $\tau \in K[\mathbf{F}]$ such that $\tau\phi_x = \phi'_x\tau$ for all $x \in \mathbf{A}$. Clearly a product of isogenies is again an isogeny. The Drinfeld $\mathbf{A}$-modules over K constitute a category $\mathrm{Drin}_{\mathbf{A}}(K)$ in which the morphisms are the isogenies.

What is an isomorphism in $\mathrm{Drin}_{\mathbf{A}}(K)$? Since only constant twisted polynomials are invertible in $K[\mathbf{F}]$, ϕ and ϕ' are isomorphic if and only if there is an element $\xi \in \mathbf{K}^\times$ such that $\xi\phi_x = \phi'_x\xi$ for all $x \in \mathbf{A}$. The two Carlitz modules are defined above as elements of $\mathrm{Drin}_{\mathbf{A}}(\mathbf{C})$. They are isomorphic via $\xi = \zeta$, where $\zeta \in \mathbf{C}$ is an $(r-1)$-st root of -1. Had we defined the Carlitz modules as elements of $\mathrm{Drin}_{\mathbf{A}}(\mathbf{K})$, then for $p \neq 2$ they would not be isomorphic because $\mathbf{K}$ does not contain an $(r-1)$-st root of -1.

The category $\mathrm{Drin}_{\mathbf{A}}(S)$ can be defined for any *scheme* S *over* $\mathbf{A}$. With two exceptions (see §§14 and 15), we consider only the case $S = \mathrm{Spec}(K)$ in these notes.

4. The action of ideals in A on $\mathrm{Drin}_{\mathbf{A}}(K)$**.** Let $\mathcal{O}$ be an order in an imaginary quadratic number field. If $\mathcal{O}$ acts on an elliptic curve E, then each invertible ideal of $\mathcal{O}$ defines an isogeny from E onto another elliptic curve E'. If E is defined over $\mathbb{C}$, this fact follows readily from the analytic theory. When E is defined over, say, a finite field, it is still true but not quite so obvious (cf. **[S]**). As we now show, the ideals of $\mathbf{A}$ act in a similar manner on $\mathrm{Drin}_{\mathbf{A}}(K)$.

Proposition 4.1. *The twisted polynomial ring* $K[\mathbf{F}]$ *is a left principal ideal ring.*

Sketch of proof. The proof goes just as in the familiar commutative theory except that one has to mind the rule (3.1). The degree of a twisted polynomial is well defined. One shows that there is a *right* division algorithm in $K[\mathbf{F}]$ by monic polynomials, and this implies that *left* ideals are principal. □

Given a Drinfeld $\mathbf{A}$-module ϕ over K and a non-zero ideal $\mathfrak{a}$ in $\mathbf{A}$, let $I_{\mathfrak{a},\phi}$ be the *left* ideal generated in $K[\mathbf{F}]$ by the twisted polynomials ϕ_a, $a \in \mathfrak{a}$. As left ideals are principal, $I_{\mathfrak{a},\phi} = K[\mathbf{F}] \cdot \phi_{\mathfrak{a}}$ for a unique monic twisted polynomial $\phi_{\mathfrak{a}}$. Clearly, $I_{\mathfrak{a},\phi}$ is carried into itself by multiplication on the *right* by the ϕ_x, $x \in \mathbf{A}$. Therefore, for every $x \in \mathbf{A}$ there is a uniquely defined $\phi'_x \in K[\mathbf{F}]$ such that

$$\phi_{\mathfrak{a}} \cdot \phi_x = \phi'_x \cdot \phi_{\mathfrak{a}}. \tag{4.2}$$

One checks easily that the map $\phi' \colon \mathbf{A} \to K[\mathbf{F}]$ defined by $x \mapsto \phi'_x$ is a Drinfeld $\mathbf{A}$-module over K except for one subtlety: $D \circ \phi' = \delta^{p^{j(\mathfrak{a})}}$, where $j(\mathfrak{a})$ is the smallest power of $\mathbf{F}$ that occurs in $\phi_{\mathfrak{a}}$ with a non-zero coefficient. So ϕ' is indeed a Drinfeld module, but possibly for a different field over $\mathbf{A}$. Actually, it is always true that $\delta^{p^{j(\mathfrak{a})}} = \delta$, as we show in §5 below, but this is not immediately obvious.

We introduce the notation

$$\phi' = \mathfrak{a} * \phi. \tag{4.3}$$

Thus, we have an operation $*$ of the non-zero ideals in $\mathbf{A}$ on $\mathrm{Drin}_{\mathbf{A}}(K)$ which we may characterize as follows: $\mathfrak{a} * \phi$ is the unique Drinfeld $\mathbf{A}$-module which is isogenous to ϕ via the isogeny $\phi_{\mathfrak{a}}$. If R is a K-algebra, the roots of $\phi_{\mathfrak{a}}(t)$ in R are the points of $\mathfrak{a}$-torsion for $\mathbf{A}$ acting on R through ϕ. In fact, much more is true. The kernel of $\phi_{\mathfrak{a}}$ as an endomorphism of $\mathbf{G}_{a/K}$ is a finite subgroup scheme of $\mathbf{G}_{a/K}$ equal to the intersection of the kernels of all the endomorphisms ϕ_a for $a \in \mathfrak{a}$.

The basic properties of the action $*$ are contained in the following lemmas. These all follow easily from the definitions and (5.8) below.

Lemma 4.4. *Let $\mathfrak{a} = w\mathbf{A}$ be a non-zero principal ideal, and let μ be the leading coefficient of ϕ_w. Then $\phi_{\mathfrak{a}} = \mu^{-1} \cdot \phi_w$, and $(\mathfrak{a} * \phi)_x = \mu^{-1} \cdot \phi_x \cdot \mu$ for all $x \in \mathbf{A}$.*

Lemma 4.5. *Let $\mathfrak{a}$ and $\mathfrak{b}$ be non-zero ideals in $\mathbf{A}$. Then*

$$\phi_{\mathfrak{a}\mathfrak{b}} = (\mathfrak{b} * \phi)_{\mathfrak{a}} \cdot \phi_{\mathfrak{b}} \tag{4.6}$$

and

$$\mathfrak{a} * (\mathfrak{b} * \phi) = (\mathfrak{a}\mathfrak{b}) * \phi. \tag{4.7}$$

Let $\mathrm{Isom}_{\mathbf{A}}(K)$ be the set of isomorphism classes of the Drinfeld $\mathbf{A}$-modules over K. It is clear from the definitions and Lemma 4.4 that $*$ also operates on $\mathrm{Isom}_{\mathbf{A}}(K)$ and that the principal ideals in $\mathbf{A}$ act trivially. It follows that the ideal class group $\mathbf{Pic}(\mathbf{A})$ also acts on $\mathrm{Isom}_{\mathbf{A}}(K)$.

5. The rank and height of a Drinfeld A-module. As we prove in §8 below, Drinfeld $\mathbf{A}$-modules over $\mathbf{C}$ exist in great profusion. Therefore, what we always knew was true for $\mathbf{k} = \mathbb{F}_r(T)$ with ∞ the unique pole of T is actually true in general—the ring $\mathbf{A}$ is a ring of (twisted) polynomials! For $\tau \in K[\mathbf{F}]$, $\tau \neq 0$, let $\deg \tau$ be the degree of τ as a polynomial in $\mathbf{F}$. If $\phi \colon \mathbf{A} \to K[\mathbf{F}]$ is a Drinfeld $\mathbf{A}$-module, then the map $\mathrm{v}_\phi \colon \mathbf{A} - \{0\} \to \mathbb{Z}$ defined by $\mathrm{v}_\phi(x) = -\deg \phi_x$ satisfies

$$\text{i) } \mathrm{v}_\phi(xy) = \mathrm{v}_\phi(x) + \mathrm{v}_\phi(y) \quad \text{and} \quad \text{ii) } \mathrm{v}_\phi(x+y) \geq \min\{\mathrm{v}_\phi(x), \mathrm{v}_\phi(y)\}.$$

It therefore extends to a valuation (also called v_ϕ) on $\mathbf{k}$. Which place of $\mathbf{k}$ does it define? Since v_ϕ is negative on $\mathbf{A}$, it can only define the distinguished place ∞. Hence, there is a rational number r_ϕ such that

$$\mathrm{v}_\phi(x) = r_\phi m d_\infty \cdot \mathrm{v}_\infty(x) = -r_\phi m \cdot \deg x \tag{5.1}$$

for all $x \in \mathbf{k}^\times$. We call r_ϕ the *rank* of ϕ. The rank r_ϕ is a positive integer, although this fact is not immediate from (5.1). To say it another way, the degree of ϕ_x as a polynomial in $\mathbf{F}^m$ is always divisible by $\deg x$.

Definition. Let $\phi \in \mathrm{Drin}_{\mathbf{A}}(K)$. The *characteristic* of ϕ is the kernel of the map $\delta \colon \mathbf{A} \to K$. It is therefore a prime ideal char(ϕ) in $\mathbf{A}$. If char(ϕ) $= 0$, we say the ϕ has *generic characteristic*.

Suppose $\mathfrak{q} = \mathrm{char}(\phi)$ is not zero, and let $\mathrm{v}_\mathfrak{q}$ be the normalized valuation on $\mathbf{k}$ at $\mathrm{v}_\mathfrak{q}$. Then the low degree ends of the twisted polynomials ϕ_x for $x \in \mathbf{A}$ define a valuation which must be a multiple of $\mathrm{v}_\mathfrak{q}$. Let $j(x)$ denote the smallest integer k such that $\mathbf{F}^k$ appears in ϕ_x with non-zero coefficient. Then there is a rational number h_ϕ, called the *height* of ϕ, such that

$$j(x) = h_\phi m \deg \mathfrak{q} \cdot \mathrm{v}_\mathfrak{q}(x) \tag{5.2}$$

for all $x \in \mathbf{A}$.

Theorem 5.3. *The rank r_ϕ of a Drinfeld $\mathbf{A}$-module ϕ is a positive integer. If* char(ϕ) $\neq 0$, *then the height h_ϕ is also a positive integer.*

Proof. Under the taking of points, the Drinfeld $\mathbf{A}$-module ϕ induces an ordinary $\mathbf{A}$-module structure on $\mathbf{G}_{a/K}(\overline{K}) = \overline{K}^+$. Since $\overline{K}$ is algebraically closed and the elements of $\mathbf{A}$ act on $\overline{K}$ via polynomials in $K[t]$, $\overline{K}^+$ is a *divisible* $\mathbf{A}$-module. For any non-zero ideal $\mathfrak{a}$ in $\mathbf{A}$, let

$$\Lambda(\mathfrak{a}) = \{t \in \overline{K}^+ : \phi_\mathfrak{a}(t) = 0\}$$

be the $\mathfrak{a}$-torsion submodule of $\overline{K}^+$ for this action of $\mathbf{A}$. Since $\phi_\mathfrak{a}(t)$ is a polynomial of degree $p^{\deg \phi_\mathfrak{a}}$, $\Lambda(\mathfrak{a})$ is finite, and we have

$$\#\Lambda(\mathfrak{a}) = p^{\deg \phi_\mathfrak{a} - j(\mathfrak{a})}. \tag{5.4}$$

Let $\mathfrak{p}$ be a non-zero prime ideal of $\mathbf{A}$, and let $g_\mathfrak{p}$ be the dimension of the $\mathbf{A}/\mathfrak{p}$-vector space $\Lambda(\mathfrak{p})$. We will show that

$$g_\mathfrak{p} = \begin{cases} r_\phi - h_\phi & \text{if } \mathfrak{p} = \mathrm{char}(\phi) \\ r_\phi & \text{if } \mathfrak{p} \neq \mathrm{char}(\phi). \end{cases} \tag{5.5}$$

Let $\pi_\mathfrak{p} \in \mathbf{A}$ be a uniformizing parameter at $\mathfrak{p}$. Because $\overline{K}^+$ is a divisible $\mathbf{A}$-module, the sequence

$$0 \to \Lambda(\mathfrak{p}) \longrightarrow \Lambda(\mathfrak{p}^{e+1}) \xrightarrow{\pi_\mathfrak{p}} \Lambda(\mathfrak{p}^e) \to 0 \tag{5.6}$$

is exact for all $e \geq 1$. As the reader can show by induction on e, any $\mathbf{A}/\mathfrak{p}$-basis for $\Lambda(\mathfrak{p})$ lifts back through the maps $\pi_\mathfrak{p}$ in (5.6) to a free basis for $\Lambda(\mathfrak{p}^e)$ as an $\mathbf{A}/\mathfrak{p}^e$-module. Therefore

$$\#\Lambda(\mathfrak{p}^e) = N(\mathfrak{p}^e)^{g_\mathfrak{p}} = p^{g_\mathfrak{p} m \cdot \deg(\mathfrak{p}^e)}. \tag{5.7}$$

Since the class number $h(\mathbf{A}) = \#(\mathbf{Pic}(\mathbf{A}))$ of the Dedekind domain $\mathbf{A}$ is finite, $\mathfrak{p}^{h(\mathbf{A})} = x\mathbf{A}$ is a principal ideal. Taking $e = h(\mathbf{A})$ in (5.7) and $\mathfrak{a} = x\mathbf{A}$ in (5.4), we learn from the equality of the right-hand sides that

$$g_{\mathfrak{p}} m \cdot \deg x = \deg \phi_x - j(x) = \begin{cases} (r_\phi - h_\phi) m \cdot \deg x & \text{if } \mathfrak{p} = \operatorname{char}(\phi) \\ r_\phi m \cdot \deg x & \text{if } \mathfrak{p} \neq \operatorname{char}(\phi) . \end{cases}$$

Therefore, we have $r_\phi = g_{\mathfrak{p}}$ for any $\mathfrak{p} \neq \operatorname{char}(\phi)$, and $r_\phi = g_{\mathfrak{p}} + h_\phi$ when $\mathfrak{p} = \operatorname{char}(\phi) \neq 0$. □

Let $\mathfrak{q} = \operatorname{char}(\phi)$. Equations (5.4)–(5.6) above imply that

$$\deg \phi_{\mathfrak{a}} = r_\phi m \cdot \deg \mathfrak{a} \quad \text{and} \quad j(\mathfrak{a}) = h_\phi m \deg \mathfrak{q} \cdot v_{\mathfrak{q}}(\mathfrak{a}) \tag{5.8}$$

for all prime power ideals $\mathfrak{a} = \mathfrak{p}^e$. Since every ideal is a product of prime powers, these equations hold for all non-zero ideals $\mathfrak{a}$ in $\mathbf{A}$.

Corollary 5.9. *Suppose ϕ has rank one and characteristic $\mathfrak{q} \neq 0$. Let $\mathfrak{b} = \mathfrak{q}^e$, $e \geq 1$, be a power of $\mathfrak{q}$. Then $\phi_{\mathfrak{b}} = \mathbf{F}^{\deg \mathfrak{b}}$.*

We may now settle the subtle point which arose in our definition of $\mathfrak{a} * \phi$ in §4.

Corollary 5.10. *For every non-zero ideal $\mathfrak{a}$ in $\mathbf{A}$, $D \circ (\mathfrak{a} * \phi) = \delta$. Therefore $\mathfrak{a} * \phi \in \mathrm{Drin}_{\mathbf{A}}(K)$.*

Proof. From the discussion in §4, $D \circ (\mathfrak{a} * \phi) = \delta^{p^{j(\mathfrak{a})}}$. If $\operatorname{char}(\phi) = 0$, then $j(\mathfrak{a}) = 0$, and we are done. If $\operatorname{char}(\phi) = \mathfrak{q} \neq 0$, then the image $\delta(\mathbf{A})$ of δ is a subfield of K which is isomorphic to $\mathbf{A}/\mathfrak{q}$. Since h_ϕ is an integer, the second equation in (5.8) above implies that exponentiation by $p^{j(\mathfrak{a})}$ is just the identity map on $\delta(\mathbf{A})$. □

The *Tate module* of ϕ at the non-zero prime ideal $\mathfrak{p} \neq \operatorname{char}(\phi)$ in $\mathbf{A}$ is the inverse limit

$$T_{\mathfrak{p}}(\phi) = \varprojlim \Lambda(\mathfrak{p}^e) \qquad (e \geq 0) \tag{5.11}$$

under the maps $\pi_{\mathfrak{p}}$ of (5.6). Let $\mathbf{A}_{\mathfrak{p}}$ be the ring of $\mathfrak{p}$-adic integers in the completion $\mathbf{k}_{\mathfrak{p}}$ of $\mathbf{k}$ at $\mathfrak{p}$. By (5.5), $T_{\mathfrak{p}}(\phi)$ is a free $\mathbf{A}_{\mathfrak{p}}$-module of rank r_ϕ. For $\phi, \phi' \in \mathrm{Drin}_{\mathbf{A}}(K)$, there is a morphism

$$0 \to \mathrm{Hom}(\phi, \phi') \longrightarrow \mathrm{Hom}_{\mathbf{A}_{\mathfrak{p}}}(T_{\mathfrak{p}}(\phi), T_{\mathfrak{p}}(\phi')) \tag{5.12}$$

of additive groups which is injective since a non-zero isogeny acting on $\overline{\mathbf{K}}$ can have only a finite set of roots. Drinfeld used the map (5.12) with $\phi = \phi'$ to prove the following analog of a result of Deuring on the endomorphism rings of elliptic curves.

Theorem 5.13. *Let $\phi \in \mathrm{Drin}_{\mathbf{A}}(K)$. The endomorphism ring $\mathrm{End}(\phi)$ of ϕ is a projective $\mathbf{A}$-module of finite rank less than or equal to r_ϕ^2.*

The proof of this theorem in **[Dr1]** is very brief. The reader will do better to adapt Silverman's proof (**[Sil]**, Chapter III, §7) to the category $\mathrm{Drin}_{\mathbf{A}}(K)$.

Corollary 5.14. *If $r_\phi = 1$, then $\mathrm{End}(\phi) = \mathbf{A}$ and $\mathrm{Aut}(\phi) = \mathbb{F}_r^\times$.*

The structure of the ring $\mathrm{End}(\phi)$ in case $r_\phi = 2$ is known in some detail. See **[Ge2]** for proofs and applications to "finite" Drinfeld modules.

We note finally, that the rank and height of ϕ are preserved under non-zero isogenies. This follows almost immediately from the definitions. In particular, both the rank and the height are invariant under the $*$ operation.

6. The leading coefficient map. The map $\mathfrak{a} \mapsto D(\phi_\mathfrak{a})$ associates an element of K to every non-zero ideal $\mathfrak{a}$ in $\mathbf{A}$. By (4.6), these elements satisfy

$$D(\phi_{\mathfrak{ab}}) = D((\mathfrak{b} * \phi)_\mathfrak{a}) \cdot D(\phi_\mathfrak{b}) \tag{6.1}$$

for all non-zero ideals $\mathfrak{a}$ and $\mathfrak{b}$ in $\mathbf{A}$. By utilizing the leading coefficients of the polynomials ϕ_x, $x \in \mathbf{A}$, we can define an element $\mu_\phi(\pi^{-1}) \in K$ at every uniformizer $\pi \in \mathbf{K}$. These elements provide an analog at ∞ for the $D(\phi_\mathfrak{a})$.

Definition. For every $x \in \mathbf{A}$, let $\mu_\phi(x)$ be the leading coefficient of ϕ_x.

The elements $\mu_\phi(x)$ satisfy the following analog of (6.1):

$$\mu_\phi(xy) = \mu_\phi(x) \cdot \mu_\phi(y)^{N(x)^{r_\phi}} = \mu_\phi(y) \cdot \mu_\phi(x)^{N(y)^{r_\phi}} \tag{6.2}$$

for all $x, y \in \mathbf{A}$.

Let $\mathfrak{m}$ be the maximal ideal of the ring of integers in the local field $\mathbf{K}$. Following Deligne **[De]**, we extend the map $x \mapsto \mu_\phi(x)$ to the homogeneous elements of grade $\ell \le 0$ in the graded ring $\mathrm{Gr}(\mathbf{K}) = \bigoplus_{\ell \in \mathbb{Z}} \mathfrak{m}^\ell/\mathfrak{m}^{\ell+1}$. For $n \gg 0$, every element of $\mathfrak{m}^{-n}/\mathfrak{m}^{-n+1}$ is the image $\bar{x}$ of an element $x \in \mathfrak{m}^{-n} \cap \mathbf{A}$ by the Riemann–Roch Theorem. For $a = \bar{x} \ne 0$ in $\mathfrak{m}^{-n}/\mathfrak{m}^{-n+1}$, we put $\mu_\phi(a) = \mu_\phi(x)$. This definition is independent of the choice of x. To prove it, note that if $\deg x = \deg y$ and $\bar{x} = \bar{y}$, then $\deg(x - y) < \deg x$. We have then $\deg \phi_{x-y} < \deg \phi_x$ which implies that $\mu_\phi(x) = \mu_\phi(y)$.

The function μ_ϕ is now defined on the homogeneous elements of sufficiently small negative grade in $\mathrm{Gr}(\mathbf{K})$. On these elements, μ_ϕ satisfies the following two properties:

$$\mathrm{grad}(a) = \mathrm{grad}(b) \implies \mu_\phi(a + b) = \mu_\phi(a) + \mu_\phi(b) \tag{6.3}$$

and

$$\mu_\phi(ab) = \mu_\phi(a) \cdot \mu_\phi(b)^{N(a)^{r_\phi}}. \tag{6.4}$$

If now $a \in \mathrm{Gr}(\mathbf{K})$ is an arbitrary homogeneous element of negative grade, we define

$$\mu_\phi(a) \stackrel{\mathrm{def}}{=} \mu_\phi(ab) \cdot \mu_\phi(b)^{-N(b)^{r_\phi}},$$

where b is an element of sufficiently small negative grade. The element is independent of the choice of b by (6.4).

If $\phi' = \xi\phi\xi^{-1}$ for $\xi \in \mathbf{K}$, then from the definitions,

$$\mu_{\phi'}(a) = \xi^{1-N(a)^{r_\phi}} \cdot \mu_\phi(a). \tag{6.5}$$

In particular, for $a = \pi^{-1}$, the inverse of a uniformizer at ∞,

$$\mu_{\phi'}(\pi^{-1}) = \xi^{1-r^{dr_\phi}} \cdot \mu_\phi(\pi^{-1}). \tag{6.6}$$

Let $\kappa(\infty) = \mathfrak{m}^0/\mathfrak{m}^1$ be the residue class field at ∞, and let

$$i_\phi : \kappa(\infty) \to K \tag{6.7}$$

be the restriction of μ_ϕ to $\kappa(\infty)$. By (6.5), the monomorphism i_ϕ depends only on the isomorphism class of ϕ.

7. Formal k-modules. We assume throughout this section that

$$\delta : \mathbf{A} \to K$$

is a monomorphism, and we denote also by 'δ' the natural extension of δ to $\mathbf{k}$. Let $K[[\mathbf{F}]]$ be the ring of *left twisted power series* generated over K by $\mathbf{F}$ subject to the relations (3.1). We define the derivative map $D: K[[\mathbf{F}]] \to \mathbf{K}$ just as we did for twisted polynomials.

Definition. Any ring morphism $\psi: \mathbf{k} \to K[[\mathbf{F}]]$ such that $D \circ \psi = \delta$ will be called a *formal* $\mathbf{k}$*-module over* K. We say that ψ is *non-trivial* if its image contains a power series which is not a constant.

If ϕ is a Drinfeld $\mathbf{A}$-module over K, then each ϕ_x for $x \in \mathbf{A} - \{0\}$ is invertible in $K[[\mathbf{F}]]$, and therefore ϕ extends to a non-trivial formal $\mathbf{k}$-module that we also denote by 'ϕ'.

Lemma 7.1. *Suppose $\sigma \in K[[\mathbf{F}]]$ has constant term $D(\sigma) = y$ with y transcendental over $\mathbb{F}_p$. Then there is a unique power series*

$$\lambda_\sigma = \sum_{i=0}^{\infty} c_i \mathbf{F}^i$$

in $K[[\mathbf{F}]]$ with $c_0 = 1$ such that $\lambda_\sigma \cdot y = \sigma \lambda_\sigma$.

Proof. If $\sigma = \sum_{j=0}^{\infty} a_j \mathbf{F}^j$, then $\lambda_\sigma \cdot y = \sigma \lambda_\sigma$ is equivalent to the recurrences

$$\left(y^{p^i} - y\right) \cdot c_i = \sum_{j=1}^{i} a_j \cdot c_{i-j}^{p^j} \tag{7.2}$$

for all $i \geq 1$. Since y is transcendental over $\mathbb{F}_p$, the coefficient on the left above is never zero. Thus, given the initial value $c_0 = 1$, (7.2) is uniquely solvable for the coefficients c_i, $i \geq 1$. □

Corollary 7.3. *If $D(\sigma)$ is transcendental over $\mathbb{F}_p$, then $\lambda_\sigma \cdot K \cdot \lambda_\sigma^{-1}$ is the centralizer of σ in $K[[\mathbf{F}]]$.*

Proof. The centralizer of $D(\sigma)$ in $K[[\mathbf{F}]]$ is clearly just K. Since σ is the image of $D(\sigma)$ under the inner automorphism through λ_σ, the centralizer of σ is the image of K under the same automorphism. □

Corollary 7.4. *If $D(\sigma)$ is transcendental over $\mathbb{F}_p$, then for each $w \in K$,*

$$\tau_w = \lambda_\sigma \cdot w \cdot \lambda_\sigma^{-1}$$

is the unique power series in $K[[\mathbf{F}]]$ with constant term w which commutes with σ.

Proposition 7.5. *Let ψ be a formal* $\mathbf{k}$ *-module over* K. *Then there is a unique power series*

$$\lambda_\psi = \sum_{i=0}^{\infty} c_i \mathbf{F}^{mi}$$

such that $c_0 = 1$ *and* $\psi_x = \lambda_\psi \cdot \delta(x) \cdot \lambda_\psi^{-1}$ *for all* $x \in \mathbf{k}$.

Proof. Choose $z \in \mathbf{k}$ so that $\delta(z) = y$ is transcendental over $\mathbb{F}_p$. Let λ_ψ be the twisted power series associated to $\sigma = \psi_z$ by Lemma 7.1. If $x \in \mathbf{k}$, then $\psi_x \cdot \psi_z = \psi_{xz} = \psi_z \cdot \psi_x$, which implies that $\psi_x = \lambda_\psi \cdot \delta(x) \cdot \lambda_\psi^{-1}$ by Corollary 7.4. Since $\delta(\zeta)$ is a constant for all $\zeta \in \mathbb{F}_r$, λ_ψ belongs to the subring $K[[\mathbf{F}^m]]$ of $K[[\mathbf{F}]]$. □

This proposition shows that a Drinfeld $\mathbf{A}$ -module ϕ over K of generic characteristic determines a unique twisted power series λ_ϕ in $K[[\mathbf{F}]]$ with $D(\lambda_\phi) = 1$. Conversely, in order to construct a Drinfeld $\mathbf{A}$ -module over K, all we have to do (!) is to construct a non-trivial twisted power series λ with $D(\lambda) = 1$ such that $\lambda \cdot \delta(x) \cdot \lambda^{-1}$ is in $K[\mathbf{F}]$ for all $x \in \mathbf{A}$. We show how to do this in the next section.

Example: The Carlitz module. For the Carlitz module $\phi: T \mapsto T - \mathbf{F}^m$ of §§1 and 3, the reader is invited to check by direct calculation that

$$\lambda_\phi = \sum_{i=0}^{\infty} \frac{(-1)^i}{F_i} \cdot \mathbf{F}^{mi}.$$

This fact also follows from (1.5) and (1.6).

8. The analytic construction of Drinfeld A -modules. Throughout this section,

$$\delta: \mathbf{A} \to \mathbf{C}$$

is the inclusion map.

The exponential functions associated with lattices in $\mathbf{C}$ provide a powerful means of constructing Drinfeld $\mathbf{A}$ -modules of any rank. We will use this analytic method to prove the existence theorem on which the whole theory is based. Except for the special case $\mathbf{A} = \mathbb{F}_r[T]$, writing down examples of Drinfeld $\mathbf{A}$ -modules by non-analytic means is not an easy task. If $\mathbf{k}$ is the function field of a hyperelliptic curve over $\mathbb{F}_r$ with ∞ chosen so that the curve has its standard equation away from ∞, then there is a fairly straightforward algebraic algorithm for computing Drinfeld $\mathbf{A}$ -modules of rank one. See **[H7]** for a description of this algorithm and some examples with $r = 2$ and $r = 3$. D. Dummit has recently used the algorithm to compute the Drinfeld modules of rank one associated to *all* elliptic curves over $\mathbb{F}_r$ for $2 \le r \le 13$. See Dummit's article in these proceedings for some examples of Drinfeld modules associated to curves of genus 2.

When this hyperelliptic algorithm is successful, it proves the existence of a Drinfeld $\mathbf{A}$ -module. We know that it is, in fact, always successful because of the analytic constructions that we will now describe.

Definition. A *lattice* is a discrete $\mathbf{A}$ -submodule Γ of $\mathbf{C}$ such that $\mathbf{K}\Gamma$ has finite $\mathbf{K}$ -dimension r_Γ. The integer r_Γ is called the *rank* of Γ. Since $\mathbf{C}$ is an infinite dimensional $\mathbf{K}$ -vector space, lattices of every positive rank exist.

The following theorem is proved in the same way as its analog over $\mathbb{R}$:

Theorem 8.1. *Let Γ be a lattice, and let x be a non-constant element of $\mathbf{A}$. Then Γ is an $\mathbb{F}_r[x]$-module of finite rank. Further, every $\mathbb{F}_r[x]$-basis of Γ is linearly independent over the closure of $\mathbb{F}_r(x)$ in $\mathbf{K}$.*

Definition. A power series $f(z)$ with coefficients in $\mathbf{C}$ is $\mathbb{F}_r$*-linear* if only monomials with exponent a power of r appear in $f(z)$ with non-zero coefficient. Let $D(f)$ denote the coefficient of z in $f(z)$. We call $D(f)$ the *tangent vector* of $f(z)$.

A basic theorem of non-archimedean analysis states that an entire function on $\mathbf{C}$ (i.e. a function defined by an everywhere convergent power series) is determined up to a multiplicative constant by its roots, multiplicities being counted. In particular, we have

Lemma 8.2. *If $f_1(z)$ and $f_2(z)$ are everywhere convergent $\mathbb{F}_r$-linear power series with $D(f_1) = D(f_2) \neq 0$ and if $f_1(z)$ and $f_2(z)$ have the same set of roots, then $f_1(z) = f_2(z)$.*

The *exponential function* associated to the lattice Γ is the function $e_\Gamma(z)$ defined for $z \in \mathbf{C}$ by the infinite product

$$e_\Gamma(z) = z \cdot \prod_{\gamma \in \Gamma - \{0\}} \left(1 - \frac{z}{\gamma}\right). \tag{8.3}$$

The series

$$\sum_{\gamma \in \Gamma - \{0\}} \frac{1}{\gamma}$$

is absolutely convergent in the ∞-adic topology on $\mathbf{C}$, as one sees without difficulty. The infinite product (8.3) therefore converges for all $z \in \mathbf{C}$ and so defines an entire function on $\mathbf{C}$. In fact, we can multiply out this product into an everywhere convergent $\mathbb{F}_r$-linear power series

$$e_\Gamma(z) = z + \sum_{i=1}^{\infty} c_i z^{r^i} \tag{8.4}$$

by considering the partial products of (8.3) over the $\gamma \in \Gamma$ with $-v_\infty(\gamma)$ less than a given bound b. Since such γ's constitute a finite dimensional $\mathbb{F}_r$-vector space, each partial product multiplies out into an $\mathbb{F}_r$-linear polynomial, and so the $\mathbb{F}_r$-linear series (8.4) appears in the limit as $b \to \infty$. All this is fairly easy because we are dealing with a non-archimedean valuation.

Theorem 8.5. *The function $e_\Gamma(z)$ is an entire function on $\mathbf{C}$ with the properties:* 1) *$e_\Gamma(z)$ is a surjective $\mathbb{F}_r$-linear endomorphism of $\mathbf{C}$, and* 2) *$e_\Gamma(z)$ is periodic with Γ as a group of periods.*

Proof. The theorem is immediate from (8.3) and (8.4) if we recall that entire functions on the non-archimedean field $\mathbf{C}$ are necessarily surjective. □

This theorem shows that $\mathbf{C}/\Gamma$ is isomorphic to $\mathbf{C}$ itself by an $\mathbb{F}_r$-linear map.

We now show that $e_\Gamma(z)$ admits complex multiplications by the elements of $\mathbf{A}$. Let Γ' and Γ be a pair of lattices with Γ of finite index in Γ'. As $e_\Gamma(z)$ induces an isomorphism of the $\mathbb{F}_r$-vector spaces Γ'/Γ and $e_\Gamma(\Gamma')$, $e_\Gamma(\Gamma')$ is a finite set. Put

$$P(\Gamma'/\Gamma;t) = t \cdot \prod_{\sigma \in e_\Gamma(\Gamma') - \{0\}} \left(1 - \frac{t}{\sigma}\right). \tag{8.6}$$

Then $P(\Gamma'/\Gamma;t)$ is an $\mathbb{F}_r$-linear polynomial of degree $\#(\Gamma'/\Gamma)$ associated to the pair Γ'/Γ in a canonical way.

Theorem 8.7. *Let $\Gamma'' \supseteq \Gamma' \supseteq \Gamma$ be lattices with $\#(\Gamma''/\Gamma)$ finite. Then*

$$e_{\Gamma'}(z) = P(\Gamma'/\Gamma; e_\Gamma(z)) \tag{8.8}$$

for all $z \in \mathbf{C}$, and

$$P(\Gamma''/\Gamma;t) = P\left(\Gamma''/\Gamma'; P(\Gamma'/\Gamma;t)\right). \tag{8.9}$$

Proof. From the definitions, both sides of (8.8) have the same set of roots in $\mathbf{C}$. Since each is defined by an everywhere convergent power series with tangent vector one, equality holds by Lemma 8.2. The identity (8.9) is an immediate consequence of (8.8) □

Given a lattice Γ and any $x \in \mathbf{A}$, the index $\#(x^{-1}\Gamma/\Gamma)$ is finite by Theorem 8.1. If we set

$$\phi_x^\Gamma(t) = x \cdot P(x^{-1}\Gamma/\Gamma;t), \tag{8.10}$$

then $\phi_x^\Gamma(t)$ is an $\mathbb{F}_r$-linear polynomial of degree $\#(x^{-1}\Gamma/\Gamma)$ and with tangent vector x. Now from the definitions, if $\Gamma' = x^{-1}\Gamma$, then $e_{\Gamma'}(z) = x^{-1} \cdot e_\Gamma(xz)$. Therefore, (8.8) implies that $e_\Gamma(z)$ admits the "complex multiplication"

$$e_\Gamma(xz) = \phi_x^\Gamma(e_\Gamma(z)). \tag{8.11}$$

The $\mathbb{F}_r$-linear polynomial ϕ_x^Γ is unique and canonically defined.

From (8.9), (8.11) and the uniqueness,

$$\phi_{x+y}^\Gamma(t) = \phi_x^\Gamma(t) + \phi_y^\Gamma(t)$$

and

$$\phi_{xy}^\Gamma(t) = \phi_x^\Gamma(\phi_y^\Gamma(t))$$

for all elements x and y in $\mathbf{A}$. These two identities show that the polynomials ϕ_x^Γ (t) provide a faithful representation of $\mathbf{A}$ as a ring of $\mathbb{F}_r$-linear endomorphisms of the $\mathbb{F}_r$-vector space $\mathbf{C}$. It is clear that this representation factors through the twisted polynomial ring $\mathbf{C}[\mathbf{F}^m]$. Thus $\phi^\Gamma : x \mapsto \phi_x^\Gamma$ is a Drinfeld $\mathbf{A}$-module over $\mathbf{C}$!

What is the rank of ϕ^Γ? Let us compute $\deg \phi_x^\Gamma(t)$ in terms of $\mathrm{v}_\infty(x)$:

$$\deg \phi_x^\Gamma(t) = \#(x^{-1}\Gamma/\Gamma) = \#(\Gamma/x\Gamma) = r^k$$

by Theorem 8.1 where

$$k = \operatorname{rank}_{\mathbb{F}_r[x]}\Gamma = r_\Gamma \cdot \operatorname{rank}_{\mathbb{F}_r[x]}\mathbf{A} = r_\Gamma \cdot [\mathbf{k} : \mathbb{F}_r(x)].$$

Now since ∞ is the only pole of x, $-d_\infty \cdot \mathrm{v}_\infty(x) = [\mathbf{k} : \mathbb{F}_r(x)]$. Thus, we have proved

Theorem 8.12. *Each lattice* Γ *defines a Drinfeld* $\mathbf{A}$*-module* ϕ^Γ *over* $\mathbf{C}$ *via the complex multiplications* (8.11). *The rank of* ϕ^Γ *equals the rank* r_Γ *of* Γ.

We may now state our fundamental result, the *analytic uniformization theorem* for Drinfeld $\mathbf{A}$-modules over $\mathbf{C}$.

Theorem 8.13. *Let* ϕ *be a Drinfeld* $\mathbf{A}$*-module over* $\mathbf{C}$. *Then there is a uniquely determined lattice* Γ *in* $\mathbf{C}$ *such that* $\phi = \phi^\Gamma$.

Sketch of proof. Let

$$\lambda_\phi = \sum_{i=0}^{\infty} c_i \mathbf{F}^{mi}$$

be the formal twisted power series associated to ϕ by Proposition 7.5. One proves first that the $\mathbb{F}_r$-linear power series

$$\lambda_\phi(z) = \sum_{i=0}^{\infty} c_i z^{r^i}$$

converges for all $z \in \mathbf{C}$. This follows from any one of the recurrences (7.2) with $\sigma = \phi_y$, y a non-constant element of $\mathbf{A}$. The identity $\lambda_\phi \cdot x = \phi_x \lambda_\phi$ is then equivalent to the complex multiplications (8.11) with $\lambda_\phi(z)$ replacing $e_\Gamma(z)$. It remains then to show that the set of roots of λ_ϕ is a lattice Γ. This follows from the fact that $\lambda_\phi(z)$ admits complex multiplications by the elements of $\mathbf{A}$. For the details, see §§4 and 5 of **[H2]**. □

We leave it to the reader to check that $\xi \in \mathbf{C}$ is an isomorphism from ϕ^Γ to $\phi^{\Gamma'}$ (i.e. $\phi_x^{\Gamma'} = \xi\phi^\Gamma\xi^{-1}$) if and only if $\Gamma' = \xi\Gamma$. In fact, there is a natural isomorphism of categories between $\mathrm{Drin}_{\mathbf{A}}(\mathbf{C})$ and the category whose objects are the lattices in $\mathbf{C}$ and whose set of morphisms from Γ to Γ' is given by $\mathrm{Hom}(\Gamma, \Gamma') = 0$ if $r_\Gamma \neq r_{\Gamma'}$ and otherwise by

$$\mathrm{Hom}(\Gamma, \Gamma') = \{z \in \mathbf{C} : z\Gamma \subseteq \Gamma'\}.$$

Finally, we ask how can one understand the action $*$ of §4 in this analytic setting? The answer is given by the following theorem, which we state without proof.

Theorem 8.14. *Let* Γ *be a lattice, and let* $\mathfrak{a}$ *be a non-zero ideal in* $\mathbf{A}$. *Put* $\phi' = \mathfrak{a} * \phi^\Gamma$, *and let* Γ' *be the lattice associated to* ϕ' *by the analytic uniformization theorem. Then*

$$\Gamma' = D(\phi_{\mathfrak{a}}^\Gamma) \cdot \mathfrak{a}^{-1}\Gamma. \tag{8.15}$$

9. Uniformizing isomorphism classes in $\mathbf{Drin_A}(\mathbf{C})$. We call two lattices Γ and Γ' *homothetic* if $\Gamma' = \xi\Gamma$ for some element $\xi \in \mathbf{C}^\times$. Let $\mathbf{Lat}_k$ be the set of lattices of rank k in $\mathbf{C}$, and let $\mathfrak{L}_k$ be the set of homothety classes in $\mathbf{Lat}_k$. The map

$$\Gamma \mapsto \mathfrak{a}^{-1}\Gamma \tag{9.1}$$

defines an action of fractional ideals $\mathfrak{a}$ on $\mathbf{Lat}_k$, and this action clearly descends to an action of $\mathbf{Pic}(\mathbf{A})$ on $\mathfrak{L}_k$. Since every rank one lattice is homothetic to a fractional ideal of $\mathbf{A}$, we have

Proposition 9.2. *The set of homothety classes* $\mathcal{L}_1$ *is a principal homogeneous space for* **Pic**$(\mathbf{A})$. *In particular,* $\mathcal{L}_1$ *is finite and* $\#(\mathcal{L}_1) = h(\mathbf{A})$.

We remind the reader that a *principal homogeneous space* for a given group G is a space on which G acts transitively and faithfully.

Let $\mathcal{D}_k$ be the set of isomorphism classes of rank k Drinfeld $\mathbf{A}$-modules over $\mathbf{C}$. The analytic uniformizing theorem implies that the map $\Gamma \mapsto \phi^\Gamma$ descends to a natural bijective correspondence from $\mathcal{L}_k$ to $\mathcal{D}_k$. Theorem 8.14 says that this map is an isomorphism from $\mathcal{L}_k$ as a representation space for **Pic**$(\mathbf{A})$ via (9.1) to $\mathcal{D}_k$ as a representation space for **Pic**$(\mathbf{A})$ under the $*$ operation. We may then restate Proposition 9.2 as follows.

Theorem 9.3. *There are exactly* $h(\mathbf{A})$ *isomorphism classes of rank one Drinfeld* $\mathbf{A}$*-modules over* $\mathbf{C}$, *and* $\mathcal{D}_1$ *is a principal homogeneous space for* **Pic**$(\mathbf{A})$ *under the* $*$ *action.*

Part II: The Rank One theory with Applications to the Class Field Theory of k

In problem twelve of his famous list, Hilbert asks for an *explicit* construction of the class field theory over *any* given number field k. Hilbert was aware of the cyclotomic theory over $k = \mathbb{Q}$, in which class fields are constructed by adjoining the special values of $e^{2\pi i z}$ at all $z \in \mathbb{Q}$. He was aware also of the construction of the class fields of an imaginary quadratic number field k by adjoining special values of elliptic functions. Nowadays we understand that, in both examples, the class fields are constructed by adjoining the division points of the action of the ring of integers $\mathcal{O}_k$ of k on some algebraic group G. In the cyclotomic theory, G is the multiplicative group G_m. Over an imaginary quadratic number field k, G is an elliptic curve which admits complex multiplications by $\mathcal{O}_k$.

For the function field $\mathbf{k}$, the analog of Hilbert's problem twelve is solved by adjoining division points of rank one Drinfeld modules acting on $G = G_a$. More precisely, the class fields of $\mathbf{k}$ in which our fixed place ∞ splits completely are constructed by adjoining division points of a suitably normalized Drinfeld $\mathbf{A}$-module over $\mathbf{C}$ of rank one (see §16). One obtains all the class fields of $\mathbf{k}$ by varying ∞.

Throughout Part II, except where explicitly stated otherwise, $\delta\colon \mathbf{A} \to \mathbf{C}$ is the inclusion map.

10. Fields of definition in C. Let ϕ be a Drinfeld $\mathbf{A}$-module over $\mathbf{C}$ of any rank. The Galois group $\mathrm{Gal}(\mathbf{C}/\mathbf{k})$ acts as a group of automorphisms of the twisted polynomial ring $\mathbf{C}[\mathbf{F}]$. Following the map $x \mapsto \phi_x$ by the action of an element $\sigma \in \mathrm{Gal}(\mathbf{C}/\mathbf{k})$ therefore defines a new Drinfeld module $\sigma\phi$ of the same rank as that of ϕ. Thus, $\mathrm{Gal}(\mathbf{C}/\mathbf{k})$ acts on $\mathrm{Drin}_{\mathbf{A}}(\mathbf{C})$, and this action induces an action of $\mathrm{Gal}(\mathbf{C}/\mathbf{k})$ on the set of isomorphism classes of Drinfeld $\mathbf{A}$-modules over $\mathbf{C}$. As the reader can check from the definitions,

$$\mathfrak{a} * \sigma\phi = \sigma(\mathfrak{a} * \phi) \tag{10.1}$$

for all $\sigma \in \mathrm{Gal}(\mathbf{C}/\mathbf{k})$ and every non-zero ideal $\mathfrak{a}$ in $\mathbf{A}$.

We say that ϕ is *defined over* a subextension $K/\mathbf{k}$ of $\mathbf{C}/\mathbf{k}$ if $\phi_x \in K[\mathbf{F}]$ for every $x \in \mathbf{A}$. A subfield K of $\mathbf{C}$ is a *field of definition* for ϕ if ϕ is isomorphic over $\mathbf{C}$ to a Drinfeld module ϕ' which is defined over K. As is clear from the definitions and (5.8), if ϕ is defined over K, then $\mathfrak{a} * \phi$ is also defined over K for any non-zero ideal $\mathfrak{a}$ in $\mathbf{A}$.

Proposition 10.2. *If ϕ has rank one, then the completion $\mathbf{K}$ of $\mathbf{k}$ at ∞ is a field of definition for ϕ.*

Proof. By Theorem 8.13, $\phi = \phi^\Gamma$ for a lattice Γ of rank one over $\mathbf{A}$. Because $r_\Gamma = 1$, Γ is isomorphic to a fractional ideal of $\mathbf{A}$, and so there is an element $\xi \in \mathbf{C}$ such that $\Gamma' = \xi\Gamma \subset \mathbf{k} \subset \mathbf{K}$. Therefore, the analytic constructions in §8 all take place within $\mathbf{K}$, which shows that $\phi^{\Gamma'} = \xi\phi\xi^{-1}$ is indeed defined over $\mathbf{K}$. □

Theorem 10.3. *There exists a field of definition K_ϕ, finitely generated over $\mathbf{k}$, which is contained in every field of definition for ϕ.*

Proof. For $x \in \mathbf{A}$, let

$$\phi_x = x + \sum_{\nu=1}^{r_\phi \deg x} c_\nu(\phi, x) \cdot \mathbf{F}^{m\nu}. \tag{10.4}$$

Then for any $\xi \in \mathbf{C}$, we have

$$c_\nu(\xi\phi\xi^{-1}, x) = \xi^{1-r^\nu} \cdot c_\nu(\phi, x). \tag{10.5}$$

Fix a non-constant element $y \in \mathbf{A}$, and let $\nu = \nu_1, \ldots, \nu_s$ be the indices of the coefficients $c_\nu(\phi, y)$ of ϕ_y which are not zero. Let g be the greatest common divisor of the set of integers $\{r^{\nu_i} - 1 : 1 \le i \le s\}$ and write

$$g = \sum_{j=1}^{s} e_j \cdot (r^{\nu_j} - 1) \qquad (e_j \in \mathbb{Z}).$$

From (10.5), it is clear that the elements

$$I_{\nu_i}(\phi, y) = c_{\nu_i}(\phi, y) \cdot \left(\prod_{j=1}^{s} c_{\nu_j}(\phi, y)^{e_j}\right)^{(1-r^{\nu_i})/g} \in \mathbf{C} \tag{10.6}$$

for $1 \le i \le s$ depend only on the isomorphism class of ϕ and hence belong to every field of definition for ϕ. Let K_ϕ be the field generated over $\mathbf{k}$ by the elements $I_{\nu_i}(\phi, y)$. We will show that K_ϕ is a field of definition for ϕ.

Choose $\xi \in \mathbf{C}$ so that

$$\xi^g = \prod_{j=1}^{s} c_{\nu_j}(\phi, y)^{e_j}. \tag{10.7}$$

For $1 \le i \le s$, we may rewrite (10.6) as

$$I_{\nu_i}(\phi, y) = \xi^{1-r^{\nu_i}} \cdot c_{\nu_i}(\phi, y),$$

so that $\xi\phi_y\xi^{-1}$ has coefficients in K_ϕ. It follows now from (7.2) and Proposition 7.5 that $\xi\phi_x\xi^{-1}$ has coefficients in K_ϕ for all $x \in \mathbf{A}$. Thus, ϕ is defined over K_ϕ. □

We call K_ϕ the *smallest field of definition* for ϕ.

For ϕ of rank one, we prove in §15 below that K_ϕ is the maximal unramified abelian extension $H_{\mathbf{A}}$ of $\mathbf{k}$ in which ∞ splits completely. For rank one $\mathbf{A}$-modules, therefore, K_ϕ is independent of the choice of ϕ. We call $H_{\mathbf{A}}$ the *Hilbert Class Field* of $\mathbf{A}$. We go further in §15 and prove the following analog of Hasse's Theorem (cf. [**S**]).

Theorem 10.8. *Let $\mathfrak{p}$ be a non-zero prime ideal in $\mathbf{A}$, and let $\mathbf{Frob}_\mathfrak{p}$ be the Frobenius automorphism of $H_{\mathbf{A}}$ at $\mathfrak{p}$. For any rank one Drinfeld module ϕ defined over $H_{\mathbf{A}}$, we have*

$$\mathbf{Frob}_\mathfrak{p}(\phi) \cong \mathfrak{p} * \phi. \tag{10.9}$$

The proof of (10.9) depends upon the theory of *reduction* of Drinfeld modules, which we introduce in the next section.

11. Reduction of Drinfeld modules over C. Let $K/\mathbf{k}$ be a subextension of $\mathbf{C}/\mathbf{k}$. A discrete valuation ring in $\mathbf{C}$ is *finite* if it contains $\mathbf{A}$. Let $\mathcal{O}_\mathfrak{P}$ be a finite discrete valuation ring in $\mathbf{C}$ with field of fractions K, let $\mathfrak{P}$ be the maximal ideal of $\mathcal{O}_\mathfrak{P}$, and let $\mathrm{v}_\mathfrak{P}$ be the normalized discrete valuation on K with valuation ring $\mathcal{O}_\mathfrak{P}$. We form the twisted polynomial subring $\mathcal{O}_\mathfrak{P}[\mathbf{F}]$ of $K[\mathbf{F}]$ and equip it with the canonical reduction morphism $\mathbf{red}_\mathfrak{P}: \mathcal{O}_\mathfrak{P}[\mathbf{F}] \to (\mathcal{O}_\mathfrak{P}/\mathfrak{P})[\mathbf{F}]$. Let $\phi \in \mathrm{Drin}_{\mathbf{A}}(\mathbf{C})$ be a Drinfeld $\mathbf{A}$-module which is defined over K.

Definition. We say that ϕ has *stable reduction modulo* $\mathfrak{P}$ (or *over* $\mathcal{O}_\mathfrak{P}$) if there is an element $\xi \in K$, $\xi \neq 0$, such that $\phi'_x = \xi\phi_x\xi^{-1} \in \mathcal{O}_\mathfrak{P}[\mathbf{F}]$ for all $x \in \mathbf{A}$ and such that $\mathbf{red}_\mathfrak{P} \circ \phi'$ is a Drinfeld $\mathbf{A}$-module over $\mathcal{O}_\mathfrak{P}/\mathfrak{P}$. We say that ϕ has *good reduction modulo* $\mathfrak{P}$ (or *over* $\mathcal{O}_\mathfrak{P}$) if, in addition, the ranks of ϕ and $\mathbf{red}_\mathfrak{P} \circ \phi'$ are equal.

If ϕ has rank one, then every stable reduction is good.

It may happen that a given ϕ has no stable (resp. good) reduction over $\mathcal{O}_\mathfrak{P}$. In that case, we seek a finite extension field K' of K in $\mathbf{C}$ and a valuation ring $\mathcal{O}_{\mathfrak{P}'}$ with field of fractions K' and maximal ideal $\mathfrak{P}'$ which sits over $\mathfrak{P}$ (i.e. $\mathcal{O}_\mathfrak{P} \cap \mathfrak{P}' = \mathfrak{P}$) such that ϕ has stable (resp. good) reduction over $\mathcal{O}_{\mathfrak{P}'}$. When this strategy succeeds, we say that ϕ has *potential stable* (resp. *potential good*) *reduction modulo* $\mathfrak{P}$.

Theorem 11.1. *A Drinfeld $\mathbf{A}$-module ϕ over K always has potential stable reduction modulo $\mathfrak{P}$.*

Proof. Let $x_1, \ldots, x_s$ generate $\mathbf{A}$ as an $\mathbb{F}_r$-algebra. For $1 \leq i \leq s$ and all $\nu > 0$, let $c_\nu(\phi, x_i)$ be the coefficient of $\mathbf{F}^{m\nu}$ in ϕ_{x_i}. Then

$$k = \max\left\{\frac{-\mathrm{v}_\mathfrak{P}(c_\nu(\phi, x_i))}{r^\nu - 1}\right\},$$

where the max is over $1 \leq i \leq s$ and $1 \leq \nu \leq \deg\phi_{x_i}$ is a rational number. There exists an extension field K' of K which has a valuation $\mathrm{v}_{\mathfrak{P}'}$ extending $\mathrm{v}_\mathfrak{P}$ with ramification number equal to the denominator of k. Clearly, ϕ has stable reduction modulo $\mathfrak{P}'$. □

Let μ_ϕ be the leading coefficient function of ϕ as defined in §6.

Definition. Let B be a subring of K. We say that ϕ is *defined over* B if for all $x \in \mathbf{A}$, $\phi_x \in B[\mathbf{F}]$ and $\mu_\phi(x)$ is a unit of B.

Proposition 11.2. *Suppose ϕ is defined over $\mathcal{O}_\mathfrak{P}$. Then for every non-zero ideal $\mathfrak{a}$ in $\mathbf{A}$, $\phi_\mathfrak{a} \in \mathcal{O}_\mathfrak{P}[\mathbf{F}]$, and $\mathfrak{a} * \phi$ is defined over $\mathcal{O}_\mathfrak{P}$.*

Proof. Good exercise for the reader. Note that the roots of $\phi_\mathfrak{a}(t)$ are all integral over $\mathcal{O}_\mathfrak{P}$. □

Proposition 11.3. *Suppose $\mu_\phi(y)$ is a unit of the finite valuation ring $\mathcal{O}_\mathfrak{P}$ for some non-constant $y \in \mathbf{A}$. If ϕ has rank one, then ϕ is defined over $\mathcal{O}_\mathfrak{P}$.*

Proof. Since $r_\phi = 1$, we know from Theorem 11.1 that there is a finite extension field K' of K and an element $\xi \in K'$ such that $\phi' = \xi\phi\xi^{-1}$ has good reduction over a valuation ring $\mathcal{O}_{\mathfrak{P}'}$ in K' which sits over $\mathcal{O}_\mathfrak{P}$. The $\mathbf{A}$-module ϕ' is therefore defined over $\mathcal{O}_{\mathfrak{P}'}$. Since $\mu_{\phi'}(y) = \xi^{1-N(y)} \cdot \mu_\phi(y)$, we see that $\mathrm{v}_{\mathfrak{P}'}(\xi) = 0$. We conclude that the coefficients of $\phi_x = \xi^{-1}\phi'_x\xi$ belong to $\mathcal{O}_\mathfrak{P}$ for all $x \in \mathbf{A}$. Further, (6.2) implies that $\mathrm{v}_\mathfrak{P}(\mu_\phi(x)) = 0$ for all $x \in \mathbf{A}$. □

Proposition 11.4. *Assume that $K/\mathbf{k}$ is finite. Suppose ϕ has rank one and is defined over a finite valuation ring $\mathcal{O}_\mathfrak{P}$ in K which is unramified over $\mathbf{k}$. Let $\mathfrak{p} = \mathfrak{P} \cap \mathbf{A}$, a non-zero prime ideal in $\mathbf{A}$. For $e \geq 1$, put $\mathfrak{a} = \mathfrak{p}^e$ and $\mathfrak{b} = \mathfrak{p}^{e-1}$. Then $\phi_\mathfrak{b}(t)$ divides $\phi_\mathfrak{a}(t)$ in $\mathcal{O}_\mathfrak{P}[t]$, and the quotient is Eisenstein at $\mathfrak{p}$.*

Proof. We first settle the case $e = 1$. The reduction of ϕ modulo $\mathfrak{P}$ has characteristic $\mathfrak{p}$, which implies by Corollary 5.9 that every coefficient of $\phi_\mathfrak{p}$ other than its leading coefficient belongs to $\mathfrak{P}$. Therefore, in order to prove that $\phi_\mathfrak{p}(t)/t$ is Eisenstein at $\mathfrak{P}$, it suffices to show that

$$\mathrm{v}_\mathfrak{P}(D(\phi_\mathfrak{p})) \leq 1.$$

Choose $x \in \mathbf{A}$ so that $\mathrm{v}_\mathfrak{P}(x) = 1$, and let $x\mathbf{A} = \mathfrak{d}\mathfrak{p}$. Then by (4.6),

$$\phi_x = \mu_\phi(x) \cdot (\mathfrak{p} * \phi)_\mathfrak{d} \cdot \phi_\mathfrak{p}$$

so that

$$1 = \mathrm{v}_\mathfrak{P}(D(\phi_x)) = \mathrm{v}_\mathfrak{P}(D(\mathfrak{p} * \phi)_\mathfrak{d}) + \mathrm{v}_\mathfrak{P}(D(\phi_\mathfrak{p})) \geq \mathrm{v}_\mathfrak{P}(D(\phi_\mathfrak{p}))$$

by (6.1) and Proposition 11.2.

For $e > 1$, we have

$$\phi_\mathfrak{a}(t) = f(\phi_\mathfrak{b}(t)) \cdot \phi_\mathfrak{b}(t)$$

where $f(t) = (\mathfrak{b} * \phi)_\mathfrak{p}(t)/t$. We know that $f(t)$ is Eisenstein at $\mathfrak{P}$, and $\phi_\mathfrak{b}(t) \equiv t^{N(\mathfrak{b})}$ (mod $\mathfrak{P}$) by Corollary 5.9. □

12. Normalization via a sign-function. Much of the theory of Drinfeld modules reminds one of the theory of elliptic curves. For this reason, Drinfeld called such modules *elliptic modules* when he introduced the concept in **[Dr1]**. The theory of elliptic curves with complex multiplication clearly motivates the development of the theory of rank one $\mathbf{A}$-modules sketched in §10. The proof of Theorem 10.8 in **[H2]** follows the classical Deuring proof of its analog over an imaginary quadratic number field.

In this and subsequent sections, we will follow a simpler path to Theorem 10.8. The idea is to introduce a sign-function sgn and use it to define an extension $\mathbf{Pic}^+(\mathbf{A})$ of $\mathbf{Pic}(\mathbf{A})$ which is a kind of *narrow* ray class group. The addition of the sign-function rigidifies the structures so that one can avoid having to deal with isomorphism classes of rank one $\mathbf{A}$-modules. One works instead with 'sgn-normalized' $\mathbf{A}$-modules, and this leads to a richer theory and simpler proofs. Of course, narrow ray class groups do not exist over an imaginary quadratic number field.

From now on, the term '$\mathbf{A}$-module' will refer to a Drinfeld $\mathbf{A}$-module over $\mathbf{C}$ of rank one. We define $W_\infty = r^d - 1$, where $d = d_\infty$ is the degree of ∞. It is convenient also to put $W_{\mathbf{k}} = r - 1$.

Let $\mathbf{U}_\infty$ be the group of units in the valuation ring of the local field $\mathbf{K}$, and let $\mathbf{U}_\infty^{(1)}$ be the subgroup of units which are congruent to one modulo ∞. Let $\mathbb{F}_\infty \cong \mathbb{F}_{r^d}$ be the field of constants in $\mathbf{K}$. The sign-function sgn: $\mathbf{K} \to \mathbb{F}_\infty$ introduced at the beginning of Part I depends upon the choice of the uniformizer π. Here is a more natural definition of sign-functions.

Definition. A *sign-function* sgn: $\mathbf{K}^\times \to \mathbb{F}_\infty^\times$ is a co-section of the inclusion map $\mathbb{F}_\infty^\times \hookrightarrow \mathbf{K}^\times$ such that $\mathrm{sgn}(\mathbf{U}_\infty^{(1)}) = 1$. In addition, we put $\mathrm{sgn}(0) = 0$. Let σ be an $\mathbb{F}_r$-automorphism of $\mathbb{F}_\infty$. The composite map $\sigma \circ \mathrm{sgn}$ is called a *twisted sign-function* or a *twisting of* sgn *by* σ.

Proposition 12.1. *Let* sgn *and* sgn$'$ *be sign-functions on* $\mathbf{K}$. *Then there is an element* $\zeta \in \mathbb{F}_\infty^\times$ *such that*

$$\mathrm{sgn}(x) = \mathrm{sgn}'(x) \cdot \zeta^{\deg x/d}$$

for all $x \in \mathbf{K}$.

Proof. The map $x \mapsto \mathrm{sgn}(x)/\mathrm{sgn}'(x)$, $x \in \mathbf{K}^\times$, is trivial on $\mathbf{U}_\infty$ and therefore factors through $\mathrm{v}_\infty\colon \mathbf{K}^\times \to \mathbb{Z}$. Now, $\deg x = -d \cdot \mathrm{v}_\infty(x)$. □

Corollary 12.2. *There are exactly* W_∞ *sign-functions on* $\mathbf{K}$.

Definition. Let $\phi \in \mathrm{Drin}_{\mathbf{A}}(\mathbf{C})$. We say that ϕ is *normalized* if the leading coefficient $\mu_\phi(x)$ of ϕ_x belongs to $\mathbb{F}_\infty$ for all $x \in \mathbf{A}$. If for some sign-function sgn, the map $x \mapsto \mu_\phi(x)$ is a twisting of sgn, then we say that ϕ is sgn-*normalized.*

The distinguished place ∞ of our function field $\mathbf{k}$ has been fixed throughout our discussion. We now choose a distinguished sign-function sgn. Our base object becomes a triple $(\mathbf{k}, \infty, \mathrm{sgn})$ that we think of as analogous to $\mathbb{Q}$ with its archimedean place together with the usual sign on the completion $\mathbb{R}$. It is this analogy that suggests the term *positive* for the elements $x \in \mathbf{A}$ such that $\mathrm{sgn}(x) = 1$.

Theorem 12.3. *Every* $\mathbf{A}$*-module* ϕ' *is isomorphic over* $\mathbf{C}$ *to a* sgn*-normalized* $\mathbf{A}$*-module* ϕ.

Proof. Let π be a uniformizer of $\mathbf{K}$ which is positive for sgn. Choose $\xi \in \mathbf{C}$ so that

$$\xi^{W_\infty} = 1/\mu_{\phi'}(\pi^{-1}),$$

where $\mu_{\phi'}$ is the leading coefficient function introduced in §6. If $\phi = \xi\phi'\xi^{-1}$, then

$$\mu_\phi(\pi^{-1}) = 1 \tag{12.4}$$

by (6.6). Since π is positive for sgn, for any $x \in \mathbf{K}^\times$ we have

$$\operatorname{sgn}(x) = c_n \in \mathbb{F}_\infty$$

where $n = \mathrm{v}_\infty(x) \leq 0$ and c_n is defined by (0.1). Therefore, for $x \in \mathbf{A}$ we have

$$\mu_\phi(x) = \mu_\phi(c_n\pi^n) = i_\phi(c_n) = i_\phi(\operatorname{sgn}(x))$$

by (6.4) and (12.4). Thus, ϕ is sgn-normalized with the isomorphism invariant map i_ϕ determining the twist of sgn. □

If ϕ is sgn-normalized, then the leading coefficient function $\mu_\phi\colon \mathbf{A} \to \mathbb{F}_\infty$ extends naturally to $\mathbf{k}$. In subsequent sections, we will write $\mu_\phi(x)$ for $x \in \mathbf{k}$ without further explanations.

13. The narrow class group Pic$^+$(A). How many sgn-normalized $\mathbf{A}$-modules are there? Let us first compute the number of sgn-normalized $\mathbf{A}$-modules in each isomorphism class in $\mathcal{D}_1$ (see §9).

Proposition 13.1. *Let* ϕ *be* sgn*-normalized. If* $\phi' = \xi\phi\xi^{-1}$ *is also* sgn*-normalized, then* $\xi \in \mathbb{F}_\infty^\times$ *and* $\mu_\phi = \mu_{\phi'}$.

Proof. Since $\mu_\phi(\pi^{-1}) = \mu_{\phi'}(\pi^{-1}) = 1$, we have $\xi \in \mathbb{F}_\infty$ by (6.6). That $\mu_\phi = \mu_{\phi'}$ is clear since $\deg\phi_x = \deg x$ is divisible by d. □

Corollary 13.2. *Let*

$$\varkappa = \frac{r^d - 1}{r - 1} = \frac{W_\infty}{W_{\mathbf{k}}}. \tag{13.3}$$

Then each isomorphism class in $\mathcal{D}_1$ *contains exactly* $\varkappa$ sgn*-normalized* $\mathbf{A}$*-modules.*

Proof. We know by Theorem 12.3 that each class in $\mathcal{D}_1$ contains a sgn-normalized $\mathbf{A}$-module ϕ. For each $\xi \in \mathbb{F}_\infty^\times$, $\xi\phi\xi^{-1}$ is clearly sgn-normalized. Since $\operatorname{Aut}(\phi) = \mathbb{F}_r^\times$, such ξ produce exactly $\varkappa$ distinct $\mathbf{A}$-modules isomorphic to ϕ. By Proposition 13.1, these are the only possibilities. □

Corollary 13.4. *There are exactly* $\varkappa \cdot h(\mathbf{A})$ sgn*-normalized* $\mathbf{A}$*-modules.*

Let Γ be any rank one lattice in $\mathbf{C}$. By Theorem 12.3 and the analytic theory, there is an element $\xi(\Gamma)$ in $\mathbf{C}$ such that $\Gamma' = \xi(\Gamma)\cdot\Gamma$ is the lattice of a sgn-normalized $\mathbf{A}$-module $\phi^{\Gamma'}$. By Proposition 13.1, the element $\xi(\Gamma)$ is determined up to multiplication by elements of $\mathbb{F}_\infty^\times$. In fact, the $\varkappa$ lattices $\zeta\Gamma'$ for $\zeta \in \mathbb{F}_\infty^\times$ are precisely the lattices associated to the sgn-normalized $\mathbf{A}$-modules which are isomorphic to $\phi^{\Gamma'}$. In spite of this ambiguity, we call $\xi(\Gamma)$ the *invariant* of Γ. The element

$$\Delta(\Gamma) = \xi(\Gamma)^{W_\infty}$$

is unambiguously determined by Γ. It is the analog of the classical discriminant function on rank two $\mathbb{Z}$-lattices in $\mathbb{C}$. Since for all $z \in \mathbf{C}$

$$\Delta(z\Gamma) = z^{-W_\infty} \cdot \Delta(\Gamma),$$

we say that Δ has *weight* W_∞. Jing Yu **[Y]** has shown that the invariants $\xi(\mathfrak{a})$ associated to fractional ideals $\mathfrak{a}$ of $\mathbf{A}$ are transcendental over $\mathbf{k}$. See **[Go1]** and **[Ge1]** for the interpretation of Δ as a cusp form on the Drinfeld upper half-plane. For the application of the invariants $\xi(\mathfrak{a})$ to the special values of L-functions at $s = 0$ and to analytic class number formulas, see **[G–R]**, **[H3]**, **[H4]** and **[Gr–R]**.

Let $\mathbf{X}$ be the set of sgn-normalized $\mathbf{A}$-modules. We leave to the reader the pleasure of checking that $\mathbf{X}$ is preserved under the $*$ action of ideals in $\mathbf{A}$ and the action of $\mathrm{Gal}(\mathbf{C}/\mathbf{k})$ (cf. §10). The one action commutes with the other by (10.1). Which ideals $\mathfrak{a}$ in $\mathbf{A}$ stabilize a given point $\phi \in \mathbf{X}$? Since $\mathbf{Pic}(\mathbf{A})$ acts transitively on $\mathcal{D}_1$, $\mathfrak{a} = x\mathbf{A}$ must be a principal ideal. Further, we see from Lemma 4.4 that $\mu_\phi(x) \in \mathrm{Aut}(\phi) = \mathbb{F}_r^\times$. Thus, the stabilizer of ϕ is the monoid of principal ideals

$$\{x\mathbf{A} : x \in \mathbf{A},\ \mu_\phi(x) \in \mathbb{F}_r^\times\} = \{x\mathbf{A} : x \in \mathbf{A},\ \mathrm{sgn}(x) = 1\}.$$

For any fractional ideal $\mathfrak{a} \in \mathfrak{I}(\mathbf{A})$, there is a positive $x \in \mathbf{A}$ such that $x\mathfrak{a} \subseteq \mathbf{A}$. The action of the monoid of integral ideals in $\mathfrak{I}(\mathbf{A})$ on $\mathbf{X}$ therefore extends to an action of $\mathfrak{I}(\mathbf{A})$ itself on $\mathbf{X}$. The stabilizer of ϕ under this extended action is, of course, the subgroup

$$\mathcal{P}^+(\mathbf{A}) = \{x\mathbf{A} : x \in \mathbf{k},\ \mathrm{sgn}(x) = 1\} \tag{13.5}$$

consisting of those fractional ideals which can be generated by a positive element of $\mathbf{k}$.

Definition 13.6. The quotient group $\mathbf{Pic}^+(\mathbf{A}) = \mathfrak{I}(\mathbf{A})/\mathcal{P}^+(\mathbf{A})$ is called the *narrow class group* of $\mathbf{A}$ relative to sgn.

The action of $\mathfrak{I}(\mathbf{A})$ on $\mathbf{X}$ descends to an action of $\mathbf{Pic}^+(\mathbf{A})$ on $\mathbf{X}$. Since sgn is a multiplicative map, the index of $\mathcal{P}^+(\mathbf{A})$ in the subgroup of all principal ideals is $\varkappa$. It follows that

$$h^+(\mathbf{A}) \stackrel{\mathrm{def}}{=} \#\left(\mathbf{Pic}^+(\mathbf{A})\right) = \varkappa \cdot h(\mathbf{A}) = \#(\mathbf{X}) \tag{13.7}$$

by Corollary 13.4. We have now proved the following fundamental result.

Theorem 13.8. *The set* $\mathbf{X}$ *of* sgn-*normalized* $\mathbf{A}$-*modules is a principal homogeneous space for* $\mathbf{Pic}^+(\mathbf{A})$ *under the* $*$ *action.*

14. The normalizing field $H_{\mathbf{A}}^{+}$. Let $\phi \in \mathbf{X}$ be a sgn-normalized $\mathbf{A}$-module. Let y be any non-constant element of $\mathbf{A}$, and let $H_{\mathbf{A}}^{+}$ be the field generated over $\mathbf{k}$ by the coefficients of ϕ_y. By (7.2) and Proposition 7.5, $H_{\mathbf{A}}^{+}$ is independent of the choice of y. All the $\mathbf{A}$-modules $\mathfrak{a} * \phi$ for non-zero ideals $\mathfrak{a}$ in $\mathbf{A}$ are defined over $H_{\mathbf{A}}^{+}$, and this set comprises *all* the sgn-normalized $\mathbf{A}$-modules by Theorem 13.8. The extension $H_{\mathbf{A}}^{+}/\mathbf{k}$ is therefore independent of the choice of ϕ. However, it does depend upon the choice of sgn.

Definition. We call $H_{\mathbf{A}}^{+}$ the *normalizing field* for rank one Drinfeld $\mathbf{A}$-modules over the base triple $(\mathbf{k}, \infty, \mathrm{sgn})$.

Since the $\mathbf{A}$-modules $\sigma\phi$ for $\sigma \in \mathrm{Gal}(\mathbf{C}/\mathbf{k})$ are sgn-normalized, they are all defined over $H_{\mathbf{A}}^{+}$. We see that $H_{\mathbf{A}}^{+}$ contains all the conjugates of its finite set of generators. The extension $H_{\mathbf{A}}^{+}/\mathbf{k}$ is therefore *finite* and *normal.*

Proposition 14.1. *The extension $H_{\mathbf{A}}^{+}/\mathbf{k}$ is Galois with an abelian Galois group isomorphic to a subgroup of* $\mathbf{Pic}^{+}(\mathbf{A})$.

Proof. We show first that $H_{\mathbf{A}}^{+}/\mathbf{k}$ is *separable.* Of course, $H_{\mathbf{A}}^{+}$ contains the smallest field of definition K_ϕ of ϕ. Now, $K_\phi/\mathbf{k}$ is an algebraic subextension of $\mathbf{K}/\mathbf{k}$ (cf. Proposition 10.2). Since every element of $\mathbf{K}$ which is algebraic over $\mathbf{k}$ is separable over $\mathbf{k}$, the extension $K_\phi/\mathbf{k}$ is separable. Let $\xi \in \mathbf{C}$ be an isomorphism which takes ϕ to an $\mathbf{A}$-module $\phi' = \xi\phi\xi^{-1}$ which is defined over K_ϕ. If $x \in \mathbf{A}$ is non-constant and positive, then by (6.5)

$$\xi^{1-N(x)} = \mu_{\phi'}(x) \in K_\phi,$$

which implies that $K_\phi(\xi)/K_\phi$ is separable. Our extension $H_{\mathbf{A}}^{+}/\mathbf{k}$ is a subextension of $K_\phi(\xi)/\mathbf{k}$ and hence is also separable. Since we already know that $H_{\mathbf{A}}^{+}/\mathbf{k}$ is finite and normal, we conclude that it is indeed Galois.

The action of $\mathrm{Gal}(H_{\mathbf{A}}^{+}/\mathbf{k})$ commutes with the action of $\mathbf{Pic}^{+}(\mathbf{A})$. By Theorem 13.8, therefore, there is a well-defined group monomorphism $\mathrm{Gal}(H_{\mathbf{A}}^{+}/\mathbf{k}) \hookrightarrow \mathbf{Pic}^{+}(\mathbf{A})$. □

According to Proposition 11.3, ϕ is defined over every finite valuation ring in $H_{\mathbf{A}}^{+}$. It is therefore defined over the integral closure B^{+} of $\mathbf{A}$ in $H_{\mathbf{A}}^{+}$. This means that ϕ may be reduced modulo any non-zero prime ideal $\mathfrak{P}$ of B^{+}. Let $\mathbf{red}_{\mathfrak{P}}\colon B^{+} \longrightarrow B^{+}/\mathfrak{P}$ be the reduction map, and let $\mathfrak{p}$ be the prime ideal of $\mathbf{A}$ which sits under $\mathfrak{P}$.

Lemma 14.2. *The reduction map $\phi \mapsto \mathbf{red}_{\mathfrak{P}} \circ \phi$ is injective on* $\mathbf{X}$.

Proof. Suppose that ϕ and ϕ' in $\mathbf{X}$ reduce modulo $\mathfrak{P}$ to the same Drinfeld $\mathbf{A}$-module over $B^{+}/\mathfrak{P}$. By Theorem 13.8, $\phi' = \mathfrak{a} * \phi$ for some ideal $\mathfrak{a}$ in $\mathbf{A}$. We may assume that $\mathfrak{a}$ is prime to $\mathfrak{P}$ as follows. We choose via the Weak Approximation Theorem an element $z \in \mathbf{k}$ such that $z \equiv 1 \pmod{\infty}$ and $\mathrm{v}_{\mathfrak{P}}(z) = -\mathrm{v}_{\mathfrak{P}}(\mathfrak{a})$. Let $z\mathfrak{a} = \mathfrak{n}/\mathfrak{d}$ with $\mathfrak{n}$ and $\mathfrak{d}$ relatively prime ideals in $\mathbf{A}$. The ideal $\mathfrak{n} \cdot \mathfrak{d}^{h^{+}(\mathbf{A})-1}$ is then an ideal prime to $\mathfrak{P}$ which belongs to the same narrow ideal class in $\Im(\mathbf{A})$ as does $\mathfrak{a}$.

Reducing the defining equation $\phi_\mathfrak{a} \cdot \phi_x = \phi'_x \cdot \phi_\mathfrak{a}$ modulo $\mathfrak{P}$, we obtain

$$\phi_\mathfrak{a} \cdot \phi_x \equiv \phi_x \cdot \phi_\mathfrak{a} \pmod{\mathfrak{P}}$$

for all $x \in \mathbf{A}$. Thus, $\mathbf{red}_{\mathfrak{P}} \circ \phi_{\mathfrak{a}} \in \mathrm{End}(\mathbf{red}_{\mathfrak{P}} \circ \phi) = \mathbf{A}$ by Corollary 5.14, and so there is a $y \in \mathbf{A}$ so that

$$\phi_{\mathfrak{a}} \equiv \phi_y \pmod{\mathfrak{P}}. \tag{14.3}$$

Since the leading coefficient of $\phi_{\mathfrak{a}}$ is one, $\mu_\phi(y) = 1$, and so y is a positive element of $\mathbf{A}$. Thus, the proof will be complete if we can show that $\mathfrak{a} = y\mathbf{A}$. Let $\mathfrak{b} = \mathfrak{a} + y\mathbf{A}$. By (14.3), the torsion modules $\Lambda(\mathfrak{a})$, $\Lambda(\mathfrak{b})$ and $\Lambda(y\mathbf{A})$ in the algebraic closure of $B^+/\mathfrak{P}$ are equal. It follows now via (5.4) and (5.8) that $\deg \mathfrak{a} = \deg \mathfrak{b} = \deg y$. □

Proposition 14.4. *The extension $H_{\mathbf{A}}^+/\mathbf{k}$ is unramified at every finite place $\mathfrak{p}$ of $\mathbf{k}$.*

Proof. A finite place of $\mathbf{k}$ is essentially a non-zero prime ideal $\mathfrak{p}$ of $\mathbf{A}$. Let $\sigma \in \mathrm{Gal}(H_{\mathbf{A}}^+/\mathbf{k})$ belong to the inertia group $I_{\mathfrak{p}}$ at $\mathfrak{p}$. From the definitions, we have

$$\sigma\phi \equiv \phi \pmod{\mathfrak{P}}$$

for any $\phi \in \mathbf{X}$. By Lemma 14.2, this implies that $\sigma\phi = \phi$. Since $H_{\mathbf{A}}^+$ is generated over $\mathbf{k}$ by the coefficients of some ϕ_y, we conclude that $\sigma = 1$. □

Lemma 14.5. *In the notation of* (10.4), *we have*

$$c_\nu(\phi, x)^{N\mathfrak{p}} \equiv c_\nu(\mathfrak{p} * \phi, x) \pmod{\mathfrak{P}} \tag{14.6}$$

for all $x \in \mathbf{A}$ and $1 \le \nu \le \deg x$.

Proof. If $\phi' = \mathfrak{p} * \phi$, then by definition

$$\phi_{\mathfrak{p}} \cdot \phi_x = \phi'_x \cdot \phi_{\mathfrak{p}}.$$

By Corollary 5.9, the reduction modulo $\mathfrak{P}$ of this equation is

$$\mathbf{F}^{m \deg \mathfrak{p}} \cdot \phi_x \equiv \phi'_x \cdot \mathbf{F}^{m \deg \mathfrak{p}} \pmod{\mathfrak{P}},$$

which implies (14.6). □

Let k be an imaginary quadratic number field, and let $\mathfrak{a}$ and $\mathfrak{b}$ be ideals in $\mathcal{O}_k$. Let $j(\mathfrak{b})$ be the j-invariant of the elliptic curve determined by $\mathfrak{b}$ viewed as a lattice in $\mathbb{C}$. Then, as is well known from the theory of complex multiplication, $k(j(\mathfrak{b}))$ is algebraic and unramified over k and one has

$$j(\mathfrak{b})^{\sigma_{\mathfrak{a}}} = j(\mathfrak{a}^{-1}\mathfrak{b})$$

where $\sigma_{\mathfrak{a}}$ is the Artin automorphism. In view of Theorem 8.14, equation (15.7) below is the $\mathbf{A}$-module analog of this classical identity. Actually, we can do much better than (15.7) for the $\phi \in \mathbf{X}$.

For a non-zero ideal $\mathfrak{a}$ in $\mathbf{A}$, let $\sigma_{\mathfrak{a}} \in \mathrm{Gal}(H_{\mathbf{A}}^+/\mathbf{k})$ be the Artin automorphism associated to $\mathfrak{a}$. If $\mathfrak{a} = \mathfrak{p}$ is prime, then $\sigma_{\mathfrak{p}} = \mathbf{Frob}_{\mathfrak{p}}$, the Frobenius automorphism at $\mathfrak{p}$. We may now prove the key result of the rank one theory.

Theorem 14.7. *For every non-zero ideal $\mathfrak{a}$ in $\mathbf{A}$, we have*

$$\sigma_{\mathfrak{a}}\phi = \mathfrak{a} * \phi \tag{14.8}$$

for every $\phi \in \mathbf{X}$. The natural map from $\mathrm{Gal}(H_{\mathbf{A}}^+/\mathbf{k})$ into $\mathbf{Pic}^+(\mathbf{A})$ is an isomorphism, and so $H_{\mathbf{A}}^+/\mathbf{k}$ has degree $h^+(\mathbf{A}) = \varkappa \cdot h(\mathbf{A})$.

Proof. If $\phi \in \mathbf{X}$ and $\mathfrak{a} = \mathfrak{p}$ is prime, then (14.6) implies that

$$\mathbf{Frob}_{\mathfrak{p}}\phi \equiv \mathfrak{p} * \phi \pmod{\mathfrak{P}}.$$

Since reduction modulo $\mathfrak{P}$ is injective on $\mathbf{X}$ by Lemma 14.2, equation (14.8) holds for the prime $\mathfrak{a}$. To prove (14.8) in general, we proceed by induction on the number of prime ideals dividing $\mathfrak{a}$, using (4.7) for the induction step.

The Galois group $\mathrm{Gal}(H_{\mathbf{A}}^{+}/\mathbf{k})$ injects naturally into $\mathbf{Pic}^{+}(\mathbf{A})$ via its action on the principal homogeneous space $\mathbf{X}$ of $\mathbf{Pic}^{+}(\mathbf{A})$. Equation (14.8) proves that this injection is surjective. □

For any $x \in \mathbf{k}^{\times}$, we write σ_x for the Artin automorphism $\sigma_{x\mathbf{A}}$ associated to the principal ideal $x\mathbf{A}$.

Corollary 14.9. *For $x \in \mathbf{k}^{\times}$, we have*

$$\sigma_x\phi = \mu_\phi(x)^{-1}\phi\mu_\phi(x). \tag{14.10}$$

Proof. If $x \in \mathbf{A}$, then equation (14.8) with $\mathfrak{a} = x\mathbf{A}$ and Lemma 4.4 imply (14.10). Let $xy^{-1} \in \mathbf{k}^{\times}$ with $x, y \in \mathbf{A}$. We have then

$$\begin{aligned}\sigma_y\sigma_{xy^{-1}}\phi &= \sigma_x\phi = \mu_\phi(x)^{-1}\phi\,\mu_\phi(x)\\ &= \mu_\phi(y)^{-1}\left(\mu_\phi(xy^{-1})^{-1}\phi\,\mu_\phi(xy^{-1})\right)\mu_\phi(y)\\ &= \sigma_y\left(\mu_\phi(xy^{-1})^{-1}\phi\,\mu_\phi(xy^{-1})\right)\end{aligned}$$

which implies the result. □

15. The Hilbert class field $H_{\mathbf{A}}$. We wish to identify the smallest field of definition K_ϕ of a given $\mathbf{A}$-module ϕ. By Theorem 12.3, we may assume that ϕ is sgn-normalized. As a first step in identifying K_ϕ, we note from (6.7) and §14 that

$$\mathbf{k}\cdot\mathbb{F}_\infty \subseteq K_\phi \subseteq H_{\mathbf{A}}^{+}. \tag{15.1}$$

Choose $\xi \in \mathbf{C}$ so that $\phi' = \xi\phi\xi^{-1}$ is defined over K_ϕ. Because $\mathrm{Aut}(\phi) = \mathbb{F}_r^{\times}$, the greatest common divisor g in the proof of Theorem 10.3 is just $r - 1 = W_{\mathbf{k}}$. It follows then from (10.7) that

$$\xi_0 \overset{\text{def}}{=} \xi^{W_{\mathbf{k}}} \in H_{\mathbf{A}}^{+}.$$

Since $\mu_\phi(\pi^{-1}) = 1$, we see via (6.6) that

$$\xi_0^{\varkappa} = \xi^{W_\infty} = 1/\mu_{\phi'}(\pi^{-1}) \in K_\phi. \tag{15.2}$$

In fact, $H_{\mathbf{A}}^{+} = K_\phi(\xi_0)$ because the coefficients of $\phi_y = \xi\phi'_y\xi^{-1}$ generate $H_{\mathbf{A}}^{+}$ for any non-constant $y \in \mathbf{A}$.

Let $\mathcal{P}(\mathbf{A})$ be the group of principal fractional ideals of $\mathbf{A}$. It is clear from the definitions that the natural sequence

$$1 \to \mathcal{P}(\mathbf{A})/\mathcal{P}^{+}(\mathbf{A}) \longrightarrow \mathbf{Pic}^{+}(\mathbf{A}) \longrightarrow \mathbf{Pic}(\mathbf{A}) \to 1 \tag{15.3}$$

is exact. We will show that, under the natural identification of $\mathbf{Pic}^{+}(\mathbf{A})$ with $\mathrm{Gal}(H_{\mathbf{A}}^{+}/\mathbf{k})$, K_ϕ is the fixed field of $\mathcal{P}(\mathbf{A})/\mathcal{P}^{+}(\mathbf{A})$. In particular, K_ϕ is independent of the choice of ϕ.

Proposition 15.4. *The subfield K_ϕ of $H_{\mathbf{A}}^+$ is the fixed field of the automorphisms σ_x, $x \in \mathbf{k}^\times$. The extension $H_{\mathbf{A}}^+/K_\phi = K_\phi(\xi_0)/K_\phi$ is a Kummer extension of degree $\varkappa$, and for any $x \in \mathbf{k}^\times$ we have*

$$\xi_0^{\sigma_x} = \mu_\phi(x)^{W_{\mathbf{k}}} \cdot \xi_0 \,. \tag{15.5}$$

Proof. By Corollary 14.9, each σ_x fixes the invariants (10.6). As these invariants generate K_ϕ over $\mathbf{k}$, each σ_x fixes K_ϕ. In order to show that K_ϕ is the fixed field of these automorphisms, we first prove (15.5). Let 'σ_x' also denote some extension of σ_x to an isomorphism from $H_{\mathbf{A}}^+(\xi)$ into $\mathbf{C}$. We have then

$$\begin{aligned}\phi' = \sigma_x\phi' &= \xi^{\sigma_x} \cdot \sigma_x\phi \cdot \xi^{-\sigma_x} = \xi^{\sigma_x} \cdot \mu_\phi(x)^{-1}\phi\, \mu_\phi(x) \cdot \xi^{-\sigma_x} \\ &= \left(\xi^{\sigma_x - 1}\mu_\phi(x)^{-1}\right)\phi' \left(\xi^{\sigma_x - 1}\mu_\phi(x)^{-1}\right)^{-1}\end{aligned}$$

by (14.10). Thus $\xi^{\sigma_x - 1}\mu_\phi(x)^{-1} \in \mathrm{Aut}(\phi') = \mathbb{F}_r^\times$ which implies (15.5). We see now from (15.5) and (15.2) that $\varkappa \le [H_{\mathbf{A}}^+/K_\phi] \le \varkappa$. Since $\mathcal{P}(\mathbf{A})/\mathcal{P}^+(\mathbf{A})$ has order $\varkappa$ by (13.7) and (15.3), K_ϕ is indeed the fixed field of the automorphisms σ_x, $x \in \mathbf{k}^\times$. □

Definition. Let $H_{\mathbf{A}}$ be the common field of definition of the Drinfeld $\mathbf{A}$-modules of rank one over $\mathbf{C}$. We call $H_{\mathbf{A}}$ the *Hilbert class field of* $\mathbf{A}$.

Theorem 15.6. *The extension $H_{\mathbf{A}}/\mathbf{k}$ is completely split over ∞ and unramified over every finite place of $\mathbf{k}$. It is an extension of degree $h(\mathbf{A})$ with Galois group isomorphic to $\mathbf{Pic}(\mathbf{A})$ under the Artin map. If ϕ is an $\mathbf{A}$-module which is defined over $H_{\mathbf{A}}$, then*

$$\sigma_{\mathfrak{a}}\phi \cong \mathfrak{a} * \phi \tag{15.7}$$

for all non-zero ideals $\mathfrak{a}$ in $\mathbf{A}$.

Proof. The infinite place ∞ splits completely because $H_{\mathbf{A}}/\mathbf{k}$ is a Galois subextension of $\mathbf{K}/\mathbf{k}$. Proposition 14.4 implies that any finite place is unramified. That the Galois group is isomorphic to $\mathbf{Pic}(\mathbf{A})$ follows from the exact sequence (15.3). Finally, (15.7) is a consequence of (14.8). In fact, the reader will find it an amusing exercise to prove that if $\xi\phi\xi^{-1}$ is sgn-normalized, then

$$\mathfrak{a} * (\xi\phi\xi^{-1}) = \sigma_{\mathfrak{a}}\left(\xi^{N(\mathfrak{a})\sigma_{\mathfrak{a}}^{-1} - 1}\phi\,\xi^{1 - N(\mathfrak{a})\sigma_{\mathfrak{a}}^{-1}}\right).$$

The element $\xi_0^{N(\mathfrak{a})\sigma_{\mathfrak{a}}^{-1} - 1}$ belongs to $H_{\mathbf{A}}$ by (15.5) because the exponent annihilates roots of unity. □

One knows from class field theory that $\mathbf{Pic}(\mathbf{A})$ is the largest group which can possibly serve as the Galois group of an unramified abelian extension of $\mathbf{k}$ in which ∞ splits completely. Therefore $H_{\mathbf{A}}$ has the requisite property of a Hilbert class field. It is the maximal unramified abelian extension of $\mathbf{k}$ in which ∞ splits completely. Class field theory also tells us that the extension $H_{\mathbf{A}}^+/H_{\mathbf{A}}$ is totally ramified over all the infinite places of $H_{\mathbf{A}}$.

Let B be the integral closure of $\mathbf{A}$ in $H_{\mathbf{A}}$. We know now that $H_{\mathbf{A}}$ is a field of definition for every $\mathbf{A}$-module ϕ. Is B a "ring of definition" for ϕ? An affirmative answer to this fundamental question is provided by the following theorem of Takahashi.

Theorem 15.8 (Takahashi). *Every* $\mathbf{A}$*-module* ϕ *is isomorphic to an* $\mathbf{A}$*-module* ϕ' *which is defined over* B.

In **[Ta]**, Takahashi develops an analog of the Néron–Ogg–Shafarevich criterion for Drinfeld modules and derives Theorem 15.8 as a corollary. We will give a more direct proof in §17 below. One beautiful application of Takahashi's Theorem is the following constructive Principal Ideal Theorem for $H_{\mathbf{A}}$.

Theorem 15.9 (Principal Ideal Theorem). *Let* ϕ *be an* $\mathbf{A}$*-module which is defined over* B. *If* $\mathfrak{a}$ *is any non-zero ideal in* $\mathbf{A}$, *then* $\mathfrak{a}B$ *is generated as an ideal in* B *by the element* $D(\phi_{\mathfrak{a}})$.

Proof. By (6.1), we may assume that $\mathfrak{a} = \mathfrak{p}$ is a prime ideal. We know from Propositions 11.4 and 14.4 that $v_{\mathfrak{P}}(D(\phi_{\mathfrak{p}})) = 1$ for all prime ideals $\mathfrak{P}$ of B which divide $\mathfrak{p}$. We must show that no other prime ideals of B divide $D(\phi_{\mathfrak{p}})$. Choose $e \geq 1$ so that $\mathfrak{p}^e = x\mathbf{A}$ is principal and put $\mathfrak{b} = \mathfrak{p}^{e-1}$. Then for any prime $\mathfrak{P} \nmid \mathfrak{p}$ of B, we have

$$v_{\mathfrak{P}}(D((\mathfrak{p} * \phi)_{\mathfrak{b}})) + v_{\mathfrak{P}}(D(\phi_{\mathfrak{p}})) = v_{\mathfrak{P}}(\mu_{\phi}(x)^{-1} \cdot x) = 0$$

by (6.1) as $\mu_{\phi}(x)$ is a unit of B. Since both ϕ and $\mathfrak{p} * \phi$ are defined over B, the two summands on the right above are non-negative. Therefore they must both be zero. □

The Principal Ideal Theorem actually holds for a large family of "Hilbert Class Fields" in function fields. See Rosen **[R]** for a very readable development of this theory.

16. The narrow ray class fields $K_{\mathfrak{m}}$**.** Let $\mathfrak{m}$ be a non-zero, proper ideal in $\mathbf{A}$. The $\mathfrak{m}$-torsion points of any given sgn-normalized $\mathbf{A}$-module generate an abelian extension $K_{\mathfrak{m}}$ of $\mathbf{k}$ which behaves like a narrow ray class field of conductor $\mathfrak{m}$. The maximal subfield $K_{\mathfrak{m}}^+$ of $K_{\mathfrak{m}}$ in which ∞ splits completely is the ray class field modulo $\mathfrak{m}$. One obtains an explicit description of the class field theory of the maximal abelian extension of $\mathbf{k}$ in which ∞ splits completely. The techniques we use to develop the theory of the extensions $K_{\mathfrak{m}}$ are familiar from §14. We will therefore sketch the main ideas, leaving the details to the reader.

For a given non-zero, proper ideal $\mathfrak{m}$ in $\mathbf{A}$, let $\mathfrak{I}_{\mathfrak{m}}(\mathbf{A})$ be the subgroup of all fractional ideals in $\mathfrak{I}(\mathbf{A})$ that are prime to $\mathfrak{m}$, and let

$$\mathcal{P}_{\mathfrak{m}}^+(\mathbf{A}) = \{x\mathbf{A} : x \in \mathbf{k}^\times,\ x \text{ positive},\ x \equiv 1 \pmod{\mathfrak{m}}\}.$$

The quotient $\mathbf{Pic}_{\mathfrak{m}}^+(\mathbf{A}) = \mathfrak{I}_{\mathfrak{m}}(\mathbf{A})/\mathcal{P}_{\mathfrak{m}}^+(\mathbf{A})$ is called the *narrow ray class group modulo* $\mathfrak{m}$ relative to sgn.

Let $\phi \in \mathbf{X}$, and let $\Lambda_{\phi}(\mathfrak{m}) \subset \mathbf{C}$ be the set of $\mathfrak{m}$-torsion points of ϕ. By (5.8) with $\mathfrak{a} = \mathfrak{m}$, $\Lambda_{\phi}(\mathfrak{m})$ is a rank one, cyclic $\mathbf{A}$-module isomorphic to $\mathbf{A}/\mathfrak{m}$. If $\Phi(\mathfrak{m})$ is the order of the multiplicative group of $\mathbf{A}/\mathfrak{m}$, then $\Lambda_{\phi}(\mathfrak{m})$ has exactly $\Phi(\mathfrak{m})$ generators as an $\mathbf{A}$-module.

Let $\mathbf{X}_{\mathfrak{m}}$ be the set of pairs (ϕ, λ), where $\phi \in \mathbf{X}$ and λ is a generator of $\Lambda_{\phi}(\mathfrak{m})$. We extend the $*$-action of the integral ideals $\mathfrak{a}$ in $\mathfrak{I}_{\mathfrak{m}}(\mathbf{A})$ to $\mathbf{X}_{\mathfrak{m}}$ as follows:

$$\mathfrak{a} * (\phi, \lambda) \stackrel{\text{def}}{=} (\mathfrak{a} * \phi, \phi_{\mathfrak{a}}(\lambda)), \tag{16.1}$$

leaving to the reader the check that $\phi_{\mathfrak{a}}(\lambda)$ generates $\Lambda_{\mathfrak{a}*\phi}(\mathfrak{m})$. As in §13, one shows that this action extends to an action of $\mathfrak{I}_{\mathfrak{m}}(\mathbf{A})$ and that the stabilizer of any point (ϕ, λ) is $\mathcal{P}^{+}_{\mathfrak{m}}(\mathbf{A})$. Since the cardinalities of $\mathbf{X}_{\mathfrak{m}}$ and $\mathbf{Pic}^{+}_{\mathfrak{m}}(\mathbf{A})$ are the same, we deduce the following analog of Theorem 13.8.

Theorem 16.2. *The set* $\mathbf{X}_{\mathfrak{m}}$ *is a principal homogeneous space for* $\mathbf{Pic}^{+}_{\mathfrak{m}}(\mathbf{A})$ *under the* $*$ *action.*

Let $K_{\mathfrak{m}} = H^{+}_{\mathbf{A}}(\Lambda_{\phi}(\mathfrak{m}))$. Proceeding in much the same way as we did in §14, we may show that the extension $K_{\mathfrak{m}}/\mathbf{k}$ is Galois and unramified away from ∞ and the prime ideals dividing $\mathfrak{m}$. That each prime divisor of $\mathfrak{m}$ ramifies in $K_{\mathfrak{m}}/\mathbf{k}$ is clear from Proposition 11.4. The key result is the following complement for (14.8). For any ideal $\mathfrak{a}$ in $\mathbf{A}$ that is prime to $\mathfrak{m}$ and any $\lambda \in \Lambda_{\phi}(\mathfrak{m})$, we have

$$\lambda^{\sigma_{\mathfrak{a}}} = \phi_{\mathfrak{a}}(\lambda) \tag{16.3}$$

where $\sigma_{\mathfrak{a}} \in \mathrm{Gal}(K_{\mathfrak{m}}/\mathbf{k})$ is the Artin automorphism. It follows that $K_{\mathfrak{m}}$ is independent of the choice of $\phi \in \mathbf{X}$. Further, $K_{\mathfrak{m}}/\mathbf{k}$ is abelian with $\mathrm{Gal}(K_{\mathfrak{m}}/\mathbf{k}) \cong \mathbf{Pic}^{+}_{\mathfrak{m}}(\mathbf{A})$. We call $K_{\mathfrak{m}}/\mathbf{k}$ the *narrow ray class extension modulo* $\mathfrak{m}$.

Since the positive elements of $\mathbf{A}$ generate $\mathbf{A}/\mathfrak{m}$, the map $x \mapsto \sigma_x = \sigma_{x\mathbf{A}}$ for $x \in \mathbf{A}$ prime to $\mathfrak{m}$ sets up an isomorphism between $(\mathbf{A}/\mathfrak{m})^{\times}$ and $\mathrm{Gal}(K_{\mathfrak{m}}/H^{+}_{\mathbf{A}})$. Further, we see from (16.3) and Lemma 4.4 that for any $\lambda \in \Lambda_{\phi}(\mathfrak{m})$ and any $x \equiv 1 \pmod{\mathfrak{m}}$ in $\mathbf{k}^{\times}$, we have

$$\lambda^{\sigma_x} = \mu_{\phi}(x)^{-1}\lambda\,. \tag{16.4}$$

Therefore, $\mathrm{Gal}(K_{\mathfrak{m}}/\mathbf{k})$ contains a natural subgroup I_{∞} isomorphic to $\mathbb{F}^{\times}_{\infty}$. In fact (cf. **[H5]**), I_{∞} equals both the decomposition group and the inertia subgroups at ∞. The infinite place therefore splits completely in the fixed field $K^{+}_{\mathfrak{m}}$ of I_{∞}. The subfield $K^{+}_{\mathfrak{m}}$ is, in fact, the ray class field of $\mathbf{k}$ of conductor $\mathfrak{m}$ that is completely split over ∞. If we think of $K_{\mathfrak{m}}$ as the analog over the triple $(\mathbf{k}, \infty, \mathrm{sgn})$ of a cyclotomic number field, then $K^{+}_{\mathfrak{m}}$ is the analog of its real subfield.

Let K^{+}_{∞} be the union of the fields $K^{+}_{\mathfrak{m}}$ over all non-zero, proper ideals $\mathfrak{m}$ in $\mathbf{A}$. Then K^{+}_{∞} is the maximal abelian extension in which ∞ splits completely. If ∞' is a place of $\mathbf{K}$ different from ∞, then one can show (cf. **[H1]**, Theorem 7.2) that the composite $K^{+}_{\infty} \cdot K^{+}_{\infty'}$ is the maximal abelian extension of $\mathbf{k}$.

Let '∞' also denote that extension of ∞ to $K^{+}_{\mathfrak{m}}$ which is specified by the inclusion of $K^{+}_{\mathfrak{m}}$ in $\mathbf{K}$. The lattice $\Gamma = \xi(\mathfrak{m}) \cdot \mathfrak{m}$ determines a sgn-normalized $\mathbf{A}$-module ϕ^{Γ} associated to the field $K^{+}_{\mathfrak{m}}$, and (by §8) the element

$$\lambda = e_{\Gamma}(1) = \xi(\mathfrak{m}) \cdot \prod_{m \in \mathfrak{m} - \{0\}} \left(1 - \frac{1}{m}\right) \tag{16.5}$$

is an $\mathfrak{m}$-torsion point for ϕ^{Γ}. As one may show (cf. **[H5]**), the norm of this λ down to $K^{+}_{\mathfrak{m}}$ is a Brumer–Stark 'unit' (see **[St]** and **[T]**) at the place ∞ of $K^{+}_{\mathfrak{m}}$. Of course, Γ is not quite canonically determined by $\mathfrak{m}$, but the norm of λ is canonical. The product expansion (16.5) is actually equivalent at ∞ to the refined Stark conjecture introduced by B. Gross in **[Gr]**. See **[H6]** for a proof of this conjecture in function fields.

17. A proof of Takahashi's Theorem. We first prove a Principal Ideal Theorem for the normalizing field $H_{\mathbf{A}}^{+}$.

Theorem 17.1. *Let $\phi \in \mathbf{X}$. If $\mathfrak{a}$ is any ideal in $\mathbf{A}$, then $\mathfrak{a}B^{+}$ is generated as an ideal of B^{+} by the element $D(\phi_{\mathfrak{a}})$.*

Proof. Since any sgn-normalized ϕ is defined over B^{+}, the techniques used in the proof of Theorem 15.9 work equally well over B^{+}. □

Corollary 17.2. *Let $\phi \in \mathbf{X}$. Then for any $\sigma \in \mathrm{Gal}(H_{\mathbf{A}}^{+}/\mathbf{k})$ and any $\mathfrak{a}$ in $\mathbf{A}$, $D(\phi_{\mathfrak{a}})^{1-\sigma}$ is a unit of B^{+}.*

We are now in a position to prove Takahashi's Theorem. Our proof is based upon the results of §16. We may assume that the $\mathbf{A}$-module ϕ of Theorem 15.8 is sgn-normalized. We will show that there is a unit $\xi_0 \in B^{+}$ such that if $\xi^{W_{\mathbf{k}}} = \xi_0$, then $\phi' = \xi\phi\xi^{-1}$ is defined over $H_{\mathbf{A}}$.

Lemma 17.3. *Let $\phi \in \mathbf{X}$, and let $\mathfrak{m}$ be a non-zero, proper ideal in $\mathbf{A}$. Then there is an element $\gamma_{\mathfrak{m}} \in K_{\mathfrak{m}}$ such that*

$$\gamma_{\mathfrak{m}}^{W_{\mathbf{k}}} = (-1)^{\frac{N(\mathfrak{m})-1}{W_{\mathbf{k}}}} \cdot D(\phi_{\mathfrak{m}}). \tag{17.4}$$

Further, for any element $x \in \mathbf{k}^{\times}$ such that $x \equiv 1 \pmod{\mathfrak{m}}$,

$$\gamma_{\mathfrak{m}}^{\sigma_x} = \mu_{\phi}(x)^{\frac{1-N(\mathfrak{m})}{W_{\mathbf{k}}}} \cdot \gamma_{\mathfrak{m}}. \tag{17.5}$$

Proof. Since $K_{\mathfrak{m}}$ is the splitting field of $\phi_{\mathfrak{m}}(t)$ over $H_{\mathbf{A}}^{+}$,

$$D(\phi_{\mathfrak{m}}) = \prod_{\lambda \neq 0} \lambda \qquad (\lambda \in \Lambda_{\phi}(\mathfrak{m}))\,.$$

Now, $\Lambda_{\phi}(\mathfrak{m})$ is an $\mathbb{F}_r$-vector space of dimension $\deg(\mathfrak{m})$ (see §5). Let λ_* run through a set of representatives for the orbits of $\Lambda_{\phi}(\mathfrak{m}) - \{0\}$ under the action of $\mathbb{F}_r^{\times}$, and put $\gamma_{\mathfrak{m}} = \prod \lambda_*$. Then

$$\begin{aligned} D(\phi_{\mathfrak{m}}) &= \prod_{\lambda_*}\left(\prod_{\zeta \in \mathbb{F}_r^{\times}} \zeta\lambda_*\right) = \prod_{\lambda_*}(-\lambda_*^{W_{\mathbf{k}}}) \\ &= (-1)^{\frac{N(\mathfrak{m})-1}{W_{\mathbf{k}}}}\left(\prod \lambda_*\right)^{W_{\mathbf{k}}} = (-1)^{\frac{N(\mathfrak{m})-1}{W_{\mathbf{k}}}} \cdot \gamma_{\mathfrak{m}}^{W_{\mathbf{k}}}. \end{aligned}$$

Further, (16.4) implies that

$$\gamma_{\mathfrak{m}}^{\sigma_x} = \prod_{\lambda_*} \lambda_*^{\sigma_x} = \prod_{\lambda_*} \mu_{\phi}(x)^{-1} \cdot \lambda_* = \mu_{\phi}(x)^{\frac{1-N(\mathfrak{m})}{W_{\mathbf{k}}}} \cdot \gamma_{\mathfrak{m}},$$

proving (17.5). □

Since $\mathbf{k}$ contains a place of degree one, we can choose ideals $\mathfrak{m}_j$ in $\mathbf{A}$ and integers n_j, $1 \le j \le \ell$, such that

$$\sum_{j=1}^{\ell} n_j \cdot (N(\mathfrak{m}_j) - 1) = W_{\mathbf{k}} . \tag{17.6}$$

For each j, $1 \le j \le \ell$, let $\gamma_{\mathfrak{m}_j}$ be the $W_{\mathbf{k}}$-th root of the right hand side of (17.4) with $\mathfrak{m} = \mathfrak{m}_j$. Put

$$\gamma_0 = \prod_{j=1}^{\ell} \gamma_{\mathfrak{m}_j}^{n_j} .$$

By (17.4) and (17.6),

$$\gamma_0^{W_{\mathbf{k}}} = - \prod_{j=1}^{\ell} D(\mathfrak{m}_j)^{n_j} . \tag{17.7}$$

Put $\mathfrak{m} = \prod \mathfrak{m}_j$, and let

$$P_{\mathfrak{m}}(\mathbf{k}) \stackrel{\text{def}}{=} \{x \in \mathbf{k}^{\times} : x \equiv 1 \pmod{\mathfrak{m}}\} .$$

We may now compute the action of the automorphisms σ_x, $x \in P_{\mathfrak{m}}(\mathbf{k})$, on γ_0. By (17.5) and (17.6), we have

$$\gamma_0^{\sigma_x} = \prod_{j=1}^{\ell} \left(\mu_\phi(x)^{\frac{1-N(\mathfrak{m})}{W_{\mathbf{k}}}} \gamma_{\mathfrak{m}_j} \right)^{n_j} = \mu_\phi(x)^{-1} \cdot \gamma_0 . \tag{17.8}$$

Choose an ideal $\mathfrak{a}$ in $\mathbf{A}$ such that $\deg \mathfrak{a} \equiv 1 \pmod{d}$, and put $\xi_0 = \gamma_0^{1-\sigma_{\mathfrak{a}}}$. By (17.7) and Corollary 17.2, $\xi_0^{W_{\mathbf{k}}}$ is a unit of B^+. We claim that, in fact, $\xi_0 \in H_{\mathbf{A}}^+$ so that ξ_0 is also a unit of B^+.

If $\tau \in \mathrm{Gal}(H_{\mathbf{A}}^+(\gamma_0)/H_{\mathbf{A}}^+)$, then

$$\gamma_0^{\tau} = \zeta \cdot \gamma_0 \qquad (\zeta \in \mathbb{F}_r^{\times})$$

by (17.7). Now, the extension $H_{\mathbf{A}}^+(\gamma_0)/\mathbf{k}$ is abelian because it is a subextension of the abelian extension $K_{\mathfrak{m}}/\mathbf{k}$. Therefore

$$\xi_0 = \gamma_0^{1-\sigma_{\mathfrak{a}}} \in H_{\mathbf{A}}^+(\gamma_0) ,$$

and for all $\tau \in \mathrm{Gal}(H_{\mathbf{A}}^+(\gamma_0)/H_{\mathbf{A}}^+)$,

$$\xi_0^{\tau} = \gamma_0^{(1-\sigma_{\mathfrak{a}})\tau} = \gamma_0^{\tau(1-\sigma_{\mathfrak{a}})} = (\zeta\gamma_0)^{1-\sigma_{\mathfrak{a}}} = \xi_0$$

as $\mathbb{F}_r \subset \mathbf{k}$. Thus, $\xi_0 \in H_{\mathbf{A}}^+$ as claimed.

Now let $\phi' = \xi\phi\xi^{-1}$, where $\xi^{W_{\mathbf{k}}} = \xi_0$. Then for any $x \in P_{\mathfrak{m}}(\mathbf{k})$, we have

$$\begin{aligned}\sigma_x\phi' = \sigma_x(\xi\phi\xi^{-1}) = \xi^{\sigma_x} \cdot \sigma_x\phi \cdot \xi^{-\sigma_x} = \xi^{\sigma_x} \cdot \mu_\phi(x)^{-1}\phi\,\mu_\phi(x) \cdot \xi^{-\sigma_x} \\ = \left(\xi^{\sigma_x - 1}\mu_\phi(x)^{-1}\right)\phi'\left(\xi^{\sigma_x-1}\mu_\phi(x)^{-1}\right)^{-1} .\end{aligned}$$

However, by (17.8)

$$\left(\xi^{\sigma_x - 1}\mu_\phi(x)^{-1}\right)^{W_{\mathbf{k}}} = \xi_0^{\sigma_x - 1}\mu_\phi(x)^{-W_{\mathbf{k}}} = \left(\gamma_0^{\sigma_x - 1}\right)^{1-\sigma_{\mathfrak{a}}} \cdot \mu_\phi(x)^{-W_{\mathbf{k}}}$$
$$= \mu_\phi(x)^{\sigma_{\mathfrak{a}} - 1 - W_{\mathbf{k}}} = \mu_\phi(x)^{N(\mathfrak{a}) - 1 - W_{\mathbf{k}}} = 1$$

as $\deg \mathfrak{a} \equiv 1 \pmod{d}$. Thus, $\sigma_x \phi' = \phi'$ for all $x \in P_{\mathfrak{m}}(\mathbf{k})$. By the Weak Approximation Theorem and Proposition 15.4, $H_{\mathbf{A}}$ is the fixed field of these σ_x. Since $\mu_{\phi'}(y) \in B^\times$ for all $y \in \mathbf{A} - \{0\}$, we conclude from Proposition 11.3 that ϕ' is defined over B.

Bibliography

[A] E. Artin, Quadratische Körper im Gebiete der höheren Kongruenzen I, II, *Math. Z.* **19** (1924) 153–246.

[C1] L. Carlitz, The arithmetic of polynomials in a Galois field, *Amer. J. of Math.* **54** (1932) 39–50.

[C2] L. Carlitz, On polynomials in a Galois field, *Bull. Amer.Math. Soc.* **38** (1932) 736–744.

[C3] L. Carlitz, On certain functions connected with polynomials in a Galois field, *Duke Math. J.* **1** (1935) 137–168.

[C4] L. Carlitz, A class of polynomials, *Trans. Amer. Math. Soc.* **43** (1938) 167–182.

[D] R. Dedekind, Abriss einer Theorie der höheren Congruenzen in Bezug auf einer reellen Primzahl–Modulus, *J. Reine Angew. Math.* **54** (1857) 1–26.

[De] P. Deligne, Private correspondence dated February 20, 1988.

[D–H] P. Deligne and D. Husemoller, Survey of Drinfeld modules, *Contemp. Math.* **67** (1987) 25–91.

[Dr1] V.G. Drinfeld, Elliptic modules (Russian), *Math. Sbornik* **94** (1974) 594–627=*Math. USSR S.* **23** (1974) 561–592.

[Dr2] V.G. Drinfeld, Elliptic modules II (Russian), *Math. Sbornik* **102** (1977) 182–194=*Math. USSR S.* **31**, (1977) 159–170.

[G–R] S. Galovich and M. Rosen, Units and class groups in cyclotomic function fields, *J. Number Theory* **14** (1982) 156–184.

[Ge1] E.-U. Gekeler, *Drinfeld Modular Curves*, Lecture Notes in Math. No. 1231. Springer–Verlag, Berlin 1986.

[Ge2] E.-U. Gekeler, On finite Drinfeld modules, *J. Algebra* **141** (1991) 187–203.

[Go1] D. Goss, The algebraist's upper half-plane, *Bull. Amer. Math. Soc. (N.S.)* **2** (1980) 391–415.

[Go2] D. Goss, The arithmetic of function fields 2: The 'cyclotomic' theory, *J. Algebra* **81** (1983) 107–149.

[Gr] B. Gross, On the values of abelian L-series at $s = 0$, *J Univ. Tokyo, Section IA Math.* **35** (1988) 177–197.

[Gr–R] B. Gross and M. Rosen, Fourier Series and Special Values of L-functions, *Adv. Math.* **69** (1988) 1–31.

[H1] D. R. Hayes, Explicit class field theory for rational function fields, *Trans. Amer. Math. Soc.* **189** (1974) 77–91.

[H2] D. R. Hayes, Explicit class field theory in global function fields, *Studies in algebra and number theory, Advances in Mathematics Supplementary Studies* **6** (1979) 173–217.

[H3] D. R. Hayes, Analytic class number formulas in global function fields, *Invent. Math.* **65** (1981) 49–69.

[H4] D. R. Hayes, Elliptic units in function fields, in *Proceedings of a Conference on Modern Developments Related to Fermat's Last Theorem*, D. Goldfeld ed., Birkhäuser, Boston 1982.

[H5] D. Hayes, Stickelberger elements in function fields, *Compositio Math.* **55** (1985) 209–239.

[H6] D. Hayes, The refined $\mathfrak{p}$-adic abelian Stark conjecture in function fields, *Invent. Math.* **94** (1988) 505–527.

[H7] D. Hayes, On the reduction of rank-one Drinfeld modules, *Math. Comp.* **57** (1991) 339–349.

[K] H. Kornblum, Über die Primfunktionen in einer arithmetischen Progression, *Math. Z.* **5** (1919) 100–111.

[R] M. Rosen, The Hilbert class field in function fields, *Exposition. Math.* **5** (1987) 365–378.

[S] J.-P. Serre, Complex multiplication, in *Algebraic Number Theory*, J.W.S. Cassels and A. Frölich, eds., Thompson, Washington 1967.

[Sil] J. Silverman, *The arithmetic of elliptic curves*. Springer–Verlag, New York 1985.

[St] H. Stark, L-functions at $s = 1$. IV. First derivatives at $s = 0$, *Adv. Math.* **35** (1980) 197–235.

[Ta] T. Takahashi, Good reduction of elliptic modules, *J. Math. Soc. of Japan* **34** (1982) 475–487.

[T] J. Tate, *Les conjectures de Stark sur les fonctions L d'Artin en $s = 0$*. Birkhäuser, Boston 1984.

[W] A. Weil, Sur les fonctions algébriques à corps de constantes fini, *Comptes Rendus* **210** (1940) 592–594.

[Y] J. Yu, Transcendence and Drinfeld modules, *Invent. Math.* **83** (1986) 507–517.

Department of Mathematics and Statistics, University of Massachusetts, Amherst MA 01003
Email: dhayes@qmodz.math.umass.edu

Galois Calculus and Carlitz Exponentials

Y. Hellegouarch

It is well known ([2], p.460) that the theory of Drinfeld modules appears under the dual aspects of t-modules and t-motives.

To the first side belongs the notion of the Drinfeld exponential

$$\exp_E\colon \mathrm{Lie}(E) \to E(\overline{K})$$

where E is an abelian t-module and $\overline{K}$ some algebraic closure of $K = \mathbb{F}_q((t^{-1}))$.

It is known that the kernel $H_*(E)$ of $\exp_E$ is a discrete A-submodule of $\mathrm{Lie}(E)$ and that the surjectivity of $\exp_E$ is equivalent to

$$\mathrm{rank}_A(H_*(E)) = \mathrm{rank}(E),$$

see [2] p. 475, and recall that $A = \mathbb{F}_q[t]$.

In a previous paper ([9]) I gave a construction of an abstract exponential belonging to the ring of formal Ore series $K\{\{\sigma\}\}$ and verifying the functional equation

$$\exp \cdot t\sigma^0 = \varphi(t) \cdot \exp$$

where $\varphi(t)$ is the Ore polynomial

$$\varphi(t) = t\sigma^0 + a_1\sigma + \cdots + a_n\sigma^n,$$

σ being an F-endomorphism of $K = F((t^{-1}))$ given by

$$\sigma(t) = t^d + \alpha_1 t^{d-1} + \cdots$$

with $d \geq 2$, and $\alpha_i \in F$ (any given field).

By the same procedure I also obtained an abstract logarithm in the same ring satisfying

$$\exp \cdot \log = \log \cdot \exp = \sigma^0.$$

Then the question arose to give a realization of our exponential as an additive map between certain groups and it seemed that a good candidate to consider was the Newton-Puiseux field P of F (see below §II, 1): P is almost algebraically closed and σ acts canonically on it. Moreover we have $\sigma(P) = P$ in the typical case where $\sigma(t) = t^d$ with $d \not\equiv 0 \bmod \mathrm{char}(F)$.

So it turns out that exp defines, through a new type of "entire" series, a local linear homeomorphism $P \to P$ and that its kernel $H_*(P)$ is a discrete A-submodule of P (see [9]).

However, many questions remain open and our original motivation was to know whether we have

$$\mathrm{rank}_A(H_*(P)) \geq \deg_\sigma(\varphi(t)).$$

The first aim of this paper is to give an affirmative answer to this question in the "Carlitz case" where

$$\varphi(t) = t\sigma^0 - \sigma .$$

In fact, we construct an element $\pi \in P\backslash\{0\}$ such that

$$H_*(P) \supset A\pi := \{\lambda\pi ;\ \lambda \in A\} .$$

The techniques we are using are partly inspired by a paper of Carlitz's (**[3]**) and partly by the calculus of "divided differences" of the Ancients.

It turns out that our notations make it easier when we consider a sequence of endomorphisms of K, the Carlitz case corresponding to the particular case of a constant sequence.

In this particular case, the "Galois derivatives" we are considering belong to the Ore domain $K\{\sigma\}$ which is similar to a "pure t-motive" (see **[2]** p. 494) except for the nature of σ which is not necessarily the Frobenius endomorphism. A central result is that the limit of the Galois derivatives, suitably normalized to begin with σ^0, is the Carlitz generalized exponential suitably combined with a scaling of ratio π (see Theorem 1).

Those were the facts I explained in the "Journées Arithmétiques de Marseille" in 1989 together with a few other results of mine (see **[5]**, **[6]**, **[7]**). Moreover, a further generalization is given here in which the sequence of endomorphisms of K is no longer constant but purely periodical of period ℓ. Here the Carlitz exponential still exists and preserves its periodicity, but then it appears as a linear combination of ℓ rational series of endomorphisms. And also the Ore polynomial ring $K\{\sigma\}$ has to be replaced by a more complex non-associative F-algebra $\mathfrak{D}$.

Those results seem to point towards the possibility of generalizations in the direction of ultrametric analysis in K, in the direction of t-motives and t-modules for a non-algebraic σ and, finally, in the direction of σ-transcendence questions (see **[9]** for a definition and some results).

I. Galois calculus

In the first paragraph we notice that Galois derivatives appear spontaneously when we look for a good concept of degree in rings of skew polynomials. But we introduce a second and more intrinsic definition of them when we have a field.

The second paragraph develops some basic properties of Galois derivatives in the more general setting arising from the second definition.

The third one introduces a reverse process, which might be thought of as a sort of primitivization.

Finally we introduce a (non-necessarily associative) algebra which partly generalizes the skew polynomial ring we started with.

1) Introduction. Let A be an integral domain. We know, **[4]** p. 10, that if we ask for a right A-module structure on the left A-module of "polynomials"

$$P(Z) = a_0 + a_1 Z + \cdots + a_m Z^m$$

such that

$$\deg(Pa) = \deg P\,; \qquad \text{when } a \in A\backslash\{0\},$$

(where $\det P = m$ means $a_m \neq 0$) then we must write

$$Za = \sigma(a)Z + \partial a \tag{1}$$

where σ is an injective endomorphism of A and ∂ is a σ-***derivation***, i.e.

$$\begin{cases} \partial \text{ is additive } A \to A \\ \partial(ab) = (\partial a)b + \sigma(a)(\partial b)\,. \end{cases} \tag{2}$$

Then it is known that this left A-module can be given a canonical ring structure: if $\partial = 0$ we could write $A[Z;\sigma]$ for this ring, as in **[4]** p. 11, but we will prefer the simpler notations $A\{\sigma\}$ and

$$P(\sigma) = a_0 + a_1\sigma + \cdots + a_m\sigma^m\,.$$

Then, a being any element in A and $P(\sigma)$ any element in $F[\sigma]$, where $F = \operatorname{Inv}\sigma$ is the ring of elements of A left fixed by σ, we have a Taylor formula

$$P(a+\sigma) = P(a) + D'_a P\cdot\sigma + \cdots + D_a^{(m)}P\cdot\sigma^m\,. \tag{3}$$

where m is (greater than) the degree of P and where the coefficients are in A.

When $a \in F$ the (ordinary) Taylor formula states that

$$D_a^{(n)}P = \frac{1}{n!}\,\frac{d^nP(a)}{da^n}\,,$$

at least when $\operatorname{char}(A)$ does not divide $n!$ In the general case we will say that

Definition. $D_a^{(n)}P$ is the $n^{\text{th}}\sigma$-derivative of $P(a)$ in a.

Example. Take $P(\sigma) = \sigma^m$, then we have by induction on n

$$D_a^{(n)}(a^m) = \sum_{\gamma_0+\cdots+\gamma_n=m-n} a^{\gamma_0}[\sigma(a)]^{\gamma_1}\ldots[\sigma^n(a)]^{\gamma_n}\,. \tag{4}$$

Now it is clear that $P \to D_a^{(n)}P$ is F-linear, so if

$$P(\sigma) = \sum_{r=0}^{m} c_r\sigma^r\,, \quad \text{with} \quad c_\gamma \in F$$

we have

$$D_a^{(n)}P = \sum_{r=n}^{m} c_r D_a^{(n)}(\sigma^r)\,.$$

Suppose that $\sigma^r(a) \neq a$ for all $r \leq n$, then it turns out (see next paragraph) that when A is a commutative ***field*** K, $D_a^{(n)}P$ can be calculated by a recursive process. To be precise, if we write $b = P(a)$ then, on the above assumptions, we have

$$\begin{cases} D_a^{(0)}P = D_a^{(0)}(b) := b \\ D_a^{(1)}P = D_a^{(1)}(b) := \frac{\sigma(b)-b}{\sigma(a)-a} \\ D_a^{(n)}P = D_a^{(n)}(b) := \frac{\sigma[D_a^{(n-1)}(b)]-D_a^{(n-1)}(b)}{\sigma^n(a)-a}\,. \end{cases} \tag{5}$$

Example. Suppose K is a commutative field and that for any $n \geq 1$ we have $\sigma^n(a) \neq a$ and $D_a^{(n)}(b) \neq 0$. Then the above considerations imply that $b \notin F[a]$, where $F = \operatorname{Inv}(\sigma)$ is the fixed field of σ.

Remark. When K is a commutative field, let us denote by K_σ the K-bimodule defined by

$$\lambda_a : x \mapsto \sigma(a)x, \qquad \rho_a : x \mapsto xa.$$

Then it can be easily seen that (**[10]** p. 373)

$$H^0(K, K_\sigma) = 0, \quad H^1(K, K_\sigma) = 0.$$

So every σ-derivation of K is $\lambda(\mathrm{id}_K - \sigma)$, with $\lambda \in K$.

2) Derivatives. From now on we suppose that K is a commutative field equipped with a sequence of (non-necessarily commuting) endomorphisms $\sigma_1, \ldots, \sigma_n$.

We denote by F the subfield of K left (pointwise) fixed by all the σ_i and we write

$$\nabla(\sigma_n, \ldots, \sigma_1) = \{x \in K\,;\ \sigma_i \ldots \sigma_1(x) \neq \sigma_j \ldots \sigma_1(x) \text{ for all } i \text{ and } j \text{ such that } 0 \leq i < j \leq n\}.$$

When $x \in \nabla(\sigma_n, \ldots, \sigma_1)$ we define F-linear operators on K by the formulae

$$\begin{cases} [\emptyset]_x := \mathrm{id} := \mathrm{id}_K \\ [\sigma_1]_x := \frac{\sigma_1 - \mathrm{id}}{\sigma_1(x) - x} \\ \vdots \\ [\sigma_n, \ldots, \sigma_1]_x := \frac{[\sigma_n, \ldots, \sigma_2]_{\sigma_1(x)} \circ \sigma_1 - [\sigma_{n-1}, \ldots, \sigma_1]_x}{\sigma_n \ldots \sigma_1(x) - x}. \end{cases} \tag{6}$$

It is clear that the right hand operators are in the K-vector space $\mathcal{D}_n$ generated by id, σ_1, $\sigma_2\sigma_1, \ldots, \sigma_n \ldots \sigma_1$ in $\mathcal{L}_F(K)$ and, since $\nabla(\sigma_n, \ldots, \sigma_1) \neq \emptyset$ by the existence of x, a result of Artin (**[11]** p. 209) shows that

$$\mathcal{D}_n = K\ \mathrm{id} \oplus K\sigma_1 \oplus \cdots \oplus K\sigma_n \ldots \sigma_1. \tag{7}$$

Remark. It is clear that $\mathcal{D}_n$ acts on the left on K and if we apply the operators in (6) to $b \in K$, we get (5) when $\sigma_i = \sigma$ for $i = 1, \ldots, n$ and $x = a$.

Accordingly we will write $[\sigma_n, \ldots, \sigma_1]_x(y)$, or more simply $[\sigma_n, \ldots, \sigma_1]_x y$ for the image of y under $[\sigma_n, \ldots, \sigma_1]_x$.

Now we will give some formulae which are reminiscent of the classical formulae of calculus.

a) Chain rule derivation.
 If x and $y \notin \operatorname{Inv}\sigma$, we have

$$[\sigma]_x z = ([\sigma]_y z)([\sigma]_x y).$$

b) Leibniz's formula.
 If $x \in \nabla(\sigma_n, \ldots, \sigma_1)$ we have have

$$[\sigma_n, \ldots, \sigma_1]_x(yz) = \sum_{i=0}^{n} ([\sigma_i, \ldots, \sigma_1]_x y)([\sigma_n, \ldots, \sigma_{i+1}]_{\sigma_i \ldots \sigma_1 x} \sigma_i \ldots \sigma_1 z). \tag{8}$$

The proof can be carried by induction on n.

c) Derivative of x^m.

If $x \in \nabla(\sigma_n, \ldots, \sigma_1)$ we have

$$[\sigma_n, \ldots, \sigma_1]_x(x^m) = \sum_{\gamma_0+\cdots+\gamma_n=m-n} x^{\gamma_0}(\sigma_1 x)^{\gamma_1} \cdots (\sigma_n \ldots \sigma_1 x)^{\gamma_n}. \tag{9}$$

The proof can be carried out by induction on m using Leibniz's formula.

Fundamental Corollary. *If $x \in \nabla(\sigma_n, \ldots, \sigma_1)$ and if $P(X) \in F[X]$ is of degree $< n$, then*

$$[\sigma_n, \ldots, \sigma_1]_x P(x) = 0. \tag{10}$$

d) Derivative of x^{-1}.

If $x \in \nabla(\sigma_n, \ldots, \sigma_1)$ we have

$$[\sigma_n, \ldots, \sigma_1]_x(x^{-1}) = \frac{(-1)^n}{x(\sigma_1 x)(\sigma_2\sigma_1 x)\cdots(\sigma_n \ldots \sigma_1 x)}.$$

This can be seen by induction on n, using Leibniz's formula with $y = x^{-1}$ and $z = x$.

e) General formula for $[\sigma_n, \ldots, \sigma_1]_x$.

Let us define $D_n \in \mathcal{D}_n$ by the determinant

$$D_n := \begin{vmatrix} 1 & x & \ldots & x^{n-1} & \mathrm{id} \\ 1 & \sigma_1 x & \ldots & \sigma_1(x^{n-1}) & \sigma_1 \\ \vdots & & & & \\ 1 & \sigma_n \ldots \sigma_1 x & \ldots & \sigma_n \ldots \sigma_1(x^{n-1}) & \sigma_n \ldots \sigma_1 \end{vmatrix}.$$

Then we have

$$[\sigma_n, \ldots, \sigma_1]_x = \frac{1}{D_n(x^n)} D_n \in \mathcal{D}_n. \tag{11}$$

This can be seen by induction on n.

f) Newton's formula.

Let $x \in \nabla(\sigma_n, \ldots, \sigma_1)$ and $y = P(x)$ where $P(X) \in F[X]$ is a polynomial in X of degree $\leq n$. Then we have

$$\begin{aligned} P(X) = {} & y + ([\sigma_1]_x y)(X - x) + \cdots \\ & + [\sigma_n, \ldots, \sigma_1]_x y (X - x)(X - \sigma_1 x) \cdots (X - \sigma_{n-1} \ldots \sigma_1 x). \end{aligned} \tag{12}$$

The proof can be carried on X^n by induction on n through the use of Leibniz's formula with $y = x$ and $z = x^{n-1}$.

Examples.

1) Let ω be a primitive root of unity in $\mathbb{C}$ of order $d > 1$ and take $K = \mathbb{Q}(2^{1/d}, \omega)$, σ a generator of $\mathrm{Gal}\,(K/\mathbb{Q}(\omega))$.

Then apply formula (9) with $n < d$, $\sigma_i = \sigma$ for $1 \leq i \leq n$, $x = 2^{1/d}$, $m = dh$, $h \in \mathbb{N}$. The left hand side is obviously null, and we have

$$\sum_{\gamma_0+\cdots+\gamma_n=dh-n} \omega^{\gamma_1+2\gamma_2+\cdots+n\gamma_n} = 0. \tag{13}$$

Now let A be a (non-commutative) ring of characteristic zero containing ω in its center and let u and v in A be such that

$$vu = \omega uv.$$

Then we have

$$(u+v)^d = u^d + v^d \tag{14}$$

as can be seen by taking $h = 1$ in (13).

Using (14) we can extend further (13) and we get

$$\begin{cases} \sum_{\gamma_0+\cdots+\gamma_0=dh-n} \omega^{\gamma_1+2\gamma_2+\cdots+n\gamma_n} = 0 & \text{if } d \text{ does not divide } n \\ \sum_{\gamma_0+\cdots+\gamma_n=d(h-i)} \omega^{\gamma_1+2\gamma_2+\cdots+n\gamma_n} = \binom{h}{i} & \text{if } d \text{ divides } n. \end{cases} \tag{15}$$

2) The same formula applied to $K = \mathbb{Q}(\omega),\ x = \omega,\ y = \omega^{dh} = 1$, gives

$$\sum_{\gamma_0+\cdots+\gamma_n=dh-n} \omega^{\gamma_0+\gamma_1\rho+\cdots+\gamma_n\rho^n} = 0 \tag{16}$$

where ρ denotes a generator of $(\mathbb{Z}/d\mathbb{Z})^*$, this group being assumed cyclical.

3) Primitivization. We are now looking for a reverse process concerning the previous operators.

If $x \in \nabla(\sigma_n, \ldots, \sigma_1)$, we denote by h_x the application x id in $\mathcal{D}_n$, further we will write σ_0 in place of id.

Then given $A(\sigma) \in \mathcal{D}_n$, we define new elements in $\mathcal{D}_n$ by the formulae

$$\begin{cases} \{\emptyset\}_x A = A \\ \{\sigma_0\}_x A = \{\emptyset\}_x A(\sigma) \circ h_x - x\{\emptyset\}_x A(\sigma) \\ \vdots \\ \{\sigma_n, \ldots, \sigma_0\}_x A \\ \quad = \{\sigma_{n-1}, \ldots, \sigma_0\}_x A(\sigma) \circ h_x - \sigma_n \ldots \sigma_1(x)\{\sigma_{n-1}, \ldots, \sigma_0\}_x A(\sigma). \end{cases} \tag{17}$$

Notice that $A \mapsto \{\sigma_1, \ldots, \sigma_0\}_x A$ is K-linear for any i.

The fundamental relations between $[\sigma_1, \ldots, \sigma_i]_x$ and $\{\sigma_n, \ldots, \sigma_0\}_x$ are the following:

$$\begin{cases} \{\sigma_{i-1}, \ldots, \sigma_0\}_x[\sigma_j, \ldots, \sigma_1]_x = [\sigma_j, \ldots, \sigma_{i+1}]_{\sigma_i \ldots \sigma_1 x} \sigma_i \cdots \sigma_1 & \text{if } i \le j \\ \{\sigma_{i-1}, \ldots, \sigma_0\}_x[\sigma_j, \ldots, \sigma_1]_x = 0 & \text{if } i > j. \end{cases} \tag{18}$$

Proof. Induction on i.

1) If $i = 0$, there is nothing to prove.

2) Suppose that the first relation in (18) is true for $i-1$, then by (17) we have

$$\begin{aligned} \{\sigma_{i-1}, \ldots, \sigma_0\}_x[\sigma_j, \ldots, \sigma_1]_x &= \{\sigma_{i-2}, \ldots, \sigma_0\}_x([\sigma_j, \ldots, \sigma_1]_x \circ h_x) \\ &\quad - \sigma_{i-1} \ldots \sigma_1 x\{\sigma_{i-2}, \ldots, \sigma_0\}_x[\sigma_j, \ldots, \sigma_1]_x. \end{aligned} \tag{19}$$

By Leibniz's formula, we have:

$$[\sigma_j, \ldots, \sigma_1]_x \circ h_x = \sigma_j \ldots \sigma_1 x[\sigma_j, \ldots, \sigma_1]_x + [\sigma_{j-1}, \ldots, \sigma_1]_x$$

and since $A \mapsto \{\sigma_{i-2}, \ldots, \sigma_0\}_x A$ is K-linear we have in the right hand side of (19)

$$\sigma_j \ldots \sigma_1 x \{\sigma_{i-2}, \ldots, \sigma_0\}_x [\sigma_j, \ldots, \sigma_1]_x + \{\sigma_{i-2}, \ldots, \sigma_0\}_x [\sigma_{j-1}, \ldots, \sigma_1]_x \\ - \sigma_{i-1} \ldots \sigma_1 x \{\sigma_{i-2}, \ldots, \sigma_0\}_x [\sigma_j, \ldots, \sigma_1]_x .$$

Then by the induction hypothesis

$$\sigma_j \ldots \sigma_1 x [\sigma_j, \ldots, \sigma_i]_{\sigma_{i-1} \ldots \sigma_1 x} \sigma_{i-1} \ldots \sigma_1 + [\sigma_{j-1}, \ldots, \sigma_i]_{\sigma_{i-1} \ldots \sigma_1 x} \sigma_{i-1} \ldots \sigma_1 \\ - \sigma_{i-1} \ldots \sigma_1 x [\sigma_j, \ldots, \sigma_i]_{\sigma_{i-1} \ldots \sigma_1 x} \sigma_{i-1} \ldots \sigma_1 .$$

Using (6) we get

$$(\sigma_j \ldots \sigma_1 x - \sigma_{i-1} \ldots \sigma_1 x)[\sigma_j, \ldots \sigma_i]_{\sigma_{i-1} \ldots \sigma_1 x} \\ = [\sigma_j, \ldots, \sigma_{i+1}]_{\sigma_i \ldots \sigma_1 x} \circ \sigma_i - [\sigma_{j-1}, \ldots, \sigma_i]_{\sigma_{i-1} \ldots \sigma_1 x}$$

which gives the required result. □

Now, from (7), we know that any $A(\sigma) \in \mathcal{D}_n$ can be written

$$A(\sigma) = a_0 \operatorname{id} + a_1 \sigma_1 + \cdots + a_d \sigma_d \cdots \sigma_1$$

with $d \le n$ and $a_i \in K$.

When $a_d \neq 0$ we say that d is the ***degree*** of $A(\sigma)$. Then we have the two following formulae:

$$\begin{cases} A(\sigma h_y) = \sum\limits_{i=0}^{\deg A} [\{\sigma_{i-1}, \ldots, \sigma_0\}_x A](y) [\sigma_i, \ldots, \sigma_1]_x \\ A(\sigma) = \sum\limits_{i=0}^{\deg A} [\{\sigma_{i-1}, \ldots, \sigma_0\}_x A](1) [\sigma_i, \ldots, \sigma_1]_x . \end{cases} \tag{20}$$

Proof.

1) Since $[\sigma_i, \ldots, \sigma_1]_x$ is of degree i, we can write

$$A(\sigma h_y) = b_0 [\emptyset]_x + b_1 [\sigma_1]_x + \cdots + b_d [\sigma_d, \ldots, \sigma_1]_x . \tag{21}$$

Applying $\{\sigma_{i-1}, \ldots, \sigma_0\}_x$ to (21) we get by (18)

$$\{\sigma_{i-1}, \ldots, \sigma_1\}_x A(\sigma h_y) = \sum_{j=i}^{\deg A} b_j [\sigma_j, \ldots, \sigma_{i+1}]_{\sigma_i \ldots \sigma_1 x} \sigma_i \ldots \sigma_1 .$$

Applying this element (of $\mathcal{D}_n$) to $1 \in K$, we get

$$\{\sigma_{i-1}, \ldots, \sigma_0\}_x A(\sigma h_y)(1) = [\{\sigma_{i-1}, \ldots, \sigma_0\}_x A(\sigma)](y) = b_i .$$

2) The second result is the particular case $y = 1$. □

Example. $[\sigma_i, \ldots, \sigma_1]_x$-development of $\sigma_d \ldots \sigma_1$.

If $A(\sigma) = \sigma_d \ldots \sigma_1$ we have

$$\begin{cases} \{\sigma_0\} A = [\sigma_d \ldots \sigma_1(x) - x] A \\ \{\sigma_1, \sigma_0\} A = [\sigma_d \ldots \sigma_1(x) - x][\sigma_d \ldots \sigma_1(x) - \sigma_1(x)] A \\ \text{etc.} \end{cases}$$

If we write

$$\begin{cases} [n]_i = \sigma_n \dots \sigma_1(x) - \sigma_i \dots \sigma_1(x) \\ F_{n,i} = [n]_0 \dots [n]_{i-1} \end{cases}$$

then we get by induction

$$\{\sigma_{i-1}, \dots, \sigma_0\}_x A = F_{d,i} A \tag{22}$$

and we have

$$\sigma_d \dots \sigma_1 = \sum_{i=1}^{d} F_{d,i}[\sigma_i, \dots, \sigma_1]_x .$$

4) Multiplication in $\mathcal{D}_n$. Our aim is to define a non-associative F-algebra structure on $\mathcal{D}_n$. The main point is the definition of the product of monomials, we write

$$\begin{cases} (\lambda\sigma_i \dots \sigma_1) \cdot (\mu\sigma_j \dots \sigma_1) = [\lambda\sigma_i \dots \sigma_1(\mu)]\sigma_{i+j} \dots \sigma_1 & \text{when } i + j \le n \\ (\lambda\sigma_i \dots \sigma_1) \cdot (\mu\sigma_j \dots \sigma_1) = 0 & \text{when } i + j > n. \end{cases} \tag{23}$$

With this notation Leibniz's formula becomes

$$\begin{aligned} &[\sigma_m, \dots, \sigma_1]_x(yz)\sigma_m \dots \sigma_1 \\ &\qquad = \sum_{i+j=m} ([\sigma_i, \dots, \sigma_1]_x(y)\sigma_i \dots \sigma_i)([\sigma_j, \dots, \sigma_1](z)\sigma_j \dots \sigma_1) \end{aligned} \tag{24}$$

when $m \le n$.

Proposition 1.

1) *The map $y \to \gamma(y) := \sum_{i=0}^{n} (-1)^i [\sigma_1, \dots, \sigma_1]_x(y)\sigma_i \dots \sigma_1$ is a ring-morphism γ of K into $\mathcal{D}_n$.*
2) *If $M \in F[x]$, then $\gamma(M)$ is nothing other than $M(x\sigma_0 - \sigma_1)$ developed along the basis σ_0, σ_1, $\sigma_2\sigma_1, \dots, \sigma_n \dots \sigma_1$ of $\mathcal{D}_n$.*

Proof. The first assertion is a direct consequence of (24). To prove the second, one need only consider the case of a monomial x^m and make an induction on m.

The result is obvious if $m = 0$.

Let it be true for $m - 1$, then one has

$$(x\sigma_0 - \sigma_1)^m = (x\sigma_0 - \sigma_1)^{m-1} \cdot (x\sigma_0 - \sigma_1).$$

By the induction hypothesis one gets

$$\left(\sum_{i=0}^{m-1} (-1)^i [\sigma_i, \dots, \sigma_1]_x(x^{m-1})\sigma_i \dots \sigma_1 \right) \cdot (x\sigma_0 - \sigma_1)$$

and by (23) we obtain

$$\sum_{i=0}^{m-1} (-1)^i \{[\sigma_i, \dots, \sigma_1]_x(x^{m-1})\sigma_i \dots \sigma_1(x) + [\sigma_{i-1}, \dots, \sigma_1]_x(x^{m-1})\}\sigma_i \dots \sigma_1 .$$

Then Leibniz's formula gives the expected result. □

Remark. Since K is a commutative field, $\operatorname{Im}\gamma$ is an associative and commutative subalgebra of $\mathcal{D}_n$. But since $\mathcal{D}_n$ is not associative and commutative in general, γ cannot be surjective in general.

II. Generalized Carlitz exponentials

A situation in which Galois calculus applies naturally is the case where $K = F((t^{-1}))$, F being ***any*** commutative field.

In this case we will choose $x = t$ (the "canonical fluent") and consider a purely periodical sequence of continuous endomorphisms of K for its t^{-1}-topology:

$$\sigma_1, \sigma_2, \ldots, \sigma_\ell, \sigma_{\ell+1} = \sigma_1, \ldots, \sigma_{2\ell+1} = \sigma_1, \text{ etc.}$$

such that $\sigma_i|_F = \mathrm{id}\,|_F$.

In the second paragraph we will see that, under adequate conditions ($\deg \sigma_i(t) \geq 2$ for all i) canonically normalized derivatives $E_n(\sigma)$ tend to a definite operator $E(\sigma)$ and that we have

$$\lim_{n\to\infty} E_n(\sigma) = E(\sigma) = \frac{e(\sigma \circ h_\pi)}{\pi}$$

for a certain "transcendental" quantity π and a certain "rational" series e.

Finally we will stress properties of $e(\sigma)$ which make it appear as a generalization of a Drinfeld exponential of rank 1.

1) Some prerequisites. We give here a shortened version of the first part of paper **[9]**.

F being any commutative field, and t an indeterminate, we set

$$A = F[t], \quad k = F(t), \quad K = F((t^{-1})).$$

If $z \in K\backslash\{0\}$ we have

$$z = a_d t^d + a_{d-1}t^{d-1} + \cdots + a_0 + a_{-1}t^{-1} + \cdots$$

with $a_d \in F^\times$ and $a_i \in F$.

We will say that d is the ***degree*** of z, a_d its ***signature*** and we will write

$$\deg z := d, \quad \operatorname{sgn}(z) := a_d, \quad |z| = \rho^d$$

where $\rho > 1$ is fixed throughout this work.

When $\operatorname{sgn}(z) = 1$ we will say that z is ***monic*** and when $z = 0$, we will write

$$\deg 0 = -\infty, \quad |0| = 0.$$

Then it is clear that any continuous F-endomorphism σ of K is given by

$$z(t) \overset{\sigma}{\mapsto} z[\sigma(t)]$$

with $\det \sigma(t) \geq 1$.

Definition. We will say that the continuous F-endomorphism $\sigma: K \to K$ is normal if and only if $\sigma(t) = t^d$ for some $d \geq 1$ and that it is ***normalizable*** if and only if there is some monic $u \in K$ of degree one such that $\sigma(u) = u^d$.

Proposition 2. *Let p be the characteristic of F and let σ be a continuous F-endomorphism of K. We suppose that $\sigma(t)$ is monic and we denote its degree by d.*

i) *If $p = 0$ and $d \geq 2$, then σ is normalizable.*

ii) *If $p > 0$, we write $d = p^h d_0$ where p does not divide d_0. Then if $d_0 \geq 2$ and $\sigma(t) \in F((t^{-p^h}))$, σ is normalizable.*

Proof.

1) We consider first the case where $p = 0$ or $h = 0$, and write

$$\begin{cases} \sigma(t) = t^d(1 + \alpha_1 t^{-1} + \alpha_2 t^{-2} + \cdots) := \tau \\ u = t(1 + \lambda_1 t^{-1} + \lambda_2 t^{-2} + \cdots) \end{cases} \tag{25}$$

with given $\alpha_i \in F$ and unknown λ_i.

The equation $\sigma(u) = u^d$ is tantamount to:

$$\begin{aligned} &(1 + \alpha_1 t^{-1} + \alpha_2 t^{-2} + \cdots)(1 + \lambda_1 \tau^{-1} + \lambda_2 \tau^{-2} + \cdots) \\ &\quad = (1 + \lambda_1 t^{-1} + \lambda_2 t^{-2} + \cdots)^d \end{aligned} \tag{26}$$

and since $d \geq 2$, we have a triangular system:

$$\begin{cases} \alpha_1 = d\lambda_1 \\ \alpha_2 = d\lambda_2 + \binom{d}{2}\lambda_1^2 - \varepsilon_2\lambda_1 \\ \text{etc.} \end{cases}$$

with $\varepsilon_2 = 0$ if $d \geq 3$, $\varepsilon_2 = 1$ if $d = 2$, etc., which gives the λ_i by induction on i.

2) If $p > 0$ and $d = p^h d_0$, then (26) can be replaced by:

$$(1 + \alpha_1 t^{-1} + \alpha_2 t^{-2} + \cdots)(1 + \lambda_1 \tau^{-1} + \lambda_2 \tau^{-2} + \cdots) = (1 + \lambda_1^q t^{-q} + \lambda_2^q t^{-2q} + \cdots)^{d_0}$$

where we put $q = p^h$.

If $\sigma(t) \in F((t^{-q}))$, then we can replace t by t^q in the first factor of the left hand side and if $d_0 \geq 2$, we can proceed exactly as in the first case. □

Definition. Let P denote the inductive limit of the fields $F((t^{-1/\nu}))$ where $\nu \in \mathbb{N}\backslash p\mathbb{N}$ are ordered by divisibility. Then we will say that P is the ***Newton-Puiseux*** field of F.

Remark. When F is algebraically closed, every polynomial in $P[X]$ whose degree is not divisible by p is reducible (see **[9]**).

Now if $\sigma(t)$ is monic, we can extend σ canonically to P by the rule

$$\sigma(t^{1/\nu}) := t^{d/\nu}(1 + \alpha_1 t^{-1} + \alpha_2 t^{-2} + \cdots)^{1/\nu}$$

where we use the notations of (25). We recall that p does not divide v and we use the binomial expansion in the right hand side. If σ is normalizable and if p does not divide $\deg \sigma(t)$, then it is clear that σ extends to an F-*automorphism of* P.

Proposition 3. *Let σ be a normalizable endomorphism of K of degree $d \geq 2$ such that $d \not\equiv 1 \bmod p$ and let a be any monic in P. Then there is a monic $z \in P\backslash\{0\}$ such that*

$$\frac{\sigma(z)}{z} = a \tag{27}$$

and any other solution z' of equation (27) can be written $z' = \lambda z$ with $\lambda \in F^\times$.

Proof. Suppose that $\sigma(u) = u^d$ and $a \in K$. Write

$$\begin{cases} a = u^s(1 + c_1 u^{-1} + c_2 u^{-2} + \cdots) \\ z = u^r(1 + \gamma_1 u^{-1} + \gamma_2 u^{-2} + \cdots) \end{cases}$$

with given $c_i \in F$ and unknown γ_i.

Necessarily we must have $r = \frac{s}{d-1} \in \mathbb{Q}$, and $u^r \in P$ since $d \not\equiv 1 \bmod p$. Then equation (27) can be written

$$1 + \gamma_1 u^{-d} + \gamma_2 u^{-2d} + \cdots = (1 + \gamma_1 u^{-1} + \gamma_2 u^{-2} + \cdots)(1 + c_1 u^{-1} + c_2 u^{-2} + \cdots)$$

and since $d \geq 2$, we have the triangular system

$$\begin{cases} \gamma_1 + c_1 = 0 \\ \gamma_1 + c_1\gamma_1 + c_2 = \varepsilon_2\gamma_1 \\ \text{etc.} \end{cases}$$

with $\varepsilon_2 = 1$ if $d = 2$ and $\varepsilon_2 = 0$ if $d > 2$, etc. So we can find the $\gamma_i \in F$ by induction on i. Now if z' is a second solution of (27) we have

$$\sigma\left(\frac{z'}{z}\right) = \frac{z'}{z}$$

and this implies that $\dfrac{z'}{z} \in F^\times$.

Finally, if $a \in F((t^{-1/\nu}))\backslash K$, the F-automorphism of P sending t onto t^ν leads us back to $a \in K$. □

2) The limit of the normalized derivatives. Let us consider first the more general situation in which we have ***any*** sequence of endomorphisms (σ_n) such that $\deg \sigma_n(t) \geq 2$ and $\operatorname{sgn}(\sigma_n(t)) = 1$ for any n.

Then it is clear that, for any n, we have $t \in \nabla(\sigma_n, \ldots, \sigma_1)$ so that $[\sigma_n, \ldots, \sigma_1]_t$ is well defined. Now we want to ***normalize*** this operator in such a way that id would have coefficient 1: we obtain the ***normalized derivative*** E_n defined by

$$E_n = \mathrm{id} - \frac{L_n}{F_1 L_{n,1}}\sigma_1 + \cdots + (-1)^i \frac{Ln}{F_i L_{n,i}}\sigma_i \ldots \sigma_1 + \cdots + (-1)^n \frac{L_n}{F_n}\sigma_n \ldots \sigma_1 \qquad (28)$$

with the notations

$$\begin{cases} [n]_i = \sigma_n \ldots \sigma_1(t) - \sigma_i \ldots \sigma_1(t) \\ F_{n,i} = [n]_0 \ldots [n]_{i-1} \\ L_{n,i} = [n]_i[n-1]_i \ldots [i+1]_i \\ F_n = F_{n,n}, \quad L_n = L_{n,0}. \end{cases} \qquad (29)$$

Now we consider the K-vector space

$$\mathcal{D} = K\,\mathrm{id} \oplus K\sigma_1 \oplus K\sigma_2\sigma_1 \oplus \cdots$$

equipped with the product topology, and we define a non-necessarily associative F-algebra structure on it by the formula

$$(\lambda\sigma_i \ldots \sigma_1) \cdot (\mu\sigma_j \ldots \sigma_1) = \lambda\sigma_i \ldots \sigma_1(\mu)\sigma_{i+j} \ldots \sigma_1.$$

Then it is clear that $\mathcal{D}$ is a graded algebra, so that we can speak of its completion $\widehat{\mathcal{D}}$ for its graduation.

Our first task will be to find the limit of E_n in $\widehat{\mathcal{D}}$ when n tends to infinity. For this end, we will introduce the notation Δ_i to indicate that all the endomorphisms are to be composed with the next i ones, for example one has

$$\Delta_i L_{n-i} = [n]_i \dots [i+1]_i = L_{n,i}. \tag{31}$$

Let us also set (following **[3]**)

$$\eta_n = [1]_0 [2]_1 \dots [n]_{n-1}. \tag{32}$$

Then we get

$$\eta_n = \eta_i \cdot \Delta_i(\eta_{n-i}). \tag{33}$$

So if we write

$$\xi_i = \frac{\eta_i}{L_i} \tag{34}$$

we get

$$\frac{L_n}{L_{n,i}} = \frac{L_n}{\Delta_i L_{n-i}} = \frac{\eta_n \Delta_i(\xi_{n-i})}{\xi_n \Delta_i(\eta_{n-i})} = \eta_i \frac{\Delta_i(\xi_{n-i})}{\xi_n}$$

and the coefficient of $\sigma_i \dots \sigma_1$ in E_n is:

$$a_{n,i} = (-1)^i \frac{\eta_i}{F_i} \frac{\Delta_i(\xi_{n-i})}{\xi_n}. \tag{35}$$

Lemma 1. *The sequence $n \mapsto \xi_n$ converges in K.*

Proof. From (34) we get

$$\frac{\xi_{i+1}}{\xi_i} = \frac{L_i}{L_{i+1}} \cdot \frac{\eta_{i+1}}{\eta_i} = \frac{\Delta_i[1]}{[i+1]} = 1 - \frac{[i]}{[i+1]}$$

with the convention

$$[n] = [n]_0 = \sigma_n \dots \sigma_1(t) - t.$$

Thus we have

$$\xi_n = \prod_{i=0}^{n-1} \left(1 - \frac{[i]}{[i+1]}\right) \tag{36}$$

and this product is obviously convergent in K since $\deg \sigma_i(t) \geq 2$. □

Definition. We will write

$$\xi_\infty = \lim_{n \to \infty} \xi_n = \prod_{i=0}^{\infty} \left(1 - \frac{[i]}{[i+1]}\right). \tag{37}$$

From now on we must make use of the assumption of periodicity for the sequence $n \mapsto \sigma_n$ and we shall write

$$\theta = \sigma_\ell \dots \sigma_1. \tag{38}$$

Lemma 2. *If we suppose that $n \mapsto \sigma_n$ has period ℓ, we get*

$$\lim_{n\to\infty} a_{n,j\ell+i} = (-1)^{j\ell+i}\frac{\eta_{j\ell+i}}{F_{j\ell+i}} \cdot \frac{\sigma_i \ldots \sigma_1\theta^j(\xi_\infty)}{\xi_\infty}.$$

Proof. This is obvious from (35), when we observe that

$$\lim_{n\to\infty} \Delta_{j\ell+i}(\xi_{n-j\ell-i}) = \sigma_i \ldots \sigma_1\theta^j(\xi_\infty).$$

□

From now on we will suppose the following

Hypotheses.
1) *θ is normalizable.*
2) $\deg\theta \not\equiv 1 \bmod p$ *where $p =$ characteristic F.*

Proposition 2 gives us sufficient conditions for θ to be normalizable. Then Proposition 3 ensures the existence of $\zeta \in P$, monic, such that

$$\frac{\theta(\zeta)}{\zeta} = \eta_\ell. \tag{39}$$

From this we deduce that

$$\frac{\theta^j(\zeta)}{\zeta} = \eta_{j\ell}, \tag{40}$$

and now we can improve on Lemma 2.

Lemma 3. *Write*

$$\pi = \zeta\xi_\infty. \tag{41}$$

Then we have

$$\lim_{n\to\infty} a_{n,\ell j+i} = \frac{(-1)^{j\ell+i}}{F_{j\ell+i}} \frac{\eta_i\zeta}{\sigma_i \ldots \sigma_1(\zeta)} \frac{\sigma_i \ldots \sigma_1\theta^j(\pi)}{\pi}.$$

Proof. From (39) and (40) we get

$$\sigma_i \ldots \sigma_1\left[\frac{\theta^j(\zeta)}{\zeta}\right] = \frac{\eta_{j\ell+i}}{\eta_i}.$$

Multiplying this equation by $\eta_i\sigma_i \ldots \sigma_1\theta^j(\xi_\infty)$ we have

$$\eta_i\frac{\sigma_i \ldots \sigma_1\theta^j(\pi)}{\sigma_i \ldots \sigma_1(\zeta)} = \eta_{j\ell+i}\sigma_i \ldots \sigma_1\theta^j(\xi_\infty) prose.$$

Then, using Lemma 2, we obtain

$$\lim_{n\to\infty} a_{n,j\ell+i} = \frac{(-1)^{j\ell+i}}{F_{j\ell+i}}\eta_i\frac{\sigma_i \ldots \sigma_1\theta^j(\pi)}{\sigma_i \ldots \sigma_1(\zeta)\cdot\xi_\infty}.$$ □

Definition. For $0 \leq i \leq \ell - 1$, we write

$$e_i = \sum_{j=0}^{\infty} \frac{(-1)^{j\ell+i}}{F_{j\ell+i}} \sigma_i \dots \sigma_1 \theta^j \in \widehat{\mathfrak{D}}.$$

Theorem 1. *When n tends to infinity in $\mathbb{N}$, E_n tends to a limit E in $\widehat{\mathfrak{D}}$ such that*

$$E = \frac{1}{\pi} e \cdot \pi \sigma_0$$

where

$$e = e_0 + \frac{\eta_1 \zeta}{\sigma_i(\zeta)} e_1 + \dots + \frac{\eta_{\ell-1}\zeta}{\sigma_{\ell-1} \dots \sigma_1(\zeta)} e_{\ell-1}. \tag{42}$$

Proof. This result follows directly from the above definition and Lemma 3. □

Remarks.

1) As in the classical case π is not uniquely determined, and all other choices are $\lambda\pi$ with $\lambda \in F^\times$ (see Proposition 3).
2) The operators e_i should be considered as ***rational***. In fact, so they are when $\sigma_i(t)$ for $i = 1, \dots, \ell$ are in $\mathbb{Z}$.
3) On the contrary π should be considered as ***transcendental***, as indeed is the case when F is a finite field and $\sigma_1, \dots, \sigma_\ell$ are certain powers of the Frobenius endomorphism of R (see **[1]** for a particular case).

3) Properties of the generalized Carlitz exponential.. Our aim is to show that the Carlitz exponential $e \in \widehat{\mathfrak{D}}$ defines an F-linear application $P \to P$ and to exhibit some of the properties of this application.

Lemma 4. *Let $i \in \{0, \dots, \ell - 1\}$. Then for any $z \in P$ the series*

$$e_i(z) = \sum_{j=0}^{\infty} \frac{(-1)^{j\ell+i}}{F_{j\ell+i}} \sigma_i \dots \sigma_1 \theta^j(z)$$

is convergent in P.

Proof. It is clear that

$$\deg F_{j\ell+i} = d_1 \dots d_i d^j (j\ell + i)$$

where $d = \deg \theta(t)$ (see (38)) and $d_h = \deg \sigma_h(t)$ for $1 \leq h \leq i$.

On the other side

$$\deg[\sigma_i \dots \sigma_1 \theta^j(z)] = d_1 \dots d_i \, d^j \deg z.$$

Then it follows that the degree of the general term of the series in the right hand side tends to $-\infty$. □

Since we gave a meaning to $e_i(z)$ for $i = 0, \dots, \ell - 1$, Theorem 1 gives us a way to define $e(z)$:

$$e(z) := e_0(z) + \frac{\eta_1 \zeta}{\sigma_1(\zeta)} e_1(z) + \dots + \frac{\eta_{\ell-1}\zeta}{\sigma_{\ell-1} \dots \sigma_1(\zeta)} e_{\ell-1}(z).$$

Lemma 5. *For any* $z \in P$, *we have*

$$\lim_{n\to\infty} E_n(z) = \frac{e(\pi z)}{\pi}.$$

Proof. The coefficient of $\sigma_i \dots \sigma_1 \theta^j$ in E_n (for n big enough) is

$$a_{n,j\ell+i} = (-1)^{j\ell+i} \frac{L_{j\ell+i}}{F_{j\ell+i}}$$

and in e it is (by Lemma 3)

$$b_{j\ell+i} = (-1)^{j\ell+i} \frac{1}{F_{j\ell+i}} \frac{\eta_i \zeta}{\sigma_i \dots \sigma_1(\zeta)} \frac{\sigma_i \dots \sigma_1 \theta^j(\pi)}{\pi}.$$

Now we will show that the two terms

$$|a_{n,j\ell+i}\sigma_i \dots \sigma_1 \theta^j(z)| \qquad \text{and} \qquad |b_{j\ell+i}\sigma_i \dots \sigma_1 \theta^j(z)|$$

both tend to zero when $j\ell + i$ tends to ∞.

Then Lemma 3 will give the desired result (since P is ultrametric).

On one hand we have

$$\begin{aligned}\deg[a_{n,j\ell+i}\sigma_i \dots \sigma_1 \theta^j(z)] &\le \ell(d + d^2 + \dots + d^{j+1}) - d_1 \dots d_i d^j (j\ell + i) \\ &\quad + d_1 \dots d_i d^j \deg(z) \\ &\le \alpha d^j [1 - \lambda(j\ell + i)], \qquad \text{with } \alpha \text{ and } \lambda > 0,\end{aligned}$$

so it does tend to $-\infty$ when j tends to ∞.

On the other hand, we have seen in Lemma 4, that $|b_{j\ell+i}\sigma_i \dots \sigma_1 \theta^j(z)|$ tends to zero. So both terms do tend to zero. □

Theorem 2.

i) *For any* $z \in P$, $e(\pi z)$ *converges in* P *and we have*

$$\lim_{n\to\infty} E_n(z) = \frac{e(\pi z)}{\pi}.$$

ii) *For any* $M \in A$ *and* $z \in P$ *we have*

$$e(\pi z + \pi M) = e(\pi z).$$

iii) *If* $m = \deg M$ *we have*

$$e_i(Mz) = \sum_{n=0}^{m} (-1)^n ([\sigma_n, \dots, \sigma_1]_t M) \sigma_n \dots \sigma_1 [e_{\text{“}i-n\text{”}}(z)]$$

where “$i - n$” *denotes the residue of* $i - n$, *modulo* ℓ, *in* $\{0, 1, \dots, \ell - 1\}$.

iv) *The map* $M \overset{\gamma}{\mapsto} \sum_{n=0}^{m} (-1)^n ([\sigma_n, \dots, \sigma_1]_t M) \sigma_n \dots \sigma_1$ *is an* F-*morphism of* A *in* $\mathfrak{D}$, *and if we write*

$$\varphi = e_0 + e_1 + \dots + e_{\ell-1} \in \widehat{\mathfrak{D}}$$

we have

$$\varphi \cdot M\sigma_0 = \gamma(M) \cdot \varphi.$$

v) *Let α be in K. Then if we write $\varepsilon = \varphi(\alpha) \in K$, we have*

$$\varphi \cdot \alpha\sigma_0 = \sum_{n=0}^{\infty}[\{\sigma_{n-1}, \ldots, \sigma_0\}\varphi](\alpha)[\sigma_n, \ldots, \sigma_1]_t$$
$$= \sum_{n=0}^{\infty}(-1)^n\sigma_n \ldots \sigma_1(\varepsilon)[\sigma_n, \ldots, \sigma_1]_t.$$

Proof.
i) Follows from Lemmas 4 and 5.
ii) Follows from (i) and from the fact that

$$E_n(M) = 0$$

if $n > \deg M$ by formula (10).
iii) Applying the first relation (20) to the partial sum $e_{i,h}(\sigma)$ of $e_i(\sigma)$ we have, for any y and $z \in P$

$$e_{i,h}(yz) = \sum_{n=0}^{h}\{\sigma_{n-1}, \ldots, \sigma_0\}_t e_{i,h}(y)[\sigma_n, \ldots, \sigma_1]_t(z). \tag{43}$$

Taking $z = M$ and $h > m = \deg M$, we have

$$e_{i,h}(My) = \sum_{n=0}^{m}[\sigma_n, \ldots, \sigma_1]_t(M)\{\sigma_n, \ldots, \sigma_0\}_t e_{i,h}(y).$$

But (22) shows that

$$\{\sigma_{n-1}, \ldots, \sigma_0\}_t e_{i,h} = (-1)^n\sigma_n \ldots \sigma_1 e_{\text{``i-n''},h-n}$$

so we get the result when h tends to ∞.
iv) The assertion about γ comes from Leibniz's rule as in Proposition 1. Summing on i the relations in iii), we get

$$\varphi \cdot M = \gamma(M) \cdot \varphi.$$

v) This is only the particular case $y = \alpha$ in (43) with summation on i and $h \to \infty$. □

Remarks.
1) When the operators e_i are considered as *rational* and $\alpha = 1$, then ε should be considered as ***transcendental***. This is the case when F is a finite field and $\sigma_1, \ldots, \sigma_\ell$ are certain powers of the Frobenius endomorphism of K (see **[12]** for a particular case and **[9]** for a more general result).
2) In classical analysis we can find different analogs for the functions e and φ (in case $\ell = 1$). The function $x \mapsto e^x$ is such an analog, and when $\alpha \in \mathbb{Q}$ is such that $1 < e^\alpha < 2$, the last formula of Theorem 2 suggests the true identity (binomial formula)

$$e^{\alpha x} - 1 = \sum_{n=1}^{\infty} \varepsilon^n \binom{x}{n}$$

where ε denotes $e^\alpha - 1$.

Of course this identity is still valid in $\mathbb{R}$ for any α such that $0 < |e^\alpha - 1| < 1$.

3) The equations in part iii) of Theorem 2 can also be written in matrix notation

$$\begin{pmatrix} e_0(Mz) \\ \vdots \\ e_{\ell-1}(Mz) \end{pmatrix} = \Gamma(M) \begin{pmatrix} e_0(z) \\ \vdots \\ e_{\ell-1}(z) \end{pmatrix}$$

where $\Gamma(M)$ is an $\ell \times \ell$ matrix with coefficients in $\mathcal{D}$, the mapping $\Gamma\colon A \to \mathrm{M}_{\ell\times\ell}(\mathcal{D})$ being a morphism.

4) In the (classical) Carlitz situation where F is a finite field $\mathbb{F}_q$, $\ell = 1$ and σ is replaced by the Frobenius endomorphism τ of K given by

$$\tau(z) = z^q$$

we get the Carlitz exponential e_τ.

Now if $z \in P$ we have

$$\tau(z) = z^q = z(t^q) = \sigma(z)$$

where σ is defined by $\sigma|\mathbb{F}_q = \mathrm{id}$ and $\sigma(t) = t^q$, so that we could also write this exponential e_σ as, in fact, e_τ and e_σ agree on P.

But if we extend σ to the Newton-Puiseux field $\widetilde{P}$ of an algebraic closure $\overline{\mathbb{F}}_q$ of $\mathbb{F}_q$ by decreeing that $\sigma|_{\overline{\mathbb{F}}_q} = \mathbb{1}_{\overline{\mathbb{F}}_q}$ then we no longer have $e_\tau = e_\sigma$ on $\widetilde{P}$ and we get two different types of Carlitz modules: an "algebraic" one and an "analytic" one.

5) When K is any commutative field, $\ell = 1$ and $\mathrm{id}, \sigma, \ldots, \sigma^n, \ldots$ are K-linearly independent, $\mathcal{D}$ is just the Ore polynomial ring $K\{\sigma\}$ of paragraph I.1.

If moreover, σ is an automorphism of P, let us define the non-commutative ring $P[T,\tau]$ by the relations

$$T\tau = \tau T, \qquad xT = Tx, \qquad \tau x = \sigma(x)\tau$$

for all $x \in K$. Then $P[T,\tau]$ can be made to operate on $P \otimes_K \mathcal{D}$ by the rule

$$T \to t\,\mathrm{id} - \sigma, \qquad \tau \to \sigma$$

and $P \otimes_K \mathcal{D}$ can be viewed as an analog of a pure T-motive **[2]** for which $\theta = t$.

References

[1] J-P. Allouche, Sur la transcendance de la série formelle Π, Sém. Théorie Nombres, Bordeaux 2, (1990), 103–116.

[2] G.W. Anderson, t-Motives, Duke Math. J. 53, (1986), 457–502.

[3] L. Carlitz, On certain functions connected with polynomials in a Galois field, Duke Math. J. 1, (1935), 137–168.

[4] P.M. Cohn, *Skew Field Constructions*, Cambridge University Press (1977).

[5] Y. Hellegouarch, Calcul différentiel galoisien, prépublication no. 42, Université de Caen (1989).

[6] Y. Hellegouarch, A generalization of Terjanian's theorem, C.R. Math. Rep. Acad. Sci. Canada, 12, (1990), 103–108.

[7] Y. Hellegouarch, Théorème de Terjanian Généralisé, Sém. Théorie Nombres, Bordeaux 2, (1990), 245–254.

[8] Y. Hellegouarch, Un analogue d'un théorème d'Euler, C.R. Acad. Sci. Paris, Sér. I Math. 313, (1991), 155–158.

[9] Y. Hellegouarch, Modules de Drinfeld généralisés, In: Approximation diophantienne et Nombres transcendants, Luminy 1990, á paraître.

[10] N. Jacobson, *Basic Algebra II*, Freeman (1989).

[11] S. Lang, *Algebra*, 1st ed, Addison-Wesley (1965).

[12] L.I. Wade, Certain quantities transcendental over $GF(p,x)$, Duke Math. J. 8, (1941), 701–720.

U.F.R. de Sciences, Université de Caen, 14032 Caen Cedex, France

A Two-Dimensional Analogue of Stickelberger's Theorem

*G. W. Anderson**

Inspired by an example due in part to R. Coleman and in part to D. Thakur, we exhibit meromorphic functions on certain surfaces of the form $X \times_{\mathbb{F}_q} X$ (X an abelian covering of the projective line over $\mathbb{F}_q$) that are closely related to (i) the abelian L-functions attached to the covering and (ii) an $\mathbb{F}_q[T]$-analogue of the gamma function. These relationships are described in our main result (Theorem 2 of §3.2).

Our main result and Stickelberger's theorem are analogous in that both specify the divisor of some explicitly given function on a regular scheme, the regular scheme in question being $X \times X$ in the former case and the spectrum of the ring of integers of a cyclotomic field in the latter. Moreover, the proofs of both results are similar in rough outline: The first step is to estimate from above the support of the divisor of the function under consideration; the second is to make a local calculation in order to find the multiplicities of the primes not already ruled out. In the Stickelberger case, the first step is accomplished by determining the complex absolute values of Gauss sums; the second is accomplished by expanding Gauss sums p-adically and evaluating the leading term. In our case, the first step is accomplished by employing (albeit in disguised fashion) a remarkable observation of Drinfeld concerning the vanishing of the cohomology of shtukas (to which we refer hereafter by the tag "Drinfeld's trick"); the second is accomplished by explicit calculations in the ring of power series in two variables over $\mathbb{F}_q$.

We begin in §1 with a review of explicit class field theory for $\mathbb{F}_q(T)$. The Coleman/Thakur example is reviewed in §2. In §3 we state our main result, explain a connection with the Brumer-Stark conjecture and give a new proof of the $\mathbb{F}_q[T]$-case of a result of D. Hayes **[6]** concerning the assembly of Brumer-Stark units into algebraic Hecke characters. §§4-6 are devoted to proving Theorem 2.

The results of this paper can be considerably generalized by systematic exploitation of Drinfeld's trick combined with methods of *soliton theory*. The utility of soliton methods is due chiefly to the fact that they provide means to study local properties of theta functions which remain valuable in characteristic p. This will be explained in future papers.

Acknowledgements. The author gratefully acknowledges helpful conversation and correspondence with R. Coleman, D. Goss, D. Hayes, T. Shiota and D. Thakur. The author also thanks the referee for useful criticism.

* Partially supported by NSF grants DMS-8957375 and DMS-8902621. The paper was presented at a conference on function field arithmetic held at The Ohio State University in June 1991 with NSF support.

1. Review of explicit CFT for $\mathbb{F}_q(T)$

1.1 The Carlitz exponential.

1.1.1. We begin with a review of some well-known results of L. Carlitz. Set

$$A =: \mathbb{F}_q[T], \qquad A_+ =: \{a \in A \mid a\colon \text{monic}\}.$$

Let $K =: \mathbb{F}_q(T)$ be the field of fractions of A. We identify the completion of K at the place $T = \infty$ with the Laurent series field $\mathbb{F}_q((1/T))$.

1.1.2. The *Carlitz exponential*

$$\exp_C(z) =: z + \sum_{n=1}^{\infty} \frac{z^{q^n}}{(T^{q^n} - T^{q^{n-1}}) \cdots (T^{q^n} - T)}$$

is uniquely characterized by the following two properties: Firstly, $\exp_C(z)$ is $\mathbb{F}_q$-linear with linear term equal to z. Secondly, $\exp_C(z)$ satisfies the functional equation

$$\exp_C(Tz) = T \exp_C(z) + \exp_C(z)^q. \tag{1}$$

1.1.3. The set of zeros of the function $\exp_C(z)$ in the algebraic closure of $\mathbb{F}_q((1/T))$ is an A-submodule of rank one and

$$\overline{\pi} =: \overline{T} T \prod_{i=1}^{\infty} (1 - T^{1-q^i})^{-1} \in \mathbb{F}_q((1/\overline{T})) \tag{2}$$

is a generator where $\overline{T}$ is a $(q-1)^{\text{st}}$ root of $-T$, fixed once and for all.

1.2 The explicit reciprocity law.

1.2.1. Next we review some results due to Carlitz, an account of which can be found in the paper **[4]** of D. Hayes. Fix $f \in A_+$. Then

$$\zeta_f =: \exp_C\left(\frac{\overline{\pi}}{f}\right)$$

generates an abelian extension of K in $\mathbb{F}_q((1/\overline{T}))$ unramified away from the infinite place and from primes dividing f. The ring $\mathbb{F}_q[T, \zeta_f]$ is the integral closure of A in $K(\zeta_f)$. For example, for all $c \in \mathbb{F}_q$,

$$(T - c) + z^{q-1} \text{ is the minimal polynomial of } \zeta_{T-c} \text{ in } K[z]. \tag{3}$$

1.2.2. For any monic irreducible polynomial $v \in A$ not dividing f, the action of the corresponding Artin symbol

$$\sigma_v \in G =: \mathrm{Gal}(K(\zeta_f)/K)$$

is given by the rule

$$\sigma_v \exp_C\left(\frac{\overline{\pi} a}{f}\right) = \exp_C\left(\frac{\overline{\pi} a v}{f}\right) \qquad (a \in A). \tag{4}$$

This last is the *explicit reciprocity law* for $\mathbb{F}_q(T)$.

1.2.3. The Artin symbol map may be extended to an isomorphism

$$a \mapsto \sigma_a : (A/f)^\times \tilde{\to} G.$$

Note that under this isomorphism, the inertia subgroup

$$\mathrm{Gal}(\mathbb{F}_q((1/\overline{T}))/\mathbb{F}_q((1/T))) \subset G$$

at ∞ corresponds to the subgroup

$$\mathbb{F}_q^\times \subset (A/f)^\times$$

and

$$\sigma_\varepsilon \overline{T} = \varepsilon \overline{T} \qquad (\varepsilon \in \mathbb{F}_q^\times). \tag{5}$$

1.3 Stickelberger elements. Set

$$\Theta_f(t) =: \sum_{\substack{a \in A_+ \\ \deg(a) < \deg(f) \\ (a,f)=1}} \sigma_a t^{\deg(f)-\deg(a)-1} \in \mathbb{Z}[G][t],$$

where $\mathbb{Z}[G][t]$ denotes the ring of polynomials in a variable t with coefficients in the integral group ring $\mathbb{Z}[G]$. The function $\Theta_f(t)$ is a variant of an L-function. More precisely, one has

$$t^{\deg(f)-1}\Theta_f(t^{-1}) + \frac{t^{\deg(f)} \sum_{\sigma \in G} \sigma}{1-qt} = \prod_{\substack{v \in A_+ \\ v:\text{irreducible} \\ v \nmid f}} (1 - \sigma_v t^{\deg(v)})^{-1}.$$

For example, in the case $f = T(T-1)$ that we are going to consider in detail in the next section,

$$\Theta_{T(T-1)}(t) = t + \sum_{0,1 \neq c \in \mathbb{F}_q} \sigma_{T-c}. \tag{6}$$

Set

$$\left(\sum_{g \in G} c_g g\right)^* =: \sum_{g \in G} c_g g^{-1} \qquad (c_g \in \mathbb{Z}).$$

We are ready to state a proposition providing a focus for our discussion.

Proposition 1. *$\Theta_f(1)^* \in \mathbb{Z}[G]$ annihilates the class group of $\mathbb{F}_q[T, \zeta_f]$.*

The proposition is well-known, as it is a consequence of the Brumer-Stark conjecture in the function field case. We note that for the latter there are at least two proofs available, cf. Tate **[9]** and Hayes **[5]**. As an application of our main result (Theorem 2 of §3.2) we give a different proof of Proposition 1.

2. An example

We present an example due in various parts to R. Coleman, to D. Thakur and the author; the rest of the paper will be devoted to generalizing it.

2.1 The geometric setting. Let $X/\mathbb{F}_q$ be the smooth projective model of the affine plane curve

$$U =: \mathrm{Spec}(\mathbb{F}_q[u,v]/(1+u^{q-1}-v^{q-1})).$$

Set

$$\mathrm{Frob} =: (u,v) \mapsto (u^q, v^q) \colon U \to U$$

and given $\alpha, \beta \in \mathbb{F}_q^\times$ set

$$[\alpha,\beta] =: (u,v) \mapsto (\alpha u, \beta v) \colon U \tilde{\to} U.$$

The maps Frob and $[\alpha,\beta]$ have unique extensions to maps $X \to X$ which are again denoted Frob and $[\alpha,\beta]$.

2.2 Coleman's function. Consider the meromorphic function

$$\phi =: -\frac{1\otimes u}{u\otimes 1} + \frac{1\otimes v}{v \otimes 1}$$

on $X \times X$. (Here and elsewhere in the paper $\times =: \times_{\mathrm{Spec}(\mathbf{F}_q)}$ and $\otimes =: \otimes_{\mathbf{F}_q}$.) Coleman's brilliant observation in the paper **[2]** was that, modulo horizontal and vertical components,

$$\begin{aligned} \mathrm{divisor}\,(1-\phi) &= \mathrm{graph}\,(\mathrm{Frob}\colon X \to X) \\ &\quad + \sum_{0,1\neq c\in\mathbf{F}_q} \mathrm{graph}\,([-c,1-c]\colon X \tilde{\to} X), \end{aligned} \tag{7}$$

and hence, by the theory of correspondences,

$$\mathrm{Frob}_* = -\sum_{0,1\neq c\in\mathbf{F}_q} [-c,1-c]_* \tag{8}$$

in the ring of endomorphisms of the Jacobian of X. Thus Coleman proved in a simple and direct way a fact the standard proof of which proceeds by way of the Lefschetz-Weil trace formula!

2.3 A proof of Proposition 1 in the case $f = T(T-1)$. In the paper **[2]** Coleman also gave a simple explicit proof of the Brumer-Stark conjecture in the special case which, in our context, coincides with Proposition 1 in the case $f = T(T-1)$. We recall the proof briefly, in language somewhat different from Coleman's. Set

$$B =: \mathbb{F}_q[T, \zeta_{T(T-1)}] \subset \mathbb{F}_q((1/\overline{T})).$$

By (3) we have

$$1 + \zeta_T^{q-1} - \zeta_{T-1}^{q-1} = 0$$

and therefore the pair

$$\tau =: (\zeta_T, \zeta_{T-1})$$

is a B-valued point of X. Note that the ring of functions regular on U is isomorphic to B under the map

$$g \mapsto g(\tau): \Gamma(U, \mathcal{O}_U) \tilde{\to} B.$$

Let $\mathbf{p}$ be a prime of B not dividing $T(T-1)$ and let d denote the degree of the residue field $B/\mathbf{p}$ over $\mathbb{F}_q$. Let

$$\xi_i =: (u_i, v_i) \qquad (i = 0, \ldots, d-1)$$

be the $\mathrm{Gal}(\overline{\mathbb{F}}_q/\mathbb{F}_q)$-orbit of $\overline{\mathbb{F}}_q$-valued points of U corresponding via τ to $\mathbf{p}$; by this we mean that for all functions g regular on U one has $g(\xi_i) = 0$ for some and hence all $i = 0, \ldots, d-1$ if and only if $g(\tau) \equiv 0 \bmod \mathbf{p}$. Finally, set

$$\chi(\mathbf{p}) =: \prod_{i=0}^{d-1} (1 - \phi(\xi_i, \tau)) = \prod_{i=0}^{d-1} (1 + u_0^{-q^i} \zeta_T - v_0^{-q^i} \zeta_{T-1}) \in B.$$

Then, by means of (4), (6) and (7) (or bare-handedly), one can check that

$$\chi(\mathbf{p}) \text{ generates the ideal } \mathbf{p}^{\Theta_{T(T-1)}(1)^*}. \tag{9}$$

This proves Proposition 1 in the case $f = T(T-1)$.

2.4 Existence of a conductor for χ. The most tantalizing feature of Coleman's idea is that it gives a formula for the Stark unit $\chi(\mathbf{p})$ that, combined with Weil reciprocity, provides natural means to show that $\mathbf{p} \mapsto \chi(\mathbf{p})$ is an "algebraic Hecke character". To be more precise, given any ideal I of B prime to $T(T-1)$, set

$$\chi(I) =: \prod_{\mathbf{p}} \chi(\mathbf{p})^{\mathrm{ord}_\mathbf{p} I},$$

the product extended over prime ideals $\mathbf{p}$ of B dividing I. Let $\mathbf{f}$ be the radical of the ideal of B generated by $T(T-1)$. We claim that

$$b \equiv 1 \bmod \mathbf{f} \Rightarrow \chi(bB) = b^{\Theta_{T(T-1)}(1)^*}, \tag{10}$$

and thus, by analogy with classical terminology, one is justified in calling χ an algebraic Hecke character of infinity type $\Theta_{T(T-1)}(1)^*$ and conductor dividing $\mathbf{f}$. Now by construction, left and right sides of equation (10) have the same valuation at all primes of B and in fact, both sides have the same valuation at all places of the fraction field of B, as is easy to check. Consequently, the desired equation holds up to an undetermined nonzero factor in $\mathbb{F}_q$. In order to fix this factor, it will be enough to show that for some prime $\mathbf{q}$ of B dividing T, one has

$$b \equiv 1 \bmod \mathbf{f} \Rightarrow \chi(bB) \equiv 1 \bmod \mathbf{q}. \tag{11}$$

In order to prove (11) we switch to geometric language.

2.5 Weil reciprocity. Fix $b \in B$ such that

$$b \equiv 1 \bmod \mathbf{f}.$$

Let g be the unique meromorphic function on X regular on U such that

$$g(\tau) = b.$$

Let D be the component of the divisor of g supported away from the poles of u and v. We view D as a $\mathrm{Gal}(\overline{\mathbb{F}}_q/\mathbb{F}_q)$-invariant element

$$D = \sum_i m_i(u_i, v_i)$$

of the free abelian group on the set of $\overline{\mathbb{F}}_q$-valued points of U with both coordinates nonvanishing. (The points (u_i, v_i) appearing here are a new batch unrelated to those appearing in §2.3 above.) Set

$$e =: \prod_i (1 + u_i^{-1}u - v_i^{-1}v)^{m_i}.$$

Then e is the unique meromorphic function on X such that

$$e(\tau) = \chi(bB).$$

Let

$$\kappa =: (0, 1),$$

an $\mathbb{F}_q$-valued point of U, and note that κ corresponds via τ to a prime $\mathfrak{q}$ of B dividing T. In order to prove (11) it will suffice to prove

$$e(\kappa) = \prod_i (1 - v_i^{-1})^{m_i} = 1. \tag{12}$$

The first equality above holds by definition; the second we prove by exploiting Weil reciprocity. Define a meromorphic function

$$h =: 1 - v^{-1}$$

on X. Noting that $h = 1$ at each pole of g, we have

$$e(\kappa) = \prod_\xi h(\xi)^{\mathrm{ord}_\xi g}$$

where the product is extended over all $\overline{\mathbb{F}}_q$-valued points ξ of X in the support of the divisor of g. But g takes the value 1 at each point in the support of the divisor of h (in particular, the functions g and h have disjoint supports) and therefore, by Weil reciprocity,

$$e(\kappa) = \prod_\eta g(\eta)^{\mathrm{ord}_\eta h} = 1$$

where the product is extended over all $\overline{\mathbb{F}}_q$-valued points η of X in the support of the divisor of h. Therefore (12) holds. The claim (10) is proved.

2.6 Thakur's formula. A remarkable characterization of Coleman's function ϕ was found by Thakur, who obtained the formula

$$1-\phi(\mathrm{Frob}^N\tau,[\alpha,\beta]) = \prod_{\substack{a\in A_+\\ \deg(a)=N-1}} \left(1+\frac{-\frac{\alpha}{T}+\frac{\beta}{T-1}}{a}\right) \tag{13}$$

$$(N>0,\ \alpha,\beta\in\mathbb{F}_q^\times).$$

Now recall the definitions. We have

$$\begin{aligned}1-\phi(\mathrm{Frob}^N\tau,[\alpha,\beta]\tau) &= 1+\alpha\zeta_T^{1-q^N}-\beta\zeta_{T-1}^{1-q^N}\\ &= 1+\alpha(-T)^{-\frac{q^N-1}{q-1}}-\beta(1-T)^{-\frac{q^N-1}{q-1}}.\end{aligned}$$

Therefore formula (13) boils down to an identity involving rational functions in one variable over $\mathbb{F}_q$; this identity and a collection of more general such were given by Thakur **[10**, §9.3]. Thakur's proof of (13) and various generalizations were based on the *Moore determinant formula*,

$$\det_{i,j=1}^{n} x_i^{q^{j-1}} = \prod_{m=1}^{n}\prod_{a_1\in\mathbb{F}_q}\cdots\prod_{a_{m-1}\in\mathbb{F}_q}\left(x_m+\sum_{i=1}^{m-1}a_ix_i\right), \tag{14}$$

an equality holding in the polynomial ring $\mathbb{F}_q[x_1,\ldots,x_n]$.

The product occurring on the right side of (13) is a reciprocal partial product of the $\mathbb{F}_q[T]$-version of the *gamma function* (cf. papers of Goss **[3]** and Thakur **[10]**)

$$\Gamma(z)=:z^{-1}\prod_{a\in A_+}\left(1+\frac{z}{a}\right)^{-1}$$

for a few special values of $z\in\mathbb{F}_q(T)$. All this seems to suggest (with hindsight...) that if one could prove an analogue of (13) in which $-\frac{\alpha}{T}+\frac{\beta}{T-1}$ were replaced by arbitrary $z\in\mathbb{F}_q(T)$, one would simultaneously get a generalization of Coleman's example.

3. Formulation and discussion of the main result

3.1 Notation and setting.

3.1.1. Fix $f\in A_+$. Set

$$B=:\mathbb{F}_q[T,\zeta_f]\subset\mathbb{F}_q((1/\overline{T})).$$

As in §1.3, let $a\mapsto\sigma_a\colon(A/f)^\times\tilde{\to}\mathrm{Gal}(K(\zeta_f)/K)$ denote the Artin symbol isomorphism. We write $\times=:\times_{\mathrm{Spec}(\mathbb{F}_q)}$ and $\otimes=:\otimes_{\mathbb{F}_q}$. For any $\mathbb{F}_q$-scheme S the qth power Frobenius endomorphism is denoted by Frob.

3.1.2. Let $X/\mathbb{F}_q$ be a smooth projective geometrically irreducible curve equipped with a point τ with values in the fraction field of B inducing an $\mathbb{F}_q$-linear isomorphism

$$g\mapsto g(\tau)\colon(\text{function field of }X)\tilde{\to}(\text{fraction field of }B).$$

Let $U/\mathbb{F}_q$ be the open subscheme of X corresponding via τ to $\mathrm{Spec}(B)$; by this we mean that for all $b \in B$ there exists a unique regular function g on U such that $g(\tau) = b$. Let U' be the open subscheme of U corresponding via τ to $\mathrm{Spec}(B[f^{-1}])$. Let ∞ denote the unique reduced closed subscheme of X complementary to U, and let V be the unique open subscheme of X containing ∞ such that $V \cap U = U'$. For each $a \in (A/f)^\times$ there exists a unique $[a] \in \mathrm{Aut}(X/\mathbb{F}_q)$ such that

$$[a]\tau = \sigma_a \tau .$$

For each nonnegative integer i and $a \in (A/f)^\times$, set

$$Z_a^i =: \mathrm{graph}\,([a] \circ \mathrm{Frob}^i \colon X \to X) \subset X \times X .$$

3.2 The main result.

Theorem 2. *There exists a unique meromorphic function ϕ on $X \times X$ regular on $V \times U$ such that for all $a \in A$ with $\deg(a) < \deg(f)$ and $(a, f) = 1$ and positive integers N,*

$$1 - \phi(\mathrm{Frob}^N \tau, [a]\tau) = \prod_{\substack{n \in A_+ \\ \deg(n) = N-1}} \left(1 + \frac{a}{fn}\right) . \tag{15}$$

One has

$$\begin{aligned}\mathrm{divisor}\,(\phi\,|_{V \times V}) &= \infty \times V - V \times \infty \\ &+ \sum_{\substack{a \in A \\ \deg(a) < \deg(f) \\ (a,f)=1}} \sum_{i=0}^{\deg(f)-\deg(a)-2} Z_a^i \cap (V \times V)\end{aligned} \tag{16}$$

and

$$\begin{aligned}\mathrm{divisor}\,((1-\phi)\,|_{V \times X}) &= -V \times \infty \\ &+ \sum_{\substack{a \in A_+ \\ \deg(a) < \deg(f) \\ (a,f)=1}} Z_a^{\deg(f)-\deg(a)-1} \cap (V \times X) .\end{aligned} \tag{17}$$

The proof of the theorem will be given in §6; in the course of the proof an explicit formula for ϕ will be developed. The uniqueness of the function ϕ is clear: If ϕ' also satisfies (15), then $\phi - \phi'$ vanishes on the transpose of the correspondence Z_a^N for all $a \in (A/f)^\times$ and $N > 0$; this is impossible unless $\phi' = \phi$.

3.3 Annihilation of the Jacobian of X. From the description of the divisor of $1 - \phi$ on $X \times X$ provided by the theorem, one deduces via the theory of correspondences that in the endomorphism ring of the Jacobian of X,

$$\Theta_f(\mathrm{Frob})_* = \sum_{\substack{a \in A_+ \\ \deg(a) < \deg(f) \\ (a,f)=1}} ([a] \circ \mathrm{Frob}^{\deg(f)-\deg(a)-1})_* = 0 . \tag{18}$$

Of course this is a consequence of the Lefschetz-Weil trace formula and the Cayley-Hamilton theorem. The point we wish to make here is that since the proof of Theorem 2 involves no use of the theory of abelian varieties and yields an explicit formula for ϕ, the proof of (18) proceeding by way of Theorem 2 is very much in the spirit of Coleman's proof of (8) by way of (7).

3.4 Construction of a family of Stark units. Let $\mathbf{p}$ be a prime of B not dividing f. Let d be the degree of the residue field $B/\mathbf{p}$ over $\mathbb{F}_q$. Let $\xi_0, \ldots, \xi_{d-1}$ be the $\mathrm{Gal}(\overline{\mathbb{F}}_q/\mathbb{F}_q)$-orbit of $\overline{\mathbb{F}}_q$-valued points of U' corresponding via τ to $\mathbf{p}$. Set

$$\chi(\mathbf{p}) = \prod_{i=0}^{d-1}(1 - \phi(\xi_i, \tau)) \in B.$$

From Theorem 2 one readily deduces the following two statements:

$$\chi(\mathbf{p}) \text{ generates the ideal } \mathbf{p}^{\Theta_f(1)^*}. \tag{19}$$

$$\sigma_a\chi(\mathbf{p}) \in \overline{T}^d\mathbb{F}_q[[1/\overline{T}]]^\times \qquad (a \in (A/f)^\times). \tag{20}$$

In view of (19), Proposition 1 becomes a corollary of Theorem 2. We think of (20) as an analogue of the fact that the complex absolute values of all conjugates of a Jacobi sum are the same. One can actually prove a result more strongly underscoring the analogy with Jacobi sums:

$$\prod_{\varepsilon \in \mathbb{F}_q^\times} \sigma_\varepsilon\chi(\mathbf{p}) = \mathbf{N}_{B/A}\mathbf{p}, \tag{21}$$

where $\mathbf{N}_{B/A}\mathbf{p} \in A_+$ is the unique monic generator of the norm from B to A of the ideal $\mathbf{p}$. The essential point of the proof of this last is the functional equation (37) appearing below; we omit further details.

3.5 Existence of a conductor for χ. Pursuing the analogy with Jacobi sums further, one expects that the Stark units $\chi(\mathbf{p})$ may be assembled into an algebraic Hecke character. This is indeed the case as the following proposition, due to Hayes **[6]**, affirms. We give a different proof. It must be noted that Hayes actually worked with a base ring $A = \Gamma(X \backslash \infty, \mathcal{O}_X)$, where $X/\mathbb{F}_q$ is a smooth projective geometrically irreducible curve and ∞ a closed point of degree one, and obtained a general result reducing in the case $A = \mathbb{F}_q[T]$ to Proposition 3.

Let $\mathbf{f}$ denote the radical of the ideal of B generated by f. For each ideal I of B prime to f put

$$\chi(I) =: \prod_{\mathbf{p}} \chi(\mathbf{p})^{\mathrm{ord}_{\mathbf{p}} I},$$

the product extended over primes $\mathbf{p}$ of B dividing I.

Proposition 3. *For all $b \in B$ such that*

$$b \equiv 1 \bmod \mathbf{f},$$

one has

$$\chi(bB) = b^{\Theta_f(1)^*}.$$

In order to prove Proposition 3, it suffices by (19) and (20) to show that for some prime $\mathbf{q}$ of B dividing $\mathbf{f}$,

$$\chi(bB) \equiv 1 \bmod \mathbf{q}. \tag{22}$$

Let g be the unique meromorphic function on X regular on U such that $g(\tau) = b$. We have

$$\chi(bB) = \prod_{\xi} (1 - \phi(\xi, \tau))^{\mathrm{ord}_\xi g}, \tag{23}$$

where the product is extended over all $\overline{\mathbb{F}}_q$-valued point ξ of U' in the support of the divisor of g. Now the divisor of g is supported in V, and $\phi(\xi, \tau) = 0$ if ξ is an $\overline{\mathbb{F}}_q$-valued points of X in the support of the divisor of g. Let κ be an $\overline{\mathbb{F}}_q$-valued point of X in the complement of U by Theorem 2. Therefore (23) remains true if we extend the product over all $\overline{\mathbb{F}}_q$-valued points ξ of X in the support of the divisor of g. Let κ be an $\overline{\mathbb{F}}_q$-valued point of U in the $\mathrm{Gal}(\overline{\mathbb{F}}_q/\mathbb{F}_q)$-orbit corresponding via τ to $\mathbf{q}$. In order to prove the congruence (22) it will be enough to prove

$$\prod_{\xi} (1 - \phi(\xi, \kappa))^{\mathrm{ord}_\xi g} = 1, \tag{24}$$

where again the product is extended over all $\overline{\mathbb{F}}_q$-valued points ξ in the support of the divisor of g.

Now extend scalars to $\overline{\mathbb{F}}_q$. Set $\overline{X} =: X \otimes \overline{\mathbb{F}}_q$ and $\overline{V} =: V \otimes \overline{\mathbb{F}}_q$. Let h denote the unique meromorphic function on $\overline{X}$ such that

$$h(\tau) = 1 - \phi(\tau, \kappa).$$

By Theorem 2, h is regular and nowhere vanishing on $\overline{V}$ and, in particular, the divisors of h and g are disjoint. By Weil reciprocity we have

$$\prod_{\xi} h(\xi)^{\mathrm{ord}_\xi g} = \prod_{\eta} g(\eta)^{\mathrm{ord}_\eta h},$$

where the product on the right is extended over all closed points η of $\overline{X}$ in the support of the divisor of h and the product on the left is extended over all closed points ξ of $\overline{X}$ in the support of the divisor of g. Since by hypothesis g takes the value 1 at each closed point in the complement of $\overline{V}$, the product on the right equals 1; the product on the left coincides with the left side of (24). Thus (24) is proved. The proof of Proposition 3 is complete.

Remark. At the Ohio conference, Hayes suggested that perhaps our proofs are not so different. In both cases, the main step is an application of a reciprocity law, namely Eisenstein reciprocity in Hayes' case, and Weil reciprocity in ours. But Eisenstein and Weil reciprocity in the function field case are more or less the same thing.

4. Formulary

4.1 An *ad hoc* residue calculus.

4.1.1. Let R be a complete regular noetherian local ring containing a field k, let L denote the fraction field of R and let $\mathcal{M}$ denote the maximal ideal of R. Let $\mathcal{H}_{R/k}$ be the functor assigning to each k-vector space W the L-vector space

$$\mathcal{H}_{R/k}W =: L \otimes_R (\varprojlim W \otimes_k R/\mathcal{M}^n).$$

Note that if W is equipped with the structure of a k-algebra, $\mathcal{H}_{R/k}W$ is equipped with the structure of an L-algebra. Let

$$\mathcal{H}_{R/k}((1/t)), \qquad \mathcal{H}_{R/k}[t], \qquad \mathcal{H}_{R/k}[[1/t]]$$

be the L-algebras obtained by applying the functor $\mathcal{H}_{R/k}$ to the k-algebras

$$k((1/t)), \qquad k[t], \qquad k[[1/t]],$$

respectively, where t is a variable.

4.1.2. The general element $f \in \mathcal{H}_{R/k}((1/t))$ may be represented as a doubly-infinite Laurent series

$$f(t) = \Delta^{-1} \sum_i a_i t^i$$

with

$$0 \neq \Delta \in R, \qquad a_i \in R, \qquad \lim_{i \to +\infty} a_i = 0,$$

where the limit is taken $\mathcal{M}$-adically. Set

$$\oint f(t)dt =: \Delta^{-1} a_{-1} \in L.$$

The ring $\mathcal{H}_{R/k}((1/t))$ contains $k((1/t))$ and also, for each $0 \neq m \in \mathcal{M}$, the Taylor expansion

$$-m \sum_{j=0}^{\infty} (mt)^j$$

of $(t - m^{-1})^{-1}$ at $t = 0$. One has a residue formula

$$\oint \frac{w(t)dt}{t - m^{-1}} = \begin{cases} -w(m^{-1}) & \text{if } w(t) \in (1/t)k[[1/t]] \\ 0 & \text{if } w(t) \in k[t] \end{cases} \tag{25}$$

which may be checked by expanding the integrand in a Laurent series.

4.1.3. In what follows we work exclusively with the following special case of the residue calculus defined above: We take R to be the ring $\mathbb{F}_q[[\overline{x}, \overline{y}, \overline{T}^{-1}]]$ where $\overline{x}$, $\overline{y}$ and $\overline{T}$ are independent variables and we set

$$x =: -\overline{x}^{q-1}, \qquad y =: -\overline{y}^{q-1}, \qquad T =: -\overline{T}^{q-1}.$$

(Note that, in particular, the fraction field of R contains the field $\mathbb{F}_q((1/\overline{T}))$ considered in §1.) We take k to be $\mathbb{F}_q$, and we write $\overline{\mathcal{H}}$ instead of $\mathcal{H}_{R/k}$. From the general formula (25) we deduce the special case

$$\oint \frac{w(t)dt}{t - T^{q^i}} = \begin{cases} -w^{q^i} & \text{if } w \in (1/T)\mathbb{F}_q[[1/T]] \\ 0 & \text{if } w \in \mathbb{F}_q[T] \end{cases} \tag{26}$$

which is of particular importance in the sequel.

4.2 The series $\mathbf{E}(y,t)$. Set

$$\mathbf{E}(y,t) =: \prod_{i=0}^{\infty}(1 - y^{q^i}t)^{-1} \in \mathbb{F}_q[[y,t]]^{\times}.$$

Note that $\mathbf{E}(y,t)^{\pm 1} \in \overline{\mathcal{H}}[t]$ because $\mathbf{E}(y,t)^{\pm 1}$ reduces modulo y^k to a polynomial in t for all positive integers k. We claim that $\mathbf{E}(y,t)$ is the unique element of $\mathbb{F}_q[[y,t]]$ satisfying

$$\mathbf{E}(0,t) = 1, \qquad \mathbf{E}(y^q,t) = (1 - yt)\mathbf{E}(y,t). \tag{27}$$

Suppose $F(y,t) \in \mathbb{F}_q[[y,t]]$ also satisfies (27). Then

$$1 \equiv F(y^{q^n},t) \equiv (1 - yt)\cdots(1 - y^{q^{n-1}}t)F(y,t) \bmod y^{q^n}$$

for all positive integers n and this is enough to prove the claim. It follows that

$$\mathbf{E}(T^{-1},t) = \sum_{n=0}^{\infty}\overline{T}^{-1}\exp_C\left(\frac{\overline{\pi}}{T^{n+1}}\right)t^n, \tag{28}$$

and in turn that

$$\overline{T}\oint \mathbf{E}(T^{-1},t)w(t)dt = \exp_C(\overline{\pi}w) \qquad (w \in \mathbb{F}_q((1/T))). \tag{29}$$

We remark that the infinite product formula (2) for $\overline{\pi}$ may be deduced by comparing the residues at $t = T$ on both sides of equation (28).

4.3 The coefficients $g_n(x,y)$. We define

$$g_n(x,y) =: \oint \mathbf{E}(x,t)^{-1}\mathbf{E}(y,t)t^{-n}\frac{dt}{t} \in \mathbb{F}_q[[x,y]] \qquad (n = 0,1,2,\ldots).$$

Given $0 \neq f,\ g \in \mathbb{F}_q[[x,y]]$, we write $f \sim g$ if there exists $u = u(x,y) \in \mathbb{F}_q[[x,y]]$ such that $u(0,0) = 1$ and $f = ug$. We claim that

$$g_n(x,y) \sim \prod_{i=0}^{n-1}(y - x^{q^i}). \tag{30}$$

There is nothing to prove if $n = 0$, so assume that $n > 0$. Since

$$\oint\left(\prod_{i=0}^{k-1}(1 - x^{q^i}t)\right)t^{-n}\frac{dt}{t} = 0 \qquad (0 \leq k \leq n-1),$$

the divisibility of the left side of (30) by the right side is clear. Now since $g_n(0,y)$ is the coefficient with which t^n appears in the expansion of $\mathbf{E}(y,t)$ in powers of t, it is clear that

$$g_n(0,y) = y^n + O(y^{n+1}),$$

and from this we deduce that the left side of (30) divided by the right side is a power series with constant term 1. The claim is proved.

4.4 The polynomials $\Psi_k(z)$. For each nonnegative integer k, set

$$\Psi_k(z) =: \left(\prod_{\substack{a\in A\\ \deg(a)<k}} z-a\right) \Big/ \left(\prod_{\substack{a\in A_+\\ \deg(a)=k}} a\right) \in K[z].$$

These polynomials were introduced long ago by Carlitz and more recently, for example, they were employed in the paper [**1**]. By the Moore determinant formula (14) one has

$$\Psi_k(z) = \begin{vmatrix} 1 & \dots & T^{k-1} & z \\ \vdots & & \vdots & \vdots \\ 1 & \dots & T^{(k-1)q^k} & z^{q^k} \end{vmatrix} \Big/ \begin{vmatrix} 1 & \dots & T^{k} \\ \vdots & & \vdots \\ 1 & \dots & T^{kq^k} \end{vmatrix}. \tag{31}$$

One also has

$$1+\Psi_k(z) = \prod_{\substack{a\in A_+\\ \deg(a)=k}} \left(1+\frac{z}{a}\right), \tag{32}$$

as can be checked by noting that the expressions on both sides of the claimed equality are polynomials of degree q^k in z with constant term 1 that vanish for all values of z of the form $-a$ with $a \in A$ monic of degree k. We claim that for all $w \in (1/T)\mathbb{F}_q[[1/T]]$,

$$(-T)^{-\frac{q^k-1}{q-1}} \oint w(t)\mathbf{E}(T^{-q^k},t)^{-1}\mathbf{E}(T^{-1},t)dt = \begin{cases} -\Psi_{k-1}(w) & (k>0) \\ \oint w(t)dt & (k=0) \end{cases}. \tag{33}$$

The case $k=0$ is trivial and so we may assume that $k>0$. In order to prove the claim, consider the partial fraction decomposition

$$\prod_{i=0}^{k}(z-T^{q^i})^{-1} =: \sum_{i=0}^{k} \left\{ {k \atop i} \right\} (z-T^{q^i})^{-1} \quad \left(\left\{ {k \atop i} \right\} \in K\right).$$

Call the expression on the left side of this equation $R(z)$. Then for all monic $a \in A$ of degree k,

$$-1 = \mathrm{Res}_{z=\infty}(a(z)R(z)dz) = -\sum_{i=0}^{k} \left\{ {k \atop i} \right\} a^{q^i},$$

whence it follows that

$$\Psi_k(z) = \sum_{i=0}^{k} \left\{ {k \atop i} \right\} z^{q^i}.$$

Since

$$(-T)^{-\frac{q^k-1}{q-1}} \mathbf{E}(T^{-q^k}, t)^{-1} \mathbf{E}(T^{-1}, t) = \prod_{i=0}^{k-1} (t - T^{q^i})^{-1},$$

the claim now follows from (26) and the definitions.

5. Solitons

5.1 Definition. For each $w \in \mathbb{F}_q((1/T))$ we define

$$\Phi_w(\overline{x}, \overline{y}) =: (\overline{x}/\overline{y}) \oint w(t)\mathbf{E}(x,t)^{-1}\mathbf{E}(y,t)dt \in (\overline{x}/\overline{y})\mathbb{F}_q[[x,y]].$$

We call such a function a *soliton*. Note the similarity of the definition to formula (29). Note that $\Phi_w(\overline{x}, \overline{y})$ depends $\mathbb{F}_q$-linearly on w. Let $\langle w \rangle$ denote the unique element of the set $(w + \mathbb{F}_q[T]) \cap (1/T)\mathbb{F}_q[[1/T]]$. Note that

$$\Phi_w(\overline{x}, \overline{y}) = \Phi_{\langle w \rangle}(\overline{x}, \overline{y}).$$

5.2 A note on terminology. It seems that, inspired by Krichever's ideas about commuting differential operators and nonlinear partial differential equations possessing *solitary wave* or *soliton* solutions such as the *Korteweg-de Vries (KdV) equation*

$$u_{xxx} + uu_x = u_t,$$

Drinfeld came up with the idea of a *shtuka* (cf. Mumford [7]). This suggests a general heuristic principle:

> Methods for solving "soliton equations", transposed to characteristic p, become methods to study function field arithmetic.

Now Sato and the Kyoto school have devised another approach to the solution of soliton equations (cf. Segal-Wilson [8], especially Chapter 8). We have found that the Kyoto school notion of the τ-function transposes nicely to characteristic p and provides the means to construct and to manipulate functions such as $\Psi_w(\overline{x}, \overline{y})$ (hence our choice of terminology). Using the Kyoto school ideas, one ought to be able to obtain generalization of all the results of this paper to the case of a base ring of the form $\Gamma(X\backslash\infty, \mathcal{O}_X)$ with $X/\mathbb{F}_q$ a smooth projective geometrically irreducible curve and ∞ a closed point. Details will appear in future papers.

5.3 An interpolation formula. The soliton $\Phi_w(\overline{x}, \overline{y})$ figures in an *interpolation formula*

$$1 - \Phi_w(\overline{T}^{-q^N}, \overline{T}^{-1}) = \prod_{\substack{a \in A_+ \\ \deg(a) = N-1}} \left(1 + \frac{\langle w \rangle}{a}\right) \qquad (N > 0). \tag{34}$$

This requires no proof, as it is merely a restatement of (33) combined with (32). Note that for $w \in \mathbb{F}_q(T)$ the interpolation formula above is a power series analogue of (15). Note also that the formula (34) uniquely characterizes $\Phi_w(\overline{x}, \overline{y})$, for were this not to be the case, one could exhibit a nonzero element of the unique factorization domain $\mathbb{F}_q[[\overline{x}, \overline{y}]]$ belonging to infinitely many prime ideals of height one. From the definitions it follows easily that

$$\Phi_w(\overline{T}^{-1}, \overline{T}^{-q^m}) = \oint t^m w(t)dt \qquad (m \geq 0,\ \langle w \rangle \in T^{-m-1}\mathbb{F}_q[[1/T]]). \quad (35)$$

5.4 A distribution relation. We claim that the following *distribution relation* holds:

$$1 - \Phi_w(\overline{x}^{q^{\deg(f)}}, \overline{y}) = \prod_{\substack{b \in A \\ \deg(b) < \deg(f)}} \frac{1 - \Phi_{w + \frac{b}{f}}(\overline{x}, \overline{y})}{1 - \Phi_{\frac{b}{f}}(\overline{x}, \overline{y})} \qquad (f \in A_+). \quad (36)$$

(Cf. [10, Theorem 6.2.1, p. 50].) But this follows via the characterization (34) from the identity

$$1 + \Psi_{N+\deg(f)}(z) = \prod_{\substack{b \in A \\ \deg(b) < \deg(f)}} \frac{1 + \Psi_N\left(\frac{z+b}{f}\right)}{1 + \Psi_N\left(\frac{b}{f}\right)},$$

which may be checked by noting that the expressions on both sides are polynomials in z with coefficients in K of the same degree with the same zeros and with constant term 1.

5.5 A functional equation. We claim that

$$1 - \Phi_w(\overline{x}, \overline{y})^{q-1} = \left(\frac{1}{y} - \frac{1}{x}\right) \frac{\Phi_w(\overline{x}^q, \overline{y})}{\Phi_w(\overline{x}, \overline{y})}. \quad (37)$$

In order to prove the claim, we begin by noting that

$$\Psi'_N(0) = \frac{1}{(T - T^q) \cdots (T - T^{q^N})},$$

as may be verified with the aid of (31) and the Vandermonde determinant formula. Then note that

$$1 - \Psi_{N-1}(z)^{q-1} = \frac{\Psi'_{N-1}(0)}{\Psi'_N(0)} \frac{\Psi_N(z)}{\Psi_{N-1}(z)} \qquad (N > 0)$$

which may be checked by observing that both sides are polynomials in z of the same degree with the same sets of zeros and the same constant terms. The claim follows now by the characterization (34).

5.6 Asymptotics. In this paragraph we make the local calculation decisive for the proof of the theorem. Given $0 \neq f,\ g$ in the fraction field of $\mathbb{F}_q[[\overline{x}, \overline{y}]]$, we write $f \sim g$ if there exists $u = u(\overline{x}, \overline{y}) \in \mathbb{F}_q[[\overline{x}, \overline{y}]]$ such that $u(0,0) = 1$ and $f = ug$. Fix $w \in \mathbb{F}_q((1/T))$ such that $\langle w \rangle \neq 0$. Let m be the unique nonnegative integer and $\varepsilon \in \mathbb{F}_q^\times$ the unique nonzero constant such that

$$\langle w \rangle = \varepsilon T^{-m-1} + \sum_{i < -m-1} w_i T^i \qquad (w_i \in \mathbb{F}_q).$$

We claim that

$$\Phi_w(\overline{x},\overline{y}) \sim \varepsilon \frac{\overline{x}}{\overline{y}} \prod_{i=0}^{m-1} (y - x^{q^i}) \tag{38}$$

and

$$1 - \Phi_w(\overline{x},\overline{y}) \sim 1 - \varepsilon \frac{\overline{x}^{q^m}}{\overline{y}}. \tag{39}$$

In order to prove the claims, note firstly that from (30) it follows that $g_m(x,y)$ divides $g_{m+i}(x,y)$ in $\mathbb{F}_q[[x,y]]$ for all positive indices i and secondly that the ratio $g_{m+i}(x,y)/g_m(x,y)$ belongs to the maximal ideal of $\mathbb{F}_q[[x,y]]$. Therefore we can write

$$\Phi_w(\overline{x},\overline{y}) = \varepsilon \frac{\overline{x}}{\overline{y}} g_m(x,y) \left(1 + \sum_{i>0} w_{-m-1-i} \frac{g_{m+i}(x,y)}{g_m(x,y)}\right),$$

where the factor on the right is a power series with unit constant term. This formula together with (30) proves (38). As for the second claim (39), we have in any case by (37) and (38) that

$$1 - \Phi_w(\overline{x},\overline{y})^{q-1} \sim 1 - \frac{x^{q^m}}{y}.$$

But by (35) one has

$$\Phi_w(\overline{T}^{-1}, \overline{T}^{-q^m}) = \varepsilon.$$

These last two relations imply the claim (39).

6. Proof of the theorem

We return to the setting of §3. Set

$$d =: \deg(f).$$

6.1 The construction of ϕ.

6.1.1. Let R be an $\mathbb{F}_q$-algebra. Set

$$H(R) =: R[t]/f(t),$$

and let $H^\times(R)$ denote the multiplicative group of the ring $H(R)$. As an R-module, $H(R)$ is free of rank d on the basis $1,\ t,\ \ldots, t^{d-1}$ and hence for each $h \in H(R)$ there exist unique coefficients

$$h_i \in R \qquad (0 \le i \le d-1)$$

such that

$$h \equiv \sum_{i=0}^{d-1} h_i t^i \bmod f(t).$$

Set

$$\theta(h) =: h_{d-1}, \qquad h^F =: \sum_{i=0}^{d-1} h_i^q t^i .$$

(We choose the notation θ because, in fact, θ is a generalized theta function on $H^\times$, the latter viewed as the Picard variety of a singular model of the projective line.) Note that

$$(g,h) \mapsto \theta(gh)\colon H(R) \times H(R) \to R \text{ is a perfect pairing.} \tag{40}$$

Given an affine $\mathbb{F}_q$-scheme Y, we write

$$H(Y) =: H(\Gamma(Y, \mathcal{O}_Y)).$$

6.1.2. Let R be an $\mathbb{F}_q$-algebra. Given $h \in H(R)$, let $\Delta(h) \in R$ denote the determinant of multiplication by h viewed as an endomorphism of the R-module underlying $H(R)$. Note that

$$\Delta(t - t_0) = (-1)^d f(t_0) \qquad (t_0 \in R), \tag{41}$$

and

$$H^\times(R) = \{h \in H(R) \mid \Delta(h) \in R^\times\}. \tag{42}$$

6.1.3. We claim that for all positive integers N and $a \in A$,

$$\theta(a(t)(t-T)^{-1} \ldots (t - T^{q^{N-1}})^{-1}) = -\Psi_{N-1}\left(\left\langle \frac{a}{f} \right\rangle\right). \tag{43}$$

Note that $t - T^{q^i}$ belongs to $H^\times(A[f^{-1}])$ by (42); this is how one makes sense of $(t - T^{q^i})^{-1}$ as it occurs on the left side of the preceding equation. We have

$$\oint \frac{a(t)dt}{f(t)(t-T)\cdots(t-T^{q^{N-1}})} = -\Psi_{N-1}\left(\left\langle \frac{a}{f} \right\rangle\right)$$

by (33). Let $P(t) \in K[t]$ be the unique polynomial of degree strictly less than d such that the rational function

$$\frac{a(t)}{(t-T)\cdots(t-T^{q^{N-1}})} - P(t)$$

of t defined over K vanishes at each zero of $f(t)$ at least as many times as does $f(t)$. Then by definition

$$\theta(P(t)) = \theta(a(t)(t-T)^{-1} \cdots (t - T^{q^{N-1}})^{-1}).$$

But

$$\oint \frac{a(t)dt}{f(t)(t-T)\cdots(t-T^{q^{N-1}})} = \oint \frac{P(t)dt}{f(t)} = \theta(P(t)).$$

The claim is proved.

6.1.4. Let R be an A-algebra. Define a functor of A-algebras

$$M(R) =: \{h \in H(R) \mid h^F = (t-T)h\}.$$

We claim that there exists a unique

$$\mu \in M(B)$$

such that

$$\theta(a(t)\mu) = \exp_C\left(\overline{\pi}\frac{a}{f}\right) \qquad (a \in A). \tag{44}$$

Uniqueness and existence of $\mu \in H(B)$ satisfying (44) follows from (40). It remains to verify that μ indeed belongs to $M(B)$. But for all $a \in A$ we have

$$\begin{aligned}\theta(a(t)(t-T)\mu) &= \exp_C(\overline{\pi}Ta/f) - T\exp_C(\overline{\pi}a/f)\\ &= \exp_C(\overline{\pi}a/f)^q\\ &= \theta(a(t)\mu)^q\\ &= \theta(a(t)\mu^F)\end{aligned}$$

which is enough by (40) to prove the claim.

6.1.5. We have

$$\begin{aligned}\Delta(\mu)^q &= \Delta(\mu^F)\\ &= \Delta((t-T)\mu)\\ &= (-1)^d f\Delta(\mu),\end{aligned}$$

hence

$$\Delta(\mu)^{q-1} = (-1)^d f,$$

hence

$$\mu \in H^\times(B[f^{-1}]). \tag{45}$$

Also note that

$$\sigma_a\mu = a(t)\mu \qquad (a \in (A/f)^\times), \tag{46}$$

as follows from (4) and the definition of μ.

6.1.6. We define $u \in H(U)$ to be the unique point specializing via τ to $\mu \in M(B)$. Note that by (45) one has $u \in H^\times(U')$. Let $u\otimes 1 \in H^\times(U' \times U)$ and $1\otimes u \in H(U'\times U)$ be defined by pulling u back under the first projection $U'\times U \to U'$ and second projection $U'\times U \to U$, respectively. Finally let ϕ be the meromorphic function on $X\times X$ regular on $U'\times U$ defined by the formula

$$\phi =: \theta\left(\frac{1\otimes u}{u\otimes 1}\right).$$

Then for all positive integers N and $a \in A$ such that $(a,f) = 1$,

$$\begin{aligned}1 - \phi(\mathrm{Frob}^N \tau, \sigma_a\tau) &= 1 - \theta(\mu^{-F^N}(\sigma_a\mu))\\ &= 1 - \theta(a(t)(t-T)^{-1}\cdots(t-T^{q^{N-1}})^{-1})\\ &= 1 + \Psi_{N-1}\left(\left\langle\frac{a}{f}\right\rangle\right)\end{aligned}$$

and hence ϕ satisfies (15) as required.

6.2 Estimation of the supports of the divisors of ϕ and $1-\phi$.

6.2.1. The following lemma and its proof were inspired by a remarkable observation of Drinfeld (cf. [7, p. 146]) which asserts the vanishing of the cohomology of certain coherent sheaves.

Lemma 4. *Let $\xi,\ \eta \in \overline{\mathbb{F}}_q$ be given such that*

$$f(\xi) \neq 0 \tag{47}$$

and

$$\xi^{q^i} \neq \eta \qquad (0 \leq i \leq d-1). \tag{48}$$

Let $h \in H(\overline{\mathbb{F}}_q)$ be given such that

$$(t-\xi)h^F = (t-\eta)h. \tag{49}$$

If $\theta(h)^q - \theta(h) = 0$ then $h = 0$.

In order to prove the lemma, set

$$R_n(t) =: \prod_{i=0}^{n-1} \frac{t-\eta^{q^i}}{t-\xi^{q^i}} \in \overline{\mathbb{F}}_q(t) \qquad (0 \leq n \leq d).$$

By (47) one has $(t-\xi^{q^i}) \in H^\times(\overline{\mathbb{F}}_q)$, hence $R_n(t)$ may be construed as an element of $H(\overline{\mathbb{F}}_q)$, and in this sense we have

$$h^{F^n} = R_n(t)h.$$

Let $P_n(t) \in \overline{\mathbb{F}}_q[t]$ be the unique polynomial of degree strictly less than d such that the rational function $R_n(t) - P_n(t)$ vanishes at each zero of $f(t)$ at least as many times as does $f(t)$; it is possible to select $P_n(t)$ because, by (47), $R_n(t)$ has no poles among the zeros of $f(t)$. Note that then

$$\theta(h^{F^n}) = \theta(R_n(t)h) = \theta(P_n(t)h).$$

Set

$$r_n(t) =: R_{n+1}(t) - R_n(t), \quad p_n(t) =: P_{n+1}(t) - P_n(t) \qquad (0 \leq n \leq d-1).$$

Then

$$0 = (\theta(h)^q - \theta(h))^{q^n} = \theta(h^{F^{n+1}} - h^{F^n}) = \theta(p_n(t)h).$$

In order to prove the lemma, it will be enough by (40) to prove that the polynomials $p_0(t), \ldots, p_{d-1}(t)$ are linearly independent. Suppose that

$$\sum_{i=0}^{d-1} c_i p_i(t) = 0 \qquad (c_i \in \overline{\mathbb{F}}_q).$$

Then the linear combination

$$r(t) =: \sum_{i=0}^{d-1} c_i r_i(t)$$

vanishes at each zero of $f(t)$ to at least the same order as does $f(t)$ and also vanishes at $t = \infty$, hence vanishes at least $d+1$ times; since $r(t)$ is of degree not exceeding d, in fact, $r(t)$ vanishes identically. But by (48), the collection $R_0(t), \ldots, R_d(t)$ is linearly independent over $\overline{\mathbb{F}}_q$, hence the collection $r_0(t), \ldots, r_{d-1}(t)$ is linearly independent, hence the coefficients c_i vanish identically. The proof of the lemma is complete.

6.2.2. We claim that

$$\text{support}\,(\text{divisor}\,(1 - \phi^{q-1})) \cap (U' \times U) \subseteq \bigcup_{i=0}^{d-1} \bigcup_{a \in (A/f)^\times} Z_a^i. \tag{50}$$

We define

$$\mu^{-1} \otimes \mu \in H(B[f^{-1}] \otimes B)$$

to be the product of the image of $\mu^{-1} \in H^\times(B[f^{-1}])$ under $b \mapsto b \otimes 1$ and the image of $\mu \in H(B)$ under $b \mapsto 1 \otimes b$. In terms of rings what we have to prove is that

$$1 - \theta(\mu^{-1} \otimes \mu)^{q-1} \in \left((B[f^{-1}] \otimes B) \left[\prod_{i=0}^{d-1} (T^{q^i} \otimes 1 - 1 \otimes T)^{-1} \right] \right)^\times .$$

Let λ be an arbitrary $\mathbb{F}_q$-linear homomorphism of the ring appearing on the right above into $\overline{\mathbb{F}}_q$. Let

$$h \in H(\overline{\mathbb{F}}_q), \quad \xi \in \overline{\mathbb{F}}_q, \quad \eta \in \overline{\mathbb{F}}_q$$

be the specializations under λ of

$$\mu^{-1} \otimes \mu, \quad T \otimes 1, \quad 1 \otimes T$$

respectively. It will be enough to prove that

$$\theta(h)^{q-1} \neq 1.$$

But h, ξ, η satisfy the hypotheses of Lemma 4, hence this last holds. The claim is proved.

6.2.3. We claim that

$$\text{support}\,(\text{divisor}\,(\phi)) \cap (U' \otimes U') \subseteq \bigcup_{i=0}^{d-1} \bigcup_{a \in (A/f)^\times} Z_a^i. \tag{51}$$

In terms of rings what we have to prove is that

$$\theta(\mu^{-1} \otimes \mu) \in \left((B[f^{-1}] \otimes B[f^{-1}]) \left[\prod_{i=0}^{d-1} (T^{q^i} \otimes 1 - 1 \otimes T)^{-1} \right] \right)^\times .$$

Let λ be an arbitrary $\mathbb{F}_q$-linear homomorphism of the ring appearing on the right above into $\overline{\mathbb{F}}_q$, and let $h \in H(\overline{\mathbb{F}}_q)$ and $\xi,\ \eta \in \overline{\mathbb{F}}_q$ be as defined above. Then $h,\ \xi,\ \eta$ satisfy the hypotheses of Lemma 4 and hence either $\theta(h) \neq 0$ or $h = 0$. But the latter possibility is excluded: Since

$$\mu^{-1} \otimes \mu \in H^{\times}(B[f^{-1}] \otimes B[f^{-1}]),$$

necessarily $h \in H^{\times}(\overline{\mathbb{F}}_q)$ and, in particular, $h \neq 0$. The claim is proved.

6.3 Computation of multiplicities.

6.3.1. Since X is complete, the point τ may be viewed as a morphism

$$\tau\colon \mathrm{Spec}(\mathbb{F}_q[[1/\overline{T}]]) \to X$$

inducing an isomorphism of $\mathbb{F}_q[[1/\overline{T}]]$ with the completion of the local ring of X at the closed point in the image. Let

$$\imath\colon \mathrm{Spec}(\mathbb{F}_q[[\overline{x}, \overline{y}]]) \to X \times X$$

be the composite

$$\mathrm{Spec}(\mathbb{F}_q[[\overline{x}, \overline{y}]]) \to \mathrm{Spec}(\mathbb{F}_q[[1/\overline{T}]])^2 \to X \times X$$

where the arrow on the right is $\tau \times \tau$, and the arrow on the left is induced by the ring homomorphism $g(\overline{T}) \otimes h(\overline{T}) \mapsto g(1/\overline{x}) \otimes h(1/\overline{y})$. For all $b,\ c \in A$ prime to f, let

$$\imath_{b,c} =: ([b]^{-1} \times [c]) \circ \imath\colon \mathrm{Spec}(\mathbb{F}_q[[\overline{x}, \overline{y}]]) \to X \times X.$$

Note that for any meromorphic function ψ on $X \times X$, the pull-back $\imath_{b,c}^{*}\psi$ may be characterized as the unique element of the fraction field of $\mathbb{F}_q[[\overline{x}, \overline{y}]]$ such that for all sufficiently large positive integers N one has

$$(\imath_{b,c}^{*}\psi)(\overline{T}^{-q^N}, \overline{T}^{-1}) = \sigma_b^{-1}(\psi(\mathrm{Frob}^N \tau, [bc]\tau)).$$

If in the light of the preceding observation we compare the interpolation formulas (15) and (34), we find that

$$\imath_{b,c}^{*}\phi = \Phi_{\frac{bc}{f}}(\overline{x}, \overline{y}). \tag{52}$$

This says that the functions

$$\Phi_{\frac{a}{f}}(\overline{x}, \overline{y}) \qquad (a \in A,\ \deg(a) < \deg(f),\ (a, f) = 1)$$

are "branches" of a single global meromorphic function on $X \times X$, namely the function ϕ.

6.3.2. Define divisors

$$D_0 =: \infty \times V - V \times \infty + \sum_{\substack{a \in A \\ \deg(a) < \deg(f) \\ (a,f)=1}} \sum_{i=0}^{\deg(f)-\deg(a)-2} Z_a^i$$

and

$$D_1 =: -V \times \infty + \sum_{\substack{a \in A_+ \\ \deg(a) < \deg(f) \\ (a,f)=1}} Z_a^{\deg(f)-\deg(a)-1}$$

of $X \times X$. In order to prove the theorem, in view of (50) and (51), it will be enough to prove that

$$(\infty \times \infty) \cap \text{support}\,(\text{divisor}\,(\phi) - D_0) = \emptyset,$$
$$(\infty \times \infty) \cap \text{support}\,(\text{divisor}\,(1-\phi) - D_1) = \emptyset.$$

6.3.3. Given any divisor E of $X \times X$, write $\imath_{b,c}^* E$ instead of $\imath_{b,c}^* \mathcal{O}_{X \times X}(E)$. By definition $\imath_{b,c}^* E$ is a rank one free $\mathbb{F}_q[[\overline{x},\overline{y}]]$-submodule of the fraction field of $\mathbb{F}_q[[\overline{x},\overline{y}]]$; we write $\imath_{b,c}^* E = (g)$ if an element g of the fraction field of $\mathbb{F}_q[[\overline{x},\overline{y}]]$ generates $\imath_{b,c}^* E$ over $\mathbb{F}_q[[\overline{x},\overline{y}]]$. Now note that every closed point lying in the closed subscheme $\infty \times \infty \subset X \times X$ appears in the image of $\imath_{b,c}$ for some b, $c \in (A/f)^\times$ and, moreover, $\imath_{b,c}$ induces an isomorphism of $\mathbb{F}_q[[\overline{x},\overline{y}]]$ with the completion of the local ring of $X \times X$ at the closed point in the image of $\imath_{b,c}$. In order to prove the theorem it will be enough to check that

$$\imath_{b,c}^* D_0 = (\Phi_{\frac{bc}{f}}(\overline{x},\overline{y})), \qquad \imath_{b,c}^* D_1 = (1 - \Phi_{\frac{bc}{f}}(\overline{x},\overline{y})) \tag{53}$$

for all $b, c \in (A/f)^\times$.

6.3.4. In any case, we have

$$\imath_{b,c}^*(\infty \times X) = (\overline{x}), \qquad \imath_{b,c}^*(X \times \infty) = (\overline{y})$$

and

$$\imath_{b,c}^* Z_{abc}^i = \imath^* Z_a^i.$$

The latter following from the consideration of the Cartesian square below.

$$\begin{array}{ccc} Z_a^i & \subset & X \times X \\ \downarrow & & \downarrow \quad [b]^{-1} \times [c] \\ Z_{abc}^i & \subset & X \times X \end{array}$$

By the observations of §1.2.3 we have

$$\imath^* Z_a^i = 1 \qquad (0 < \deg(a) < \deg(f), \quad (a,f) = 1),$$

and

$$\imath^* Z_\varepsilon^i = (\varepsilon^{-1}\overline{x}^{q^i} - \overline{y}) \qquad (\varepsilon \in \mathbb{F}_q^\times).$$

Now write

$$\left\langle \frac{bc}{f} \right\rangle = \varepsilon T^{-m-1} + \text{lower powers of } T \qquad (\varepsilon \in \mathbb{F}_q^\times, \ m \in \mathbb{Z}_{\geq 0}).$$

We have

$$\imath_{b,c}^* D_0 = \left(\frac{\overline{x}}{\overline{y}} \prod_{i=0}^{m-1} (y - x^{q^i}) \right) \tag{54}$$

and

$$\imath_{b,c}^{*} D_1 = \left(1 - \varepsilon \frac{\overline{x}^{q^m}}{\overline{y}}\right) . \tag{55}$$

Comparison with (38) and (39) now proves (53). The proof of Theorem 2 is complete.

References

[1] Anderson, G. and Thakur, D.: Tensor powers of the Carlitz module and zeta values, *Ann. of Math.* **132** (1990), 159–191.

[2] Coleman, R.: On the Frobenius endomorphisms of Fermat and Artin-Schreier curves. *Proc. Amer. Math. Soc.* **102**, (1988), 463–466.

[3] Goss, D.: The Γ-function in the arithmetic of function fields. *Duke Math. J.* **56**, (1988), 163–191.

[4] Hayes, D.: Explicit class field theory for rational function fields. *Trans. Amer. Math. Soc.* **189**, (1974), 77–91.

[5] Hayes, D.: The refined $\mathfrak{p}$-adic abelian Stark conjecture in function fields. *Invent. Math.* **94**, (1988), 505–527.

[6] Hayes, D.: Hecke characters and Eisenstein reciprocity in function fields. Preprint 1991.

[7] Mumford, D.: An algebro-geometric construction of commuting operators and of solutions to the Toda lattice equation, Korteweg-de Vries equation and related nonlinear equations. *Intl. Symp. Alg. Geom.*, Kyoto, (1977), 115–153.

[8] Segal, G. and Wilson, G.: Loop groups and equations of KdV type. *Inst. Hautes Études Sci. Publ. Math.* **61**, (1985), 5–65.

[9] Tate, J.: *Les Conjectures de Stark sur les Fonctions L d'Artin en* $s = 0$. Boston, Birkhäuser 1984.

[10] Thakur, D.: Gamma functions for function fields and Drinfeld modules. *Ann. of Math.* **134**, (1991), 25–64.

Department of Mathematics, University of Minnesota, Minneapolis, MN 55455

On Gamma Functions for Function Fields

D. S. Thakur

This expository account of gamma functions for function fields is divided into two parts. The first part (Sections I-III) deals with the work done by Carlitz, Goss and myself. We will not follow the chronological path and hence will not give any detailed references. (See **[C]**, **[G]**, **[T4]** and references given there.) The second part (Sections IV-V) is about some new results. (To appear in **[T6]**.) For various interesting aspects of the classical gamma function, the reader can refer to **[Ar]**, **[D]**, **[G-K]**.

Notation.

q: a power of a prime p
K: a function field of one variable with field of constants $\mathbb{F}_q$
∞: a fixed place of K, called the infinite place (other classical uses of the symbol ∞ should be clear from the context)
A: the ring of integers outside ∞
K_∞: the completion of K at ∞
Ω: the completion of an algebraic closure of K_∞
v: a place of K other than ∞, also called a finite place
K_v: the completion of K at v

For simplicity, we will restrict to the case where ∞ has degree one: In general, the nature of the results is expected to be the same after the right generalization of the definitions.

Let us fix a uniformizing parameter t at ∞, so that we have, for $x \in K_\infty^*$, the Laurent expansion $x = \operatorname{sgn}(x)t^{-\deg(x)}+$ lower order terms, with $\operatorname{sgn}(x) \in \mathbb{F}_q^*$ and $\deg(x) \in \mathbb{Z}$. We call x monic or positive, if $\operatorname{sgn}(x) = 1$. Let A_+ (resp. $\mathbb{Z}_+$) be the set of positive elements of A (resp. $\mathbb{Z}$) and let $A_{\geq 0} =: A_+ \cup \{0\}$ and $\mathbb{Z}_{\geq 0} := \mathbb{Z}_+ \cup \{0\}$.

Section I. Constructions and basic properties

We will consider three gamma functions. The first one is the classical gamma function and the other two are connected with function fields. To stress the analogies, we will use the same notation, but we will distinguish three cases.

Case I: 'Characteristic 0'. For $n \in \mathbb{Z}_{\geq 0}$, put $n! := n(n-1)\cdots 1$ and for $n \in \mathbb{Z}_+$, put $\Gamma(n) := (n-1)!$. Then, as is well-known, Euler's gamma function $\Gamma(z) := \int_0^\infty e^{-r} r^z \, dr/r$ (for $z \in \mathbb{C} - \mathbb{Z}_{\leq 0}$) interpolates the values above at the integers, continues to satisfy $\Gamma(z+1) = z\Gamma(z)$ and is logarithmically convex. Put $z! := \Gamma(z+1)$. We can think of Γ as a gamma function at infinity.

For a prime p and $n \in \mathbb{Z}_+$, let $\Gamma_p(n) := (-1)^n \prod_{j<n,(p,n)=1} j$. Then for $z \in \mathbb{Z}_p$, Morita's p-adic gamma function $\Gamma_p(z)$ is defined as the limit of $\Gamma_p(n)$, as n tends to z, p-adically. We get a continuous function $\Gamma_p: \mathbb{Z}_p \to \mathbb{Z}_p^*$. The p-adic factorial is defined by the shift of the variable by one. If you have not seen the p-adic gamma function before, some remarks are in order here.

Remarks.
(1) The reason for removing the factors divisible by p is that for large n, the factorial of n is p-adically close to zero, no matter what n is close to p-adically.
(2) The reason for the interpolation and the minus signs is essentially the generalization $\prod_{k \le j < k+p^n} j \equiv -1 \bmod p^n$ of Wilson's theorem $(p-1)! \equiv -1 \bmod p$. The general principle is: in the product of all elements of a finite abelian group ($(\mathbb{Z}/p^n\mathbb{Z})^*$ in this case), the elements not of order 2 get paired off with their inverses and cancel, so we have to count the number of order 2 elements only.
(3) Morita's Γ_p is the most well-known and successful p-adic interpolation of the gamma function, but there is nothing sacred about it. There are other candidates; e.g. Diamond's p-adic log-gamma function defined on $\mathbb{C}_p - \mathbb{Z}_p$ as against $\log_p \Gamma_p$ which makes sense only on $\mathbb{Z}_p$. Also note that Γ_p is not rigid analytic.

Case II. 'Mixed characteristic'. Let D_i be the product of the monic elements of A of degree i and let $D_{i,v}$ be the product of monic elements, prime to v, of A of degree i. For $n \in \mathbb{Z}_{\ge 0}$, with base q expansion $n = \sum n_i q^i$ $(0 \le n_i < q)$, define

$$n! := \Pi(n) := \Gamma(n+1) := \prod D_i^{n_i}, \qquad \Pi_v(n) := \Gamma_v(n+1) := \prod (-D_{i,v})^{n_i}.$$

Example. $A = \mathbb{F}_q[T]$, $n = q^2 + q + 1$. Then

$$\Pi(n) = D_2 D_1 D_0 = (T^{q^2} - T)(T^{q^2} - T^q) \times (T^q - T) \times 1.$$

Applying the remark (2) above to $(A/v^n A)^*$ rather than $(\mathbb{Z}/p^n\mathbb{Z})^*$, we can see that $-D_{i,v} \to 1$ v-adically as i tends to infinity. Hence for $z \in \mathbb{Z}_p$, $z = \sum z_i q^i$, $(0 \le z_i < q)$, we can define v-adic factorial value as

$$\Pi_v(z) := \Gamma_v(z+1) := \prod (-D_{i,v})^{z_i}.$$

This gives a continuous function $\Gamma_v: \mathbb{Z}_p \to A_v^*$. Note that for any v, the domain of Γ_v is just $\mathbb{Z}_p$.

To get ∞-adic interpolation, we remove the '∞-factor'. Put $\overline{D_i} := D_i / t^{-\deg(D_i)}$. It can be shown that as i tends to infinity, $\overline{D_i}$ tends to 1, ∞-adically. Hence we get $\overline{\Gamma}: \mathbb{Z}_p \to K_\infty^*$ by defining

$$\overline{\Pi}(z) := \overline{\Gamma}(z+1) := \prod \overline{D_i}^{z_i}$$

for $z \in \mathbb{Z}_p$, with $z = \sum z_i q^i$ $(0 \le z_i < q)$. The bar over Γ is a reminder that we are interpolating the 'unit part' and not the original values. Following a suggestion of E.-U. Gekeler, we can make sense out of '$\deg(\Gamma)$' (as explained in the following paragraph), and with a little sacrifice, we can do much better.

Using the Riemann-Roch theorem, if we count the number of monic elements of A of degree i, we see that $\deg(D_i)$ tends to zero, q-adically as i tends to infinity. Hence if we look at the p-adic completion $\widehat{K}_\infty^* := \lim K_\infty^*/(K_\infty^*)^{p^n}$, where we can make sense out of p-adic powers, we can define $\Gamma: \mathbb{Z}_p \to \widehat{K}_\infty^*$ via

$$\Gamma(z) := \overline{\Gamma}(z) t^{-\sum z_i \deg(D_i)}.$$

This way we recover the 'degree part', but K_∞^* does not inject into $\widehat{K}_\infty^*$, unless $q = 2$, because the elements in the finite field project to one.

Remarks.
(1) The factorial is again defined by shifting the variable by one.
(2) $\widehat{K}_\infty^*$ has a natural Galois group interpretation via the local class field theory.

This finishes the construction. Why these are good analogues of gamma functions will become clear as we go along. Here we just note an analogy, observed by W. Sinnott, which works for $A = \mathbb{F}_q[T]$, but not in general.

The prime factorization of the factorial (of Case I or II) of n is given by

$$\prod_{p \text{ positive}} p^{n_p}, \qquad n_p = \sum_{e>0} \left[\frac{n}{Np^e}\right].$$

Note that the gamma function of Case II has domain of characteristic zero and range of characteristic p. By the analogy with the Case I, we might have hoped for a function from Ω to Ω, apart from the poles. This hope is realized in

Case III. 'Characteristic p'. Define

$$\Gamma(z) := \frac{1}{z} \prod_{n \in A_+} \left(1 + \frac{z}{n}\right)^{-1} \in \Omega$$

for $z \in \Omega - A_{\leq 0}$, where $A_{\leq 0} := \{-n : n \text{ monic or } 0\}$ ('the negative integers'). (Note that $A_{\leq 0} = A$, when $q = 2$.) This Γ has simple poles at $A_{\leq 0}$ and no other poles or zeros. In other words, its divisor is similar to that of Γ of the Case I. From the point of view of divisors, it is appropriate to define the factorial as $z! := \Pi(z) := z\Gamma(z)$ rather than by the shift of variable by one, as before.

Note that in Case I, we have $z\Gamma(z) = z! = \Gamma(z+1)$; whereas the first equality holds for Case III and the second for Case II. In fact, in Case III, the domain is of characteristic p, so adding one p-times brings you back, so the recursion with respect to $z \to z+1$ would not have been of much use anyway!

To get v-adic interpolation (see the remarks at the end also), we remove the 'v-factors'. For $a \in A_v$, let $u(a)$ be a or 1 according to whether v is prime to a or not. For $z \in A$, put

$$\Pi_v(z) := \prod_{n \in A_+} \frac{u(n)}{u(z+n)}.$$

This is rational, i.e., lies in K, because the partial product over n's of degree j is one, for j sufficiently large. It is easy to see that Π_v interpolates to $\Pi_v: A_v \to A_v^*$. Put $\Gamma_v(z) := \Pi_v(z)/u(z)$.

We are finally finished with the constructions. Let us make a table to keep track.

Case I	Case II	Case III
$\Gamma(\mathbb{Z}_+) \subset \mathbb{Z}_+$	$\Gamma(\mathbb{Z}_+) \subset A_{>0}^*$	$\Gamma(A - A_{\leq 0}) \subset K^*$
$\Gamma: \mathbb{C} - \mathbb{Z}_{\leq 0} \to \mathbb{C}^*$	$\Gamma: \mathbb{Z}_p \to \widehat{K}_\infty^*$	$\Gamma: \Omega - A_{\leq 0} \to \Omega^*$
$\Gamma_p: \mathbb{Z}_p \to \mathbb{Z}_p^*$	$\Gamma_v: \mathbb{Z}_p \to A_v^*$	$\Gamma_v: A_v \to A_v^*$

Table 1

Curiously, in Case III for $A = \mathbb{F}_q[T]$, if $a \in A_{>0}$, then $1/\Gamma(a) \in A$, rather than $\Gamma(a) \in A$. For example $\Gamma(T) = 1/(2T(T+1))$.

So far we were dealing with simple arithmetic in A and Drinfeld modules were, at least, not visible.

Section II. Drinfeld modules and analogies

In Case I, the period $2\pi i$ of $\mathbb{G}_m$ and the torsion (i.e. roots of unity) of $\mathbb{G}_m$ show up in

Theorem I. *In Case I,*
(1) *Reflection formula:* $\Gamma(z)\Gamma(1-z) = \pi/\sin \pi z$;
(2) *Multiplication formula:*

$$\Gamma(z)\Gamma(z+1/n)\cdots\Gamma(z+(n-1)/n)/\Gamma(nz) = (2\pi)^{(n-1)/2} n^{1/2-nz};$$

(3) $\Gamma(1/2) = \sqrt{\pi}$

We have similar formulas with Γ_p, except that π gets replaced by roots of unity.

In the function field case, the place of the multiplicative group is taken by ρ, a rank one sgn-normalized Drinfeld A-module of generic characteristic, and $2\pi i$ is replaced by the period $\tilde{\pi}$ of ρ. (For simplicity, we assume that the lattice corresponding to ρ is $\tilde{\pi}A$.) We have

Theorem II. *In Case II,*
(1) $\Gamma(z)\Gamma(1-z) = \Gamma(0)$;
(2) $\Gamma(z)\Gamma(z+1/n)\cdots\Gamma(z+(n-1)/n)/\Gamma(nz) = \Gamma(0)^{(n-1)/2}$, *for n prime to p;*
(3) $\Gamma(0)^{q-1} = -\tilde{\pi}^{q-1}$, $\Gamma(1/2) = \sqrt{\tilde{\pi}}$ *times a root of unity, when p is not* 2.

(1) and (2) also hold for Γ replaced by Γ_v and in place of (3), we have $\Gamma_v(0) = (-1)^{1-\deg v}$. Note that powers of the periods are analogous in both the theorems.

Proof of (1). Note that $-1 = \sum(q-1)q^i$, so if $z = \sum z_i q^i$ $(0 \leq z_i < q)$, then the q-base expansion of $-1-z$ is just $-1-z = \sum(q-1-z_i)q^i$. Hence

$$\begin{aligned}\Gamma(z+1)\Gamma(-z) = \Pi(z)\Pi(-1-z) &= \prod d_i^{z_i+q-1-z_i} \\ &= \prod D_i^{q-1} = \Pi(-1) = \Gamma(0).\end{aligned}$$

Replacing z by $z+1$ gives (1). □

Since the value of D_i was not used at all in the proof, the same argument works for Γ_v also. The proof of (2) is also a formal q-base digit manipulation. In other words, the reflection and multiplication formulae are 'explained' by digit manipulations.

Now the Case I reflection formula can be rephrased as $z!(-z)! = \pi z/\sin \pi z$. Observe that, $\sin \pi z$ is zero for $z \in \mathbb{Z}$, whereas $e(\tilde{\pi} z)$ is zero for $z \in A$. Now the left hand side of the formula can be rewritten as $\prod_{\theta \in \mathbb{Z}^*} (\theta z)!$. We have an analogue:

Theorem III. *In Case III,*
(1) $\prod_{\theta \in A^* = \mathbb{F}_q^*} (\theta z)! = \tilde{\pi} z / e(\tilde{\pi} z)$;
(2) *If* $n \in A_+$ *has degree* d, *then* $\prod_{a \bmod n} \Gamma(z + a/n)/\Gamma(nz)$ *is* $\tilde{\pi}^{(q^d-1)/(q-1)}$ *multiplied by a rational function of* z *and an algebraic constant.*

Remarks.
(1) There are similar results for Γ_v. For this, and for detailed statements and proofs, see Sections 2, 3, 6 of **[T4]**.
(2) The $\tilde{\pi}$ power appearing, namely $(q^d-1)/(q-1) = (Nm-1)/(q-1)$, is analogous to $(n-1)/2$ of Cases I and II, since Nm or n are just the number of relevant residue classes.
(3) For z a fraction, $e(\tilde{\pi} z)$ gives torsion of ρ.

Proof of Part (1) *of Theorem III.* In the nonarchimedean analysis here, zeros and poles determine the functions up to multiplication by a constant, hence we have a simple product formulae for Γ and the exponential. Hence (1) follows by straightforward comparison: $\tilde{\pi} z/e(\tilde{\pi} z) = z/e_A(z) = \prod(1 + z/a)^{-1}$, whereas the factorial has a similar product expansion except that a is restricted to monic elements.

Similarly, part (2) is also proved by 'matching of poles'. □

Section III. Special values

Now we look at the questions of algebraicity, transcendence, relations to the periods of the special values of gamma functions at fractional arguments.

First note how special value combinations occurring in the reflection and multiplication formulae, for z a fraction, introduce cyclotomic ('$\sin \pi a/b$') and Kummer ('$n^{1/2-nz}$') extensions and keep in mind the vague connection $\Gamma(1/a) \leftrightarrow e(\tilde{\pi}/a) \leftrightarrow a$-torsion of ρ.

We want to describe an unified treatment, which requires some unified notation. Let us make a table (see Table 2). In this table, $f = a/N$, with a, $N \in I$ and r was defined by $\mathbb{F}_{q^r} = \mathbb{F}_q(\mu_N)$.

The theorems above take care of the values of Γ at integers, so we restrict to proper fractions. In all the three cases, if $f \in F - I$, $i \in I$, then $\Gamma(f+i)/\Gamma(f) \in B \subset \mathcal{A}$. So for our question of algebraicity, it is enough to look at $f \in (F-I)/I$. To extract this information, we define for $f \in F - I$, a rational number $\langle f \rangle \in \mathbb{Q}$ such that $\langle f \rangle$ depends only on $f \bmod I$:

Case I and II. For unique $n \in \mathbb{Z}$, $0 < f - n < 1$. Put $\langle f \rangle := f - n$.

	Case I	Case II	Case III
I : integers in domain	$\mathbb{Z}$	$\mathbb{Z}$	A
F : fractions in domain	$\mathbb{Q}$	$\{a/b \in \mathbb{Q}$: $(p,b)=1\}$ $= \{a/(q^m-1)\}$	K
$\mathcal{A}$: algebraic numbers in range	$\overline{\mathbb{Q}}$	$\overline{K}$	$\overline{K}$
underlying objects	$\mathbb{G}_m$, CM elliptic curves	$\mathbb{G}_m$, CM ρ's	CM ρ's
exponential e	classical e	classical for domain, e_ρ for range	e_ρ
period θ	$2\pi i$	$2\pi i$ for domain, $\tilde{\pi}$ for range	$\tilde{\pi}$
B : base field	$\mathbb{Q}$	K	K
$\mathcal{O}$: base ring	$\mathbb{Z}$	A	A
extension $B(e(\theta f))$	usual cyclotomic	constant field	Drinfeld cyclotomic
Galois group G identified with	$\mathrm{Gal}(\mathbb{Q}(\mu_N)/\mathbb{Q})$ $= (\mathbb{Z}/N\mathbb{Z})^*$	$\mathrm{Gal}(K(\mu_N)/K)$ $= q^{(\mathbb{Z}/r\mathbb{Z})}$	$\mathrm{Gal}(K(\Lambda_N)/K)$ $= (A/NA)^*$

Table 2

Case III. For unique $a \in A$, $f - a = -a_1/a_2$, where $a_1, a_2 \in A$, $\deg a_1 < \deg a_2$, a_2 monic. Put $\langle f \rangle := 1$ or 0, according as a_1 is monic or not.

Motivation for this definition due to Anderson will be provided later. Note that for $q = 2$, $\langle f \rangle = 1$.

Consider a finite formal sum $\underline{f} = \oplus m_i[f_i]$, $m_i \in \mathbb{Z}$, $f_i \in F - I$. This is nothing but an integral linear combination of symbols $[f_i]$, i.e., a divisor. Let N be a common denominator for the f_i's. Put $\Gamma(\underline{f}) := \prod \Gamma(f_i)^{m_i}$ and make similar definitions for Π, Γ_v, Π_v, etc. Also put $m(\underline{f}) := \sum m_i \langle -f_i \rangle \in \mathbb{Q}$.

For $\sigma \in G$, let $\underline{f}^{(\sigma)} := \oplus m_i[f_i^{(\sigma)}]$, where $f_i^{(\sigma)}$ is just multiplication of f_i by σ as an element of the identification of the Galois group given in the table. For example, in Case II, if σ corresponds to q^j, then $f_i^{(\sigma)} = q^j f_i$.

For a finite place v, let $\mathrm{Frob}_v \in G$ be the Frobenius, so that it makes sense to talk of Frob_v-orbits of f_i or $\underline{f}$.

Consider the hypothesis/recipe/conjecture:

H1. If $m(\underline{f}^{(\sigma)})$ is independent of $\sigma \in G$, then $\Gamma(\underline{f})/\theta^{m(\underline{f})}$ and $\Gamma_v(\underline{f})$ belong to $\mathcal{A}$.

H2. If $\underline{f}$ is a linear combination of Frob_v-orbits, then $\Gamma_v(\underline{f}) \in \mathcal{A}$.

Examples. We leave it as an exercise to check that the cases of the hypothesis with $\underline{f} = \langle f\rangle + \langle 1-f\rangle$, $m(\underline{f}) = 1$ and with $\underline{f} = \oplus_{j=0}^{n-1}\langle z+j/n\rangle - \langle nz\rangle$, $m(\underline{f}) = (n-1)/2$ are compatible with the reflection and multiplication formulae mentioned above in all three cases.

What is known about H1 and H2? H1 is true in all the three cases, but still we have written it as an hypothesis, because we will consider more situations later, which fit in this framework, but for which it is not known.

In Case I, $\underline{f}$ satisfying the condition in H1 is a linear combination of the examples above (i.e., the conclusion follows from the known reflection and the multiplication formulae) by a result of Koblitz and Ogus. With a proof analogous to that of Koblitz and Ogus, the same can be proved for Case III. Even though the analogous statement is not true for the Case II, H1 can be proved directly then.

H2 is known for the Case I, by the Gross-Koblitz theorem and for Case II, with $A = \mathbb{F}_q[T]$ by the analogue of the Gross-Koblitz result in this case. (We will see more about this in the second half of the paper.) There is some evidence in other cases.

In the complex multiplication situation, we also have a Chowla-Selberg type formula (up to multiplication by elements of $\mathcal{A}$) for the period, in terms of the gamma values at appropriate fractions. This formula is predicted via a simple calculation involving the 'brackets' introduced above. It works as follows.

Let E be a Drinfeld A-module (or elliptic curve for the Case I) over B^{sep} with complex multiplication by the integral closure of $\mathcal{O}$ in the appropriate abelian extension L of B as in the table, e.g., a constant field extension for Case II and Drinfeld cyclotomic extension for Case II, etc. (In fact, we can get much more flexibility using higher dimensional A-motives of Anderson, e.g., the solitons in **[A]** should give rise to higher dimensional A-motives with Drinfeld cyclotomic CM whose periods are values of $\Gamma(z)$, for $z \in K$, in Case III, with $A = \mathbb{F}_q[T]$. But here we will be content with this simple case.)

For $f \in F - I$, let $h(f)\colon \mathrm{Gal}(B^{\mathrm{sep}}/B) \to \mathbb{Q}$ be defined by $h(f)(\sigma) := \langle -f'\rangle$, where $e(\theta f)^\sigma = e(\theta f')$. (Note that in Case II, the exponential used here is the classical exponential according to the table.)

Let $\chi_{L/B}$ be the characteristic function of $\mathrm{Gal}(L/B)$. In other words

$$\chi_{L/B}\colon \mathrm{Gal}(B^{\mathrm{sep}}/B) \to \mathbb{Z}$$

is such that $\chi_{L/B}(\sigma)$ is 1, if σ is the identity on L and is 0 otherwise.

We then have the hypothesis/recipe/conjecture:

H3. If $\chi_{L/B} = \sum m_f h(f)$, $m_f \in \mathbb{Z}$, then $\mathrm{Period}_E / \prod \Gamma(f)^{m_f} \in \mathcal{A}$.

This is known in Case I. In Cases II and III, there is only some evidence.

Example. $\mathcal{O} = A = \mathbb{F}_q[T]$, $K = B = \mathbb{F}_q(T)$, $L = \mathbb{F}_{q^k}(T)$. Then the rank k Drinfeld A-module: $u \mapsto Tu + u^{q^k}$ is the rank one Carlitz module over $\mathbb{F}_{q^k}[T]$. Hence we want to express its period $\tilde{\pi}_k$ in terms of the gamma values at fractions with the gamma function built up from the base. If F is the q-power Frobenius, then $\chi_{L/B}(F^n)$

is 1, if k divides n and is 0 otherwise. The same is true for

$$\left[qh\left(\frac{q^{k-1}}{1-q^k}\right)-h\left(\frac{1}{1-q^k}\right)\right](F^n)=q\left\langle\frac{q^{k+n-1}}{q^k-1}\right\rangle-\left\langle\frac{q^n}{q^k-1}\right\rangle.$$

Hence H3 predicts that $\Gamma(q^{k-1}/(1-q^k))^q/\Gamma(1/(1-q^k))$ is an algebraic multiple of $\tilde{\pi}_k$. Let us verify this, using the theorems above. For simplicity, we will disregard the convergence questions and consider equalities only up to algebraic multiples.

We have observed

$$\begin{aligned}\tilde{\pi}=(-1)!&=\left(\frac{1}{1-q}\right)!^{q-1}=(D_0D_1D_2\ldots)^{q-1}\\&=\left(\frac{D_0^q}{D_1}\frac{D_1^q}{D_2}\cdots\right)=(T^q-T)(T^{q^2}-T)\cdots.\end{aligned}$$

Hence

$$\tilde{\pi}_k=(T^{q^k}-T)(T^{q^{2k}}-T)\cdots=\prod\frac{D_{kn-1}^q}{D_{kn}}=\left(\frac{q^{k-1}}{1-q^k}\right)!^q\Big/\left(\frac{1}{1-q^k}\right)!$$

Remarks.

(1) If we use transcendence results concerning periods or algebraic independence results concerning different periods, then the hypotheses Hi give transcendence results for some special gamma values. For instance, combining the example above with the theorems quoted before, one can express $(1/(1-q^2))!$, which can be thought of as a rough analogue of $\Gamma(1/3)$, as a monomial in $\tilde{\pi}$ and $\tilde{\pi}_2$. This allows us to conclude transcendence of these values from transcendence results about different periods due to Thiery or Yu (announced at this conference). As for the question of a period expressible in terms of a single gamma value, apart from the relation between $\tilde{\pi}$ and $\Gamma(0)$, it is easy to show that for the gamma function of Case III, and $A=\mathbb{F}_q[T]$, $\Gamma(1/T)$ is the period of the Carlitz module for the ring $\mathbb{F}_q[\Lambda_T]$ and is hence transcendental.

(2) It is natural to ask whether the Hi's are best possible. In Case I, the conjecture that H1 is best possible follows, by work of Deligne, from some very general algebraic geometrical conjectures. Such a link has not yet been developed in the other cases.

(3) In Case I, the Γ function is closely connected with ζ-functions and L-functions, it is in some sense a factor at infinity for the ζ-function and hence appears in functional equations. Such connections are missing in the other cases. We only have $\Pi'(x)/\Pi(x)=\zeta(x,1)$ in Case III.

(4) The Hurwitz formula $\zeta(x,s)=\langle -x\rangle-1/2+s\log\Gamma(x)+o(s^2)$ around $s=0$ connects partial zeta, 'brackets' and logarithm of Γ. For Case III, the partial zeta value for $x=a_1/a_2$ (with $a_2\in A_+$) is obtained by summing $q^{-s\deg a}$ over monic a congruent to a_1 modulo a_2 and hence it equals $1-1/(q-1)$ or $-1/(q-1)$ according as a_1 is monic or not. This is the motivation for the definition of the brackets as above.

(5) There is some evidence for a v-adic Chowla-Selberg-Ogus formula.

(6) How about generalizations? First, there is a two variable gamma function of Goss on the domain $\mathbb{Z}_p\times\Omega$, which handles both constant fields and Drinfeld cyclotomic extensions at once and gamma functions of Cases II and III occur as its specializations. Analogues of Hi can be formulated for it and they are open. In another direction,

Anderson has some ideas on adelic gamma function theory for function fields parallel to his hyperadelic gamma theory in the classical case. His soliton theory, sketched at this conference, gives tools to address most of the questions above. Anderson also suggests that the study of analogue of Barnes multiple gamma function would be very fruitful.

Section IV. Construction of still another gamma function

We now introduce a new gamma function more intimately connected with Drinfeld modules and explain some results about its interpolations and special values. It is obtained from a different kind of generalization of the gamma function of Case II for $A = \mathbb{F}_q[T]$.

Construction. The key analogy is that if $A = \mathbb{F}_q[T]$ and $\rho_T = T + F$, then $e_\rho(z) = \sum z^{q^i}/D_i$ parallel to $e^z = \sum z^n/n!$ in the classical case, because D_i is just $q^i!$ in Case II. Let A be arbitrary now and let ρ be a sgn-normalized rank one (of generic characteristic) Drinfeld A-module. It can be shown that $e_\rho(z) = \sum z^{q^i}/d_i$ with $d_i \in H$, where H is the ('totally real') Hilbert class field. Now one can define new factorials as in Case II by replacing D_i by d_i.

This analogy can be rephrased as follows: In Case I, closely related to Euler's famous integral representation of the gamma function, which we used to define Γ in Case I, there is Hankel's formula

$$\frac{1}{n!} = \int_C e^z z^{-n} \frac{dz}{z}$$

where C is a contour starting at $-\infty$ on the real axis, encircling the origin once in positive sense and returning to $-\infty$. Interpreting the right hand side as the residue of the integrand, we have the same formula for the new factorial, when $n = q^i$.

Another reason, for general n, is that $\zeta(\text{even}) = B_{\text{even}}\tilde{\pi}^{\text{even}}/(\text{even})!$, since this relation is obtained by looking at $z/e_\rho(z)$.

Better justification comes from the special values results below.

Interpolation at ∞. Remove the ∞-factor. Put $\overline{d_i} := d_i/t^{-\deg d_i}$.

Theorem IV. *As $i \to \infty$, we have*

$$\overline{d_i} \to 1, \quad (\infty - adically) \quad and \quad \deg d_i \to 0, \quad (q - adically).$$

Hence we can define the unit part interpolation $\overline{\Gamma}$ and the completed interpolation Γ just as in Case II, but with d_i's in place of D_i's. These functions satisfy the same functional equations, because the same proofs involving the digit manipulation trick go through. It is interesting to note that in **[T1]**, I had suggested studying this gamma function and had the theorem and interpolations in special cases, but then Case II became more successful and I gave up on this!

Interpolation at v. Now in general, the d_i's are fractions in H rather than integers in A. They do not have a natural interpretation as a product of some integers and so we can not imitate the classical interpolation procedure of removing the factors divisible by v from the product. To explain how the interpolation is achieved, we assume for simplicity that the class number of A is 1 and that v is a prime monic element of A of degree l. An element of degree i, divisible by v, is v times an element of degree $i-l$. So in Case II, $D_{i,v} = D_i/D_{i-l}v^k$, where k is chosen so that $D_{i,v}$ is a unit at v. We take this clue and define $d_{i,v}$ in exactly analogous fashion. (In general, we can use a principal v, which then splits in H and hence $H \subset K_v$ or use a local parameter w at a prime of H above v and look at the completion of the Hilbert class field. For simplicity, we restrict to the class number one case and refer to **[T6]** for the general case.)

Theorem V. *As* $i \to \infty$, $-d_{i,v} \to 1$, *v-adically.*

Hence imitating the procedure in Case II, but with $d_{i,v}$ in the place of $D_{i,v}$, we get the new Γ_v.

Section V. Special values and Gauss sums

For this section we will assume that the reader is familiar with **[G-K]**, **[T2]**, **[T3]**. In **[T2]**, **[T3]** analogues of Gauss sums, g_j's, were defined and studied. In the $A = \mathbb{F}_q[T]$ case, various analogues of classical theorems were proved about them in **[T2]**. But in **[T3]** it was shown that their behavior including factorization is weird in general: They need not lie above a prime. I attributed the failure to find a Gross-Koblitz formula in general for Case II to this strange behavior, but the following analogue of the Gross-Koblitz theorem shows that the Gauss sums are perfectly fine, but the matching gamma function is the one defined in the previous section.

Theorem VI. *For suitable* μ, $v^{\mu}/\Pi_v(q^j/(1-q^l)) = g_j$.

The proof consists of using the connection between Drinfeld modules and Shtukas and establishing that the Jacobi sums of **[T2]**, **[T3]** are essentially specializations of Shtukas. Once this is done, the rest of the proof is same as that in **[T2]** for $A = \mathbb{F}_q[T]$ after the reinterpretation of the notation!

Remarks.
(1) For the case of class number more than one, Theorems V and VI are still valid, essentially after 'replacing v by w' and 'dealing with sign'.
(2) We can again ask whether the Hi's are true and best possible for this gamma function. H2 is true by Theorem VI. H1 would be true if $\Gamma(0)$ is an algebraic multiple of $\tilde{\pi}$. But the results of **[T3]** imply that the naive analogue of Deligne's theorem (see **[G-K]**) would not hold in this case.
(3) Using the connection between Shtukas and Jacobi sums mentioned above, we can generalize the result in **[T3]** about the factorization of Gauss sums for the rational function field with ∞ being any place. We have also computed factorization for some examples of class number more than one. Anderson has explained the Shtukas and thus the factorizations beautifully, in terms of the theta divisor.
(4) Gamma functions and Gauss sums go hand in hand. The Gross-Koblitz formula follows exactly the same way for the Case II gamma function with Shtukas replaced by

some special functions found by Anderson and Gauss sums defined as their appropriate specializations. These Gauss sums are the analogues of Brumer-Stark units due to Tate-Deligne-Hayes (see **[H]**) for constant field extensions. After my talk, Anderson suggested that similar results should hold for Case III gamma functions, with his results in **[A]**, which give still other functions whose specializations are 'Brumer-Stark units for the geometric extensions'. To carry out this suggestion, we need the following change in the v-adic interpolation in Case III:
Let $e_i := \prod(1+z/n)^{-1}$, where the product is taken over all $n \in A_+$ of degree i. These are nothing but the partial gamma products. Put $e_{i,v} := e_i/e_{i-l}v^k$ where, as before v is of the degree l and k is chosen so that $e_{i,v}$ is unit at v. Then define $\Pi_v(z) := \prod_i e_{i,v}$. This works with the proof described above!
(5) In view of 5.8 and **[T4]** and the analogies explained in **[T5]**, it seems worthwhile to study the factorial defined by $\Pi(z) := \prod \left(1 + \left\{ \begin{matrix} z \\ q^i \end{matrix} \right\}\right)^{-1}$, in the notation of these references. It bears the same logical relationship to the factorial of Case III, as does the factorial of Section IV to that of Case II. We can show that it has interpolations at all the places. The special values are under investigation at present.
(6) We take this opportunity to correct some errors in **[T4]**: The last exponent $1 - q^d$, in the first formula in 3.4, should be replaced by q^d. The reference **[Y1]** in 1.6 should be changed to **[Y]** or **[Th]** of the present paper.

References

[A] G. Anderson, A two dimensional analogue of Stickelberger's theorem. This volume.

[Ar] E. Artin, The Gamma Function, Holt, Rinehart and Winston, Inc., New York, 1964.

[C] L. Carlitz, An analogue of the von-Staudt-Clausen theorem, Duke Math. J. **3** (1937), 503–517.

[D] P. Davis, Leonhard Euler's integral: A historical approach to the gamma function, Amer. Math. Monthly **66** (1959), 849–869.

[G] D. Goss, The Γ-function in the arithmetic of function fields, Duke Math. J. **56** (1988), 163–191.

[G-K] B. Gross and N. Koblitz, Gauss sums and the p-adic Γ function, Ann. Math. **109** (1979), 569–581.

[H] D. Hayes, Stickelberger elements in function fields, Compositio Math. **55** (1985), 209–239.

[T1] D. Thakur, Number fields and function fields (zeta and gamma functions at all primes), In: Proc. Conf. on p-adic analysis, Hengelhoef 1986, pp. 149–157. Eds.: N. De Grande-De Kimpe, L. Van Hamme.

[T2] D. Thakur, Gauss sums of $\mathbb{F}_q[T]$, Invent. Math. **94** (1988), 105–112.

[T3] D. Thakur, Gauss sums for function fields, J. Number Theory **37** (1991), 242–252.

[T4] D. Thakur, Gamma functions for function fields and Drinfeld modules, Ann. Math. **134** (1991), 25–64.

[T5] D. Thakur, Drinfeld modules and arithmetic in the function fields, Preprint.

[T6] D. Thakur, Shtukas and Jacobi sums, Preprint.

[Th] A. Thiery, Indépendence Algébrique des Périodes et Quasi-Périodes d'un Module de Drinfeld. This volume.

[Y] J. Yu, Transcendence in finite characteristic. This volume.

School of Mathematics, Tata Institute of Fundamental Research, Homi Bhabha Road, Colaba, Bombay, India

Groups of Elliptic Units in Global Function Fields

H. Oukhaba

Introduction

Let k be a global function field and $\mathbb{F}_q$ be its field of constants. Fix a place ∞ of k. Let $\mathcal{O}_k$ be the Dedekind ring of elements of k regular outside of ∞, and let k_∞ be the completion of k at ∞.

For each finite abelian extension F of k we let $\mathcal{O}_F$ be the integral closure of $\mathcal{O}_k$ in F. We know that $\mathcal{O}_F$ is a Dedekind ring with a finite ideal class number $h(\mathcal{O}_F)$.

As usual we denote by $\mathcal{O}_F^\times$ the group of units of $\mathcal{O}_F$, and $\mu(F) \subset \mathcal{O}_F^\times$ the finite multiplicative group of nonzero constants of F. We have $\mu(k) = \mathcal{O}_k^\times$, and in general the quotient group $\mathcal{O}_F^\times/\mu(F)$ is a free abelian group of rank $r_F - 1$, where r_F is the exact number of places of F sitting over ∞.

Now suppose that $F \subset k_\infty$, meaning that the place ∞ splits completely in F/k. Then the structure of $\mathcal{O}_F^\times$ is very close to that of the group of units of a finite abelian extension of an imaginary quadratic field, especially from the point of view of the elliptic units theory.

In this paper we give, under the hypotheses specified in Theorem 1, a method to construct subgroups of $\mathcal{O}_F^\times$, cf. §3. The description of these subgroups is rather easy. Their elements are special values of elliptic functions, cf. §2. In addition they verify a nice index formula, cf. §4, generalizing those established by Hayes and by Galovich and Rosen respectively in **[6]** and **[1]**.

When the extension F/k is unramified we take our inspiration from the recent work of G. Robert, cf. **[12]**, **[13]** and **[14]**, where he suggests an alternative definition of the elliptic units group of an unramified abelian extension of an imaginary quadratic field. This group, as it was proved by D. Kersey in **[8]**, verifies Theorem 1 too.

We use here the fundamental concept of a Drinfeld $\mathcal{O}_k$-module, especially its applications proposed by D. Hayes, cf. **[5]** and **[7]**.

We have:

Theorem 1. *Let F be a finite extension of k such that $F \subset k_\infty$. Suppose in addition that one of the following two conditions occurs*

1. *The extension F/k is unramified.*
2. *One prime ideal of $\mathcal{O}_k$ ramifies in F/k and* $\deg(\infty) = 1$.

Then $\mathcal{O}_F^\times$ contains a subgroup $\mathcal{E}_F$ of finite index whose elements are elliptic units, i.e., special values of elliptic functions, and we have

$$[\mathcal{O}_F^\times : \mathcal{E}_F] = \frac{h(\mathcal{O}_F)}{[H_{(1)} : F \cap H_{(1)}]}, \tag{1}$$

where $H_{(1)}$ is the maximal unramified abelian extension of k such that $H_{(1)} \subset k_\infty$.

For short, we shall give the proof of (1) only when the extension F/k is unramified. Recall that the proof in general can be found in the author's thesis, cf. **[10]**.

1. Some preliminaries

It is well known that the place ∞ extends in a unique manner to the algebraic closure $\overline{k}_\infty$ of k_∞. Call it ∞ again, and let Ω be the completion of $\overline{k}_\infty$ at ∞. Then we know by the general theory that Ω is also algebraically closed.

Definition 1.1. By a lattice Γ of Ω we mean a discrete $\mathcal{O}_k$-module of finite rank.

One can associate to each lattice Γ of Ω its exponential function e_Γ defined for all $z \in \Omega$ by

$$e_\Gamma(z) \stackrel{\text{dfn}}{=} z \prod_{\substack{\gamma \in \Gamma \\ \gamma \neq 0}} (1 - z/\gamma) \tag{1.1}$$

where γ runs over the set of nonzero elements of Γ. The product (1.1) converges uniformly on every bounded set of Ω and hence defines an entire function, surjective and $\mathbb{F}_q$-linear, with Γ as its set of roots. The derivative $e'_\Gamma(z) = 1$ for all $z \in \Omega$. On the other hand, if $x \in \Omega^\times$ and $z \in \Omega$, then we have the homogeneity relation

$$e_{x\Gamma}(xz) = x e_\Gamma(z). \tag{1.2}$$

Let $\Gamma \subset \overline{\Gamma}$ be two lattices of Ω such that the index $[\overline{\Gamma} : \Gamma]$ is finite. Then $e_{\overline{\Gamma}}$ and e_Γ are such that

$$e_{\overline{\Gamma}}(z) = P(\overline{\Gamma}/\Gamma;\, e_\Gamma(z)) \tag{1.3}$$

where $P(\overline{\Gamma}/\Gamma; t)$ is the linear polynomial whose set of roots is the finite group $e_\Gamma(\overline{\Gamma})$, and whose leading coefficient is

$$\delta(\Gamma, \overline{\Gamma}) \stackrel{\text{dfn}}{=} \prod_{\substack{\rho \in \overline{\Gamma}/\Gamma \\ \rho \neq 0}} e_\Gamma(\rho)^{-1}, \tag{1.4}$$

where ρ runs over a complete system of nonzero representatives of $\overline{\Gamma}$ modulo Γ.

If $\overline{\Gamma} = \mathfrak{d}^{-1}\Gamma$ where $\mathfrak{d}$ is an integral ideal of $\mathcal{O}_k$, then we define $\Phi_\mathfrak{d}^\Gamma(t)$ to be the monic polynomial

$$\Phi_\mathfrak{d}^\Gamma(t) \stackrel{\text{dfn}}{=} \delta(\Gamma, \overline{\Gamma})^{-1} P(\overline{\Gamma}/\Gamma; t).$$

Consider the twisted polynomial ring $\Omega\{\tau\}$, where τ is the Frobenius endomorphism of Ω which maps z to z^p, with the action $\tau a = a^p \tau$ if $a \in \Omega$. The positive prime p being the characteristic of Ω. Let Γ be a lattice of Ω. Then the linear polynomials

$$P_x^\Gamma = x P(x^{-1}\Gamma/\Gamma; t), \qquad x \in \mathcal{O}_k \backslash \{0\}$$

can be viewed as elements of $\Omega\{\tau\}$. Moreover, we obtain a ring homomorphism

$$P^{\Gamma}: \mathcal{O}_k \to \Omega\{\tau\}$$
$$x \mapsto P_x^{\Gamma}$$

and, in fact, P^{Γ} is a Drinfeld $\mathcal{O}_k$-module of rank equal to the rank of the lattice Γ. See **[5]** for more details.

Remark 0. The exponential function e_{Γ} can be considered as the analogous of the Klein form K_L associated to a complex lattice L because these two functions have very similar properties which we shall investigate in a future paper in common with G. Robert, see also **[10]** and **[11]**.

Definition 1.2. Let $\mathfrak{a}$ be a fractional ideal of $\mathcal{O}_k$. Then we define $N\mathfrak{a}$ to be the ideal norm of $\mathfrak{a}$. In particular if $\mathfrak{a} \subset \mathcal{O}_k$, then $N\mathfrak{a}$ is the exact number of elements of the quotient $\mathcal{O}_k/\mathfrak{a}$. If $x \in \mathcal{O}_k$, then we shall write Nx instead of $Nx\mathcal{O}_k$.

Lemma 1.3. *Let $x \in \Omega^{\times}$ and $\Gamma_1 \subset \Gamma_2 \subset \Gamma_3$ be lattices of Ω such that the index $[\Gamma_3:\Gamma_1]$ is finite. Then we have*

$$\delta(x\Gamma_1, x\Gamma_2) = x^{1-[\Gamma_2:\Gamma_1]}\delta(\Gamma_1, \Gamma_2) \tag{1.5}$$
$$\delta(\Gamma_1, \Gamma_3) = \delta(\Gamma_1, \Gamma_2)^{[\Gamma_3:\Gamma_2]}\delta(\Gamma_2, \Gamma_3). \tag{1.6}$$

Proof. The lemma is an easy consequence of (1.2) and (1.3). □

Corollary 1.4. *Let $x \in \mathcal{O}_k\backslash\{0\}$ and $\Gamma \subset \overline{\Gamma}$ be two lattices of Ω which have the same rank r as $\mathcal{O}_k$-modules. Then*

$$(x\delta(\Gamma, x^{-1}\Gamma))^{[\overline{\Gamma}:\Gamma]} = x\delta(\overline{\Gamma}, x^{-1}\overline{\Gamma})\delta(\Gamma, \overline{\Gamma})^{(Nx)^r-1}. \tag{1.7}$$

Proof. We have

$$\delta(\Gamma, x^{-1}\Gamma)^{[x^{-1}\overline{\Gamma}:x^{-1}\Gamma]}\delta(x^{-1}\Gamma, x^{-1}\overline{\Gamma}) = \delta(\Gamma, \overline{\Gamma})^{[x^{-1}\overline{\Gamma}:\overline{\Gamma}]}\delta(\overline{\Gamma}, x^{-1}\overline{\Gamma})$$

since by (1.6) both terms of this equation are equal to $\delta(\Gamma, x^{-1}\overline{\Gamma})$. But $[x^{-1}\overline{\Gamma}: x^{-1}\Gamma] = [\overline{\Gamma}: \Gamma]$ and $[x^{-1}\overline{\Gamma}: \overline{\Gamma}] = (Nx)^r$. Now use Formula (1.5) to find (1.7). □

Let $x, y \in \mathcal{O}_k\backslash\{0\}$ and Γ be a lattice of Ω of rank r. If we put $\overline{\Gamma} = y^{-1}\Gamma$ in Corollary 1.4, then Formula (1.7) becomes

$$(x\delta(\Gamma, x^{-1}\Gamma))^{(Ny)^r-1} = (y\delta(\Gamma, y^{-1}\Gamma))^{(Nx)^r-1}. \tag{1.8}$$

Remark 1. It is easy to prove that the rational integers $(Nx)^r - 1$, where x runs over the nonzero elements of $\mathcal{O}_k$, have g.c.d. equal to

$$w_\infty^{(r)} \stackrel{\text{dfn}}{=} q^{r\deg(\infty)} - 1.$$

In particular, if $r = 1$, then one can find a finite subset $\mathcal{H}$ of $\mathcal{O}_k\backslash\{0\}$ and rational integers m_y, $y \in \mathcal{H}$ such that

$$w_\infty \stackrel{\text{dfn}}{=} w_\infty^{(1)} = \sum_{y\in\mathcal{H}} m_y(Ny - 1). \tag{1.9}$$

Remark 2. Let $K(\infty)$ be the constant field of k_∞. Then one knows that $K(\infty)$ is a finite extension of $\mathbb{F}_q$ with degree $[K(\infty):\mathbb{F}_q] = \deg(\infty)$. Hence $\#K(\infty)^\times = w_\infty$.

Proposition 1.5. *There is a unique sign function s, i.e., a co-section of the inclusion morphism $K(\infty)^\times \hookrightarrow k_\infty^\times$ such that $s(z) = 1$ if $|z-1|_\infty < 1$, characterized by*

$$s\left(\prod_{y\in\mathcal{H}} y^{m_y}\right) = 1. \tag{1.10}$$

Moreover, for every lattice Γ of rank 1 the number

$$\Delta_s(\Gamma) \stackrel{\text{dfn}}{=} \prod_{y\in\mathcal{H}} (y\delta(\Gamma, y^{-1}\Gamma))^{m_y} \tag{1.11}$$

depends only on Γ and s; and we have the formula

$$\Delta_s(\Gamma)^{\frac{1-Nx}{w_\infty}} x\delta(\Gamma, x^{-1}\Gamma) = s(x)^{\sigma_\Gamma} \tag{1.12}$$

where σ_Γ is an $\mathbb{F}_q$-automorphism of $K(\infty)$ uniquely determined by Γ.

Proof. See **[10]**, chap. II, §3 or **[11]**. □

Remark 3. In fact, one can prove more precise results. In particular, how to construct all the sign functions using the Formulas (1.9), (1.11) and (1.12), cf. **[10]** or **[11]**.

The analytic calculus of E. U. Gekeler, cf. **[2]**, chap. IV, pp. 34-36, gives an explicit formula for the automorphism σ_Γ of $K(\infty)/\mathbb{F}_q$ when Γ is homothetic to a fractional ideal of $\mathcal{O}_k$.

Lemma 1.6. *Let τ be the Frobenius automorphism $x \mapsto x^q$ of $K(\infty)/\mathbb{F}_q$. Let Γ be a lattice of Ω homothetic to a fractional ideal $\mathfrak{a}$ of $\mathcal{O}_k$. Then we have*

$$\sigma_\Gamma = \tau^{1-g-\deg\mathfrak{a}}$$

where g is the genus of k and $\deg\mathfrak{a}$ is defined by $N\mathfrak{a} = q^{\deg\mathfrak{a}}$.

Proof. See **[2]**, chap. IV, pp. 34-36. □

The invariants $\Delta_s(\Gamma)$ have many properties, especially those given by the following Lemma.

Lemma 1.7. *Let $x \in \mathcal{O}_k\backslash\{0\}$ and $\Gamma \subset \overline{\Gamma}$ be two lattices of Ω of rank 1. Then we have*

$$\Delta_s(\Gamma)^{Nx-1} = (x\delta(\Gamma, x^{-1}\Gamma))^{w_\infty}; \tag{1.12}$$

$$\Delta_s(z\Gamma) = z^{-w_\infty}\Delta_s(\Gamma), \qquad z \in \Omega^\times; \tag{1.13}$$

$$\Delta_s(\Gamma)^{[\overline{\Gamma}:\Gamma]} = \Delta_s(\overline{\Gamma})\delta(\Gamma, \overline{\Gamma})^{w_\infty}. \tag{1.14}$$

Proof. Formula (1.12) is an immediate consequence of the definition of $\Delta_s(\Gamma)$, Property (1.8) and Formula (1.9). The equality (1.13) can be deduced from (1.5) and (1.9). On the other hand, one gets (1.14) using (1.9) and Corollary 1.4. □

Let $H_{(1)}$ be the maximal unramified abelian extension of k such that $H_{(1)} \subset k_\infty$. Then we know by Class Field Theory that $H_{(1)}$ is associated to the idèle group

$$\mathfrak{J}_{(1)} \overset{\text{dfn}}{=} k^\times \cdot \mathfrak{U}_{(1)}$$

where $\mathfrak{U}_{(1)}$ contains all idèles t of k such that the component t_v of t at each place $v \neq \infty$ of k verifies $t_v \in \mathcal{O}_{k,v}^\times$. (See **2-2** for the definition of $\mathcal{O}_{k,v}$.) Moreover, we have $h(\mathcal{O}_k) = [H_{(1)}: k]$. We shall write h instead of $h(\mathcal{O}_k)$.

Proposition 1.8. *Let $\mathfrak{a}_1$, $\mathfrak{a}_2$, and $\mathfrak{a}$ be fractional ideals of $\mathcal{O}_k$. Then the quotient $\Delta_s(\mathfrak{a}_1)/\Delta_s(\mathfrak{a}_2) \in H_{(1)}$ and generates in $\mathcal{O}_{H_{(1)}}$ the ideal*

$$(\mathfrak{a}_2\mathfrak{a}_1^{-1}\mathcal{O}_{H_{(1)}})^{w_\infty}.$$

Moreover, if $(\mathfrak{a}, H_{(1)}/k)$ is the automorphism of $H_{(1)}/k$ associated to $\mathfrak{a}$ by the Artin map , then we have

$$\left(\frac{\Delta_s(\mathfrak{a}_1)}{\Delta_s(\mathfrak{a}_2)}\right)^{(\mathfrak{a}_1, H_{(1)}/k)} = \frac{\Delta_s(\mathfrak{a}^{-1}\mathfrak{a}_1)}{\Delta_s(\mathfrak{a}^{-1}\mathfrak{a}_2)}.$$

Proof. See **[10]**, chap. II, §4 or **[11]**. □

2. The elliptic function Ψ

2-1. Let $\Gamma \subset \overline{\Gamma}$ be two lattices of Ω such that $[\overline{\Gamma}:\Gamma]$ is finite. We let

$$\Psi(z;\Gamma,\overline{\Gamma}) \overset{\text{dfn}}{=} \delta(\Gamma,\overline{\Gamma})\frac{e_\Gamma(z)^{[\overline{\Gamma}:\Gamma]}}{e_{\overline{\Gamma}}(z)}.$$

The function $z \mapsto \Psi(z;\Gamma,\overline{\Gamma})$ is well defined on the complement of $\overline{\Gamma}\backslash\Gamma$ in Ω. It vanishes on Γ and, in fact, is elliptic (i.e. periodic) with respect to Γ. Moreover, as a rational fraction in $e_\Gamma(z)$ its divisor is

$$[\overline{\Gamma}:\Gamma](0)_\Gamma - \sum_{\rho\in\overline{\Gamma}/\Gamma} (\rho)_\Gamma.$$

On the other hand, we have the homogeneity formula

$$\Psi(\lambda z;\lambda\Gamma,\lambda\overline{\Gamma}) = \Psi(z;\Gamma,\overline{\Gamma}), \qquad \text{for all } \lambda \in \Omega^\times.$$

Also if $\Gamma_1 \subset \Gamma_2 \subset \Gamma_3$ are lattices of Ω such that the index $[\Gamma_3:\Gamma_1]$ is finite, then we have

$$\Psi(z;\Gamma_1,\Gamma_3) = \Psi(z;\Gamma_1,\Gamma_2)^{[\Gamma_3:\Gamma_2]}\Psi(z;\Gamma_2,\Gamma_3). \tag{2.1}$$

Proposition 2.1. *Let Γ, Γ', $\overline{\Gamma}$, $\overline{\Gamma}'$ be lattices of Ω such that*

1) $\Gamma' \subset \Gamma \subset \overline{\Gamma}$ *and* $\Gamma' \subset \overline{\Gamma}' \subset \overline{\Gamma}$;
2) $\Gamma \cap \overline{\Gamma}' = \Gamma'$;
3) $\overline{\Gamma}/\Gamma \simeq \overline{\Gamma}'/\Gamma'$ *and* $\Gamma/\Gamma' \simeq \overline{\Gamma}/\overline{\Gamma}'$;
4) *The index* $[\overline{\Gamma}:\Gamma']$ *is finite.*

Then we have

$$\Psi(z;\Gamma,\overline{\Gamma}) = \prod_{i=1}^{[\Gamma:\Gamma']} \Psi(z+\rho_i;\Gamma',\overline{\Gamma}')$$

$$\frac{\delta(\Gamma,\overline{\Gamma})}{\delta(\Gamma',\overline{\Gamma}')} = \prod_{i=2}^{[\Gamma:\Gamma']} \Psi(\rho_i;\Gamma',\overline{\Gamma}')$$

where $\{\rho_1, \ldots, \rho_{[\Gamma:\Gamma']}\}$ *is a complete system of representatives of* $\Gamma \bmod \Gamma'$ *such that* $\rho_1 \in \Gamma'$.

Proof. These distribution formulas are an immediate consequence of the fundamental relation (1.3). □

2-2. Let $\mathfrak{m} \neq (1)$ be an integral ideal of $\mathcal{O}_k$, and let $H_\mathfrak{m}$ be the maximal abelian extension of k whose conductor divides $\mathfrak{m}$ and such that $H_\mathfrak{m} \subset k_\infty$. The existence of $H_\mathfrak{m}$ is well assured by Class Field Theory. In addition, if we denote by $\mathfrak{m}_v$ (resp. $\mathcal{O}_{k,v}$) the completion of $\mathfrak{m}$ (resp. $\mathcal{O}_k$) at each place $v \neq \infty$ of k, then we know by Class Field Theory that $H_\mathfrak{m}$ is associated to the idèle group

$$\mathcal{J}_\mathfrak{m} \overset{\text{dfn}}{=} k^\times \cdot \mathcal{U}_\mathfrak{m}$$

where $\mathcal{U}_\mathfrak{m}$ contains all idèles t of k such that the component t_v of t at each place $v \neq \infty$ of k verifies $t_v \in \mathcal{O}_{k,v}^\times$ and $t_v \equiv 1 \bmod \mathfrak{m}_v$.

Theorem 2.2. *Let* $\mathfrak{a}$, $\mathfrak{b}$ *be two integral ideals of* $\mathcal{O}_k$ *prime to* $\mathfrak{m}$. *Then we have*

$$\Psi(1;\mathfrak{m},\mathfrak{a}^{-1}\mathfrak{m}) \in H_\mathfrak{m}$$

and if $(\mathfrak{b}, H_\mathfrak{m}/k)$ *is the automorphism of* $H_\mathfrak{m}/k$ *associated to* $\mathfrak{b}$ *by the Artin map, then*

$$\Psi(1;\mathfrak{m},\mathfrak{a}^{-1}\mathfrak{m})^{(\mathfrak{b},H_\mathfrak{m}/k)} = \Psi(1;\mathfrak{b}^{-1}\mathfrak{m},\mathfrak{a}^{-1}\mathfrak{b}^{-1}\mathfrak{m}).$$

Moreover, let $T(\mathfrak{m})$ *be the set of prime ideals of* $\mathcal{O}_k$ *dividing* $\mathfrak{m}$. *Then* $\Psi(1;\mathfrak{m},\mathfrak{a}^{-1}\mathfrak{m})$ *generates in* $\mathcal{O}_{H_\mathfrak{m}}$ *the ideal*

$$\begin{cases} \mathfrak{q}_{H_\mathfrak{m}}^{\frac{N\mathfrak{a}-1}{w_k}} & \text{if } T(\mathfrak{m}) = \{\mathfrak{q}\} \\ \mathcal{O}_{H_\mathfrak{m}} & \text{if } \#T(\mathfrak{m}) \geq 2 \end{cases}$$

where $w_k = \#\mu(k)$ *and* $\mathfrak{q}_{H_\mathfrak{m}}$ *is the product in* $\mathcal{O}_{H_\mathfrak{m}}$ *of the prime ideals dividing* $\mathfrak{q}$.

Proof. We shall see that $\Psi(1;\mathfrak{m},\mathfrak{a}^{-1}\mathfrak{m})$ can be expressed as function of a primitive $\mathfrak{m}$ division point. This allows us to use Theorem 4.12 and Lemma 4.19 of **[7]**. Indeed, since Ψ is homogeneous, we have

$$\Psi(1;\mathfrak{m},\mathfrak{a}^{-1}\mathfrak{m}) = \frac{(\xi e_\mathfrak{m}(1))^{N\mathfrak{a}}}{\Phi_\mathfrak{a}^{\xi\mathfrak{m}}(\xi e_\mathfrak{m}(1))} \tag{2.2}$$

where ξ is any nonzero element of Ω.

Now choose $\xi \in \Omega^\times$ such that $\xi^{w_\infty} = \Delta_s(\mathfrak{m})$, where s is the sign function of Proposition 1.5 and $\Delta_s(\mathfrak{m})$ is defined by Formula (1.13). Then we know, cf. **[7]**, §4, that

$\lambda \overset{\text{dfn}}{=} \xi e_{\mathfrak{m}}(1)$ belongs to the abelian extension $H_{\mathfrak{m},s}$ of k associated to the idèle group

$$\mathfrak{J}_{\mathfrak{m},s} \overset{\text{dfn}}{=} k^{\times} \cdot \mathfrak{U}_{\mathfrak{m},s}$$

where $\mathfrak{U}_{\mathfrak{m},s}$ contains those idèles t such that the component t_v of t at each place $v \neq \infty$ of k verifies $t_v \in \mathcal{O}_{k,v}^{\times}$ and $t_v \equiv 1 \mod \mathfrak{m}_v$ with $s(t_\infty) = 1$.

Moreover, if $\mathfrak{d}$ is an integral ideal of $\mathcal{O}_k$ prime to $\mathfrak{m}$ then we have, cf. [7], Theorem 4.12,

$$\lambda^{(\mathfrak{d},H_{\mathfrak{m},s}/k)} = \Phi_{\mathfrak{d}}^{\Gamma}(\lambda) \tag{2.3}$$

where $\Gamma = \xi \cdot \mathfrak{m}$.

Remark 4. The lattice $\Gamma = \xi \cdot \mathfrak{m}$ is called special because the associated Drinfeld $\mathcal{O}_k$-module P^{Γ} is s-normalized, i.e.

$$x\delta(\Gamma, x^{-1}\Gamma) = s(x)^{\sigma_\Gamma}, \qquad \text{for all } x \in \mathcal{O}_k \backslash \{0\}. \tag{2.4}$$

So, if $\mathfrak{d}$ is principal and generated by some $x \in \mathcal{O}_k$, $x \equiv 1 \mod \mathfrak{m}$, then

$$\Phi_{\mathfrak{d}}^{\Gamma}(\lambda) = \Phi_{\mathfrak{d}}^{\Gamma}(e_\Gamma(\xi)) = s(x)^{-\sigma_\Gamma} \cdot \lambda$$

by Formulas (1.3) and (2.4). In this case (2.3) becomes

$$\lambda^{(\mathfrak{d},H_{\mathfrak{m},s}/k)} = s(x)^{-\sigma_\Gamma} \cdot \lambda. \tag{2.3'}$$

Now using (2.2) and (2.3) we find the following expression

$$\Psi(1; \mathfrak{m}, \mathfrak{a}^{-1}\mathfrak{m}) = \lambda^{N\mathfrak{a} - (\mathfrak{a}, H_{\mathfrak{m},s}/k)}. \tag{2.5}$$

But $H_{\mathfrak{m}}$ is just the fixed field of all automorphisms $(\mathfrak{d}, H_{\mathfrak{m},s}/k)$ of $H_{\mathfrak{m},s}/k$, where $\mathfrak{d}$ is any principal ideal of $\mathcal{O}_k$ generated by some $x \equiv 1 \mod \mathfrak{m}$. Hence the first assertion is easily deduced from (2.3) and (2.5).

On the other hand, Lemma 4.19 of [7] gives the prime factorization of λ in $\mathcal{O}_{H_{\mathfrak{m},s}}$. We have

$$\lambda \mathcal{O}_{H_{\mathfrak{m},s}} = \begin{cases} \mathfrak{q}_{H_{\mathfrak{m},s}} & \text{if } T(\mathfrak{m}) = \{\mathfrak{q}\} \\ \mathcal{O}_{H_{\mathfrak{m},s}} & \text{if } \#T(\mathfrak{m}) \geq 2 \end{cases} \tag{2.6}$$

where $\mathfrak{q}_{H_{\mathfrak{m},s}}$ is the product in $\mathcal{O}_{H_{\mathfrak{m},s}}$ of the prime ideals dividing $\mathfrak{q}$. But in the case $T(\mathfrak{m}) = \{\mathfrak{q}\}$, each prime factor of $\mathfrak{q}_{H_{\mathfrak{m}}}$ has ramification number w_k in $H_{\mathfrak{m},s}/H_{\mathfrak{m}}$. So we can deduce the prime factorization of $\Psi(1; \mathfrak{m}, \mathfrak{a}^{-1}\mathfrak{m})$ in $\mathcal{O}_{H_{\mathfrak{m}}}$. □

Remark 5. In fact, one can give an alternative proof of Theorem 2.2, see [10] for instance.

Using classical techniques and Theorem 2.2, one can interpret the distribution formulas of Proposition 2.1 as norm formulas. Indeed we have

Proposition 2.3. *Let* $\mathfrak{q}$ *be a prime ideal of* $\mathcal{O}_k$, *and let* $\mathfrak{b}$ *be an integral ideal prime to* $\mathfrak{m}\mathfrak{q}$. *Then*

$$N_{H_{\mathfrak{m}\mathfrak{q}}/H_{\mathfrak{m}}}(\Psi(1; \mathfrak{m}\mathfrak{q}, \mathfrak{b}^{-1}\mathfrak{m}\mathfrak{q}))^{w_{\mathfrak{m}}} = \begin{cases} \Psi(1, \mathfrak{m}, \mathfrak{b}^{-1}\mathfrak{m}) & \text{if } \mathfrak{q} \mid \mathfrak{m} \\ \Psi(1; \mathfrak{m}, \mathfrak{b}^{-1}\mathfrak{m})^{1-(\mathfrak{q}^{-1}, H_{\mathfrak{m}}/k)} & \text{if } \mathfrak{q} \nmid \mathfrak{m} \\ \frac{\delta(\mathcal{O}_k, \mathfrak{b}^{-1})}{\delta(\mathfrak{q}, \mathfrak{b}^{-1}\mathfrak{q})} & \text{if } \mathfrak{m} = (1) \end{cases}$$

where $w_{\mathfrak{m}} = 1$ *if* $\mathfrak{m} \neq (1)$ *and* $w_{\mathfrak{m}} = w_k$ *if* $\mathfrak{m} = (1)$.

Proof. See **[10]**, chap. IV, Theorem 5.3. □

3. The group $\mathcal{E}_F$

In this section the field F is always a finite abelian extension of k such that $F \subset k_\infty$. We begin with the following remark.

Remark 6. Let $\mathfrak{a}$ be an integral ideal of $\mathcal{O}_k$ prime to the conductor of F/k. Then we have

$$\alpha^{(\mathfrak{a},F/k)} = \alpha^{N\mathfrak{a}}, \qquad \text{for all } \alpha \in \mu(F),$$

where $(\mathfrak{a}, F/k)$ is the automorphism of F/k associated to $\mathfrak{a}$ by the Artin map. In particular if $(\mathfrak{a}', F/k) = (\mathfrak{a}, F/k)$, then $N\mathfrak{a}' \equiv N\mathfrak{a} \bmod w_F$, where w_F is the exact number of roots of unity in F, i.e., $w_F = \#\mu(F)$. This means that we have a well defined Dirichlet character

$$\varphi_F \colon \mathrm{Gal}(F/k) \to \mathbb{Z}(w_F)^\times$$
$$\tau \mapsto \varphi_F(\tau),$$

given by the condition

$$\varphi_F(\tau) \equiv N\mathfrak{b} \bmod w_F,$$

$\mathfrak{b}$ being any integral ideal of $\mathcal{O}_k$ prime to the conductor of F/k and such that $\tau = (\mathfrak{b}, F/k)$.

Moreover one has the following useful result.

Lemma 3.1. *Let* $J_{F/k}$ *be the annihilator of the* $\mathbb{Z}[\mathrm{Gal}(F/k)]$*-module* $\mu(F)$. *Let* T *be a finite set of primes of* $\mathcal{O}_k$ *containing at least all those ramified in* F/k. *Then* $J_{F/k}$ *is generated over* $\mathbb{Z}$ *by the elements* $N\mathfrak{q} - (\mathfrak{q}, F/k)$, *where* $\mathfrak{q}$ *runs over the prime ideals of* $\mathcal{O}_k$ *such that* $\mathfrak{q} \notin T$. *In particular, the integer* $w_F \overset{\text{dfn}}{=} \#\mu(F)$ *is the g.c.d. of* $N\mathfrak{q} - 1$, $\mathfrak{q} \notin T$ *and* $(\mathfrak{q}, F/k) = 1$.

Proof. See **[15]**, chap. IV, §1. □

3-1. In this section we suppose that the extension F/k is unramified. In order to have a pleasant definition of $\mathcal{E}_F$ we need to prove the following result.

Theorem 3.2. *Let* $a > 0$ *be a natural number. Let* $\mathfrak{q}$ *be a prime ideal of* $\mathcal{O}_k$ *and* $\sigma = (\mathfrak{q}, F/k)$ *be the automorphism of* F/k *associated to* $\mathfrak{q}$ *by the Artin map. Then the intersection with* $\mathcal{O}_F^\times$ *of the subgroup of* $F^\times$ *generated by* $\mu(F)$ *and by*

$$N_{H_{\mathfrak{q}^a}/F}(\Psi(1; \mathfrak{q}^a, \mathfrak{b}^{-1}\mathfrak{q}^a))$$

where $\mathfrak{b}$ *runs over the set of integral ideals of* $\mathcal{O}_k$ *prime to* $\mathfrak{q}$, *depends only on* σ. *(We will denote this intersection by* $\mathcal{E}_{F,\sigma}$.*)*

Proof. The theorem is an immediate corollary of Proposition 3.6. □

Definition 3.3. The group $\mathcal{E}_F$ of Theorem 1 is the group generated by all $\mathcal{E}_{F,\sigma}$, $\sigma \in \mathrm{Gal}(F/k)$, i.e.

$$\mathcal{E}_F \overset{\mathrm{dfn}}{=} \prod_{\sigma \in \mathrm{Gal}(F/k)} \mathcal{E}_{F,\sigma}. \tag{3.1}$$

Now it remains to give Proposition 3.6. For this we need some definitions and remarks.

Definition 3.4. Let $\tau \in \mathrm{Gal}(F/k)$.

i) If $F = H_{(1)}$, then we define

$$\partial_{H_{(1)}}(\tau) \overset{\mathrm{dfn}}{=} \alpha^{w_\infty} \Delta_s(\mathfrak{b})^h \tag{3.2}$$

where $\mathfrak{b}$ is any fractional ideal of $\mathcal{O}_k$ such that $(\mathfrak{b}, H_{(1)}/k) = \tau^{-1}$ and $\mathfrak{b}^h = \alpha\mathcal{O}_k$, with $\alpha \in k$.

ii) In general, we define

$$\partial_F(\tau) \overset{\mathrm{dfn}}{=} \prod_{\substack{\widetilde{\tau} \in \mathrm{Gal}(H_{(1)}/k) \\ \widetilde{\tau}|_F = \tau}} \partial_{H_{(1)}}(\widetilde{\tau}).$$

Remark 7. The expression defined by (3.2) depends only on τ since the discriminant function Δ_s is homogeneous of weight $-w_\infty$, cf. Property (1.13).

Lemma 3.5. *Let τ_1, τ_2 and τ be automorphisms of F/k. Then we have*

$$\frac{\partial_F(\tau_1)}{\partial_F(\tau_2)} \in \mathcal{O}_F^\times$$

and

$$\left(\frac{\partial_F(\tau_1)}{\partial_F(\tau_2)}\right)^\tau = \frac{\partial_F(\tau_1\tau)}{\partial_F(\tau_2\tau)}.$$

In particular if $\widetilde{\tau}_1$, and $\widetilde{\tau}_2$ are automorphisms of $H_{(1)}/k$ such that $\widetilde{\tau}_i\,|_F = \tau_i$, $i = 1, 2$, then

$$N_{H_{(1)}/F}\left(\frac{\partial_{H_{(1)}}(\widetilde{\tau}_1)}{\partial_{H_{(1)}}(\widetilde{\tau}_2)}\right) = \frac{\partial_F(\tau_1)}{\partial_F(\tau_2)}.$$

Proof. This lemma is easily derived from Proposition 1.8, above. □

Proposition 3.6. *The elements of $\mathcal{E}_{F,\sigma}^{w_k w_\infty h}$ are all the products*

$$\prod_{\tau \in \mathrm{Gal}(F/k)} \left(\frac{\partial_F(1)\partial_F(\tau\sigma^{-1})}{\partial_F(\sigma^{-1})\partial_F(\tau)}\right)^{m_\tau} \tag{3.3}$$

where m_τ, $\tau \in \mathrm{Gal}(F/k)$, are rational integers such that

$$\sum_{\tau \in \mathrm{Gal}(F/k)} m_\tau(\varphi_F(\tau) - 1) \equiv 0 \bmod w_F. \tag{3.4}$$

Proof. First recall, cf. Theorem 2.2, that $N_{H_{\mathfrak{q}^a}/F}(\Psi(1;\mathfrak{q}^a,\mathfrak{b}^{-1}\mathfrak{q}^a))$ generates in $\mathcal{O}_F$ the ideal

$$\mathfrak{q}_F^{[H_{(1)}:F]\frac{N\mathfrak{b}-1}{w_k}}$$

where $\mathfrak{q}_F$ is the product in $\mathcal{O}_F$ of the prime ideals dividing $\mathfrak{q}$. It follows that the elements of $\mathcal{E}_{F,\sigma}$ are, up to roots of unity in F, all the products

$$\prod_{\mathfrak{b}\in B} N_{H_{\mathfrak{q}^a}/F}(\Psi(1;\mathfrak{q}^a,\mathfrak{b}^{-1}\mathfrak{q}^a))^{m_\mathfrak{b}} \tag{3.5}$$

where B is any finite set of integral ideals of $\mathcal{O}_k$ prime to $\mathfrak{q}$, and $m_\mathfrak{b}$, $\mathfrak{b}\in B$ are rational integers, with the condition

$$\sum_{\mathfrak{b}\in B} m_\mathfrak{b}(N\mathfrak{b}-1)=0 \tag{3.6}$$

necessary to get units.

On the other hand, if we adopt the notation $\tau_\mathfrak{b} \overset{\text{dfn}}{=} (\mathfrak{b}, F/k)$, then Lemma 3.1 allows us to write the product (3.3) as

$$\prod_{\mathfrak{b}\in\widetilde{B}} \left(\frac{\partial_F(1)\partial_F(\tau_\mathfrak{b}\sigma^{-1})}{\partial_F(\tau_\mathfrak{b})\partial_F(\sigma^{-1})}\right)^{m_\mathfrak{b}} \tag{3.3'}$$

where $\widetilde{B}$ is a finite set of integral ideals of $\mathcal{O}_k$ prime to $\mathfrak{q}$ and $m_\mathfrak{b}$, $\mathfrak{b}\in\widetilde{B}$ are rational integers satisfying condition (3.6).

Now we have the formula

$$N_{H_{\mathfrak{q}^a}/F}(\Psi(1;\mathfrak{q}^a,\mathfrak{b}^{-1}\mathfrak{q}^a))^{w_k} = N_{H_{(1)}/F}\left(\frac{\delta(\mathcal{O}_k,\mathfrak{b}^{-1})}{\delta(\mathfrak{q},\mathfrak{b}^{-1}\mathfrak{q})}\right)$$

which we deduce from Proposition 2.3. Consequently

$$N_{H_{\mathfrak{q}^a}/F}(\Psi(1;\mathfrak{q}^a,\mathfrak{b}^{-1}\mathfrak{q}^a))^{w_k w_\infty h} = \gamma^{[H_{(1)}:F]w_\infty(N\mathfrak{b}-1)}\left(\frac{\partial_F(1)}{\partial_F(\sigma^{-1})}\right)^{N\mathfrak{b}}\frac{\partial_F(\tau_\mathfrak{b}\sigma^{-1})}{\partial_F(\tau_\mathfrak{b})}$$

where $\gamma\in\mathcal{O}_k$ is chosen such that $\mathfrak{q}^h=\gamma\mathcal{O}_k$.

This proves that the power $w_k w_\infty h$ of (3.5) is exactly equal to (3.3′). □

Remark 8. In fact, we see by Proposition 3.6 that $\mathcal{E}_{F,\sigma}$ depends only on the cyclic subgroup of $\mathrm{Gal}(F/k)$ generated by σ, i.e., if $\sigma'\in\mathrm{Gal}(F/k)$ is such that $\sigma'\in\langle\sigma\rangle$ then we have

$$\mathcal{E}_{F,\sigma'}\subset\mathcal{E}_{F,\sigma}.$$

In general one can ask, what about the intersection $\mathcal{E}_{F,\sigma'}\cap\mathcal{E}_{F,\sigma}$? In any case this group contains $\mathcal{E}_{F,\sigma''}$, if σ'' is a generator of the cyclic group $\langle\sigma\rangle\cap\langle\sigma'\rangle$.

Remark 9. We deduce from Theorem 2.2 and Formula (2.1) the relation

$$\Psi(1;\mathfrak{q}^a,\mathfrak{b}^{-1}\mathfrak{q}^a)^{(\mathfrak{d},H_{\mathfrak{q}^a}/k)} = \frac{\Psi(q;\mathfrak{q}^a,\sigma^{-1}\mathfrak{b}^{-1}\mathfrak{q}^a)}{\Psi(1;\mathfrak{q}^a,\mathfrak{d}^{-1}\mathfrak{q}^a)^{N\mathfrak{b}}}. \tag{3.7}$$

In particular the group $\mathcal{E}_{F,\sigma}$ is stable under the Galois action.

Remark 10. There is no difficulty to see that the behavior of the groups $\mathcal{E}_{F,\sigma}$ under the norm map is given by the relation

$$\mu(F)N_{\widetilde{F}/F}(\mathcal{E}_{\widetilde{F},\widetilde{\sigma}}) = \mathcal{E}_{F,\sigma},$$

where $\widetilde{F} \subset k_\infty$ is any unramified abelian extension of k such that $F \subset \widetilde{F}$, and $\widetilde{\sigma} \in \mathrm{Gal}(\widetilde{F}/k)$ is an extension of σ to $\widetilde{F}$.

In particular we have

$$\mu(F)N_{\widetilde{F}/F}(\mathcal{E}_{\widetilde{F}}) = \mathcal{E}_F.$$

Remark 11. Let us suppose, just a moment, that $\deg(\infty) = 1$. Then D. R. Hayes introduces, cf. **[6]**, Definition 1.3, a subgroup E of $\mathcal{O}_{H_{(1)}}^\times$ whose elements are called elliptic units of level $G = \mathrm{Gal}(H_{(1)}/k)$. He proves, cf. Proposition 3.7, that there is a maximal subgroup $\overline{E}$ of $\mathcal{O}_{H_{(1)}}^\times$ such that

$$\overline{E}^{w_k} = E.$$

He then shows, cf. Corollary 4.5, that the index of $\mathbb{F}_q^\times N_{H_{(1)}/F}(\overline{E})$ in $\mathcal{O}_F^\times$ is equal to $h(\mathcal{O}_F)/[H_{(1)}: F]$.

In fact the groups $\overline{E}$ and $\mathcal{E}_{H_{(1)}}$ are equal. Consider a prime ideal $\mathfrak{q}$ of $\mathcal{O}_k$ and let $\lambda_\mathfrak{q}$ be any $\mathfrak{q}$-division point. Call $I_\mathfrak{q}$ the augmentation ideal of the group ring $\mathbb{Z}[\mathrm{Gal}(H_{\mathfrak{q},s}/k)]$, and let $\overline{E}_\mathfrak{q}$ be the subgroup of $\mathcal{O}_{H_{(1)}}^\times$ defined as follows:

$$\overline{E}_\mathfrak{q} \overset{\text{dfn}}{=} \mathbb{F}_q^\times N_{H_\mathfrak{q}/H_{(1)}}(\lambda_\mathfrak{q}^{I_\mathfrak{q}}).$$

Then we deduce from Corollary 3.6, Proposition 2.6, and Formula (1.9) of loc. cit. that $\overline{E}$ is generated by all the groups $\overline{E}_\mathfrak{q}$.

But since $\deg(\infty) = 1$ we see that the elements of $I_\mathfrak{q}$ are all the finite sums

$$\sum_\mathfrak{b} m_\mathfrak{b}(N\mathfrak{b} - (\mathfrak{b}, H_{\mathfrak{m},s}/k))$$

where $\mathfrak{b}$ are integral ideals of $\mathcal{O}_k$ prime to $\mathfrak{q}$ and $n_\mathfrak{b}$ are rational integers with Condition (3.6). Hence Formula (2.5) and Theorem 3.2 imply that

$$\overline{E}_\mathfrak{q} = \mathcal{E}_{H_{(1)},\sigma}$$

where $\sigma = (\mathfrak{q}, H_{(1)}/k)$. In particular we have $\overline{E} = \mathcal{E}_{H_{(1)}}$.

3-2. Now suppose that the conductor of F/k is a positive power of some prime ideal $\mathfrak{q}$ of $\mathcal{O}_k$, i.e., $\mathrm{cond}(F/k) = \mathfrak{q}^e$, $e > 0$.

Definition 3.7. Let $A_F^0 = A_F \cap \mathcal{O}_F^\times$, where A_F is the subgroup of $F^\times$ generated by all

$$N_{H_{\mathfrak{q}^e}/F}(\Psi(1; \mathfrak{q}^e, \mathfrak{b}^{-1}\mathfrak{q}^e))$$

where $\mathfrak{b}$ runs over the set of integral ideals of $\mathcal{O}_k$ prime to $\mathfrak{q}$.

Lemma 3.8. *The group A_F^0 is formed by all the products*

$$\prod_{\mathfrak{b}\in B} N_{H_{\mathfrak{q}^e}/F}(\Psi(1;\mathfrak{q}^e,\mathfrak{b}^{-1}\mathfrak{q}^e))^{m_\mathfrak{b}}$$

where B is any finite set of integral ideals of $\mathcal{O}_k$ prime to $\mathfrak{q}$, and $m_\mathfrak{b}$, $\mathfrak{b}\in B$, are rational integers with Condition (3.6).

Proof. Clear because we know by Theorem 2.2 that $N_{H_{\mathfrak{q}^e}/F}(\Psi(1;\mathfrak{q}^e,\mathfrak{b}^{-1}\mathfrak{q}^e))$ generates in $\mathcal{O}_F$ the ideal

$$\mathfrak{q}_F^{[H_{(1)}:F\cap H_{(1)}]\frac{N\mathfrak{b}-1}{w_k}}. \qquad \square$$

Now we are ready to give an ad-hoc definition of the group $\mathcal{E}_F$ of Theorem 1 when the degree of the place ∞ is equal to 1. In fact, one can compute the index in $\mathcal{O}_F^\times$ of the group $\mathcal{E}_F$, defined by (3.8), even if $\deg(\infty)\neq 1$, cf. **[10]**. In that case, the group $\mathcal{E}_F$ is not big enough to give the desired index formula.

Definition 3.9. When the degree of ∞ is equal to 1, i.e., $\deg(\infty)=1$, we let $\mathcal{E}_F$ be the subgroup of $\mathcal{O}_F^\times$ generated by $\mathcal{E}_{F\cap H_{(1)}}$ and A_F^0, so that

$$\mathcal{E}_F \stackrel{\text{dfn}}{=} \mathcal{E}_{F\cap H_{(1)}}\cdot A_F^0. \tag{3.8}$$

Remark 12. One can choose $k=\mathbb{F}_q(t)$ to be a rational function field, and ∞ the prime divisor corresponding to the pole of t. Thus we have $\mathcal{O}_k=\mathbb{F}_q[t]$. Then it is well known that every prime ideal of $\mathcal{O}_k$ is generated by a unique monic prime polynomial of $\mathbb{F}_q[t]$. Moreover, we have $H_{(1)}=k$ because $\deg(\infty)=1$ and the class number of k equals 1 also.

Now let F be as in the above Definition 3.9. Then the group $\mathcal{E}_F$ is just equal to $F_q^\times A_F^0$. But we see, thanks to Formula (2.5), that A_F^0 can be described as the group

$$N_{H_{\mathfrak{q}^e}/F}(\lambda_{\mathfrak{q}^e}^{I_{\mathfrak{q}^e}})$$

where $\lambda_{\mathfrak{q}^e}$ is any primitive $\mathfrak{q}^e$-division point and $I_{\mathfrak{q}^e}$ is the augmentation ideal of $\mathbb{Z}[\mathrm{Gal}(H_{\mathfrak{q}^e,s}/k)]$. This shows, in particular, that $\mathcal{E}_{H_{\mathfrak{q}^e}}$ and the group $\mathcal{E}$, introduced by S. Galovich and M. Rosen in **[1]**, Theorem 3, are equal.

4. The index formula

As in §3, the field F is a finite abelian extension of k such that $F\subset k_\infty$.

4-1. Consider the logarithm map

$$\ell_F\colon F^\times \to \mathbb{Z}[\mathrm{Gal}(F/k)]$$
$$x\mapsto \sum_{\sigma\in\mathrm{Gal}(F/k)} v_\infty(x^\sigma)\sigma^{-1},$$

where v_∞ is the normalized valuation of k_∞ associated to ∞. It is obvious that ℓ_F is a group morphism. Its kernel in $\mathcal{O}_F^\times$ is $\mu(F)$ since if $z\in\mathcal{O}_F^\times$ is such that $\ell_F(z)=0$, then z is a unit at each place of F. Moreover, the image $\ell_F(\mathcal{O}_F^\times)$ is contained in the augmentation ideal $I_{F/k}$ of the group ring $\mathbb{Z}[\mathrm{Gal}(F/k)]$.

Let h_F be the class number of F and d_F be the common degree of the places of F sitting over ∞. Then one can prove, cf. **[10]**, that the index $[I_{F/k}\colon \ell_F(\mathcal{O}_F^\times)]$ is finite and verifies the formula

$$[I_{F/k}\colon \ell_F(\mathcal{O}_F^\times)] = \frac{d_F h_F}{h(\mathcal{O}_F)}. \tag{4.1}$$

4-2. Let χ be a character of $\mathrm{Gal}(F/k)$. Then we can define $\chi(v)$ for each place v of k in a well known manner. In particular $\chi(\infty) = 1$ because ∞ splits completely in F/k. One can associate to χ the L-function

$$L(s;\chi,F/k) \stackrel{\mathrm{dfn}}{=} \prod_v (1-\chi(v)Nv^{-s})^{-1}$$

defined on the complex half-plane $\mathrm{Re}(s) > 1$. Actually it has a meromorphic continuation to the whole complex plane, with simple poles at 0 and 1 when the extension F/k is unramified and $\chi = 1$. In this case we have $L(s;\chi,F/k) = \zeta_F(s)$, the Dedekind zeta function of the global field F.

We call χ primitive if it is not obtained from a character $\widetilde{\chi}$ of the Galois group $\mathrm{Gal}(\widetilde{F}/k)$ for some sub-extension $\widetilde{F}$ of F. The L-functions associated to primitive characters verify a well known functional equation. In addition, we have the following decomposition

$$\zeta_F(s) = \zeta_k(s) \prod_{\substack{\chi \text{ primitive} \\ \chi\neq 1}} L(s;\chi,F/k) \tag{4.2}$$

of the Dedekind zeta function of F, cf. **[15]**, chap. XIII, Théorème 10. Hence by the evaluation of the principal parts of (4.2) at $s = 0$ we get the analytic class number formula

$$\frac{h(\mathcal{O}_F)R(\mathcal{O}_F)}{w_F} = \frac{h}{w_k} \prod_{\substack{\chi \text{ primitive} \\ \chi\neq 1}} L(0;\chi,F/k) \tag{4.3}$$

where $R(\mathcal{O}_F)$ is the regulator of the Dedekind ring $\mathcal{O}_F$, in particular $R(\mathcal{O}_k) = 1$. Also we know that $R(\mathcal{O}_F)$ can be expressed as the index of $\ell_F(\mathcal{O}_F^\times)$ in $I_{F/k}$, cf. **[10]**, chap I, §3, i.e.

$$R(\mathcal{O}_F) = [I_{F/k}\colon \ell_F(\mathcal{O}_F^\times)]. \tag{4.4}$$

4-3. Now suppose that the extension F/k is unramified. Then it is easy to see that

$$\ell_F\left(\frac{\partial_F(\sigma)}{\partial_F(1)}\right) = \omega_F((\sigma)-(1)) \tag{4.5}$$

where ω_F is the element of $I_{F/k}$ given by

$$\omega_F \stackrel{\mathrm{dfn}}{=} \sum_{\tau\in\mathrm{Gal}(F/k)} v_\infty\left(\frac{\partial_F(\tau)}{\partial_F(1)}\right)\tau^{-1}.$$

We can consider ω_F as an endomorphism of $I_{F/k}$ and then calculate $\det \omega_F$ by the Frobenius determinant formula

$$\det \omega_F = \prod_{\chi \neq 1} \left(\sum_{\tau \in \mathrm{Gal}(F/k)} \chi(\tau) v_\infty(\partial_F(\tau)) \right),$$

the product being taken over all non-trivial characters of $\mathrm{Gal}(F/k)$. But if χ is a nontrivial character of the Galois group of F/k then we have the limit formula

$$L(0;\chi, F/k) = \frac{1}{w_k w_\infty h} \sum_{\sigma \in \mathrm{Gal}(F/k)} \chi(\sigma) v_\infty(\partial_F(\sigma)) \tag{4.6}$$

deduced from Proposition 7.19 of **[3]**. So, using (4.3) we find

$$\det \omega_F = (w_k w_\infty h)^{[F:k]-1} \frac{h(\mathcal{O}_F) R(\mathcal{O}_F)}{h} \frac{w_k}{w_F}.$$

Hence $\det \omega_F \neq 0$. This implies that $\omega_F I_{F/k}$ has finite index in $I_{F/k}$ given by

$$[I_{F/k} : \omega_F I_{F/k}] = \det \omega_F .$$

Now let $M(\varphi_F)$ be the ideal of the group ring $\mathbb{Z}[\mathrm{Gal}(F/k)]$ generated by all the finite sums

$$\left[\sum_{\sigma \in \mathrm{Gal}(F/k)} n_\sigma((1) - (\sigma)) \right] ((1) - (\sigma')), \qquad \sigma' \in \mathrm{Gal}(F/k)$$

where n_σ, $\sigma \in \mathrm{Gal}(F/k)$, are rational integers satisfying the Condition (3.4).

It is obvious that $M(\varphi_F) \subset I_{F/k}^2$. In fact $M(\varphi_F)$ has finite index in $I_{F/k}^2$ since we know by Théorème 2 of **[9]** that $M(\varphi_F)$ is formed of all the finite sums

$$\sum_{\sigma, \sigma' \in \mathrm{Gal}(F/k)} n_{\sigma,\sigma'}((1) - (\sigma))((1) - (\sigma'))$$

where $n_{\sigma,\sigma'}$, σ, $\sigma' \in \mathrm{Gal}(F/k)$, are rational integers such that

$$\sum n_{\sigma,\sigma'} \left(\frac{\varphi_F(\sigma) - 1}{w_k} \right) \left(\frac{\varphi_F(\sigma') - 1}{w_k} \right) \equiv 0 \bmod \frac{w_F}{w_k}.$$

Hence Théorème 1 of **[9]** gives

$$[I_{F/k}^2 : M(\varphi_F)] = \frac{w_F}{w_k}.$$

Proposition 4.1. *The image of $\mathcal{E}_F^{w_k w_\infty h}$ by the logarithm map is*

$$\ell_F(\mathcal{E}_F^{w_k w_\infty h}) = \omega_F M(\varphi_F). \tag{4.7}$$

It follows that the index of $\mathcal{E}_F$ in $\mathcal{O}_F^\times$ is finite. It is equal to

$$[\mathcal{O}_F^\times : \mathcal{E}_F] = \frac{h(\mathcal{O}_F)}{[H_{(1)} : F]}.$$

Proof. Formula (4.7) is an immediate consequence of (4.5), Proposition 3.6 and Definition 3.3. It implies the equality

$$[\mathcal{O}_F^\times : \mu(F)\mathcal{E}_F^{w_k w_\infty h}] = [\ell_F(\mathcal{O}_F^\times) : \omega_F M(\varphi_F)] .$$

Moreover, Formula (4.4) gives the following decomposition:

$$\begin{aligned} &R(\mathcal{O}_F)[\ell_F(\mathcal{O}_F^\times) : \omega_F M(\varphi_F)] \\ &\qquad = [I_{F/k} : \omega_F I_{F/k}][\omega_F I_{F/k} : \omega_F I_{F/k}^2][\omega_F I_{F/k}^2 : \omega_F M(\varphi_F)] . \end{aligned}$$

But since $\det \omega_F \neq 0$ we have

$$[\omega_F I_{F/k} : \omega_F I_{F/k}^2] = [I_{F/k} : I_{F/k}^2] = [F : k]$$

$$[\omega_F I_{F/k}^2 : \omega_F M(\varphi_F)] = [I_{F/k}^2 : M(\varphi_F)] = \frac{w_F}{w_k}$$

so that

$$[\mathcal{O}_F^\times : \mu(F)\mathcal{E}_F^{w_k w_\infty h}] = (w_k w_\infty h)^{[F:k]-1} \frac{h(\mathcal{O}_F)}{[H_{(1)} : F]} .$$

In particular $\mathcal{E}_F^{w_k w_\infty h}$ is a free abelian group of rank $[F:k] - 1$. Hence we have

$$[\mathcal{E}_F : \mu(F)\mathcal{E}_F^{w_k w_\infty h}] = (w_k w_\infty h)^{[F:k]-1}$$

which gives the index formula of the proposition. □

References

[1] S. Galovich, M. Rosen, The class number of cyclotomic function fields, J. Number Theory 13, (1981), 363–375.

[2] E.- U. Gekeler, Drinfeld modular curves, Lecture Notes in Math. 1231, (1986).

[3] B. Gross, M. Rosen, Fourier series and the special values of L-functions, Adv. Math. 69, (1988), 1–31.

[4] D. Goss, Modular forms for $\mathbb{F}_r[T]$, J. Reine Angew. Math. 317, (1980), 16–39.

[5] D. R. Hayes, Explicit class field theory for global function fields, Studies in Algebra and Number Theory (G. C. Rota ed.), New York, Academic Press, (1979), 173–217.

[6] D. R. Hayes, Elliptic units in function fields, In: Proc. of a conference on modern developments related to Fermat's last theorem, (D. Goldfeld ed.), Birkhäuser, Boston, 1982.

[7] D. R. Hayes, Stickelberger elements in function fields, Compositio. Math. 55 (1985), 209–239.

[8] D. Kubert, S. Lang, Modular units, Grundlehren Math. Wiss. 244, Springer (1981).

[9] H. Oukhaba, G. Robert, Etude d' un idéal particulier associé à un caractère de Dirichlet d' un groupe fini, Séminaire de Théorie des Nombres, Bordeaux 3 (1991), 117–127.

[10] H. Oukhaba, Fonctions discriminant, formules pour le nombre de classes, et unités elliptiques; le cas des corps de fonctions (associé à des courbes sur des corps finis), Thèse (Grenoble, Institut Fourier, 10 juin 1991).

[11] H. Oukhaba, On discriminant functions associated to Drinfeld Modules of rank 1, Prépublication de l'Instut Fourier, à paraître.

[12] G. Robert, Concernant la relation de distribution satisfaite par la fonction φ associée à un réseau complexe, Invent. Math. 100, (1990), 231–257.

[13] G. Robert, La racine douzième canonique de $\Delta(L)^{[\underline{L}:L]}/\Delta(\underline{L})$, Prog. in Math,(1991), Birkhäuser, à paraître .

[14] G. Robert, A propos des unités de Stark (dans les extensions abéliennes ramifiées) d' un corps quadratique imaginaire, à paraître.

[15] J. Tate, Les conjectures de Stark sur les fonctions L d' Artin en $s = 0$, Cours à Orsay, Notes de D. Bernardi et N. Schappacher, Prog. Math.47, Birkhäuser (1984).

[16] A. Weil, Basic number theory, Grundlehren. Math. Wiss. 144, Springer (1974).

Institut Fourier, Université de Grenoble 1, B.P.74, 38402 St Martin D'Héres Cedex, France

Class Number "Parity" for Cyclic Function Fields

*Keqin Feng**

Abstract

Let $k = \mathbb{F}_q(T)$, M be a monic polynomial in $\mathbb{F}_q[T]$, $k(\Lambda_M)$ and $k(\Lambda_M)^+$ be the cyclotomic function field with conductor M and its maximal "real" subfield respectively. For a cyclic extension K/k, $K \subset k(\Lambda_M)^+$ with degree $n = [K:k]$, $(n, q-1) = 1$, a criterion for $l \mid h(O_K)$ and $l \mid h(K)$ is presented in terms of the cyclotomic unit group of K where $h(O_K)$ and $h(K)$ are the class numbers for the ring O_K and the field K respectively, and l is any prime divisor of $q-1$. As applications of this elementary criterion, we determine the divisibility of $h(O_K)$ and $h(K)$ by l for several cyclic function fields K.

Key Words. Class number, cyclotomic function field, cyclic function field, cyclotomic units.

1. Introduction

There are many papers concerned with the parity of the class number $h(K)$ of a number field K which is related to the signature of the unit group U_K or of the cyclotomic unit group C_K if K is abelian (see **[1]**, **[3]**, **[6]**, **[7]**, **[8]**, et al.). As a simple and typical example, G. Gras **[7]** proved that if K is an abelian number field with odd degree n, and the order of $2 \pmod{n}$ is even, then $h(K)$ is odd if and only if each totally positive cyclotomic unit in K is a square of a cyclotomic unit of K. Combining Gras' results and the method in Davis **[2]** for computing the signature of cyclotomic units, Feng **[3]** gives an elementary criterion of $2 \mid h(K)$ for several kinds of cyclic number fields K. The purpose of this paper is to present their analogies for cyclic function fields.

Let $k = \mathbb{F}_q(T)$ be the rational function field over the finite field $\mathbb{F}_q$, $R = \mathbb{F}_q[T]$, M be a monic polynomial in R, $k(\Lambda_M) = k(\lambda)$ be the cyclotomic function field with conductor M where λ is a primitive M-torsion element in the Carlitz R-module $\overline{k}$ ($=$ the algebraic closure of k), $k(\Lambda_M)^+ = k(\lambda^{q-1})$ be the maximal "real" subfield of $k(\Lambda_M)$ (see **[5]** for details on cyclotomic function fields). In this paper a cyclic function field K always means a cyclic extension K/k with degree $n = [K:k]$ satisfying the following conditions:

(I) $(n, q-1) = 1$;

* Granted by National Scientific Foundation 1991-5.

(II) Conductor $(K) = M$ and $K \subseteq k(\Lambda_M)^+$;

(III) Either n is a prime number or M is a power of an irreducible polynomial in R.

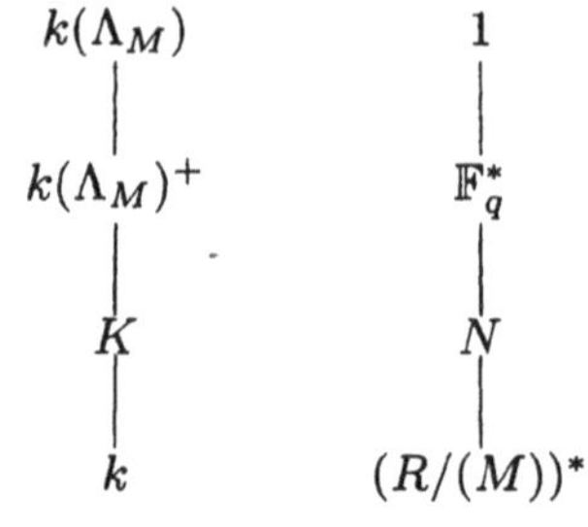

The Galois group $G(k(\Lambda_M)/k)$ is canonically isomorphic to $(R/(M))^*$ by $\sigma_A \mapsto A \pmod{M}$ for $A \in R$, $(A, M) = 1$ where σ_A is the automorphism determined by $\sigma_A(\lambda) = \lambda^A$. With this isomorphism we have $G(k(\Lambda_M)^+/k) \cong (R/(M))^*/\mathbb{F}_q^*$. Under the condition (II), the Galois group $G(k(\Lambda_M)/K)$ is isomorphic to a certain subgroup N of $(R/(M))^*$, $\mathbb{F}_q^* \subset N$, the cyclic group $G = G(K/k)$ with order n is isomorphic to $(R/(M))^*/N$. We choose a set N' of monic polynomials such that N' is a representative set of $N/\mathbb{F}_q^*$. Then we have

$$N_{k(\Lambda_M)^+/K}(\alpha) = \prod_{B \in N'} \sigma_B(\alpha) \qquad \text{for } \alpha \in k(\Lambda_M)^+. \tag{1.1}$$

Let O_K be the ring of integers of K (= the integral closure of R in K), $h(O_K)$ and $h(K)$ be the class numbers of the ring O_K and the field K respectively. It is well known that

$$h(K) = h(O_k)R_K$$

where R_K is a positive integer called the regulator of K.

In this paper, the "parity" of class numbers means the divisibility of $h(O_K)$ and $h(K)$ by a prime divisor l of $q-1$. From now on we fix a prime divisor l of $q-1$. In Section 2, we present a criterion for $l \mid h(O_K)$ in terms of the cyclotomic unit group of K. Then there are even more elementary criterions for $l \mid h(O_K)$ and $l \mid h(K)$ in Section 3. As applications of these criterions we determine the divisibility of $h(O_K)$ and $h(K)$ by l for several cyclic function fields K in the last section.

I would like to thank Professor Armand Brumer and Mike Rosen for their useful conversations and suggestions. I am obliged to Professor David Goss for organizing such a successful conference and providing a warm atmosphere.

2. Cyclotomic units and class number parity

Let $U_K = (O_K)^*$ (the unit group of K). The cyclotomic unit group C_K of K is the subgroup of U_K generated by $\mathbb{F}_q^*$ and the cyclotomic units $\eta_A((A, M) = 1)$ of K where

$$\eta_A = N_{k(\Lambda_M)^+/k}(\lambda^A/\lambda)$$

$$= \prod_{B \in N'} (\lambda^{BA}/\lambda^{B}) . \tag{2.1}$$

Lemma 2.1. *Let K/k be a cyclic extension satisfying the condition* (I), (II) *and* (III). *Then*

$$[U_K : C_K] = h(O_K) .$$

Proof. For any intermediate field K of $k(\Lambda_M)^+/k$, Feng and Yin **[4]** proved that

$$[U_K : C_K] = h(O_K) i(K) , \tag{2.2}$$

$$i(K) = \prod_{\substack{\chi \neq 1 \\ \chi(N)=1}} \prod_{P|M} (1 - \chi^*(P))$$

where χ runs through the nontrivial characters of R mod M, such that $\chi(N) = 1$, χ^* is the primitive character associated with χ, and P runs through the irreducible factors of M. The proof of formula (2.1) is very like the proof of Corollary 8.8 in **[10]** for the number field case. Under the condition (III), we have $\chi^*(P) = 0$ for all irreducible factors P of M. Therefore $i(K) = 1$ and Lemma 2.1 follows. □

We introduce several subgroups of U_K and C_K. The infinite prime divisor $\infty = \left(\frac{1}{T}\right)$ of k splits completely in $k(\Lambda_M)^*$, so does in K: $\infty = \wp_1, \ldots, \wp_n$. For each infinite prime divisor $\wp_i$ of K, $K_{\wp_i} = k_\infty = \mathbb{F}_q\left(\left(\frac{1}{T}\right)\right)$. An element $\alpha \in K^*$ is called an l-th power at infinity if α is an l-th power in $K_{\wp_i}$, for each i ($1 \leq i \leq n$). We consider the following subgroups of U_K:

$$\begin{aligned} U_K^1 &= \{\varepsilon \in U_K | N_{K/k}(\varepsilon) = 1\}; \\ U_K^{(l)} &= \{\varepsilon) \in U_K | N_{K/k}(\varepsilon) \in (\mathbb{F}_q^*)^l\}; \\ U_K^+ &= \{\varepsilon \in U_K | \varepsilon \text{ is } l\text{-th power at infinity}\}; \\ (U_K)^l &= \{\varepsilon^l | \varepsilon \in U_K\} . \end{aligned}$$

Similarly we have the subgroups C_K^1, $C_K^{(l)}$, C_K^+, and $(C_K)^l$ of C_K.

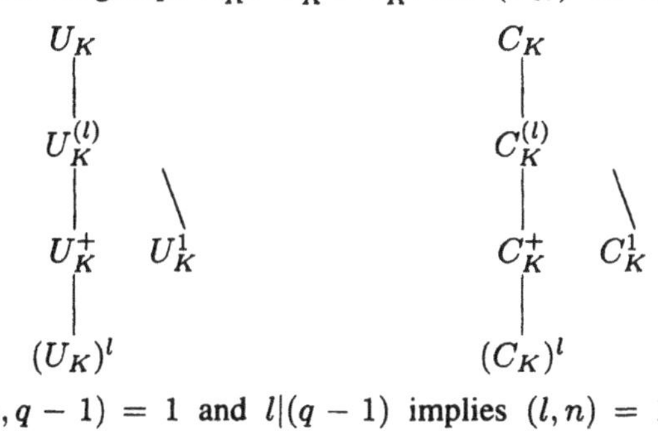

The conditions $(n, q-1) = 1$ and $l|(q-1)$ implies $(l, n) = 1$ and the following decompositions

$$\begin{aligned} U_K &= U_K^1 \times \mathbb{F}_q^* , & C_K &= C_K^1 \times \mathbb{F}_q^* \\ U_K^{(l)} &= U_K^1 \times (\mathbb{F}_q^*)^l , & C_K^{(l)} &= C_K^1 \times (\mathbb{F}_q^*)^l \end{aligned}$$

from which we can restate the Lemma 2.1 as

$$h(O_K) = [U_K^1 : C_K^1] = [U_K^{(l)} : C_K^{(l)}]. \tag{2.3}$$

Next, we need a criterion for $l | h(O_K)$ from class field theory.

Lemma 2.2. *Under the assumptions of Lemma 2.1, $l \mid h(O_K)$ if and only if there exists an unramified extension $K(\sqrt[l]{\alpha})/K$ ($\alpha \in K^*$) of degree l such that each infinite prime divisor $\wp_i (1 \le i \le n)$ of K splits in $K(\sqrt[l]{\alpha})$.*

Proof. The Hilbert class field of K introduced by Rosen **[9]** means the maximal unramified abelian extension $H(K)$ of K such that each infinite prime divisor of K splits completely in $H(K)$. Rosen proved that $H(K)/K$ is a finite extension and the Galois group $G(H(K)/K)$ is isomorphic to the class group of O_K. Lemma 2.2 follows from this fact since $l \mid q-1$ and any abelian extension of K with degree l is a Kummer extension. □

Now we present a criterion for $l \mid h(O_K)$ in terms of cyclotomic units which is an analogy of Gras' result **[7]**, **[8]** for the number field case.

Theorem 2.3. *Under the assumptions of Lemma 2.1, the following conditions are equivalent to each other*
(a) $l \nmid h(O_K)$;
(b) $C_K^+ \cap C_K^1 = (C_K^1)^l$;
(c) $C_K^+ = (C_K)^l$.

Proof. (a) $\Leftrightarrow$ (b): If $l \mid h(O_K) = [U_K^1 : C_K^1]$ (by (2.3)), we have an element $\varepsilon \in U_K^1 - C_K^1$ such that $\varepsilon^l \in C_K^1$. It is easy to see that $\varepsilon^l \in (C_K^+ \cap C_K^1) - (C_K^1)^l$. Therefore $C_K^+ \cap C_K^1 \neq (C_K^1)^l$. On the other hand, suppose that $u \in (C_K^+ \cap C_K^1) - (C_K^1)^l$. If $u = \alpha^l$, $\alpha \in K^*$, then $\alpha \in U_K$. We can choose $a \in \mathbb{F}_q^*$, $a^l = 1$ such that $a\alpha \in U_K^1$. Then $a\alpha \notin C_K^1$ (otherwise $u = (a\alpha)^1 \in (C_K^1)^l$) and $(a\alpha)^l = u \in C_K^1$. Therefore, $l \mid [U_K^1 : C_K^1] = h(O_K)$. If $u \notin (K^*)^l$, $K(\sqrt[l]{u})/K$ isa Kummer extension with degree l. $u \in U_K$ implies that each finite prime divisor of K is unramified in $K(\sqrt[l]{u})$. $u \in C_K^+$ implies that each infinite prime divisor of K splits completely in $K(\sqrt[l]{u})$. We also have $l \mid h(O_K)$ from Lemma 2.2.

(b) $\Leftrightarrow$ (c): Since

$$C_K^+ = (C_K^+ \cap C_K^1) \times (\mathbb{F}_q^*)^l$$
$$(C_K)^l = (C_K^1)^l \times (\mathbb{F}_q^*)^l.$$

Thus the lemma is established. □

3. Elementary Criterions for $l \mid h(O_K)$ and $l \mid h(K)$

Theorem 2.3 says that $l \nmid h(O_K)$ if and only if $C_K^+ = (C_K)^l$. In this section we present an elementary method to determine if each l-th power cyclotomic unit at infinity is an l-th power of a cyclotomic unit in K.

Let $\wp$ be a fixed infinite prime divisor of K, $K_\wp = \mathbb{F}_q\left(\left(\frac{1}{T}\right)\right)$. Each element $\alpha \in K_\wp^*$ has its $\left(\frac{1}{T}\right)$-adic expansion

$$\alpha = \sum_{\lambda=m}^{\infty} c_\lambda \left(\frac{1}{T}\right), \qquad c_\lambda \in \mathbb{F}_q, \qquad c_m \neq 0, \qquad m = v_\wp(\alpha).$$

We denote $\operatorname{sgn}(\alpha) = c_m$. From the Hensel lemma we know that

$$\alpha \in (K_\wp^*)^l \text{ if and only if } \operatorname{sgn}(\alpha) \in (\mathbb{F}_q^*)^l \text{ and } l \mid v_\wp(\alpha).$$

Therefore $\alpha \in K^*$ is an l-th power at infinity if and only if $\operatorname{sgn}(\tau(\alpha)) \in (\mathbb{F}_q^*)^l$ and $l \mid v_\wp(\tau(\alpha))$ for each $\tau \in G = G(K/k)$.

We denote

$$\Omega = \mathbb{F}_l[G]/(1 + \sigma + \cdots + \sigma^{n-1})$$

where σ is a generator of the cyclic Galois group G with order n. One knows that Ω is a semisimple algebra (since $(n, l) = 1$) and $\overline{C}_K^{(l)} = C_K^{(l)}/(C_K)^l$ is an Ω-module. Let N and N' be defined as in Section 1. We can choose a polynomial $A \in R = \mathbb{F}_q[T]$ such that $\{1, A, A^2, \ldots, A^{n-1}\}$ is a representative set of $(R/(M))^*$ mod N. Then we can choose $\sigma = \sigma_{A|K}$ as a generator of G, and C_K is generated by $\mathbb{F}_q^*$ and

$$\begin{aligned} \eta_i &= N_{k(\Lambda_M)^+/k}(\sigma^i(\lambda^A/\lambda)) \\ &= N_{k(\Lambda_M)^+/k}(\lambda^{A^{i+1}}/\lambda^{A^i}) \\ &= \prod_{B \in N'} (\lambda^{A^{i+1}B}/\lambda^{A^i B}) \qquad (1 \le i \le n-2). \end{aligned} \tag{3.1}$$

We can choose $a \in \mathbb{F}_q^*$ such that $\eta = a\eta_0 \in C_K^{(l)}$. It is easy to see that

$$\overline{C}_K^{(l)} = \overline{\eta}^\Omega$$

by considering that the $\mathbb{F}_q$-dimensions of both sides are $n-1$. Thus $\overline{C}_K^{(l)}$ is a faithful cyclic Ω-module generated by $\overline{\eta}$.

At last, we fix a nontrivial homomorphism ϕ from the multiplicative group $\mathbb{F}_q^*$ to the additive group $\mathbb{F}_l = \mathbb{Z}/l\mathbb{Z}$. Now we consider the mapping

$$S' : \overline{C}_K^{(l)} \to \mathbb{F}_l[X]/(X^n - 1);$$

$$(\overline{\varepsilon}) \mapsto S'(\varepsilon, X) = \sum_{j=0}^{n-1} s(\varepsilon^{\sigma^j}) X^{n-j-1} (\operatorname{mod} X^n - 1) \text{ for } \varepsilon \in C_K^{(l)},$$

where

$$s(\varepsilon) = \phi(\operatorname{sgn}(\varepsilon)).$$

Since

$$S'(\varepsilon, 1) = \sum_{j=0}^{n-1} s(\varepsilon^{\sigma^j}) = s\left(\prod_{\tau \in G} \varepsilon^{\tau}\right) = s(N_{K/k}(\varepsilon)) = 0$$

for $\varepsilon \in C_K^{(l)}$, we know that $(X-1) \mid S'(\varepsilon, X)$. Therefore, we have the mapping

$$S: \overline{C_K^{(l)}} \to \mathbb{F}_l[X]/(1+X+\cdots+X^{n-1});$$
$$\bar{\varepsilon} \mapsto S(\varepsilon, X) = S'(\varepsilon, X)/(X-1) \text{ for } \varepsilon \in C_K^{(l)}.$$

We have $S(\overline{\varepsilon^{\sigma}}) = XS(\bar{\varepsilon})$. If we make $\mathbb{F}_l[X]/(1+X+\cdots+X^{n-1})$ into an Ω-module by $f(X)^{\sigma} = Xf(X)$, then S is a homomorphism of Ω-modules.

Similarly, we have another homomorphism of Ω-modules

$$T: \overline{C_K^{(l)}} \to \mathbb{F}_l[X]/(1+X+\cdots+X^{n-1});$$
$$\bar{\varepsilon} \mapsto T(\varepsilon, X) = \sum_{j=0}^{n-1} t(\varepsilon^{\sigma^j}) X^{n-j-1}/(X-1) \text{ for } \varepsilon \in C_K^{(l)},$$

where

$$t(\varepsilon) = v_{\mathfrak{p}}(\varepsilon) \pmod{l}.$$

It is easy to see that

$$\overline{C_K^+}(= C_K^+/(C_K)^l) = \operatorname{Ker}(S) \cap \operatorname{Ker}(T). \tag{3.2}$$

The following theorem gives a criterion for $l \mid h(O_K)$ by using formula (3.2).

Theorem 3.1. *Suppose that K/k is a cyclic extension satisfying the conditions* (I), (II) *and* (III). *Let*

$$f_i(X) = S(\eta_i, X), \qquad g_i(X) = T(\eta_i, X) \qquad (0 \le i \le n-2) \tag{3.3}$$

$$f(X) = (1+X+\cdots+X^{n-1}, f_0(X), \ldots, f_{n-2}(X))$$
$$g(X) = (1+X+\cdots+X^{n-1}, g_0(X), \ldots, g_{n-2}(X)).$$

Then

$$l \mid h(O_K) \quad \text{if and only if} \quad (f(X), g(X)) = 1.$$

Proof. $Im(S)$ is the ideal of $\mathbb{F}_l[X]/(1+X+\cdots+X^{n-1})$ generated by $f_i(X)$ $(0 \le i \le n-2)$. Therefore $Im(S)$ is the principal ideal generated by $f(X)$. Let

$$h(X) = (1+X+\cdots+X^{n-1})/f(X).$$

Then

$$S(\overline{C}_K^{(l)^{h(\sigma)}}) = h(X)f(X)\mathbb{F}_l[X]/(1+X+\cdots+X^{n-1}) = (0)$$

which means that $(\overline{C}_K^{(l)})^{h(\sigma)} \subseteq \mathrm{Ker}(S)$. Since

$$\dim_{\mathbf{F}_l}(\overline{C_K^{(l)}})^{h(\sigma)} = \dim_{\mathbf{F}_l}(\overline{\eta})^{h(\sigma)\Omega} = \dim_{\mathbf{F}_l} h(\sigma)\Omega = \deg f(X),$$

$$\begin{aligned}\dim_{\mathbf{F}_l} \mathrm{Ker}(S) &= \dim_{\mathbf{F}_l}(\overline{C}_K^{(l)}) - \dim_{\mathbf{F}_l} Im(S) \\ &= (n-1) - (n-1-\deg f(X)) = \deg f(X).\end{aligned}$$

We know that $\mathrm{Ker}(S) = \overline{\eta}^{h(\sigma)\Omega}$. Similarly we have $\mathrm{Ker}(T) = \overline{\eta}^{r(\sigma)\Omega}$ where $r(X) = (1 + X + \cdots + X^{n-1})/g(X)$, and

$$\overline{C_K^+} = \mathrm{Ker}(T) \cap \mathrm{Ker}(S) = \overline{\eta}^{l(\sigma)\Omega}$$

where

$$\begin{aligned} l(X) = [h(X), r(X)] &= \left[\frac{1+X+\cdots+X^{n-1}}{f(X)}, \frac{1+X+\cdots+X^{n-1}}{g(X)}\right] \\ &= \frac{1+X+\cdots+X^{n-1}}{(f(X), g(X))}. \end{aligned}$$

Therefore

$$\dim_{\mathbf{F}_l} \overline{C_K^+} = \deg(f(X), g(X))$$

and

$$\begin{aligned} l \mid h(O_K) &\Leftrightarrow C_K^+ = C_K^l \qquad \text{(Theorem 2.3)} \\ &\Leftrightarrow \overline{C_K^+} = (1) \\ &\Leftrightarrow (f(X), g(X)) = 1. \end{aligned}$$

The result follows. □

The criterion for $l \mid h(O_K)$ presented in Theorem 3.1 is quite elementary except that the polynomials $f_i(X)$ and $g_i(X)$ in formula (3.3) involve the cyclotomic units. Now we show that $f_i(X)$ and $g_i(X)$ can be computed in very elementary way, namely, by some operations on polynomials without any cyclotomic units. Our method is stimulated by Davis' **[2]** method for the number field case.

The infinite prime divisor $\wp$ of K splits completely in $k(\Lambda_M^+)$, so we have prime divisor β^+ in $k(\Lambda_M)^+$ such that $\beta^+ \| \wp$. On the other hand, β^+ is totally ramified in $k(\Lambda_M)$, so we have a prime divisor β in $k(\Lambda_M)$ such that $\beta^+ = \beta^{q-1}$. According to Galovich and Rosen **[5**, Proposition 1.10**]**, there exists a primitive M-torsion element λ such that

$$v_\beta(\lambda) = (d-1)(q-1) - 1$$

where $d = \deg M$. For any $A' \in R$, $(A', M) = 1$, $a = \deg A'$, $0 \le a < d$, we have the following Carlitz action law:

$$\sigma_{A'}(\lambda) = \lambda^{A'} = \lambda^{q^a} + \begin{bmatrix} A' \\ a-1 \end{bmatrix} \lambda^{q^{a-1}} + \cdots + \begin{bmatrix} A' \\ 1 \end{bmatrix} \lambda^q + \begin{bmatrix} A' \\ 0 \end{bmatrix} \lambda$$

where $\begin{bmatrix} A' \\ i \end{bmatrix}$ is a polynomial of degree $(a-i)q^i$ in R, $\begin{bmatrix} A' \\ 0 \end{bmatrix} = A$. Since $v_\beta(T) = -(q-1)$, we have

$$v_\beta\left(\begin{bmatrix} A' \\ i \end{bmatrix} \lambda^{q^i}\right) = -q^i(a-i)(q-1) + q^i[(d-1)(q-1)-1]$$
$$= q^i[(q-1)(d+i-a-1)-1].$$

The right-hand side of the above equality is an increasing function of i, the main term of $\lambda^{A'}$ is $\begin{bmatrix} A' \\ 0 \end{bmatrix} \lambda = A'\lambda$, and

$$v_\beta(\lambda^{A'}) = v_\beta(A'\lambda),$$
$$v_\beta(\lambda^{A'}/\lambda) = v_\beta(A') = -a(q-1),$$
$$v_{\beta^+}(\lambda^{A'}/\lambda) = -a = -\deg A'.$$

Therefore by formula (3.1) we have

$$\begin{aligned} t(\eta_i^{\sigma^j}) &= t\left(\prod_{B\in N'} \lambda^{BA^{i+j+1}}/\lambda^{BA^{i+j}}\right) \\ &= \sum_{B\in N'} t\left(\frac{\lambda^{BA^{i+j+1}}/\lambda}{\lambda^{BA^{i+j}}/\lambda}\right) \\ &= \sum_{B\in N'} (-\deg[BA^{i+j+1}] + \deg[BA^{i+j}]) \end{aligned} \tag{3.4}$$

where $[B]$ (for $B \in R$, $(B, M) = 1$) denotes the polynomial satisfying

$$[B] \equiv B \pmod{M}, \qquad 0 \le \deg[B] < \deg M.$$

Similarly, we have (for $A' \in R$, $(A', M) = 1$, $0 \le \deg A' < \deg M$)

$$\begin{aligned} \operatorname{sgn}(\lambda^{A'}/\lambda) &= \operatorname{sgn}(A'\lambda/\lambda) = \operatorname{sgn}(A') \\ &= \text{the leading coefficient of polynomial } A' \\ s(\lambda^{A'}/\lambda) &= s(A') = \phi(\text{the leading coefficient of } A') \\ s(\eta_i^{\sigma^j}) &= \sum_{B\in N'} (s[BA^{i+j+1}] - s[BA^{i+j}]). \end{aligned} \tag{3.5}$$

Therefore we get the following complete elementary expressions for $f_i(X)$ and $g_i(X)$.

Theorem 3.2. *Suppose that K/k is a cyclic extension satisfying the conditions* (I), (II) *and* (III). *A is a polynomial in R, such that $G = G(K/k)$ is generated by $\sigma = \sigma_{A|K}$, $G(k(\Lambda_M)/K) \cong N$, $\mathbb{F}_q^* \subseteq N \subseteq (R/(M))^*$, N' is a representative system of polynomials for $N/\mathbb{F}_q^*$. Then*

$$(X-1)f_i(X) = \sum_{j=0}^{n-1} \sum_{B\in N'} (s[BA^{i+j+1}] - s[BA^{i+j}])X^{n-j-1} \tag{3.6}$$
$$(0 \le i \le n-2)$$

$$g_0(X) = \sum_{j=0}^{n-1} \sum_{B \in N'} \deg[BA^j] X^{n-j-1}$$

$$g_i(X) = X^i g_0(X).$$

In particular, $g(X) = (1 + x + \cdots + X^{n-1}, g_0(X))$.

Proof. The formula (3.6) comes from the formula (3.5) and the definition of $f_i(X) = S(\eta_i, X)$. Similarly, from formula (3.4) we have

$$(X-1)g_i(X) = \sum_{j=0}^{n-1} \sum_{B \in N'} (-\deg[BA^{i+j+1}] + \deg[BA^{i+j}]) X^{n-j} X^{n-j-1}.$$

Since $\deg[A'] = \deg[aA]$ for $a \in \mathbb{F}_q^*$, $A' \in R$, we have

$$\sum_{B \in N'} \deg[BA^i] = \sum_{B \in N'} \deg[BA^{i+n}]$$

and

$$\begin{aligned}(X-1)g_i(X) &= (-X^{i+1} + X^i) \sum_{j=0}^{n-1} \sum_{B \in N'} \deg[BA^j] X^{n-j-1} \\ &= (X^{i+1} - X^i) g_0(X).\end{aligned}$$

Therefore

$$g_i(X) = X^i g_0(X) \qquad (0 \le i \le n-2)$$

and

$$\begin{aligned}g(X) &= (1 + X + \cdots + X^{n-1}, g_0(X), \ldots, g_{n-2}(X)) \\ &= (1 + X + \cdots + X^{n-1}, g_0(X)).\end{aligned}$$

The result is now established. □

Remark 3.3. If we can choose A such that $A^n \equiv 1 \pmod{M}$, then $s[BA^{i+n}] = s[BA^i]$. By the same argument as for $g_i(X)$ we have

$$f_0(X) = \sum_{j=0}^{n-1} \sum_{B \in N'} s[BA^j] X^{n-j-1},$$

$$f_i(X) = X^i f_0(X) \qquad (0 \le i \le n-2)$$

and

$$f(X) = (1 + X + \cdots + X^{n-1}, f_0(X)).$$

Theorem 3.1 tells us that $l \mid h(O_K)$ if and only if $\mathrm{Ker}(S) \cap \mathrm{Ker}(T) \neq (1)$. The next theorem shows us that $\mathrm{Ker}(T) \neq (1)$ just gives the same information for $h(K)$.

Theorem 3.4. *Suppose that K/k is a cyclic extension satisfying the conditions* (I), (II) *and* (III). *Then*

$$l \mid h(K) \quad \text{if and only if} \quad \mathrm{Ker}(T) \neq (1).$$

Proof. Galovich and Rosen **[5]** showed an analytic formula for the class number of the fields $k(\Lambda_M)$ and $h(k(\Lambda_M)^+)$. Using the same argument we also have a similar formula for $h(K)$:

$$h(K) = \prod_{i=1}^{n-1} \sum_{\substack{C \in R \\ \mathrm{sgn} C = 1 \\ (C,M)=1 \\ 0 \leq C < \deg M}} -\chi^i(C) \deg C$$

where the character χ of $(R/(M))^*$ satisfying $\chi(N) = 1$ is a generator of the character group of $G = G(K/k)$. Let $\chi(A) = \zeta$ (a primitive n-th root of 1), then

$$h(K) = \prod_{i=1}^{n-1} \left(-\sum_{j=0}^{n-1} \sum_{B \in N'} \deg[BA^j] \zeta^{i(n-\lambda)} \right).$$

Therefore

$$\begin{aligned} l \mid h(K) \Leftrightarrow {} & \text{there exists } i (1 \leq i \leq n-1) \text{ such that} \\ & \sum_{j=0}^{n-1} \sum_{B \in N'} \deg[BA^j] \zeta^{i(n-j)} = 0 \pmod{\mathfrak{L}} \\ & \text{for some prime ideal } \mathfrak{L} \text{ in } \mathbb{Z}[\zeta], \mathfrak{L} \mid l \\ \Leftrightarrow {} & g_0(w) = 0 \text{ for some primitive } n\text{-th root} \\ & w \text{ of } 1 \text{ in the extension field } \mathbb{Z}[\zeta]/\mathfrak{L} \text{ of } \mathbb{F}_l \\ \Leftrightarrow {} & g(X) = (1 + X + \cdots + X^{n-1}, g_0(X)) \neq 1 \\ \Leftrightarrow {} & \mathrm{Ker}(T) \neq (1). \end{aligned}$$

The proof of the theorem is now complete. □

4. Applications

In this section we give some applications of our elementary criterion for $l \mid h(O_K)$ and $l \mid h(K)$ for several particular cases of cyclic function fields K. First, we consider the case that $M = P$ is a monic irreducible polynomial, $d = \deg P$. Then $(R/(P))^* = \mathbb{F}_{q}^*$. Let $\alpha \in R$ be a generator of the cyclic group $(R/(P))^*$, then $\alpha^{\frac{q^d-1}{q-1}}$ is a generator of $\mathbb{F}_q^*$. We can choose

$$A = \alpha, \qquad N' = \{\alpha^{nr} \mid 0 \leq r \leq m-1\}$$

$$\phi(a) = 1 \pmod{l}$$

where $m = [k(\Lambda_P)^+ : K)]$, $mn = [k(\Lambda_P)^+ : k] = \frac{q^d-1}{q-1}$. The following result is an analogy of Feng **[3]**.

Theorem 4.1. *Suppose that* K/k *is a cyclic extension satisfying the conditions* (I) *and* (II), *and* $M = P$ *is a monic irreducible polynomial. Let*

$$f'(X) = \sum_{j=0}^{n-1} \sum_{r=0}^{m-1} s[\alpha^{nr+j}] X^{n-j-1}.$$

Then

$$f(X) = (1 + X + \cdots + X^{n-1}, (X-1)f'(X) + 1)$$

and

$$g(X) = (1 + X + \cdots + X^{n-1}, g_0(X))$$

where

$$g_0(X) = -\sum_{j=0}^{n-1} \sum_{r=0}^{m-1} \deg[\alpha^{nr+j}] X^{n-j-1}.$$

Proof. From Theorem 3.2 we know that

$$(X-1)f_i(X) = \sum_{j=0}^{n-1} \sum_{r=0}^{m-1} (s[\alpha^{nr+j+i+1}] - s[\alpha^{nr+j+i}]) X^{n-j-1}.$$

From $nm \le t \le 2nm - 1$, we have $s[\alpha^t] = s[a\alpha^{t-mn}] = s[a] + s[\alpha^{t-nm}] = \phi(a) + s[\alpha^{t-mn}] = s[\alpha^{t-mn}] + 1$. Therefore

$$\begin{aligned}(X-1)f_i(X) &= (X^{i+1} - X^i) \left(\sum_{j=0}^{n-1} \sum_{r=0}^{m-1} s[\alpha^{nr+j}] X^{n-j-1} \right) \\ &\quad + \sum_{j=n-i-1}^{n-1} X^{n-j-1} - \sum_{j=n-i}^{n-1} X^{n-j-1} \\ &= X^i (X-1) f'(X) + X^i = X^i((X-1)f'(X) + 1)\end{aligned}$$

and

$$\begin{aligned}f(X) &= (1 + X + \cdots + X^{n-1}, f_0(X), \ldots, f_{n-2}(X)) \\ &= (1 + X + \cdots + X^{n-1}, (X-1)f'(X) + 1).\end{aligned}$$

The equality $g(X) = (1 + X + \cdots + X^{n-1}, g_0(X))$ comes from Theorem 3.2. □

From now on we assume the following condition instead of condition (III):

(III)' $n = p$ is an odd prime number and l is a generator of the cyclic multiplicative group $\mathbb{F}_q^*$.

Lemma 4.2. *Suppose that* K/k *is a cyclic extension satisfying the conditions* (I), (II) *and* (III)'.

(a) *Let*

$$t_j = -\sum_{B \in N'} \deg[BA^j] \qquad (0 \le j \le p-1).$$

Then $l \mid h(K)$ *if and only if* $t_0 \equiv t_1 \equiv \ldots \equiv t_{p-1} \pmod{l}$.

(b) *If* $M = P$ *is a monic irreducible polynomial in* $R = \mathbb{F}_q[T]$, $\frac{\Phi(P)}{q-1} = pm$, $\alpha \in R$ *is a generator of* $(R/(P))^*$, $f'(X)$ *and* $g_0(X) \in \mathbb{F}_l[X]$ *defined in Theorem 4.1 are given by*

$$f'(X) = \sum_{j=0}^{p-1} s_j X^{p-j-1}, \qquad g_0(X) = \sum_{j=0}^{p-1} t_j X^{p-j-1}$$

where

$$s_j = \sum_{r=0}^{m-1} s[\alpha^{pr+1}], \qquad t_j = -\sum_{r=0}^{m-1} \deg[\alpha^{pr+j}],$$

then

$$l \mid h(K) \Leftrightarrow t_0 = t_1 = \cdots = t_{p-1} \pmod{l}$$

$$l \mid h(O_K) \Leftrightarrow l \mid h(K) \text{ and } s_j \equiv s_0 \frac{j}{p} \pmod{l} \quad (1 \le j \le p-1). \tag{4.1}$$

Proof. (a) We have $g_0 = \sum_{j=0}^{p-1} t_j X^{p-j-1}$. The condition (III)' implies that $1 + X + \cdots + X^{p-1}$ is irreducible in $\mathbb{F}_l[X]$. Therefore

$$\begin{aligned} l \mid h(K) &\Leftrightarrow (1 + X + \cdots + X^{p-1}, g_0(X)) \neq 1 \\ &\Leftrightarrow 1 + X + \cdots + X^{p-1} \mid g_0(X) \\ &\Leftrightarrow t_0 \equiv t_1 \equiv \ldots \equiv t_{p-1} \pmod{l}. \end{aligned}$$

(b) From Theorem 4.1 we know that

$$\begin{aligned} l \mid h(O_K) \Leftrightarrow\ & 1 + X + \cdots + X^{p-1} \mid g_0(X) \text{ (which is equivalent to } l \mid h(K)) \\ & \text{and } 1 + X + \cdots + X^{p-1} \mid (X-1)f'(X) + 1. \end{aligned}$$

But

$$\begin{aligned} (X-1)f'(X) + 1 &= (X-1)(s_{p-1} + s_{p-2}X + \cdots + s_0 X^{p-1}) + 1 \\ &\equiv (1 + s_0 - s_{p-1}) + (s_{p-1} - s_{p-2})X + \cdots + (s_1 - s_0)X^{p-1} \\ &\quad \pmod{X^p - 1}. \end{aligned}$$

Therefore

$$\begin{aligned} & 1 + X + \cdots + X^{p-1} \mid (X-1)f'(X) + 1 \\ & \Leftrightarrow 1 + s_0 - s_{p-1} \equiv s_{p-1} - s_{p-2} \equiv \ldots \equiv s_1 - s_0 \pmod{l} \\ & \Leftrightarrow s_j \equiv s_0 + \frac{j}{p} \pmod{l} \qquad (1 \le j \le p-1) \end{aligned}$$

by a simple computation. □

Example 4.3. $q = 7$, $P = P(T) = T^3 + 5$ is irreducible in $\mathbb{F}_7[T]$. Let γ be a root of P, $\gamma^3 = 2$, then $\alpha = \gamma + 1$ is a generator of $\mathbb{F}_7(\gamma)^*$. Let $k = \mathbb{F}_7(T) \subset K \subset k(\Lambda_P)^+$,

$[K:k]=3,\ l=2$. A calculation shows that $(s_0,s_2,s_2)\equiv(1,0,1)$ and $(t_0,t_1,t_2)\equiv(1,1,1)\,(\mathrm{mod}\ 2)$. From Lemma 4.2 we know that $2\mid h(O_K)$. The analytic formula gives $h(K)=12$.

Example 4.4. We show several examples of cyclic function fields K with even $h(K)$ and odd $h(O_K)$. Let $k=\mathbb{F}_q(T)$, $P=P(T)$ be irreducible in $\mathbb{F}_q[T]$, $P(\gamma)=0$, α be a generator of $\mathbb{F}_q(\gamma)^*$, $k\subset K\subset k(\Lambda_P)^+$, $p=[K:k]$ be an odd prime number and let $l=2$ be a generator of $\mathbb{F}_p^*$. For the following cyclic function fields K, we have $t_0\equiv t_1\equiv\ldots\equiv t_{p-1}\,(\mathrm{mod}\ 2)$, but the condition (4.1) (which is $s_0\equiv s_2\equiv\ldots\equiv s_{p-1}\equiv s_{p-2}+1\equiv\ldots\equiv s_3+1\equiv s_1+1\,(\mathrm{mod}\ 2)$ for $l=2$) is not satisfied. Therefore $2\mid h(K)$ and $2\nmid h(O_K)$.

q	$P(T)$	α	p	$(t_0t_1\ldots t_{p-1})$	$(s_0,s_1\ldots s_{p-1})$ (mod 2)
7	T^3+6T+5	γ	3	(111)	(011)
7	T^3+6T+2	γ	3	(111)	(000)
3	T^4+T^2+2T+1	$2\gamma^2+1$	5	(00000)	(00000)
3	T^4+T^2+T+1	$2\gamma^2+1$	5	(00000)	(00000)
3	$T^4+2T^3+T^2+2T+1$	$\gamma+1$	5	(00000)	(01001)
3	$T^4+T^3+T^2+T+1$	$\gamma+2$	5	(00000)	(01001)
3	$T^4+T^3+T^2+2T+2$	γ	5	(00000)	(01001)
3	$T^4+2T^3+T^2+T+2$	γ	5	(00000)	(01001)

Example 4.5. $q=4$, $\mathbb{F}_4=\{0,1,\alpha,\alpha^2\}$, $\alpha^2=\alpha+1$, $M=(1+\alpha T+T^2)^2$, $k=\mathbb{F}_4(T)$, $n=2$, $l=q-1=3$. Then

$$G(k(\Lambda_M)^+/k)\cong(\mathbb{F}_4[T]/(M))^*/\mathbb{F}_4^*)\cong C_5\times C_2^4,$$

where C_5 is the cyclic group of order 5 generated by $1+T^2$, C_2^4 is the elementary 2-group with basis

$$a_1=T^2+T^3,\quad a_2=1+T^2+T^3,\quad a_3=T+T^2+\alpha T^3,\quad a_4=\alpha T+T^3.$$

There are exactly 15 quadratic extensions K/k, $K\subset k(\Lambda_P)^+$, corresponding to 15 subgroups $N'(\cong G(k(\Lambda_M)^+/K)$ with order 40 of $C_5\times C_2^4$. For $N'=C_5\times\langle a_1,a_2,a_3\rangle$, we choose $A=a_4$, $G(K/k)=\{1,\sigma_A\}$. A calculation shows that $(t_0,t_1)\equiv(s_0,s_1)\equiv(1,1)\,(\mathrm{mod}\ 3)$ where

$$t_j=-\sum_{B\in N'}\deg[BA^j],\qquad s_j=\sum_{B\in N'}s[BA^j];$$

$$g_0(X)=s_1+s_0X,\qquad f_0(X)=t_1+t_0X\in\mathbb{F}_3[X].$$

From Theorem 3.2 and Remark 3.3 we know that $3\mid h(O_K)$. The following table shows that there are 3 quadratic extensions K with $3\mid h(O_K)$, another 3 quadratic extensions K with $3\mid h(K)$ and $3\nmid h(O_K)$ (the analytic formula gives $h(K)=2^2\cdot3\cdot107$ for all of these 6 fields K), the remaining 9 quadratic extensions K with $3\nmid h(K)$ (the analytic formula gives $h(K) = 8 \cdot 107$ or $16 \cdot 107$).

cases	N'	A	(t_0, t_1) (mod 3)	(s_0, s_1) (mod 3)
	$C_5 \times \langle a_1, a_2, a_3 \rangle$	a_4	$(1,1)$	$(1,1)$
$3 \mid h(O_K)$	$C_5 \times \langle a_2, a_1a_3, a_4 \rangle$	a_1	$(1,1)$	$(1,1)$
	$C_5 \times \langle a_1, a_2a_3, a_3a_4 \rangle$	a_2	$(1,1)$	$(1,1)$
$3 \mid h(K)$	$C_5 \times \langle a_1a_2, a_3, a_4 \rangle$	a_1	$(1,1)$	$(0,2)$
$3 \nmid h(O_K)$	$C_5 \times \langle a_2, a_1a_3, a_1a_4 \rangle$	a_1	$(1,1)$	$(2,0)$
	$C_5 \times \langle a_1, a_4, a_2a_3 \rangle$	a_2	$(1,1)$	$(2,0)$

References

[1] J. V. Armitage; A. Fröhlich, Class number and unit signatures, *Mathematika*, **14** (1967), 94–98.

[2] D. Davis, Computing the number of totally positive circular units which are squares, *J. Number Theory* **10** (1978), 1–9.

[3] Keqin Feng, An elementary criterion on parity of class number of cyclic number fields, *Scientia Sinica* **25** (1982), 1032–1041.

[4] Keqin Feng; Linsheng Yin, On maximal independent systems of cyclotomic units in cyclotomic function fields, *Scientia Sinica* **34** (1991), 252–261.

[5] S. Galovich; M. Rosen, Units and class groups in cyclotomic function fields, *J. Number Theory* **14** (1982), 156–184.

[6] D. A. Garbanati, Unit signatures and even class number, and relative class numbers, *J. Reine Angew. Math.* **274** (1975), 376–384.

[7] G. Gras, Signature des unités cyclotomiques et parité du nombre de class des extensions abéliennes de Q de degré impair, *Bull. Soc. Math. France* **103** (1977), 177–190.

[8] G. Gras; M.-N. Gras, Signature des unités cyclotomiques et parité du nombre de classes des extensions cycliques de Q de degré premier impair, *Ann. Inst. Fourier, Grenoble* **25** (1975), 1–22.

[9] M. Rosen, The Hilbert class field in function fields, *Exposition. Math.* (1987), 365–378.

[10] L. C. Washington, *Introduction to cyclotomic fields*, Springer-Verlag, New York, 1982.

Department of Mathematics, China University of Science and Technology, Hefei, 230026, P. R. China

Genus Two Hyperelliptic Drinfeld Modules over $\mathbb{F}_2$

*D. S. Dummit**

The purpose of this paper is to provide some explicit examples of (rank one) Drinfeld modules, and in particular to describe all the rank one Drinfeld modules where the infinite prime has degree one for hyperelliptic curves of genus $g = 2$ over the finite field $\mathbb{F}_2$ of two elements. In addition, the elliptic Drinfeld modules computed by Hayes **[H1]** over $\mathbb{F}_2$ and $\mathbb{F}_3$ are listed in the Appendix (the elliptic Drinfeld modules over $\mathbb{F}_q$ for $q \leq 13$ will be found in **[D-H]**). The first section describes the determination of the genus 2 curves in question and the reader may wish to skip directly to the description of the explicit computation of the corresponding Drinfeld modules in Section 2.

§1. Genus two curves over $\mathbb{F}_2$

Let C be a curve defined over the finite field $\mathbb{F}_2$ of genus $g = 2$ with a Weierstraß point $\mathcal{O}$ rational over $\mathbb{F}_2$. Then C is a hyperelliptic curve (i.e., the function field of C is a separable quadratic extension of a (unique) rational function field) of genus two over $\mathbb{F}_2$ having an $\mathbb{F}_2$-rational Weierstraß point. Let x denote a function on C regular except for a double pole at $\mathcal{O}$ (this is the definition of a Weierstraß point) defining C as a 2-fold covering of $\mathbb{P}^1$. The field of functions $\mathbb{F}_2(C)$ is then a quadratic extension of the rational function field $\mathbb{F}_2(x)$ that is ramified at the prime defined by $1/x$ in $\mathbb{F}_2(x)$. By the Riemann-Roch Theorem, the dimension of the space $L(3\mathcal{O})$ of functions on C with no poles except for a pole of order at most 3 at $\mathcal{O}$ is $l(3\mathcal{O}) = \deg(3\mathcal{O}) - g + 1 = 3 - 2 + 1 = 2$. Hence 1, x give a basis for $L(3\mathcal{O})$ over $\overline{\mathbb{F}_2}$. Similarly, the dimension of $L(4\mathcal{O})$ is 3, with $1, x, x^2$ as basis. Since $L(5\mathcal{O})$ has dimension 4, there is a function y on C defined over $\mathbb{F}_2$ with a pole of order precisely 5 at $\mathcal{O}$ and having no other poles. If $\mathbb{F}_2(C)$ denotes the function field of functions on C defined over $\mathbb{F}_2$, then $[\mathbb{F}_2(C) : \mathbb{F}_2(x)] = 2$ and $[\mathbb{F}_2(C) : \mathbb{F}_2(y)] = 5$, so $\mathbb{F}_2(C) = \mathbb{F}_2(x, y)$.

Using the Riemann-Roch Theorem it is easy to determine the following $\mathbb{F}_2$-rational bases for the corresponding vector spaces:

$$L(5\mathcal{O}) = \langle 1, x, x^2, y \rangle$$
$$L(6\mathcal{O}) = \langle 1, x, x^2, x^3, y \rangle$$
$$L(7\mathcal{O}) = \langle 1, x, x^2, x^3, y, xy \rangle$$
$$L(8\mathcal{O}) = \langle 1, x, x^2, x^3, x^4, y, xy \rangle$$
$$L(9\mathcal{O}) = \langle 1, x, x^2, x^3, x^4, y, xy, x^2y \rangle,$$

* Research partially sponsored by grants from the NSA and the NSF

since in each case the added function is the only function in the list with a pole of that order, hence cannot be linearly dependent on the previous functions. Because $L(10\mathcal{O})$ has dimension 9 and contains the functions y^2 and x^5 in addition to the functions in $L(9\mathcal{O})$, it follows that there is some linear relation

$$a_0y^2 + a_1x^2y + a_2xy + a_3y = b_0x^5 + b_1x^4 + b_2x^3 + b_3x^2 + b_4x + b_5\,, \qquad (1)$$

with coefficients in $\mathbb{F}_2$. If either a_0 or b_0 were 0, then by comparing the order of the pole at $\mathcal{O}$ on both sides of this relation it would follow that both a_0 and b_0 would be 0 and the remaining terms would define a linear dependence among the basis elements for $L(9\mathcal{O})$, a contradiction. Hence we must have $a_0 = b_0 = 1$.

Because $L(2\mathcal{O}) = \langle 1, x\rangle$ and $L(5\mathcal{O}) = \langle 1, x, x^2, y\rangle$, the possible coordinate transformations on such a model for C are given by:

$$\begin{aligned} x &\mapsto x + r \\ y &\mapsto y + sx^2 + tx + u, \qquad r, s, t, u \in \mathbb{F}_2 . \end{aligned} \qquad (2)$$

Under the action of these coordinate transformations, there are 30 orbits of separable curves having Weierstraß form as in equation (1). Each of the affine curves defined by (1) are singular at the point at infinity, and 12 of the 30 affine curves in addition have an affine singular point, so are in fact of genus at most one. For example, the curve defined affinely by the equation $y^2+x^2y = x^5+x^4$ is singular at (0,0). Dividing by x^4 gives the equation $(y/x^2)^2+(y/x^2) = x+1$, which shows that the function field is birationally isomorphic under the change of variables $s = x$, $t = y/x^2$ to the genus 0 curve $t^2 + t = s + 1$. Similar computations apply to the remaining curves. One class, represented by the curve $y^2 + x^2y + xy + y = x^5$ is of genus 1 over $\mathbb{F}_4$ (if $\mathbb{F}_4 = \mathbb{F}_2(\alpha)$, then the transformation $x \mapsto x + \alpha$, $y \mapsto y + \alpha$ gives the curve $y^2 + x^2y + xy = x^5 + \alpha x^4 + \alpha x^2$; dividing by x^2 shows that the birational transformation $t = y/x$, $s = x$ defines a birational equivalence to the (genus 0) curve $t^2 + st + t = s^3 + \alpha s^2 + \alpha$). The remaining 11 classes are birationally equivalent over $\mathbb{F}_2$ to curves of genus less than 2.

This leaves a total of 18 classes of genus 2 curves over $\mathbb{F}_2$ with an $\mathbb{F}_2$-rational Weierstraß point:

$$\begin{aligned}
C_1&: y^2 + y = x^5\\
C_2&: y^2 + y = x^5 + x\\
C_3&: y^2 + y = x^5 + x + 1\\
C_4&: y^2 + y = x^5 + x^3\\
C_5&: y^2 + y = x^5 + x^3 + 1\\
C_6&: y^2 + y = x^5 + x^3 + x\\
C_7&: y^2 + xy = x^5 + 1\\
C_8&: y^2 + xy = x^5 + x^2 + 1\\
C_9&: y^2 + xy = x^5 + x^3 + 1\\
C_{10}&: y^2 + xy = x^5 + x^3 + x^2 + 1\\
C_{11}&: y^2 + x^2y = x^5 + x
\end{aligned}$$

$$C_{12}: y^2 + x^2y = x^5 + x^4 + x$$
$$C_{13}: y^2 + x^2y = x^5 + x + 1$$
$$C_{14}: y^2 + x^2y = x^5 + x^4 + x + 1$$
$$C_{15}: y^2 + x^2y + xy = x^5 + 1$$
$$C_{16}: y^2 + x^2y + xy + y = x^5 + 1$$
$$C_{17}: y^2 + x^2y + xy + y = x^5 + x$$
$$C_{18}: y^2 + x^2y + xy + y = x^5 + x + 1.$$

Each of these curves is nonsingular affinely and the point at infinity in each model is singular. The nonsingular model is the projective closure of the image of the affine points under the map taking the point (x, y) to the point $(1, x, x^2, x^3, y)$ in projective 4-space, $\mathbb{P}^4$. In terms of projective coordinates x_0, x_1, x_2, x_3 and x_4 for $\mathbb{P}^4$, the nonsingular model is given by the homogeneous equations

$$x_0x_2 = x_1^2, \qquad x_0x_3 = x_1x_2, \qquad x_1x_3 = x_2^2 \tag{3}$$

together with the equation obtained from the affine model above by homogenizing (1) with respect to x_0, replacing y by x_4 and x by x_1, using the relations above and cancelling a factor of x_0^3. This has the effect of replacing the elements 1, x, x^2, x^3, x^5, y^2, x^2y, xy, y in each C_i by x_0^2, x_1x_0, x_2x_0, x_3x_0, x_2x_3, x_4^2, x_4x_2, x_4x_1, x_4x_0, respectively. For example, the projective nonsingular model for C_{18} is the curve in $\mathbb{P}^4$ defined by the equations (3) together with the homogeneous equation

$$x_4^2 + x_4x_2 + x_4x_1 + x_4x_0 = x_2x_3 + x_1x_0 + x_0^2.$$

It is easy to check that each of the nonsingular projective models have exactly one point at infinity, namely $(0, 0, 0, 1, 0) \in \mathbb{P}^4$, in addition to their affine points.

Remark 1. The requirement that the curve C have an $\mathbb{F}_2$-rational Weierstraß point at infinity was made to simplify the Drinfeld module of the corresponding affine coordinate ring, since in this case there is a unique prime of degree one at infinity. It is certainly possible to find curves C of genus two over $\mathbb{F}_2$ with no $\mathbb{F}_2$-rational Weierstraß points. For example, the curve defined by $C : y^2 + y = 1/(x^3 + x + 1)$ is of genus two with no $\mathbb{F}_2$-rational Weierstraß points; $\mathbb{F}_2(C)$ is a quadratic extension of $\mathbb{F}_2(x)$ ramified only at the prime $(x^3 + x + 1)$ of degree 3 (the prime at infinity splits in this case). The Weierstraß points for C over $\overline{\mathbb{F}_2}$ lie over the primes $x - \alpha$ of $\overline{\mathbb{F}_2}(x)$, where $\alpha^3 + \alpha + 1 = 0$, and are not $\mathbb{F}_2$-rational.

Remark 2. If Q is a prime in $\mathbb{F}_2(C)$ ramified in the extension $\mathbb{F}_2(C)/\mathbb{F}_2(x)$, then the degree, $V(Q)$, of the ramification divisor at Q_2 is the (finite) sum of the number of non-identity elements in the ith higher ramification group for Q in the extension:

$$V(Q) = \sum_{i=0}^{\infty} (|V_i| - 1).$$

If Q is not ramified in $\mathbb{F}_2(C)/\mathbb{F}_2(x)$ then $V(Q) = 0$, and since the characteristic is 2, $V(Q)$ is even for every Q.

The Riemann-Hurwitz genus formula in this case gives

$$\sum_Q V(Q)\deg(Q) = 6.$$

It follows that there are precisely one, two, or three $\mathbb{F}_2$-rational Weierstraß points:

(a) if there is precisely one $\mathbb{F}_2$-rational Weierstraß point Q, then $V(Q) = 6$ if Q is the only Weierstraß point over $\overline{\mathbb{F}_2}$ and $V(Q) = 2$ if there are three Weierstraß points over $\overline{\mathbb{F}_2}$ (the other two being rational over $\mathbb{F}_4$);

(b) if there are precisely two $\mathbb{F}_2$-rational Weierstraß points, then the ramification divisor has degrees $\{2, 4\}$ at these points;

(c) if there are three $\mathbb{F}_2$-rational Weierstraß points, then the ramification divisor has degrees $\{2, 2, 2\}$ at these points.

The curves C_1 through C_6 each have a unique Weierstraß point (at infinity), rational over $\mathbb{F}_2$.

The curves C_7 through C_{14} each have two Weierstraß points (the point at infinity and the point with $x = 0$), both $\mathbb{F}_2$-rational. For C_7 through C_{10} the corresponding degrees in the ramification divisor are $\{4, 2\}$, respectively, and for C_{11} through C_{14} they are $\{2, 4\}$, respectively.

The curve C_{15} has three Weierstraß points (the point at infinity and (0,1), (1,0)), all $\mathbb{F}_2$-rational.

The curves C_{16} through C_{18} each have only one $\mathbb{F}_2$-rational Weierstraß point (at infinity), but have a total of three Weierstraß points over $\overline{\mathbb{F}_2}$ (the point at infinity and the points with $x = \alpha$ where $\alpha^2 + \alpha + 1 = 0$).

Remark 3. The pairs of curves $\{C_7, C_{11}\}$, $\{C_8, C_{12}\}$, $\{C_9, C_{13}\}$, $\{C_{10}, C_{14}\}$, have an interesting duality, already suggested by their Weierstraß points. The curves in each pair are birationally isomorphic (by the map $x \mapsto 1/s$, $y \mapsto t/s^3$), so have the same function field. The corresponding affine coordinate rings therefore have the same class numbers and the same number of closed points of any degree (i.e., the number of points on the two curves over any $\mathbb{F}_{2^n}$ are the same), yet the pair of affine coordinate rings are not isomorphic (this can be seen from the Weierstraß point computations above, for example).

§2. Explicit Computation of Drinfeld Modules

Let $\mathbf{k} = \mathbb{F}_2(x, y)$ be the function field over $\mathbb{F}_2$ of one of the curves in Section 1, and let $\mathbf{A} = \mathbb{F}_2[x, y]$ be the ring corresponding to the choice of the infinite prime ∞ in $\mathbf{k}$ as the unique prime in $\mathbf{k}$ lying above the prime in $\mathbb{F}_2(x)$ for which $1/x$ is a uniformizer. This ∞ is ramified from $\mathbb{F}_2(x)$ to $\mathbf{k}$ (the point at infinity in the nonsingular model is a Weierstraß point).

A Drinfeld $\mathbf{A}$-module defined over the algebraic closure of $\mathbf{k}$ is defined by a homomorphism ϕ of $\mathbf{A}$ into the twisted polynomial ring* $\overline{\mathbf{k}}\{\mathbf{F}\}$ (polynomials in the Frobenius

* This notation differs slightly from that used in Hayes' expository article **[H2]**—the curly brackets are used to emphasize the non-commutativity.

element $\mathbf{F}$ with $\mathbf{F}c = c^2\mathbf{F}$ for $c \in \overline{\mathbf{k}}$). In particular, a (sgn-normalized) rank one Drinfeld module in this case is determined by its values on x and y:

$$\begin{aligned} \phi_x &= x + a\mathbf{F} + \mathbf{F}^2 \\ \phi_y &= y + c_1\mathbf{F} + c_2\mathbf{F}^2 + c_3\mathbf{F}^3 + c_4\mathbf{F}^4 + \mathbf{F}^5 . \end{aligned} \tag{4}$$

Here it is known that the coefficients a, c_1, c_2, c_3, c_4 lie in the Hilbert class field H of $\mathbf{A}$ (the maximal abelian unramified extension of $\mathbf{k}$ in which the prime ∞ splits completely), in fact generating the integral closure of $\mathbf{A}$ in H.

The key to the explicit computation of the module in (3) is the observation of D. Hayes **[H1]** that the relation $\phi(xy) = \phi(yx)$, i.e., $\phi_x\phi_y = \phi_y\phi_x$ actually determines the Drinfeld module. In other words, the fact that ϕ_x and ϕ_y satisfy all of the relations that x and y satisfy ($\phi_y^2 + \ldots = \phi_x^5 + \ldots$) is already a consequence of the commutativity relation. This reduces the computation of the Drinfeld module to the determination of two twisted polynomials as in (4) that commute. The proof uses the following Lemma (stated for the q-Frobenius over the finite field $\mathbb{F}_q$; $q = 2$ in our case), cf. **[H2**, Section 6**]**.

Lemma. (1) *Let $P = a_0 + a_1\mathbf{F} + \ldots \in \overline{\mathbf{k}}\{\{\mathbf{F}\}\}$ be any element in the twisted power series ring with non-constant constant term: $a_0 \notin \overline{\mathbb{F}_q}$. Then there is a unique twisted power series $\psi = 1 + b_1\mathbf{F} + b_2\mathbf{F}^2 + \ldots \in \overline{\mathbf{k}}\{\{\mathbf{F}\}\}$ with $P\psi = \psi a_0$.*
(2) *Given any $c_0 \in \overline{\mathbf{k}}$ there is a unique twisted power series $Q = c_0 + c_1\mathbf{F} + \ldots \in \overline{\mathbf{k}}\{\{\mathbf{F}\}\}$ commuting with P: $PQ = QP$.*

Proof. (1) By comparing coefficients in the relation $P\psi = \psi a_0$ we obtain the system of equations:

$$\begin{aligned} a_0 &= a_0 \\ a_0 b_1 + a_1 &= b_1 a_0^q \\ a_0 b_2 + a_1 b_1^q + a_2 &= b_2 a_0^{q^2} \\ &\vdots \end{aligned}$$

and since a_0 is non-constant it follows that $a_0^{q^n} \neq a_0$ for any n, so these equations uniquely define $b_1, b_2, b_3, \ldots$ recursively.
(2) Similarly, if $Q = c_0 + c_1\mathbf{F} + \ldots \in \overline{\mathbf{k}}\{\{\mathbf{F}\}\}$ commutes with P, then we obtain the equations

$$\begin{aligned} a_0 c_0 &= c_0 a_0 \\ a_0 c_1 + a_1 c_0^q &= c_0 a_1 + c_1 a_0^q \\ a_0 c_2 + a_1 c_1^q + a_2 c_0^{q^2} &= c_0 a_2 + c_1 a_1^q + c_2 a_0^{q^2} \\ &\vdots \end{aligned}$$

which uniquely define $c_1, c_2, \ldots$ recursively.

Proposition. *Suppose ϕ_x, ϕ_y are twisted polynomials as in (4) with $\phi_x\phi_y = \phi_y\phi_x$. Then the map ϕ defined by mapping x to ϕ_x and y to ϕ_y is an injection of $\mathbf{A}$ into the twisted polynomial ring $\overline{\mathbf{k}}\{\mathbf{F}\}$, i.e., defines a (sgn-normalized) rank one Drinfeld $\mathbf{A}$-module.*

Proof. By the first part of the lemma, there is a unique twisted power series $\psi = 1 + \ldots$ with $\phi_x \psi = \psi x$, i.e., with $\phi_x = \psi x \psi^{-1}$. The map $\mathbf{A} \rightarrow \overline{\mathbf{k}}\{\{\mathbf{F}\}\}$ defined by $z \mapsto \psi z \psi^{-1}$ is evidently an injective ring homomorphism of $\mathbf{A}$ into $\overline{\mathbf{k}}\{\{\mathbf{F}\}\}$ that maps x to ϕ_x. Further, $\psi y \psi^{-1}$ is a twisted power series with constant term y that commutes with $\psi x \psi^{-1} = \phi_x$. Since ϕ_y is also such a power series by assumption, the uniqueness in the second part of the lemma shows that $\psi y \psi^{-1} = \phi_y$. Hence the ring homomorphism defined by conjugation by ψ in fact maps $\mathbf{A}$ into the twisted *polynomial* ring, and maps x to ϕ_x, y to ϕ_y, completing the proof. □

To determine the rank one Drinfeld modules for the curves C_i, it therefore suffices to determine all commuting twisted polynomials as in (4). Comparing coefficients in the relation $\phi_x \phi_y = \phi_y \phi_x$ gives the equations ($q = 2$ here; similar relations hold for any such hyperelliptic curve, cf. **[H1]**):

$$\begin{aligned} xy &= yx \\ xc_1 + ay^2 &= ay + c_1 x^2 \\ xc_2 + ac_1^2 + y^4 &= y + c_1 a^2 + c_2 x^4 \\ xc_3 + ac_2^2 + c_1^4 &= c_1 + c_2 a^4 + c_3 x^8 \\ xc_4 + ac_3^2 + c_2^4 &= c_2 + c_3 a^8 + c_4 x^{16} \\ x + ac_4^2 + c_3^4 &= c_3 + c_4 a^{16} + x^{32} \\ a + c_4^4 &= a^{32} + c_4 . \end{aligned} \tag{5}$$

(The situation is even easier in the elliptic case—see the Appendix.) The first of these equations is trivial and the next four can be used to solve recursively for c_1, c_2, c_3, c_4 in terms of x, y, a:

$$\begin{aligned} c_1 &= \frac{1}{x^2 - x} a(y^2 - y) \\ c_2 &= \frac{1}{x^4 - x}(y^4 - y + ac_1^2 - c_1 a^2) \\ c_3 &= \frac{1}{x^8 - x}(ac_2^2 - c_2 a^4 + c_1^4 - c_1) \\ c_4 &= \frac{1}{x^{16} - x}(ac_3^2 - c_3 a^8 + c_2^4 - c_2) . \end{aligned} \tag{6}$$

When these expressions for c_1, c_2, c_3, c_4 are substituted into the remaining two relations in (5), one obtains two equations in a with coefficients in $\mathbf{k}$. Clearing the (nonzero) denominators gives two polynomials with coefficients in $\mathbf{A}$ that a must satisfy, hence a must be a root of the greatest common divisor of these two polynomials. Conversely, any root a of this greatest common divisor (with corresponding c_1, c_2, c_3, c_4 defined by (6)) defines two commuting twisted polynomials as in (4), hence by the proposition defines a rank one Drinfeld module. Hence the explicit determination of the Drinfeld modules for each of the curves C_i is reduced to the computation of the greatest common divisor of two polynomials (of degrees 60 and 31) with coefficients in $\mathbb{F}_2[x, y]$.

As a check on the computations, note that since a generates the Hilbert class field H, the degree of the (separable part of the) polynomial satisfied by a is h, the class number of the ring $\mathbf{A}$, which here is the same as the class number of the field $\mathbf{k}$. In all these examples, the greatest common divisor is in fact separable (in the elliptic case one can prove this always happens). The class number of $\mathbf{k} = \mathbb{F}_2(C)$ can be computed by counting the $\mathbb{F}_{2^n}$ points on the projective nonsingular model $\tilde{C}$ of the curve C ($n = 1, 2$ actually suffice); the trace formula states

$$|\tilde{C}(\mathbb{F}_{2^n})| = 1 + 2^n - [\alpha^n + (\frac{p}{\alpha})^n + \beta^n + (\frac{p}{\beta})^n]$$

where $\alpha, p/\alpha, \beta, p/\beta$ are the $2g = 4$ roots of the L-function for $\tilde{C}$. As mentioned in Section 1, in this case

$$|\tilde{C}(\mathbb{F}_{2^n})| = 1 + |C(\mathbb{F}_{2^n})|,$$

so we may determine α and β by counting the number of points on the affine curve C over $\mathbb{F}_2$ and $\mathbb{F}_4$ (a check on the computations is then provided by computing the number of points defined over $\mathbb{F}_8, \ldots$). The class number h is given by evaluating the L-function at 1, i.e.,

$$h = (1 - \alpha)(1 - p/\alpha)(1 - \beta)(1 - p/\beta).$$

In this case, this gives

$$h = \frac{|C(\mathbb{F}_2)|^2 + |C(\mathbb{F}_4)|}{2} + |C(\mathbb{F}_2)| - 1.$$

The number of points over the finite fields $\mathbb{F}_{2^n}$, $n = 1, 2, 3, 4$ and the computed class number h (agreeing with the degree of the polynomial satisfied by a) are listed along with the curves in the next section. Note that for a given curve C the number of points over $\mathbb{F}_{2^n}$ is either always even or always odd. This can be seen from the action of the hyperelliptic involution (the map $(x, y) \mapsto (x, y + a_1x^2 + a_2x + a_3)$ in the model (1)) on the $\mathbb{F}_{2^n}$-rational points by considering the number of fixed points.

§3. Explicit Examples

Curve $C_1 : y^2 + y = x^5$ has class number $h = 5$, with $|C(\mathbb{F}_{2^1})| = 2$, $|C(\mathbb{F}_{2^2})| = 4$, $|C(\mathbb{F}_{2^3})| = 4$, and $|C(\mathbb{F}_{2^4})| = 32$. Its (sgn-normalized) rank one Drinfeld modules are determined by the elements a satisfying

$$a^5 + (x^4 + x^2)a^4 + (x^4 + x^2)a^3 + (x^8 + x^6 + x^5 + x^2 + x + 1)a^2 + (x^5 + x)a + (x^{10} + x^9 + x^7 + x^6 + x^5 + x^4 + x^3 + 1) = 0.$$

Curve C_2: $y^2 + y = x^5 + x$ has class number $h = 15$, with $|C(\mathbb{F}_{2^1})| = 4$, $|C(\mathbb{F}_{2^2})| = 8$, $|C(\mathbb{F}_{2^3})| = 4$, and $|C(\mathbb{F}_{2^4})| = 16$. Its (sgn-normalized) rank one Drinfeld modules are determined by the elements a satisfying

$$\begin{aligned}
a^{15} &+ (x^4+x^2)a^{14} + (x^6+x^5+x^4+x^3+x^2+x+1)a^{12} \\
&+ (x^8+x)a^{11} + (x^{12}+x^{10}+x^8+x^6+x^4+x^2+1)a^{10} \\
&+ (x^8+x^6+x^5+x^3+1)a^9 + (x^{14}+x^{13}+x^{12}+x^{11}+x^8+x^7+x^6+x^5)a^8 \\
&+ (x^8+x)a^7 + (x^{12}+x^9+x^8+x^6+x^4+x^3+1)a^6 \\
&+ (x^8+x^6+x^5+x^4+x^3+x^2+1)a^5 \\
&+ (x^{14}+x^{13}+x^{11}+x^{10}+x^7+x^5+x^2+x+1)a^4 \\
&+ (x^{16}+x^{12}+x^9+x^8+x^6+x^4+x^3+x+1)a^3 \\
&+ (x^{20}+x^{18}+x^{14}+x^{13}+x^{12}+x^{11}+x^8+x^7+x^6+x^4+x^3+x)a^2 \\
&+ (x^{16}+x^{14}+x^{13}+x^{11}+x^{10}+x^8+x^7+x^6+x^4+x^3+x^2+x)a \\
&+ (x^{22}+x^{21}+x^{20}+x^{19}+x^{18}+x^{17}+x^{16}+x^{14}+x^{13}+x^{11}+x^9+x^8 \\
&+ x^6+x^4+x^2+x+1) = 0.
\end{aligned}$$

Curve C_3 : $y^2+y = x^5+x+1$ has class number $h = 3$, with $|C(\mathbb{F}_{2^1})| = 0$, $|C(\mathbb{F}_{2^2})| = 8$, $|C(\mathbb{F}_{2^3})| = 12$, and $|C(\mathbb{F}_{2^4})| = 16$. Its (sgn-normalized) rank one Drinfeld modules are determined by the elements a satisfying

$$a^3 + (x^4+x^2)a^2 + (x^6+x^5+x^4+x^3) = 0.$$

Curve C_4 : $y^2+y = x^5+x^3$ has class number $h = 13$, with $|C(\mathbb{F}_{2^1})| = 4$, $|C(\mathbb{F}_{2^2})| = 4$, $|C(\mathbb{F}_{2^3})| = 16$, and $|C(\mathbb{F}_{2^4})| = 8$. Its (sgn-normalized) rank one Drinfeld modules are determined by the elements a satisfying

$$\begin{aligned}
a^{13} &+ (x^4+x^2)a^{12} + (x^2+x+1)a^{10} + (x^8+x+1)a^9 \\
&+ (x^{12}+x^{10}+x^8+x^6+1)a^8 + (x^8+x+1)a^7 \\
&+ (x^{10}+x^9+x^4+x^3+x^2+x+1)a^6 \\
&+ (x^{10}+x^9+x^3+x^2+1)a^5 + (x^{10}+x^9+x^6+x^5+1)a^4 \\
&+ (x^{12}+x^{10}+x^5+x^4+x^3+x^2)a^3 \\
&+ (x^{14}+x^{13}+x^{12}+x^{11}+x^{10}+x^9+x^8+x^7+x^2+x)a^2 \\
&+ (x^{16}+x^{14}+x^{13}+x^{12}+x^{11}+x^{10}+x^9+x^7+x^6+x^5+x^3+x)a \\
&+ (x^{20}+x^{18}+x^{16}+x^{14}+x^{13}+x^{11}+x^{10}+x^7+x^3+x^2+1) = 0.
\end{aligned}$$

Curve C_5 : $y^2+y = x^5+x^3+1$ has class number $h = 1$, with $|C(\mathbb{F}_{2^1})| = 0$, $|C(\mathbb{F}_{2^2})| = 4$, $|C(\mathbb{F}_{2^3})| = 0$, and $|C(\mathbb{F}_{2^4})| = 8$. Its (sgn-normalized) rank one Drinfeld modules are determined by the elements a satisfying

$$a + (x^4+x^2) = 0.$$

Curve C_6 : $y^2+y = x^5+x^3+x$ has class number $h = 7$, with $|C(\mathbb{F}_{2^1})| = 2$, $|C(\mathbb{F}_{2^2})| = 8$, $|C(\mathbb{F}_{2^3})| = 8$, and $|C(\mathbb{F}_{2^4})| = 24$. Its (sgn-normalized) rank one Drinfeld modules are determined by the elements a satisfying

$$a^7+(x^4+x^2)a^6+(x^4+x^2)a^5+(x^8+x^4+x^3+x^2+x+1)a^4+ (x^5+x^4+x+1)a^3+(x^8+x^7+x^6+x^3+x^2+1)a^2+ (x^8+x^7+x^6+x^5+x^4+x^3+x^2+x)a+ (x^{12}+x^{10}+x^9+x^8+x^4+x^2+x+1)=0.$$

Curve C_7 : $y^2+xy=x^5+1$ has class number $h=8$, with $|C(\mathbb{F}_{2^1})|=3$, $|C(\mathbb{F}_{2^2})|=3$, $|C(\mathbb{F}_{2^3})|=15$, and $|C(\mathbb{F}_{2^4})|=23$. Its (sgn-normalized) rank one Drinfeld modules are determined by the elements a satisfying

$$a^8+(x^4+x)a^7+(x^6+x^5+x^2+x)a^6+(x^8+x^6+x^5+x^4+x^2)a^5+ (x^{10}+x^9+x^7+x^6+x^5+x^4+x^3)a^4+(x^7+x^5+x^3)a^3+ (x^{10}+x^9+x^8+x^6+x^5+x^4+x^3)a^2+ (x^{12}+x^{10}+x^9+x^8+x^7+x^6+x^4)a+ (x^{14}+x^{13}+x^{12}+x^{10}+x^9+x^7+x^4)=0.$$

Curve C_8 : $y^2+xy=x^5+x^2+1$ has class number $h=2$, with $|C(\mathbb{F}_{2^1})|=1$, $|C(\mathbb{F}_{2^2})|=3$, $|C(\mathbb{F}_{2^3})|=1$, and $|C(\mathbb{F}_{2^4})|=23$. Its (sgn-normalized) rank one Drinfeld modules are determined by the elements a satisfying

$$a^2+(x^4+x)a+(x^6+x^5+x^4+x)=0.$$

Curve C_9 : $y^2+xy=x^5+x^3+1$ has class number $h=4$, with $|C(\mathbb{F}_{2^1})|=1$, $|C(\mathbb{F}_{2^2})|=7$, $|C(\mathbb{F}_{2^3})|=7$, and $|C(\mathbb{F}_{2^4})|=23$. Its (sgn-normalized) rank one Drinfeld modules are determined by the elements a satisfying

$$a^4+(x^4+x)a^3+(x^6+x^5+x^4+x)a^2+(x^6+x^2)a+(x^8+x^6+x^4+x^2)=0.$$

Curve C_{10} : $y^2+xy=x^5+x^3+x^2+1$ has class number $h=10$, with $|C(\mathbb{F}_{2^1})|=3$, $|C(\mathbb{F}_{2^2})|=7$, $|C(\mathbb{F}_{2^3})|=9$, and $|C(\mathbb{F}_{2^4})|=23$. Its (sgn-normalized) rank one Drinfeld modules are determined by the elements a satisfying

$$a^{10}+(x^4+x)a^9+(x^6+x^5+x^2+x)a^8+(x^8+x^5+x^4)a^7+ (x^{10}+x^9+x^8+x^7+x^5+x^3+x^2)a^6+ (x^{10}+x^8+x^7+x^6+x^5+x^4+x^3)a^5+ (x^{12}+x^{10}+x^9+x^7+x^5+x^4+x^3)a^4+(x^{12}+x^{10}+x^9+x^8+x^7)a^3+ (x^{14}+x^{13}+x^{10}+x^9+x^6+x^5+x^4)a^2+ (x^{14}+x^{12}+x^{11}+x^9+x^8+x^7+x^5)a+ (x^{16}+x^{14}+x^{12}+x^{11}+x^8+x^7+x^5)=0.$$

Curve C_{11} : $y^2+x^2y=x^5+x$ has class number $h=8$, with $|C(\mathbb{F}_{2^1})|=3$, $|C(\mathbb{F}_{2^2})|=3$, $|C(\mathbb{F}_{2^3})|=15$, and $|C(\mathbb{F}_{2^4})|=23$. Its (sgn-normalized) rank one

Drinfeld modules are determined by the elements a satisfying

$$\begin{aligned} &a^8 + (x^4 + x^2)a^7 + (x^6 + x^3 + x^2 + x)a^6 + (x^8 + x^6 + x^5 + x^3)a^5 \\ &\quad + (x^{10} + x^8 + x^7 + x^6 + x^4)a^4 + (x^9 + x^5 + x^4)a^3 \\ &\quad + (x^9 + x^7 + x^6 + x^4 + x^3)a^2 + (x^{12} + x^{11} + x^8 + x^7 + x^6 + x^5)a \\ &\quad + (x^{14} + x^{10} + x^9 + x^8 + x^4) = 0. \end{aligned}$$

Curve C_{12} : $y^2 + x^2y = x^5 + x^4 + x$ has class number $h = 2$, with $|C(\mathbb{F}_{2^1})| = 1$, $|C(\mathbb{F}_{2^2})| = 3$, $|C(\mathbb{F}_{2^3})| = 1$, and $|C(\mathbb{F}_{2^4})| = 23$. Its (sgn-normalized) rank one Drinfeld modules are determined by the elements a satisfying

$$a^2 + (x^4 + x^2)a + (x^6 + x^3 + x^2 + x) = 0.$$

Curve C_{13} : $y^2 + x^2y = x^5 + x + 1$ has class number $h = 4$, with $|C(\mathbb{F}_{2^1})| = 1$, $|C(\mathbb{F}_{2^2})| = 7$, $|C(\mathbb{F}_{2^3})| = 7$, and $|C(\mathbb{F}_{2^4})| = 23$. Its (sgn-normalized) rank one Drinfeld modules are determined by the elements a satisfying

$$a^4 + (x^4 + x^2)a^3 + (x^6 + x^4 + x^3 + x)a^2 + (x^6 + x^5 + x^4 + x^3)a + (x^8 + x^6 + x^4 + x^2) = 0.$$

Curve C_{14} : $y^2 + x^2y = x^5 + x^4 + x + 1$ has class number $h = 10$, with $|C(\mathbb{F}_{2^1})| = 3$, $|C(\mathbb{F}_{2^2})| = 7$, $|C(\mathbb{F}_{2^3})| = 9$, and $|C(\mathbb{F}_{2^4})| = 23$. Its (sgn-normalized) rank one Drinfeld modules are determined by the elements a satisfying

$$\begin{aligned} a^{10} + &(x^4 + x^2)a^9 + (x^6 + x^4 + x^3 + x)a^8 + (x^8 + x^4)a^7 + \\ &(x^{10} + x^8 + x^7 + x^4 + x^2)a^6 + (x^{10} + x^6 + x^4)a^5 + \\ &(x^{12} + x^{10} + x^9 + x^8 + x^7 + x^4 + x^3)a^4 + (x^{12} + x^8 + x^7 + x^6)a^3 + \\ &(x^{14} + x^{12} + x^{11} + x^6 + x^5 + x^4)a^2 + (x^{14} + x^{12} + x^9 + x^6)a + \\ &(x^{16} + x^{13} + x^9 + x^8 + x^7 + x^6 + x^5) = 0. \end{aligned}$$

Curve C_{15} : $y^2 + x^2y + xy = x^5 + 1$ has class number $h = 4$, with $|C(\mathbb{F}_{2^1})| = 2$, $|C(\mathbb{F}_{2^2})| = 2$, $|C(\mathbb{F}_{2^3})| = 8$, and $|C(\mathbb{F}_{2^4})| = 30$. Its (sgn-normalized) rank one Drinfeld modules are determined by the elements a satisfying

$$a^4 + (x^4 + x)a^3 + (x^5 + x^3 + x^2 + x)a^2 + (x^8 + x^2)a + (x^9 + x^8 + x^5 + x^2) = 0.$$

Curve C_{16} : $y^2 + x^2y + xy + y = x^5 + 1$ has class number $h = 6$, with $|C(\mathbb{F}_{2^1})| = 2$, $|C(\mathbb{F}_{2^2})| = 6$, $|C(\mathbb{F}_{2^3})| = 8$, and $|C(\mathbb{F}_{2^4})| = 30$. Its (sgn-normalized) rank one Drinfeld modules are determined by the elements a satisfying

$$\begin{aligned} a^6 + &(x^4 + x)a^5 + (x^5 + x^3)a^4 + (x^8 + x^6 + x^5 + x^3)a^3 + (x^9 + x^8 + x^6 + x^4)a^2 + \\ &(x^{10} + x^8 + x^6)a + (x^{11} + x^{10} + x^8) = 0. \end{aligned}$$

Curve C_{17} : $y^2 + x^2y + xy + y = x^5 + x$ has class number $h = 14$, with $|C(\mathbb{F}_{2^1})| = 4$, $|C(\mathbb{F}_{2^2})| = 6$, $|C(\mathbb{F}_{2^3})| = 10$, and $|C(\mathbb{F}_{2^4})| = 14$. Its (sgn-normalized) rank one Drinfeld modules are determined by the elements a satisfying

$$\begin{aligned}
&a^{14} + (x^4 + x)a^{13} + (x^5 + x^4 + x^3 + x^2)a^{12} + (x^8 + x^4 + x^2 + x)a^{10} \\
&\quad + (x^{12} + x^9 + x^5 + x^4 + 1)a^9 + (x^{13} + x^{12} + x^{11} + x^{10} + x^8 + x^4 + x^3 + x)a^8 \\
&\quad + (x^9 + x^7 + x^6 + x)a^7 + (x^{10} + x^6 + x^5 + x^3 + 1)a^6 \\
&\quad + (x^{12} + x^{11} + x^{10} + x^9 + x^8 + x^5 + x^3)a^5 + (x^{13} + x^9 + x^8 + x^3 + 1)a^4 \\
&\quad + (x^{13} + x^{11} + x^{10} + x^9 + x^8 + x^6 + x^2 + x + 1)a^3 \\
&\quad + (x^{16} + x^{14} + x^{10} + x^9 + x^7 + x^5 + x)a^2 \\
&\quad + (x^{20} + x^{17} + x^{15} + x^{14} + x^{13} + x^{12} + x^{11} + x^9 + x^7 + x^3 + x^2 + x + 1)a \\
&\quad + (x^{21} + x^{20} + x^{19} + x^{18} + x^{15} + x^{14} + x^6 + x^5 + x^4 + x^3 + 1) = 0.
\end{aligned}$$

Curve C_{18} : $y^2+x^2y+xy+y = x^5+x+1$ has class number $h = 2$, with $|C(\mathbb{F}_{2^1})| = 0$, $|C(\mathbb{F}_{2^2})| = 6$, $|C(\mathbb{F}_{2^3})| = 6$, and $|C(\mathbb{F}_{2^4})| = 14$. Its (sgn-normalized) rank one Drinfeld modules are determined by the elements a satisfying

$$a^2 + (x^4 + x)a + (x^5 + x^4 + x^3 + x^2) = 0.$$

§4. Some questions

Question 1. As mentioned before, it is possible to prove in the elliptic case ($g = 1$) that the greatest common divisor of the polynomials computed in Section 2 is in fact a separable polynomial (over any $\mathbb{F}_q$). In all the computed examples above, this polynomial was again separable. Is this true in general for all hyperelliptic curves over arbitrary $\mathbb{F}_q$?

Question 2. One obvious regularity about the examples in the previous section is their trace terms: always $x^4 + x^2$ or $x^4 + x$ and completely regular within families of curves. The same regularity occurs in the elliptic examples computed over $\mathbb{F}_2$, $\mathbb{F}_4$, and $\mathbb{F}_8$, where the trace terms are again additive polynomials over $\mathbb{F}_2$ (see **[D-H]**). Is this true over $\mathbb{F}_{2^n}$?

Appendix: Elliptic Drinfeld modules over $\mathbb{F}_2$ and $\mathbb{F}_3$

It is perhaps worthwhile to include here the (sgn-normalized) rank one Drinfeld modules computed by Hayes **[H1]** for all of the *elliptic* function fields over $\mathbb{F}_2$ and $\mathbb{F}_3$. In this case,

$$\begin{aligned}
\phi_x &= x + a\mathbf{F} + \mathbf{F}^2 \\
\phi_y &= y + c_1\mathbf{F} + c_2\mathbf{F}^2 + \mathbf{F}^3
\end{aligned}$$

and the equations of Section 2 in this case are simply

$$\begin{aligned}
c_1 &= \frac{1}{x^q - x}a(y^q - y) \\
c_2 &= \frac{1}{x^{q^2} - x}(y^{q^2} - y + ac_1^q - c_1a^q)
\end{aligned}$$

defining c_1 and c_2 in terms of a, and the polynomial satisfied by a is the greatest common divisor of the polynomials obtained from the two equations

$$c_1^{q^2} - c_1 = c_2 a^{q^2} - a c_2^q + x^{q^3} - x$$
$$c_2^{q^2} - c_2 = a^{q^3} - a\,.$$

In the following, we list the curve and the equation that a satisfies that determines the (sgn-normalized) rank one Drinfeld modules for the curve as in the hyperelliptic examples. In **[H1]** one can also find the explicit values of c_1 and c_2, as well as some comments regarding the ring of integers in the Hilbert class field H for some of these examples.

Example 1. $y^2 + y = x^3 + x + 1$ over $\mathbb{F}_2$:

$$a + (x^2 + x) = 0\,.$$

Example 2. $y^2 + xy = x^3 + x^2 + 1$ over $\mathbb{F}_2$:

$$a^2 + (x^2 + x)a + (x^3 + x) = 0\,.$$

Example 3. $y^2 + y = x^3$ over $\mathbb{F}_2$:

$$a^3 + (x^2 + x)a^2 + (x + 1)^2 a + (x + 1)^4 = 0\,.$$

(Note that Example 3 and Example 4 in **[H1]** are the same curve, which explains the discrepancy in the enumeration here from that in **[H1]**.)

Example 4. $y^2 + xy = x^3 + 1$ over $\mathbb{F}_2$:

$$a^4 + (x^2 + x)a^3 + (x^3 + x)a^2 + (x^4 + x^3 + x^2)a + (x^5 + x^4 + x^2) = 0\,.$$

Example 5. $y^2 + xy = x^3 + x$ over $\mathbb{F}_2$:

$$a^5 + (x^2 + x)a^4 + a^3 + a^2 + (x^4 + x + 1)a + (x^3 + x + 1)(x^3 + x^2 + 1) = 0\,.$$

Example 6. $y^2 = x^3 + 2x + 2$ over $\mathbb{F}_3$:

$$a + 2y(x^3 + 2x) = 0\,.$$

Example 7. $y^2 = x^3 + 2x^2 + 2x$ over $\mathbb{F}_3$:

$$a^2 + (2x^3 + x^2 + x + 2)ya + (x^6 + x^5 + x^4 + x^3 + x^2 + x) = 0\,.$$

Example 8. $y^2 = x^3 + x^2 + 2$ over $\mathbb{F}_3$:

$$a^3 + (2x^3 + 2x^2)ya^2 + (2x^6 + 2x^5 + 2x^3 + 2x^2)a + (2x^6 + 2x^3)y = 0\,.$$

Example 9. $y^2 = x^3 + 2x$ over $\mathbb{F}_3$:

$$a^4 + (2x^3 + x + 2)ya^3 + (2x^3 + x)a^2 + (x^3 + 2x)ya + (x^9 + x^6 + x^4 + 2x^3 + x^2) = 0\,.$$

Example 10. $y^2 = x^3 + x$ over $\mathbb{F}_3$:

$$a^4 + (2x^3 + 2x + 2)ya^3 + (x^4 + 2x)a^2 + (x^4 + 2x)ya +$$
$$(x^9 + x^7 + x^6 + 2x^5 + 2x^4 + x^3 + x^2) = 0\,.$$

Example 11. $y^2 = x^3 + 2x^2 + 1$ over $\mathbb{F}_3$:

$$a^5+(2x^3 + x^2 + 1)ya^4 + (x^6 + x^5 + x^2 + 2)a^3 + (2x^3 + 2)ya^2+ (x^9 + 2x^4 + 2x^3 + 2x)a + (2x^9 + x^7 + 2x^6 + 2x^4 + x^3 + x + 1)y = 0.$$

Example 12. $y^2 = x^3 + x^2 + 2x$ over $\mathbb{F}_3$:

$$\begin{aligned} a^6&+(2x^3 + 2x^2 + x + 2)ya^5 + (2x^6 + 2x^5 + x^4 + 2x^2 + x)a^4+ \\ &(2x^6 + 2x^3)ya^3 + (x^9 + x^6 + 2x^5 + 2x^4 + x^2)a^2+ \\ &(x^9 + 2x^8 + x^7 + x^4 + 2x^3 + x^2)ya+ \\ &(x^{12} + 2x^{11} + 2x^{10} + x^8 + 2x^6 + x^5 + x^3) = 0. \end{aligned}$$

Example 13. $y^2 = x^3 + 2x + 1$ over $\mathbb{F}_3$:

$$\begin{aligned} a^7&+(2x^3 + x + 1)ya^6 + (x^3 + 2x + 2)a^5 + (2x^3 + x)ya^4+ \\ &(2x^9 + 2x^6 + 2x^4 + 2x^3 + 2x^2 + 2x + 2)a^3+ \\ &(2x^6 + 2x^4 + 2x^3 + 2x^2 + x + 1)ya^2+ \\ &(x^6 + x^4 + x^3 + x^2 + 2x + 2)a+ \\ &(2x^{12} + x^{10} + x^9 + 2x^6 + x^2 + 2x + 1) = 0. \end{aligned}$$

References

[D-H] D. Dummit and D. Hayes, Elliptic rank-one Drinfeld modules, in preparation.

[H1] D. Hayes, On the reduction of rank-one Drinfeld modules, *Math. Comp.* **57** (1991), 339–49.

[H2] D. Hayes, A brief introduction to Drinfeld modules, this volume.

Department of Mathematics and Statistics, University of Vermont, 16 Colchester Ave., Burlington, VT 05405
Email: dummit@griffin.uvm.edu

A Short Introduction to Rigid Analytic Spaces

David Goss

1. Introduction

This paper is a *slightly* edited version of the appendix to the author's 1977 Harvard thesis *π-adic Eisenstein series for Function Fields*. It is being included here on the request of members of the workshop and is to be viewed as an *aid* in understanding rigid spaces. For more information, the reader can consult the original articles listed in the references, or the books **[B]**, **[BGR]**, and **[FP]** which have appeared since this piece was first written.

Let K be a field of arbitrary characteristic complete with respect to a discrete valuation $|?|$. We shall denote its maximal compact subring by R and we let M be the associated maximal ideal. Let $\overline{K}$ be a fixed algebraic closure of K equipped with the canonical extension of $|?|$. Finally let $k = R/M$ be the residue field.

1.1. Definition. We let $D^n \subseteq \mathbb{A}^n$ be the nth polydisc, i.e., $D^n = \{(z_1, \ldots, z_n) \mid |z_i| \leq 1 \text{ all } i\}$.

Given a function $f: D^n \to \mathbb{A}^1$ there are two (at least) notions of "analyticity": The first requires that f is analytic if and only if it is locally expandable in a power series. The second notion requires, instead, that f be expanded in a power series around the origin converging in *all* of D^n. In complex analysis, through analytic continuation, these two concepts are *equivalent*. However, in our case the following example shows that the second notion is really much stronger than the first.

1.2. Example. Let $U_1 = \{z \mid |z| = 1\}$, $U_2 = \{z \mid |z| < 1\}$. Clearly, $U_1 \cup U_2 = D^1$. The function f with $f(U_1) = 0$, $f(U_2) = 1$ is analytic in the first sense but not in the second.

The global functions on D^n (i.e., those that satisfy the second sense of analyticity) are the ones of interest to us. Thus the first step towards repairing the fact that both senses of analyticity for K are not equivalent is to study their properties. Once one understands the theory of these functions in polydiscs (i.e, the "affine theory"), the next step will be to globalize into general "rigid spaces". Having accomplished this, we will then have put some sort of "analytic continuation" into analysis over K. This analytic continuation will be sufficient in that the theorems of G.A.G.A. type will then be true.

During the original writing of this work, the author was supported by an NSF graduate fellowship.
Key Words: rigid analytic spaces, Tate algebras, Grothendieck topology

2. Affine Theory

(See [T].) The most fruitful way to attempt such a study is to follow closely the affine theory of algebraic geometry. In particular, the spaces D^n will be represented as maximal ideal spaces of certain algebras.

Let $f = \sum a_\alpha T^\alpha$, $\{a_\alpha\} \subset K$, $\alpha \in N^n$, be a power series. It is easy to see that a necessary and sufficient condition for f to converge in D^n is for $|a_\alpha| \to 0$ as $\alpha \to \infty$.

2.1. Definition. The subring of all formal power series $\sum\limits_\alpha a_\alpha T^\alpha$ with $\{a_\alpha\} \subset K$, $a_\alpha \to 0$ as $\alpha \to \infty$, is called the *algebra of restricted formal power series* over K. This ring contains the subring of all restricted power series with coefficients in R, (being the ring of restricted formal power series over R).

Notation. The algebra of restricted formal power series in K, (resp. R) will be denoted $K\{T_1, \ldots, T_n\}$, (resp. $R\{T_1, \ldots, T_n\}$).

The ring $K\{T_1, \ldots, T_n\}$ has a metric on it given by $\|f\| = \max\limits_\alpha\{|a_\alpha|\}$. Thus $R\{T_1, \ldots, T_n\} = \{f \mid \|f\| \leq 1\}$. This ring is complete and separated for the topology induced by this metric. Moreover, the topology is given by considering $R\{T_1, \ldots, T_n\}$ as a neighborhood of zero. Finally, note that $K\{T_1, \ldots, T_n\} = R\{T_1, \ldots, T_n\} \otimes_R K$.

2.2. Proposition. *Let* $f, g \in K\{T_1, \ldots, T_n\}$. *Then*

$$\|f \cdot g\| = \|f\| \cdot \|g\|.$$

Proof. Write $f = c \cdot f_0$ with $\|f\| = |c|$ and similarly $g = d \cdot g_0$. Thus $f_0, g_0 \in R\{T_1, \ldots, T_n\}$. It is enough to show that $\|f_0 \cdot g_0\| = \|f_0\| \cdot \|g_0\| = 1 \cdot 1 = 1$.

Let $\overline{?}$ denote reduction with respect to the ideal $M \cdot R\{T_1, \ldots, T_n\}$. Clearly $\overline{f}$ and $\overline{g}$ are nonzero. As $R\{T_1, \ldots, T_n\}/M \cdot R\{T_1, \ldots, T_n\}$ is a polynomial ring over k, we see that $\overline{f} \cdot \overline{g} = \overline{f \cdot g}$ is also nonzero. The result follows. □

2.3. Theorem. *The algebra* $K\{T_1, \ldots, T_n\}$ *is Noetherian.*

Proof. Since

$$K\{T_1, \ldots, T_n\} = R\{T_1, \ldots, T_n\} \otimes K$$

it is enough to see that $R\{T_1, \ldots, T_n\}$ is Noetherian. But $R\{T_1, \ldots, T_n\}$ is the completion of $R[T_1, \ldots, T_n]$ at M. As $R[T_1, \ldots, T_n]$ is Noetherian its completion is well known to *also* be Noetherian. □

2.4. Definition. Let $J \subseteq K\{T_1, \ldots, T_n\}$ be an ideal. Any algebra of the form $B = K\{T_1, \ldots, T_n\}/J$ is called a *Tate algebra*. The ring B is said to be *topologically of finite type*.

Remarks.

1) Actually Tate is his original paper does not require that $|?|$ be discrete. Theorem 2.3 is still true in this more general situation.

2) Tate shows that any ideal J in $K\{T_1, \ldots, T_n\}$ is *closed*. Thus B inherits a topology from $K\{T_1, \ldots, T_n\}$. This topology is given by taking the image of $R\{T_1, \ldots, T_n\}$ in B as a neighborhood of zero.

The following lemma is crucial for the study of Tate algebras. We shall not attempt to prove it but will point out that it is an analogue of the classical *Noether's normalization lemma.*

2.5. Lemma. *Let B be topologically of finite type. Then B contains a subring H, with B finite over H and $H \simeq K\{T_1, \ldots, T_n\}$.* □

Let B be a Tate algebra and let $J \subset B$ be a maximal ideal; thus B/J is a field over K. The above lemma gives the existence of an $H \simeq K\{T_1, \ldots, T_n\}$ with B/J finite over H. As B/J is a field, one sees easily that H must also be a field; i.e., H contains *no* transcendental elements. We thus have

2.6. Theorem. *Every maximal ideal in B is of finite codimension.* □

Note. The above "Nullstellensatz" is the analogue of Hilbert's on one hand and that of Gelfand-Mazur in commutative Banach algebras on the other.

2.7. Definition. For any topologically finite B, $\mathrm{Max}(B) \subset \mathrm{Spec}(B)$ is the set of all maximal ideals.

Theorem 2.6 is the key to Tate's approach to rigid spaces. To see this one has but to understand what the maximal ideals of $K\{T_1, \ldots, T_n\}$ look like. Let $L \subset \overline{K}$ be a finite extension and let $a \in L$ with $|a| \leq 1$. Then, as L is complete, a determines a maximal ideal in $K\{T\}$ by evaluating T at a. In general, given L and $(a_1, \ldots, a_n) \in L^n$ with $|a_i| \leq 1$ for all i, we obtain a maximal ideal of $K\{T_1, \ldots, T_n\}$ as above. If $\sigma \in \mathrm{Gal}(\overline{K}/K)$, then $(\sigma(a_1), \ldots, \sigma(a_n))$ determines the same ideal as $(a_1, \ldots, a_n)$. By Theorem 2.6, this construction gives *all* the maximal ideals of $K\{T_1, \ldots, T_n\}$. Thus the maximal ideal space looks like $\{(z_1, \ldots, z_n) \mid |z_i| \leq 1,\ z_i \in \overline{K}\}$ up to Galois action. In other words, we have represented D^n as the space of maximal ideals of $K\{T_1, \ldots, T_n\}$.

Let B now be an arbitrary Tate algebra; thus B is of the form $K\{T_1, \ldots, T_n\}/J$ for some ideal $J \subset B$. Consequently $\mathrm{Max}(B)$ represents some closed subspace of D^n.

Let B_1, B_2 be Tate algebras. If $\varphi\colon B_1 \to B_2$ is a homomorphism then we deduce an obvious map map $\widetilde{\varphi}$ on maximal ideal spaces, $\widetilde{\varphi}\colon \mathrm{Max}(B_2) \to \mathrm{Max}(B_1)$. This follows easily from Theorem 2.6. As a consequence of the closedness of ideals, one sees that φ is a *continuous* map.

2.8. Definition. Let B be a Tate algebra. We define

$$B_{00} = \{f \in B \mid |f(x)| \leq 1 \text{ all } x \text{ in } \mathrm{Max}(B)\}.$$

B_{00} is clearly a subring of B. Moreover, if $B = K\{T_1, \ldots, T_n\}$ then

$$R\{T_1, \ldots, T_n\} \subseteq B_{00};$$

and one can show that in this case $B_{00} = R\{T_1, \ldots, T_n\}$. For more general algebras B, we know the existence of some subring $H = K\{T_1, \ldots, T_n\}$ of B with B finite over H. Tate shows then that B_{00} is integral over $R\{T_1, \ldots, T_n\}$. If B is reduced then B_{00} is also shown to be bounded. This implies the following very important proposition:

2.9. Proposition. *Let B be reduced and let $\{f_i\}$, $i \in N$, be a sequence of elements of B which converges uniformly on $\mathrm{Max}(B)$. Then the limit exists as an element of B (i.e., a sequence of functions which converges uniformly determines a rigid analytic function).*

Proof. By uniform convergence we can find an N so that for $i, j > N$

$$|f_i(x) - f_j(x)| < \varepsilon .$$

Let π be a uniformizer of K. The fact that B_{00} is bounded means that for some n we have $(f_i - f_j) \in \pi^n B_{00}$. As $\varepsilon \to 0$, $n \to \infty$ and we are done. □

Finally we mention:

1) B is a Jacobson ring. This implies that any function f with $f(x) = 0$, for all geometric points x, is nilpotent.
2) Any f with $f(x) \neq 0$ on geometric points is invertible.

3. Affine subspaces of Affine Space

Let B be a Tate algebra. We say that $\mathrm{Max}(B)$ is the *affine space* associated to B. Morphisms of affine spaces are just dual to morphisms of the associated Tate algebras. In order to build up more general spaces we now need to define affine subspaces of an affine space.

For the rest of this paper, T shall denote the category of Tate algebras.

3.1. Definition. Let $B \in T$ and $X = \mathrm{Max}(B)$. Let $Y \subseteq X$. We say Y is an *affine subspace* of X if and only if the functor which associates to each $C \in T$ the set of morphisms $B \xrightarrow{\varphi} C$, with $\tilde{\varphi}(\mathrm{Max}(C)) \subseteq Y$, is representable. If this functor is representable, the we let B_Y denote the corresponding Tate algebra.

There is an obvious map $B \to B_Y$.

Example. Let $b \in K$ and $f \in B$. Then the set of points $x \in \mathrm{Max}(B)$ with $|f(x)| \leq |b|$ is represented by $B\{T\}/(f - bT)$, where $B\{T\}$ has the obvious meaning. The set of points where $|f(x)| \geq |b|$ is given by $B\{T\}/(b - fT)$.

In general the most important affine subspaces will be constructed in a manner similar to the example.

3.2. Proposition. *Let Y be an affine subset of* $\mathrm{Max}(B)$. *Then the map* $\mathrm{Max}(B_Y) \to Y$ *is bijective. Moreover, if $N \in \mathrm{Max}(B_Y)$ and $M \in Y \subseteq \mathrm{Max}(B)$ are corresponding ideals, then the morphism $B/M^n \to B_Y/N^n$ is bijective for all n.*

Proof. The difficult part is the last. Let M be as in the hypothesis. We have the following diagram:

$$\begin{array}{ccc} B & \xrightarrow{\theta} & B_Y \\ \varphi\downarrow & \overset{\psi}{\swarrow} & \downarrow\varphi' \\ B/M^n & \underset{\theta'}{\longrightarrow} & B_Y/M^nB_Y; \end{array}$$

where θ is the canonical map, θ' comes from θ, φ and φ' are the canonical surjections, and, finally, ψ is the unique arrow given by the representability of Y.

Since the square is commutative, the bottom triangle is also by the universality of θ. Then the surjectivity of θ' follows from that of φ'. The injectivity of θ' is easily deduced. Thus θ' is bijective. In particular, when $n = 1$ we obtain $N = MB_Y$. The result follows. □

Let $\widehat{}$ denote the completion of a Tate algebra B at some point $y \in \operatorname{Max}(B)$.

3.3. Corollary. *Let Y be an affine subspace of* $\operatorname{Max}(B)$ *and let $y \in Y$. Then* $\widehat{B}_y = (\widehat{B}_Y)_y$. □

3.4. Corollary. *Let $Y \subseteq D^n$ be an open affine subspace, then Y is reduced.*

Proof. The algebra $B = K\{T_1, \ldots, T_n\}$ is regular and therefore B_Y is also by the corollary. This clearly implies that B_Y is reduced. □

4. Intersection and Fiber Product of Affines

4.1. Proposition. *Let $A, B, C \in T$ and set $X = \operatorname{Max}(A)$, $Y = \operatorname{Max}(B)$, and $Z = \operatorname{Max}(C)$. Let*

$$\begin{array}{ccc} X & & Y \\ & \searrow \quad \swarrow & \\ & Z & \end{array}$$

be morphisms. Then the fiber product of this diagram exists and is an affine space. Its algebra is

$$A\widehat{\otimes}_C B$$

where $\widehat{}$ denotes the completion under the obvious topology.

Proof. All formal consequences of the definitions. □

4.2. Corollary. *The intersection of a finite number of affines is affine and the inverse image of an affine is affine under a map dual to an algebra homomorphism.* □

5. Types of Coverings

5.1. Lemma. *Let $B \in T$ and $T_1, \dots, T_n$ be free variables. Then*

$$B\{T_1, \dots, T_{n-1}\}\{T_n\} \simeq B\{T_1, \dots, T_n\}.$$

(This may be stated as "topologically free over topologically free over is topologically free".) □

5.2. Corollary. *The product of D^1 with itself is D^2, etc.* □

5.3. Definition. An affine subspace Y of $X = \mathrm{Max}(B)$ is called *special* if there are two finite families (f_i), (g_j) in B so that

$$Y = \{x \in X \mid |f_i(x)| \leq 1 \text{ for all } i \text{ and } |g_j(x)| \geq 1 \text{ for all } j\}.$$

(Such a set is clearly affine as it is represented by

$$B\{T_i, S_i\}/(f_i - T_i, 1 - g_jS_j).)$$

5.4. Lemma. *The intersection of a finite number of special affines is special and the inverse image of a special affine is special.*

Proof. Follows from definitions. □

In his paper, Tate proved that a finite covering of an affine by special affines has a Čech cohomology complex (defined in an obvious manner) which is acyclic. In particular with such coverings we would be able to build functions locally.

Grauert and Gerritzen **[G]** established that *all* finite coverings of an affine by affines are acyclic. This is done by using the following type of covering:

5.5. Definition. Let $f_1, \dots, f_n \in B$ with B in T. We suppose that these functions have no common zero (and thus generate the unit ideal). Let $1 \leq i \leq n$. Then the set of points $x \in \mathrm{Max}(B)$ where $|f_i(x)| \geq |f_j(x)|$, $j = 1, \dots, n$ is called a *standard affine*. This affine is represented by $B_i = B\{T_1, \dots T_n\}/(f_j - T_jf_i)$ for $j = 1, \dots, n$.

Of course $\mathrm{Max}(B) = \bigcup \mathrm{Max}(B_i)$. In B_i, the function f_i generates the ideal $(f_1, \dots, f_n)$. Also one can check that the intersection of standard affines is a standard affine and the inverse image of a standard affine is standard.

Grauert and Gerritzen show that every covering by a finite number of affines can be refined to a standard covering. Using this and results on standard coverings they show that *all* coverings by a finite number of affines are acyclic.

Let $\mathrm{Max}(B) \xrightarrow{\widetilde{\varphi}} \mathrm{Max}(A)$ be a morphism dual to the map $\varphi: A \to B$. Assume that $\widetilde{\varphi}$ is étale and bijective. Then, in **[G]** it is shown that $\mathrm{Max}(B)$ is an open affine of $\mathrm{Max}(A)$ via $\widetilde{\varphi}$.

5.6. Example. In $\mathbb{A}^1$ we may build functions locally as follows: Let $a, b \in K$ $|a| \leq |b|$. Suppose we have a function on $\{x \mid |x| \leq |a|\}$ and on $\{x \mid |a| \leq |x| \leq |b|\}$ which agrees on $\{x \mid |x| = |a|\}$. Then as these spaces are affine we deduce a function on the whole disc of radius $|b|$, i.e., a power series around the origin converging in the disc of radius $|b|$.

5.7. Example. As a final example of an affine space we show how to represent *over* K affine spaces of the form $\{x \mid |x| \le |a|\}$ for any a in $\overline{K}$. We can of course assume that $a = \pi^n$, $n \in \mathbb{Q}$ where π is the uniformizer of k. Let $n = \frac{u}{v}$ for $u, v \in \mathbb{Z}$. We shall let $n > 0$, the case $n < 0$ being similar. Then the set $\{x \mid |x| \le |a|\}$ is represented by

$$K\{T_1, T_2\}/(\pi^v T_1 - \pi^u T_2).$$

6. Kiehl's approach to Rigid Analytic Spaces

(See **[K1]**, **[K2]**.) The problem of globalizing the affine theory was pursued by Kiehl. The approach taken is to restrict the number of open sets that may be used until one has the *bare minimum* left for patching. This is accomplished with the notion of a Grothendieck Topology.

6.1. Definition. A *Grothendieck Topology* Ω is given by a category, $\mathrm{Cat}(\Omega)$, and a set, $\mathrm{Cov}(\Omega)$, of families $\{U_i \xrightarrow{\phi_i} U\}_{i \in I}$ of maps in $\mathrm{Cat}(\Omega)$ (called *coverings of* U) with the following properties:

(1) If ϕ is an isomorphism, then $\{\phi\} \in \mathrm{Cov}(\Omega)$.

(2) If $\{U_i \to U\} \in \mathrm{Cov}(\Omega)$ and $\{V_{ij} \to U_i\} \in \mathrm{Cov}(\Omega)$ for each i, then the family $\{V_{ij} \to U\}$ obtained by composition is in $\mathrm{Cov}(\Omega)$.

(3) If $\{U_i \to U\} \in \mathrm{Cov}(\Omega)$ and $V \to U \in \mathrm{Cat}(\Omega)$ is arbitrary, then $U_i \times_U V$ exists and $\{U_i \times_U V \to V\} \in \mathrm{Cov}(\Omega)$.

Suppose on a topological space X we are given a Grothendieck Topology Ω. We will assume the objects in the category come from open subsets of X and that morphisms are inclusion. We will also suppose the existence of a collection W of open subsets of X which will play the role of a *subbasis*. We shall have $U \in \mathrm{Cat}(\Omega)$ if and only if there is a covering of U with elements of W of a specified sort. The elements of $\mathrm{Cat}(\Omega)$ will be the *admissible subsets* of X. We suppose $\{\emptyset, X\} \in \mathrm{Cat}(\Omega)$. Moreover, a covering $\{U_i\}$ of U with U_i, all i, and U in $\mathrm{Cat}(\Omega)$ is to be in $\mathrm{Cov}(\Omega)$ if it may be refined to a covering of our special sort. We shall call the elements of $\mathrm{Cov}(\Omega)$ the *admissible coverings*. Finally, our topology is to possess the following completeness axiom: Let $U \in \mathrm{Cat}(\Omega)$, and $V \subseteq U$ an open subset. Then $V \in \mathrm{Cat}(\Omega)$ if and only if there is an admissible covering $\{U_i\}$ of U so that $U_i \cap V \in \mathrm{Cat}(\Omega)$ for all i.

To apply this to our situation we first topologize our affine spaces $X = \mathrm{Max}(B)$ by taking the open affine subsets as a subbasis. We shall say that an open set $U \subseteq X$ is *admissible* (i.e., in $\mathrm{Cat}(\Omega)$) if and only if there is a covering with affine subsets $\{U_i\}_{i \in I}$ with the following property: Suppose we have a morphism $\mathrm{Max}(C) \xrightarrow{\pi} \mathrm{Max}(B)$ (dual to an algebra map) *with image in* U. Then $\{\pi^{-1}(U_i)\}_{i \in I}$ can be refined to a *finite* affine covering. The affines thus play the role of W.

On a general X, with a Grothendieck Topology, we have the notion of sheaves of rings, sets, etc. In particular on $X = \mathrm{Max}(B)$ we have a structure sheaf O_X. That this is a sheaf is exactly the reason for the whole setup. In particular for $\mathrm{Max}(B)$ we can easily define the stalk of a sheaf and one sees that $\mathrm{Max}(B)$ is a locally ringed space.

6.2. Definition. A morphism $f = (\varphi, \psi): X \to Y$ of spaces of the type discussed above is a pair of mappings (φ, ψ) such that:

(1) φ a continuous map of the underlying spaces and continuous in the Grothendieck Topology (thus the inverse of an admissible set (cover) is an admissible set (cover)).

(2) $\psi: O_Y \to O_X$ is a homomorphism, i.e., a system of homomorphisms for all admissible $U \in Y$ so that

$$O_Y(U) \to O_X(\varphi^{-1}(U)).$$

(3) The map on sheaves should commute with restriction, and on stalks it should be a local homomorphism.

6.3. Definition. The local ringed space $\mathrm{Max}(B)$ for $B \in T$ with its Grothendieck Topology is called the *affine rigid analytic space* associated to B.

Any morphism of affine spaces arises from a morphism of global rings (and of course the reverse is true).

6.4. Definition. A locally ringed space with a Grothendieck Topology is called *rigid analytic* if and only if it has an admissible covering through open affines.

6.5. Examples.

(1) $\mathbb{A}^1$ may be given the structure of a rigid space. Indeed we simply take a covering by closed discs of increasing radii. Note that there is no finite covering by affines.

(2) Any K-algebraic scheme may be given a rigid structure in an obvious fashion. If the variety is projective then we can find a covering by a finite number of affines. It is easy to see that open affine subschemes are admissible.

(3) $\mathbb{P}^1$ has the covering by $\{t \mid |t| \geq 1\}$ and $\{t \mid |t| \leq 1\}$. These meet in $\{t \mid |t| = 1\}$. Note that $\{t \mid |t| = 1\}$ is represented by $K\{T, T^{-1}\}$.

Note. In his paper Kiehl used only special coverings. However by Grauert and Gerritzen we can use any cover by a finite number of affines.

7. Raynaud's approach via Formal Schemes and G.A.G.A.

(See **[R1]**.) The basic idea here is to connect rigid spaces with formal schemes. Having done this one then has the use of the machinery of formal geometry.

Over $\mathrm{Spf}(R)$, we have the formal scheme defined by $R\{T_1, \ldots, T_n\}$. The fiber over the closed point is, of course, $\mathrm{Spec}\,((R/M)[T_1, \ldots, T_n])$. Since $K\{T_1, \ldots, T_n\} = R\{T_1, \ldots, T_n\} \otimes K$, the rigid space associated to $K\{T_1, \ldots, T_n\}$ appears as the "generic fiber" of the structure map. *However*, as the category of formal schemes over R has *no* generic fiber, one needs to develop a theory for them. This theory is essentially the theory of rigid spaces.

More precisely, let π be a uniformizer of K. By multiplying by a high enough power of π, for any Tate algebra B we can find an R-algebra C of the form

$$R\{T_1, \ldots, T_n\}/(f_1, \ldots, f_n)$$

with $C \otimes K = B$. (Such R-algebras are also called topologically of finite type.) If $X = \mathrm{Max}(B)$, then $\mathcal{X} = \mathrm{Spf}(C)$ will be called an R-*model* for X. It is obvious how to generalize this construction to a rigid space with a finite covering by affines.

Any affine open of $\mathrm{Spf}(C)$ gives an admissible open subset of the associated rigid space. However, these are not the most general coverings. In order to get the most general covering we shall use the fact that every covering of the rigid space may be refined to a standard covering.

Recall that a standard covering is given by elements $f_1, \ldots, f_n$ so that these functions generate the unit ideal of B. By using a high enough power of π we can assume that $f_1, \ldots, f_n$ are in C. In C, however, they do not generate necessarily the unit ideal but, at least, the ideal they generate contains a power of π. Call this ideal J.

Let X_i be the locus where $|f_i|$ is maximal among the $|f_j|$, $j = 1, \ldots, n$. Then we know X_i comes from the R-model $C\{T_1, \ldots, T_n\}/(f_j - T_j f_i) = C_i$. Since we are interested in the rigid space X_i we may divide C_i by the ideal of all elements who are killed by a power of π. Call the resulting ring $\overline{C_i}$. In $\overline{C_i}$, the ideal J is generated by a nonzero divisor. Indeed if $f_i \cdot x = 0$, then as f_i divides π^m some m, we have $\pi^m \cdot x = 0$, and so $x = 0$. Therefore J is an invertible sheaf. The observation of Raynaud is that the rings $\overline{C_i}$ correspond to *blowing up* J in the category of formal schemes.

If we start off with C and blow up J over $\mathrm{Spec}(C)$ and then complete, we end up with a formal scheme which has a covering by $\mathrm{Spf}(\overline{C_i})$ for all i. This motivates Raynaud's definition of rigid spaces.

Let $\mathcal{F}$ be the category of R-formal schemes of finite type.

7.1. Definition. The category E of K-rigid analytic spaces, of finite type and quasi-separated (i.e., the intersection of affines has a finite affine cover), is the category deduced from $\mathcal{F}$ by making invertible those morphisms which are blow-ups of ideals containing a power of π.

Remark. The definition says the category of K-rigid spaces has objects, the objects of $\mathcal{F}$, and an arrow $\mathcal{X} \to \mathcal{Y}$ in E is a diagram

$$\begin{array}{ccc} \mathcal{X}' & & \\ \mu\downarrow & \searrow & \\ \mathcal{X} & & \mathcal{Y} \end{array}$$

with μ a blow-up of the type discussed.

This category is equivalent to the category of rigid spaces in Kiehl's sense which are given by a finite union of affines and are quasi-separated. In particular Kiehl has a notion of proper map which will correspond to proper maps of formal schemes.

The use of R-modules allows one to utilize Grothendieck's existence theorem of formal schemes (E.G.A. III 3.4.21). Through this, one has the theorems of G.A.G.A. type. Thus we have for instance, on a complete variety, all rigid maps are algebraic, all line bundles are algebraic, all rigid sections of line bundles are algebraic, etc. These theorems were known also to Kiehl.

8. Connectivity

8.1. Definition. A rigid analytic space X is *connected* if and only if it is not possible to find an admissible cover $X = U_1 \cup U_2$, with $U_1 \cap U_2 = \emptyset$. It is *geometrically connected* if it is connected for each finite extension L over K.

An affine rigid space is connected if it is not possible to write the ring of global functions as a product of algebras. It is not true that any affine subspace of a connected affine space is connected.

8.2. Proposition. *Let X be a K-variety. If X is geometrically connected as a rigid analytic space, then it is geometrically connected as a variety. As a corollary to G.A.G.A., a variety which is complete and connected is connected as a rigid space.* □

9. $\mathbb{A}^1$

A rigid function on $\mathbb{A}^1$ is a power series with an infinite radius of convergence.

9.1. Theorem. *Let $f(x)$ be an entire function, then:*

1) *The roots $\{a_i\}$ of f are discrete and $a_i \in \overline{K}$.*
2) *Let n be the multiplicity of zero as a root, then*
$$f(x) = Cx^n \prod_{a_i \neq 0} (1 - x/a_i),$$
where the roots are counted with multiplicity.
3) *Let $K^s \subset \overline{k}$ be the separable closure. Any discrete sequence $\{a_i\} \subset K^s$ which is Galois stable gives rise to an entire function.*

Proof. See (**[Ro]**). □

9.2. Corollary. *Any entire function is determined up to a constant by its divisor.* □

References

[B] V. Berkovich: *Spectral Theory and Analytic Geometry over Non-Archimedean Fields*, Providence: American Mathematical Society (1990).

[BGR] S. Bosch, U. Güntzer, and R. Remmert: *Non-Archimedean analysis: A systematic approach to rigid analytic geometry*, Grundlehren Math. Wiss. 261, Springer, Berlin-Heidelberg-New York (1984).

[FP] J. Fresnel, M. van der Put: *Géométrie Analytique Rigide et Applications*, Birkhäuser, Boston-Basel-Stuttgart (1981).

[G] H. Grauert, L. Gerritzen: Die Azyklizität der affinoiden Überdeckungen, In: *Global Analysis*, Edited by D.C. Spencer and S. Iyanaga, Univ. of Tokyo Press (1969), 159–184.

[Gr] A. Grothendieck: *Éléments de Géométrie Algébrique III*, (rédigé avec la collaboration de J. Dieudonné), Inst. Hautes Études Sci. Publ. Math. (1961).

[K1] R. Kiehl: Der Endlichkeitsatz für eigentliche Abbildungen in der nichtarchimedischen Funktionentheorie, *Invent. Math.* **2**, (1967), 191–214.

[K2] R. Kiehl: Theorem A und Theorem B in der nichtarchimedischen Funktionentheorie, *Invent. Math.* **2**, (1967), 256–273.

[R1] M. Raynaud: Géométrie Analytique Rigide d'aprés Tate, Kiehl,... *Bull. Soc. Math. France. Mémoire* **39/40** (1974), 319–327.

[R2] M. Raynaud: Unpublished Notes.

[Ro] A. Robert: *Elliptic Curves*, Lecture Notes in Math. 326, Springer, Berlin-Heidelberg-New York (1973), II.70–II.85.

[T] J. Tate: Rigid Analytic Spaces, *Invent. Math.* **12**, (1971), 257–258.

Department of Mathematics, Ohio State University, Columbus, Ohio 43210
Email: goss@coltrane.mps.ohio-state.edu

Lecture on Rigid Geometry

Marc Reversat

The purpose of this lecture, through the example of the Drinfeld-upper-half-plane and its quotients, is to describe how rigid geometry can be used in the theory of Drinfeld modules and what sort of tools are coming with the rigid analytic structure. So, we do not give an account about the foundation of rigid geometry and its main applications ; the bibliography gives all the material to study rigid geometry, its main results and methods.

Of course, nothing is new in this paper and all is well known by the specialists.

1. Some Notation

Let $\mathcal{C}$ be a smooth projective curve over $\mathbb{F}_q$ (the field with q element) and ∞ be a point of $\mathcal{C}$ rational over $\mathbb{F}_q$. We set:

$A = \mathcal{O}_{\mathcal{C}}(\mathcal{C} - \{\infty\})$: ring of rational function over $\mathcal{C}$, regular outside ∞.

$K =$ quot (A) : quotient field of A.

K_∞ : completion of K at the place ∞.

C : completion of the algebraic closure of K_∞ (which is always algebraically closed).

π : uniformizing parameter of K at ∞.

$|?|$: absolute value of C, coming from the place ∞, with $|\pi| = 1$.

Let K_∞^0, C^0 be the valuation rings of K_∞ and C for the absolute value $|?|$, K_∞^{00} and C^{00} the ideals of valuations, $\bar{K}_\infty = {}^{K_\infty^0}/_{K_\infty^{00}}$ and $\bar{C} = {}^{C^0}/_{C^{00}}$ the residue fields (note that $\bar{K}_\infty = \mathbb{F}_q$ and that $\bar{C}$ is an algebraic closure of $\mathbb{F}_q$). If t belongs to K_∞^0 or C^0, we write $\bar{t}$ for its image in the residue field.

The Drinfeld-upper-half-plane is

$$\Omega = \mathbb{P}_C^1 \setminus \mathbb{P}_C'(K_\infty) = C \setminus K_\infty.$$

2. The rigid analytic structure of the Drinfeld-upper-half-plane

For all $n \in \mathbb{Z}$ set

$$D_n = \{z \in C \mid |\pi|^{n+1} \leq |z| \leq |\pi|^n \text{and, for all } \rho \in \mathbb{F}_q^*,$$
$$|z - \pi^n \rho| \geq |\pi|^n, \ |z - \pi^{n+1}\rho| \geq |\pi|^{n+1}\}$$

and for all $\lambda \in K$, let

$$D_{n,\lambda} = \lambda + D_n.$$

One has

$$\Omega = \bigcup_{\substack{\lambda\in K\\ n\in\mathbb{Z}}} D_{n,\lambda} \tag{2.1}$$

(note that: $(D_{n,\lambda} = D_{n,\lambda'}) \iff (n = n'$ and $|\lambda - \lambda'| \le |\pi|^{n+1}))$.

The $D_{n,\lambda}$ are affinoid spaces : let $A_{n,\lambda}$ be the algebra of holomorphic functions on $D_{n,\lambda}$, then we have

$$A_{n,\lambda} = C\Big\langle \frac{z-\lambda}{\pi^n}, \frac{\pi^{n+1}}{z-\lambda}, \frac{1}{\frac{z-\lambda}{\pi^n}-\rho}, \frac{1}{\frac{z-\lambda}{\pi^{n+1}}-\rho}; \rho \in \mathbb{F}_q^* \Big\rangle; \tag{2.2}$$

this means that $A_{n,\lambda}$ is the set of series of the form

$$f = \sum_{k\ge 0} a_k \Big(\frac{z-\lambda}{\pi^n}\Big)^k + \sum_{k>0} b_k \Big(\frac{z^{n+1}}{z-\lambda}\Big)^k + \sum_{\rho\in\mathbb{F}_q^*}\sum_{k>0} c_{k,\rho}\Big(\frac{1}{\frac{z-\lambda}{\pi^n}-\rho}\Big)^k$$
$$+ \sum_{\rho\in\mathbb{F}_q^*}\sum_{k>0} d_{k,\rho}\Big(\frac{1}{\frac{z-\lambda}{\pi^{n+1}}-\rho}\Big)^k$$

where a_k, b_k, $c_{k,\rho}$, $d_{k,\rho}$ are in C and with

$$\lim_{k\to+\infty} a_k = \lim_{k\to+\infty} b_k = \lim_{k\to+\infty} c_{k,\rho} = \lim_{k\to+\infty} d_{k,\rho} = 0.$$

The algebra $A_{n,\lambda}$ is equipped with a norm, the spectral norm, which is defined by

$$\| f \|^{sp}_{A_{n,\lambda}} = \sup_{z\in D_{n,\lambda}} |f(z)|.$$

In fact, this supremum is a maximum, and we have

$$\| f \|^{sp}_{A_{n,\lambda}} = \max_{\rho,k_1,k_2,k_3,k_4} \Big\{ |a_{k_1}|, |b_{k_2}|, |c_{k_3,\rho}|, |d_{k_4,\rho}| \Big\} \tag{2.3}$$

(see the expression of f above).

Of course, we have isomorphisms of normed algebras $A_{n,\lambda} \simeq A_{n',\lambda'}$ for all n, $n' \in \mathbb{Z}$ and $\lambda, \lambda' \in K$, and also isomorphisms of affinoid spaces $D_{n,\lambda} \simeq D_{n',\lambda'}$.

If $D_{n,\lambda} \ne D_{n',\lambda'}$ and $D_{n,\lambda} \cap D_{n',\lambda'} \ne \emptyset$, the intersection $D_{n,\lambda} \cap D_{n',\lambda'}$ is a circumference "in the boundary" of $D_{n,\lambda}$ (and $D_{n',\lambda'}$), and is of the form

$$\Big\{ z \in C \mid |z - \lambda - \rho\pi^n| = |\pi|^n \ \forall \rho \in \mathbb{F}_q \Big\} \tag{2.4}$$

or the same set with $n+1$ in place of n (for instance, $D_{n,\lambda} \cap D_{n-1,\lambda}$ or $D_{n,\lambda} \cap D_{n,\lambda+\pi^n}$ are equal to the above set). The analytic space defined by (2.4) is affinoid, its algebra of holomorphic functions is $C\Big\langle \frac{z-\lambda}{\pi^n}, \frac{\pi^n}{z-\lambda-\rho\pi^n}; \rho \in \mathbb{F}_q \Big\rangle$.

Consider again $D_{n,\lambda}$ and $D_{n',\lambda'}$ such that $D_{n,\lambda} \ne D_{n',\lambda'}$ and $D_{n,\lambda} \cap D_{n',\lambda'} \ne \emptyset$; such $D_{n,\lambda}$ and $D_{n',\lambda'}$ can be glued on their affinoid subspace $D_{n,\lambda} \cap D_{n',\lambda'}$; this gives the analytic space $D_{n,\lambda} \cup D_{n',\lambda'}$. With this process, one constructs an analytic structure on the Drinfeld-upper-half-plane Ω, which is the gluing of the analytic structure of the $D_{n,\lambda}$ (cf. (2.1)). So, Ω is equipped with a sheaf, $\mathcal{O}_\Omega$, the sheaf of holomorphic functions on Ω. One has

$$\mathcal{O}_\Omega(D_{n,\lambda}) = A_{n,\lambda}\,. \tag{2.5}$$

3. An analytic reduction and the Bruhat-Tits tree

With the help of the spectral norm, one can associate to each $D_{n,\lambda}$ an (algebraic) affine variety over $\bar{C}$, denoted $\bar{D}_{n,\lambda}$.

For instance, consider $D = \{z \in C \mid |\pi| \le |z| \le 1\}$ and its algebra

$$A = \Big\{f = \sum_{k\ge 0} a_k z^k + \sum_{k>0} b_k \big(\frac{\pi}{z}\big)^k \mid a_k, b_k \in C, a_k \to 0, b_k \to 0\Big\}$$
$$= C < z, \frac{\pi}{z} > .$$

The spectral norm on A is

$$\| f \|_D^{sp} = \sup_{z\in D} |f(z)| = \max_{k_1,k_2}\{|a_{k_1}|, |b_{k_2}|\}. \tag{3.1}$$

Let

$$A^o = \{f \in A \mid \| f \|_D^{sp} \le 1\};$$

this is a C^o-algebra, and let

$$A^{oo} = \{f \in A \mid \| f \|_D^{sp} < 1\}$$

which is an ideal of A^o. We set $\bar{A} = {}^{A^o} /_{A^{oo}}$.

We have (see (3.1))

$$\bar{A} = \bar{C}[x, y] = \frac{\bar{C}[X, Y]}{(XY)}$$

where x (resp. y) is the image of z (resp. $\frac{\pi}{z}$). Let

$$\bar{D} := \mathrm{Spm}(\bar{A}). \tag{3.2}$$

The space $\bar{D}$ is an affine variety over $\bar{C}$. It has two irreducible components, which are affine lines over $\bar{C}$ and which meet in an ordinary double point rational over $\mathbb{F}_q$. The affine variety $\bar{D}$ is called *the canonical reduction of the affinoid space* D and we have the reduction map

$$D \xrightarrow{r_D} \bar{D}, \quad t \mapsto \Big(x - \bar{t}, y - \overline{\big(\tfrac{\pi}{t}\big)}\Big).$$

(Note that D is identified with the maximal spectrum of A by $t \mapsto (z-t)A + (\frac{\pi}{z} - \frac{\pi}{t})A$; more generally, every affinoid space is the maximal spectrum of its algebra of holomorphic functions on itself.) By r_D, the circumference $\{|z| = 1\}$ is mapped on an irreducible component minus the double point, the circumference $\{|z| = |\pi|\}$ is mapped on the others, and $\{|\pi| < |z| < 1\}$ on the double point.

The same method allows to construct the canonical reduction of the $D_{n,\lambda}$:

$$\bar{D}_{n,\lambda} = \mathrm{Spm}\Big({}^{A^o_{n,\lambda}} /_{A^{oo}_{n,\lambda}}\Big) = \mathrm{Spm}\,(\bar{A}_{n,\lambda})\,.$$

$\bar{D}_{n,\lambda}$ is an affine variety over $\bar{C}$, it has two irreducible components, which are projective lines over $\bar{C}$ minus the rational points over $\mathbb{F}_q$, except one, which is the intersection of the two components and which is an ordinary double point (see (2.2) and (2.3)). We have also the reduction map

$$D_{n,\lambda} \xrightarrow{r_{D_{n,\lambda}}} \bar{D}_{n,\lambda}, \quad t \mapsto \left(x - \overline{(\frac{t-\lambda}{\pi^n})}, y - \overline{(\frac{\pi^{n+1}}{t-\lambda})}, \cdots\right)$$

where x, y are the images in $A_{n,\lambda}$ of $\frac{z-\lambda}{\pi^n}, \frac{\pi^{n+1}}{z-\lambda}$.

The covering $(D_{n,\lambda})_{n,\lambda}$ of Ω has the following main property : for all n, n', λ, λ', we have

$$r_{n,\lambda}^{-1}\left(r_{n,\lambda}(D_{n,\lambda} \cap D_{n',\lambda'})\right) = D_{n,\lambda} \cap D_{n',\lambda'}$$

(in general $r_{n,\lambda}^{-1}\left(r_{n,\lambda}(\ )\right)$ is bigger than the initial subset of $D_{n,\lambda}$). We say that $(D_{n,\lambda})_{n,\lambda}$ is a pure covering of Ω (M. van der Put) or that Ω equipped with the covering $(D_{n,\lambda})_{n,\lambda}$ is a formal analytic space (S. Bosch). With this property, we can glue the $\bar{D}_{n,\lambda}$ on the canonical reduction of the $D_{n,\lambda} \cap D_{n',\lambda'}$. Finally, we obtain a reduction $\Omega \to \bar{\Omega}$ of Ω such that $r \mid D_{n,\lambda} = r_{D_{n,\lambda}}$.

When we glue $\bar{D}_{n,\lambda}$ and $\bar{D}_{n',\lambda'}$, with $D_{n,\lambda} \neq D_{n',\lambda'}$ and $D_{n,\lambda} \cap D_{n',\lambda'} \neq \emptyset$, we glue an irreducible component of $\bar{D}_{n,\lambda}$ with one of $\bar{D}_{n',\lambda'}$. So we fill up the "missing point" of the irreducible component of $\bar{D}_{n,\lambda}$, or $\bar{D}_{n',\lambda'}$, which corresponds to a hole in the boundary of $D_{n,\lambda}$ in which is $D_{n',\lambda'}$, and also to a hole in the boundary of $D_{n',\lambda'}$ in which is $D_{n,\lambda}$ (see also (2.4)). Then $\bar{D}_{n,\lambda} \cup \bar{D}_{n',\lambda'}$ is an algebraic variety over $\bar{C}$ with three irreducible components, two, L_1 and L_2, are projective lines over $\bar{C}$ minus rational points over $\mathbb{F}_q$ plus one point, rational over $\mathbb{F}_q$, which is the intersection with the third component, L_3. This third component L_3 is also a projective line over $\bar{C}$, minus the rational points over $\mathbb{F}_q$ except for two of them which are the intersections with the two other components L_1 and L_2 and which are ordinary double points for good (and different) choices of $D_{n,\lambda}, D_{n',\lambda'}$ and $D_{n'',\lambda''}$ (the first picture high on the left of Figure 1 is $\bar{D}_{n,\lambda} \cup \bar{D}_{n',\lambda'}$, the others $\bar{D}_{n,\lambda} \cup \bar{D}_{n',\lambda'} \cup \bar{D}_{n'',\lambda''}$).

Then, we see that $\bar{\Omega}$ is a scheme over $\bar{C}$, locally of finite type, each of its irreducible components is a $\mathbb{P}^1_{\bar{C}}$ which meets exactly $(q+1)$ other irreducible components, each of them in one point which is ordinary and rational over $\mathbb{F}_q$.

The intersection graph of $\bar{\Omega}$ is obtained by associating to each irreducible component a vertex, two vertices being connected by an edge if and only if the corresponding irreducible components meet. *The intersection graph of $\bar{\Omega}$ is canonically the Bruhat-Tits tree of* $\mathrm{GL}(2, K_\infty)$. To see this (briefly) note that each vertex of the intersection graph corresponds by the reduction to a circumference of the form $\{z \in C \mid |z-\lambda| = |\pi|^n\}$ (see (2.4) and the discussion above about the construction of $\bar{\Omega}$); then, to each such vertex we associate the vertex in the Bruhat-Tits tree given by the module $K_\infty^o(e_1 - \lambda e_2)\pi^{-n} \oplus K_\infty^o e_2$, where $\{e_1, e_2\}$ is the canonical basis of $K_\infty \oplus K_\infty$. This gives a bijection between the vertices of the two graphs, which is in fact an isomorphism of graphs (the idea here is in fact from D. Mumford, who to each vertex $[M]$ of the Bruhat-Tits tree associates the scheme Proj (Sym. M) ; M is a K_∞^o-module of rank 2 ; **[M]**).

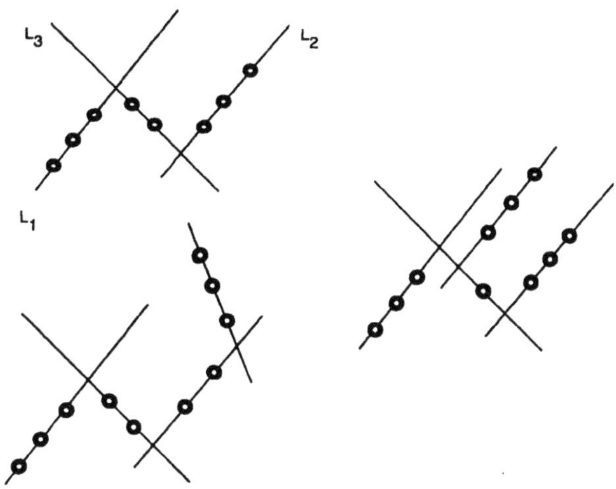

Figure 1 – Pictures with $q = 3$

4. On some quotients of Ω

Now we denote G the intersection graph of $\bar{\Omega}$ and by τ the Bruhat-Tits tree of the group $\mathrm{GL}(2, K_\infty)$ (see § 3).

Consider the group $\mathrm{GL}(2, A)$; it acts on Ω, on the covering $(D_{n,\lambda})_{n,\lambda}$, hence on $\bar{\Omega}$ and on the intersection graph G. Via its action on the set of K_∞^o-lattices in $K_\infty \oplus K_\infty$, it acts also on the Bruhat-Tits tree τ, and the preceding isomorphism, between G and τ, is compatible with these two actions. Take $\Gamma \subset \mathrm{GL}(2, A)$ to be an arithmetic subgroup (for instance $\Gamma \supset \mathrm{Ker}\big(\mathrm{GL}(2, A) \to \mathrm{GL}(2, A/_I)\big)$, where I is an ideal of A). One sees easely that $\Gamma \setminus \Omega$ has a natural rigid analytic structure (quotient of the one of Ω) and a natural covering $\Gamma \setminus (D_{n,\lambda})_{n,\lambda}$, which gives a reduction of $\Gamma \setminus \Omega$ which is $\Gamma \setminus \bar{\Omega}$. On the other hand one knows * that $\Gamma \setminus \Omega = M_\Gamma(C)$, the C-valued points of an affine curve M_Γ (**[Dr1]**, **[D-H]**, **[G2]**). The associated graph of the reduction $\overline{M_\Gamma(C)} := \Gamma \setminus \bar{\Omega}$ is $\Gamma \setminus G \simeq \Gamma \setminus \tau$. This last graph has a finite number of ends and is finite outside the ends (**[S]**).

There exists a canonical compactification of M_Γ, $\widetilde{M_\Gamma}$, so that the difference $\widetilde{M_\Gamma} \setminus M_\Gamma$ is a finite number of points. There exists a bijection between the set of these missing

* It is possible to prove the existence of $M_\Gamma(C)$ by rigid analytic geometry. First we consider an analytic smooth completion of $M_\Gamma(C)$ and then we use the following result (non published, but well-known by the specialists) : every rigid analytic space of dimension 1, connex, proper and non singular is an algebraic curve.

points and the set of ends of $\Gamma \setminus \tau$. One can give here a concrete description of this bijection.

A subgraph of $\Gamma \setminus G = \Gamma \setminus \tau$ of the form •—• corresponds, via the analytic reduction, to an annulus in $\Gamma \setminus \Omega$ (note that $\Gamma \setminus \Omega$ is locally isomorphic to $\mathbb{P}^1_C$); so, an end, which is a subgraph of the form •—•—•—•—•—•—··· corresponding to a sequence $(\mathbf{A}_i)_{i \in \mathbb{N}}$ of annuli in $\Gamma \setminus \Omega$ such that $\mathbf{A}_i \cap \mathbf{A}_{i+1}$ will be a circumference for all i (see Figure 2).

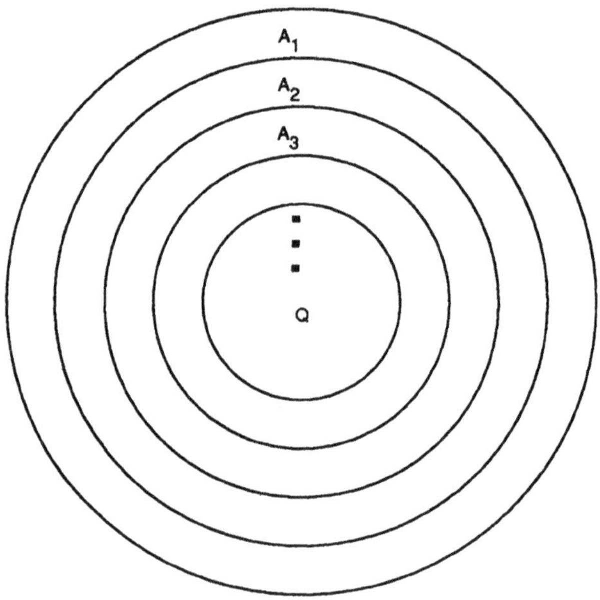

Figure 2

$\bigcup_{j \geq i} \mathbf{A}_i$ is for all i a disk with a missing point (Q in Figure 2) which is the point in $\widetilde{M_\Gamma}(C) - M_\Gamma(C)$ corresponding to the end considered here. Now, let $\mathfrak{U}$ be the covering of $\widetilde{M_\Gamma}(C)$ obtained by replacing for all ends the collection of the $\mathbf{A}_i$ by one disk, which is the union of the $\mathbf{A}_i$ plus the missing point. It is not difficult to see that $\mathfrak{U}$ is a finite pure covering of $\widetilde{M_\Gamma}(C)$, so that $\mathfrak{U}$ gives an analytic reduction $\overline{\widetilde{M_\Gamma}(C)}$ (which, as $\bar{\Omega}$ and $\overline{M_\Gamma(C)}$, are built by $\mathbb{P}'_{\bar{C}}$ with ordinary double points rational over $\mathbb{F}_q$) and that the intersection graph of $\overline{\widetilde{M_\Gamma}(C)}$ is just the finite part of $\Gamma \setminus G$, that is the subgraph of $\Gamma \setminus G$ outside the ends.

In fact, we have more precise information: because of the functoriality of the analytic reduction, we have a canonical, surjective, morphism $\overline{M_\Gamma(C)} \longrightarrow \overline{\widetilde{M_\Gamma}(C)}$ of schemes over $\bar{C}$, which comes from the inclusion $M_\Gamma(C) \subset \widetilde{M_\Gamma}(C)$, and which gives a surjective map between the intersection graph $\Gamma \setminus G \simeq \Gamma \setminus \tau$ of $\overline{M_\Gamma(C)}$ and the intersection graph of $\overline{\widetilde{M_\Gamma}(C)}$. This map consists in effacing the ends.

5. The use of rigid analytic structures and reductions, an example

So, we have described analytic structures and reductions of Ω, $M_\Gamma(C)$, $\widetilde{M_\Gamma}(C)$ and the relations between those of $M_\Gamma(C)$ and $\widetilde{M_\Gamma}(C)$. Much information concerning the geometry of $\widetilde{M_\Gamma}$ can be deduced from this ; here we give just one example. (Recall that by GAGA theorems, **[K]**, many of the algebraic and analytic objects attached to $\widetilde{M_\Gamma}$ are the same or have related properties.)

There exists general methods to study the cohomology of analytic sheaves with the help of reduction (**[G-v]** p.141, **[B]**). For instance, one has the following result:

Let X be an analytic space over C with a finite pure covering $\mathfrak{U}$, which gives a reduction $\bar{X}$. We suppose that $\bar{X}$ is projective (over $\bar{C}$) and denote $\mathcal{O}_X$ the sheaf of holomorphic functions on X, $\mathcal{O}_X^o$ the subsheaf of functions with spectral norms not greater than 1. Then

$$\mathcal{O}_X(X) = C, \mathcal{O}_X^o(C) = C^o$$

$H^1(\mathfrak{U}, \mathcal{O}_X^o)$ is a free C^o-module of finite type and

$$H^1(X, \mathcal{O}_X) \simeq H^1(\mathfrak{U}, \mathcal{O}_X^o) \bigotimes_{C^o} C, H^1(\bar{X}, \mathcal{O}_{\bar{X}}) \simeq H^1(\mathfrak{U}, \mathcal{O}_X^o) \bigotimes_{C^o} \bar{C}$$

($\mathcal{O}_{\bar{X}}$ is the structural sheaf of the algebraic variety $\bar{X}$).

All the preceding hypotheses are true for $\widetilde{M_\Gamma}(C)$, and we have:

$$\begin{aligned}\dim_C H^1\left(\widetilde{M_\Gamma}(C), \mathcal{O}_{\widetilde{M_\Gamma}(C)}\right) &= \text{number of loops in the intersection graph} \\ &\quad\ \text{of the reduction of } \widetilde{M_\Gamma}(C) \\ &= \text{number of loops in } \Gamma \setminus \tau, \text{ where } \tau \text{ is the} \\ &\quad\ \text{Bruhat-Tits tree of } \mathrm{GL}(2, K_\infty).\end{aligned}$$

On the other hand, by GAGA, $H^1(\tilde{M}_\Gamma(C), \mathcal{O}_{\tilde{M}_\Gamma(C)})$ does not depend on whether $\mathcal{O}_{\tilde{M}_\Gamma(C)}$ is the sheaf of holomorphic functions or the sheaf of (algebraic) regular functions. So,we find: genus of $\tilde{M}_\Gamma$ = number of loops in $\Gamma \setminus \tau$. This is well known (see **[Dr1]**, **[D-H]**, **[G1]** and **[G2]**), and comes easily from general properties of rigid geometry.

6. Concluding remarks

We hope that in these few lines, we have motivated the reader to use rigid geometry in the study of Drinfeld modules, in particular of their modular objects. We have treated just some aspects, in dimension one. In the higher dimensional case, the analogues of Ω are (**[Dr1]**, **[D-H]**]) $\Omega_d = \mathbb{P}_C^d -$ rational hyperplanes over K_∞.

There exists on Ω_d a rigid analytic structure and a reduction map $\Omega_d \to \bar{\Omega}_d$ which gives, as in the case $d = 1$, the Bruhat-Tits building of $\mathrm{GL}(d+1, K_\infty)$ (**[Mu]**, **[V]**). Here also, one can deduce many properties of Ω_d and its quotients.

Bibliographie

There exist now three books to study rigid analytic geometry.

L. Gerritzen, M. van der Put, "Schottky groups and Mumford curves", Lecture Notes in Math. 817, Springer-Verlag, 1980.

J. Fresnel, M. van der Put, "Géométrie analytique rigide et applications", Progr. Math. 18, Birkhäuser P M 18, 1981.

S. Bosch, U. Güntzer, R. Remmert "Non-archimedean analysis", Springer-Verlag, 1984.

And, of course, one can also read the main articles of the foundations:

J.Tate, *Rigid analytic spaces*, Private notes, 1962; and Invent. Math. 12, (1971), 257–289.

R. Kiehl, *Der Endlichkeitssatz für eigentliche Abbildungen in der nichtarchimedischen Funktionentheorie*, Invent. Math. 2, (1967), 191–214.

R. Kiehl, *Theorem A und Theorem B in der nichtarchimedischen Funktionentheorie*, Invent. Math. 2, (1967), 256–273.

L. Gerritzen, H. Grauert *Die Azyklität der affinoiden Überdeckungen*, In: Global Analysis "Papers in honor of K. Kodaira", 159–184, University of Tokyo Press, Princeton University Press (1969).

The connexion between rigid geometry and formal schemes is clarified in

M. Raynaud *Géométrie analytique rigide d'après Tate, Kiehl...*, Bull. Soc. Math. France, mémoire 39/40, (1974), 319-327.

The author knows examples of uses of rigid geometry in the theory of Drinfeld modules (of course, this small list is not exhaustive !) in the two papers of V.G. Drinfeld himself (**[Dr1]**, **[Dr2]**, see also the survey of Deligne-Husemoller **[D-H]**) and in some works of E.U. Gekeler (see the Springer Lecture Notes 1231 and its list of references, **[G2]**). A forthcoming paper of E.U. Gekeler and the author uses rigid geometry to study the Jacobians of Drinfeld-modular curves (see also the review in this volume).

References

[B] S. Bosch, Orthonormalbasen in der nichtarchimedischen Funktionentheorie, Manuscripta Math. 1, (1969), 35–57.

[D-H] P. Deligne, D. Husemoller, Survey on Drinfeld modules, Notes IHES, Bures-sur- Yvette, 1977.

[Dr1] V.G. Drinfeld, Elliptic modules, (russian) Mat. Sbornik 94, (1974), 594–627; English translation, Math. USSR-Sb. 23 (1976), 561–592.

[Dr2] V.G. Drinfeld, Elliptic Modules II, (russian) Mat. Sbornik 102, 1972; English translation, Math USSR-Sb. 31, (1977), 159–170.

[G1] E.U. Gekeler, Über Drinfeld'sche Modulkurven vom Hecke typ., Compositio Math. 57, (1986), 219–236.

[G2] E.U. Gekeler, Drinfeld Modular Curves, Springer Lecture Notes in Math. 1231, Springer 1986.

[G-v] L.Gerritzen, M. van der Put, Schottky groups and Mumford curves, Springer Lecture Notes in Math. 817, Springer 1980.

[K] U. Köpf, Über eigentlische Familien algebraischer Varietäten über affinoiden Räumen, Schriftenreihe Math. Inst. Univ. Münster 2. Serie, 7, (1974).

[M] D. Mumford, An analytic construction of degenerating curves over complete local fields, Compositio Math. 24, (1972), 129–174.

[Mu] G.A. Mustafin, Non archimedean uniformization, Math. USSR-Sb. 34, (1978), 187–214.

[S] J.P. Serre, Arbres, amalgames, SL_2, Astérisque 26, Soc. Math. France, 1977.

[V] H. Voskuil, Ultrametric uniformization and symmetric spaces, Thesis, Groningen, Holland, 1990.

Laboratoire de Topologie et Géométrie, Université Toulouse III, URA 1408, 31062 Toulouse Cedex, France
Email: AsV@frcict81.bitnet

Moduli for Drinfeld Modules

E.-U. Gekeler

The goal of the present paper is to describe the set of isomorphism classes of Drinfeld modules of some fixed rank r. We will neither present new results nor even full proofs, but rather try to explain, in down-to-earth terms, the ideas and methods of some groundlaying articles like **[DR]**, **[Dr]**, **[DH]**, and to show the arithmetical significance of the results so obtained. As a heuristical principle, we will use the analogy between Drinfeld modules and elliptic curves (or abelian varieties), the latter being easier accessible, or at least better known, than Drinfeld modules. Thus we shall mimic, in most cases without special mention, some constructions familiar from the theory for elliptic curves. We hope however to convince the reader that, from a moduli point of view, Drinfeld modules are easier to treat than abelian varieties or even elliptic curves, the reason being that Drinfeld modules basically are linear objects.

0. Notation

We will use the following notation:

$\mathbb{F}_q =$ finite field of characteristic p with q elements;
$\tau =$ the endomorphism $(x \mapsto x^q)$ of the additive group scheme $\mathbb{G}_a$ in characteristic p;
$K =$ function field in one variable with $\mathbb{F}_q$ as its field of constants;
$\infty =$ some place of K, fixed once for all, with associated normalized absolute value $|?|$;
$A =$ ring of elements of K regular outside of ∞;
$K_\infty =$ completion of K with respect to $|?|$;
$C =$ completed algebraic closure of K_∞ with respect to the unique extension, also denoted by $|?|$, of $|?|$.

Note that A fully determines the data K, K_∞, C. Divisors of K coprime with ∞ will be identified with (non-zero) ideals of A. In general, we shall not distinguish between "places", and "prime ideals".

"deg" will either denote the degree of a divisor or the degree function on K. For $0 \neq f$ in A, we have $|f| = q^{\deg f} = \#(A/f)$.

The *basic example* will be given by the triple (K, ∞, A), where K is the rational function field $\mathbb{F}_q(T)$, ∞ the usual place at infinity, and A the polynomial ring $\mathbb{F}_q[T]$. Then $\deg f$ is the usual degree of f as a polynomial, and K_∞ turns out to be the field of formal Laurent series $\mathbb{F}_q((T^{-1}))$ in the reciprocal T^{-1} of T.

L always stands for some A-field provided with a structure $\gamma: A \to L$ of A-algebra. The A-characteristic $\mathrm{char}_A(L)$ of L is the kernel of γ, which is either a place of A or 0. In the latter case, one often writes $\mathrm{char}_A(L) = \infty$ or "generic".

The endomorphism ring of the additive group scheme $\mathbb{G}_{a|L}$ is generated over L by the Frobenius morphism $\tau_p = (x \mapsto x^p)$, i.e., is the twisted polynomial ring $L\{\tau_p\}$, where the braces indicate the commutation rule $\tau_p \circ x = x^p \circ \tau_p$ for "constants" x. We let $L\{\tau\}$ be the subalgebra generated by $\tau = (x \mapsto x^q)$, and $\deg_\tau f$ the well-defined degree of a "polynomial" f in τ.

It follows from Krasner's theorem that C is algebraically closed, thus an algebraically closed extension of K complete with respect to $|?|$. In the obvious analogy of

$$\begin{aligned} K &\text{ with the field } \mathbb{Q} \text{ of rational numbers}, \\ A &\text{ with the ring } \mathbb{Z} \text{ of integers}, \\ K_\infty &\text{ with the field } \mathbb{R} \text{ of real numbers}, \\ C &\text{ plays the role of the field of complex numbers}. \end{aligned}$$

Note that in contrast with the classical case, C contains discrete A-lattices of arbitrarily large ranks.

1. The Moduli Problem

(1.1). Recall that a *Drinfeld A-module of rank $r \in \mathbb{N}$ over L* is an A-module structure on the additive group scheme $\mathbb{G}_{a|L}$, given by a ring homomorphism

$$\begin{aligned} \Phi: A &\to L\{\tau\} \subset \mathrm{End}_L(\mathbb{G}_a) \\ f &\mapsto \Phi_f \end{aligned}$$

such that the following holds: Given $f \in A$, write Φ_f in the form

$$\Phi_f = \sum_{0 \le i \le N} a_i \tau^i \qquad a_N \neq 0, \ N = N(f). \tag{1.1.1}$$

Then for any f,

$$\text{(i) } a_0 = \gamma(f) \qquad \text{and} \qquad \text{(ii) } N = r \cdot \deg f, \text{ if } f \neq 0. \tag{1.1.2}$$

A *morphism* $u: \Phi \to \Phi'$ of Drinfeld modules over L is some element u of $L\{\tau\}$ such that for f in A,

$$u \circ \Phi_f = \Phi'_f \circ u \tag{1.1.3}$$

holds. We write symbolically $u \circ \Phi = \Phi' \circ u$. Correspondingly, one defines endo-, iso-, and automorphisms of Drinfeld modules.

(1.2) Basic example. Let A be the polynomial ring $\mathbb{F}_q[T]$. Then some Φ of rank r is given by the operator

$$\Phi_T = \gamma(T) + \sum_{1 \le i \le r} g_i \tau^i, \tag{1.2.1}$$

subject only to $g_r \neq 0$. Two such, Φ and Φ', are isomorphic if and only if there exists some invertible element $u \in L\{\tau\}$, i.e., $u \in L^*$, such that $u \circ \Phi = \Phi' \circ u$. This says that for $1 \leq i \leq r$,

$$u^{q^i-1} g_i' = g_i . \qquad (1.2.2)$$

If the rank r equals two and L is algebraically closed, it is easy to see that Φ and Φ' are isomorphic if and only if $j(\Phi) = j(\Phi')$, where the j-invariant is defined as

$$j(\Phi) = g_1^{q+1}/g_2 . \qquad (1.2.3)$$

(1.3). We define $\mathcal{M}^r(1)(L)$ to be the set of isomorphism classes of rank-r Drinfeld A-modules over L. It follows from the above that $\mathcal{M}^r(1)(L)$ corresponds to the set of points $(g_1 : \ldots : g_r)$ with $g_r \neq 0$ of the weighted projective space $\operatorname{Proj} L[g_1, \ldots, g_r]$, provided that $A = \mathbb{F}_q[T]$ and L is algebraically closed. Here, the weight of g_i is $q^i - 1$, and the Proj equals the quotient of $\mathbb{A}^r \backslash \{0\}$ by the action of the multiplicative group $\mathbb{G}_{m|L}$ described in (1.2.2). For $r = 2$, we get

$$\begin{array}{ccccc} \mathcal{M}^2(1)(L) & \xrightarrow{\cong} & \{x \in \operatorname{Proj} L[g_1, g_2] \mid g_2 \neq 0\} & \xrightarrow{\cong} & \mathbb{A}^1(L) \\ \Phi & \mapsto & (g_1 : g_2) & \mapsto & j(\Phi) = g_1^{q+1}/g_2 . \end{array} \qquad (1.3.1)$$

In order to describe $\mathcal{M}^r(1)(L)$ in the general case, we have to take two difficulties into account:

(a) A might be more complicated than a polynomial ring. Even writing down one single Drinfeld module Φ will then become a highly non-trivial problem (see **[Ha2]**, **[Ha3]**). If $\{f\}$ is a system of generators of A over $\mathbb{F}_q$, Φ is given by the coefficients of the Φ_f. These satisfy a complicated system of equations, whose set of solutions in general is not accessible by elementary methods.

(b) If L fails to be algebraically closed, Drinfeld modules may become isomorphic over the algebraic closure without being so over L. Hence a problem of "L-Forms" arises.

(1.4). It turns out that for a systematic approach, we have to generalize our problem and deal with Drinfeld modules over arbitrary A-schemes S (instead of $S = \operatorname{Spec} L$ only). So let S be an A-scheme. Intuitively, a Drinfeld module over S is a continuous family of Drinfeld modules over the residue class fields of S. The precise definition is as follows:

A *Drinfeld module* $(\mathcal{L}, \Phi)$ of rank r over S consists of a line bundle $\mathcal{L}$ over S and a ring homomorphism

$$\begin{array}{ccc} \Phi : A & \longrightarrow & \operatorname{End}_S(\mathcal{L}, +) \\ f & \longmapsto & \Phi_f \end{array}$$

into the endomorphism ring of the additive group scheme $(\mathcal{L}, +)$ underlying $\mathcal{L}$, subject to the following condition:

There exists a trivialization of $\mathcal{L}$ by open affine subschemes $\operatorname{Spec} B$ of S, such that over each piece, $\Phi_f|\operatorname{Spec} B$ has the form

$$\Phi_f|\operatorname{Spec} B = \sum_{0 \le i \le N(f)} a_i \tau^i, \tag{1.4.1}$$

with $a_i \in B$ and

$$\begin{aligned} &\text{(i) } a_0 = \text{image of } f \text{ under the canonical morphism from } A \text{ to } B; \\ &\text{(ii) } N(f) = r \cdot \deg f, \text{ and the highest coefficient } a_{N(f)} \text{ is a unit}. \end{aligned} \tag{1.4.2}$$

We briefly write "Φ" for the object $(\mathcal{L}, \Phi)$. For $S = \operatorname{Spec} L$, we find the previous definition. Morphisms are morphisms of the corresponding group schemes (*but not:* morphisms of line bundles) compatible with the A-actions. We define $\mathcal{M}^r(1)(S)$ as the set of isomorphism classes of rank-r Drinfeld A-modules over S. Since a Drinfeld module over S may be pulled back to any S-scheme $S' \to S$, $\mathcal{M}^r(1)(?)$ is a contravariant set-valued functor on the category Sch_A of A-schemes.

(1.5). We would like to find some scheme $M^r(1)$ whose set $M^r(1)(S)$ of S-valued points ($=$ morphisms $S \to M^r(1)$ of A-schemes) corresponds bijectively and functorially to $\mathcal{M}^r(1)(S)$. More generally, let M be an A-scheme, and define the contravariant functor

$$\underline{M}\colon \operatorname{Sch}_A \to \operatorname{Sets}$$

by

$$S \mapsto \underline{M}(S) = \operatorname{Hom}_A(S, M) = M(S).$$

By Yoneda's lemma, $\underline{M}$ determines M, since $M \mapsto \underline{M}$ is a fully faithful embedding of Sch_A into the category $\operatorname{Hom}(\operatorname{Sch}_A, \operatorname{Sets})$ of contravariant functors. Functors of type $\underline{M}$ are called *representable.*

(1.6). Suppose that $\mathcal{M}^r(1)$ was representable by the A-scheme $M^r(1)$. Then to the identity morphism of $M^r(1)$ there corresponds a universal object, i.e., a rank-r Drinfeld module Φ over $M^r(1)$ such that any rank-r Drinfeld module over any A-scheme S is the pull-back of Φ through a unique morphism from S to $M^r(1)$. This implies, however, that the objects classified by $\mathcal{M}^r(1)$ ($=$ rank-r Drinfeld modules) have no non-trivial automorphisms. But the multiplicative group $\mathbb{F}_q^*$ is contained in $\operatorname{Aut}(\Phi)$ for any Drinfeld module Φ. So we cannot expect $\mathcal{M}^r(1)$ to be representable. In order to overcome this obstacle and to obtain a representable moduli problem, we shall "rigidify" Drinfeld modules by imposing level structures that reduce to 1 the automorphism group of the resulting objects.

(1.7). Let $\Phi = (\mathcal{L}, \Phi)$ be defined over S, and let $0 \neq f \in A$. The scheme ${}_f\Phi$ of *f-division points* (or f-torsion points) of Φ is the subscheme $\ker(\Phi_f)$ of $(\mathcal{L}, +)$. For a not necessarily principal ideal $\mathfrak{n}$ of A, we set ${}_\mathfrak{n}\Phi = \bigcap {}_f\Phi$ $(f \in \mathfrak{n})$. It is an easy consequence of (1.4.2) that ${}_\mathfrak{n}\Phi$ is a flat and finite group scheme of degree $\#(A/\mathfrak{n})^r$ over S. It is étale (i.e., reduced) outside of the support ($=$ inverse image in S of the primes that divide $\mathfrak{n}$) of $\mathfrak{n}$. If S is the spectrum of an algebraically closed field L, as a discrete

A-module,

$$_{\mathfrak{n}}\Phi(L) = \{x \in L \mid \Phi_f(x) = 0 \text{ for all } f \in \mathfrak{n}\}$$

is isomorphic with

$$\begin{aligned} &(A/\mathfrak{n})^r, \text{ if } \mathrm{char}_A(L) \text{ does not divide } \mathfrak{n}, \\ &(A/\mathfrak{n})^s \text{ for some } s < r, \text{ if } \mathrm{char}_A(L) \text{ divides } \mathfrak{n}. \end{aligned} \tag{1.7.1}$$

Intuitively, a level-$\mathfrak{n}$ structure on Φ should be the choice of an isomorphism of $(A/\mathfrak{n})^r$ with $_{\mathfrak{n}}\Phi$. This works well outside of the support of $\mathfrak{n}$ (in that case, $_{\mathfrak{n}}\Phi$ is locally isomorphic with $(A/\mathfrak{n})^r$ in the étale topology), but fails if $\mathfrak{n}$ is divisible by a residue characteristic of S. The "correct" definition is as follows:

(1.7.2). A *level-$\mathfrak{n}$-structure* on Φ is the choice of a morphism α of the constant scheme of A-modules $(\mathfrak{n}^{-1}/A)^r$ to $_{\mathfrak{n}}\Phi$ that identifies the sum $\sum \alpha(i)$ $(i \in (n^{-1}/A)^r)$ with $_{\mathfrak{n}}(\Phi)$ as Cartier divisors on $\mathcal{L}$.

There is an obvious notion of morphism of objects (Φ, α), and it turns out that the automorphism group of (Φ, α) is trivial if $\mathfrak{n}$ is divisible by at least two different primes. Call such divisors $\mathfrak{n}$ *admissible*. We set

$$\mathcal{M}^r(\mathfrak{n})(S) = \left\{ \begin{array}{l} \text{set of isomorphism classes of Drinfeld modules } (\Phi, \alpha) \\ \text{of rank } r \text{ over } S \text{ provided with a structure } \alpha \text{ of level } \mathfrak{n}. \end{array} \right. \tag{1.7.3}$$

Then $\mathcal{M}^r(n)$ is a contravariant functor from Sch_A to Sets that under certain circumstances will be representable. For $\mathfrak{n} = (f)$ principal, we write $\mathcal{M}^r(f) = \mathcal{M}^r(\mathfrak{n})$, which is compatible with the notation of (1.5).

(1.8) Theorem (Drinfeld [Dr, Sect. 4/5]). *Let $\mathfrak{n}$ be an admissible divisor, i.e., an ideal of A divisible by at least two different primes. Then*
(i) *The functor $\mathcal{M}^r(\mathfrak{n})$ is representable by an A-scheme $M^r(\mathfrak{n})$.*
(ii) *$M^r(\mathfrak{n})$ is affine, smooth, of finite type and dimension $r-1$ over A. (So its global dimension is $r - 1 + \dim A = r$.)*
(iii) *If $\mathfrak{m}$ is an ideal contained in $\mathfrak{n}$, the forgetful morphism from $M^r(\mathfrak{m})$ to $M^r(\mathfrak{n})$ is finite and flat, and even étale outside of the support of $\mathfrak{m}\mathfrak{n}^{-1}$.*

The existence of $M^r(\mathfrak{n})$ is not difficult to prove: It may be written down as the spectrum of a certain A-algebra given explicitly through generators and relations. The hard part is (ii), notably the smoothness, which results from a careful analysis **[Dr]** of the deformation properties of Drinfeld modules.

(1.9). We have need for another set of definitions. Let

$$\begin{aligned} \mathcal{A} &= \mathcal{A}_f \times K_\infty \text{ be the adèle ring of } K, \text{ where} \\ \mathcal{A}_f &= \text{ring of adèles without } \infty - \text{component, and} \\ \widehat{A} &= \varprojlim A/\mathfrak{n} \ (\mathfrak{n} \text{ running through the ideals of } A), \end{aligned}$$

which is a compact open subring of $\mathcal{A}_f$.

We further let G be the group scheme $GL(r)$ with center Z, and we define the congruence subgroup $G(\widehat{A}, \mathfrak{n})$ by the exactness of the sequence

$$1 \to G(\widehat{A}, \mathfrak{n}) \to G(\widehat{A}) \to G(A/\mathfrak{n}) \to 1.$$

There is a canonical action of $G(\mathcal{A}_f)/Z(K)$ on the scheme

$$M^r = \varprojlim M^r(\mathfrak{n}), \tag{1.9.1}$$

which we briefly describe. First note that each element of $G(\mathcal{A}_f)$ may be written as the product of an element of $Z(K)$ and an element of

$$\mathrm{W} = \text{semi-group of matrices in } G(\mathcal{A}_f) \text{ with entries in } \widehat{A}.$$

Now the points of M^r "are" Drinfeld modules of rank r with a system of compatible level structures for all levels $\mathfrak{n}$. W acts through matrix multiplication on the points of M^r, thus on M^r itself. Since

$$Z(K) \cap W = Z(\mathbb{F}_q) \cong \mathbb{F}_q^*,$$

which is contained in the automorphism group of any Drinfeld module, $Z(K) \cap W$ acts trivially, and there exists a unique extension to an action of $G(\mathcal{A}_f)/Z(K)$ on M^r.

(1.10). For any *admissible* divisor $\mathfrak{n}$, we have canonically

$$G(\widehat{A}, \mathfrak{n}) \backslash M^r \xrightarrow{\cong} M^r(\mathfrak{n}).$$

For arbitrary ideals $\mathfrak{n}$ of A, including the trivial ideal $(1) = A$, we *define* the scheme $M^r(\mathfrak{n})$ as the quotient of M^r by $G(\widehat{A}, \mathfrak{n})$. By (1.8), there exists an isomorphism of functors (see (1.5))

$$\varphi_{\mathfrak{n}}: \mathcal{M}^r(\mathfrak{n}) \xrightarrow{\cong} \underline{M}^r(\mathfrak{n}), \tag{1.10.1}$$

i.e., for any A-scheme S a bijection $\varphi_{\mathfrak{n},S}$, functorial in S, from $\mathcal{M}^r(\mathfrak{n})(S)$ to

$$M^r(\mathfrak{n})(S) = G(\widehat{A}, \mathfrak{n}) \backslash (M^r(S)),$$

provided that $\mathfrak{n}$ is admissible. If $\mathfrak{n}$ is not admissible, choose $\mathfrak{m}$ admissible and divisible by $\mathfrak{n}$. Then $\varphi_{\mathfrak{m}}$ descends to a unique morphism

$$\varphi_{\mathfrak{n}}: \mathcal{M}^r(\mathfrak{n}) \to \underline{M}^r(\mathfrak{n})$$

that fits into the obvious commutative diagram and is independent of the choice of $\mathfrak{m}$. It is an easy exercise to verify that

(i) $\varphi_{\mathfrak{n}}$ is universal for morphisms of $\mathcal{M}^r(\mathfrak{n})$ to representable functors, i.e., any morphism from $\mathcal{M}^r(\mathfrak{n})$ to (the functor represented by) an A-scheme factors through $\varphi_{\mathfrak{n}}$; (1.10.2)

and

(ii) $\varphi_{\mathfrak{n},S}$ is bijective, if S is the spectrum of an algebraically closed field.

The above conditions state the $M^r(\mathfrak{n})$ is "the best possible scheme" that may be constructed from the functor $\mathcal{M}^r(\mathfrak{n})$; it is called the *coarse modular scheme* for the moduli problem

" Drinfeld modules of rank $r+$ level-$\mathfrak{n}$ structures".

In (1.3), we have described $M^r(1)$ in the special case where A is the polynomial ring $\mathbb{F}_q[T]$.

Finally, for an arbitrary open subgroup $\mathcal{K}$ of $G(\widehat{A})$, we set

$$M^r_{\mathcal{K}} = \mathcal{K}\backslash M^r. \tag{1.10.3}$$

Clearly, $M^r_{\mathcal{K}} = M^r(\mathfrak{n})$ if $\mathcal{K} = \mathcal{K}(\mathfrak{n}) = G(\widehat{A}, \mathfrak{n})$. In general, $M^r_{\mathcal{K}}$ is a coarse modular scheme (the assertions analogous with (1.10.2) hold) for a moduli problem that depends on $\mathcal{K}$. The coarse modular scheme is a *fine modular scheme* (i.e., it represents the corresponding functor) if and only if over algebraically closed fields, the associated objects have no non-trivial automorphisms.

(1.11) Example. Let $\mathcal{K}$ be the *Hecke congruence subgroup*

$$\mathcal{K}_0(\mathfrak{n}) = \left\{ \begin{pmatrix} a & b \\ c & d \end{pmatrix} \in G(\widehat{A}) \mid c \equiv 0 \bmod \mathfrak{n} \right\}$$

of $GL(2, \widehat{A})$. It corresponds to a certain moduli problem

"Drinfeld modules of rank $2+$ level-$\mathfrak{n}$ Hecke structures".

For reasons discussed in (1.7), it is somewhat technical to write down the precise definitions of what a level-$\mathfrak{n}$ Hecke structure on a rank-2 Drinfeld module Φ should be. If Φ is defined over an A-scheme S whose residue characteristics are coprime with $\mathfrak{n}$, it is simply the choice of an $A/\mathfrak{n}$-submodule scheme of ${}_{\mathfrak{n}}\Phi$ isomorphic with $A/\mathfrak{n}$. The *Hecke modular scheme* $M^2_0(\mathfrak{n}) = M^2_{\mathcal{K}}$ ($\mathcal{K}$ as above) is especially important for arithmetical reasons (**[Ja-La]**, **[Ma]**, **[Ge2]**, **[Ge3]**, **[Ge8]**). It is not a fine modular scheme, but may be studied geometrically through its covers by the fine modular schemes associated with $\mathcal{K} = \mathcal{K}_0(\mathfrak{n}) \cap \mathcal{K}(\mathfrak{m})$, where $\mathfrak{m}$ is an admissible divisor coprime with $\mathfrak{n}$ **[DR]**.

2. Analytical Description of Modular Schemes

(2.1). Let us look, for a moment, to the elliptic curve situation. The Weierstraß parameterization of complex elliptic curves yields a description of the complex points of the elliptic modular scheme $M^{\text{ell}}(1)$. Namely, to a lattice $\Lambda = \mathbb{Z}\omega_1 + \mathbb{Z}\omega_2$ in $\mathbb{C}$, one first associates its Weierstraß function (see e.g. **[La]** for the following)

$$\wp(z) = z^{-2} + \sum_{0 \neq \lambda \in \Lambda} \left((z-\lambda)^{-2} - \lambda^{-2}\right). \tag{2.1.1}$$

From its well-known functional equation

$$\wp'(z)^2 = 4\wp(z)^3 - g_2\wp(z) - g_3, \tag{2.1.2}$$

one obtains the isomorphism of Riemann surfaces

$$\begin{aligned}\mathbb{C}/\Lambda &\xrightarrow{\cong} E(\mathbb{C}),\\ z &\mapsto (\mathfrak{p}(z), \mathfrak{p}'(z))\end{aligned}$$

where E is the elliptic curve with affine equation

$$Y^2 = 4X^3 - g_2X - g_3.$$

The coefficients are the so-called *Eisenstein series*

$$\begin{aligned} g_i(\Lambda) &= c_i \cdot G_{2i}(\Lambda) \\ G_{2i}(\Lambda) &= \sum_{0 \neq \lambda \in \Lambda} \lambda^{-2i} \qquad (i = 2, 3), \end{aligned} \tag{2.1.3}$$

with $c_2 = 60$ and $c_3 = 140$. The discriminant

$$\Delta = g_2^3 - 27g_3^2 \tag{2.1.4}$$

never vanishes. Each elliptic curve over $\mathbb{C}$ is obtained this way, and two of them are isomorphic if and only if their j-invariants

$$j(E) = g_2^3/\Delta \tag{2.1.5}$$

are equal, if and only if the associated lattices are similar (i.e., agree up to a scalar factor). There results the chain of canonical bijections (in fact, isomorphisms of complex manifolds)

$$\begin{aligned} SL(2,\mathbb{Z})\backslash H &\xrightarrow{\cong} \left\{ \begin{array}{c} \text{similarity} \\ \text{classes of} \\ \text{lattices in } \mathbb{C} \end{array} \right\} \xrightarrow{\cong} M^{\text{ell}}(1)(\mathbb{C}) \xrightarrow{\cong} \mathbb{A}^1(\mathbb{C}) = \mathbb{C}. \\ \omega &\mapsto \text{class of } \mathbb{Z}\omega + \mathbb{Z} \mapsto \begin{array}{c}\text{class of} \\ E = \mathbb{C}/(\mathbb{Z}\omega + \mathbb{Z})\end{array} \mapsto j(\omega) = j(E). \end{aligned} \tag{2.1.6}$$

Here, the modular group $SL(2,\mathbb{Z})$ acts through fractional linear transformations

$$\begin{pmatrix} a & b \\ c & d \end{pmatrix} \omega = (a\omega + b)/(c\omega + d)$$

on the complex upper half-plane H. Attaching to $\omega \in H$ the values on the lattice $\mathbb{Z}\omega + \mathbb{Z}$ of the functions g_2, g_3, Δ, j, etc., we may regard these as function on H. They are modular forms of weights $4, 6, 12, 0, \ldots$ respectively, and enjoy a number of famous well-known (but less understood) properties. Similar descriptions exist for the elliptic modular schemes "with level structures".

(2.2). The precise form of equations (2.1.1)-(2.1.5) is not important for our purposes. But let us abstract the essential facts:

- With a lattice Λ, one can associate a Λ-periodic function, $\mathfrak{p}_\Lambda$:
- $\mathfrak{p}_\Lambda$ satisfies an algebraic functional equation that supplies $\mathbb{C}/\Lambda$ with a certain algebraic structure (elliptic curve = abelian variety of dimension one). That latter object makes sense algebraically, without reference to the way it has been constructed;

- We have a bijection
  $\{\text{classes of lattices}\} \xrightarrow{\cong} \{\text{classes of elliptic curves over } \mathbb{C}\}$;
- Lattices, thus modular schemes may be described through modular forms.

We will see that the same feature applies to Drinfeld modules.

(2.3). Next, we briefly resume the "Weierstraß parametrization" for Drinfeld modules (see **[Dr]**, **[DH]**). An *A-lattice of rank r* in C (an *r-lattice* for short) is a finitely generated (thus projective) A-submodule Λ of C of projective rank r that satisfies the discreteness condition:

$$\Lambda \text{ intersected with each bounded subset of } C \text{ is finite.} \tag{2.3.1}$$

With Λ, we associate its *exponential* (or *lattice*) *function*

$$e_\Lambda(z) = z\prod{}'(1 - z/\lambda), \tag{2.3.2}$$

where the primed product is always extended over the non-zero points of Λ. The product converges and defines a surjective, entire and $\mathbb{F}_q$-linear function $e_\Lambda\colon C \to C$. For $n \in A$, e_Λ satisfies an equation

$$e_\Lambda(nz) = \Phi_n^\Lambda(e_\Lambda(z)) \tag{2.3.3}$$

with some operator $\Phi_n^\Lambda \in C\{\tau\}$. The map $n \mapsto \Phi_n^\Lambda$ is a ring homomorphism from A to $C\{\tau\}$ and, in fact, a Drinfeld module Φ^Λ of rank r. Moreover, each rank-r Drinfeld module Φ over C arises as Φ^Λ from a uniquely determined r-lattice Λ, and Φ^Λ, $\Phi^{\Lambda'}$ are isomorphic as Drinfeld modules if and only if Λ and Λ' are similar. We therefore have the set-theoretical description

$$\begin{aligned}\{\text{similarity classes of } r\text{-lattices in } C\} &\xrightarrow{\cong} M^r(1)(C) \\ \Lambda &\mapsto \Phi^\Lambda.\end{aligned} \tag{2.3.4}$$

(2.4). Let Y be a projective A-module of rank r with automorphism group Γ_Y. When expressed with respect to a K-basis $y_1, \ldots, y_r$ of $Y \otimes K$, Γ_Y is a subgroup of $GL(r, K)$ commensurable with $GL(r, A)$.

An injection $i\colon K \otimes Y \hookrightarrow C$ maps Y to an A-lattice Λ (i.e., (2.3.1) is satisfied) if and only if $y_1, \ldots, y_r$ are mapped to K_∞-independent points of C. Such an injection is given by a vector $\tilde{\omega} = (\omega_1, \ldots, \omega_r)$ in C^r which is contained in no hyperplane defined over K_∞. Two points $\tilde{\omega}$, $\tilde{\omega}'$ give rise to the same lattice if and only if they satisfy $\tilde{\omega}' = \gamma\tilde{\omega}$ with some $\gamma \in \Gamma_Y$. (Here we assume that Γ_Y acts as a matrix group from the right on Y and from the left on C^r, see **[Ge4**, pp. 10–12**]** for more details.) Therefore, letting

$$\tilde{\Omega}^r = C^r \text{ minus the hyperplanes defined over } K_\infty,$$

we get

$$\Gamma_Y \backslash \tilde{\Omega}^r \xrightarrow{\cong} \{A\text{-lattices isomorphic with } Y\}. \tag{2.4.1}$$

Put $\Omega^r = \widetilde{\Omega}^r / C^*$, which is a subset of projective $r-1$-space over C. Dividing out the action of C^*, we arrive at the bijection

$$\dot{\bigcup_{Y \in \mathcal{P}^r_A}} \Gamma_Y \backslash \Omega^r \overset{\cong}{\to} \left\{ \begin{matrix} \text{similarity classes of} \\ \text{rank}-r\ A-\text{lattices in } C \end{matrix} \right\}, \tag{2.4.2}$$

where Y runs through a system of representatives for the (finite) set $\mathcal{P}^r_A$ of isomorphism classes of projective A-modules of rank r. Combined with (2.3.4), this yields

$$\dot{\bigcup_{Y \in \mathcal{P}^r_A}} \Gamma_Y \backslash \Omega^r \overset{\cong}{\to} M^r(1)(C) \tag{2.4.3}$$

canonically. Note that Ω^r has a natural structure of rigid analytic space over C. (See **[Re]**. In fact, it is a manifold of dimension $r-1$ even defined over K_∞.) Now all the groups Γ_Y act on Ω^r with finite stabilizers, so the quotients exist as (possibly singular) analytic spaces, and (2.4.3) is biholomorphic **[Dr]**. Corresponding descriptions exist for the C-valued points of the modular schemes $M^r(\mathfrak{n})$ with arbitrary level-$\mathfrak{n}$-structures. For details, including the canonical adèlic description, we refer to **[Ge4]**.

(2.5) Example. Let again A be the polynomial ring $\mathbb{F}_q[T]$. Since every finitely generated projective A-module is free, we may suppose without restriction that $Y = A^r$ and $\Gamma = \Gamma_Y = GL(r, A)$. We then get

$$GL(r, A) \backslash \Omega^r \overset{\cong}{\to} M^r(1)(C). \tag{2.5.1}$$

Recall that we have described $M^r(1)$ in (1.3) as an open subvariety in a certain weighted projective space. The coordinates of the corresponding mapping from Ω^r to that space are given by modular forms.

Assume further that $r = 2$. This is the case closest to the one discussed in (2.1). Here,

$$\Omega := \Omega^2 = \mathbb{P}^1(C) \setminus \mathbb{P}^1(K_\infty) = C \setminus K_\infty, \tag{2.5.2}$$

upon which $GL(2, K_\infty)$ acts through fractional linear transformations. For $\omega \in C$ not contained in K_∞, let Λ be the 2-lattice $A\omega + A$, $\Phi = \Phi^\Lambda$ the corresponding Drinfeld module, and write

$$\Phi_T = T + g(\omega)\tau + \Delta(\omega)\tau^2. \tag{2.5.3}$$

g and Δ are functions on Ω that behave nicely under $\Gamma = GL(2, A)$; they are modular forms of weight $q-1$ and q^2-1, respectively. Now the j-invariant $j(\omega) = j(\Lambda) = g(\omega)^{q+1}/\Delta(\omega)$ defines a map from Ω to C that induces an isomorphism of analytic spaces

$$\begin{aligned} j: \Gamma \backslash \Omega &\overset{\cong}{\to} M^2(1)(C) = \mathbb{A}^1(C) = C \\ \omega &\mapsto j(\omega). \end{aligned} \tag{2.5.4}$$

It turns out (see below) that functions like g, Δ and their generalizations can be expressed through easily accessible series in ω, also called Eisenstein series. They may be studied both from a function theoretic and an arithmetic point of view **[Go1]**, **[Go2]**,

[Ge6]. As in the classical case sketched in (2.1), the knowledge so obtained on the modular schemes will provide information on number theoretical properties of K and A.

3. Drinfeld Modular Forms

(3.1). Following Katz' approach **[Ka]**, **[KM]**, Drinfeld modular forms for arbitrary ranks and levels may be defined in a purely algebraical fashion **[Go1]**. However, since most of the features relevant for the present article are most easily explained analytically, we will take the naive point of view and regard modular forms simply as functions on the "symmetric spaces" Ω^r which behave well under certain group actions. Moreover, for fixing the ideas, we restrict to the case where the rank r equals two, which corresponds to classical elliptic modular forms. So we will be dealing with functions on the "upper half plane" $\Omega = \Omega^2 = C \backslash K_\infty$. Some hints for the higher rank case may be found in **[Ge5]** and **[Kap]**.

(3.2). As usual, we need some definitions. In the whole section, G denotes the group scheme $GL(2)$. Let Y be an A-lattice in K^2, i.e., Y is a projective submodule with $Y \otimes K = K^2$. Necessarily, Y has rank two. A *congruence subgroup* of Γ_Y (see (2.4)) is a subgroup that contains the kernel of the "reduction mod $\mathfrak{n}$" mapping for some ideal $\mathfrak{n}$ of A. An *arithmetic subgroup* Γ of $G(K)$ will be a congruence subgroup of some group Γ_Y.

(3.3). Let k be a non-negative integer and m a class mod $q-1$. A *modular form of weight k and type m* for the arithmetic group Γ is a function $f: \Omega \to C$ that satisfies

$$f\left(\frac{az+b}{cz+d}\right) = (\det \gamma)^{-m}(cz+d)^k f(z) \qquad \left(\gamma = \begin{pmatrix} a & b \\ c & d \end{pmatrix} \in \Gamma\right); \quad (3.3.1)$$

$$f \text{ is holomorphic in the rigid analytic sense;} \quad (3.3.2)$$

$$f \text{ is holomorphic at the cusps of } \Gamma. \quad (3.3.3)$$

(3.4). We briefly explain the meaning of the last condition. More details (and proofs of the statements to follow) may be found in **[Ge4**, pp. 45/46**]**. As for classical modular curves, the "open Riemann surface" $\Gamma \backslash \Omega$ is smoothly compactified by adjoining a finite number of "cusps". The result equals the space $\overline{M}_\Gamma(C)$ of C-points of a smooth projective algebraic curve $\overline{M}_\Gamma$ over C. $\overline{M}_\Gamma$ is the algebraic compactification of a component of $M_{\mathcal{K}} \times C$ for some $\mathcal{K}$ as in (1.10.3). The set

$$\operatorname{cusp}(\Gamma) = \overline{M}_\Gamma(C) \setminus M_\Gamma(C)$$

of *cusps* of Γ is in canonical bijection with the set of orbits $\Gamma \backslash \mathbb{P}^1(K)$ of Γ on $\mathbb{P}^1(K)$.

Let s_∞ be the cusp that corresponds to ∞, and let Γ_∞^u be the unipotent part of the intersection of Γ with the stabilizer in $G(K)$ of ∞. Γ_∞^u consists of the matrices $\begin{pmatrix} 1 & b \\ 0 & 1 \end{pmatrix}$, where b runs through a certain subgroup H of K commensurable with A.

If e.g. Γ is a full congruence subgroup, H will be a fractional ideal of A. Now set

$$e_H(z) = z \prod_{b\in H}{}'(1 - z/b), \tag{3.4.1}$$

whose properties are similar to those described in (2.3.2). It is the reciprocal

$$t_H(z) = e_H^{-1}(z) \tag{3.4.2}$$

that will serve as the substitute of classical $q(z) = \exp(2\pi i z)$, i.e., as a uniformizer at the cusp s_∞. If the function f on Ω satisfies (3.3.1), it is invariant under shifts $z \mapsto z + b$ with $b \in H$, and may therefore be expanded in a Laurent series in $t_H(z)$. Now the holomorphy of f at s_∞ means: f has a Laurent expansion (convergent for $|t_H|$ small) without polar terms

$$f(z) = \sum_{i\geq 0} a_i t_H^i(z). \tag{3.4.3}$$

This also makes clear how to define the vanishing order at s_∞ of f. Holomorphy properties of f at the other cusps are defined similarly.

(3.5). We set $M_{k,m}(\Gamma)$ for the C-vector space of modular forms of weight k and type m for Γ. As in the elliptic modular case, there is a close connection to multi-differentials on the curve $\overline{M}_\Gamma$. Namely, the identity

$$\frac{d(\gamma z)}{dz} = \frac{\det \gamma}{(cz+d)^2}, \qquad \gamma = \begin{pmatrix} a & b \\ c & d \end{pmatrix}, \tag{3.5.1}$$

implies that the differential $f(z)dz$ on Ω is Γ-invariant if f is in $M_{2,1}(\Gamma)$. More generally, the multi-differential $f(z)(dz)^k$ is Γ-invariant if f is a form of weight $2k$ and type k for Γ. Now since $e_H(z)$ has constant derivative $de_H(z)/dz = 1$, $dt_H(z)/dz = -t_H^2$, i.e.,

$$dz = -t_H^2(z)dt_H, \tag{3.5.2}$$

which means that f must vanish of order $2k$ at a cusp s in order that $f(z)(dz)^k$ be holomorphic at s. Let $M_{k,m}^i(\Gamma) \subset M_{k,m}(\Gamma)$ be the subspace of forms vanishing of order i at all the cusps. We obtain the isomorphism $f \mapsto f(z)(dz)^k$ of $M_{2k,k}^{2k}(\Gamma)$ with the space of k-differentials of $\overline{M}_\Gamma$. As is easy to see, this implies that all the spaces $M_{k,m}(\Gamma)$ are finite dimensional.

(3.6) Examples. In what follows, let Γ be the arithmetic group $GL(2, A)$ and $\Gamma(\mathfrak{n})$ its $\mathfrak{n}$-th principal congruence subgroup ($\mathfrak{n} \subsetneq A$ an ideal). Let k be a non-negative integer and

$$E^{(k)}(z) = \sum_{(a,b)\in A^2}{}'(az+b)^{-k}. \tag{3.6.1}$$

(Recall that primed sums or products are always over the non-zero elements of a group.)

The sum converges for $z \in \Omega$ and defines a modular form of weight k, type 0 for Γ **[Go1]**, the *Eisenstein series of weight* k of Γ. It is non-zero if and only if

$k \equiv 0 \bmod q-1$. Accordingly, let u, v be elements of A, and

$$E_{u,v}^{(k)}(z) = \sum{}'(az+b)^{-k}, \tag{3.6.2}$$

the sum now extended over pairs $(a,b) \equiv (u,v) \bmod \mathfrak{n}$ only. Then $E_{u,v}^{(k)}$ is modular of weight k for $\Gamma(\mathfrak{n})$. (Here we omit the type since $\Gamma(\mathfrak{n})$ is contained in $SL(2,K)$.) Considering suitable congruence conditions, we may tailor Eisenstein series for any given arithmetic subgroup of $G(K)$.

Next, let $f \in A$, and write

$$\Phi_f = f + a_1(z)\tau + \cdots + a_N(z)\tau^N \tag{3.6.3}$$

for the Drinfeld module $\Phi = \Phi^\Lambda$ attached to the rank-two lattice $\Lambda = Az + A$ in C. The *coefficient form* $z \mapsto a_i(z)$ is a modular form of weight $q^i - 1$, type 0 for Γ. These forms generalize the forms g and Δ of (2.5.3). The highest coefficient $a_N(z)$, ($N = 2 \cdot \deg f$) is a cusp form: It vanishes at the cusps, but is non-zero on all of Ω.

(3.6.4). Let x, y be elements of $\mathfrak{n}^{-1}$, not both contained in A, $\Lambda = Az + A$ as above. By (2.3.2) and (2.3.3), $e_\Lambda(xz+y)$ is a non-zero $\mathfrak{n}$-torsion point of Φ^Λ. The function $z \mapsto e_\Lambda^{-1}(xz+y)$ is modular of weight one for $\Gamma(\mathfrak{n})$, some kind of *Weber* or *Fricke function*.

(3.6.5). It is slightly more laborious to write down modular forms of non-zero type. A typical example is obtained as follows (see **[Ge6]** for details):

Let N be the subgroup of elements $\begin{pmatrix} * & * \\ 0 & 1 \end{pmatrix}$ of Γ and set

$$\alpha(\gamma, z) = (\det \gamma)^{-m}(cz+d)^k \quad \text{for } \gamma = \begin{pmatrix} a & b \\ c & d \end{pmatrix} \in \Gamma. \tag{3.6.5.1}$$

Then

$$\alpha(\gamma\delta, z) = \alpha(\gamma, \delta z)\alpha(\delta, z); \tag{3.6.5.2}$$

$$\begin{aligned} N\backslash\Gamma &\xrightarrow{\cong} \{(c,d) \mid c, d \in A,\ (c,d) = 1)\}; \\ \gamma &\mapsto (c,d) \end{aligned} \tag{3.6.5.3}$$

$$\alpha^{-1}(\gamma, z)t_A^m(\gamma z) \text{ depends only on the class of } \gamma \text{ in } N\backslash\Gamma. \tag{3.6.5.4}$$

Hence the *Poincaré series*

$$P_{k,m}(z) = \sum_{\gamma \in N\backslash\Gamma} \alpha^{-1}(\gamma, z)t_A^m(\gamma z) \tag{3.6.5.5}$$

is well-defined as a function on Ω, provided the sum converges. For $k > 0$, this is not hard to verify, applying easy estimates on t_A. Furthermore, $P_{k,m}$ behaves well at cusps and does not vanish identically if $k \equiv 2m \bmod q-1$ and $m \leq k/(q+1)$. Under these conditions, $P_{k,m}$ is modular for Γ of weight k and type $m \bmod q-1$.

(3.7). All of the above examples may easily be generalized to cover the case of arbitrary arithmetic subgroups. In fact, despite the analytical definition of e.g. Eisenstein and

Poincaré series, all of them are essentially algebraic in nature. We just mention some of the relations they satisfy. First, letting $a_i = a_i(f, z)$ be the i-th coefficient form as in (3.6.3), we have for $k \geq 0$

$$f \cdot E^{(q^k-1)} = \sum_{i+j=k} E^{(q^i-1)} \cdot a_j^{q^i}. \tag{3.7.1}$$

([**Ge4**, II, 2.10] Here $a_0 = a_0(f, z)$ is the constant $f \in A$ and $E^{(0)} = -1$.) Thus we can recursively determine the a_i from the $E^{(q^i-1)}$ and vice versa. For example, in the case $A = \mathbb{F}_q[T]$ as in (2.5), we have the relations [**Go3**]

$$\begin{aligned} g &= (T^q - T) \cdot E^{(q-1)} \qquad \text{and} \\ \Delta &= (T^{q^2} - T^q) \cdot E^{(q-1)^{q+1}} + (T^{q^2} - T)E^{(q^2-1)} \end{aligned} \tag{3.7.2}$$

between the coefficient forms g, Δ and the Eisenstein series of weights $q-1$ and q^2-1. Next, all the Eisenstein series $E^{(k)}$ may be calculated from those whose indices have the special form $q^i - 1$ [**Ge6**, Prop. 3.4]. Furthermore, the Poincaré series appear in a purely algebraic manner as the "Serre derivatives" of coefficient forms [**Ge9**]. If e.g. $A = \mathbb{F}_q[T]$, the space $M_{q+1,1}(\Gamma)$ has dimension one. A non-zero form may (up to scalars) be described alternatively [**Ge6**] as

- a Poincaré series $P_{q+1,1}$,
- the Serre derivative of $g(z)$, or
- the $(q-1)$-th root of Δ.

(3.8). The relation between Drinfeld modular forms and modular curves is obvious from (3.5). This is used to derive geometric properties of $\overline{M}_\Gamma$ from function theoretic properties of modular forms (see Section 4). Let us first note that unlike the classical case, it is a difficult problem to calculate the genera or other invariants of the curves $\overline{M}_\Gamma$. Namely, apart from the easy special case $A = \mathbb{F}_q[T]$ described in (1.3.1) and (2.5), the usual Hurwitz formula argument is not available, by lack of a priori knowledge of the genus of the curve "without level".

We define an *elliptic point* of the arithmetic group Γ as some $z \in \Omega$ whose stabilizer Γ_z is strictly larger than the "usual" $\Gamma \cap Z(K)$. In our basic example $A = \mathbb{F}_q[T]$, those z with $j(z) = 0$ are elliptic for $\Gamma = GL(2, A)$. If now $\Gamma = \Gamma_Y$ is a maximal arithmetic group, we have [**Ge4**, p. 53]: let k be divisible by $q^2 - 1$, and let $0 \neq f \in M_{2k,0}(\Gamma)$ be a form whose divisor on $\overline{M}_\Gamma$ has degree d. Then

$$d = k[2g(\overline{M}_\Gamma) - 2 + qc/(q-1) + qe/(q+1)], \tag{3.8.1}$$

where $g(\overline{M}_\Gamma)$ is the genus and c (resp. e) the number of cusps (resp. elliptic points) of $\overline{M}_\Gamma$. Similar formulas exist for subgroups of Γ_Y. In view of the arithmetic meaning of the numbers c and e (loc. cit.), this yields a method to determine $g(\overline{M}_\Gamma)$ once the divisor of *one* f as above is known.

4. An Assortment of Results

There are many direct implications of the moduli theory of Drinfeld modules for the arithmetic of K. We list and discuss a few of them.

(4.1). (Class field theory) First, the case where the rank r equals one is more or less the same as abelian class field theory for K. Namely, the scheme M^1 of (1.9.1) equipped with the action of the idèle class group of K is canonically isomorphic with the spectrum of the ring of A-integers of the maximal abelian extension of K that splits completely at ∞ **[Dr**, Theorem 1**]**, **[Ha1]**, **[Ha2]**. Class fields of K may explicitly be constructed as the fields generated by torsion points of distinguished rank-1 Drinfeld modules. So the statement is a simultaneous analogue of the Kronecker-Weber theorem and the main theorem of Complex Multiplication.

(4.2). (Langlands correspondence) The 1-adic cohomology of the schemes M^r is provided with natural actions of $GL(r, \mathcal{A}_f)$ and the local Galois group at ∞. These actions commute; the precise nature of their common irreducible constituents is described by a conjectural reciprocity law, whose general validity would imply the Langlands correspondence over K. The reciprocity law is proved for $r = 2$ by Drinfeld **[Dr**, Theorem 2**]**; for $r > 2$, it is the subject of recent work of Laumon, Rapoport, Schneider, Stuhler **[SS]**, **[LRS]**.

(4.3). (Cohomology of arithmetic groups) Let us now focus on the rank-2 case. As is proved in **[Dr]** and **[Ge4**, VA**]**, the first Betti number $b_i(\Gamma)$ of an arithmetic group Γ equals the genus of the corresponding curve $\overline{M}_\Gamma$. Following the strategy indicated at the end of the last section, we can calculate $b_1(\Gamma)$. More precisely, in **[Ge4]**, the expansions of modular forms at the cusps of $\overline{M}_\Gamma$ are examined. Besides properties of independent interest on the arithmetic nature of the "Fourier coefficients", the zero order of certain coefficient forms is expressed through the zeta function of the function field K (loc. cit., VI, Sections 4, 5). For a maximal arithmetic group $\Gamma = \Gamma_Y$, the result is

$$\begin{aligned} b_1(\Gamma) &= g(\overline{M}_\Gamma) \qquad (4.3.1) \\ &= 1 + (q^2-1)^{-1}\left[\frac{q^{\deg\infty}-1}{q-1}P(q) - \frac{q(q+1)}{2}\deg\infty\cdot P(1) + \eta\right]. \end{aligned}$$

Here $\deg\infty$ is the degree of the place ∞ over $\mathbb{F}_q$, $P(X)$ is the numerator polynomial of the zeta function of K, and η equals $-\frac{q(q-1)}{2}P(-1)$ for odd and 0 for even values of $\deg\infty$. Some values for $\Gamma = GL(2, A)$ and $g =$ genus of $K = 0$ or 1 are given in the table extracted from (loc. cit.):

g = genus of K	deg ∞	$b_1(\Gamma) = g(\overline{M}_\Gamma)$
0	1, 2, 3	0
0	4	q
0	5	$q(q+1)$
0	6	$q(q^2+q+2)$
1	1	0
1	2	q^2
1	3	$q(q^2-1+\#)$

(4.3.2)

The symbol $\#$ stands for the number of $\mathbb{F}_q$-rational points of the elliptic curve associated with K. It is a remarkable fact that $g(\overline{M}_\Gamma) = 0$ independently of $\#$, if $g = \deg\infty = 1$. Similarly explicit formulas exist for the Hecke modular curves (1.11).

(4.4). (Representation theory of $GL(2, \mathbb{F}_q)$) Just recently, J. Teitelbaum **[Te]** found an application of the description of the ring of modular forms for $GL(2, \mathbb{F}_q[T])$ to the theory of modular representations of $GL(2, \mathbb{F}_q)$. He used it to prove multiplicity formulas for the occurrence of the Steinberg module in the symmetric powers of the canonical representation of $GL(2, \mathbb{F}_q)$ on $\mathbb{F}_q^2$. These formulas have been found earlier by Kuhn, Mitchell, and Priddy, using different methods.

(4.5). (Cuspidal divisor class groups) The classical Manin-Drinfeld theorem states that each degree zero divisor supported by the cusps on an elliptic modular curve has finite order in the Picard group. The obvious analogue in the Drinfeld modular context is equally true: **[Ge1]**, **[Ge4**, VI, 5.12**]**. This fact even extends to higher rank modular schemes **[Ge5]**, **[Kap]**.

(4.6). (Class numbers of division algebras) Fix $r \in \mathbb{N}$ and some finite place $\mathfrak{p}$ of K, and let $D = D(r, \mathfrak{p})$ be the central division algebra over K uniquely determined by its Brauer invariants $\mathrm{inv}_v(D) = 1/r$ if $v = \mathfrak{p}$, $\mathrm{inv}_v(D) = -1/r$ if $v = \infty$, and $\mathrm{inv}_v(D) = 0$ if v is a place of K different from $\mathfrak{p}$ and ∞. The arithmetic of D is closely related to the geometry of certain modular schemes, notably their supersingular points modulo $\mathfrak{p}$. In some cases (see **[Ge8]** for $r = 2$, **[Ge7]** for r arbitrary but $A = \mathbb{F}_q[T]$), this yields class number formulas for maximal A-orders in D.

(4.7). (Analogue of Shimura-Taniyama-Weil conjecture) Last but not least, we should mention the connection with the classification of elliptic curves over K. As will be explained in **[GR]**, decomposing the l-adic cohomology of a Drinfeld modular curve $\overline{M}_0(\mathfrak{n})$ of Hecke type (1.11) as a module under the Hecke algebra yields a classification up to isogeny of elliptic curves over K with split multiplicative reduction at ∞ and conductor $\mathfrak{n} \cdot \infty$. Thus for instance, the "genus $= 0$" statements of (4.3.2) imply that over ground fields K of genus 0 (resp. 1) with a place "∞" of degree ≤ 3 (resp. of degree 1), there is no elliptic curve $E|_K$ with split multiplicative reduction at ∞ and good reduction elsewhere.

References

[DH] P. Deligne and D. Husemöller: Survey of Drinfeld modules. Contemp. Math. 67, (1987), 25–91.

[DR] P. Deligne and M. Rapoport: Les schémas de modules de courbes elliptiques. In: P. Deligne, W. Kuijk (Ed.): Modular Forms of One VariableII, Lecture Notes in Mathematics 349, Springer (1973).

[Dr] V. G. Drinfeld: Elliptic modules. Math. USSR-Sb. 23(4), (1976), 561–592.

[Ge1] E.-U. Gekeler: Modulare Einheiten für Funktionenkörper. J. Reine Angew. Math. 348, (1984), 94–115.

[Ge2] E.-U. Gekeler: Über Drinfeld'sche Modulkurven vom Hecke-Typ. Compositio Math. 57, (1986), 219–236.

[Ge3] E.-U. Gekeler: Automorphe Formen über $\mathbb{F}_q[T]$ mit kleinen Führer. Abh. Math. Sem. Hamburg 55, (1985), 111–146.

[Ge4] E.-U. Gekeler: Drinfeld Modular Curves. Lecture Notes in Math. 1231, Springer (1986).

[Ge5] E.-U. Gekeler: Satake compactification of Drinfeld modular schemes. Proceedings of the Conference on P-adic Analysis, Hengelhoef, 1986.

[Ge6] E.-U. Gekeler: On the coefficients of Drinfeld modular forms. Invent. Math. 93, (1988), 667–700.

[Ge7] E.-U. Gekeler: Sur les classes d'idéaux des ordres de certains corps gauches. C. R. Acad. Sc. Paris, Sér. I Math. 309, (1989), 577–580.

[Ge8] E.-U. Gekeler: Sur la géométrie de certaines algèbres de quaternions. Sém. Théor. Nombres Bordeaux 2, (1990), 143–153.

[Ge9] E.-U. Gekeler: DeRham cohomology and the Gauss-Manin connection for Drinfeld modules. Conference on P-adic Analysis, Trento 1989. In: F. Baldassari, S. Bosch, B. Dwork (Ed.): p-adic Analysis, Lecture Notes in Math. 1454.

[GR] E.-U. Gekeler and M. Reversat: Some results on the Jacobians of Drinfeld modular curves. This volume.

[Go1] D. Goss: π-adic Eisenstein series for function fields. Compositio Math. 41, (1980), 3–38.

[Go2] D. Goss: Modular forms for $\mathbb{F}_r[T]$. J. Reine Angew. Math. 317, (1980), 16–39.

[Go3] D. Goss: The algebraist's upper half-plane. Bull. Amer. Math. Soc. (N.S.) 2, (1980), 391–415.

[Ha1] D. Hayes: Explicit class field theory for rational function fields. Trans. Amer. Math. Soc. 189, (1974), 77–91.

[Ha2] D. Hayes: Explicit class field theory in global function fields. In: G. C. Rota (Ed.), Studies in Algebra and Number Theory, Academic Press, New York, 1979.

[Ha3] D. Hayes: On the reduction of rank-one Drinfeld modules. Math. of Comp. 57, (1991), 339–349.

[JL] H. Jacquet and R. P. Langlands. Automorphic forms on $GL(2)$. Lecture Notes in Math. 114.

[Kap] M. M. Kapranov: On cuspidal divisors on the modular varieties of elliptic modules. Math. USSR-Izv. 30(3), (1988).

[Ka] N. Katz: P-adic properties of modular schemes and modular forms. In: W. Kuijk, J.-P. Serre (Ed.): Modular Forms of One Variable III, Lecture Notes in Math. 350.

[KM] N. Katz and B. Mazur: Arithmetic moduli of elliptic curves. Ann. of Math. Stud. 108, Princeton 1985.

[La] S. Lang: Elliptic Functions. Addison-Wesley, Reading 1973.

[LRS] G. Laumon, M. Rapoport, and U. Stuhler: $\mathfrak{D}$-elliptic sheaves and the Langlands correspondence. Preprint, Orsay 1991.

[Ma] B. Mazur: Modular curves and the Eisenstein ideal. Inst. Hautes Études Sci. Publ. Math. 47, 1977.

[Re] M. Reversat: Lecture on rigid geometry. This volume.

[SS] P. Schneider and U. Stuhler: The cohomology of p-adic symmetric spaces. Invent. Math. 105, (1991), 47–122.

[Te] J. Teitelbaum: The Poisson Kernel for Drinfeld Modular Curves. J. Amer. Math. Soc. 4, (1991), 491–511.

FB 9 - Mathematik, Universität des Saarlandes, Saarbrücken, D-6600 Germany

Ramifications Arising from Drinfeld Modules

Yuichiro Taguchi

Introduction

In this paper, we study various ramifications arising from division points of Drinfeld modules, abelian T-modules, formal modules, etc. A motivation for this is to know how many isogeny classes and isomorphism classes of Drinfeld A-modules exist over a finite extension of the fraction field of A. We will see (cf. Remark (3.4)) that, modulo the isogeny conjecture, an isogeny class can contain infinitely many isomorphism classes and, without any restriction on ramification at the infinite places, there can be infinitely many isogeny classes.

To explain some of the results, let F be a function field in one variable over a finite field, ∞ a fixed place of F, A the ring of elements of F which are regular outside ∞, and K a finite extension of F. Given a Drinfeld A-module ϕ over K and a prime v of A, we denote by $K(\phi; v^n)$ the field of v^n-division points of ϕ. Then it turns out (Corollary (1.6)) that the ramification at various primes in the tower $(K(\phi; v^n)/K)_{n\geq 1}$ is bounded at the places over ∞ by a divisor depending only on ϕ, and at the finite places, it is controlled in a fairly precise way in terms of the "discriminant" $\Delta(\phi)$. Roughly speaking, $\Delta(\phi)$ is the coefficient of the leading term of the defining equation of ϕ. For finite places, this result is analogous to the case of abelian varieties over number fields. (At least one has the Hermite-Minkowski theorem for number fields, which assures the existence of an estimate of discriminants.) But at infinite places, there occur new phenomena, which we describe by example in §2. We construct explicitly an infinite family of Drinfeld modules with everywhere good reduction and with ramification at infinity becoming arbitrarily large (Example (2.1)), as well as an infinite family of mutually non-isomorphic Drinfeld modules with everywhere good reduction and with bounded ramification at infinity (Example (2.2)). In §3, we give a proposition on v-adic Galois representations (a positive characteristic version of a theorem of Faltings), and discuss how many isomorphism and isogeny classes can exist. §4 and §5 are generalizations of §1 to the cases of finite submodules of higher dimensional formal modules. Theorem (4.6) is an A-module version of Théorème 1 of [5].

Acknowledgement. This paper is an extended version of my talk at the workshop "The arithmetic of function fields" held at The Ohio State University in June, 1991. I would like to thank the organizer David Goss for his efforts and his hospitality during my stay at The Ohio State University after the workshop (I found Example (2.1) there). I am also grateful to Greg W. Anderson for asking a question, a partial answer to which is the content of §5.

Notation. Throughout this article, A is a "basic" Dedekind ring and F is its fraction field; in §§1 – 3 (global context), F is a function field in one variable over a finite field, ∞ is a fixed place of F, and A is the ring of elements of F regular outside ∞, whereas in §§4 and 5 (local context), A is a complete discrete valuation ring with finite residue field.

In either context, K will mainly be used to denote a finite extension of F, and then $\mathfrak{O}_K$ will denote the integral closure of A in K.

If we are in the global context, a *prime* of $\mathfrak{O}_K$ means a non-zero prime ideal of $\mathfrak{O}_K$, which is identified with a place of K, and called a *finite place* of K. A place of K is called *infinite* if it extends the place ∞ of F. A place of K will often be identified with a *normalized* valuation of K. A non-zero fractional ideal of $\mathfrak{O}_K$ is often regarded as a divisor of K and denoted additively. So we use notations like, e.g., $\mathfrak{a} \leq \mathfrak{b}$ for such $\mathfrak{a}$ and $\mathfrak{b}$. If w is a prime of $\mathfrak{O}_K$ or a place of K, $\mathfrak{O}_{K,w}$ and K_w denote respectively the completions of $\mathfrak{O}_K$ and K with respect to w.

For a field K, K^{sep} denotes a fixed separable closure of K, and G_K denotes the absolute Galois group Gal(K^{sep}/K). For a finite separable extension L/K, $\mathfrak{D}(L/K)$ (resp. $\mathfrak{d}(L/K)$) denotes the different (resp. discriminant) of L/K if it can be defined at all.

If G is a group scheme and v is a non-zero element or a non-zero ideal of End(G), then ${}_vG$ denotes the subgroup scheme Ker(v) of G.

1. Finite places

In this section, we estimate the differents at finite places of the field extensions arising from division points of Drinfeld modules.

Let F be a function field in one variable over a finite field, ∞ a fixed place of F, and A the ring of elements of F regular outside ∞. We assume that the field of constants is $\mathbb{F}_q$, the finite field with $q = p^f$ elements. For $a \in A - 0$, we define $\deg(a)$ by $\mathrm{Card}(A/aA) = q^{\deg(a)}$.

Let K be a field of characteristic $p > 0$, and $\mathbb{G}_a$ the additive group scheme over K. After choosing a coordinate X of $\mathbb{G}_a$, we can identify $\mathrm{End}_K(\mathbb{G}_a)$ with the non-commutative ring of additive polynomials of X with coefficients in K, where the product is the composition of maps. So in the following, if

$$\phi : A \longrightarrow \mathrm{End}_K(\mathbb{G}_a)\,; \quad a \mapsto \phi_a$$

is a Drinfeld module over K, we think of ϕ_a as a polynomial $\phi_a(X) \in K[X]$ via this identification. If ϕ is of rank r, the degree of $\phi_a(X)$ as a polynomial of X is $q^{r\deg(a)}$ for all $a \in A - 0$.

Let R be, for example, a Dedekind ring over A, and K its fraction field. For a Drinfeld module ϕ over K, we have a minimal model $(\tilde{\phi}, \mathfrak{m})$ of ϕ over R ([10], §2), where $\mathfrak{m}$ is a fractional ideal of R. If R is a discrete valuation ring, we can take $\mathfrak{m} = R$, and $(\tilde{\phi}, R)$ is characterized as the unique (up to isomorphisms) Drinfeld module $\tilde{\phi}$ over K such that $\tilde{\phi}_a(X) \in R[X]$ for all $a \in A$ and the valuations of the coefficients of $\tilde{\phi}_a(X)$ are minimal.

Lemma (1.1). *Let R be a discrete valuation ring over A, and K its fraction field. Let ϕ be a Drinfeld module over K of rank r, and $(\tilde{\phi}, R)$ the minimal model of ϕ over R. Then there exists an ideal $\mathfrak{n}$ of R such that the leading coefficient of $\tilde{\phi}_a(X) \in R[X]$ divides $\mathfrak{n}^{\delta(r,a)}$ for any $a \in A - 0$, where $\delta(r,a) := (q^{r\deg(a)} - 1)/(q-1)$.*

Proof. Clearly, we may assume $\tilde{\phi} = \phi$. Take non-constant elements x and $y \in A$. Since y is algebraic over $\mathbb{F}_q[x]$, we have a non-trivial relation

$$\sum \alpha_{ij} x^i y^j = 0, \qquad \alpha_{ij} \in \mathbb{F}_q^*.$$

Let $z = \sum \beta_{ij} x^i y^j$ be the sum of the $\alpha_{ij} x^i y^j$'s with degree

$$d := \max_{i,j} \{ \deg(\alpha_{ij} x^i y^j) \}$$

in the left side of this equality. We must have then

$$\deg(z) < d.$$

If $\phi_x(X) = xX + \cdots + x_m X^{q^m}$ and $\phi_y(X) = yX + \cdots + y_n X^{q^n}$ with $m = r\deg(x)$, $n = r\deg(y)$, and x_m, $y_n \in R$, then we have

$$\phi_{x^i y^j}(X) = x^i y^j X + \cdots + x_m^{1+q^m+\cdots+q^{(i-1)m}} y_n^{q^{im}(1+q^n+\cdots+q^{(j-1)n})} X^{q^{im+jn}}.$$

So the coefficient of $X^{q^{rd}}$ in $0 = \sum \alpha_{ij} \phi_{x^i y^j}$ is

$$\sum \beta_{ij} x_m^{\frac{q^{im}-1}{q^m-1}} y_n^{q^{im}\frac{q^{jn}-1}{q^n-1}},$$

where the sum is over i and j with $im + jn = rd$. Since this sum must be zero, there exist two terms in the sum (say, of indices (i,j) and (i',j'), with $i \neq i'$ and $j \neq j'$) with the same valuation. Denoting by v the valuation of R, we have

$$\frac{q^{im}-1}{q^m-1} v(x_m) + q^{im}\frac{q^{jn}-1}{q^n-1} v(y_n) = \frac{q^{i'm}-1}{q^m-1} v(x_m) + q^{i'm}\frac{q^{j'n}-1}{q^n-1} v(y_n).$$

Noticing the relation $rd = im + jn = i'm + j'n$, we see from this that

$$v(y_n) = \frac{q^n-1}{q^m-1} v(x_m).$$

Hence, if $x_m \mid \mathfrak{n}^{\frac{q^m-1}{q-1}}$, then $y_n \mid \mathfrak{n}^{\frac{q^n-1}{q-1}}$. Now the proof is complete. □

Definition (1.2). Let K be a finite extension of F. For a Drinfeld module ϕ over K and a prime w of $\mathfrak{O}_K$, consider its minimal model over $\mathfrak{O}_{K,(w)}$, and define $\Delta_w(\phi)$ to be the smallest ideal $\mathfrak{n}$ of $\mathfrak{O}_{K,(w)}$ with the property stated in Lemma (1.1). Define also $\Delta(\phi) := \sum_w \Delta_w(\phi)$, where the sum is over all primes of $\mathfrak{O}_K$ and $\Delta_w(\phi)$ is regarded as a divisor of K.

$\Delta(\phi)$ measures in a sense the "badness" of the reductions of ϕ at finite places.

To estimate the difference, we begin with

Lemma (1.3). *Let R be a complete discrete valuation ring, K the fraction field, and $\phi(X) \in R[X]$ a separable polynomial with coefficients in R. Assume the coefficient of the leading term of ϕ is a unit. Let α be a root of ϕ in K^{sep}. Then the different $\mathfrak{D}(K(\alpha)/K)$ divides the principal ideal $(\phi'(\alpha))$.*

Note that $\phi'(\alpha) = a_0$ if ϕ is of the form $\phi(X) = \sum_i a_i X^{p^i}$ and K is of positive characteristic p.

Proof. Since the minimal polynomial of α divides $\phi(X)$ and all the roots of $\phi(X)$ are integral over R, this follows from Cor. 2 (p. 66) to Prop. 11 of §6, Chap. III of [8]. □

Lemma (1.4). *Let R be a complete discrete valuation ring of characteristic $p > 0$ and K the fraction field. Let $\phi(X) = \sum_{i=0}^{N} a_i X^{p^i}$ be a separable polynomial in $R[X]$ with $a_0 a_N \neq 0$, and α a root of ϕ in K^{sep}. Then the different $\mathfrak{D}(K(\alpha)/K)$ divides the principal ideal $(a_0 a_N^{p^N-2})$.*

Proof. $\tilde{\phi}(X) := a_N^{p^N-1}\phi(X/a_N) = \sum_{i=0}^{N} a_i a_N^{p^N-1-p^i} X^{p^i}$ is a separable monic polynomial in $R[X]$ and $a_N\alpha$ is a root of $\tilde{\phi}$. The assertion now follows from the previous lemma. □

For a finite extension K of F and $a \in A - 0$, let $K(\phi; a) = K({}_a\phi(K^{sep}))$ denote the finite separable extension of K obtained by adding the a-division points of ϕ. Let $\mathfrak{D}_f(/)$ denote the finite part of $\mathfrak{D}(/)$, i.e., the sum of the components of $\mathfrak{D}(/)$ not lying over ∞. Since the extension $K(\phi; a)/K$ is obtained by adding r roots of $\phi_a(X)$ which form an (A/aA)-base of ${}_a\phi(K^{sep}) \simeq (A/aA)^r$, we see from Lemmas (1.1) and (1.4) the following

Proposition (1.5). *Let ϕ be a Drinfeld module over a finite extension K of F of rank r. For $a \in A - 0$, we have*

$$\mathfrak{D}_f(K(\phi; a)/K) \leq r\big[(a) + \delta(r, a)(q^{r\deg(a)} - 2) \cdot \Delta(\phi)\big].$$

For an infinite place w of K, let $\Lambda_w(\phi)$ denote the A-lattice in K_w^{sep} corresponding to $\phi \otimes_K K_w$ ([3], §3). This is a G_{K_w}-stable projective A-module of rank r. In particular, it is finitely generated over A, and the fixed subfield $K_w(\Lambda_w(\phi))$ of K_w^{sep} by the kernel of the natural representation $G_{K_w} \longrightarrow \mathrm{Aut}(\Lambda_w(\phi))$ is a finite extension of K_w. We have for $a \in A - 0$, ${}_a\phi(K_w^{sep}) \simeq \Lambda_w(\phi)/a\Lambda_w(\phi)$ as G_{K_w}-modules, which is rational over $K_w(\Lambda_w(\phi))$. Hence we have:

Corollary (1.6). *Let r be a positive integer, v a prime of A, and $\mathfrak{n}$ a non-zero ideal of $\mathfrak{O}_K$. Let S be the set of finite places of K consisting of the finite places lying above v or dividing $\mathfrak{n}$. Then there exists a family $(N(w, n))_{w \in S, n \in \mathbb{N}}$ of non-negative integers which has the following property:*

For any Drinfeld module ϕ over K of rank r with $\Delta(\phi) \leq \mathfrak{n}$ and for any $n \in \mathbb{N}$, we have

$$\mathfrak{d}(K(\phi; v^n)/K) \leq \sum_{w \in S} N(w, n) \cdot (w) \; + \; M(\phi) \cdot \infty,$$

where $M(\phi)$ is an integer ≥ 0 depending on ϕ but not on n.

2. Infinite places — Examples

In contrast to the classical case (where we have only $\mathbb{C}/\mathbb{R}$), we have more complicated field extensions at infinity in the Drinfeld module case, if the rank r is bigger than one. In this section, we give two typical examples which clarify this contrast.

Let A, F, and ∞ be as in §1. For a Drinfeld module ϕ over a finite extension K of F and an infinite place w of K, let $\Lambda_w(\phi)$ denote, as in §1, the A-lattice corresponding to $\phi \otimes_K K_w$.

Example (2.1). Let $A = \mathbb{F}_q[T]$, $F = \mathbb{F}_q(T)$, $\infty = (\frac{1}{T})$, and r an integer ≥ 2. Then there exists an infinite family $(\phi^{(n)})_{n\geq 1}$ of Drinfeld modules over F of rank r which has the following properties:

(i) $\phi^{(n)}$ has everywhere good reduction over A;

(ii) the ramification of the corresponding lattice $\Lambda^{(n)} = \Lambda_\infty(\phi^{(n)})$ at ∞ becomes arbitrarily large, i.e., $\mathrm{ord}_\infty(\mathfrak{D}(F_\infty(\Lambda^{(n)})/F_\infty))$ tends to infinity as n does.

Especially, for any finite place v of F, there arise infinitely many isomorphism classes of v-adic G_F-representations $T_v(\phi) \otimes_{A_v} F_v$ from Drinfeld modules ϕ of rank r over F with everywhere good reduction.

Construction. Consider a Drinfeld module $\phi^{(n)}$ over A defined by

$$\phi_T^{(n)}(X) = TX + a_1 X^q + \cdots + a_r X^{q^r}, \qquad a_i \in A,$$

where we assume:

(1) $a_r \in A^\times = \mathbb{F}_q^\times$;

(2) $\mathrm{ord}_\infty(a_{r-1}) = -n(q^r - q^{r-1}) + 1$;

(3) $\mathrm{ord}_\infty(a_{r-1}) \leq \mathrm{ord}_\infty(a_i)$ for $1 \leq i \leq r-1$.

Let v denote the normalized valuation $\mathrm{ord}_\infty(\cdot)$ extended uniquely to a fixed separable closure F_∞^{sep} of F_∞. The Newton polygon of $\phi_T(X) \in F_\infty[X]$ shows that

$$\phi_T(X) \text{ has } (q^r - q^{r-1}) \text{ roots } \lambda \text{ with } v(\lambda) = -n + \frac{1}{q^r - q^{r-1}} \tag{4}$$

(take and fix one such λ), and the other non-zero roots have non-negative valuations. Consequently, $V := \{\lambda' \in {}_T\phi(F_\infty^{sep})\ ;\ v(\lambda') \geq 0\}$ forms an $(r-1)$-dimensional $\mathbb{F}_q$-vector space.

By (4), the degree of the minimal polynomial of λ over F_∞ cannot exceed $(q^r - q^{r-1})$. On the other hand, the denominator of $v(\lambda)$, expressed as a reduced fraction, is $(q^r - q^{r-1})$. Hence the extension $F_\infty(\lambda)/F_\infty$ is totally ramified of degree $(q^r - q^{r-1})$, and in particular, it is wildly ramified.

Let L be the Galois closure in F_∞^{sep} of $F_\infty(\lambda)/F_\infty$. Since L/F_∞ is also wildly ramified, there exists an element $\sigma \in \mathrm{Gal}(L/F_\infty)$ of order p. Then $\sigma(\lambda)$ is of the form

$$\sigma(\lambda) = \alpha\lambda + \lambda', \qquad \text{for some } \alpha \in \mathbb{F}_q^\times \text{ and } \lambda' \in V.$$

Since $v(\sigma(\lambda')) = v(\lambda') > v(\lambda)$, we again have $\sigma(\lambda') \in V$. Hence $\sigma^p = 1$ implies $\alpha^p = 1$. Thus

$$\sigma(\lambda) = \lambda + \lambda', \qquad \lambda' \in V.$$

Set $\pi := \frac{\lambda}{T^n}$, so that $v(\pi) = \frac{1}{q^r - q^{r-1}}$ and π is a uniformizer of $F_\infty(\lambda)$. Since

$$v(\sigma(\pi) - \pi) = v((\sigma(\lambda) - \lambda)/T^n) = n + v(\lambda') \geq n,$$

and

$$\mathfrak{D}(F_\infty(\lambda)/F_\infty) = \prod_{\tau \in \mathrm{Gal}(L/F_\infty)-1} (\tau(\pi) - \pi),$$

this different, and hence $\mathfrak{D}(F_\infty(\Lambda)/F_\infty)$, can become arbitrarily large, as asserted before.

Example (2.2). Let A, F, ∞, and r be as in Example (2.1). Then there exists an infinite family $(\phi^{(n)})_{n \geq 0}$ of mutually non-isomorphic Drinfeld modules of rank r over a finite extension L of F which has the following properties:

(i) $\phi^{(n)}$ has everywhere good reduction over the integral closure $\mathfrak{O}_L$ of A in L.

(ii) Let w be an infinite place of L, and set $\Lambda^{(n)} := \Lambda_w(\phi^{(n)})$. Then there are, in fact, only finitely many field extensions in the set $\{L_w(\Lambda^{(n)})/L_w \ ; \ n \in \mathbb{N}\}$.

Construction. Let $K = \mathbb{F}_q(t)$, $\mathfrak{O}_K = \mathbb{F}_q[t]$, and C the Carlitz $\mathfrak{O}_K$-module defined by

$$C_t(X) = tX + X^q.$$

Take an irreducible element $T = f(t) \in \mathfrak{O}_K$ of degree r, and let A be the subring $\mathbb{F}_q[T]$ of $\mathfrak{O}_K$ ($\infty := (\frac{1}{T}) = (\frac{1}{t^r})$ in K). Define a Drinfeld A-module ϕ over $\mathfrak{O}_K$ by

$$\phi_T := C_{f(t)}.$$

Then ϕ has rank r, and everywhere good reduction over $\mathfrak{O}_K$. By explicit class field theory ([7]), the field $L := K({}_{f(t)}C(K^{sep})) = K({}_T\phi(K^{sep}))$ is an abelian extension of K with Galois group $(\mathfrak{O}_K/(f(t)))^\times \simeq \mathbb{F}_{q^r}^\times$, and the prime $(f(t))$ ramifies totally in L. In particular, the polynomial $C_{f(t)}(X)/X = \phi_T(X)/X$ over $\mathfrak{O}_K$ is Eisenstein at (T); if we write

$$\phi_T(X) = TX + a_1 X^q + \cdots + a_{r-1} X^{q^{r-1}} + X^{q^r},$$

then we have

$$\mathrm{ord}_T(a_i) \geq 1, \quad 1 \leq i \leq r-1. \tag{1}$$

Write

$$\phi_T(X) = TX \prod_{\lambda \in {}_T\phi - 0} (1 - \frac{X}{\lambda}) \qquad \text{in } K^{sep}[X].$$

Looking at the Newton polygons of $\phi_T(X)$ at various finite places of K, we see that, for any $\lambda \in {}_T\phi(K^{sep}) - 0$, (λ) is the unique prime ideal of $\mathfrak{O}_L$ lying over (T). So we can take a prime element τ (say $\tau :=$ one of the λ's) of $\mathfrak{O}_L$ and write

$$\lambda = \tau\lambda_1, \quad \lambda_1 \in \mathfrak{O}_L^\times$$

for each $\lambda \in {}_T\phi(K^{sep}) - 0$. Define Drinfeld modules $\phi^{(n)}$ $(n = 0, 1, 2, \cdots)$ over L by

$$\phi_T^{(n)}(X) = TX \prod_{\lambda \in {}_T\phi - 0} (1 - \frac{X}{\tau\lambda_1^{p^n}}).$$

If $\phi_T^{(n)}(X) = TX + a_1^{(n)} X^q + \cdots + a_r^{(n)} X^{q^r}$, then

$$a_i^{(n)} = T \sum \prod_{(q^i-1)} \frac{1}{-\tau\lambda_1^{p^n}} = T\tau^{1-q^i} (\sum \prod_{(q^i-1)} \frac{1}{\lambda_1})^{p^n},$$

where $\sum\prod_{(q^i-1)}$ denotes the sum of the products of $(q^i - 1)$ elements, the sum taken over all possible choices of $(q^i - 1)$ λ's from ${}_T\phi(K^{sep}) - 0$. Set $b_i := \sum\prod_{(q^i-1)} \frac{1}{\lambda_1}$. Since $(T) = (\tau^{q^r-1})$ and, for $1 \leq i \leq r-1$,

$$a_i = a_i^{(0)} = T\tau^{1-q^i} b_i$$

is by (1) an element of $\mathfrak{O}_K$ divisible by T, b_i is integral; $b_i \in \mathfrak{O}_L$. Note that

$$\mathrm{ord}_\tau(b_i) = \mathrm{ord}_\tau(a_i) - (q^r - q^i) > 0. \tag{2}$$

Since

$$a_i^{(n)} = T\tau^{1-q^i} b_i^{p^n} = a_i b_i^{p^n-1},$$

we have for a prime w of $\mathfrak{O}_L$,

$$\mathrm{ord}_w(a_i^{(n)}) = \begin{cases} \mathrm{ord}_\tau(a_i) + (p^n - 1)\mathrm{ord}_\tau(b_i) & \text{if } w \mid T \\ p^n \mathrm{ord}_w(a_i) & \text{if } w \nmid T \end{cases}. \tag{3}$$

Moreover we have for $i = r$,

$$a_r^{(n)} = T\tau^{1-q^r} (\prod_{(q^r-1)} \frac{1}{\lambda_1})^{p^n} \in \mathfrak{O}_L^\times.$$

We have thus obtained an infinite family $(\phi^{(n)})_{n\in\mathbb{N}}$ of Drinfeld modules over $\mathfrak{O}_L$ with everywhere good reduction. (2) and (3) imply that these are mutually non-isomorphic, and yet the T-division points ${}_T\phi^{(n)}(L^{sep})$ are rational over L. Further, it will be shown in §3 (Cor. (3.2), (ii)) that there arise, in fact, only finitely many extensions $L_w(\Lambda_w(\phi^{(n)}))/L_w$ for all $w \mid \infty$.

3. Some finiteness and infiniteness

In this section, A, F and ∞ are as in §1. Let v be a prime of A, K a finite extension of F, and $\mathfrak{n}$ a positive divisor of K. Let V be a finite dimensional F_v-vector space, and $\rho : G_K \longrightarrow \mathrm{GL}(V)$ an F_v-linear continuous representation of G_K. Consider the following condition for ρ:

(*) ρ is unramified outside $\mathrm{Supp}(\mathfrak{n})$, and there exists in V a G_K-stable A_v-lattice T such that $\mathfrak{d}(K'/K) \leq \mathfrak{n}$, where K' is the fixed subfield of K^{sep} by the kernel of the map $G_K \longrightarrow \mathrm{Aut}(T/vT)$ induced by ρ.

Proposition (3.1). *Let r be a positive integer and $\mathfrak{n}$ a positive divisor of K. Then there exists a finite set S of finite places of K disjoint from* $\mathrm{Supp}(\mathfrak{n})$ *which has the following property:*

Let $\rho_i : G_K \longrightarrow \mathrm{GL}(V_i)$, $i = 1, 2$, be two r-dimensional v-adic representations which are semi-simple and satisfy the above condition (*). If the characteristic polynomials of $\rho_1(\mathrm{Frob}_w)$ and $\rho_2(\mathrm{Frob}_w)$ coincide for all $w \in S$, then we have $\rho_1 \simeq \rho_2$.

Proof. (Cf. Proof of Theorem 5 of [4].) Since there are only finitely many separable extensions of K with given degree and discriminant, there exists a finite Galois extension $K_{\mathfrak{n}}/K$ which contains all separable extensions K'/K such that $[K' : K] \leq \mathrm{Card}(\mathrm{GL}_r(\mathbb{F}_{q_v}))$ and $\mathfrak{d}(K'/K) \leq \mathfrak{n}$, where $q_v := \mathrm{Card}(A/vA)$. By Čebotarev, there exists a finite set S of finite places of K disjoint from Supp($\mathfrak{n}$) such that Gal($K_{\mathfrak{n}}/K$) is filled with the images of the conjugacy classes of Frob_w for $w \in S$. We will show that this S has the required property. Choose G_K-stable A_v-lattices T_i of V_i for $i = 1, 2$, with the property as in the assumption (*), and let M_j, $1 \leq j \leq r$, be the A_v-subalgebra of $\mathrm{End}_{A_v}(\wedge^j T_1 \times \wedge^j T_2)$ generated by the image of $\wedge^j \rho_1 \times \wedge^j \rho_2$. By the assumption of semi-simplicity and by a version of the Brauer-Nesbitt theorem (cf. [9]), it suffices to show $\mathrm{Tr}(m_1; \wedge^j V_1) = \mathrm{Tr}(m_2; \wedge^j V_2)$ for all $m = (m_1, m_2) \in M_j$, $1 \leq j \leq r$, which is already true by assumption if m is conjugate to the image of Frob_w for some $w \in S$. It remains to show that these images together with their conjugates generate M_j over A_v. They generate the A_v/vA_v-module M_j/vM_j, because $G_{K_{\mathfrak{n}}}$ acts trivially on $T_1/vT_1 \times T_2/vT_2$ according to our choice of $K_{\mathfrak{n}}$. By Nakayama's lemma, they generate the A_v-module M_j. □

Hereafter in §3, w denotes an infinite place of K.

Since the Galois representations $T_v(\phi) \otimes_{A_v} F_v$ are semi-simple ([10]), Corollary (1.6) and Proposition (3.1) imply

Corollary (3.2). *Let v be a prime of A, r a positive integer, and $\mathfrak{n}$ a positive divisor of K. Then:*
(i) *There arise only finitely many isomorphism classes of Galois representations $T_v(\phi) \otimes_{A_v} F_v$ from Drinfeld modules ϕ over F of rank r such that $\Delta(\phi) \leq \mathfrak{n}$ and*

$$\mathfrak{d}(K_w({}_v\phi(K_w^{sep})/K_w) \leq \mathfrak{n}$$

for all infinite places w of K.
(ii) *For a Drinfeld module ϕ over K and an infinite place w of K, let $\Lambda_w(\phi)$ be the A-lattice ($\subset K_w^{sep}$) corresponding to $\phi \otimes_K K_w$. Then there arise only finitely many field extensions $K_w(\Lambda_w(\phi))/K_w$ from Drinfeld modules ϕ as in* (i). *Especially, $\mathfrak{d}(K_w(\Lambda_w(\phi))/K_w)$ is bounded.*

Remark (3.3). Over K_w, there may exist an infinite family $(\phi^{(n)})_{n \in \mathbb{N}}$ of Drinfeld modules of rank $r \geq 2$ such that ${}_v\phi^{(n)}(K_w^{sep})$ is rational over K_w but the ramification of the corresponding lattices $\Lambda^{(n)}$ is not bounded. For example, let $A := \mathbb{F}_q[T]$ and $\lambda^{(n)}$ a root of the Artin-Schreier equation $X^q - X = T^n$. Consider the rank two Drinfeld module $\phi^{(n)}$ over F_∞ corresponding to the A-lattice $\Lambda^{(n)} := A \cdot \lambda^{(n)} + A \cdot \frac{1}{T}$ ($\subset F_\infty^{sep}$). Then G_{F_∞} acts trivially on ${}_T\phi^{(n)}(F_\infty^{sep}) \simeq \Lambda^{(n)}/T\Lambda^{(n)}$ (since $\sigma(\lambda^{(n)}) - \lambda^{(n)} \in \mathbb{F}_q \subset A \subset T\Lambda^{(n)}$ for $\sigma \in G_{F_\infty}$), but $\mathfrak{d}(F_\infty(\lambda^{(n)})/F_\infty)$ is not bounded.

Remark (3.4). It is conjectured that the isogeny classes of Drinfeld modules ϕ over K are in one to one correspondence with the Galois representations $T_v(\phi) \otimes_{A_v} F_v$. If this is true, then Example (2.1), Example (2.2) and Corollary (3.2) imply the following:

For a Drinfeld module ϕ over K, consider the positive divisor of K

$$\overline{\Delta}(\phi) := \Delta(\phi) + \sum_{w|\infty} \mathfrak{d}(K_w(\Lambda_w(\phi))/K_w).$$

Suppose we are given a positive divisor $\mathfrak{n}$ of K. Then:
(i) There exist only finitely many isogeny classes of Drinfeld modules ϕ of rank $r \geq 1$ with $\overline{\Delta}(\phi) \leq \mathfrak{n}$ (Corollary (3.2)).
(ii) Let $A = \mathbb{F}_q[T]$ and $F = \mathbb{F}_q(T)$. Then
(ii-1) there exist infinitely many isogeny classes of Drinfeld modules ϕ of rank $r \geq 2$ over F with $\Delta(\phi) \leq \mathfrak{n}$ (Example (2.1));
(ii-2) there exists an isogeny class of Drinfeld modules of rank $r \geq 2$ over some finite extension of F which contains infinitely many isomorphism classes (Example (2.2) + Corollary (3.2)).

Remark (3.5). The above definition of $\overline{\Delta}(\phi)$ is not good. We hope to find a definition of the infinite component of $\overline{\Delta}(\phi)$ which is calculated directly from the defining equation of ϕ and with which we can bound $\sum \mathfrak{d}(K_w(\Lambda_w(\phi))/K_w)$.

4. Higher dimensional cases

The content of this section is an A-module version of a theorem of Fontaine (Théorème 1 of [5]), which can be regarded as a higher dimensional generalization of Lemma (1.3).

First we give a preliminary on Taylor expansions.

Let R be a commutative ring and $R[[X]] = R[[X_1, \cdots, X_h]]$ the ring of formal power series over R in h variables. For a multi-index $n = (n_1, \cdots, n_h) \in \mathbb{N}^h$ ($\mathbb{N}$ is the set of natural numbers including 0), we define a "differential operator" $\frac{\delta^n}{\delta X^n}$ as follows:

If $f(X) = \sum a_m X^m = \sum a_{m_1,\cdots,m_h} X_1^{m_1} \cdots X_h^{m_h} \in R[[X]]$, then

$$\frac{\delta^n}{\delta X^n} f(X) := \sum a_m \binom{m}{n} X^{m-n}$$

$$= \sum a_{m_1,\cdots,m_h} \binom{m_1}{n_1} \cdots \binom{m_h}{n_h} X_1^{m_1-n_1} \cdots X_h^{m_h-n_h},$$

where $\binom{m}{n} = \binom{m_1}{n_1} \cdots \binom{m_h}{n_h}$ is the "multi-binomial coefficient" with $\binom{m_i}{n_i} := 0$ if $n_i > m_i$.

Remarks (4.1).
(1) $\frac{\delta^n}{\delta X^n}$ is R-linear.
(2) $\frac{\partial^n}{\partial X^n} = n! \frac{\delta^n}{\delta X^n}$ (where $n! := n_1! \cdots n_h!$) is the usual differential operator, and $\frac{\delta^n}{\delta X^n} = \frac{1}{n!}(\frac{\delta}{\delta X})^n$ if $n!$ is invertible in R. In particular, we have $\frac{\partial}{\partial X_j} = \frac{\delta}{\delta X_j}$.
(3) For $f(X) \in R[[X]]$, put $f_Y(X) := f(X+Y) \in R[[X,Y]] = R[[X]][[Y]]$. We have

$$\frac{\delta^n}{\delta X^n} f_Y(X) = (\frac{\delta^n}{\delta X^n} f)(X+Y) \quad \text{in } R[[X,Y]].$$

(4) We have

$$\frac{\delta^n}{\delta X^n}(fg) = \sum_{k+l=n} (\frac{\delta^k}{\delta X^k} f)(\frac{\delta^l}{\delta X^l} g) \quad \text{for } f,\ g \in R[[X]].$$

(5) Let S be an R-algebra and I an ideal of S. Assume S is complete with respect to the I-adic topology. If $f(X) \in R[[X]]$ has the value $f(x) \in S$ at a point $x = (x_1, \cdots, x_h) \in S^h$, then $\frac{\delta^n}{\delta X^n} f(X)$ also has the value $\frac{\delta^n}{\delta X^n} f(x)$ at x for any $n \in \mathbb{N}^h$.

Proposition (4.2). *For $f(X) \in R[[X]]$, we have the formal Taylor expansion (or rather, the binomial expansion)*

$$f(X+Y) = \sum_{|n| \geq 0} \frac{\delta^n}{\delta X^n} f(X) \cdot Y^n \quad \text{in } R[[X, Y]]. \tag{4.2.1}$$

Let S and I be as in Remark (4.1), (5). If $f(X)$ has the value $f(x) \in S$ at $x \in S^h$ and y is an element of I^h, then $f(x+y) \in S$ also exists and we have

$$f(x+y) = \sum_{|n| \geq 0} \frac{\delta^n}{\delta X^n} f(x) \cdot y^n \quad \text{in } S. \tag{4.2.2}$$

Proof. Write $f(X+Y) = \sum a_n(X) Y^n$ with $a_n(X) \in R[[X]]$. Applying $\frac{\delta^n}{\delta Y^n}$ to both sides and reducing modulo Y, we obtain (cf. Remark (4.1), (3))

$$\frac{\delta^n}{\delta X^n} f(X) = a_n(X)$$

and hence (4.2.1).

The latter half of the Proposition is obvious. □

Next we recall Fontaine's numbering of the ramification groups of a local field and some of his results ([5], §1). In the rest of this section, if L is a discrete valuation field, $\mathfrak{O}_L$ (resp. $\mathfrak{m}_L$, resp. k_L) denotes the integer ring of L (resp. the maximal ideal of $\mathfrak{O}_L$, resp. the residue field $\mathfrak{O}_L/\mathfrak{m}_L$).

In the following, K is a complete discrete valuation field with perfect residue field k of characteristic $p \neq 0$. Let v_K denote the valuation on K normalized by $v_K(K^\times) = \mathbb{Z}$, and also its unique extension to any algebraic extension of K. If $\mathfrak{a}$ is a subset of an algebraic extension of K, we put $v_K(\mathfrak{a}) := \inf\{v_K(x); x \in \mathfrak{a}\}$.

For a finite Galois extension L/K, Fontaine defines a filtration with lower (resp. upper) numbering $G_{(i)}$ (resp. $G^{(u)}$) ($i, u \in \mathbb{R}$) on the Galois group $G = \mathrm{Gal}(L/K)$, which is connected with the usual filtration G_i (resp. G^u) defined in Chapitre IV of [8] by

$$G_i = G_{((i+1)/e)}, \qquad \text{resp.} \quad G^u = G^{(u+1)},$$

where $e = e_{L/K}$ is the ramification index of L/K.

He also defines a real number $i_{L/K}$ (resp. $u_{L/K}$), which is characterized as the largest real number i (resp. u) such that $G_{(i)} \neq 1$ (resp. $G^{(u)} \neq 1$). $i_{L/K}$ and $u_{L/K}$

are connected by

$$u_{L/K} = \int_0^{i_{L/K}} (G_{(x)} : 1)dx.$$

Then he proves the following

Proposition (4.3). *Let L be a finite Galois extension of K.*
(1) ([**5**, 1.3]) $v_K(\mathfrak{D}(L/K)) = u_{L/K} - i_{L/K}$.
(2) ([**5**, 1.5]) *For a real number $m \geq 0$, consider the following property (P_m) on the extension L/K:*

$$(P_m)\left\{\begin{array}{c} For\ any\ algebraic\ extension\ E\ of\ K,\ if\ there\ exists \\ an\ \mathfrak{O}_K-algebra\ homomorphism:\ \mathfrak{O}_L \to \mathfrak{O}_E/\mathfrak{a}^m_{E/K} \\ (where\ \mathfrak{a}^m_{E/K} := \{x \in \mathfrak{O}_E; v_K(x) \geq m\}), \\ then\ there\ exists\ a\ K-embedding:\ L \hookrightarrow E. \end{array}\right\}$$

Then
(i) *if $m > u_{L/K}$, L/K has the property (P_m);*
(ii) *if L/K has the property (P_m), we have $m > u_{L/K} - e^{-1}_{L/K}$.*

Now we shall refine Fontaine's Proposition 1.7 of [**5**] as follows. The main point is that it works, *mutatis mutandis*, even in positive characteristics.

Proposition (4.4). *Let B be a finite flat $\mathfrak{O}_K$-algebra which is locally of complete intersection over $\mathfrak{O}_K$. Suppose that there exists an element $a \in \mathfrak{O}_K$ such that $\Omega^1_{B/\mathfrak{O}_K}$ is a flat (B/aB)-module.*
(i) *Let S be a finite flat $\mathfrak{O}_K$-algebra and I an ideal of S. Suppose either the S-submodule $a^{-1}I^{p-1}$ of $K \otimes_{\mathfrak{O}_K} S$ is topologically nilpotent (i.e., $\bigcap_{n\geq 1}(a^{-1}I^{p-1})^n = 0$), or I has a PD-structure such that $\bigcap_{n\geq 1} I^{[n]} = 0$.*
(a) *For any $\mathfrak{O}_K$-algebra homomorphism $u : B \longrightarrow S/aI$, there exists an $\mathfrak{O}_K$-algebra homomorphism $\hat{u} : B \longrightarrow S$ which is uniquely determined by u (mod I) and makes the following diagram commutative:*

$$\begin{array}{ccc} B & \stackrel{u}{\longrightarrow} & S/aI \\ {\scriptstyle \hat{u}}\downarrow & & \downarrow \\ S & \longrightarrow & S/I \end{array}$$

(b) *The canonical map of sets*

$$\mathrm{Hom}_{\mathfrak{O}_K-\mathrm{alg}}(B, S) \longrightarrow \mathrm{Hom}_{\mathfrak{O}_K-\mathrm{alg}}(B, S/I)$$

is injective.
(ii) *The K-algebra $B_K := K \otimes_{\mathfrak{O}_K} B$ is étale. Let L be the smallest subfield of a separable closure K^{sep} of K which contains the images $u(B)$ for all*

$$u \in \mathrm{Hom}_{K-\mathrm{alg}}(B_K, K^{sep}).$$

Then L/K is a finite Galois extension and $u_{L/K} \leq v_K(a) + \frac{1}{p-1} \cdot \min\{v_K(a), v_K(p)\}$.

The proof is essentially the same as the original one due to Fontaine, but here we reproduce his proof of (i) to make clear the meaning of the condition in (i).

Proof. We may and do suppose B is a local ring, because B is the product of a finite number of local rings. Let $\mathfrak{m}_B$ be the maximal ideal of B. Replacing K by an unramified extension if necessary, we may also suppose $B/\mathfrak{m}_B = k$, the residue field of $\mathfrak{O}_K$.

Then $\Omega^1_{B/\mathfrak{O}_K}$ is a free (B/aB)-module. Let $x_1, \cdots, x_h$ be elements of $\mathfrak{m}_B$ the images of which form a k-base of $\mathfrak{m}_B/(\mathfrak{m}_B^2 + \mathfrak{m}_K B)$. We see from the definition of differential modules that $dx_1, \cdots, dx_h$ generate $\Omega^1_{B/\mathfrak{O}_K}$, and further, they form a (B/aB)-base of $\Omega^1_{B/\mathfrak{O}_K}$ because of the canonical isomorphisms

$$\Omega^1_{B/\mathfrak{O}_K} \otimes_B B_o \xrightarrow{\sim} \Omega^1_{B_o/k} \qquad (B_o := B/\mathfrak{m}_K B),$$
$$\mathfrak{m}_B/(\mathfrak{m}_B^2 + \mathfrak{m}_K B) \xrightarrow{\sim} \mathfrak{m}_{B_o}/\mathfrak{m}_{B_o}^2 \xrightarrow{\sim} \Omega^1_{B_o/k} \otimes_{B_o} k,$$

where $\mathfrak{m}_{B_o} = \mathfrak{m}_B/\mathfrak{m}_K B$ is the maximal ideal of B_o.

Now let

$$\alpha : \mathfrak{O}_K[[X_1, \cdots, X_h]] \longrightarrow B$$

be the unique continuous $\mathfrak{O}_K$-algebra homomorphism such that $\alpha(X_j) = x_j$, and let $J := \mathrm{Ker}(\alpha)$. Since B is finite of complete intersection over $\mathfrak{O}_K$, J is generated by h elements, say $P_1, \cdots, P_h \in \mathfrak{O}_K[[X_1, \cdots, X_h]]$.

For each i, we have $\sum_j \frac{\delta P_i}{\delta X_j}(x_1, \cdots, x_h)dx_j = 0$ (note $\frac{\delta}{\delta X_j} = \frac{\partial}{\partial X_j}$), which implies $\frac{\delta P_i}{\delta X_j}(x_1, \cdots, x_h) \in aB$. Hence there are $p_{ij} \in B$ such that $\frac{\delta P_i}{\delta X_j}(x_1, \cdots, x_h) = ap_{ij}$. The fact that $\Omega^1_{B/\mathfrak{O}_K}$ is a free (B/aB)-module means that the free B-submodule of $\oplus_{j=1}^h B dX_j$ generated by $\sum_j \frac{\delta P_i}{\delta X_j}(x_1, \cdots, x_h)dX_j$, $1 \le i \le h$, coincides with the one generated by adX_j, $1 \le j \le h$. We can therefore find $q_{li} \in B$ such that

$$adX_l = \sum_i q_{li}\Big(\sum_j \frac{\delta P_i}{\delta X_j}(x_1, \cdots, x_h)dX_j\Big), \quad 1 \le l \le h,$$

i.e., $a1_h = (q_{li})(ap_{ij})$. (1_h is the unit matrix of degree h.) Since B is a free $\mathfrak{O}_K$-module, we can divide both sides by a. Thus the matrix (p_{ij}) is invertible in $\mathrm{M}_h(B)$ and $(q_{li}) = (p_{ij})^{-1}$.

The case of PD-ideals is proved in [5], so we suppose $a^{-1}I^{p-1}$ is topologically nilpotent. Then the ideal $a^{-1}I^{p-1} + I$ is also topologically nilpotent. Set $I_n := (a^{-1}I^{p-1} + I)^{n-1}I$, $n \ge 1$ (so that $a^{-1}I_n^{p-1}$ is again topologically nilpotent, and S is canonically isomorphic to the projective limit of the system $(S/I_n)_{n\ge1}$). It is easily seen that $I_n^p \subset aI_{2n}$ and $I_n^2 \subset I_{2n}$. To show the assertion, it is enough to verify:

For any integer $n \ge 1$ and an $\mathfrak{O}_K$-algebra homomorphism $u : B \longrightarrow S/aI_n$, there exists an $\mathfrak{O}_K$-algebra homomorphism $u' : B \longrightarrow S/aI_{2n}$ such that $u' \pmod{I_{2n}}$ is uniquely determined by $u \pmod{I_n}$ and u' makes the following diagram commutative:

$$\begin{array}{ccc} B & \xrightarrow{u} & S/aI_n \\ {\scriptstyle u'}\downarrow & & \downarrow \\ S/aI_{2n} & \longrightarrow & S/I_n \end{array}$$

In other words, writing I for I_n and I_2 for I_{2n} :

For any elements $u_1, \cdots, u_h$ of S such that

$$P_i(u_1, \cdots, u_h) = a\lambda_i \quad \text{with some } \lambda_i \in I \quad (1 \leq i \leq h),$$

there exist $\mu_1, \cdots, \mu_h \in I$ such that $\mu_j \pmod{I_2}$ are uniquely determined by $u_j \pmod{I}$ and

$$P_i(u_1 + \mu_1, \cdots, u_h + \mu_h) \in aI_2 \quad (1 \leq i \leq h). \tag{4.4.1}$$

If $\mu_j \in I$, we have the Taylor expansion (4.2.2)

$$P_i(u_1 + \mu_1, \cdots, u_h + \mu_h) = a\lambda_i + \sum_j \frac{\delta P_i}{\delta X_j}(u_1, \cdots, u_h)\mu_j + R_i \tag{4.4.2}$$

with $R_i := \sum_{|r| \geq 2} \frac{\delta^r P_i}{\delta X^r}(u_1, \cdots, u_h)\mu^r$.

For any element $P \in J$, we have $\frac{\delta P}{\delta X_j}(x_1, \cdots, x_h) \in aB$, i.e.,

$$\frac{\delta P}{\delta X_j}(X_1, \cdots, X_h) \in a\mathfrak{O}_K[[X_1, \cdots, X_h]] + J.$$

If $|r| \geq 1$ and $r!$ is invertible in $\mathfrak{O}_K$, we see inductively (cf. Remark (4.1), (2))

$$\frac{\delta^r P}{\delta X^r}(X_1, \cdots, X_h) \in a\mathfrak{O}_K[[X_1, \cdots, X_h]] + J,$$

so

$$\frac{\delta^r P}{\delta X^r}(u_1, \cdots, u_h) \in aS + aI = aS.$$

Since $I^2 \subset I_2$, we have

$$\frac{\delta^r P}{\delta X^r}(u_1, \cdots, u_h) \cdot \mu^r \in aI_2,$$

if $|r| \geq 2$ and $r!$ is invertible in $\mathfrak{O}_K$.

On the other hand, we have

$$\mu^r \in I^{|r|} \subset I^p \subset aI_2$$

if p divides $r!$, and $\frac{\delta^r P}{\delta X^r}(u_1, \cdots, u_h)$ are always in S (Remark (4.1), (5)). Thus we have

$$R_i \in aI_2. \tag{4.4.3}$$

Take an element $P_{ij} \in \mathfrak{O}_K[[X_1, \cdots, X_h]]$ such that $\alpha(P_{ij}) = p_{ij} \in B$ for each (i, j). We have

$$\frac{\delta P_i}{\delta X_j}(x_1, \cdots, x_h) = a p_{ij},$$

i.e., $\frac{\delta P_i}{\delta X_j} = a P_{ij} + R_{ij}$ with some $R_{ij} \in J$, from which follows the congruence

$$\frac{\delta P_i}{\delta X_j}(u_1, \cdots, u_h) \equiv a P_{ij}(u_1, \cdots, u_h) \pmod{aI},$$

and

$$\frac{\delta P_i}{\delta X_j}(u_1, \cdots u_h) \cdot \mu_j \equiv a P_{ij}(u_1, \cdots, u_h) \cdot \mu_j \pmod{aI_2}. \tag{4.4.4}$$

Putting (4.4.3) and (4.4.4) into (4.4.2), we have

$$P_i(u_1 + \mu_1, \cdots, u_h + \mu_h) \equiv a(\lambda_i + \sum_j P_{ij}(u_1, \cdots, u_h) \cdot \mu_j) \pmod{aI_2}.$$

Since S is flat over $\mathfrak{O}_K$, the condition (4.4.1) for μ_j is now equivalent to

$$\lambda_i + \sum_j P_{ij}(u_1, \cdots, u_h) \cdot \mu_j \equiv 0 \pmod{I_2}, \quad 1 \leq i \leq h.$$

Since the matrix $(p_{ij}) = (P_{ij}(x_1, \cdots, x_h))$ is invertible, the matrix $(P_{ij}(u_1, \cdots, u_h))$ is invertible modulo aI. Now the existence of $\mu_j \in I$ satisfying (4.4.1) is clear. Moreover $u_j \pmod{I}$, $1 \leq j \leq h$, determine $\mu_j \pmod{I_2}$, $1 \leq j \leq h$, uniquely, because they determine $\lambda_i \equiv 0 \pmod{I}$ and $P_{ij}(u_1, \cdots, u_h) \pmod{I}$ uniquely and $I^2 \subset I_2$.

Part (b) of (i) follows immediately from Part (a).

The proof of (ii) is exactly the same as in [5]. □

Corollary (4.5). *Let the notation and hypothesis be as in Proposition (4.4). Then we have* $v_K(\mathfrak{D}(L/K)) < v_K(a) + \frac{1}{p-1}\min\{v_K(a), v_K(p)\}$ *unless* $v_K(\mathfrak{D}(L/K)) = 0$.

Proof. If L/K is unramified, then $v_K(\mathfrak{D}(L/K)) = 0$. If not, we have $i_{L/K} > 0$ and by (Prop. (4.3), (1))

$$v_K(\mathfrak{D}(L/K)) = u_{L/K} - i_{L/K} < u_{L/K} \leq v_K(a) + \frac{1}{p-1}\min\{v_K(a), v_K(p)\}. \quad \square$$

Theorem (4.6). *Let A be a complete discrete valuation ring with finite residue field, and fix a prime element π of A. Let K be a local field of "mixed characteristic" over A, i.e., a complete discrete valuation field K with perfect residue field which is endowed with an injective ring homomorphism $A \longrightarrow K$ inducing a local homomorphism $A \longrightarrow \mathfrak{O}_K$. Let $n \geq 1$ be an integer and J a finite flat π-module scheme over $\mathfrak{O}_K$* ([10], §1) *such that the invariant differential module ω_J of J is a free $(\mathfrak{O}_K/\pi^n\mathfrak{O}_K)$-module. (A typical example of such a π-module is the kernel of π^n on a π-divisible group (loc. cit.)). Let $u_o := n v_K(\pi) + \frac{1}{p-1}\min\{n v_K(\pi), v_K(p)\}$, H the kernel of the action of $G_K = \mathrm{Gal}(K^{sep}/K)$ on $J(K^{sep})$, and $L := (K^{sep})^H$. Then we have $G_K^{(u)} \subset H$ for all $u > u_o$, and $v_K(\mathfrak{D}(L/K)) < u_o$.*

Proof. As in the proof of Théorème 1 of [5], the affine ring B of J is locally of complete intersection. Since $\Omega^1_{B/\mathfrak{O}_K} = B \otimes_{\mathfrak{O}_K} \omega_J$ is a free $(B/\pi^n B)$-module, we can apply Prop. (4.4) and Corollary (4.5) with $a = \pi^n$ and obtain the theorem. □

Remark (4.7). In some simple cases, direct calculations yield sharper results. For example, let A and π be as above, F the fraction field of A, and F_n, $n \geq 0$, the field of π^n-division points of a Lubin-Tate group over A associated with π. If $L/K = F_m/F_n$ with $m > n$, we have

$$u_{L/K} = \begin{cases} m, & \text{if } n = 0 \\ q^n + (m-n-1)(q^n - q^{n-1}), & \text{if } n \geq 1, \end{cases}$$

$$i_{L/K} = \begin{cases} \frac{1}{q-1}, & \text{if } n = 0, \\ q^{n-1}, & \text{if } n \geq 1, \end{cases}$$

$$v_K(\mathfrak{D}(L/K)) = \begin{cases} m - \frac{1}{q-1}, & \text{if } n = 0 \\ (m-n)(q^n - q^{n-1}), & \text{if } n \geq 1. \end{cases}$$

5. The case of non-scalar A-actions on tangent spaces

Important classes of abelian T-modules ([1]), such as higher Carlitz modules $C^{\otimes n}$ ([2]) and tensor products of Drinfeld modules ([1], [6]), are such that the actions of T on the tangent spaces are not just multiplication by T, but T plus nilpotent linear maps. In this section, we study the ramification arising from division points of such objects.

Let A, π and K be as in Theorem (4.6); A a complete discrete valuation ring with finite residue field, π a uniformizer of A, and K a local field of "mixed characteristic" over A. Consider a smooth connected commutative formal group J over $\mathfrak{O}_K$ with an A-action

$$\phi : A \longrightarrow \mathrm{End}_{\mathfrak{O}_K}(J)\,; \qquad a \mapsto \phi_a$$

such that, for all $a \in A$, ϕ_a induces a linear map $\mathrm{Lie}(\phi_a)$ on $\mathrm{Lie}(J)$ of the form (multiplication by a) + (nilpotent map). If J is, for example, the tensor product of abelian T-modules with scalar T-actions on their tangent spaces, Theorem (4.6) for $J_n = \mathrm{Ker}(\phi_{T^n})$ would be valid because $\otimes$ and $T_v(\cdot)$ should be compatible (this is shown in [6] at least for the tensor products of two Drinfeld modules). What can be said on the ramification of the geometric points of $J_n := \mathrm{Ker}(\phi_{\pi^n})$ in other cases ?

(5.1). First assume that the nilpotent map $\mathrm{Lie}(\phi_{\pi^n}) - \pi$ is divisible by π in $\mathrm{End}_{\mathfrak{O}_K}(\mathrm{Lie}(J))$. Then, since the image of $\mathrm{Lie}(\phi_{\pi^n})$ is $\pi^n\,\mathrm{Lie}(J)$, we have

$$\Omega^1_{J_n} = \Omega^1_J / \pi^n \Omega^1_J,$$

which is a flat $B/\pi^n B$-module (B is the affine ring of J_n). So we can apply Proposition (4.4), and Theorem (4.6) remains valid for such J_n.

(5.2). We now return to a general J. Let $d := \dim(J) = \operatorname{rank}_{\mathfrak{O}_K}(\operatorname{Lie}(J))$, and let p^k be the smallest power of p, the residue characteristic of A, such that $p^k \geq d$. Then for any multiple m of p^k, the nilpotent map $\operatorname{Lie}(\phi_{\pi^m}) - \pi^m$ is divisible by π^m in $\operatorname{End}_{\mathfrak{O}_K}(\operatorname{Lie}(J))$, as is easily seen by looking at the binomial expansion of $[\pi + (\operatorname{Lie}(\phi_\pi) - \pi)]^{p^k}$. In view of (5.1), we have

Theorem (5.3). *Let J and p^k be as above. For a positive integer n, let n' be the smallest multiple of p^k not less than n, $u_o := n'v_K(\pi) + \frac{1}{p-1}\min\{n'v_K(\pi),\ v_K(p)\}$, H the kernel of the action of G_K on $J_n(K^{sep})$, and $L := (K^{sep})^H$. Then we have $G_K^{(u)} \subset H$ for all $u > u_o$, and $v_K(\mathfrak{D}(L/K)) < u_o$.*

This theorem reduces to Theorem (4.6) when n is divisible by p^k. But if n is not divisible by p^k, the ramification can be bigger than expected from Theorem (4.6), as the following example shows:

Example (5.4). Let $A = \mathbb{F}_q[[T]]$ and $K = \mathbb{F}_q((T))$, and consider the d-dimensional formal group $\widehat{\mathbb{G}}_a^{\oplus d}$ over A with an A-action defined by

$$\phi_T\begin{pmatrix} X_1 \\ \vdots \\ X_d \end{pmatrix} = \begin{pmatrix} T & -1 & & \\ & \ddots & \ddots & \\ & & \ddots & -1 \\ & & & T \end{pmatrix}\begin{pmatrix} X_1 \\ \vdots \\ X_d \end{pmatrix} + \begin{pmatrix} X_1^q \\ \vdots \\ X_d^q \end{pmatrix}.$$

Then we have

$$K(J_1(K^{sep})) = K({}_{T^d}C(K^{sep})),$$

where C is the Carlitz module defined by $C_T(X) = TX + X^q$. This means that $\mathfrak{D}(K(J_1(K^{sep}))/K)$ can become arbitrarily large according to d.

References

[1] G. W. Anderson, t-motives, *Duke Math. J.* **53** (1986), 457 – 502.

[2] G. W. Anderson and D. S. Thakur, Tensor powers of the Carlitz module and zeta values, *Ann. of Math.* **132** (1990), 159 – 191.

[3] V. G. Drinfeld, Elliptic modules, *Math. USSR-Sb.* **23** (1974), 561 – 592.

[4] G. Faltings, Endlichkeitssätze für Abelsche Varietäten über Zahlkörpern, *Invent. Math.* **73**, (1983), 349 – 366.

[5] J-M. Fontaine, Il n'y a pas de variété abélienne sur $\mathbb{Z}$, *Invent. Math.* **81** (1985), 515 – 538

[6] Y. Hamahata, On the Tate module associated to the tensor product of two Drinfeld modules I, II, preprint.

[7] D. Hayes, Explicit class field theory for rational function fields, *Trans. Amer. Math. Soc.* **189** (1974), 77 – 91.

[8] J-P. Serre, *Corps locaux* (3ème édition), Hermann, Paris (1980)

[9] J-P. Serre, A letter to D. Goss, dated April 14, 1990.

[10] Y. Taguchi, Semi-simplicity of the Galois representations attached to Drinfeld modules over fields of "infinite characteristic", *J. Number Theory* (to appear).

Tokyo Metropolitan University, Hachioji, Tokyo, 192–03, Japan
Email: taguchi@math.metro-u.ac.jp

Rigid Analytic Modular Forms: An Integral Transform Approach

Jeremy Teitelbaum

The theory of parameter spaces for Drinfeld modules leads to a construction of "characteristic p modular forms". First studied by Goss, these modular forms are functions on the non–archimedean upper half plane over a complete local field of characteristic p, but nevertheless they are similar in many ways to the classical forms for $\mathrm{SL}_2(\mathbb{Z})$. I hope in this paper to give a brief introduction to these modular forms based on a technique which views them as a kind of integral transform. This technique, based on ideas of Drinfeld and Schneider, reduces the construction of such modular forms to combinatorial considerations involving the Bruhat–Tits tree $\mathcal{T}$ of PGL_2.

I will begin with a general introduction to the upper half plane, its relation to the Bruhat–Tits tree of PGL_2, and its role in the theory of Drinfeld modules. Then I will discuss the measures and integrals which enable one to construct modular forms. I will conclude with a very brief survey of some of the technical issues in the theory.

This paper is a written version of my talk at the Ohio State Conference on Function Fields. For details, verifications, justifications, and proofs in general, I refer the reader to my paper **[T2]**.

Part I. An Introduction to the Theory

Notation. Let us begin by fixing a complete local field k_∞ of characteristic p with residue field the finite field $\mathbb{F}_q$ with q elements. We may view this field as the field of Laurent series

$$k_\infty = \mathbb{F}_q((\frac{1}{T}))$$

where the parameter is written $\frac{1}{T}$ for historical reasons. Let $|\cdot|$ be the absolute value on k_∞ with $|T| = q$, and let $\mathcal{O}$ be the valuation ring. The field k_∞ plays the role of the real numbers in our discussion; the role of the complex numbers is taken by the completion of an algebraic closure of k_∞, which we will call K.

The upper half plane. We'll begin this discussion with a description of the rigid analytic upper half plane Ω over k_∞. The easiest way to define Ω is simply to set

$$\Omega = K - k_\infty,$$

that is, Ω is the complement of k_∞ in K. Notice that, if we view K as the complex and k_∞ as the real numbers, then the classical version of Ω is the set $\mathbf{C} - \mathbf{R}$, which is the union of the upper and lower complex half planes.

The group $\mathrm{GL}_2(k_\infty)$ acts on Ω. In fact, let z be a coordinate function on Ω; then we arrange things so that $\mathrm{GL}_2(k_\infty)$ acts by linear fractional transformations on z. In other words, if $P \in \Omega$, and

$$\gamma = \begin{pmatrix} a & b \\ c & d \end{pmatrix} \in \mathrm{GL}_2(k_\infty)$$

then

$$z(\gamma P) = \frac{az(P)+b}{cz(P)+d}.$$

Of course, this is all well–defined, since $\mathrm{GL}_2(k_\infty)$ preserves k_∞.

The classical upper half plane is a Riemann surface in a rather trivial way; we must work somewhat harder to put enough structure on Ω to support the existence of a theory of analytic functions. In technical terms, we must endow Ω with the structure of a "rigid analytic space." In practice, this means that we must cover Ω with relatively simple sets on which we have a well understood notion of analytic function. These sets should overlap on subsets which are "large enough" that we can carry out some form of analytic continuation. This goal is complicated by the fact that K is a totally disconnected space, so that we cannot, for example, simply cover Ω by open sets in the topology induced on it from the natural p–adic topology on K. Two open sets in this topology are either disjoint, or one is included in the other!

In order to describe a good covering of Ω, we begin by selecting a model set as a starting point. Our model set is the set U defined by the inequalities

$$\frac{1}{q} < |z(P)| < q$$

and

$$|z(P) - \lambda| > \frac{1}{q}, \quad \lambda \in \mathbb{F}_q.$$

The open set U is constructed by cutting disks centered at points with coordinates in $\mathbb{F}_q \cup \{\infty\}$ out of $\mathbb{P}^1_K$. U should be thought of as an inner region (where $|z(P)| = 1$) with $q+1$ bounding annuli (of the form $1 < |z| < 1/q$) attached. The inner region is an example of an "affinoid domain" in the terminology of rigid analytic geometry, while the annuli are what Coleman calls "wide open spaces." Figure 1 shows a picture of U.

I will leave to the reader the job of checking that the translates of U by $\mathrm{GL}_2(k_\infty)$ cover Ω. In addition, the stabilizer of U is the subgroup $\mathrm{GL}_2(\mathcal{O})$ of $\mathrm{GL}_2(k_\infty)$. The complete set of rules for matching up the open sets in the covering is given by the following lemma, which is easy to prove by calculation.

Lemma. *Any two translates γU and τU of U are either identical, disjoint, or intersect along a boundary annulus.*

Make a simplicial complex by taking one vertex for each distinct set in the covering $\{\gamma U\}$ of Ω, and connect two vertices by an edge if the two corresponding sets overlap. It is not hard to see that the resulting complex (the *nerve* of the covering by the γU) is an infinite, homogeneous tree with $q+1$ edges leaving every vertex. Let us call this tree

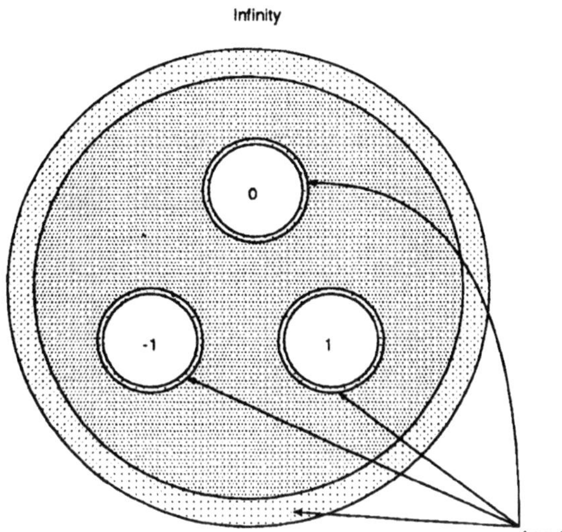

Figure 1.

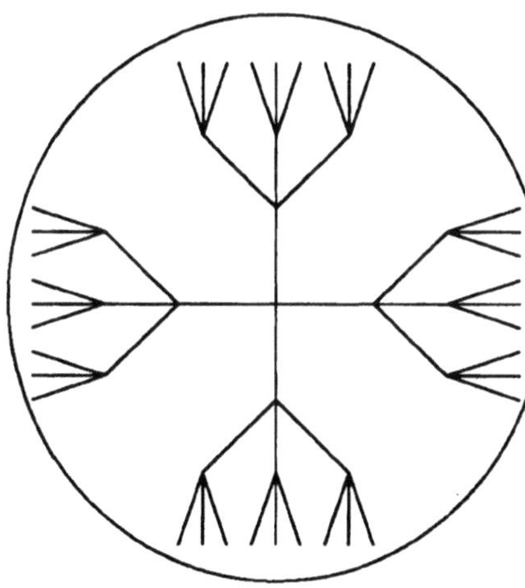

Figure 2.

$\mathcal{T}$. (See Figure 2.) Clearly, the action of $\mathrm{GL}_2(k_\infty)$ on Ω leads to an action of $\mathrm{GL}_2(k_\infty)$ on $\mathcal{T}$.

Although we have constructed $\mathcal{T}$ as the nerve of a special covering of Ω, it has many other interpretations. For example, it arises as a "Bruhat–Tits building" for $\mathrm{PGL}_2(k_\infty)$, as explained in Serre's book *Trees* (**[Se]**). It may also be viewed as a "tree of norms," as explained in Drinfeld's original paper on Elliptic Modules (**[D1]**).

Suppose P is a point of Ω. Either P belongs to exactly one set γU (if P belongs to the interior region of γU) or else P belongs to exactly two translates of U. This dichotomy means that we may define a map from $r : \Omega \to \mathcal{T}$; if P belongs to precisely one γU then $r(P)$ is the corresponding vertex; otherwise, $r(P)$ is the edge joining the two translates of U containing P.

This "reduction" map can be refined considerably (Drinfeld's tree of norms construction is particularly well–suited for such a refinement) but our *ad hoc* definition should serve us well enough.

To get a clearer picture of the reduction map, consider Figure 3, which is drawn for $q = 3$. As shown there, the four ($= q+1$) arms radiating out of a given vertex correspond to the four boundary annuli on a particular set γU in our covering. In general, the $q+1$ boundary annuli on the basic set U are naturally identified with the points of $\mathbb{P}^1_{\mathbb{F}_q}$, and the $\mathrm{GL}_2(k_\infty)$ invariance of the whole picture means that the $q+1$ edges leaving any vertex may similarly be thought of as a copy of $\mathbb{P}^1_{\mathbb{F}_q}$. Obviously the construction of the reduction map guarantees that it will commute with the action of $\mathrm{GL}_2(k_\infty)$ on $\mathcal{T}$ and Ω.

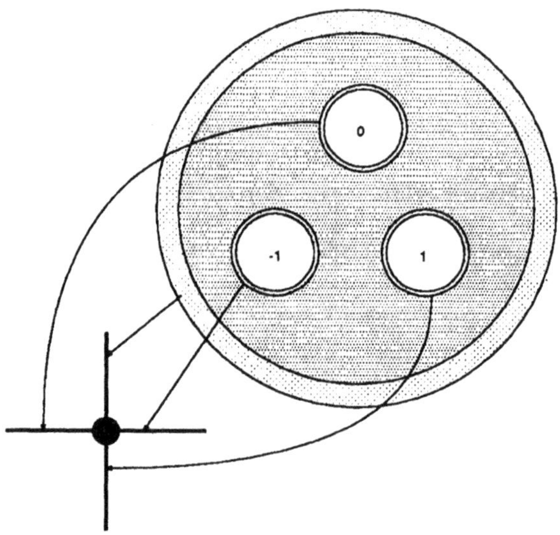

Figure 3.

There is one more feature of the tree $\mathcal{T}$ which will be essential for our investigation of analytic functions and modular forms on Ω. This is the set of *ends* of $\mathcal{T}$. To reach an end of $\mathcal{T}$, simply travel along an infinite path in $\mathcal{T}$ without back–tracking. At the end of your trip you reach an end of $\mathcal{T}$. Of course, this definition is rather imprecise. To formalize it, fix a vertex v_0 of $\mathcal{T}$ and define an end of $\mathcal{T}$ to be an infinite path leaving v_0 without back–tracking. It isn't hard to fix up this definition and make it independent of v_0. The authoritative source on all these matters is Serre's book (**[Se]**).

We put a topology on the set of ends by declaring that an open set of ends consists of all paths passing through a particular oriented edge of $\mathcal{T}$. This makes $\mathrm{Ends}(\mathcal{T})$ into a compact space. It also gives a correspondence between edges of $\mathcal{T}$ and compact open subsets of the ends of $\mathcal{T}$ which we will make heavy use of later on. (This is depicted in Figure 4.)

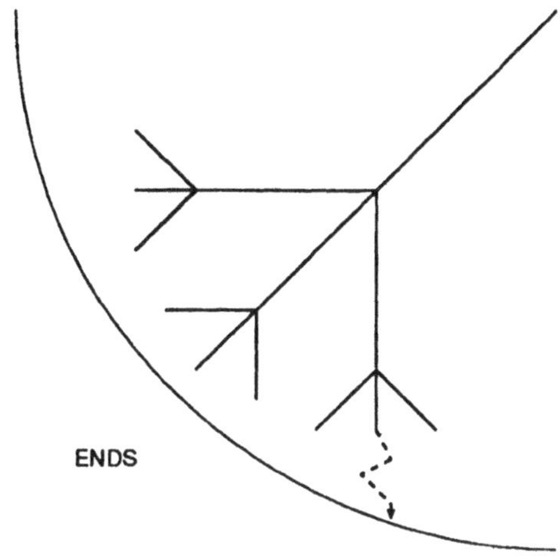

Figure 4.

Let's think for a moment about what $\text{Ends}(\mathcal{T})$ really is by returning to our interpretation of $\mathcal{T}$ as the nerve of the standard covering of Ω. Fix a sequence $e_1, e_2, \ldots$ of edges in $\mathcal{T}$ leading out to an end x. Each of these edges corresponds to an annulus in Ω; most of the time they are nested one inside the next, getting smaller and smaller (what is the exception? see below for help). It is not hard to believe that there is a unique point y in K which is surrounded by all of these nested annuli. In addition, this point y cannot belong to Ω – after all, a point in Ω must be carried, under the reduction map, either to an edge of $\mathcal{T}$ or to a vertex of $\mathcal{T}$. Therefore the point y must belong to $K - \Omega$, or k_∞. (In the exceptional case I alluded to above, the point y must be the point at infinity on $\mathbb{P}^1_{k_\infty}$.) This argument is supposed to convince the reader that there is a natural identification of the ends of $\mathcal{T}$ with $\mathbb{P}^1_{k_\infty}$ – indeed, this identification is a homeomorphism. Even more, the reduction map extends from Ω to $\mathbb{P}^1_K$, with points of Ω reducing to $\mathcal{T}$ and the points of $\mathbb{P}^1_{k_\infty}$ reducing to the ends of $\mathcal{T}$. Finally, all of this is compatible with the action of $\text{GL}_2(k_\infty)$.

Arithmetic groups and Drinfeld modules

Now that we have described the apparatus of the upper half plane, we may introduce some discrete groups; our modular forms will be functions with nice transformation properties with respect to these groups. These discrete groups are arithmetic in origin. The simplest situation arises when we choose $k = \mathbb{F}_q(T)$ and embed k into k_∞ using the "place at infinity." Then the ring $A = \mathbb{F}_q[T]$ is the ring of functions with poles only at infinity, and we have an embedding

$$\text{GL}_2(A) \hookrightarrow \text{GL}_2(k_\infty).$$

This gives us an action of $\mathrm{GL}_2(A)$ on Ω.

More generally, we may start with any global field k of characteristic p, choose a place

$$\text{“}\infty\text{”}: k \hookrightarrow k_\infty,$$

and let A be the ring of functions with poles only at infinity. Once again, we have $\mathrm{GL}_2(A)$ acting on Ω.

Of course, we needn't stop with $\Gamma = \mathrm{GL}_2(A)$ (which is the function field version of the modular group $\mathrm{SL}_2(\mathbb{Z})$). We may choose an ideal I of A and use it to construct congruence groups $\Gamma(I)$ and $\Gamma_0(I)$ by specifying congruence conditions on the matrices in the same way as we do for the modular group.

The significance of Ω in the theory of moduli of Drinfeld modules is easier to explain when we have the field k and the ring A at hand. Indeed, it is a consequence of the general theory of Drinfeld modules (which is explained in other places in this volume) that a "rank two Drinfeld A-module over K" is determined by giving an injection

$$\iota : Ax_1 + Ax_2 \hookrightarrow K.$$

The image of such a map is an A-lattice in K. Two such lattices yield the same Drinfeld module precisely when they are scalar multiples of each other. Therefore we may specify a Drinfeld A-module over K by giving $z = \iota(x_1)/\iota(x_2)$ (which must lie in Ω), and two different z's in Ω correspond to the same module precisely when they are $\mathrm{GL}_2(A)$ equivalent. In this way, the quotient space $\mathrm{GL}_2(A)\backslash\Omega$ becomes a parameter space for rank two A-modules. It is not hard to accept, by analogy with elliptic curves, that congruence subgroups yield parameter spaces for modules with "level structure." This theory is described in detail in other papers in this volume, and in **[D]**.

Quotients. As it turns out, the quotient of Ω by $\mathrm{GL}_2(A)$ or one of its congruence subgroups is an (affine) algebraic curve. This takes a fair amount of work to prove, including some information on algebraization of rigid analytic spaces (non–archimedean GAGA type theorems, as in **[GVdP]**). However, even if we can't cover all that ground in this article, we can at least say a few words about the much simpler, but related problem of studying the quotient of the tree $\mathcal{T}$ by a congruence subgroup of $\mathrm{GL}_2(A)$. Working out the structure of $\Gamma\backslash\mathcal{T}$ is a combinatorial problem and in general the resulting graphs are quite complicated. A few cases are worked out in Serre (for some interesting choices of A) and a few more cases are worked out in Gekeler's paper (**[Gek]**) for some congruence subgroups of $\mathrm{GL}_2(\mathbb{F}_q[T])$. If we don't ask for too much explicit information, however, we can make a simple general statement.

Proposition (see [Se])). *If Γ is a congruence subgroup of some $\mathrm{GL}_2(A)$, then the quotient graph $\Gamma\backslash\mathcal{T}$ is a finite graph with a finite number of ends attached.*

As we know, the ends of $\mathcal{T}$ are homeomorphic to $\mathbb{P}^1_{k_\infty}$, so the finitely many ends of $\Gamma\backslash\mathcal{T}$ correspond to the finitely many elements of $\Gamma\backslash\mathbb{P}^1_k$.

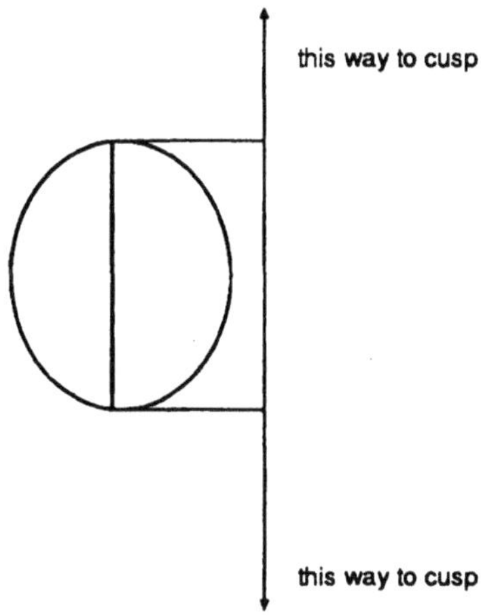

Figure 5.

Boundary Measures. As you may recall, the goal of this discussion is to construct modular forms on Ω as integral transforms of certain kinds of measures. We now have enough machinery constructed to describe the measures that we need.

Suppose first of all that M is an abelian group. An M–valued harmonic function on $\mathcal{T}$ is a function f on the edges of $\mathcal{T}$ which takes values in M and satisfies the following two properties:

$$f(e^{\mathrm{op}}) = -f(e)$$

and

$$\sum_{e \mapsto v} f(e) = 0.$$

Here e^{op} is the edge e with the opposite orientation, and the sum is the sum over the edges leaving v.

A harmonic function f has another incarnation. It may be viewed as a finitely additive, M–valued distribution (let us call it μ_f) on the ends of $\mathcal{T}$. This interpretation is based on the correspondence between oriented edges of $\mathcal{T}$ and compact opens in the ends of $\mathcal{T}$. Simply, if e is an edge of $\mathcal{T}$ and U is the corresponding open set, then $\mu_f(U) = f(e)$. If we look at Figure 3, it is easy to see that a function f being harmonic translates into the distribution μ_f being finitely additive. Furthermore, the alternating property of f guarantees that $\mu_f(\mathbb{P}^1_{k_\infty}) = 0$ – this is because the integral over $\mathbb{P}^1_{k_\infty}$ is the sum of f over the edges leaving a vertex, which is zero.

We are particularly interested in harmonic functions f taking values in a particular representation space for $\mathrm{GL}_2(K)$. Let V be a two dimensional vector space over K with the standard action of $\mathrm{GL}_2(K)$. Let X and Y be the dual basis for the dual space V^*. For each integer n and each i mod $q-1$, define

$$V(n,i) = (\det)^i \otimes \mathrm{Symm}^{n-1}(V^*)$$

and $V(-n,-i) = \mathrm{Hom}(V(n,i), K)$.

If Γ is a congruence subgroup of $GL_2(k)$, a harmonic function with values in $V(1-n, 1-i)$ which is Γ–equivariant will be called a *harmonic cocycle of weight* n *and type* i *for* Γ. Here Γ–equivariance means that $f(\gamma e) = \gamma \cdot f(e)$.

The description of $V(n,i)$ above may seem a bit tortured. In particular, tracing through it reveals that $V(n,i)$ is defined as the dual of the symmetric tensor power of the dual. In characteristic zero, some of these dualities cancel out, and one may describe $V(n,i)$ more simply. Unfortunately, in characteristic p, this isn't so, and we are stuck with this rather painful sequence of adjectives. The representation we have constructed is sometimes called the divided power representation.

We will construct modular forms by integrating against the distributions derived from harmonic cocycles. In Part II of this paper, I will say a few words about the technical details of this integration theory, which was developed by Vishik (**[V]**). For now, let me give a few of its properties. Suppose f is a harmonic cocycle of weight n, and type i. Also, if e is an edge, let $U(e)$ be the compact open in the ends of $\mathcal{T}$ corresponding to e. Since $f(e)$ takes values in $V(1-n, 1-i)$, we may evaluate $f(e)$ on elements of the symmetric tensor power $\mathrm{Symm}^{n-2}(V^*)$, and in particular on monomials X^iY^{n-2-i}. The values of $f(e)$ on these monomials determine the integrals of polynomials of degree less than $n-2$ according to the rule

$$\int_{U(e)} x^j \, d\mu_f(x) = f(e)(X^jY^{n-2-j}). \tag{1}$$

This definition implies that

$$\int_{\mathbb{P}^1_{k_\infty}} x^j \, d\mu_f(x) = 0$$

since the sum of the values of f on the edges leaving a vertex is zero. Also, the Γ–equivariance of f means that, when integrating polynomials of degree at most $n-2$, we have the change of variables formula

$$\int_{U(\gamma e)} h(x) \, d\mu_f(x) = \int_{U(e)} \det(\gamma)^{1-i}(cx+d)^{n-2} h(\gamma x) \, d\mu_f. \tag{2}$$

Modular Forms. Since the whole point of our discussion is the construction of modular forms, I ought to take this opportunity to say what modular forms are. Naturally enough, they are analytic functions on Ω which transform nicely under the action of a congruence group Γ. To say that F is analytic means that the restriction of F to any of the open sets γU in our standard covering is uniformly approximable by rational functions with poles outside of γU. In practice, this definition means that F can be written

$$F(z) = a_0 + \sum_{j=1}^{q+1} \sum_{i \geq 1} \frac{a_{ij}}{(z-e_j)^i}$$

where $e_1, \ldots, e_{q+1}$ are centers for the "missing disks" cut from γU, and the a_{ij} go to zero fast enough to insure convergence (one of the e_j is probably ∞, in which case $1/(z-\infty)$ should be interpreted as z). This representation of F follows from standard results on the rigid geometry of $\mathbb{P}^1$; you may read about it for example in **[GVdP]**.

Now we define a modular form F of weight n and type i for a congruence group Γ to be an analytic function on Ω which satisfies

$$F(\gamma z) = \det(\gamma)^{-i}(cz+d)^n F(z)$$

and which is in some sense "holomorphic at the cusps."

The cusp condition on an analytic function requires a certain amount of further explanation. For details, see Gekeler's book (**[Gek]**). Suppose for example that we consider the point where $z = \infty$. This point is stabilized by a translation group of the form $z \mapsto z + b$. The function

$$t(z) = \sum_b \frac{1}{z-b}$$

is invariant under these translations, and one can prove that F (which is also translation invariant) has an expansion as a power series in t. To say that F is holomorphic at the cusp at infinity is to say that this power series really is a power series; in other words, it has no negative powers of t. The stronger condition on F that the series have no constant term makes F a *cusp form*. The other cusps impose other, similar conditions on F.

There are several "famous" examples of modular forms for the group $\mathrm{GL}_2(\mathbb{F}_q[T])$. For example, Goss' Eisenstein Series

$$E_n(z) = \sum_{(c,d)\neq(0,0)} \frac{1}{(cz+d)^n},$$

which is a form of weight n, and his "delta" function

$$\Delta(z) = (T^q - T)^q E_{q-1}^{q+1} + (T^{q^2} - T)E_{q^2-1},$$

which is a cusp form of weight $q^2 - 1$. These functions are important for the study of Drinfeld modules, as explained in Goss' papers (and many other places).

We will have occasion to consider another important modular form, Gekeler's P function (**[Gek]**), which is of weight $q+1$ and type 1. This form is characterized by the relation

$$P^{q-1} = -\Delta$$

although it is constructed by Gekeler as a Poincaré series.

Residues. Modular forms are related to harmonic functions on $\mathcal{T}$ by means of the residue map. This map produces a harmonic cocycle of weight n and type i from a modular form of this weight and type. To understand how it is defined, let us suppose that we are given a modular form $F(z)$ of weight n and type i. Let e_0 denote the edge of $\mathcal{T}$ which corresponds to the basic annulus

$$V = \{P : 1 < |z(P)| < q\}.$$

If we restrict $F(z)$ to V, then we may expand $F(z)dz$ in a "two-sided" convergent series

$$F(z) = \sum_j a(j) z^j dz$$

and we define

$$\mathrm{Res}_{e_0} F(z)\,dz = a(-1).$$

More generally, if $e = \gamma e_0$ is any edge of $\mathcal{T}$, and $V(e) = \gamma V$ the corresponding annulus in Ω, then we may expand $F(z)dz$ as

$$F(z) = \sum_j a(j) u^j du$$

where $u = \gamma^{-1}(z)$. Then as above we set

$$\mathrm{Res}_e F(z)\,dz = a(-1).$$

Notice that we didn't use anything about the modular property of F in the above definition (yet), so it makes sense to take the residue of $zFdz$, z^2Fdz, and so on.

The crucial fact about the residue map is the residue theorem:

Proposition. *The sum of the residues of $z^j F(z)dz$ on the edges leaving a vertex is zero.*

Proof. This is true in general for holomorphic 1–forms on Ω, (and in many more cases as well). See **[GVdP]**.

Now return to our modular form F. Define a harmonic cocycle of weight n and type i on $\mathcal{T}$ by the formula

$$\mathrm{Res}_e(F)(X^j Y^{n-2-j}) = \mathrm{Res}_e\, z^j F dz.$$

Remember that harmonic cocycles of weight n are determined by giving their values on the monomials $X^j Y^{n-2-j}$; so this definition makes some kind of sense. The function on $\mathcal{T}$ defined in this way is harmonic because of the residue theorem, and it transforms appropriately, as we may verify by computing:

$$\begin{aligned}\mathrm{Res}_{\gamma e}\, z^j F dz &= \det(\gamma)^{1-i}\,\mathrm{Res}_e (az+b)^j (cz+d)^{n-2-j} F dz \\ &= (\mathrm{Res}_e F)((aX+bY)^j (cX+dY)^{n-2-j} \det(\gamma)^{1-i}) \\ &= (\gamma \cdot \mathrm{Res}_e(F))(X^j Y^{n-2-j}).\end{aligned}$$

The residue map that we have described was found by Drinfeld (in weight two) (**[D]**) and later Schneider (in higher weight) (**[Sch]**). It takes us from modular forms to harmonic cocycles. It is natural to hope that this map is an isomorphism. At first, this may seem questionable, since an easy computation shows that Eisenstein series have zero residues. However, if we restrict our consideration to cusp forms, we have all that we might wish for.

The Poisson Integral. The passage back from harmonic functions on $\mathcal{T}$ to modular forms requires interpreting the harmonic functions as measures, and then integrating against them. As I mentioned earlier, this type of integration is somewhat technical. If we grant that one can in fact do the necessary integrals, we may state the following theorem which sums up the relationship we are trying to describe.

Theorem 1. *For $n \geq 2$, the residue map is an isomorphism between the space $S(n,i)$ of cusp forms of weight n and type i for a group Γ, and the space of harmonic cocycles of weight n and type i for Γ. Moreover, let $F \in S(n,i)$ and let μ_F be the associated measure. Then*

$$F(z) = \int_{\mathbb{P}^1_{k_\infty}} \frac{1}{z-x}\, d\mu_F.$$

For now, let's just check the transformation properties of the integral, assuming that the change of variables law in equation 2 holds.

$$\begin{aligned}
F(\gamma z) &= \int \frac{1}{\gamma z - x}\, d\mu(x) \\
&= \int \frac{1}{\gamma z - \gamma x}\, d\mu(\gamma x) \\
&= \det(\gamma)^{-i} \int \frac{(cz+d)(cx+d)^{n-1}}{z-x}\, d\mu(x) \\
&= \det(\gamma)^{-i} \int (\frac{(cz+d)^n}{z-x} + P_{n-2}(x,z))\, d\mu(x) \\
&= \det(\gamma)^{-i}(cz+d)^n F(z)
\end{aligned}$$

The crucial point here is that $P_{n-2}(z,x)$ is a polynomial in x of degree $n-2$ – but the integral over all of $\mathbb{P}^1$ of such a polynomial is automatically zero, as I pointed out much earlier.

I'll conclude Part I of this paper with an example.

Example. Suppose that $k = \mathbb{F}_q(T)$ and that the selected place of k is the infinite one, so that

$$k \hookrightarrow \mathbb{F}_q((1/T))$$

is the obvious injection. Let us look for a modular form of weight $q+1$ and type 1. By the Theorem 1, this amounts to finding a harmonic cocycle of weight $q+1$ and type 1. We'll do this by descent, starting with the principal congruence group $\Gamma = \Gamma((T))$ associated to the ideal (T) of $A = \mathbb{F}_q[T]$. The quotient graph $\Gamma\backslash\mathfrak{T}$ is depicted in Figure 6; the cusps correspond in an evident way to the points of $\mathbb{P}^1_{\mathbb{F}_q}$ and the quotient group $\mathrm{GL}_2(A)/\Gamma = \mathrm{GL}_2(\mathbb{F}_q)$ acts on them in the usual way.

The edges labeled "stable" in the figure have trivial stabilizer in Γ. The other edges (the "unstable" ones) are stabilized by p–groups, and this in fact forces a Γ-equivariant harmonic function to vanish on them. I'll say a bit more about this in Part II when I talk about cuspidal measures, but for now it means we may forget about the unstable edges and just look at the stable ones. To find a cocycle for Γ, then, we need only assign elements of $V(-q,0)$ to the $q+1$ stable edges at random, insuring that the sum of our $q+1$ elements is zero (to take care of harmonicity).

If we want a form on $\mathrm{GL}_2(A)$, however, we need more. Let us label a stable edge according to the point of $\mathbb{P}^1_{\mathbb{F}_q}$ which it points towards. We need a function satisfying

$$\gamma f(e_x) = f(e_{\gamma x})$$

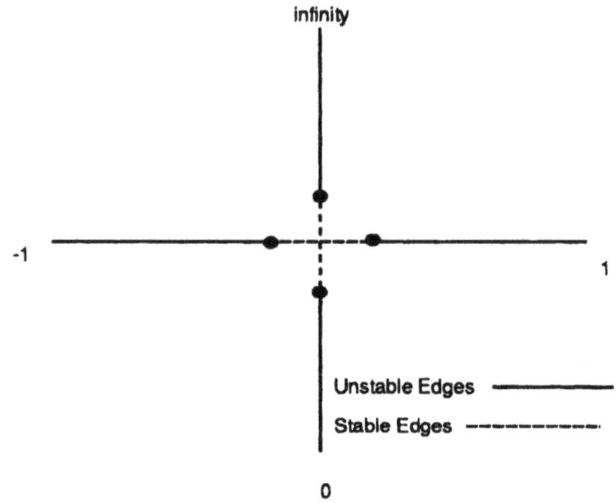

Figure 6.

for $\gamma \in \mathrm{GL}_2(A)$, and in addition

$$\sum_x f(e_x) = 0.$$

As it turns out, these conditions are easy to understand in terms of the representation theory of $\mathrm{GL}_2(\mathbb{F}_q)$. In fact, our space $V(-q,0)$ is isomorphic to $K \otimes \mathrm{Div}^0(\mathbb{F}_q)$ as a representation of $\mathrm{GL}_2(\mathbb{F}_q)$. Furthermore, $\mathrm{Div}^0(\mathbb{F}_q)$ is a projective, irreducible, self-dual representation of $\mathrm{GL}_2(\mathbb{F}_q)$ (it is the "Steinberg representation", see **[Alp]**). The two conditions imposed on our map f amount to saying that we have a $\mathrm{GL}_2(\mathbb{F}_q)$ map

$$\mathrm{Div}^0(\mathbb{F}_q) \to \mathrm{Div}^0(\mathbb{F}_q),$$

and by Schur's lemma the space of such maps is one dimensional. It follows that (up to scalar multiple) there is a unique cusp form of weight $q+1$ and type 1 for $\mathrm{GL}_2(\mathbb{F}_q)$. This characterizes Gekeler's P form, up to scalar multiplication.

Expansion Coefficients. As I mentioned earlier, modular forms have expansions at the cusps in terms of a translation invariant parameter $t(z)$:

$$F(z) = \sum_j a_j t^j.$$

We can give a simple formula for the coefficients of this expansion in terms of the associated measures. This formula was initially worked out by the author for the case of $\mathbb{F}_q(T)$, using a different method; I am grateful to Sinnott and Goss for pointing out the following argument which works more generally. Suppose that F is modular for Γ and that Γ contains the group of translations $\{z \mapsto z+b\}$ for b in some ideal I in A. Then we may split up the integral giving F:

$$\begin{aligned} F(z) &= \int_{\mathbb{P}^1_{k_\infty}} \frac{1}{z-x}\, d\mu(x) \\ &= \int_{k_\infty} \frac{1}{z-x}\, d\mu(x) \\ &= \sum_{b\in I} \int_{k_\infty/I} \frac{1}{z-(x+b)}\, d\mu(x). \end{aligned}$$

Changing the domain of integration from $\mathbb{P}^1_{k_\infty}$ to k_∞ requires a little analysis; the essential point is that the measure μ is "cuspidal" and so there is no contribution to the integral from ∞. I will discuss these issues further later on. Returning to the formulae, we interchange summation and integration to yield

$$F(z) = \int_{k_\infty/I} \sum_{b\in I} \frac{1}{z-x-b}\, d\mu(x)$$

so that

$$F(z) = \int_{k_\infty/I} t(z-x)\, d\mu(x).$$

Now, as it happens, the function t is related to the "Drinfeld Exponential" function e_I associated to the (rank 1) lattice I (See Hayes' paper in this volume, for example) by the formula $t = 1/e_I$. Further, the function e_I is additive – in other words, $e(z) - e(x) = e(z-x)$. Therefore

$$\begin{aligned} F(z) &= \int_{k_\infty/I} \frac{1}{e(z)-e(x)}\, d\mu(x) \\ &= \int_{k_\infty/I} e(z)^{-1} \sum_j (\frac{e(x)}{e(z)})^j\, d\mu(x) \\ &= \sum_j (\int_{k_\infty/I} e(x)^{j-1}\, d\mu(x)) t(z)^j \end{aligned}$$

from which we obtain the formula

$$a_j = \int_{k_\infty/I} e(x)^{j-1}\, d\mu(x),$$

which is a sort of "Cauchy's Formula" for the coefficients a_j.

Part II. Some Technical Issues

I hope in this part of the paper to touch on some of the technical issues which are important in the proof of Theorem 1. These issues are: the effect of the cusps on harmonic cocycles, the theory of integration, and the relationship between the residue map and the Poisson Integral. It is not my intention to present a complete proof of Theorem 1, for which I refer the reader to the "official" paper **[T2]**.

Cuspidal Measures. The quotient graph of $\mathcal{T}$ by a congruence group Γ is an infinite graph, and one might *a priori* expect that the space of harmonic functions on this graph would be infinite dimensional. In fact, this is not the case. Let us say that a harmonic cocycle f is *cuspidal* if it vanishes off of a finite set of edges mod Γ. Then we have the following remarkable fact:

> If f takes values in an abelian group killed by p, and f is Γ-invariant, then f is AUTOMATICALLY cuspidal.

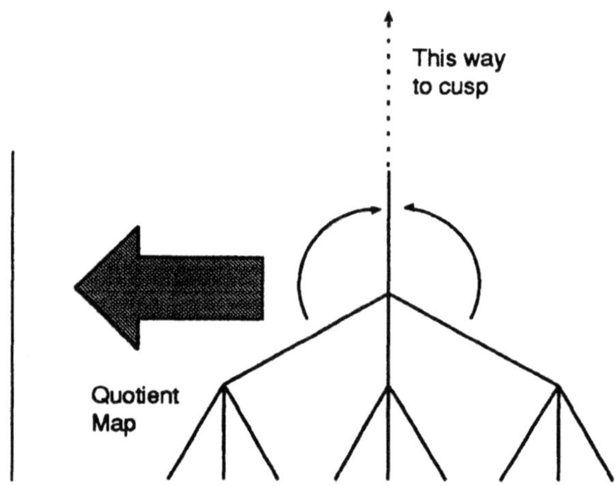

Figure 7.

To see why this is true, consider Figure 7. This figure shows (for $q = 3$) the manner in which the group stabilizing a cusp compresses a part of $\mathcal{T}$ onto the end in the quotient graph. Two facts conspire to force a harmonic function to vanish on the cusp – first, since f is equivariant, and the q edges in the diagram are glued together by Γ as indicated by the arrows, f must take the same value on each of these edges. Second, since f is harmonic, its value on the unaffected edge must be the sum of its value on the glued edges; this means that

$$f(\text{lower}) = \sum f(\text{upper}) = qf(\text{something}) = 0.$$

Integration. When f is a harmonic cocycle of weight 2, then the associated distribution μ_f takes values in K. Furthermore, since f is cuspidal, it takes only finitely many values, and consequently f is bounded. This means that μ_f is a measure, similar in form to the p-adic measures used to construct p-adic L-functions in the $\mathbb{Q}_p$-case. In this situation, one may integrate a continuous function $u(x)$ against μ_f over a compact open set of the form $a + \pi^m \mathcal{O}$ (where π is a uniformizing parameter for $\mathcal{O}$, such as $1/T$) by setting

$$\int_{a+\pi^m \mathcal{O}} u(x)\, d\mu(x) = \lim_{N\to\infty} \sum_{\substack{b \mapsto a \\ b \bmod \pi^{m+N}}} u(b)\mu_f(b + \pi^{m+N}\mathcal{O}).$$

This form of integration suffices to construct modular forms of weight 2 by the Poisson integral, since $1/(z-x)$ is continuous on $\mathbb{P}^1_{k_\infty}$ for fixed z – of course, some work to show that the integral is analytic in z is still necessary, but this is not difficult.

Suppose however that we wish to integrate μ_f when f is of weight greater than 2. We have a linear functional defined on polynomials of degree at most $n-2$, defined by

$$\int_{a+\pi^m \mathcal{O}} x^j\, d\mu_f(x) = f(X^j Y^{n-2-j}).$$

In general, the integral $\int_{a+\pi^m \mathcal{O}} d\mu_f(x)$ will be unbounded; however, a theorem of Schneider (**[Sch]**), using the cuspidal property of f in an essential way, shows that the growth may be controlled so that one may apply Vishik's integration theory to integrate sufficiently nice (i.e., locally polynomial or locally analytic) functions.

For a more complete discussion of the integration technology which is necessary in this setting, I refer the interested reader to **[MTT]** and **[Sch]**. However, I will at least offer some explanation of how one actually carries out an integral against $d\mu_f$. Suppose that u is an analytic function on a compact open $B = a + \pi^m \mathcal{O}$, with a power series expansion

$$u(x) = \sum_j r_j (x-a)^j.$$

We assume that this series converges, so that $r_j \pi^{mj} \to 0$. Approximate u by a polynomial of degree $n-2$ on B by truncating the higher order terms, yielding

$$u_a(x) = \sum_{j=0}^{n-2} r_j (x-a)^j.$$

Since u_a is a polynomial, we know how to integrate it (using equation 1). Then the integral of u over B can be computed as the following limit:

$$\int_B u(x)\, d\mu_f(x) = \lim_{N\to\infty} \sum_{\substack{b \equiv a \pmod{\pi^m} \\ b \bmod \pi^{m+N}}} \int u_b(x)$$

where each u_b is an approximation to u obtained by truncation on each sub–ball $b + \pi^{m+N}\mathcal{O}$.

The Poisson Integral is Inverse to the Residue Map. In this section I would like to indicate why the Poisson integral is the inverse of the residue map, or, in other words, why if

$$F(z) = \int_{\mathbb{P}^1_{k_\infty}} \frac{1}{z-x}\, d\mu_f(x)$$

then $\mathrm{Res}(F) = f$. Consider an edge of $\mathcal{T}$ and the corresponding annulus, as depicted in Figure 8.

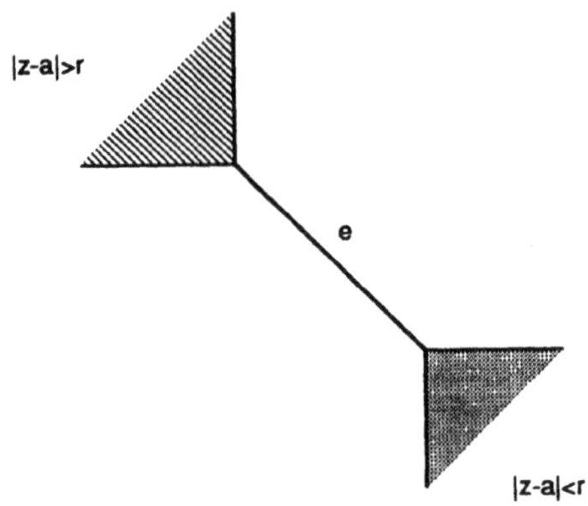

Figure 8.

Expand $F(z) = \int \frac{1}{z-x}\, d\mu(x)$ on this annulus. Write $F = F_+ + F_-$ where

$$F_+ = \int_{B_+} \frac{1}{z-a+a-x}$$

and

$$F_- = \int_{B_-} \frac{1}{z-a+a-x}.$$

Expanding in a geometric series, we obtain

$$F_+ = \sum \frac{\int (x-a)^j\, d\mu(x)}{(z-a)^{j+1}}$$

and

$$F_- = -\sum (z-a)^j \int_{B_-} (x-a)^{-j-1}\, d\mu(x).$$

Now we can calculate:

$$\begin{aligned} Res_e z^j F\,dz &= \mathrm{Res}(z-a+a)^j F\,dz \\ &= \sum \binom{j}{k} a^k \,\mathrm{Res}(z-a)^{j-k} F\,dz \\ &= \sum \binom{j}{k} a^j \int (x-a)^{j-k}\,d\mu(x) \\ &= \int x^j\,d\mu(x) \end{aligned}$$

which is precisely the relation between the residue map and the integral of $d\mu(x)$.

The Injectivity of the Residue Map. We have a pair of maps

$$C_{har}(\Gamma, n, i) \overset{\int}{\underset{\mathrm{Res}}{\leftrightarrow}} S_{n,i}(\Gamma).$$

We know that

$$\mathrm{Res} \circ \int = \mathbf{Id}.$$

To prove that these are mutually inverse isomorphisms, we must show that

$$\dim\, C_{har}(\Gamma, n, i) \geq \dim\, S_{n,i}(\Gamma).$$

The level is irrelevant here; it suffices to prove the result for groups $\Gamma(I)$. Since $\Gamma(I) \subset \mathrm{SL}_2$, the type is also irrelevant.

Let $g(I)$ be the genus of the algebraic curve $X(I) = \Gamma\backslash\Omega$, and $h(I)$ be the number of cusps of $\Gamma(I)$. By Riemann–Roch (see Gekeler's book **[Gek]**) we know that, for $n \geq 2$,

$$\dim\, S_n(\Gamma(I)) = (n-1)(g(I) + h(I) - 1).$$

Thus our task is to find at least this many harmonic cocycles. To do this, we rely on Chapter 9 of Serre's book *Trees*.

Following Serre, we say that an edge or vertex of $\mathcal{T}$ is *stable* if it has trivial stabilizer in Γ. Let S_0 and S_1 be the sets of stable vertices and edges. Serre proves that (when Γ has no torsion of order prime to p, which is true for $\Gamma(I)$):

$$\#(\Gamma\backslash S_1) - \#(\Gamma\backslash S_0) = g(I) + h(I) - 1.$$

Suppose we try to construct a harmonic cocycle f of weight n on the stable edges and vertices of $\mathcal{T}$. For each stable edge $\bmod \Gamma$, we have an $n-1$ dimensional space of possible values. This gives a total of $(n-1)\#(\Gamma\backslash S_1)$ choices for the values of f.

However, we must satisfy the condition that f be harmonic. Each stable vertex imposes $n-1$ linear conditions on f. Assuming that the conditions imposed by the vertices are independent, we see that we have at least $(n-1)(\#(\Gamma\backslash S_1) - \#(\Gamma\backslash S_0))$ harmonic functions *restricted to the stable region*.

To complete the proof of our theorem, we need to show that every function on the stable region extends uniquely to $\mathcal{T}$. This follows from Serre's result that the unstable subgraph of $\mathcal{T}$ is a union of trees.

Conclusion. This concludes our discussion of modular forms on Ω from the point of view of integral transforms. I should point out that recent work of Gekeler and Reversat (see their paper in this volume) has clarified the relationship of weight two modular forms of the type we have been discussing here, and automorphic forms for GL_2 over k. However, the arithmetic significance of the higher weight forms is still quite mysterious. I refer the reader to Goss' paper in this volume, which suggests a number of possible constructions of L-functions.

References

[Al] J. L. Alperin, *Local Representation Theory*, Cambridge Stud. Adv. Math. 11, Cambridge University Press, (1986)

[D] V. G. Drinfeld, Elliptic Modules (Russian), *Mat. Sb.* **94**, (1974) 594–627. English Translation: *Math. USSR-Sb.* **23**, (1976), 561–592.

[H] D. Hayes, A Brief Introduction to Drinfeld Modules, this volume.

[GVdP] L. Gerritzen and M. van der Put, *Schottky Groups and Mumford Curves*, Lecture Notes in Math. 817, Springer, (1980).

[Gek1] E. U. Gekeler, On the coefficients of Drinfeld modular forms, *Invent. Math.* **93**, (1988), 667–700.

[Gek2] E. U. Gekeler, *Drinfeld Modular Curves*, Lecture Notes in Math. 1231 Springer, (1986).

[Gek3] E. U. Gekeler and M. Reversat, Some Results on the Jacobians of Drinfeld Modular Curves, this volume.

[Go1] D. Goss, Modular forms for $\mathbb{F}_r[T]$, *J. Reine Angew. Math.* **317**, (1980), 16–39.

[Go2] D. Goss, π-adic Eisenstein series for Function Fields, *Compositio Math.* **41**, (1980), 3–38.

[Go3] D. Goss, On a new type of L-function for algebraic curves over a finite field, *Pacific J. Math.* **105**, (1983), 143–181.

[KM] N. Kuhn and S. Mitchell, The multiplicity of the Steinberg Representation of $\mathrm{GL}_2(\mathbb{F}_q)$ in the symmetric algebra, *Proc. Amer. Math. Soc.* **96**, (1986), 1–6.

[MTT] B. Mazur, J. Tate, and J. Teitelbaum, On p-adic analogues of the conjecture of Birch and Swinnerton–Dyer, *Invent. Math.* **84**, (1986), 1–48.

[Sch] P. Schneider, Rigid analytic L-transforms, In: *Number Theory, Noordwijkerhout 1983, Proceedings*, Lecture Notes in Math. 1068, Springer, (1983), 216–230.

[Se] J. P. Serre, *Trees*, Springer, (1980).

[T1] J. Teitelbaum, Values of p-adic L-functions and a p-adic Poisson Kernel, *Invent. Math.* **101**, (1990), 395–410.

[T2] J. Teitelbaum, The Poisson kernel for Drinfeld Modular Curves, *J. Amer. Math. Soc.* **4**, (1991), 491–511.

[V] M. Vishik, Non–archimedean measures connected with Dirichlet series, *Math. USSR-Sb.* **28**, (1976), 216–228.

[W] A. Weil, On the analogue of the modular group in characteristic p, *Collected Papers*, Volume III, Springer, (1979), 201–213.

Department of Mathematics, University of Illinois at Chicago, P. O. Box 4348, Chicago, IL 60680
Email: jeremy@math.uic.edu

Some Results on the Jacobians of Drinfeld Modular Curves

E.-U. Gekeler and M. Reversat

0. Introduction

Drinfeld modules are some kind of strange-looking analogues in positive characteristic of elliptic curves. Drinfeld introduced them, under the title of "elliptic modules", in order to prove his reciprocity law **[Dr**, Th. 2**]** on automorphic forms over function fields.

As we learned in the meantime, Drinfeld modules, even when considered as individual objects, are extremely useful in a variety of questions in the arithmetic of function fields, including e.g., explicit class field theory and transcendence theory, but their first raison d'être is that they have the "right" moduli.

Besides the case of rank 1, which corresponds to the classical cyclotomic and complex multiplication theory, it is the moduli theory of Drinfeld modules of rank 2 which most neatly accords with its number theoretic model.

Classically, there is no difference in studying (elliptic) modular forms (e.g., Hecke decompositions and eigenvalues, Fourier coefficients) and (elliptic) modular curves. That is, modular forms are multi-differentials on the compactified modular curves; conversely, the Hecke action on modular forms fully determines the action on various cohomologies, the Jacobian, etc., of the corresponding modular curve. In the rank-2 Drinfeld module situation, there still exist modular curves that can be described analytically through Drinfeld modular forms (see e.g., **[Ge2]**). Let M be such a (compactified) Drinfeld modular curve, of genus g, and with Jacobian J. By Drinfeld, the l-adic cohomology $H^1(M, \mathbb{Q}_l)$ is a space of automorphic forms in the sense of Jacquet and Langlands (**[JL]**. For a down-to-earth description in a special case, see **[Ge1]**). Clearly, there are actions of the Hecke algebra on J, thus on $H^1(M, \mathbb{Q}_l)$ and on the associated spaces of Drinfeld modular forms, and these actions agree with the ones defined naively by means of double cosets. But in contrast with the elliptic case, the action on modular forms doesn't suffice to determine the action of J. In fact, any sort of unpleasant phenomena occur:

- eigenforms of different weights may share the same set of eigenvalues;
- "multiplicity one" fails to hold even for eigenforms of weight two;
- the Hecke action on modular forms is not even semi-simple ...

So the question arises of describing the relationship between the Hecke modules of *automorphic forms* (i.e., $H^1(M, \mathbb{Q}_l)$, values in characteristic 0) and *modular forms* (values in characteristic p), notably modular forms of weight two and type one, which "are" differentials on M. In naive terms, the latter should be some kind of reduction $\bmod p$ of the former. This is best understood in the following example: Let E be an elliptic curve uniformized by M (i.e., a "Weil curve").

To E, there corresponds an automorphic eigenform φ_E with rational integral eigenvalues, but also a modular eigenform f_E of weight two, type one: the pull-back of a Neron differential of E, considered as a modular form. Both φ_E and f_E are defined up to scalars, and one would expect the eigenvalues of f_E to be those of φ_E reduced mod p.

We will give a satisfactory answer to these questions. It is related with the description of J, thus the title of the paper. From the structure of the Galois module $H^1(M, \mathbb{Q}_l)$, one knows a priori that J has split multiplicative reduction, or, what is the same **[Mu]**, may be described as a torus

$$J(C) = (C^*)^g/\Lambda,$$

where C is the field of "complex numbers", and $\Lambda \cong \mathbb{Z}^g$ is a lattice in the g-th power of C^*.

Using techniques developed by Gerritzen and van der Put in the case of Schottky groups (**[GvdP]**, **[vdP]**, see also **[Ra]**), we describe the lattice Λ through theta functions.

Our main result is expressed in the commutative diagram (4.14). Among others, it says that in the above example, f_E *is* the reduction of φ_E (after possible scaling, and after identifying f_E via Teitelbaum's transform with a certain measure). Moreover, it yields an explicit construction as a Tate curve $C^*/u^{\mathbb{Z}}$ of the elliptic curve $E|_C$ that corresponds to an automorphic eigenform φ with rational eigenvalues. The construction mimics the procedure of cycle integration which classically is used to obtain a Weil curve from its newform.

We will omit full proofs (to appear elsewhere), but present all the definitions and constructions needed to state the results, and discuss the ideas behind them. We hope this will enable the reader to understand the connections with the theory of automorphic forms and notably with the function field analogue of the Shimura-Taniyama-Weil conjecture.

1. Drinfeld modular curves

We recall some facts on the analytic theory of Drinfeld modular curves, assuming the reader to be familiar with **[Ge2]**, which will serve as a general reference. Notations and the basic set-up are as in **[Ge4]**. G will always denote the group scheme $GL(2)$ with center $Z = \{\text{scalar matrices in } G\}$.

(1.1). The *Drinfeld upper half-plane* Ω is the rigid analytic space $C \setminus K_\infty$ **[Re]**, upon which $G(K_\infty)$ acts through fractional linear transformations. An *arithmetic group* will be a congruence subgroup Γ of $\Gamma_Y = GL(Y) \subset G(K)$, where $Y \subset K^2$ is a rank-2 projective A-submodule. The quotient $\Gamma\backslash\Omega$ is the analytic space $M_\Gamma(C)$ associated with a certain smooth affine algebraic curve M_Γ. The "Riemann surfaces" M_Γ and their canonical compactifications $\overline{M}_\Gamma$ are what we here denote by *Drinfeld modular curves*; we will not be concerned with their scheme structure over A or other rationality properties.

(1.2). The *Bruhat-Tits tree* $\mathcal{T}$ of $PGL(2, K_\infty)$ is a connected homogeneous tree upon which $G(K_\infty)$ acts via its quotient $PGL(2, K_\infty)$. We let O_∞ be the integers in K_∞

with residue class field k of cardinality $q_\infty = q^{\deg \infty}$. The set $V(\mathcal{T})$ of vertices of $\mathcal{T}$ consists of the similarity classes of O_∞-lattices in K_∞^2; two vertices are adjacent if and only if they are represented by lattices $\Lambda_1 \subsetneq \Lambda_2$ such that Λ_2/Λ_1 has length one. The set of oriented edges of $\mathcal{T}$ is denoted by $E^*(\mathcal{T})$. For $e \in E^*(\mathcal{T})$, $o(e)$, $t(e)$, e^- is the origin, the terminus, the edge e inversely oriented, respectively. $E^*(\mathcal{T})$ has a simple description through $G(K_\infty)$. Namely, defining the *Iwahori subgroup* $\mathfrak{I}_\infty$ by

$$\mathfrak{I}_\infty = \left\{ \begin{pmatrix} a & b \\ c & d \end{pmatrix} \in G(O_\infty) \mid c \equiv 0 \bmod \infty \right\}, \qquad (1.2.1)$$

we have the identification

$$E^*(\mathcal{T}) = G(K_\infty)/Z(K_\infty) \cdot \mathfrak{I}_\infty \qquad (1.2.2)$$

compatible with left $G(K_\infty)$-actions.

(1.3). There is a canonical map λ from Ω to the realization $\mathcal{T}(\mathbb{R})$ of $\mathcal{T}$. Interpreting the latter as the set of similarity classes of norms on K_∞^2 **[GI]**, λ associates to $z \in \Omega$ the class of the norm

$$\|?\|_z : (u, v) \mapsto |zu + v|,$$

where $|?|$ is the absolute value on C. The inverse images of a) vertices, b) "open" edges under λ are isomorphic as analytic spaces with

a) $\mathbb{P}^1(C)$ minus $(q_\infty + 1)$ disjoint open balls;
b) an annulus $\{z \in C \mid q_\infty^{-1} < |z| < 1\}$, conveniently glued together **[DH**, Ch. III], **[Re]**.

(1.4). The map λ is equivariant with respect to the left actions of $G(K_\infty)$ on Ω and $\mathcal{T}$. Hence we may compare, for each arithmetic group Γ, the geometry of $\Gamma\backslash\Omega$ with that of the graph $\Gamma\backslash\mathcal{T}$. It is known that:

(1.4.1). $\Gamma\backslash\mathcal{T}$ is the union of a finite graph with a finite number of ends (= infinite half-lines —o—o—o—o $\cdots$) **[Se**, II, Th. 9]. A typical example is given by the graph

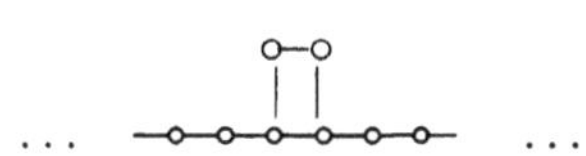

(1.4.2). λ induces a bijection of the set of ends occurring in (1.4.1) with $\overline{M}_\Gamma(C)\backslash M_\Gamma(C)$. These sets may be canonically identified with the set cusp $(\Gamma) = \Gamma\backslash\mathbb{P}^1(K)$ of *cusps* of Γ **[Se**, II, 2.3].

(1.4.3). The genus $g(\overline{M}_\Gamma)$ of $\overline{M}_\Gamma$ equals the first Betti number $b_1(\Gamma\backslash\mathcal{T})$ of $\Gamma\backslash\mathcal{T}$, which is also the rank $\dim_\mathbb{Q} \Gamma^{ab} \otimes \mathbb{Q}$ of the factor commutator group Γ^{ab} of Γ **[DH**, IV, Section 4], **[Ge2**, V, App.]. It may explicitly be calculated **[Ge2]**.

(1.5). A (Drinfeld) *modular form of weight* k *and type* m for Γ is a holomorphic function $f\colon \Omega \to C$ that under $\begin{pmatrix} a & b \\ c & d \end{pmatrix} = \gamma \in \Gamma$ transforms like

$$f\left(\frac{az+b}{cz+d}\right) = \frac{(cz+d)^k}{(\det\gamma)^m} f(z), \tag{1.5.1}$$

and is holomorphic at the cusps **[Ge2, V, Section 3]**. It is called *cuspidal (double cuspidal)* if it vanishes (of order two) at all the cusps. We let

$$M_{k,m}^{0,0}(\Gamma) \subset M_{k,m}^{0}(\Gamma) \subset M_{k,m}(\Gamma)$$

be the C-spaces of double cuspidal, cuspidal, arbitrary modular forms of weight k and type m, respectively. From

$$\frac{d\gamma z}{dz} = \frac{\det\gamma}{(cz+d)^2},$$

we see that the differential $f(z)dz$ is Γ-invariant for $f \in M_{2,1}(\Gamma)$. But unlike the classical case, we must require f double cuspidal to yield a holomorphic differential on $\overline{M}_\Gamma$, as D. Goss first observed. This is because the uniformizers $t(z)$ at cusps satisfy

$$\frac{dt(z)}{dz} = \text{const} \cdot t^2(z), \tag{1.5.2}$$

in contrast with

$$\frac{dq(z)}{dz} = \text{const} \cdot q(z)$$

for $q(z) = e^{2\pi i z}$. Therefore, simply cuspidal forms produce differentials with simple poles.

It follows that

$$\dim_C M_{2,1}^{0,0}(\Gamma) = g(\overline{M}_\Gamma) = b_1(\Gamma\backslash\mathfrak{T}). \tag{1.5.3}$$

Typical examples for modular forms are given by Eisenstein series, coefficient forms, and Poincaré series **[Go1]**, **[Go2]**, **[Ge3]**.

2. Harmonic cochains on $\mathfrak{T}$

(2.1). A *harmonic cochain* on $\mathfrak{T}$ with values in an abelian group X is a map $\varphi\colon E^*(\mathfrak{T}) \to X$ that satisfies

(2.1.1).

(i) $\varphi(e) + \varphi(e^-) = 0 \qquad (e \in E^*(\mathfrak{T}))$

(ii) $\sum_{t(e)=v} \varphi(e) = 0 \qquad (v \in V(\mathfrak{T}))$.

We put $\underline{H}(\mathfrak{T}, X)$ for the group of all harmonic cochains. Consider the additional conditions

(iii) $\varphi(\gamma e) = \varphi(e) \qquad (\gamma \in \Gamma)$;

(iv) φ has compact support modulo Γ;

(iv') φ vanishes at the cusps of $\Gamma\backslash\mathcal{T}$.
We put $\underline{H}(\mathcal{T},X)^\Gamma$, $\underline{H}_!(\mathcal{T},X)^\Gamma$, $\underline{H}_{!!}(\mathcal{T},X)^\Gamma$ for the group of $\varphi \in \underline{H}(\mathcal{T},X)$ that also satisfy (iii), (iii)+(iv), (iii)+(iv'), respectively. Note that by (iii), the elements of $\underline{H}(\mathcal{T},X)^\Gamma$ may be regarded as alternating functions on the oriented edges of $\Gamma\backslash\mathcal{T}$. Condition (ii) is then replaced by a sum condition "with multiplicities" for every vertex of $\Gamma\backslash\mathcal{T}$. Therefore, (iv) and (iv') make sense. (iv) says that f has compact support in the space $\Gamma\backslash G(K_\infty)/Z(K_\infty)\cdot\mathcal{I}_\infty$, i.e., is supported by finitely many double cosets. By (1.4.1), (iv') implies (iv), thus

$$\underline{H}_{!!}(\mathcal{T},X)^\Gamma \subseteq \underline{H}_!(\mathcal{T},X)^\Gamma, \tag{2.1.2}$$

with equality if X is torsion-free.

(2.2). The last statement is best explained with a picture. Consider an end on $\Gamma\backslash\mathcal{T}$ (see Figure 1):

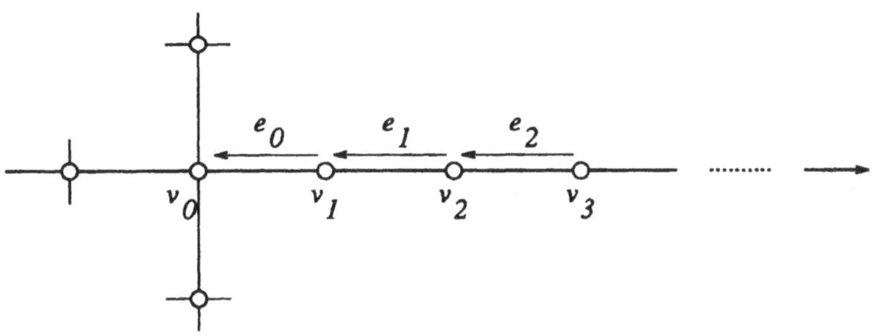

Figure 1

Any function $\varphi \in \underline{H}_{!!}(\mathcal{T},X)^\Gamma$ has to vanish on $e_0 = (v_1, v_0)$, $e_1 = (v_2, v_1)$ etc. For $\varphi \in \underline{H}_!(\mathcal{T},X)^\Gamma$, we only know that it eventually vanishes on each end, thus on $e_N, e_{N+1}\dots$ for some large N. Let $\tilde{e}_1 = (\tilde{v}_2, \tilde{v}_1)$ be an edge of $\mathcal{T}$ above e_1. All the q_∞ edges $\tilde{e} \neq \tilde{e}_1$ with $t(\tilde{e}) = \tilde{v}_1$ are identified under Γ, thus map to $\overline{e_0} = (v_0, v_1)$ in $\Gamma\backslash\mathcal{T}$ **[Se, II, Prop. 8,9]**. Therefore, equation (ii) at v_1 is equivalent with $q_\infty\varphi(e_0) = \varphi(e_1)$. Correspondingly, $q_\infty\varphi(e_1) = \varphi(e_{i+1})$, $i = 1, 2, \dots$, so (iv) implies $\varphi(e_0) = 0$, if q_∞ is no zero-divisor on X, i.e., under that condition, (iv) implies (iv'). But let X be a field of characteristic p (the crucial case). Then automatically, $\varphi \in \underline{H}(\mathcal{T},X)^\Gamma$ vanishes on $e_1, e_2, \dots$, but not necessarily on e_0.

(2.3). As we have seen, these groups strongly depend on the structure of X. Nevertheless, if the argument X is a commutative ring B, we always have:

(2.3.1). The modules $\underline{H}(\mathcal{T},?)^\Gamma$, $\underline{H}_!(\mathcal{T},?)^\Gamma$, and $\underline{H}_{!!}(\mathcal{T},?)^\Gamma$ commute with flat ring extensions $B'|B$.

We shall apply this to the extensions $\mathbb{Z} \hookrightarrow \mathbb{Q} \hookrightarrow \mathbb{C}$ and $\mathbb{F}_p \hookrightarrow C$.

(2.4). Harmonic C-valued cochains come naturally into the game by considering residues of differentials. With any holomorphic function f of Ω, we associate the map

$$\operatorname{Res} f\colon E^*(\mathcal{T}) \to C \tag{2.4.1}$$

$$e \mapsto \mathrm{Res}_e f(z)dz.$$

Here $\mathrm{Res}_e f(z)dz$ is the residue of $f(z)dz$ in the oriented annulus $\lambda^{-1}(e) \subset C$. (Choose an isomorphism of $\lambda^{-1}(e)$ with $\{w \in C \mid q_\infty^{-1} < |w| < 1\}$ such that the exterior corresponds to $o(e)$.) With respect to the coordinates w, write $f(z)dz$ as a Laurent series $\sum a_i w^i dw$. Then $\mathrm{Res}_e f(z)dz := a_{-1}$ is independent of choices made and behaves as it should, see **[GvdP**, pp. 93–95**]**. It follows from residue calculus and the description of the building map λ that $\mathrm{Res}\, f$ is always a harmonic C-valued cochain. If, furthermore, $f \in M_{2,1}(\Gamma)$, $\mathrm{Res}\, f$ is Γ-invariant.

(2.5) Theorem. *The map* $f \mapsto \mathrm{Res}\, f$ *induces isomorphisms*
(i) $M_{2,1}^0(\Gamma) \xrightarrow{\cong} \underline{H}_!(\mathcal{T}, C)^\Gamma$
(ii) $M_{2,1}^{0,0}(\Gamma) \xrightarrow{\cong} \underline{H}_{!!}(\mathcal{T}, C)^\Gamma$.

(i) is a special case of a theorem of J. Teitelbaum **[Te]**, who also characterizes the residues of higher weight modular forms. (ii) comes out by considerations as in (2.2).
Together with (2.3.1), this yields

(2.5.1) Corollary. *The set* $M_{2,1}^0(\Gamma, \mathbb{F}_p)$ *of forms that have all their residues in* $\mathbb{F}_p$ *defines a canonical* $\mathbb{F}_p$*-structure on the* C*-space* $M_{2,1}^0(\Gamma)$ *(and the same statement for* $M_{2,1}^{0,0}$*).*

From (1.5.3) and some easy graph theoretic considerations, we get

(2.5.2) Corollary.

$$\dim \underline{H}_{!!}(\mathcal{T}, C)^\Gamma = g$$
$$\dim \underline{H}_!(\mathcal{T}, C)^\Gamma = g + c - 1,$$

where g *is the genus and* c *the number of cusps of* $\overline{M}_\Gamma$.

3. Automorphic forms

In the present section, we briefly explain how $\mathbb{Q}$- or $\mathbb{C}$-valued harmonic cochains may be regarded as automorphic forms of a certain type. Our general reference on automorphic forms is **[Gel]**.

(3.1). We let $\mathcal{A}$ be the adèle ring of K with ring of integers $\mathcal{O}$ and I the idèle group. They decompose into "finite" and "infinite" parts

$$\begin{aligned} \mathcal{A} &= \mathcal{A}_f \times K_\infty \\ \mathcal{O} &= \mathcal{O}_f \times O_\infty \\ I &= I_f \times K_\infty^*. \end{aligned} \tag{3.1.1}$$

Note that $\mathcal{O}_f$ equals the completion $\widehat{A}$ of A with respect to the ideal topology. We regard K as embedded into $\mathcal{A}_f$ and $\mathcal{A}$, which induces embeddings of $G(K)$ into $G(\mathcal{A}_f)$ and $G(\mathcal{A})$. Recall that G is the group scheme $GL(2)$ with center Z.

(3.2). As explained, e.g., in **[Ha**, 2.3**]**, there is a well-known construction of an A-lattice $Y = Y(\underline{g})$ in K^2 from $\underline{g} \in G(\mathcal{A}_f)$, characterized by $Y \cdot \widehat{A} = \widehat{A}^2 \cdot \underline{g}^{-1}$ in $(\mathcal{A}_f)^2$. It identifies the double coset space

$$G(K)\backslash G(\mathcal{A}_f)/G(\widehat{A}) \xrightarrow{\cong} \mathcal{P}_A^2 \qquad (3.2.1)$$

with the set $\mathcal{P}_A^2$ of isomorphism classes of projective rank-2 A-modules. In particular, if $\mathcal{K}_f$ is an open subgroup of $G(\widehat{A})$ (i.e., of finite index), the associated $G(K)\backslash G(\mathcal{A}_f)/\mathcal{K}_f$ is finite. Furthermore, the determinant map gives a bijection

$$G(K)\backslash G(\mathcal{A}_f)/\mathcal{K}_f \xrightarrow{\cong} K^*\backslash I_f/\det \mathcal{K}_f. \qquad (3.2.2)$$

(3.3). Jacquet-Langlands automorphic forms for K are certain $\mathbb{C}$-valued functions on double coset spaces $G(K)\backslash G(\mathcal{A})/\mathcal{K}$, where $\mathcal{K}$ is an open subgroup of $G(\mathcal{O})$ **[Gel**, 3C**]**. We are going to describe this set through edges of graphs of type $\Gamma\backslash\mathcal{T}$.

Suppose that $\mathcal{K}$ decomposes as $\mathcal{K}_f \times \mathcal{K}_\infty$, and let S be a system of representatives for $G(K)\backslash G(\mathcal{A}_f)/\mathcal{K}_f$. Write $\underline{g} \in G(\mathcal{A})$ as

$$\underline{g} = \gamma \underline{x}\, \underline{k}\, g_\infty$$

with $\gamma \in G(K)$, $\underline{k} \in \mathcal{K}_f$, $g_\infty \in G(K_\infty)$, and some uniquely determined $\underline{x} \in S$. Let

$$\Gamma_{\underline{x}} := G(K) \cap \underline{x}\mathcal{K}_f\underline{x}^{-1}, \qquad (3.3.1)$$

the intersection taking place in $G(\mathcal{A}_f)$. It is an arithmetic subgroup (see (1.1)) of $GL(Y(\underline{x}))$. One easily verifies that the map

$$\begin{aligned} G(K)\backslash G(\mathcal{A})/\mathcal{K}_f \times Z(K_\infty)\mathcal{K}_\infty &\to \dot{\bigcup_{\underline{x}\in S}} \Gamma_{\underline{x}}\backslash G(K_\infty)/Z(K_\infty)\cdot\mathcal{K}_\infty \\ \underline{g} &\mapsto \begin{array}{c}\text{class of } g_\infty \text{ in} \\ \Gamma_{\underline{x}}\backslash G(K_\infty)/Z(K_\infty)\cdot\mathcal{K}_\infty\end{array} \end{aligned} \qquad (3.3.2)$$

is well-defined and bijective. If now $\mathcal{K}_\infty$ is the Iwahori group $\mathcal{I}_\infty$ defined in (1.2.1), we have the special case

$$\begin{aligned} G(K)\backslash G(\mathcal{A})/\mathcal{K}_f \times Z(K_\infty)\cdot\mathcal{I}_\infty &\xrightarrow{\cong} \dot{\bigcup_{\underline{x}\in S}} \Gamma_{\underline{x}}\backslash G(K_\infty)/Z(K_\infty)\cdot\mathcal{I}_\infty \qquad (3.3.3) \\ &= \dot{\bigcup_{\underline{x}\in S}} E^*(\Gamma_{\underline{x}}\backslash\mathcal{T}). \end{aligned}$$

Thus the question arises:

(3.4). Which kind of automorphic forms correspond under the above isomorphism to harmonic cochains?

(3.5). Automorphic forms φ are usually $\mathbb{C}$-valued, since over number fields, smoothness and growth conditions for φ at the archimedean places are needed. In our case, however, no archimedean places occur, and the domains (3.3.3) of automorphic forms are discrete sets. We may therefore take any field of characteristic zero as the coefficient field, e.g., $\mathbb{Q}$.

The answer to question (3.4), in very down-to-earth terms, is as follows:

(3.6) Theorem (Drinfeld [Dr, 10.3]). *Let $\mathcal{K}$ be an open subgroup of $G(\mathcal{O})$ of the form $\mathcal{K} = \mathcal{K}_f \times \mathcal{I}_\infty$, and let F be a field of characteristic zero. Under the bijection* (3.3.3), *the module of harmonic cochains*

$$\bigoplus_{\underline{x}\in S} \underline{H}_!(\mathcal{T}, F)^{\Gamma_{\underline{x}}}$$

corresponds to the space $W_{sp}(\mathcal{K}, F)$ of F-valued cuspidal automorphic forms φ on $G(K)\backslash G(\mathcal{A})/\mathcal{K}^f \times Z(K_\infty)\cdot \mathcal{I}_\infty$ that transform like π_{sp} under $G(K_\infty)$.

Here π_{sp} is the so-called *special representation* of $G(K_\infty)$, i.e., the irreducible representation of $G(K_\infty) = GL(2, K_\infty)$ on the space of locally constant F-valued functions on $\mathbb{P}^1(K_\infty)$, modulo the constants. The condition says that the space of right $G(K_\infty)$-translates of φ is isomorphic with a finite number of copies of π_{sp}. The property of being cuspidal is usually expressed as the vanishing of certain integrals of φ [**Ge1**, 3.17(vi)]. In our situation, it just means that φ has compact support.

(3.7). There are some additional features in the theory of automorphic forms, namely the Petersson product and Hecke operators. The Petersson product is a scalar product on automorphic forms derived from a Haar measure on the locally compact group $PGL(2, \mathcal{A}) = G(\mathcal{A})/Z(\mathcal{A})$. For a systematic treatment, see [**JL**, Section 10] or [**Gel**, Section 5]. In our context, it is the product that comes from the following measure μ on the set (3.3.3):

If the double class $cl(\underline{g})$ of $\underline{g}$ corresponds under (3.3.3) to the double class of $g_\infty \in G(K_\infty)$, then

$$\mathrm{vol}_\mu(cl(\underline{g})) = \frac{\#(\Gamma \cap Z(K))}{\#(\Gamma_{\underline{x}} \cap g_\infty \mathcal{I}_\infty g_\infty^{-1})}. \tag{3.7.1}$$

Thus each edge e of $\Gamma\backslash\mathcal{T}$ is counted with the reciprocal of the order of the stabilizer of a pre-image $\tilde{e}$ of e in $\mathcal{T}$.

Then vol_μ is finite [**Se**, p. 103, Ex. 5], and μ induces a scalar product

$$(\varphi, \psi)_\mu = \int \varphi(x)\psi(x)d\mu(x) \tag{3.7.2}$$

on the space of $\mathbb{Q}$-valued functions with compact support. Clearly, the "integral" is actually a finite sum. This product (or its euclidean real extension, or its hermitian complex extension) will be called the *Petersson product*.

(3.8) Hecke operators. There are many different ways to introduce them, e.g., [**We**, Ch. VI] or [**Gel**, p. 88]. We limit ourselves to just summarize their basic properties. Let $\mathfrak{p}$ be any prime of A unramified for $\mathcal{K}$, i.e., $G(A_\mathfrak{p}) \subset \mathcal{K}_f$ (this holds for almost all $\mathfrak{p}$). The $\mathfrak{p}$-th *Hecke operator* $T_\mathfrak{p}$ satisfies:

(3.8.1).
(i) $T_{\mathfrak{p}}$ is an endomorphism of the $\mathbb{Q}$-vector space $W_{sp}(\mathcal{K}, \mathbb{Q})$;
(ii) $T_{\mathfrak{p}}$ is a self-adjoint with respect to $(\ , \ )_{\mu}$;
(iii) any two $T_{\mathfrak{p}}$, $T_{\mathfrak{q}}$ commute.
Hence, there exists a basis $\{\varphi\}$ of simultaneous eigenforms of $W_{sp}(\mathcal{K}, \overline{\mathbb{Q}})$. Let $\varepsilon(\varphi, \mathfrak{p})$ be the eigenvalue. Then
(iv) $\varepsilon(\varphi, \mathfrak{p})$ is an algebraic integer, and
(v) $|\varepsilon(\varphi, \mathfrak{p})| \leq 2q^{1/2 \deg \mathfrak{p}}$ for each embedding into $\mathbb{C}$.

Properties (i)-(iii) are trivial, (iv) holds since $T_{\mathfrak{p}}$ comes from a correspondence on $\dot{\bigcup} E^*(\Gamma_{\underline{x}} \backslash \mathcal{T})$ with integral coefficients, but (v) is deep. It is the analogue of Ramanujan's conjecture, and has been proven by Drinfeld in **[Dr]**. The algebra generated by the $T_{\mathfrak{p}}$ is called the *Hecke algebra*.

(3.9). Now let $M_{\mathcal{K}} \times C$ be the modular scheme associated with

$$\mathcal{K} = \mathcal{K}_f \times \mathcal{I}_\infty$$

[Ge2] base extended to C and $\overline{M}_{\mathcal{K}} \times C$ its canonical compactification as a "Riemann surface". Quite analogous with (3.3.3), their C-points are given by

$$M_{\mathcal{K}}(C) = \dot{\bigcup_{\underline{x} \in S}} \Gamma_{\underline{x}} \backslash \Omega = \dot{\bigcup} M_{\Gamma_{\underline{x}}}(C), \qquad \text{and}$$

$$\overline{M}_{\mathcal{K}}(C) = \dot{\bigcup} \overline{M}_{\Gamma_{\underline{x}}}(C). \tag{3.9.1}$$

$M_{\mathcal{K}}(C)$ corresponds bijectively to classes of rank-2 A-lattices in C with some level structure, depending on $\mathcal{K}$. This is used to define, for any unramified $\mathfrak{p}$, a Hecke correspondence on $\overline{M}_{\mathcal{K}}(C)$ **[Ge2**, VIII, Section 1**]**. There result operators on many objects related to $\overline{M}_{\mathcal{K}}(C)$, e.g., its l-adic cohomology, or its Jacobian. By abuse of notation, we label them all as "Hecke operators".

(3.10). Let $l \neq p$ be a prime and sp the *special representation* of the local Galois group $\mathrm{Gal}(K_\infty^{\mathrm{sep}}/K_\infty)$ on $V = \mathbb{Q}_l^2$. In a somewhat disguised form, it is described in **[De1**, 3.1**]**. It is a tamely ramified representation that factors over the group $\widehat{\mathbb{Z}} \ltimes \mathbb{Z}_l$, and is then given by $(1,0) \mapsto \begin{pmatrix} 1 & 0 \\ 0 & q_\infty^{-1} \end{pmatrix}$, $(0,1) \mapsto \begin{pmatrix} 1 & 1 \\ 0 & 1 \end{pmatrix}$.

We have now collected all the ingredients to formulate

(3.11) Theorem. *The space $W_{sp}(\mathcal{K}, \mathbb{Q}_l)$ of $\mathbb{Q}_l$-valued cuspidal automorphic forms, tensored with (V, sp), is canonically isomorphic with the l-adic cohomology module $H^1(\overline{M}_{\mathcal{K}} \times C, \mathbb{Q}_l)$. The isomorphism is compatible with the actions of the Hecke algebra and the local Galois group* $\mathrm{Gal}(K_\infty^{\mathrm{sep}}/K_\infty)$.

Taken together, and combined with some well-known results from automorphic forms theory, Theorems 3.6 and 3.11 constitute the famous Drinfeld reciprocity law, of which we will present an application in the last section.

(3.12). In what follows, we do not investigate the "global" structure of the modules that occur in (3.6) and (3.11), but restrict attention to $\underline{H}_!(\mathcal{T},\mathbb{Q})^\Gamma$ (Γ any arithmetic group). After tensoring with (V, sp), it yields the cohomology of the curve $\overline{M}_\Gamma$, which is a component of $\overline{M}_\mathcal{K} \times C$ for a suitable open subgroup $\mathcal{K} = \mathcal{K}_f \times \mathcal{I}_\infty$ of $G(\mathcal{O})$.

4. Theta functions

(4.1). Throughout, Γ denotes an arbitrary arithmetic subgroup of $G(K)$. We set

$$\overline{\Gamma} = \Gamma^{ab}/\text{torsion} \tag{4.1.1}$$

for the factor commutator group of Γ modulo its torsion. Recall that $\overline{\Gamma}$ is a free abelian group in g generators, where g is the genus of the modular curve $\overline{M}_\Gamma$.

Let $\alpha \in \Gamma$ and $w \in \Omega$. For $z \in \Omega$ not in the orbit Γw of w, we put

$$u_{\alpha,w}(z) = \prod_{\gamma\in\Gamma}\left(\frac{z-\gamma\alpha w}{z-\gamma w}\right). \tag{4.1.2}$$

The next theorm has been proved, in the case where Γ equals the group $GL(2,A)$, by W. Radtke in **[Ra]**. The proof in the general case is essentially the same.

(4.2) Theorem.
(i) *The product for $u_{\alpha,w}$ converges locally uniformly on $\Omega\setminus\Gamma_w$. The resulting function extends to a holomorphic function without zeros on Ω, also denoted by $u_{\alpha,w}$.*
(ii) *$u_\alpha := u_{\alpha,w}$ is independent of the choice of the base point $w \in \Omega$.*
(iii) *For any $\beta \in \Gamma$, there exists a constant $c_\alpha(\beta) \in C^*$ such that u_α satisfies the functional equation*

$$u_\alpha(\beta z) = c_\alpha(\beta) u_\alpha(z).$$

(iv) *$c_\alpha(\beta)$ depends only on the classes of α and β in $\overline{\Gamma}$. The association $(\alpha,\beta) \mapsto \langle\alpha,\beta\rangle := c_\alpha(\beta)$ is a symmetric bilinear map from $\overline{\Gamma}\times\overline{\Gamma}$ to C^*.*

The u_α are called (holomorphic) *theta functions* for Γ. We set

$$(\alpha,\beta) := -\log_{q_\infty}|\langle\alpha,\beta\rangle|, \tag{4.2.1}$$

which is a symmetric bilinear $\mathbb{Q}$-valued pairing on $\overline{\Gamma}$.

(4.3). Next, we describe a map from Γ to $\underline{H}_!(\mathcal{T},\mathbb{Z})^\Gamma$. For any two vertices v, w of $\mathcal{T}$, let $s(v,w)$ be the unique geodesic (= path without backtracking) from v to w. Choose a base vertex v of $\mathcal{T}$. For $e \in E^*(\mathcal{T})$ and $\gamma \in \Gamma$, we let

$$i(e,\alpha,\gamma,v) = 1,-1,0, \tag{4.3.1}$$

if $\gamma(e)$ belongs to $s(v,\alpha(v))$, $s(\alpha(v),v)$, neither one, respectively. Note that Γ acts on $\mathcal{T}$ via its quotient by the finite subgroup $\Gamma\cap Z(K)$. Therefore, and since $i(e,\alpha,\gamma,v)$ as a function of γ has finite support,

$$\varphi_{\alpha,v}(e) = \#(\Gamma\cap Z(K))^{-1}\sum_{\gamma\in\Gamma} i(e,\alpha,\gamma,v) \tag{4.3.2}$$

is well-defined and $\mathbb{Z}$-valued.

(4.4) Lemma. *The function* $\varphi_{\alpha,v}$ *on* $E^*(\mathcal{T})$ *has the following properties:*
(i) $\varphi_{\alpha,v} \in \underline{H}_!(\mathcal{T},\mathbb{Z})^\Gamma$;
(ii) $\varphi_\alpha := \varphi_{\alpha,v}$ *does not depend on the choice of the base vertex* v;
(iii) $\varphi_{\alpha\beta} = \varphi_\alpha + \varphi_\beta$, *thus* $\alpha \mapsto \varphi_\alpha$ *induces a homomorphism* j *from* $\overline{\Gamma}$ *to* $\underline{H}_!(\mathcal{T},\mathbb{Z})^\Gamma$;
(iv) j *is injective.*

Some hints on the proof: $\varphi_{\alpha,v} \in \underline{H}(\mathcal{T},\mathbb{Z})^\Gamma$, i.e., conditions (i)–(iii) of (2.1.1), is straightforward. For (ii), it suffices to show that $\varphi_{\alpha,v} = \varphi_{\alpha,w}$ if v, w are neighbors. This is seen by comparing the geodesics $s(v,\alpha(v))$ and $s(w,\alpha(w))$ on $\mathcal{T}$. The fact that φ_α vanishes on a cusp c of Γ, i.e., on an end of $\Gamma\backslash\mathcal{T}$, results from choosing the base point v on c "sufficiently close to ∞" such that the image of $s(v,\alpha(v))$ in $\Gamma\backslash\mathcal{T}$ is null-homotopic. (iii) follows from (ii) and the formal identity $\varphi_{\alpha\beta,v} = \varphi_{\alpha,\beta v} + \varphi_{\beta,v}$. For (iv), it suffices to verify that after tensoring with $\mathbb{Q}$, j becomes surjective, both groups being isomorphic with $\mathbb{Z}^g$. This follows by standard graph theoretic constructions.

(4.5) Theorem ([GR]). $j:\overline{\Gamma} \xrightarrow{\cong} \underline{H}_!(\mathcal{T},\mathbb{Z})^\Gamma$ *is always bijective.*

We state this separately since it is much more subtle than the injectivity of j. If Γ has no non-p torsion, one applies **[Se, II, 2.9]**. In the general case, the methods of **[GvdP**, Ch. VIII**]** may be adapted to derive the surjectivity of j from this special case. Note, however, that we do not need the surjectivity to state our next results.

(4.6) Proposition. *For* $\alpha,\beta \in \overline{\Gamma}$, *we have*

$$2(\alpha,\beta) = (\varphi_\alpha,\varphi_\beta)_\mu,$$

where the right hand side is the Petersson product.

This follows from a direct calculation **[GR]**.

(4.6.1) Corollary. *The* $\mathbb{Q}$*-valued pairing* $(\ ,\ )$ *on* $\overline{\Gamma}\times\overline{\Gamma}$ *is positive definite.*

(4.6.2) Corollary. *The map* $c:\alpha \mapsto c_\alpha$ *defines an injection of* $\overline{\Gamma}$ *into the group of homomorphisms* $\mathrm{Hom}(\overline{\Gamma},C^*)$ *of* $\overline{\Gamma}$ *with values in* C^*.

The latter group is (non-canonically) isomorphic with $(C^*)^g$. As is known from **[MD]** and **[GvdP]** (see also **[My]**), the quotient of a torus $(C^*)^g = \mathbb{G}_m^g(C)$ by a lattice like $\Lambda = c(\overline{\Gamma})$ with the properties stated in the corollaries defines an abelian variety over C. Not much surprisingly, this turns out to be the Jacobian of our curve $\overline{M}_\Gamma$:

(4.7) Theorem ([GR]). *The Jacobian* $\mathrm{Jac}(\overline{M}_\Gamma)$ *of* $\overline{M}_\Gamma$ *has the analytic uniformization given by the exact sequence*

$$1 \to \overline{\Gamma} \xrightarrow{c} \mathrm{Hom}(\overline{\Gamma},C^*) \to \mathrm{Jac}(\overline{M}_\Gamma)(C) \to 0. \qquad (4.7.1)$$

(4.8). By means of theta functions, we may write down holomorphic differentials for the curves $\overline{M}_\Gamma$. Clearly, differentiating the functional equation (4.2)(iii) yields that u'_α/u_α as a function on Ω is modular of weight two and type one. Furthermore, since u_α

is invariant under parabolic elements of Γ, it has expansions around the cusps of Γ. Therefore, its vanishing orders are defined, which turn out to be zero **[Ra]**. In view of (1.5.2), this implies that u'_α/u_α is an element of $M_{2,1}^{0,0}(\Gamma)$.

(4.9) Theorem. *For $\alpha \in \Gamma$, we have $u'_\alpha/u_\alpha \in M_{2,1}^{0,0}(\Gamma, \mathbb{F}_p)$. If $\{\alpha\}$ is a $\mathbb{Z}$-basis of $\overline{\Gamma}$, the set $\{u'_\alpha/u_\alpha\}$ forms an $\mathbb{F}_p$-basis of $M_{2,1}^{0,0}(\Gamma, \mathbb{F}_p)$, thus a C-basis of $M_{2,1}^{0,0}(\Gamma)$.*

It is clear that $\operatorname{Res}(u'_\alpha/u_\alpha)$ has its values in the prime field $\mathbb{F}_p$. The second statement follows from the proof of (4.5).

(4.10). As we learned from Marius van der Put, any invertible holomorphic function f on Ω defines a $\mathbb{Z}$-valued harmonic cochain $r(f)$ on $\mathcal{T}$. Following **[vdP]**, we briefly describe how $r(f)$ is obtained.

Let $v, w \in V(\mathcal{T})$ be neighbors and e the oriented edge (v, w). With respect to an appropriate coordinate z on $\Omega \subset \mathbb{P}^1(C)$, we have

$$\lambda^{-1}(v) = \{z \in C \mid |z| = 1,\ |z - x_i| \geq 1\}, \tag{4.10.1}$$

where the $x_i \in C$ are q_∞ points with $|x_i| = 1$ and such that the balls $B(x_i, 1) = \{z \in C \mid |z - x_i| < 1\}$ are disjoint. Similarly, the inverse image under λ of the "closed" edge $[v, w]$ is given by

$$\lambda^{-1}([v,w]) = \{z \in C \mid q_\infty^{-1} \leq |z| \leq 1,\ |z - x_i| \geq 1,\ |z - y_i| \geq q_\infty^{-1}\} \tag{4.10.2}$$

with $|x_i| = 1$, $|y_i| = q_\infty^{-1}$ $(i = 1, \ldots, q_\infty)$, and the $B(y_i, q_\infty^{-1})$ mutually disjoint.
The main point is the observation:

(4.10.3). Every invertible holomorphic function f on Ω has constant absolute value on $\lambda^{-1}(v)$.

Call this $|f|_v$. From the properties of the Laurent expansion of f in the annulus $\{z \in C \mid q_\infty^{-1} \leq |z| \leq 1\}$, it follows that f can be written

$$f(z) = c \cdot z^n \cdot g(z) \tag{4.10.4}$$

with $c \in C^*$ and g invertible with constant absolute value $|g| = 1$ on $\lambda^{-1}([v, w])$. In particular

$$|f|_v = |c|, \quad |f|_w = q_\infty^{-n}|c|. \tag{4.10.5}$$

Now the value of $r(f)$ on $e = (v, w)$ is defined as

$$r(f)(e) = n = \log_{q_\infty}(|f|_v/|f|_w). \tag{4.10.6}$$

Clearly, $r(f)$ is alternating, and property (ii) of (2.1.1) follows from studying the reduction of $\lambda^{-1}(v)$ as the algebraic curve over $k = O_\infty/\pi$.

(4.11) Theorem ([vdP]). *Let $\mathcal{O}(\Omega)^*$ be the group of invertible holomorphic functions on Ω. The sequence*

$$1 \to C^* \to \mathcal{O}(\Omega)^* \xrightarrow{r} \underline{H}(\mathcal{T}, \mathbb{Z}) \to 0$$

is exact.

By construction, this sequence is compatible with the action of $G(K_\infty)$ on Ω and on $\mathcal{T}$. Hence the functions u_α give rise to Γ-invariant cochains $r(u_\alpha)$, which, moreover, have compact support modulo Γ. The last statement follows since u_α does not vanish at cusps, so its absolute value becomes constant in the neighborhood of a cusp.

Now we dispose of two different constructions of Γ-invariant integer-valued harmonic cochains from $\alpha \in \Gamma$:

(a) φ_α (see (4.4)) and

(b) $r(u_\alpha)$.

It is somewhat lengthy but straightforward to check the

(4.12) Proposition. $\varphi_\alpha = r(u_\alpha)$.

Furthermore, we dispose of two constructions of elements of $\underline{H}_{!!}(\mathcal{T}, \mathbb{F}_p)^\Gamma$ from $\alpha \in \Gamma$:

(c) reduction mod p of $\varphi_\alpha = r(u_\alpha)$

(d) $\mathrm{Res}(u'_\alpha/u_\alpha)$.

The fact that (c) takes its values in $\underline{H}_{!!}$ results from (2.2). The proof of (4.12) simultaneously yields:

(4.13) Proposition. $\mathrm{Res}(u'_\alpha/u_\alpha) \equiv \varphi_\alpha \bmod p$.

We may summarize our results in a large commutative diagram compatible with Hecke actions, see Diagram 4.14 (next page). Note that the lower rectangle is a base extension from $\mathbb{F}_p$ to C. Identifying left and right hand sides by the indicated isomorphisms, and having (3.6) in mind, the diagram says:

(4.15) "Drinfeld modular forms (double cuspidal, of weight two and type one, and with $\mathbb{F}_p$-residues) are the reductions of automorphic forms (of Jacquet-Langlands-Drinfeld type)."

Also:

(4.16) "Lifting a double cuspidal Γ-invariant $\mathbb{F}_p$-valued harmonic cochain $\mathrm{Res}\, f$ corresponds to integrating $f = u'/u$ to a theta function u for Γ."

5. Application to the Shimura-Taniyama-Weil Conjecture

In order to avoid complications arising from ideal classes, we now make the

(5.1) Assumption. A is the polynomial ring $\mathbb{F}_q[T]$.

This is not essential: The results to be presented generalize to arbitrary A of the type considered in **[Ge4]**, but then require additional notational and technical efforts.

(5.2). We first describe the Shimura-Taniyama-Weil correspondence, i.e., the characteristic p analogue of the "classical" Shimura-Taniyama-Weil conjecture, which happens to be a theorem.

Let $E|_K$ be an elliptic curve with split multiplicative reduction at the place ∞ of K. Equivalently, E has a Tate parameterization

$$E(K_\infty) = K_\infty^*/u^{\mathbb{Z}}, \qquad (5.2.1)$$

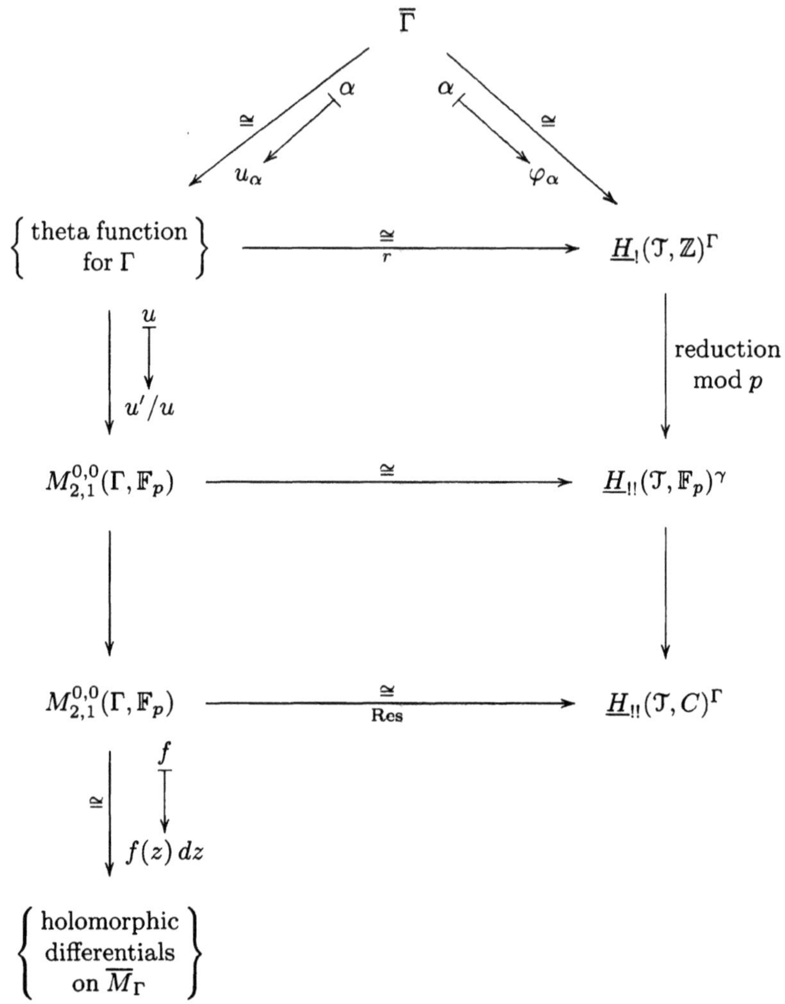

Diagram 4.14

where $u \in K_\infty^*$ has absolute value strictly less than 1.

The condition implies that the j-invariant $j(E)$ has a pole at ∞; in particular, E is not a "twisted constant curve".

Let $\mathfrak{f} = \mathfrak{f}(E)$ be the conductor of E. As a divisor of K, it decomposes

$$\mathfrak{f} = \mathfrak{n} \cdot \infty, \tag{5.2.2}$$

where $\mathfrak{n}$ is coprime with ∞, i.e., corresponds to an ideal of A, also denoted by $\mathfrak{n}$. Let

$$\Gamma = \Gamma_0(\mathfrak{n}) = \left\{ \begin{pmatrix} a & b \\ c & d \end{pmatrix} \middle| c \equiv 0(\mathfrak{n}) \right\} \subset GL(2, A) \tag{5.2.3}$$

be the *Hecke congruence subgroup* associated with $\mathfrak{n}$. We put $X_0(\mathfrak{n})$ for the curve M_Γ and $J_0(\mathfrak{n})$ for the Jacobian of $\overline{X}_0(\mathfrak{n}) = \overline{M}_\Gamma$.

With E as above, we consider the system of complex-valued L-functions $L_E(\omega, s)$ of E, where ω runs through the set of idèle class characters of K. Now there are the following very deep results, which we indicate only in key words. For precise statements and explanations (not to mention proofs!), it is indispensable to consult the references cited.

(5.3). (Grothendieck) The system $L_E(\omega, s)$ has the "correct" analytical properties (holomorphic continuation plus functional equations: **[De2**, Section 9]).

(5.4). (Jacquet-Langlands) Let $\mathcal{K}$ be the following subgroup of $G(\mathcal{O})$:

$$\mathcal{K} = \mathcal{K}_{f,0}(\mathfrak{n}) \times \mathcal{I}_\infty, \tag{5.4.1}$$

where $\mathcal{K}_{f,0}(\mathfrak{n}) = \left\{ \begin{pmatrix} \underline{a} & \underline{b} \\ \underline{c} & \underline{d} \end{pmatrix} \in G(\widehat{A}) \mid \underline{c} \equiv 0(\mathfrak{n}) \right\}$. It is some kind of global version of $\Gamma_0(\mathfrak{n})$. In the space $W_{sp}(\mathcal{K}, \mathbb{C})$ defined in (3.6), there exists a newform $\varphi = \varphi_E$, uniquely determined up to constants, with the properties:

$$\varphi \text{ is a Hecke eigenform with rational eigenvalues;} \tag{5.4.2}$$

$$\text{the system of } L\text{-functions } L_\varphi(\omega, s) \text{ agrees with } L_E(\omega, s). \tag{5.4.3}$$

There are different ways of how to attach L_φ to φ, differing by shifts $s \mapsto s \pm 1/2$ in the argument s. We use what is called the "Hecke correspondence" in **[De1**, 3.2].

(5.5). Let $\pi_E \in \mathrm{End}(W_{sp}(\mathcal{K}, \mathbb{C}))$ be the projection to $\mathbb{C}\varphi_E$. Viewed as an endomorphism of $J_0(\mathfrak{n})$, π_E maps $J_0(\mathfrak{n})$ to a one-dimensional abelian sub-variety, i.e., an elliptic curve E', which is K-isogeneous with E.

(5.6). The rules $E \mapsto \varphi_E \mapsto E'$ induce bijections between the sets of

(a) K-isogeny classes of elliptic curves E over K with conductor $\mathfrak{n} \cdot \infty$, and which are Tate curves at ∞;

(b) normalized Hecke eigenforms with rational (necessarily integral) eigenvalues in the new part of $W_{sp}(\mathcal{K}, \mathbb{C}) = \underline{H}_!(\mathcal{T}, \mathbb{C})^\Gamma$;

(c) one-dimensional abelian sub-varieties in the new part of $J_0(\mathfrak{n})$ (necessarily defined over K).

(5.5) is implicitly in **[Dr]**, whereas (5.6) results from combining (5.3) and (5.4) with Drinfeld's reciprocity law (3.11).

(5.7). The crucial point is that (b) is (relatively) easily calculated numerically. Thus one would like to have a procedure to construct an elliptic curve within the isogeny class that corresponds to the newform φ. Besides the sub-variety of $J_0(\mathfrak{n})$, there is another distinguished curve in the isogeny class, the *strong Weil curve* E_φ attached to φ. It is a *quotient* of $J_0(\mathfrak{n})$, characterized up to sign by the fact that the induced morphism

$$\pi_\varphi: \overline{X}_0(\mathfrak{n}) \to E_\varphi \tag{5.7.1}$$

is maximal **[SDB**, Section 5]. We show how to construct E_φ from φ.

(5.8). Let now $\varphi \in \underline{H}_!(\mathcal{T}, \mathbb{C})^\Gamma$ be a newform with rational eigenvalues. We may suppose that it actually takes its values in $\mathbb{Z}$ but not in $n\mathbb{Z}$ for any $n > 1$. Identifying $\underline{H}_!$ with $\overline{\Gamma}$ by means of (4.5), we regard φ as an element of $\overline{\Gamma}$.

(5.9) Proposition. *Let* $\mathrm{ev}: g \mapsto g(\varphi)$ *be the evaluation map from* $\mathrm{Hom}(\overline{\Gamma}, C^*)$ *to* C^*. *Under our conditions,* $\mathrm{ev}(c(\overline{\Gamma}))$ *is isomorphic with* $\mathbb{Z}$, *generated by some* $u \in K_\infty^*$ *with* $|u| < 1$.

Therefore, the lattice $\mathrm{ev}(c(\overline{\Gamma})) = u^{\mathbb{Z}}$ in K_∞^* uniformizes an elliptic curve: The one we have been heading for.

(5.10) Proposition. *The commutative diagram*

$$\begin{array}{ccccccccc} 1 & \to & \overline{\Gamma} & \stackrel{c}{\to} & \mathrm{Hom}(\overline{\Gamma}, C^*) & \to & J_0(\mathfrak{n})(C) & \to & 0 \\ & & \downarrow & & \downarrow \mathrm{ev} & & \downarrow & & \\ 1 & \to & u^{\mathbb{Z}} & \to & C^* & \to & C^*/u^{\mathbb{Z}} & \to & 0 \end{array}$$

induces a canonical bijection of $K_\infty^*/u^{\mathbb{Z}}$ *with* $E_\varphi(K_\infty)$ *compatible with extensions of* K_∞ *in* C.

The map π_φ is described by

$$\begin{array}{ccc} X_0(\mathfrak{n})(C) & \longrightarrow & E_\varphi(C) \\ \| & & \| \\ \Gamma\backslash\Omega & \longrightarrow & C^*/u^{\mathbb{Z}}. \\ z & \longmapsto & u_\varphi(z) \end{array} \tag{5.11}$$

If $\omega = dw/w$ is the invariant differential on E_φ coming from a coordinate w of C^*,

$$\pi_\varphi^*(\omega) = u_\varphi'(z)/u_\varphi(z)\, dz. \tag{5.11.1}$$

Thus the study of E_φ may now be replaced by the study of Drinfeld modular curves and forms, which are rather accessible objects. We expect this will lead to progress in the following set of questions (among others):

(a) conjectures of Birch and Swinnerton-Dyer on E_φ: analogue of Kolyvagin's results;
(b) conjectures of type "Gross-Zagier";
(c) analogue of Manin's conjecture: properties of the Fourier coefficients of φ.

References

[De1] P. Deligne: Formes modulaires et représentations de $GL(2)$. In: P. Deligne, W. Kuijk (Ed.): Modular Functions of One Variable II, Lecture Notes in Math. 349, Springer 1973.

[De2] P. Deligne: Les constantes des équations fonctionnelles des fonctions L. Lecture Notes in Math. 349, Springer 1973.

[DH] P. Deligne and D. Husemöller: Survey of Drinfeld Modules. Contemp. Math. 67, (1987), 25–91.

[Dr] V. G. Drinfeld: Elliptic Modules. Math. USSR-Sb. 23(4), (1976), 561–592.

[Ge1] E-U. Gekeler: Automorphe Formen über $\mathbb{F}_q(T)$ mit kleinem Führer. Abh. Math. Sem. Univ. Hamburg 55, (1985), 111–146.

[Ge2] E-U. Gekeler: Drinfeld Modular Curves. Lecture Notes in Math. 1231, Springer 1986.

[Ge3] E-U. Gekeler: On the coefficients of Drinfeld modular forms. Invent. Math. 93, (1988), 667–700.

[Ge4] E-U. Gekeler: Moduli for Drinfeld Modules. This volume.

[GR] E-U. Gekeler and M. Reversat: Jacobians of Drinfeld Modular curves. In preparation.

[Gel] S. Gelbart: Automorphic Forms on Adèle Groups. Ann. Math. Studies 83, 1975.

[GvdP] L. Gerritzen and M. van der Put: Schottky Groups and Mumford Curves. Lecture Notes in Math. 817, Springer 1980.

[GI] O. Goldman and N. Iwahori: The space of p-adic norms. Acta Math. 109, (1963), 137–177.

[Go1] D. Goss: π-adic Eisenstein Series for Function Fields. Compositio Math. 41, (1980), 3–38.

[Go2] D. Goss: The Algebraist's Upper Half-Plane. Bull. Amer. Math. Soc. (N.S.) 2, (1980), 391–415.

[Ha] G. Harder: Halbeinfache Gruppenschemata über Dedekindringen. Invent. Math. 4, (1967), 165–191.

[JL] H. Jacquet and R. P. Langlands: Automorphic Forms on $GL(2)$. Lecture Notes in Math. 114, Springer 1970.

[MD] Y. Manin and V. G. Drinfeld: Periods of p-adic Schottky groups. J. Reine Angew. Math. 262/263, (1973), 239–247.

[Mu] D. Mumford: An analytic construction of degenerating curves over local fields. Compositio Math. 24, (1972), 129–174.

[My] J. F. Myers: p-adic Schottky groups. Thesis: Harvard Univ. 1973.

[vdP] M. van der Put: Les fonctions thêta d'une courbe de Mumford. Groupe d'étude d'Analyse ultramétrique 1981/82.

[Ra] W. Radtke: Diskontinuierliche arithmetische Gruppen im Funktionenkörperfall. Thesis: Bochum 1984.

[Re] M. Reversat: Lecture on rigid geometry. This volume.

[Se] J-P. Serre: Trees. Springer 1980.

[SDB] H. P. F. Swinnerton-Dyer and B. J. Birch: Elliptic Curves and Modular Functions. Lecture Notes in Math. 476, Springer 1975.

[Te] J. Teitelbaum: The Poisson Kernel for Drinfeld Modular Curves. Journal of the Amer. Math. Soc. 4, (1991), 491–511.

[We] A. Weil: Dirichlet Series and Automorphic Forms. Lecture Notes in Math. 189, Springer 1971.

FB 9 - Mathematik, Universität des Saarlandes, Saarbrücken, D-6600 Germany
Email: gekeler@math.uni-sb.de

Laboratoire de Topologie et Géométrie, Université Toulouse III, URA 1408, 31062 Toulouse Cedex, France
Email: ASV@frcict81.bitnet

Some Integrals Attached to Modular Forms in the Theory of Function Fields

David Goss

Abstract

Let $\mathbf{A}$ be the polynomial ring in T over the finite field $\mathbb{F}_r$, $r = p^n$, p prime. Let $\mathbf{k}$ be its quotient field with Ξ denoting the prime of $\mathbf{k}$ "at infinity". We use the symbol "$\mathbf{K}$" to denote the completion of $\mathbf{k}$ at Ξ. We use the symbols "$\overline{\mathbf{K}}$" to denote a fixed algebraic closure of $\mathbf{K}$, and "∞" to denote the point "at infinity" on $\mathbb{P}^1(\mathbf{K})$. We let $|?|$ be the absolute value on $\overline{\mathbf{K}}$ which is the extension of the canonical normalized one on $\mathbf{K}$. We let $\mathcal{O} = \{x \in \mathbf{K} \mid |x| \leq 1\}$ and $\mathcal{M} \subset \mathcal{O}$ the maximal ideal.

Let Ω be the algebraic upper half plane over $\mathbf{K}$; so Ω is a rigid analytic space defined over $\mathbf{K}$ whose set of geometric points is $\{\overline{\mathbf{K}} - \mathbf{K}\}$. One has an action of $GL_2(\mathbf{A})$ on Ω which is very similar to the action of $SL_2(\mathbb{Z})$ on the complex upper half plane $\mathcal{H}$. Moreover, as in the complex theory, this action of $GL_2(\mathbf{A})$ leads naturally to "modular forms"; i.e., rigid functions on Ω with classical style invariance properties (see Definition 2.1.4), and "cusp forms". Quite recently, in **[T1]**, Jeremy Teitelbaum has associated to such cusp forms non-archimedean measures (in the sense of Vishik, Amice and Vélu) on $\mathbb{P}^1(\mathbf{K})$. These measures possess symmetries reflecting those of the cusp form. It is our purpose here to attach to these measures certain integrals with functional equations and to relate these integrals to the theory of Fourier Transforms which appears in **[G7]**. Along the way, we shall give an integral expression for the action of the Hecke operators which is quite reminiscent of the classical kernel formula (see the appendix by D. Zagier in **[L1]**).

The analogy with classical theory leads us to expect that much more than is established here will eventually be found. At various points in the paper, questions will be presented as they naturally occur.

Partially supported by NSF Grant DMS-890110 and NSA Grant MDA 904-89-H-2059. **AMS Classification**: 12A90; **Key Words**: Mellin Transforms, non-archimedean measures, modular forms.

1. Introduction

Let $f(z)$ be a cusp form of weight k for the modular group $SL_2(\mathbb{Z})$ acting on $\mathcal{H} = \{z = x + iy \in \mathbb{C} \mid y > 0\}$ with q-expansion

$$f(z) = \sum_{n\geq 1} b_n q^n = \sum_{n\geq 1} b_n e^{2n\pi iz}.$$

Let $\Gamma(s) = \int\limits_0^\infty t^s e^{-t} \frac{dt}{t}$ be Euler's gamma function. Finally, put

$$L_f(s) = \sum_{n\geq 1} b_n n^{-s}.$$

The function

$$R_f(s) = \Gamma(s)(2\pi)^{-s} L_f(s)$$

is called the "Mellin Transform of $f(z)$". A fundamental result of arithmetic asserts the analytic continuation of $R_f(s)$ and its functional equation

$$R_f(s) = i^k R_f(k-s).$$

(Since k must be even for f to be nontrivial, i^k is a real number.) These results are established by first using the integral for $\Gamma(s)$ to write

$$R_f(s) = \int_0^\infty t^s f(it)\frac{dt}{t}.$$

One then uses the symmetry properties of $f(z)$ to rewrite the integral in a form visibly symmetric under $s \mapsto k - s$.

Now turn to the polynomial theory. Here one replaces $\mathbb{Z}$ by $\mathbf{A}$, $\mathcal{H}$ by the rigid analytic space Ω and $SL_2(\mathbb{Z})$ by $GL_2(\mathbf{A})$. The notions of modular forms and q-expansions have analogs on Ω and so on. It is therefore a natural question to inquire if integrals analogous to those given above exist for Ω. It is our purpose here to show that they do exist, though the analogy with classical theory leaves us with a wealth of unanswered questions. We are able to find these integrals because of the recent seminal work of J. Teitelbaum in **[T1]**. In this paper Teitelbaum constructs non-archimedean measures μ_f from cusp forms f on Ω. The measures μ_f seem to play the role for $\mathbf{K}$ played classically by $f(it)\,dt$ as given above. In particular, μ_f possesses symmetries which mirror exactly those of the original cusp form; indeed, the space of cusp forms turns out to be *isomorphic* to the space of such measures.

As an example of the naturality of his construction, Teitelbaum gives an *integral* for the q-expansions of $f(z)$ which is very similar to *Cauchy's Theorem*, see Remarks 2.3.2.c). Teitelbaum's original derivation is valid only for the polynomial case. At the end of Section 2, however, we show that this integral actually holds in complete generality.

Once one has the Teitelbaum construction, one is able to take a form $f(z)$ of weight k and associate to it a continuous function

$$L_f(s): \mathbb{Z}_p \to \mathbf{K}_f$$

with

$$L_f(s) = \pm L_f(k-s),$$

where $\mathbf{K}_f$ is the finite extension of $\mathbf{K}$ containing the q-expansion coefficients of f. We will determine the exact sign in the above equation in Section 5.

One way to obtain $L_f(s)$ is by integration of the function $u \mapsto u^s$ over the group, $\mathcal{U}_1$, of 1-units of $\mathcal{O}$ against the measure $\frac{d\mu_f}{u}$; i.e.,

$$L_f(s) = \int_{\mathcal{U}_1} t^s \frac{d\mu_f(t)}{t}.$$

From the properties of Teitelbaum's measures, the functional equation will follow without much difficulty. Moreover, the symmetry properties of μ_f also allow us to express $L_f(s)$ as an integral over $\mathcal{M}$. Avner Ash has explained to the author how, from the point of view of *modular symbols*, integration over $\mathcal{M}$ (or the 1-units) corresponds to integration over $[0, i\infty]$ classically.

However, there is another method of obtaining $L_f(s)$ which no longer involves integration over $\mathcal{M}$ but rather integration over $\mathbb{Z}_p$. It is this second method which gives the precise relationship between $L_f(s)$ and the q-expansion of f. Roughly the idea is as follows (see Section 4 for the full story): Let

$$f(z) = \sum_{j \geq 1} b_j q^j.$$

Set

$$T_f(s) = \sum_{j \geq 1} b_j \binom{s}{j-1},$$

which is a continuous $\mathbf{K}$-valued function on $\mathbb{Z}_p$ and which can also be obtained as an integral (Lemma 4.2). Then we show

$$L_f(s) = \widehat{T_f}(s-1),$$

where "$\widehat{\ }$" denotes Fourier Transform in the sense of **[G7]**. It is this formula that allows us to prove that $L_f(s)$ satisfies a certain universal formula (Theorem 4.4). Moreover, the invertibility of the Fourier Transform allows one to pass back and forth between L_f and the q-expansion of f.

In our last section (Section 6) we will present examples of $L_f(s)$ and its functional equation. We will use Teitelbaum's measure to give an integral representation for the action of Hecke operators on the space of cusp forms. The similarity with the classical kernel for the Hecke operators **[L1]** leads us to believe that much more may be said about this integral. We also propose a possible "two-variable" L-series for f. Finally, we briefly discuss some recent work of M. Kapranov **[K1]** on the compactification of moduli spaces of Drinfeld modules of arbitrary rank. This work also allows us to conclude the finite dimensionality of spaces of modular forms of arbitrary rank **[G1]**; thus raising the exciting possibility of finding Teitelbaum measures in complete generality.

Many questions naturally arise from this work:

1) To what extent does the functional equation of $L_f(s)$ allow us (in line with classical theory) to recapture $f(z)$? One apparent difficulty lies with the functions $T_f(s)$: In order for the series given for $T_f(s)$ to converge to a continuous function, one only needs for the coefficients b_j to approach 0; which, in turn, implies that the power series for $f(z)$ converges on the closed unit ball. However, as will be seen in Section 2.1, the power series for $f(z)$ actually converges in a somewhat larger region. How can this overconvergence be expressed using $L_f(s)$?
2) Where are the Γ-functions of **[G4]** and **[Th1]**? On the surface, they are nowhere to be found. In fact, these Γ-functions are shown to be related in **[G7]** to $\mathbf{q}_C(z)^{-1}$ (where $\mathbf{q}_C$ is "q normalized" as explained in Section 2) via Fourier Transforms. *However* there one seems to need a *different parameter for the theory* (as explained in Section 3) than the one used here. No obvious explanation for this is apparent. It may happen that gamma functions will be needed for a satisfactory solution to Question 1 and the overconvergence problem; perhaps through a connection with the two-variable function of our last section.
3) Why does $L_f(s)$ appear as a Fourier Transform and not as a Dirichlet series summed over monic elements of $\mathbf{A}$?
4) Let $p > 2$. Then there is a "center" to $\mathbb{Z}_p$ under the involution $\iota(s) = k - s$. Indeed the congruence class of $\frac{k}{2} \pmod{p}$ is stable under $\iota(s)$. Is this stability somehow reflected in the location of the zeroes of $L_f(s)$?
5) Do there exist good "one-variable" L-series associated to cusp forms for *arbitrary* congruence subgroups and *general* $\mathbf{A}$? The spaces of 1-units do *not* possess the correct symmetries in general.
6) How can one give a functional equation to the two-variable L-series of Section 6?

The formalism of this paper leads to a fascinating idea which seems to have occurred to J. Teitelbaum and E.-U. Gekeler about the same time: Let E be an elliptic curve over $\mathbf{k}$ with split multiplicative reduction at Ξ. Then Drinfeld's reciprocity law implies that E is isogenous to a factor of some rank 2 moduli curve of Drinfeld modules (i.e., a quotient of Ω by an appropriate congruence subgroup of $GL_2(\mathbf{A})$). Thus E gives rise to a differential form on the modular curve which, in turn, comes from a cusp form. By associating to this cusp form its Teitelbaum measure and using the techniques of this paper, we deduce the existence of a characteristic p L-series associated to E. Does it have a functional equation? How is it related to the classical L-series of E, etc.? As of the writing of this paper, one has far more questions than answers about such functions.

It is my pleasure to thank Jeremy Teitelbaum for his essential help in this paper and Warren Sinnott for his usual keen insight into non-archimedean integration. In fact, after a preliminary version of this paper was written, Sinnott suggested to the author that a simple change of variable would give the L-series studied here functional equations of the classical form "$s \mapsto k - s$" as opposed to the form "$k \mapsto k - 2 - s$" which was deduced originally. I also thank Avner Ash for his comments on the use of modular symbols. Finally, I wish to thank M. Kapranov for kindly explaining his work to me.

2. Modular Forms and Teitelbaum's Construction

2.1. A Brief Look at Modular Forms for $GL_2(\mathbf{A})$. Our purpose in this section is to describe the results of **[T1]** needed to perform integration. In order to do so, we recall first in this subsection the basic definitions of modular forms. For more information, we refer the reader to **[G1]**, **[G3]**, **[Ge1]**, **[Ge2]**, and **[Ge3]**.

2.1.1. Definition. We put $\Omega = \overline{\mathbf{K}} - \mathbf{K}$, and $\Gamma = GL_2(\mathbf{A})$.

As is now well-known, Ω inherits a $\mathbf{K}$-rigid analytic structure as a connected analytic subspace of $\mathbb{P}^1(\overline{\mathbf{K}})$. The natural action of $GL_2(\mathbf{K})$ on $\mathbb{P}^1(\overline{\mathbf{K}})$ extends to an analytic action on Ω as follows: Let $\gamma \in GL_2(\mathbf{K})$ be written as

$$\gamma = \begin{pmatrix} a & b \\ c & d \end{pmatrix}.$$

Then, if $x \in \Omega$, we have

$$\gamma(x) = \frac{ax+b}{cx+d}.$$

One knows that this action of Γ on Ω is discontinuous. The resulting quotient $\Gamma\backslash\Omega$ is a 1-dimensional rigid analytic space that is isomorphic to $\mathbb{P}^1$ minus a point (which we denote by "∞" and call "the cusp at infinity"). One has a natural interpretation of $\Gamma\backslash\Omega$ as the coarse moduli space of Drinfeld $\mathbf{A}$-modules of rank 2.

There is a very beautiful geometry underlying this quotient construction. For our purposes we need only examine more closely the behavior of Γ "near ∞". In classical theory, a point $\tau \in \mathcal{H}$ is far from $\mathbb{R}$ (and "near ∞") if its imaginary part, $\Im(\tau)$, has large absolute value. On Ω this notion is replaced by:

2.1.2. Definition. Let $z \in \Omega$. We define the *imaginary absolute value*, $|z|_i$, of z by

$$|z|_i = \min\{|z-\alpha| \mid \alpha \in \mathbf{K}\}.$$

The reader will easily see that, in the complex situation, the obvious analog of the above definition *is* $|\Im(\tau)| = \Im(\tau)$. Let t be a positive rational number. The analogy with classical theory thus leads us to view the spaces $\Omega_t = \{z \in \Omega \mid |z|_i \geq t\}$ as "neighborhoods of ∞". In fact, these open subsets of Ω are also *analytic* neighborhoods of ∞ as they may easily be seen to possess a natural structure as analytic space.

Let Γ_∞ denote the group of translations

$$z \mapsto z + a,$$

for $a \in \mathbf{A}$. The reader may easily check that the imaginary absolute value does not change under the action of Γ_∞. Therefore, Ω_t is *stable* under the action of Γ_∞.

Now let $f(z)$ be a rigid function on Ω which is invariant under translations by elements of Γ_∞. Then $f(z)$ may be given an expansion about ∞ in the following fashion: Form the function

$$\mathbf{e}(z) = z \prod_{0 \neq a \in \mathbf{A}} \left(1 - \frac{z}{a}\right).$$

One checks easily that $\mathbf{e}(z)$ is entire and $\mathbb{F}_r$-linear. It is visibly invariant under translation by elements of $\mathbf{A}$. Standard non-archimedean function theory readily gives an isomorphism of groups

$$\widetilde{\mathbf{e}}(z)\colon \overline{\mathbf{K}}/\mathbf{A} \to \overline{\mathbf{K}},$$

which is also an isomorphism of the underlying rigid spaces. Now put

$$\mathbf{q}(z) = \mathbf{e}(z)^{-1}.$$

Since $\mathbf{e}(z)$ is additive, its derivative can be computed to be identically 1. Thus logarithmic differentiation gives immediately

$$\mathbf{q}(z) = \sum_{a \in \mathbf{A}} \frac{1}{z-a}. \tag{2.1.3}$$

By definition, $\mathbf{q}$ is invariant under the action of Γ_∞ and approaches 0 as $|z|_i$ approaches ∞ (this is immediately seen from 2.1.3.). Moreover, this fact, and estimates of a similar type involving $\mathbf{e}(z)$, allow us to see that $\mathbf{q}$ maps

$$\Gamma_\infty \setminus \{z \in \Omega \mid |z|_i > r\}$$

isomorphically to the pointed open disc

$$\mathcal{V} = \{0 \neq z \in \overline{\mathbf{K}} \mid |z| < r\}.$$

Thus $\mathbf{q}$ is a *uniformizing parameter* at ∞. As such, we have the expansion

$$f(z) = \sum_{j=-\infty}^{\infty} b_j \mathbf{q}(z)^j,$$

which converges in $\mathcal{V}$. We call this Laurent series the *expansion of f at the cusp* ∞, or (in standard terminology) the *q-expansion of $f(z)$*. We say that f is *holomorphic at* ∞ if the above expansion contains no negative terms.

Finally, let $\gamma \in \Gamma$ be written as above. An elementary argument shows that

$$|\gamma z|_i = |cz + d|^{-2} |z|_i.$$

We conclude that, for t sufficiently large, the stabilizer of Ω_t in $SL_2(\mathbf{A})$ is Γ_∞. (In $GL_2(\mathbf{A})$ it is $\Gamma_\infty \times S$ where S is the group of scalar matrices associated to elements of $\mathbb{F}_r^* = \mathbf{A}^*$.)

All of the above motivates the following basic definition. In it we let k be an integer and m a residue class of $\mathbb{Z}/(r-1)$.

2.1.4. Definition. Let f be a rigid analytic function on Ω. We call f a *modular form of weight k and type m* if

a) for $\gamma \in \Gamma$ (γ given as above)

$$f(\gamma z) = \det(\gamma)^{-m}(cz + d)^k f(z);$$

b) f is holomorphic at infinity.

We say that $f(z)$ is a *cusp form* if it is holomorphic at infinity and the constant term of its expansion is 0.

Through the use of the Riemann-Roch theorem, one finds that the space $\mathcal{M}(k,m)$ of modular forms of weight k and type m is finite dimensional. (For a computation of the dimensions of the spaces of forms of weight 0, see **[G3]**). We denote the space of cusp forms of weight k and type m by $\mathcal{S}(k,m)$.

We finish this section by explaining a subtlety involving q-expansions. Let C be the Carlitz module for $\mathbf{A}$ (see **[C1]**, **[G2]**); so C is the simplest rank-one Drinfeld $\mathbf{A}$-module and is defined by the equation

$$C_T(z) = Tz + z^r.$$

We let $\mathbf{e}_C(z)$ be the exponential of C. So $\mathbf{e}_C(z)$ is an entire $\mathbb{F}_r$-linear power series which satisfies the functional equation

$$\mathbf{e}_C(Tz) = T\mathbf{e}_C(z) + \mathbf{e}_C(z)^r.$$

From this we deduce that

$$\mathbf{e}_C(z) = \sum_{j=0}^{\infty} \frac{z^{r^j}}{D_j},$$

where

$$D_j = (T^{r^j} - T)(T^{r^j} - T^r)\cdots(T^{r^j} - T^{r^{j-1}})$$

and can be seen to be the product of all monics of degree j. The zeroes of $\mathbf{e}_C(z)$ form a rank one module over $\mathbf{A}$ which is written as $\mathbf{A}\xi$. Here ξ plays the role for $\mathbf{A}$ which is played by $2\pi i$ classically and is transcendental (by Wade **[W1]**). Since entire functions are determined up to a constant by their divisors, we find

$$\mathbf{e}(z) = \frac{1}{\xi}\mathbf{e}_C(\xi z).$$

2.1.5. Definition. We put

$$\mathbf{q}_C(z) = \mathbf{e}_C(\xi z)^{-1} = \frac{1}{\xi \mathbf{e}(z)}.$$

Thus if

$$f(z) = \sum b_j \mathbf{q}_C^j,$$

then

$$f(z) = \sum \xi^{-j} b_j \mathbf{q}^j,$$

and so on. However, from the Tate theory (**[G1]**, **[G3]**) we see that $\mathbf{q}_C$ is the "correct" parameter in the following sense: The "standard" modular forms (see next section) have $\mathbf{q}_C$-expansion coefficients which are, up to a fixed power of ξ, *rational*. Thus from now on we use only $\mathbf{q}_C$-adic expansions. In particular, we shall always express the coefficients of $\mathbf{q}$-expansions in terms of the $\mathbf{q}_C$-adic ones as above.

It also follows from the Tate theory that, as for $SL_2(\mathbb{Z})$, there exists bases for the spaces of modular forms with $\mathbf{q}_C$-expansion coefficients generating *finite* extensions of $\mathbf{k}$. It easily follows that these same bases will have $\mathbf{q}$-expansions which generate finite extensions of $\mathbf{K}$.

The reader can easily establish the obvious analog of Equation 2.1.3 for $\mathbf{q}_C(z)$.

2.1.6. Definition. Let f be a modular form. We denote by $\mathbf{K}_f$ the finite extension of $\mathbf{K}$ generated by its $\mathbf{q}$-expansion coefficients.

To finish with this circle of ideas, we denote by $\mathbf{log}(z)$ the composition inverse of $\mathbf{e}(z)$ and by $\mathbf{log}_C(z)$ the composition inverse of $\mathbf{e}_C(z)$. One finds with a little work that

$$\mathbf{log}(z) = \frac{1}{\xi}\mathbf{log}_C(\xi z),$$

and

$$\mathbf{log}_C(z) = \sum_{j=0}^{\infty}(-1)^j \frac{z^{r^j}}{L_j},$$

where

$$L_j = (T^{r^j} - T)(T^{r^{j-1}} - T)\cdots(T^r - T).$$

(For more on $\{D_j, L_j\}$, we refer the reader to **[Th1]**, **[G4]**.)

2.2. Examples of Modular Forms. We begin with the Eisenstein series:

2.2.1. Definition. Let k be a positive integer. Let $\mathcal{X}$ be the set of all ordered pairs (a, b) with $\{a, b\} \subset \mathbf{A} \times \mathbf{A}$ and $(a, b) \neq (0, 0)$. We put

$$E_k(z) = \sum_{(a,b)\in\mathcal{X}} \frac{1}{(az+b)^k}.$$

It is easy to see both that $E_k(z)$ is rigid analytic on Ω and that it is a modular form for Γ of weight k and type 0. Moreover, for $k \not\equiv 0 \pmod{r-1}$, $E_k(z)$ is identically 0.

Now let $k \equiv 0 \pmod{r-1}$. We begin by discussing the constant term of $E_k = \sum_{0\neq b\in\mathbf{A}} b^{-k}$. We abuse language and denote this sum by "$E_k(\infty)$". One sees easily that

$$E_k(\infty) = -\sum_{b\in\mathbf{M}} b^{-k},$$

where $\mathbf{M} \subset \mathbf{A}$ is the subset of *monic* elements. Denote the sum on the right by "$\zeta(k)$". It is trivial to see that $\zeta(k) \neq 0$ and thus that E_k is non-trivial. Moreover, Carlitz has shown (see **[C1]**, **[G2]**) that

$$\xi^{-k}\zeta(k) \in \mathbf{k},$$

and so the same result obviously holds for $E_k(\infty)$. We will use the symbol "$\mathcal{B}_k$" to denote the rational element

$$\xi^{-k}E_k(\infty) = -\xi^{-k}\zeta(k).$$

Upon multiplying $\mathcal{B}_k$ by the "Carlitz factorials" (**[Th1]**, **[G4]**), we obtain (-1 times) the "Bernoulli-Carlitz numbers"; i.e., $\mathbf{A}$-analogs of classical Bernoulli numbers.

In order to discuss the non-constant terms we need some results on certain "lattice" sums; for our purposes a *lattice* will be a discrete $\mathbb{F}_r$-submodule of $\overline{\mathbf{K}}$. A lattice is said to be an $\mathbf{A}$-*lattice* if it is also an $\mathbf{A}$-module. Our discussion here is based on that of E.-U. Gekeler's paper **[Ge1]** (but note that we have changed notation slightly; we use "$\mathcal{S}_k$" where "G_k" is used by Gekeler).

2.2.2. Definition. Let L be a lattice and k a positive integer. Put

$$s_{k,L}(z) = \sum_{\alpha \in L} \frac{1}{(z-\alpha)^k}.$$

So, for instance, if $L = \mathbf{A}$ then Equation 2.1.3 tells us that

$$s_{1,L}(z) = \mathbf{q}(z).$$

Newton's formula on power sums of zeroes of polynomials implies the existence of a polynomial $\mathcal{S}_k(u) = \mathcal{S}_{k,L}(u)$ such that

$$\mathcal{S}_k(s_{1,L}(z)) = s_{k,L}(z).$$

With a little work, one finds:

1) $\mathcal{S}_k(u)$ is monic of degree k;
2) $\mathcal{S}_k(0) = 0$;
3) Let $k \leq r$, then $\mathcal{S}_k(u) = u^k$.

(For more such relations, see **[Ge1**, 3.4**]**.)

Let $\mathbf{e}_L(z) = z \cdot \prod_{0 \neq \alpha \in L} (1 - \frac{z}{\alpha})$. Following **[Ge1]** we form the generating function:

$$\mathcal{S}(u,z) = \sum_{k=0}^{\infty} \mathcal{S}_k(u) z^k.$$

Through the use of the recurrence relations coming from Newton's rules, Gekeler shows

$$\mathcal{S}(u,z) = \frac{u}{1 - u\mathbf{e}_L(z)}. \tag{2.2.3}$$

This formula allows one to compute some of the functions $\mathcal{S}_k$. However, they still remain poorly understood in general.

In Remarks 3.4, we shall give a very simple derivation of Equation 2.2.3 through the use of hyper-derivatives.

Let $\alpha \in \mathbf{A}$ have degree d and let $C_\alpha(z)$ be the additive polynomial in z associated to α by the Carlitz module. The degree of $C_\alpha(z)$ in z is r^d. Let $c_\alpha(z)$ be the polynomial in z defined by

$$c_\alpha(z) = z^{r^d} C_\alpha\left(\frac{1}{z}\right),$$

and put (by abuse of language)

$$\mathbf{q}_C^{(\alpha)} = \mathbf{q}_C(\alpha z) = \mathbf{q}_C^{r^d} c_\alpha(\mathbf{q}_C).$$

Finally, put $L = \mathbf{A}\xi$ and $\mathcal{S}_k(u) = \mathcal{S}_{k,L}(u)$. From Equation 2.2.3 (for instance) one finds that $\mathcal{S}_k(u)$ has $\mathbf{k}$-rational coefficients. Thus, for $\alpha \in \mathbf{A}$, elementary algebra also shows that $\mathcal{S}_k(\mathbf{q}_C^{(\alpha)})$ has $\mathbf{k}$-rational $\mathbf{q}_C$-expansion coefficients.

We now have:

2.2.4. Proposition. *Let* $k \equiv 0 \pmod{r-1}$. *Then*

$$\xi^{-k} E_k(z) = \mathcal{B}_k - \sum_{a \in \mathbf{M}} \mathcal{S}_k(\mathbf{q}_C^{(a)}) . \qquad \square$$

In particular, the $\mathbf{q}_C$-expansion coefficients of E_k are rational multiples of ξ^k.

Next we turn to the form Δ. There are a number of ways to define this form and we choose a simple one in terms of Eisenstein series:

2.2.5. Definition. We put

$$\Delta = (T^r - T)^r E_{r-1}^{r+1} + (T^{r^2} - T) E_{r^2-1} .$$

Clearly Δ is a form of weight $r^2 - 1$ and type 0 which is readily seen to be a *cusp form*. Moreover, the computations for Eisenstein series now immediately imply that the $\mathbf{q}_C$-expansion coefficients of $\xi^{1-r^2}\Delta$ are rational. In fact, one knows **[G3]** that many of the expansion coefficients vanish.

Following classical notation, we write the $\mathbf{q}_C$-expansion of Δ as

$$-\xi^{1-r^2}\Delta = \sum_{j=1}^{\infty} \tau(j) \mathbf{q}_C^j ,$$

with $\{\tau(j)\} \subseteq \mathbf{k}$. On functorial grounds (**[G3]**), one finds that $\tau(j) \in \mathbf{A}$ for all j. This fact (and the reason for the minus sign in the definition of $\tau(j)$) will also follow immediately from the Theorem 2.2.6.

In classical $SL_2(\mathbb{Z})$ theory, the modular form Δ possesses a well-known product formula for its q-expansion coefficients. In our case, an elegant analog of this result was established by E.-U. Gekeler in **[Ge3]**. His result is:

2.2.6. Theorem 2.2.6 (Gekeler).

$$-\xi^{1-r^2}\Delta = \mathbf{q}_C^{r-1} \prod_{a \in \mathbf{M}} c_a(\mathbf{q}_C)^{(r^2-1)(r-1)} . \qquad \square$$

The algebra of forms of arbitrary weight, type 0, and with *rational* $\mathbf{q}_C$-expansion coefficients is

$$\mathbf{k}[\xi^{1-r} E_{r-1}, \xi^{1-r^2}\Delta] .$$

From Theorem 2.2.6, it seems reasonable to expect that an (r-1)-st root of $-\Delta$ is rigid analytic on Ω. To see this, Gekeler, in **[Ge1]**, constructs a "Poincaré series", h, with $h^{r-1} = -\Delta$. The form h is a cusp form of weight $r+1$ and type 1 with $\mathbf{q}_C$-expansion beginning with $-\mathbf{q}_C$ (thus the exact root of Δ is specified). We write

$$-\xi^{-1-r} h = \sum_{j=1}^{\infty} \phi(j) \mathbf{q}_C^j .$$

Clearly 2.2.6 gives a product formula for the expansion of h and shows that $\{\phi(j)\} \subseteq \mathbf{A}$. Moreover it can be shown that the algebra of *all* (i.e., arbitrary type) forms with rational coefficients is

$$\mathbf{k}[\xi^{1-r}E_{r-1}, \xi^{-1-r}h].$$

Finally, let $k \equiv 0 \pmod{r-1}$ and let $f \in \mathcal{M}(k,0)$. Then f is easily written as a linear combination of $E_k(z)$ and a cusp form.

2.3. Teitelbaum's Construction. Let $\mathcal{K}$ be a totally disconnected local field (e.g., $\mathbf{K}$ or $\mathbb{Q}_p$). The usual notion of a "*$\mathcal{K}$-valued measure μ*" on a totally-disconnected space X (as in, say, **[G6]**) is a finitely additive $\mathcal{K}$-valued function μ on the compact open subsets of X such that the set of values

$$\{\mu(U) \mid U \subseteq X,\ U \text{ compact open}\}$$

is *bounded.* With such a measure, one is able to integrate compactly supported, continuous functions from X to $\mathcal{K}$. When we talk about "measures on $\mathbb{Z}_p$" we shall *always* have this notion of measure in mind.

However, there are many ways to weaken the boundedness condition. The most obvious one is to simply eliminate it; one thus obtains the notion of *$\mathcal{K}$-valued distribution on X*. Given a distribution, one can only integrate (in the obvious fashion) compactly supported, locally constant functions.

Another way to weaken the boundedness requirement is to replace it with a *growth condition* as one decomposes a given compact open into a union of sub-compact opens (see **[MTT1]**). One can now no longer integrate *all* continuous functions but must restrict to those that are *locally analytic*. This notion of "measure" is due to Vishik, following Amice and Vélu, and we shall have it in mind when we discuss "measures on $\mathbf{K}$ (or $\mathbb{P}^1(\mathbf{K})$)". Vishik's measures are the type of objects produced by Teitelbaum in **[T1]**. The reader may refer there for more information.

Let μ be a measure on X and $g(t)$ a function that may be integrated against μ. We write

$$\int_X g(t)\, d\mu(t)$$

for the integral.

For the rest of this chapter we set $X = \mathbb{P}^1(\mathbf{K})$.

We define an action of Γ on the measures on $\mathbb{P}^1(\mathbf{K})$ as follows: Let μ be a measure on $\mathbb{P}^1(\mathbf{K})$ and let $\gamma \in \Gamma$. Then the measure $\gamma^*\mu$ is defined by

$$\int_U g(t)\, d(\gamma^*\mu)(t) = \int_{\gamma U} g(\gamma^{-1}t)\, d\mu(t).$$

Teitelbaum's idea, building on earlier papers of P. Schneider, **[S1]**, and V. Drinfeld, is to linearly associate to a modular form $f \in \mathcal{M}(k,m)$ a measure μ_f on $\mathbb{P}^1(\mathbf{K})$. This is an intricate construction which involves the *tree* of $PGL_2(\mathbf{K})$; we refer the reader to **[T1]**, **[T2]**, and **[S1]**, for a more complete discussion. We only recall here those results that are needed to actually compute integrals:

2.3.1. Theorem (Teitelbaum). *Let $f \in \mathcal{M}(k,m)$. Then the transformation $f \mapsto \mu = \mu_f$ has the following properties:*

1) *Let E be an Eisenstein series. Then μ_E is identically 0.*
2) *Let $\gamma \in \Gamma$ with*

$$\gamma = \begin{pmatrix} a & b \\ c & d \end{pmatrix}.$$

Then,

$$\gamma^* \mu = \det(\gamma)^{(1-m)}(cz+d)^{(k-2)}\mu.$$

3)

$$\int_{\mathbb{P}^1(\mathbf{K})} g(x)\, d\mu(x) = 0,$$

for $g(x)$ a polynomial of degree $k-2$.

4) *Let $f \in \mathcal{S}(k,m)$. Then we have the Poisson Kernel Formula*

$$f(z) = \int_{\mathbb{P}^1(\mathbf{K})} \frac{1}{z-x}\, d\mu(x).$$

5) *Every measure with the symmetries of Part 2) arises uniquely from a cusp form.*
6) *Let $U \subset \mathbb{P}^1(\mathbf{K})$ be compact open which is written as a disjoint union of closed balls. Let $g_j(x)$, j a non-negative integer, be a sequence of analytic functions on U which are given by convergent power series in each ball and which converge uniformly on geometric points (in the sense of rigid analysis) to a function $g(x)$. Then*

$$\int_U g(x)\, d\mu(x) = \lim_j \int_U g_j(x)\, d\mu(x).$$

7) *Let*

$$f(z) = \sum_{j=1}^{\infty} b_j \mathbf{q}_C^j = \sum_{j=1}^{\infty} \xi^{-j} b_j \mathbf{q}^j.$$

Then,

$$\xi^{-j} b_j = \int_{\mathcal{M}} \mathbf{e}(x)^{j-1}\, d\mu(x),$$

and

$$\xi^{-1} b_j = \int_{\mathcal{M}} \mathbf{e}_C(\xi x)^{j-1}\, d\mu(x). \qquad \square$$

2.3.2. Remarks .

a) 2.3.1.2) is a slight generalization of what is found in **[T1]**. It can be derived quickly from the techniques of, say, Lemma 6 of **[T1]**, and the following fact:

$$\frac{1}{\gamma z - \gamma x} = \det(\gamma)^{(-1)}(cx+d)(cz+d)/(z-x).$$

Thus, integration against a Γ-invariant measure will produce forms of type 1 and so on. (I am indebted to Teitelbaum for pointing this out to me.)

b) By standard rigid analytic geometry, the convergence condition in 2.3.1.6) forces convergence as elements of Tate algebras.
c) One can see the analogy of 2.3.1.7) with the classical *Cauchy's Theorem* by playing a little loose in the following fashion:. Let $f(z) = \sum_{j=1}^{\infty} b_j \mathbf{q}_C^j$. Then 2.3.1.7) can be restated as

$$\frac{b_j}{\xi} = \int_{\mathcal{M}} \mathbf{q}_C(\xi z)^{1-j}\, d\mu(z).$$

Now,

$$d\mathbf{q}_C = -\xi \mathbf{q}_C^2\, dz,$$

and "morally"

$$d\mu_f = f(z)\, dz.$$

Substituting, we "find"

$$-b_j = \int_{\mathcal{M}} \mathbf{q}_C(z)^{-1-j} f(z)\, dz,$$

which certainly *looks* like Cauchy's integral.

2.3.3. Corollary. *Let* $f(z)$ *be as in Theorem 2.3.1. Then*

$$f(z) = \int_{\mathcal{M}} \frac{1}{\mathbf{e}(z) - \mathbf{e}(x)}\, d\mu(x) = \int_{\mathcal{M}} \frac{1}{\mathbf{e}(z-x)}\, d\mu(x).$$

Proof. We shall give two proofs of this. The first uses 2.3.1.7) and the second (due to Warren Sinnott) uses 2.3.1.4).

Let $z \in \mathcal{V}$. Then 2.3.1.7) may be written as

$$f(z) = \sum_{j=1}^{\infty} \int_{\mathcal{M}} \mathbf{e}(x)^{j-1} \mathbf{q}(z)^j\, d\mu(x).$$

Upon changing the order of summation and integration and using the geometric series, we find

$$f(z) = \int_{\mathcal{M}} \frac{\mathbf{q}(z)}{1 - \mathbf{e}(x)\mathbf{q}(z)}\, d\mu(x),$$

which is equivalent to the result to be obtained except that we have restricted z to be in $\mathcal{V}$. On the other hand, the integral can be seen to be rigid analytic for *all* $z \in \Omega$. Thus, the result follows by the connectivity of Ω as a rigid analytic space.

For the second proof, since $f(z)$ is a cusp form, the integral in 2.3.1.4) may be computed by integration over $\mathbf{K}$ (see the remark following Lemma 13 of **[T1]**). As such, we find

$$f(z) = \int_{\mathbf{K}} \frac{1}{z - x}\, d\mu(x). \qquad (2.3.4)$$

Now, 2.3.1.2) immediately implies that μ is invariant under the elements of Γ_∞. Therefore the integral in equation 2.3.4 can be shifted over to $\mathcal{M}$. We find

$$f(z) = \int_{\mathcal{M}} \sum_{\alpha \in \mathbf{A}} \frac{1}{z - x + \alpha}\, d\mu(x).$$

Equation 2.1.3 now gives the result. □

As Warren Sinnott has pointed out, Corollary 2.3.3 can also be used to show the existence of $\mathbf{q}$-expansions for cusp forms. Indeed, one need only reverse the steps of our first proof of the corollary and then integrate to get the coefficients! We note that this procedure gives the correct expansion coefficients since they are unique.

The techniques of proof of Corollary 2.3.3 actually allows us to fill a gap in Teitelbaum's paper **[T1]**. Indeed, in **[T1]** an integral for $\mathbf{q}$-expansions is only given in the polynomial case, which is what we have been working with here. But let $\mathbf{A}$ now denote the general affine algebra of **[T1]**; so Spec($\mathbf{A}$) is a complete curve minus a closed point. Let J be a fractional ideal of $\mathbf{A}$ and let Γ_J be the group of translations by elements of J. With a little thought, one can construct a compact and open fundamental domain, $\mathcal{F}$, for the action of Γ_J on $\mathbf{K}$. Now the second proof of 2.3.3 works in complete generality to express $f(z)$ as $\displaystyle\int_{\mathcal{F}} \frac{1}{\mathbf{e}(z) - \mathbf{e}(x)}\, d\mu_f(x)$. We have seen that this formula then transforms into an integral formula for the $\mathbf{q}$-expansion coefficients of $f(z)$ and the result is established.

3. Fourier Transforms

In this section we put $X = \mathbb{Z}_p$ (in the discussion beginning our last section) and we examine *continuous* functions and measures from $\mathbb{Z}_p \to \mathbf{K}$. We will discuss the relevant portions of **[G7]**. We refer the reader there for proofs and for results not mentioned here (such as Poisson Summation).

Let $g(s)\colon \mathbb{Z}_p \to \mathbf{K}$ be a continuous function and ν a $\mathbf{K}$-*valued* measure on $\mathbb{Z}_p$. Then (see **[G5]**) one knows that $g(s)$ can be written as

$$g(s) = \sum_{j \geq 0} a_j \binom{s}{j},$$

where the $a_j \in \mathbf{K}$ and $a_j \to 0$ as $j \to \infty$. Conversely any such sequence $\{a_j\}$ defines a continuous function. This development of $f(s)$ is called its "Mahler expansion". Moreover, to ν there is a canonically associated formal power series $h(X) = h_\nu(X)$ in $\mathbf{K}[[X]]$ with

$$h(X) = h_\nu(X) = \sum_{j \geq 0} c_j X^j,$$

and

$$\int_{\mathbb{Z}_p} g(s)\, d\nu(s) = \sum_{j \geq 0} a_j c_j.$$

Conversely, this last formula also shows how *any* power series can be used to create a measure. We often abuse notation and write $\int g(s)\,dh(X)(s)$ for $\int g(s)\,d\nu(s)$, etc.

Choose now an *additive* parameter $\mathcal{Z}(X) \in \mathcal{O}[[X]]$. Thus $\mathcal{Z}'(X)$ is a *constant* in $\mathcal{O}^*$. As $\mathcal{Z}$ is a parameter, it has a composition inverse $\mathcal{Z}^{(-1)}(X)$ which has the same properties as $\mathcal{Z}(X)$. Thus, we can write

$$h(X) = \widehat{h}(\mathcal{Z}(X)).$$

and

$$h(X) = \widetilde{h}(\mathcal{Z}^{(-1)}(X)).$$

We call the measure associated to $\widehat{h}(X)$ the *Fourier Transform of* ν and the one associated to $\widetilde{h}(X)$ the *inverse Fourier Transform of* ν. Clearly,

$$\widetilde{\widehat{h}}(X) = h(X),$$

etc.

Since $\mathcal{Z}(X) = \alpha X +$higher terms, the binomial theorem allows us to raise $1+\mathcal{Z}(X)$ to the s-th power, for arbitrary $s \in \mathbb{Z}_p$. Indeed,

$$(1+\mathcal{Z}(X))^s = \sum_{t=0}^{\infty} \binom{s}{t} \mathcal{Z}(X)^t$$

and the sum on the right converges X-adically to an element of $\mathcal{O}[[X]]$. A similar statement also holds for $\mathcal{Z}^{(-1)}(X)$.

3.1. Definition. Let $g(s)\colon \mathbb{Z}_p \to \mathbf{K}$ be continuous.

a) We define the *Fourier Transform of* g, $\widehat{g}(s)\colon \mathbb{Z}_p \to \mathbf{K}$, by

$$\widehat{g}(s) = \int_{\mathbb{Z}_p} g(t)\, d(1+\mathcal{Z}(X))^s(t).$$

b) We define the *inverse Fourier Transform of* g to be

$$\widetilde{g}(s) = \int_{\mathbb{Z}_p} g(t)\, d(1+\mathcal{Z}^{(-1)}(X))^s(t).$$

It is trivial to see that both $\widehat{g}$ and $\widetilde{g}$ are continuous functions from $\mathbb{Z}_p \to \mathbf{K}$.

3.2. Theorem. *Let* $g(s)\colon \mathbb{Z}_p \to \mathbf{K}$ *be continuous. Then we have*

1) $\widetilde{\widehat{g}}(s) = \widehat{\widetilde{g}}(s) = g(s)$.

2) *("The Parseval Formula")*

$$\int_{\mathbb{Z}_p} g(t)\, dh(X)(t) = \int_{\mathbb{Z}_p} \widetilde{g}(t)\, d\widehat{h}(X)(t). \qquad \square$$

Upon substituting $\widetilde{h}(X)$ for $h(X)$ in the theorem, the next corollary results immediately.

3.3. Corollary. *With the notation of 3.2, we have*

$$\int_{\mathbb{Z}_p} g(t)\, d\tilde{h}(X)(t) = \int_{\mathbb{Z}_p} \hat{g}(t)\, dh(X)(t). \qquad \square$$

3.4. Remarks. Let j be a non-negative integer. We set $D = \frac{d}{dX}$ and

$$\frac{D^j}{j!} = \frac{d^j}{dX^j} \Big/ j!$$

as operators on $\mathcal{O}[[X]]$. (Since $\frac{D^j}{j!} X^n = \binom{n}{j} X^{n-j}$, and $\binom{n}{j} \in \mathbb{Z}$, there is no problem with the characteristic of $\mathbf{K}$.) Linear combinations (which may be infinite) of the operators $\frac{D^j}{j!}$ are called *hyper-derivatives*. Theorem 3.2 is then established by a *change of variable* formula for hyper-derivatives based *precisely* on the additivity of our parameter $\mathcal{Z}$.

Hyper-derivatives can also be used to give a quick proof of Equation 2.2.3. Indeed, let D now denote d/dz and note that

$$\frac{D^k}{k!} s_{1,L}(z) = \frac{D^k}{k!} \mathbf{e}_L(z)^{-1} = (-1)^k s_{k,L}(z).$$

Moreover, *by definition* for any function $f(z)$, $\frac{D^j}{j!} f(z)$ is the coefficient of t^j in $f(z+t)$. Thus, $(-1)^j s_{j,L}(z)$ is the coefficient of t^j in

$$\frac{1}{\mathbf{e}_L(z+u)}.$$

Upon substituting $\log_L(u)$ for u the above formula is readily seen to be equivalent to that of 2.2.3.

Finally, note that while we have worked with the field $\mathbf{K}$ in this section, *everything* carries over word for word to finite extensions of $\mathbf{K}$; e.g., the fields $\mathbf{K}_f$ for f a modular form.

4. Mellin Transforms

For this section, *fix* a form $f(z) \in \mathcal{S}(k, m)$ with expansions

$$f(z) = \sum_{j \geq 1} b_j \mathbf{q}_C^j = \sum_{j \geq 1} \xi^{-j} b_j \mathbf{q}^j.$$

We denote its Teitelbaum measure by "$\mu = \mu_f$" and, as above, by "$\mathbf{K}_f$" the field generated by its $\mathbf{q}$-expansion coefficients.

Let $h(X) \in \mathcal{O}[[X]]$. Clearly the power series for h converges for any $x \in \mathcal{M}$ and defines an analytic $\mathbf{K}$-valued function there. We therefore have a *linear operator* $\mathcal{L}$ on $\mathcal{O}[[X]]$ defined by

$$\mathcal{L}(h) = \int_{\mathcal{M}} h(x)\, d\mu(x).$$

4.1. Theorem. *There exists a unique continuous function* $L_f(s)\colon \mathbb{Z}_p \to \mathbf{K}_f$ *such that*

$$\mathcal{L}(h) = \int_{\mathbb{Z}_p} L_f(t+1)\, dh(X)(t).$$

Proof. Recall that at the beginning of Section 3, we presented a formula for computing

$$\int_{\mathbb{Z}_p} L_f(t+1)\, dh(X)(t)$$

in terms of the expansions of $L_f(s+1)$ and $h(X)$. This formula, and the linearity of the integral over $\mathcal{M}$, formally force $L_f(s+1)$ to be uniquely written as

$$L_f(s+1) = \sum_{j\geq 0} \left(\int_{\mathcal{M}} x^j \, d\mu(x) \right) \binom{s}{j}.$$

The only question is that of convergence; i.e, does the coefficient of $\binom{s}{j}$ tend to 0. However, on $\mathcal{M}$ (viewed as the $\mathbf{K}$-rational points of the closed ball of radius $\frac{1}{r}$) the functions x^j, $j > 0$, converge uniformly to 0 on geometric points. Thus the same is true for their integrals by 2.3.1.6). □

We call the function $L(s) = L_f(s)$ the *Mellin Transform* of $f(z)$. It is the goal of this section to give a simple formula (Theorem 4.4) for it. We begin by examining a simpler transform of f.

4.2. Lemma. *Put*

$$T(s) = T_f(s) = \sum_{j\geq 1} \xi^{-j} b_j \binom{s}{j-1}.$$

Then $T(s)$ *is continuous and*

$$T(s) = \int_{\mathcal{M}} (1 + \mathbf{e}(x))^s \, d\mu_f(x).$$

Proof. Since the $\mathbf{q}$-expansion of $f(z)$ converges on a disc of radius > 1 as in Section 2.1, one deduces that $b_j/\xi^j \to 0$ as $j \to \infty$. This is enough to show that T is continuous.

Now expand $(1 + \mathbf{e}(x))^s$ by the binomial theorem to obtain

$$(1 + \mathbf{e}(x))^s = \sum_{j\geq 0} \binom{s}{j} \mathbf{e}(x)^j.$$

The lemma now follows by 2.3.1.6) and 2.3.1.7).

Continuity can also be derived from the integral formula. □

Our next goal is to show how one can compute integrals over $\mathcal{M}$ in terms of integrals over $\mathbb{Z}_p$ by using the function $T(s)$ and Fourier Transforms. In order to do so, we finally fix our uniformizer $\mathcal{Z}(X)$ to be

$$\mathcal{Z}(X) = \mathbf{log}(X) = \frac{1}{\xi} \mathbf{log}_C(\xi X).$$

In fact, it follows trivially from the product expansions of the associated exponential functions that both $\mathbf{log}(X)$ and $\mathbf{log}_C(X)$ have the properties required for $\mathcal{Z}$.

4.3. Theorem. *Let f be as in Theorem 4.1, and let $h(X) \in \mathcal{O}[[X]]$. Then we have the integral formula*

$$\int_{\mathcal{M}} h(x)\, d\mu(x) = \int_{\mathbb{Z}_p} T_f(t)\, d\widetilde{h}(X)(t).$$

Proof. By definition

$$x = \mathbf{log}(\mathbf{e}(x)).$$

Thus, the power series in x can be expressed as power series in $\mathbf{e}(x)$. Teitelbaum's 2.3.1.6) and 2.3.1.7) now tell us how to read off the integral over $\mathcal{M}$ in terms of $\mathbf{q}$-expansions.

On the other hand, note that by our definitions

$$\widetilde{h}(x) = h(\mathbf{log}(x)).$$

Thus, one sees that computing the integral over $\mathcal{M}$ in terms of $\mathbf{q}$-expansions is exactly the same as computing the integral of $T_f(s)$ over $\mathbb{Z}_p$ against the measure $d\widetilde{h}(X)$ and the result is established. □

4.4. Theorem. *Let $f(z)$ be as above. Then*

$$L_f(s) = \widehat{T_f}(s-1).$$

Proof. By Corollary 3.3,

$$\int_{\mathbb{Z}_p} T_f(t)\, d\widetilde{h}(X)(t) = \int_{\mathbb{Z}_p} \widehat{T_f}(t)\, dh(X)(t).$$

Thus the theorem follows from 4.3 and the uniqueness of $L_f(s)$. □

4.5. Corollary. *We have*

$$\widetilde{L_f(s+1)} = T_f(s).$$

Proof. We know that

$$\widetilde{\widehat{T_f}}(s) = T_f(s).$$

The result follows directly. □

The importance of the corollary lies in that it provides a formal method of passing from the functions $L_f(s)$ to $\mathbf{q}_C$-expansions via $T_f(s)$.

4.6. Question. Is there any means of making the above transformation, which is really analogous to the classical *Mellin Transform*, any less formal?

5. Functional Equations

We retain the notation of the last section. Our goal now is to establish that the functions $L_f(s)$ possess *functional equations*.

Let

$$\mathcal{U}_1 \subseteq \mathcal{O}^*$$

be the subgroup of 1-units; i.e., those units $\equiv 1 \pmod{\mathcal{M}}$.

5.1. Theorem. *We have*

$$L_f(s) = \int_{\mathcal{U}_1} t^s \frac{d\mu(t)}{t}.$$

Proof. Let $s \in \mathbb{Z}_p$ and

$$h_s(X) = (1+X)^s = \sum_{j\geq 0} \binom{s}{j} X^j,$$

which we view as an element of $\mathcal{O}[[X]]$.

Recall now that, by 2.3.1.2), μ is invariant under translation by elements of Γ_∞. Thus,

$$\int_{\mathcal{U}_1} t^s \, d\mu(t) = \int_{\mathcal{M}} h_s(x)\, d\mu(x).$$

In turn, by 4.3, the right hand integral is

$$\int_{\mathbb{Z}_p} T_f(t) d\widehat{h_s}(X)(t),$$

and this is easily seen to be $\widehat{T_f}(s)$ by definition of the Fourier Transform. The proof now follows by using Theorem 4.4. □

The advantage of shifting the integral over to $\mathcal{U}_1$ is that $\mathcal{U}_1$ is stable under non-trivial elements of Γ. In particular the element $\eta \in \Gamma$ defined by

$$\eta(x) = \frac{1}{x}$$

has

$$\eta(\mathcal{U}_1) = \mathcal{U}_1.$$

Note that

$$\det(\eta) = -1,$$

and

$$\mathbb{P}^1(\mathbf{K}) = \mathcal{M} \cup \eta(\mathcal{M}) \cup \mathcal{O}^*.$$

5.2. Theorem.

$$L_f(s) = (-1)^{1-m} L_f(k-s).$$

Proof. Part 2) of Theorem 2.3.1 applied to η tells us that

$$\int_{\mathcal{U}_1} (1/t)^s \, d\mu(t) = (-1)^{1-m} \int_{\mathcal{U}_1} t^{k-2} t^s \, d\mu(t).$$

The left hand side of the equation is $L_f(-s+1)$, and the right hand side is $(-1)^{1-m}$ $L_f(k-1+s)$. The equation is seen to be equivalent to the statement of the theorem by the change of variables $w = s-1$. □

Let $p \neq 2$ (so $r-1$ is even). Then there is a "center" to $\mathbb{Z}_p$ under the involution

$$\iota(s) = k - s;$$

namely the class

$$\frac{k}{2} + (p)$$

is stable under ι. In fact, we have

5.3. Corollary. *Let* $p \neq 2$ *and let the class of* $m \pmod{r-1}$ *contain even elements. Then*

$$L_f\left(\frac{k}{2}\right) = 0.$$

Proof. The theorem says that $L_f(\frac{k}{2}) = -L_f(\frac{k}{2})$. □

6. Complements

6.1 Examples. Our first result is of a computational nature. Let $f(z)$ be as in the previous two chapters. We formally put

$$t_f(s) = \sum_{j \geq 0} b_j \binom{s}{j-1},$$

so we have gotten rid of the powers of ξ. We are not guaranteed the convergence of $t_f(s)$ and so will *only* use it as a computational device. Let $h(X) = \sum_{j \geq 0} c_j X^j \in \mathcal{O}[[X]]$. We put

$$\int_{\mathbb{Z}_p} t_f(v) \, dh(X)(v) = \sum_j b_j c_j$$

provided that this series converges.

6.1.1. Proposition.

$$L_f(s) = \xi^{-1} \int_{\mathbb{Z}_p} t_f(v) \, d\left(1 + \xi^{-1} \log_C(X)\right)^{s-1} (v).$$

Proof. Recall that

$$\mathbf{log}(X) = \xi^{-1}\mathbf{log}_C(\xi X).$$

The result then follows by expanding out the power series

$$(1 + \mathbf{log}(X))^s = \left(1 + \xi^{-1}\mathbf{log}_C(\xi X)\right)^s$$

by the binomial theorem, using the answer in the integral for the Fourier Transform of $T_f(s)$, and then keeping track of the powers of ξ that occur. □

The proposition thus allows us to do our calculations with $\mathbf{q}_C$-expansions and the function $\mathbf{log}_C(z)$.

Our first example is the form Δ of weight $r^2 - 1$ and type 0. Since constants commute with the formation of L_f we can ignore the factor of $-\xi^{r^2-1}$ appearing in Δ's $\mathbf{q}_C$-expansion. With this in mind, we set

$$t_\Delta(s) = \sum_{j\geq 1} \tau(j) \binom{s}{j-1},$$

and then calculate $L_\Delta(s)$ by using 6.1.1. From Theorem 5.2, we see

$$L_\Delta(s) = -L_\Delta(r^2 - 1 - s)$$

and that, when $p > 2$,

$$L_\Delta\left(\frac{r^2-1}{2}\right) = 0.$$

Next we discuss the form h. Here the weight is $r + 1$ and the type is 1. We put

$$t_h(s) = \sum_{j\geq 0} \phi(j) \binom{s}{j-1},$$

etc. We find

$$L_h(s) = L_h(r + 1 - s).$$

6.1.2. Remarks. Let $f \in \mathcal{S}(k, m)$. The functional equation, Theorem 5.2, determines *many* relations satisfied by the Mahler coefficients of $L_f(s)$ (and thus the $\mathbf{q}_C$-expansions of $f(z)$). Indeed, one can use standard results on binomial coefficients to also expand the right hand side of 5.2 in terms of $\binom{s}{j}$; as the Mahler coefficients are *unique* we can equate coefficients; etc. What relations arise in this fashion?

Along the same lines, where do these relations fail for Eisenstein series (if indeed they fail)? Indeed, let $E_k(z)$ be a non-trivial Eisenstein series with $\mathbf{q}_C$-expansion (as given in 2.2.4)

$$\xi^{-k} E_k = \mathcal{B}_k + \sum_{j\geq 1} e_j \mathbf{q}_C^j.$$

Let

$$T_{E_k}(s) = \sum_{j\geq 1} e_j \binom{s}{j-1},$$

and let

$$L_{E_k}(s) = \widehat{T_{E_k}}(s-1)$$

as in Section 4. Does $L_{E_k}(s)$ satisfy the same formal relations as happens for cusp forms? If this turns out to be so, then $L_{E_k}(s)$ has a functional equation (and conversely). What would this functional equation then correspond to in terms of measures as we know it cannot come directly from one of Teitelbaum's measures?

6.2. Hecke Operators. Let h be a monic element of $\mathbf{A}$. Following the classical formula, one can associate to h an operator T_h on the spaces of cusp forms $\mathcal{S}(k,m)$. However, unlike the classical $SL_2(\mathbb{Z})$ situation, $T_{gh} = T_g T_h$ for *any* g and h; one is thus reduced to studying T_h for h prime.

6.2.1. Definition. Let $f \in \mathcal{S}(k,m)$ and $h \in \mathbf{A}$ be a monic prime of degree d. We set

$$T_h f(z) = h^k f(hz) + \sum_{\deg(b)<d} f\left(\frac{z+b}{h}\right).$$

It is not difficult to see that the above definition gives an endomorphism of $\mathcal{S}(k,m)$.

6.2.2. Theorem. *Let f and h be as in 6.2.1. Let $\mu = \mu_f$ be the Teitelbaum measure associated to f. Then*

$$T_h f(z) = \int_{\mathcal{M}} \frac{h^k}{\mathbf{e}(hz-x)} + \frac{h}{\mathbf{e}(z-hx)}\, d\mu(x).$$

Proof. Using Corollary 2.3.3 and the definition of T_h, one finds

$$T_h f(z) = \int_{\mathcal{M}} \frac{h^k}{\mathbf{e}(hz-x)} + \sum_{\deg(b)<d} \frac{1}{\mathbf{e}(\frac{z+b}{h}-x)}\, d\mu(x).$$

Let ρ be the twist of the Carlitz module which is associated to the lattice $\mathbf{A}$ and exponential $\mathbf{e}(z)$. Then the set $\{\mathbf{e}(\frac{z+b}{h}-x) \mid \deg b < d\}$ is precisely the set of roots of the equation in u,

$$0 = \rho_h(u) - \mathbf{e}(z-hx).$$

The result follows upon taking the logarithmic derivative (with respect to u) of $\rho_h(u) - \mathbf{e}(z-hx)$ and setting $u=0$. □

6.2.3. Question. Can the analogy with the formulae presented in **[L1]** be pushed to give a formula for the *trace* of T_h on the space $V = \mathcal{S}(k,m)$? If so, can this be generalized to the trace of T_h on the exterior powers $\wedge^i V$ for all i?

In private communication with the author **[Se1]** (see also **[G8]**), J.-P. Serre has clarified the use of the *Brauer-Nesbitt Theorem* in the present situation. In particular, an affirmative answer to the above question would precisely *characterize* the semisimplification of the Hecke operators up to isomorphism!

6.3. Two-variable L-series. In this subsection we will use the measure associated to a cusp form $f(z) \in \mathcal{S}(k,m)$ to obtain, at least provisionally, a "two-variable" L-series. This function will converge only on a "half-plane"; it remains to be seen whether it can be analytically continued and given a functional equation, perhaps with the use of an auxiliary object such as a Γ-function. Still, from the point of view of **[G2]**, **[G6]**, **[G7]**, and **[G8]**, some such construction seems natural.

Let $\alpha \in \mathbf{K}^*$. We say that α is *positive* (or *monic*) if the highest coefficient in its expansion in $1/T$ is 1. The set of such α forms a multiplicative group which we denote by $\mathbf{P}$; clearly the set $\mathbf{M}$ of monic polynomials is contained in $\mathbf{P}$. Let $\deg(\alpha)$ be the canonical (and obvious) extension of the degree function on $\mathbf{A}$ to $\mathbf{K}$. We then denote by $\mathbf{P}_+$ the subset of $\mathbf{P}$ consisting of those positive α with $\deg(\alpha) > 0$.

6.3.1. Definition.
1) We put $\mathbf{S}_\infty = \mathbf{K}^* \times \mathbb{Z}_p$.
2) Let $\alpha \in \mathbf{P}$. We put $\langle \alpha \rangle = T^{-\deg(\alpha)}\alpha \in \mathcal{U}_1$.

Now let $\alpha \in \mathbf{P}$ and let $s = (x,y) \in \mathbf{S}_\infty$. We put

$$\alpha^s = x^{\deg(\alpha)} \langle \alpha \rangle^y .$$

This definition of exponentiation satisfies all the usual rules of "n^s"; in particular, the integer powers sit inside the group $\mathbf{S}_\infty$ as (T^i, i). We shall freely denote (T^i, i) by "i".

6.3.2. Definition. We define the zeta function of $\mathbf{A}$ by

$$\zeta(\mathbf{A}, s) =: \zeta(s) =: \sum_{n \in \mathbf{M}} n^{-s} .$$

It is easy to see that $\zeta(s)$ converges absolutely for all those $s = (x,y) \in \mathbf{S}_\infty$ with $|x| > 1$. In what follows we shall restrict s to be in this "half-plane". *All* L-series arising from Drinfeld modules (see **[G8]**) are naturally defined on $\mathbf{S}_\infty$; i.e., are two-variable. (N.B., $\zeta(s)$ arises from the Carlitz module by a simple shift of variables.)

Let $s = (x,y) \in \mathbf{S}_\infty$. We define $\binom{s}{j}$ to be $\binom{y}{j}$ with the usual meaning (y is a p-adic integer).

6.3.3. Definition. Let s be as above.
1) We define $\mathcal{L}_f(s)$ by

$$\mathcal{L}_f(s) = \int_{\mathbf{P}_+} t^{-s}\, d\mu_f(t) .$$

2) Define the formal power series $\mathcal{H}_s(X) \in \mathcal{O}[[X]]$ by

$$\mathcal{H}_s(X) = \sum_{j \geq 0} \zeta(s+j) \binom{-s}{j} X^j .$$

It is not difficult to use the properties of μ_f to show that $\mathcal{L}_f(s)$ is a continuous function of s in the above mentioned half-plane. In fact we have:

6.3.4. Proposition. *We have*

$$\mathcal{L}_f(s) = \int_{\mathbb{Z}_p} L_f(t+1)\, d\mathcal{H}_s(X)(t).$$

Proof. Note first that $\mathbf{P}_+ = \bigcup_{n \in \mathbf{M}} (n + \mathcal{M})$. Thus the invariance properties of μ_f allow us to compute the integral for $\mathcal{L}_f(s)$ over $\mathcal{M}$; we deduce

$$\mathcal{L}_f(s) = \int_{\mathcal{M}} \sum_{n \in \mathbf{M}} (n+t)^{-s}\, d\mu(t).$$

Next write $(n+t)^{-s}$ as $n^{-s}(1+t/n)^{-s}$. The result now follows from Theorem 4.1 and a little algebra. □

6.4. Higher Ranks. Teitelbaum's constructions are based on the finite dimensionality of spaces of modular forms on Ω. As classically for the complex upper half plane, this finite dimensionality is deduced via a compactification of the modular curves.

On the other hand, one knows how to define modular forms on the general $(d-1)$-dimensional rigid spaces Ω^d (where $\Omega^2 = \Omega$) **[G1]**; the definition mimics the case discussed here with $d = 2$. Moreover, the rank 2 case also shows how to associate to such modular forms $\mathbf{q}$-expansions. Indeed, in the notation of **[G1]**, one can use the Tate theory to obtain rank d Drinfeld modules over $M_I^{d-1}((\mathbf{q}))$. This gives rise to the algebraic version of the $\mathbf{q}$-expansion; the associated analytic expansion follows easily as for forms on Ω.

The difficulty in simply concluding that the spaces of modular forms are finite dimensional arises as, for $d > 1$, one does **not** obtain a compactification of the modular variety M_I^d simply by adjoining copies of M_I^{d-1}. However, in **[K1]**, Kapranov *does* compactify the modular varieties (into a complete normal variety) by first adjoining the above spaces and then adding pieces of higher codimension. Let us denote by $\widetilde{M}_I^d$ the space formed by adjoining the copies of M_I^{d-1} and by $\overline{M}_I^d$ the compactified space; one has the open immersion

$$j\colon \widetilde{M}_I^d \to \overline{M}_I^d.$$

The finite dimensionality can now be deduced as follows (I thank M. Kapranov for his explanation of the proof): Let $\mathcal{M}(k,I)$ be the coherent sheaf of modular forms of weight k on M_I^d; as in the rank 2 case, $\mathcal{M}(k,I)$ is the invertible sheaf associated to a line bundle. Moreover, as in the rank 2 case **[G1]**, this line bundle extends to a line bundle on $\widetilde{M}_I^d$; its sections are precisely the space of modular forms of weight k with *holomorphic* $\mathbf{q}$-expansion. We will also denote this space of sections by $\mathcal{M}(k,I)$. Finally, as

$$\overline{M}_I^d - \widetilde{M}_I^d$$

has codimension 2, we conclude that $j_*\mathcal{M}(k,I)$ is a *coherent* sheaf on $\overline{M}_I^d$ (see Corollaire 5.11.4 of **[Gr1]**). The result now follows from the finite dimensionality of the cohomology (in particular H^0) of coherent sheaves on complete varieties.

References

[C1] L. Carlitz: On certain functions connected with polynomials in a Galois field, *Duke Math. J.* **1** (1934), 137–168.

[Ge1] E.-U. Gekeler: On the coefficients of Drinfeld modular forms, *Invent. Math.* **93**, (1988), 667–700.

[Ge2] E.-U. Gekeler: *Drinfeld Modular Curves*, (Lecture Notes in Math. 1231), Springer, Berlin-Heidelberg-New York (1986).

[Ge3] E.-U. Gekeler: A Product Expansion for the Discriminant Function of Drinfeld Modules of Rank Two, *J. Number Theory* **21** (1985), 135–140.

[G1] D. Goss: π-adic Eisenstein Series for Function Fields, *Compositio Math.* **41**, (1980), 3–38.

[G2] D. Goss: The Arithmetic of Function Fields 2: The 'Cyclotomic' Theory, *J. Algebra* **81**(1), (1983), 107–149.

[G3] D. Goss: Modular Forms for $\mathbb{F}_r[T]$, *J. Reine Angew. Math.* **317**, (1980), 16–39.

[G4] D. Goss: The Γ-function in the Arithmetic of Function Fields, *Duke Math. J.* **56**(1), (1988), 163–191.

[G5] D. Goss: Fourier Series, Measures and Divided Power Series in the Theory of Function Fields, *K-Theory* **1**, (1989), 533–555.

[G6] D. Goss: A Formal Mellin Transform in the Theory of Function Fields, *Trans. Amer. Math. Soc.* **327**, (1991), 567–582.

[G7] D. Goss: Harmonic Analysis and the Flow of a Drinfeld Module, *J. Algebra* **146**, No. 1, (1992), 219–241.

[G8] D. Goss: L-series of t-motives and Drinfeld Modules, (this volume).

[Gr1] A. Grothendieck, (with J. Dieudonné): EGA IV: Ètude locale des schémas et des morphismes des schémas, *Inst. Hautes Études Sci. Publ. Math.* **24**, (1965), 5–231.

[K1] M. Kapranov: On Cuspidal Divisors on the Modular Varieties of Elliptic Modules, *USSR-Izv.* **30**(3), (1988), 533–547.

[L1] S. Lang: *Introduction to Modular forms*, Springer, Berlin-Heidelberg-New York (1976).

[MTT1] B. Mazur, J. Tate, and J. Teitelbaum: On p-adic analogues of the conjecture of Birch and Swinnerton-Dyer, *Invent. Math.* **84**, (1986), 1–48.

[S1] P. Schneider: Rigid Analytic L-functions, in *Number Theory, Noordwijkerhout 1983*, Lecture Notes in Math. 1068, Springer, Berlin-Heidelberg-New York, (1983), 216–230.

[Se1] J.-P. Serre: Letter to the author dated April 14, 1990.

[T1] J. Teitelbaum: The Poisson Kernel for Drinfeld Modular Curves. *J. Amer. Math. Soc.* **4**, (1991), 491–511.

[T2] J. Teitelbaum: Rigid Analytic Modular Forms: An Integral Transform Approach, (this volume).

[Th1] D. Thakur: Gamma functions for function fields and Drinfeld modules, *Ann. of Math.* **134**, (1991), 25–64.

[W1] L. Wade: Certain quantities transcendental over $GF(p^n, x)$, *Duke Math. J.* **8**, (1941), 701–720.

Department of Mathematics, Ohio State University, Columbus, Ohio 43210
Email: goss@coltrane.mps.ohio-state.edu

Transcendence in Finite Characteristic

*Jing Yu**

Classical transcendence theory deals with numbers transcendental over $\mathbb{Q}$. One starts with certain commutative algebraic groups defined over a number field. The parameterizing functions for these algebro-geometric objects then enjoy nice transcendence properties which give rise to natural transcendental numbers. Parallel to this beautiful classical theory there is also a transcendence theory over function fields of finite characteristic. The role of commutative algebraic groups is played by the Drinfeld modules. Owing to the existence of the Frobenius map, plain vector groups here become fascinating algebro-geometric objects. One has again parameterizing functions possessing nice transcendence properties. Quantities transcendental over the rational function field $\mathbb{F}_q(T)$ then arise.

In this article we shall give a survey of this aspect of finite characteristic transcendence theory. A little bit of history is in Section 1 and a brief introduction to t-modules as well as formulations of fundamental theorems is in Section 2. Some highlights of the theory are given in the final Section 3.

§1. Origins and elaborations

It all started with L. Carlitz during the 1930's, cf. **[5]**. He considered the additive group $\mathbb{G}_a$ with $\mathbb{F}_q$-linear action of $A =: \mathbb{F}_q[T]$ given by

$$\phi_C(T) : x \longmapsto Tx + x^q .$$

The parameterizing function of this simplest Drinfeld module is the $\mathbb{F}_q$-linear Carlitz exponential function satisfying

$$e_C : \overline{\mathbb{F}_q\left(\left(\frac{1}{T}\right)\right)} \longrightarrow \overline{\mathbb{F}_q\left(\left(\frac{1}{T}\right)\right)}, \quad e_C(Tz) = Te_C(z) + e_C(z)^q .$$

The kernel of this function is $\mathbb{F}_q[T]\tilde{\pi}_q$, where $\tilde{\pi}_q$ is a fundamental period of the Carlitz module given by

$$\tilde{\pi}_q = (T - T^q)^{\frac{1}{q-1}} \prod_{i=1}^{\infty} \left(1 - \frac{T^{q^i} - T}{T^{q^{i+1}} - T}\right) =: (T - T^q)^{\frac{1}{q-1}} \pi_q .$$

The $1/T$-adic expansion of such period can be written down easily, e.g. in the case $q = 3$, π_3 is

10000020100000201020002020000020100000201000202010100010100010101000010002...

---

* Supported by the National Science Council of R.O.C.

The first result in transcendence theory of finite characteristic is obtained by Carlitz's student L.I. Wade **[21]**. He proved an analogue of the Hermite-Lindemann theorem for Carlitz exponential functions thereby getting the transcendence of the fundamental periods $\tilde{\pi}_q$. Later on he also proved an analogue of the Gelfond-Schneider theorem for Carlitz modules, as well as a number of other miscellaneous transcendence results. Most of Wade's works are done via ad-hoc methods, relying on explicit expansions of the quantities in question. A very good exposition of these pioneering results can be found in **[11]**.

The fact that the Carlitz module is a perfect analogue of the multiplicative group $\mathbb{G}_m$ can be seen either algebraically by way of class field theory or transcendentally through introducing the following zeta values

$$\zeta_C(n) = \sum_{\substack{a\in A\\ a \text{ monic}}} \frac{1}{a^n} \in \mathbb{F}_q\left(\left(\frac{1}{T}\right)\right), \qquad n = 1, 2, \ldots .$$

Carlitz evaluated $\zeta_C(n)$ for those positive integers n which are divisible by $q-1$:

$$\zeta_C(n) = \frac{\tilde{\pi}_q^n B_n}{\Gamma_{n+1}},$$

where Γ_n are Carlitz factorials defined by first setting $D_i =: (T^{q^i} - T^{q^{i-1}}) \cdots (T^{q^i} - T)$ for $i \geq 1$, $D_0 =: 1$, writing down the q-adic expansion $\sum_{i=0}^{\infty} n_i q^i$ of n, and then let

$$\Gamma_{n+1} =: \prod_{i=0}^{\infty} D_i^{n_i}.$$

The B_n are the Bernoulli-Carlitz numbers defined by the series expansion

$$\frac{x}{e_C(x)} = \sum_{n=0}^{\infty} B_n \frac{x^n}{\Gamma_{n+1}}.$$

In this context, the integers divisible by $q-1$ are considered as **even** because there are now $q-1$ signs in our integral ring $A = \mathbb{F}_q[T]$.

In the beginning of 1970's, V. G. Drinfeld **[9]** introduced the Drinfeld modules which vastly extended the scope of Carlitz theory. The function field in question can be any global function field k over its field of constants $\mathbb{F}_q$, and the ring A is just the ring of functions in k regular away from a fixed place ∞. A Drinfeld module is simply an A-action on $\mathbb{G}_a$ which linearizes to the scalar action of A on $\operatorname{Lie}\mathbb{G}_a$. The algebraic closure of the completion of k at ∞, denoted by $\bar{k}_\infty$, is the universal domain for parameterizing functions of the modules. By adapting the methods of Gelfond, Schneider, and Siegel from classical transcendence theory, we have obtained an analogue of the Hermite-Lindemann theorem for arbitrary Drinfeld modules defined over $\bar{k}$, cf. **[32]**.

Theorem 1.1. *Let $e_\phi(z)$ be the exponential function parameterizing a Drinfeld A-module ϕ defined over $\bar{k}$. Then $e_\phi(\alpha)$ is transcendental over k for all $\alpha \neq 0$ in $\bar{k} \subset \bar{k}_\infty$. In particular, all nonzero elements (periods) in the kernel of $e_\phi(z)$ are transcendental.*

Notice that the exponential functions here are always $\mathbb{F}_q$-linear. This property forces them to behave like classical functions satisfying algebraic differential equations. If they happen to assume algebraic values at one point, all the higher coefficients of their Taylor expansions at that point are also algebraic.

To see the significance of this theorem for function fields arithmetic, let H be the Hilbert classfield of A, i.e. the maximal abelian unramified extension of k which is totally split at ∞. In **[15]**, D. Hayes shows there always exists a Drinfeld A-module ϕ defined over the integral closure of A in H with period lattice $\ker e_\phi = \mathfrak{A}\pi_\infty$, where $\mathfrak{A}$ is an ideal in A and π_∞ is a fundamental period for (k, ∞). Theorem 1.1 then allows us to generalize the Carlitz-Wade transcendence result for Carlitz zeta values to the zeta values associated with general pairs (k, ∞) first introduced by D. Goss, cf. **[13]**, **[19]**, and **[32]**.

As is well-known from Hilbert's 7-th problem, the classical Gelfond-Schneider theorem is just a special case of the qualitative Baker's theorem concerning linear forms in logarithms of algebraic numbers. A natural question therefore is to establish an analogue of the qualitative Baker's theorem for Drinfeld modules. This question is answered affirmatively in **[33]**. To state it, let K_ϕ denote the field of multiplications of ϕ, i.e. the field generated over k by all $\beta \in \bar{k}_\infty$ satisfying $\beta \ker e_\phi \subset \ker e_\phi$. This K_ϕ is an algebraic extension of k with degree $[K_\phi : k]$ dividing $\operatorname{rank} \phi =: \operatorname{rank}_A \ker e_\phi$.

Theorem 1.2. *Let $e_\phi(z)$ be as in Theorem 1.1. Let $\{\alpha_1, \ldots, \alpha_n\} \in \bar{k}_\infty$ such that $e_\phi(\alpha_i) \in \bar{k}$ for all i. If $\{\alpha_1, \ldots, \alpha_n\}$ are linearly independent over K_ϕ, then*

$$\{1, \alpha_1, \ldots, \alpha_n\}$$

are linearly independent over $\bar{k}$.

This theorem is proved in **[33]** with an extra separability hypothesis on the linear coefficients. This hypothesis is indeed dispensable, cf. **[8]**, **[37]**. Note also that Theorem 1.1 is a very special case of Theorem 1.2.

Transcendence theory also yields informations on the fields of definitions of Drinfeld modules. Call a finitely generated discrete A-submodule contained in $\bar{k}_\infty$ a lattice. According to Drinfeld's theory, every such lattice M is the kernel of the exponential function parameterizing a uniquely determined Drinfeld module ϕ_M. Homothetic lattices correspond to isomorphic Drinfeld modules. One is interested in the following question: Given lattice $M \subset \bar{k}$ whether its corresponding Drinfeld module ϕ_M is isomorphic to one defined over $\bar{k}$. This turns out to be the case only if ϕ_M has the biggest possible field of multiplications. In the following we say a lattice is an ideal if it is an $\mathfrak{o}$-ideal of an order $\mathfrak{o}$ in a finite extension field L/k and there is only one place of L lying above ∞.

Theorem 1.3 ([32]). *Let M be a lattice in $\bar{k}_\infty$ homothetic to one contained in $\bar{k}$. If the corresponding Drinfeld module ϕ_M is defined over $\bar{k}$ then M is homothetic to an ideal.*

If M is an ideal, then ϕ_M is not only defined over $\bar{k}$ but over a suitable (Hilbert) class field of k. Moreover, ϕ_M has potentially everywhere good reduction. The case $A = \mathbb{F}_q[T]$ and M is of rank 2 is most illuminating. It reduces to assertions completely analogous to the well-known theorems about j-invariants in classical complex multiplications theory, cf. **[15]**, **[32]**.

In recent developments of transcendence theory over global function fields one is driven to the higher dimensional Drinfeld modules. These are vector groups, i.e. $\mathbb{G}_a^n$, equipped with reasonable A-actions. Groundwork for these modules is laid in G. Anderson **[2]**. There he introduces in particular the t-modules. This leads to the Anderson-Thakur theory of tensor powers of Carlitz modules **[3]**, and finally to the transcendence of all the Carlitz zeta values, **even** as well as **odd** ones. On the other hand, by considering extensions of Drinfeld modules by trivial linear 1-dimensional module, Deligne and Anderson(cf. **[12]**, **[35]**) are led to a rich theory of periods of the second kind for Drinfeld modules. We will say more about these developments in Section 3.

§2. Fundamental theorems

In this section we shall let $A = \mathbb{F}_q[t]$, $k = \mathbb{F}_q(T)$, $k_\infty = \mathbb{F}_q((1/T))$, and $\bar{k}_\infty$ is a fixed algebraic closure of k_∞. We set $d_\infty(a)$ to be the degree in T of $a \in k$. $\mathbb{G}_a^n$ is the n-dimensional vector group. By a t-module of dimension n over $\bar{k}_\infty$, we mean a t-action given by a ring homomorphism $\phi : A \to \operatorname{End} \mathbb{G}_a^n$ satisfying the following conditions:

(i) For constants $\zeta \in \mathbb{F}_q$, $\phi(\zeta)$ is the scalar multiplication by ζ.
(ii) $(d\phi(t) - TI)^N \operatorname{Lie} \mathbb{G}_a^n = 0$, for some integer $N > 0$.

One can always parameterize these t-modules $G = (\mathbb{G}_a^n, \phi)$ analytically through exponential maps

$$\exp_G : \operatorname{Lie} \mathbb{G}_a^n(\bar{k}_\infty) = \bar{k}_\infty^n \longrightarrow \bar{k}_\infty^n = \mathbb{G}_a^n(\bar{k}_\infty).$$

These are entire $\mathbb{F}_q$-linear maps defined on $\bar{k}_\infty^n$ and satisfying

$$\exp_G(d\phi(t)(\mathbf{z})) = \phi(t)(\exp_G(\mathbf{z}))\text{ , for all } \mathbf{z} \in \bar{k}_\infty^n.$$

To each t-module G, there is a unique exponential map $\exp_G$ which depends functorially on G. A t-module $(\mathbb{G}_a^n, \phi)$ is said to be defined over $\bar{k}$ if $\phi(t)$ is an endomorphism of $\mathbb{G}_a^n$ defined over $\bar{k}$.

Examples.
(1) The one dimensional "trivial" t-module with t acting as scalar multiplication T. This t-module will be denoted by G_L. The exponential map here is just $\exp_{G_L}(z) = z$.
(2) The Carlitz module $C = (\mathbb{G}_a, \phi_C)$. More generally all Drinfeld $\mathbb{F}_q[T]$-modules are 1-dimensional t-modules under the identification of T with t.
(3) Higher Carlitz modules, $C^{\otimes n} = (\mathbb{G}_a^n, \phi_C^{\otimes n})$ (cf.**[2]**, **[36]**). These are tensor powers of the one-dimensional Carlitz module. The t-action $\phi_C^{\otimes n}$ is explicitly given by

$$\phi_C^{\otimes n}(t) : \begin{pmatrix} x_1 \\ \vdots \\ x_n \end{pmatrix} \longmapsto \begin{pmatrix} T & 1 & & 0 \\ & & \ddots & 1 \\ 0 & & & T \end{pmatrix} \begin{pmatrix} x_1 \\ \vdots \\ x_n \end{pmatrix} + \begin{pmatrix} 0 & \cdots & 0 \\ \vdots & & \vdots \\ 1 & \cdots & 0 \end{pmatrix} \begin{pmatrix} x_1^q \\ \vdots \\ x_n^q \end{pmatrix},$$

where the first square matrix on the right-hand side is the standard Jordan block, the second square matrix is the elementary one with the lower left corner entry equals 1.

The endomorphism rings of these t-modules coincide with $\phi(A)$. We shall let $\exp_n(\mathbf{z})$ denote the exponential map of $C^{\otimes n}$.

Let $G = (\mathbb{G}_a^n, \phi)$ be a t-module. Let $\mathrm{Hom}_{\mathbb{F}_q}(\mathbb{G}_a^n, \mathbb{G}_a)$ be the $\bar{k}_\infty$-vector space of all $\mathbb{F}_q$-linear morphisms from $\mathbb{G}_a^n$ to $\mathbb{G}_a$. Allow t to act on this set by $f \mapsto f \circ \phi(t)$. Then $\mathrm{Hom}_{\mathbb{F}_q}(\mathbb{G}_a^n, \mathbb{G}_a)$ becomes a $\bar{k}_\infty[t]$-module. If this $\bar{k}_\infty[t]$-module is finitely generated, we call G an abelian t-module. All Drinfeld A-modules are abelian. The higher Carlitz modules are also abelian. In fact, for $C^{\otimes n}$, the $\bar{k}_\infty[t]$-module $\mathrm{Hom}_{\mathbb{F}_q}(\mathbb{G}_a^n, \mathbb{G}_a)$ is always free of rank one, generated by the first coordinate projection on $\mathbb{G}_a^n$. On the other hand the linear module G_L is not abelian.

A given t-module $(\mathbb{G}_a^n, \phi)$ is said to be a regular t-module if there is an integer r such that for all $a \in A$ the kernel of $\phi(a)$ in $\mathbb{G}_a^n(\bar{k})$ is a free $A/(a)$-module of rank r. Abelian t-modules are regular, cf. **[2]**. Thus all examples above are regular. Moreover, they are simple t-modules in the sense that no proper connected algebraic subgroup $H \neq \{0\}$ in $\mathbb{G}_a^n$ is invariant under $\phi(A)$. A connected algebraic subgroup of $\mathbb{G}_a^n$ invariant under $\phi(A)$ will be called a t-submodule of $(\mathbb{G}_a^n, \phi)$.

An analogue of the Hermite-Lindemann theorem for general t-modules is

Theorem 2.1. *Let $G = (\mathbb{G}_a^m, \phi)$ be a simple t-module defined over $\bar{k}$, and not isomorphic to G_L. Suppose $\mathbf{u} \neq \mathbf{0}$ is a point in* $\mathrm{Lie}\, G(\bar{k}_\infty)$ *such that* $\exp_G(\mathbf{u}) \in G(\bar{k})$. *Then the last coordinate of* $\mathbf{u}$ *must be transcendental.*

We shall call the last coordinate of such a logarithmic vector $\mathbf{u}$ a last coordinate logarithm of an algebraic point on the simple t-module G. It also follows from Theorem 2.1 that both the logarithmic vector itself and the corresponding algebraic point are uniquely determined by this last coordinate logarithm. If the t-module is not simple, one has

Corollary 2.2. *Let $G = (\mathbb{G}_a^m, \phi)$ be an abelian t-module defined over $\bar{k}$. Suppose $\mathbf{u} \neq \mathbf{0}$ is a point in* $\mathrm{Lie}\, G(\bar{k}_\infty)$ *such that* $\exp_G(\mathbf{u}) \in G(\bar{k})$. *Then at least one coordinate component of* $\mathbf{u}$ *is transcendental.*

More generally, reasonable linear relations satisfied by a logarithmic vector of an algebraic point must come from algebraic relations inside the t-module in question. This is formulated as (cf. **[37]**)

Theorem 2.3. *Let $G = (\mathbb{G}_a^m, \phi)$ be a regular t-module defined over $\bar{k}$. Let $\mathbf{u}$ be a point in* $\mathrm{Lie}\, G(\bar{k}_\infty)$ *such that* $\exp_G(\mathbf{u}) \in G(\bar{k})$. *Then the smallest vector subspace in* $\mathrm{Lie}\, G(\bar{k}_\infty)$ *defined over $\bar{k}$ which is invariant under $d\phi(t)$ and contains $\mathbf{u}$ is the tangent space at the origin of a t-submodule of G.*

This is an analogue of a well-known theorem of G.Wüstholz **[28]** for commutative algebraic groups. For the proof one needs to exploit arithmetic properties of the exponential maps restricted to $d\phi(t)$-invariant vector subspaces in $\mathrm{Lie}\, G(\bar{k}_\infty)$ defined over $\bar{k}$. These maps are analytic homomorphisms in the sense below.

By an n-parameter analytic homomorphism into $G = (\mathbb{G}_a^m, \phi)$, we mean an $\mathbb{F}_q$-linear analytic map $\Phi\colon \bar{k}_\infty^n \to \mathbb{G}_a^m(\bar{k}_\infty)$ with the following properties

(i) There exists an $n \times n$ matrix $[t]$ with entries in $\bar{k}_\infty$, satisfying

$$([t] - TI)^n = 0$$

and

$$\Phi([t]\mathbf{z}) = \phi(t)\Phi(\mathbf{z}) \quad , \quad \text{for all } \mathbf{z} \in \bar{k}_\infty^n .$$

(ii) It has a Taylor expansion at 0, which is of the form

$$\Phi(\mathbf{z}) = \sum_{h=0}^{\infty} \beta_h \mathbf{z}^{(h)} \quad , \quad \text{for all } \mathbf{z} \in \bar{k}_\infty^n ,$$

where β_h are $m \times n$ matrices with entries in $\bar{k}_\infty$, and $\mathbf{z}^{(h)}$ is the column vector obtained from the column vector $\mathbf{z}$ by raising all the coordinates to their q^h-th power.

Given an n-parameter analytic homomorphism, we shall regard $\bar{k}_\infty^n$ as $\mathbb{F}_q[t]$-module by letting $t \mapsto [t]$, where $[t]$ is a fixed choice of a matrix satisfying the condition (i) above. We say that Φ is normalized if all its partial derivatives at 0 are in $\bar{k}$. In that case, it follows that all the matrix Taylor coefficients β_h have entries in K which is a finite extension field of k. The exponential maps parameterizing t-modules defined over $\bar{k}$ are always normalized analytic homomorphisms.

Fix an analytic homomorphism Φ as above. Given a polynomial Q in m variables on G, we say that Q vanishes to order at least U at $\mathbf{x}_0$ along Φ if for all $\mathbf{j} = (j_1, \ldots, j_n)$, with $|\mathbf{j}| < U$, one has $\Delta_{\mathbf{j}}^{\Phi} Q(\mathbf{x}_0) = 0$, where the hyperderivatives are defined by

$$Q(\mathbf{x} + \Phi(\mathbf{z})) = \sum_{\mathbf{j} \in \underline{\mathbb{N}}^n} \Delta_{\mathbf{j}}^{\Phi} Q(\mathbf{x}) \mathbf{z}^{\mathbf{j}} .$$

Example. Let G be a direct sum of m copies of the Carlitz module. Fix $\alpha_1, \ldots, \alpha_m$ in $\bar{k}_\infty$. We are interested in the one-parameter analytic homomorphism

$$\Phi \colon z \longmapsto (e_C(\alpha_1 z), \ldots, e_C(\alpha_m z)) .$$

Suppose Q is an $\mathbb{F}_q$-linear nonzero polynomial of (total) degree q^d in m variables which vanishes to order at least q^U along Φ at the origin. Let $\underline{\alpha}$ be the vector $(\alpha_1, \ldots, \alpha_m)$. Consider the following $(1+d)m \times (1+d)m$ determinant

$$\begin{vmatrix} \underline{\alpha} & 0 & 0 & \cdots & 0 \\ \frac{\underline{\alpha}^{(1)}}{D_1} & \underline{\alpha}^{(1)} & 0 & \cdots & 0 \\ \vdots & \vdots & \vdots & \ddots & \vdots \\ \vdots & \vdots & \vdots & \cdots & \underline{\alpha}^{(d)} \\ \vdots & \vdots & \vdots & \cdots & \frac{\underline{\alpha}^{(d+1)}}{D_1^{q^d}} \\ \vdots & \vdots & \vdots & \ddots & \vdots \\ \frac{\underline{\alpha}^{(m(d+1)-1)}}{D_{m(d+1)-1}} & \frac{\underline{\alpha}^{(m(d+1)-1)}}{D^q_{m(d+1)-2}} & \cdots & \cdots & \frac{\underline{\alpha}^{(m(d+1)-1)}}{D^{q^d}_{(m-1)(d+1)}} \end{vmatrix} .$$

Multiply each row of the above determinant by its denominator and then adding suitable columns, we arrive at Moore's determinant

$$\begin{vmatrix} \underline{\alpha} & T\underline{\alpha} & \cdots & T^d\underline{\alpha} \\ \underline{\alpha}^{(1)} & (T\underline{\alpha})^{(1)} & \cdots & (T^d\underline{\alpha})^{(1)} \\ \vdots & \vdots & \ddots & \vdots \\ \underline{\alpha}^{(m(d+1)-1)} & (T\underline{\alpha})^{(m(d+1)-1)} & \cdots & (T^d\underline{\alpha})^{(m(d+1)-1)} \end{vmatrix}.$$

If $\alpha_1, \ldots, \alpha_m$ are linearly independent over $\mathbb{F}_q(T)$, then this determinant is nonzero. Also $e_C(\alpha_1 z), \ldots, e_C(\alpha_m z)$ are algebraically independent functions, cf. [33]. Writing down the Taylor expansion of $Q(e_C(\alpha_1 z), \ldots, e_C(\alpha_m z))$ at 0, we deduce immediately that $U \leq m(d+1) - 1$. This gives an explicit measure of algebraic independence of the functions $e_C(\alpha_1 z), \ldots, e_C(\alpha_m z)$. On the other hand, if $U > m(d+1) - 1$, the vanishing of Moore's determinant implies there are linearly dependence relations among $\alpha_1, \ldots, \alpha_m$ over $\mathbb{F}_q[T]$ with coefficients $a_1, \ldots, a_m$ in T of degree bounded by d. Hence the whole image of the analytic homomorphism Φ must be contained in the t-submodule which is the connected component of $\mathbf{0}$ of the subgroup described by

$$Q'(\mathbf{x}) = \phi_C(a_1)x_1 + \cdots + \phi_C(a_m)x_m .$$

This is the simplest nontrivial case of multiplicity zero estimate for Drinfeld modules. From the high order of vanishing of a polynomial along a given analytic homomorphism, one concludes the existence of a certain t-submodule.

Given $G = (\mathbb{G}_a^n(\bar{k}_\infty), \phi)$, we shall always fix a projective envelope of $\mathbb{G}_a^n(\bar{k}_\infty)$ so that the degree of any algebraic subgroup means the degree of its Zariski closure inside the ambient projective space. With a given analytic homomorphism Φ into G, we shall also let $r(\Phi, H)$ denote the analytic codimension of $\Phi^{-1}(H)$ for a t-submodule $H \subset G$. We want to estimate zeros of a given polynomial over finite abelian groups of the following form inside G:

$$\Gamma(S) =: \Gamma(\gamma_1, \cdots, \gamma_l; S) =: \{\phi(a_1)\gamma_1 + \cdots + \phi(a_l)\gamma_l \mid a_i \in A, d_\infty(a_i) < S\}.$$

The following fundamental zero estimate theorem (cf. [37]) is among the key ingredients for proving Theorem 2.3.

Theorem 2.4. *Let Φ be an n-parameter analytic homomorphism into the t-module G of dimension m. Suppose Q is a polynomial in m variables which is of total degree $\leq D$ and vanishes to order at least $mU + 1$ along Φ at all points of $\Gamma(S)$. Then there exists a proper $\mathbb{F}_p[t]$-submodule $H \subset G$ such that*

$$\binom{U + r(\Phi, H)}{r(\Phi, H)} \operatorname{card}(\Gamma(S - m + 1) + H/H) \deg H \leq C(G) D^{\dim G/H},$$

where $C(G)$ is a positive constant depending only on G.

We do zero estimate only at those finite sets of points which form finite additive subgroups. Since such kind of sets are abundant in the finite characteristic situation, our estimate is as useful for transcendence theory as the classical multiplicity estimate treated by Masser-Wüstholz **[27]** and Philippon **[16]**. Commutative algebra together with Bézout's Theorem is all that is needed to establish this fundamental result.

§3. Some highlights

The higher Carlitz module $C^{\otimes n}$ was first introduced by G. Anderson, cf. **[2]**, **[3]**. It is indeed the n-th tensor power of the original Carlitz module C in the category of t-motives. Set $L_0 =: 1$, and $L_i =: \prod_{j=1}^{i}(T - T^{q^j})$ for $i \geq 1$. The Taylor expansion at 0 of the logarithmic function for the Carlitz module is

$$\log_C(x) = \sum_{h=0}^{\infty} \frac{x^{q^h}}{L_h}.$$

Checking the expansion of the logarithmic map for $C^{\otimes n}$ one finds

$$\log_n \begin{pmatrix} 0 \\ \vdots \\ 0 \\ x \end{pmatrix} = \begin{pmatrix} \vdots \\ \vdots \\ \sum_{h=0}^{\infty} \frac{x^{q^h}}{L_h^n} \end{pmatrix}.$$

Thus, from Theorem 2.1, one deduces that the n-th multilogarithm function of Carlitz takes transcendental values at all algebraic $\alpha \neq 0$. Noting that $\zeta_C(1)$ is actually $\log_C(1)$, Anderson-Thakur (**[3]**) made the striking discovery that all the values $\Gamma_n\zeta_C(n)$ are last coordinate logarithms of algebraic points on $C^{\otimes n}$. Hence Theorem 2.1 implies that $\zeta_C(n)$ is transcendental for all positive integers n.

The fact that higher Carlitz modules are tensor powers of C also has bearings on the periods. In **[3]**, it is shown that the kernel of the exponential map $\exp_n$ is a rank one $\mathbb{F}_q[t]$-module $d\phi_C^{\otimes n}(\mathbb{F}_q[t])\lambda_n$, where the last coordinate of the fundamental period vector λ_n can be chosen to be equal to $\tilde{\pi}_q^n$. As a consequence, all rational multiples of $\tilde{\pi}_q^n$ are last coordinate logarithms of torsion points on $C^{\otimes n}$. The corresponding torsion point can then be read off from the denominator of the rational in question.

Now both the values $\Gamma_n\zeta_C(n)$ and $\tilde{\pi}_q^n$ are last coordinate logarithms of specific algebraic points on $C^{\otimes n}$. If the integer n is **odd**, i.e. not divisible by $q-1$, then $\Gamma_n\zeta_C(n)$ is "real" while $\tilde{\pi}_q^n$ is "imaginary", and they are obviously linearly independent over k. We contend that this forces the **ratio** $\zeta_C(n)/\tilde{\pi}_q^n$ to be transcendental.

Suppose $\zeta_C(n)/\tilde{\pi}_q^n$ is algebraic. Take $G = (\mathbb{G}_a^{2n}, \phi) = C^{\otimes n} \times C^{\otimes n}$ and consider $\mathbf{u} = (\mathbf{z}, \lambda_n)$ viewed as vector in Lie G. Since $\mathbf{u}$ falls in a $d\phi(t)$-invariant hyperplane defined over $\bar{k}$, Theorem 2.3 says it is a tangent vector to the origin of a proper t-submodule inside G. As $C^{\otimes n}$ is simple with endomorphism ring $\phi(\mathbb{F}_q[t])$, a proper t-submodule is necessarily contained in the zero set of an equation of the form

$$d\phi_C^{\otimes n}(a_1)\mathbf{x}_1 + d\phi_C^{\otimes n}(a_2)\mathbf{x}_2 = \mathbf{0}.$$

This leads to a contradictory linear relation $a_1\Gamma_n\zeta_C(n) + a_2\tilde{\pi}_q^n = 0$.

Thus the value $\zeta_C(n)$ is a rational multiple of $\tilde{\pi}_q^n$ if and only if the integer n is **even**, i.e. divisible by $q-1$. In that case the corresponding algebraic point on $C^{\otimes n}$ is a **torsion** point with torsion given by the denominator of the Bernoulli-Carlitz number B_n/Γ_{n+1}. In fact one can even use Theorem 2.3 to prove the following pleasant fact: The only linear relations over $\bar{k}$ of the values $\zeta_C(n)$, $n = 1, 2, \ldots$, and $\widetilde{\pi}_q^m$, $m = 0, 1, 2, \ldots$, are the Bernoulli-Carlitz ones between $\zeta_C(n)$ and $\widetilde{\pi}_q^n$ when n is **even**.

All these results about zeta values attached to rational function fields should be expendable to the zeta values of Goss associated with general pairs (k, ∞). However this is still a very open problem in general. We refer to **[19]** for more discussions.

Fix the characteristic p. Another interesting question about these zeta values and the fundamental period $\widetilde{\pi}_q$ is to compare the values when q varies. Since $\phi_C^{\otimes n}(t)$ commutes with the scalars from $\mathbb{F}_q$, the simple t-module $C^{\otimes n}$ remains simple when one descends its field of constants from $\mathbb{F}_q$ to $\mathbb{F}_p$. Using Theorem 2.3 one can then show that there are no linear relations between the values associated to different q. In particular, if $q_1 \neq q_2$ then the ratio $\widetilde{\pi}_{q_1}^{n_1}/\widetilde{\pi}_{q_2}^{n_2}$ is transcendental for all positive integers n_1, n_2.

Let $C_q^{\otimes n}$ denote the higher Carlitz module with field of constants $\mathbb{F}_q$. Let $G = C_{q_1}^{\otimes n_1} \times C_{q_2}^{\otimes n_2}$ be considered as an $\mathbb{F}_p[t]$-module. Let $\mathbf{u} = (\lambda_{n_1}, \lambda_{n_2})$. If $\widetilde{\pi}_{q_1}^{n_1}/\widetilde{\pi}_{q_2}^{n_2} \in \bar{k}$, $\mathbf{u}$ must be contained in some $d\phi(t)$-invariant hyperplane defined over $\bar{k}$. Then $\mathbf{u} \in \operatorname{Lie} H$ for some $\mathbb{F}_p[t]$-submodule $H \subset G$. Since both λ_{n_1} and λ_{n_2} are nonzero, the projection maps from G to $C_{q_1}^{\otimes n_1}$ or $C_{q_2}^{\otimes n_2}$ are surjective when restricted to H. It follows that the connected component of $\mathbf{0}$ in H is isomorphic to both $C_{q_1}^{\otimes n_1}$ and $C_{q_2}^{\otimes n_2}$. This is impossible for $q_1 \neq q_2$.

In the case $q_2 = q_1^2$, introducing periods of second kind for the Carlitz module over $\mathbb{F}_{q_2}$, one can even adapt Chudnovsky's method to prove the algebraic independence of $\widetilde{\pi}_q$ and $\widetilde{\pi}_{q^2}$, cf. Thiery **[20]**. These varying q results are precisely those needed by Thakur **[18]** to prove the transcendence of Gamma values at $1/(1-q^2)$ and $q/(1-q^2)$ for the rational function field $\mathbb{F}_q(T)$.

Finally we come again to general Drinfeld A-modules to briefly expound the theory of periods of the second kind. So A is the ring of elements in a function field k regular away from a fixed place ∞. The field of constants of k is $\mathbb{F}_q$ and the completion of k at ∞ is k_∞. We shall write $\bar{k}\{\tau\}$ for the ring $\operatorname{End}_{\bar{k}} \mathbb{G}_a$ consisting of all $\mathbb{F}_q$-linear endomorphisms of $\mathbb{G}_a$ defined over $\bar{k}$, where τ denotes the q-th power Frobenius map. Following P. Deligne we want to consider (Lie split) extensions of the one-dimensional Drinfeld module $(\mathbb{G}_a, \phi)$ by G_L defined over $\bar{k}$:

$$\mathbf{0} \longrightarrow G_L \longrightarrow G \longrightarrow (\mathbb{G}_a, \phi) \longrightarrow \mathbf{0}.$$

The A-action ϕ' on such G can be given explicitly as

$$\phi'(a) : \begin{pmatrix} x_1 \\ x_2 \end{pmatrix} \longmapsto \begin{pmatrix} \phi(a)x_1 \\ ax_2 + \delta(a)x_1 \end{pmatrix}, \quad \text{for } a \in A,$$

where δ is an $\mathbb{F}_q$-linear biderivation from A to $\bar{k}\{\tau\}\tau$ in the sense that it satisfies $\delta(ab) = a\delta(b) + \delta(a)\phi(b)$ for all a, b in A. Writing down the exponential map for this two dimensional A-module then amounts to solving the "algebraic differential equation" given below

$$F(az) - aF(z) = \delta(a)(e_\phi(z)),$$
$$F(z) \equiv 0 \bmod z^q.$$

Denoting the unique $\mathbb{F}_q$-linear analytic solution by $F = F_\delta(z)$. Then the exponential map of G is

$$\exp_G : (z_1, z_2) \longmapsto (e_\phi(z_1), F_\delta(z_1) + z_2).$$

This analytic function $F_\delta(z)$ is easily seen to be quasi-periodic with respect to the period lattice of the Drinfeld A-module $(\mathbb{G}_a, \phi)$, i.e.
(i) $F_\delta(z + \omega) = F_\delta(z) + F_\delta(\omega)$, for $z \in \bar{k}_\infty$ and $\omega \in \ker e_\phi$,
(ii) $F_\delta(\omega)$ is A-linear in $\omega \in \ker e_\phi$.

If the biderivation δ is of the form $a \mapsto P\phi(a) - aP$ for some $P \in \bar{k}\{\tau\}$ then the corresponding module G is a trivial extension and the quasi-periodic function $F_\delta(z)$ is actually periodic. According to the de Rham Theorem of Anderson-Deligne (cf. **[12]**) nontrivial extensions exist when the Drinfeld A-module $(\mathbb{G}_a, \phi)$ has rank > 1.

Thus assume G is given as a nontrivial extension. Observe that G_L is the only proper A-submodule of G while all the vector subspaces of $\operatorname{Lie} G$ are $d\phi'(A)$-invariant. We can view the A-modules in question as regular t-modules by a choice of non-constant T in A. Applying Theorem 2.3 we conclude that if $\exp_G(u_1, u_2) \neq 0 \in \bar{k}^2$ then either $u_1 = 0$ and $u_2 \in \bar{k}$ or else the ratio of u_1 and u_2 is transcendental. In particular, let $u_1 = \omega \in \ker e_\phi$ be a period and $u_2 = 0$ then the quasi-period $F_\delta(\omega)$ is transcendental. Changing u_2 to $-F_\delta(\omega)$ gives also the transcendence of the ratio $F_\delta(\omega)/\omega$.

In the special case that $(\mathbb{G}_a, \phi)$ is a rank two $\mathbb{F}_q[T]$-module with field of multiplications quadratic over $\mathbb{F}_q(T)$, Thiery **[20]** has managed to show the algebraic independence of period ω with quasi-period $F_\delta(\omega)$. Taking the rank one Carlitz module C over $\mathbb{F}_{q^2}$ viewed as a rank two $\mathbb{F}_q[t]$-module, this results in the algebraic independence of $\tilde{\pi}_q$ and $\tilde{\pi}_{q^2}$. The algebraic independence of periods with quasi-periods is again a very interesting open problem in general. It amounts to the analogue of a well-known conjecture of Grothendieck for Drinfeld modules.

References

[1] J.-P. Allouche, Sur la transcendance de la série formelle. II, Sém. Théorie Nombres Bordeaux 2, (1990), 103–116.

[2] G. Anderson, t-motives, Duke Math. J. 53, (1986), 457–502.

[3] G. Anderson and D. Thakur, Tensor powers of the Carlitz modules and zeta values, Ann. of Math. 132, (1990), 159–191.

[4] R.Apéry, Interpolation de fractions continues et irrationalité de certaines constantes, Bull. section de sciences du C.T.H.S. 3, (1978), 37–53.

[5] L. Carlitz, On certain functions connected with polynomials in a Galois field, Duke Math. J. 1, (1935), 137–168.

[6] G.Damamme and Y.Hellegouarch, Propriétés de transcendance de la fonction zéta de Carlitz, C.R.A.S. Série I, 307, (1988), 635–637.

[7] P. Deligne and D. Husemoller, Survey of Drinfeld modules, Contemporary Math. 67, (1987), 24–91.

[8] L. Denis, Théorème de Baker et modules de Drinfeld, Preprint, (1990).

[9] V. G. Drinfeld, Elliptic modules, Math. USSR-Sb.**23**, (1974), 561–592.

[10] N.B. Dubovitskaya, Transcendency of analytic parameters associated to elliptic modules, Math. USSR-Sb.**55**, 133–143.

[11] J.M. Geijsel, Transcendence in fields of finite characteristic, Mathematical centre tracts **91**, Mathematisch Centrum, Amsterdam, 1979.

[12] E.-U. Gekeler, On the de Rham isomorphism for Drinfeld modules, J. Reine Angew. Math. **401**, (1989), 188–208.

[13] D. Goss, v-adic zeta functions, l-series and measures for function fields, Invent. Math. **55**, (1979), 107–119.

[14] D. Goss, On a new type of L-functions for algebraic curves over finite fields, Pacific J. Math. **105**, (1983), 143–181.

[15] D. Hayes, Explicit class field theory in global function fields, Studies in Algebra and Number Theory, G.-C. Rota, Academic Press, New York, 1979.

[16] P. Philippon, Lemma de zéros dans les groupes algébriques commutatifs, Bull. Soc. Math. France **114**, (1986), 355–383.

[17] D. Thakur, Gamma functions and Gauss sums for function fields and periods of Drinfeld modules, Ph.D. thesis, Harvard Univ., Cambridge, 1987.

[18] D. Thakur, Gamma functions for function fields and Drinfeld modules, Ann. of Math. **134**, (1991), 25–64.

[19] D. Thakur, Drinfeld modules and arithmetic in the function fields, Preprint, (1991).

[20] A. Thiery, Indépendance algébrique des périodes et quasi-périodes d'un module de Drinfeld, Preprint, (1991).

[21] L.I. Wade, Certain quantities transcendental over $GF(p^n, x)$, Duke Math. J. **8**, (1941), 707–729.

[22] L.I. Wade, Certain quantities transcendental over $GF(p^n, x)$ (2), Duke Math. J. **10**, (1943), 587–594.

[23] L.I. Wade, Transcendence properties of the Carlitz ψ-functions, Duke Math. J. **13**, (1946), 79–85.

[24] M. Waldschmidt, Nombres transcendants et groupes algébriques, Soc. Math. France, Astérisque, **69-70**, 1979.

[25] M. Waldschmidt, Transcendance en caractéristique finie et modules de Drinfeld, Cours IHP, 1989.

[26] M. Waldschmidt, Transcendence problems connected with Drinfeld modules, Preprint, (1991).

[27] G. Wüstholz, Multiplicity estimates on group varieties, Ann. of Math. **129**, (1989), 471–500.

[28] G. Wüstholz, Algebraische Punkte auf analytischen Untergruppen algebraischer Gruppen, Ann. of Math. **129**, (1989), 501–517.

[29] J. Yu, Transcendental numbers arising from Drinfeld modules, Mathematika **30**, (1983), 61–66.

[30] J. Yu, Transcendence theory over function fields, Duke Math. J. **52**, (1985), 517–527.

[31] J. Yu, A six exponentials theorem in finite characteristic, Math. Ann. **272**, (1985), 91–98.

[32] J. Yu, Transcendence and Drinfeld modules, Invent. Math. **83**, (1986), 507–517.

[33] J. Yu, Transcendence and Drinfeld modules: several variables, Duke Math. J. **58**, (1989), 559–575.

[34] J. Yu, Algebraic independence and Drinfeld modules, Preprint, (1989).

[35] J. Yu, On periods and quasi-periods of Drinfeld modules, Compositio Math. **74**, (1990), 235–245.

[36] J. Yu, Transcendence and special zeta values in characteristic p, Ann. of Math. **134**, (1991), 1–23.

[37] J. Yu, Analytic homomorphisms into Drinfeld modules, Preprint, (1991).

Institute of Mathematics, Academia Sinica, Taipei, Taiwan, R.O.C

Indépendance Algébrique des Périodes et Quasi-périodes d'un Module de Drinfeld

Alain Thiery

Résumé. Soit Φ un module de Drinfeld de rang 2 sur $\mathbb{F}_q[T]$ et soit η une dérivation associée à Φ. On considère la fonction entière F_η construite par P.Deligne et J.Yu (qui joue un rôle analogue à celui joué par la fonction zêta de Weierstraß pour un réseau du plan complexe). Pour tout λ, ω éléments non nuls du réseau associé à Φ tels que

$$F_\eta(\omega) - \frac{F_\eta(\lambda)}{\lambda}\omega$$

soit non nul, les deux nombres

$$F_\eta(\omega) - \frac{F_\eta(\lambda)}{\lambda}\omega \quad \text{et} \quad \frac{F_\eta(\lambda)}{\lambda}$$

sont algébriquement indépendants sur $\mathbb{F}_q(T)$. Cet énoncé est l'analogue en caractéristique finie d'un résultat démontré par Chudnovsky dans le cas complexe.

Table des matières:

I). Introduction

Le but de ce papier est de donner un équivalent en caractéristique non nulle d'un théorème de Chudnovsky qui entraîne la transcendance des nombres $\Gamma(1/4)$ et $\Gamma(1/3)$, (voir **[C]**).

Ce théorème de Chudnovsky s'énonce ainsi:

Soient $\wp$ une fonction elliptique de Weierstraß d'invariants g_2, g_3 algébriques, ζ la fonction de Weierstraß associée à $\wp$, λ une période de $\wp$ et ω un point rationnel de $\wp$ qui ne soit pas un pôle, tels que λ et ω soient $\mathbb{Q}$-linairement indépendants. Alors les deux nombres

$$\frac{2\zeta(\lambda/2)}{\lambda} \quad \text{et} \quad \zeta(\omega) - \frac{2\zeta(\lambda/2)}{\lambda}\omega$$

sont algébriquement indépendants sur $\mathbb{Q}$.

Commençons par introduire quelques notations afin de pouvoir énoncer précisément le théorème que nous nous proposons de démontrer. Ces notations seront conservées et utilisées tout au long de cet article.

Soit p un nombre premier, $q = p^n$ une puissance de p et $\mathbb{F}_q$ le corps fini à q éléments.

Notons $A = \mathbb{F}_q[T]$ l'anneau des polynômes à coefficients dans $\mathbb{F}_q$, $k = \mathbb{F}_q(T)$ le corps des fractions de A, $k_\infty = \mathbb{F}_q((\frac{1}{T}))$ le complété de k pour la valuation $-\deg$, $\overline{k}_\infty$ une clôture algébrique de k_∞ et $\overline{k}$ la clôture algébrique de k dans $\overline{k}_\infty$. On prolonge la valuation $-\deg$ à $\overline{k}_\infty$.

Le Frobenius $x \longrightarrow x^q$ sera noté τ. Soit $\Phi : A \longrightarrow k\{\tau\}$ un module de Drinfeld de rang 2, Λ le réseau de $\overline{k}_\infty$ associé à Φ et $e_\Lambda \in k\{\{\tau\}\}$ la fonction exponentielle de Drinfeld associée à Φ. On rappelle que $\Lambda = \operatorname{Ker} e_\Lambda$ et que, pour tout $a \in A$ et $z \in \overline{k}_\infty$, on a $\Phi_a(e_\Lambda(z)) = e_\Lambda(az)$.

Soit $\eta : A \longrightarrow k\{\tau\} \circ \tau$ une dérivation, c'est-à-dire une application $\mathbb{F}_q$-linéaire telle que, pour tout $a, b \in A$, on ait $\eta_{ab} = a\eta_b + \eta_a \circ \Phi_b$. D'après Gekeler **[G]**, il existe une unique fonction entière $F_\eta \in k\{\{\tau\}\} \circ \tau$ telle que pour tout $a \in A$ et $z \in \overline{k}_\infty$, on ait

$$F_\eta(az) - aF_\eta(z) = \eta_a \circ e_\Lambda(z).$$

Soient λ et ω des éléments non nuls de Λ tels que $F_\eta(\omega) - \frac{F_\eta(\lambda)}{\lambda}\omega$ soit non nul. On notera $\theta = F_\eta(\lambda)/\lambda$. Le nombre θ est alors transcendant (voir Appendice).

Notre principal résultat est le suivant:

Théorème. *Les deux nombres*

$$F_\eta(\omega) - \frac{F_\eta(\lambda)}{\lambda}\omega \qquad \textit{et} \qquad \frac{F_\eta(\lambda)}{\lambda}$$

sont algébriquement indépendants sur $\mathbb{F}_q(T)$.

Il s'agit bien d'un analogue du théorème de Chudnovsky cité plus haut, la fonction e_Λ jouant le rôle de la fonction elliptique associée à un réseau du plan complexe et F_η celui de la fonction ζ de Weierstraß correspondante.

II). Lemme de Zéros

Nous reprenons les notations introduites dans le premier paragraphe. De plus nous posons $F(z) = F_\eta(z) - \theta z$. On a alors par hypothèse $F(\omega) \neq 0$.

Proposition 1. *Soient s, m, N trois entiers rationnels avec $m \geq 2$, L un sous-corps de $\overline{k}_\infty$ contenant k et $P \in L[X,Y]$ un polynôme irréductible sur L. On désigne par $x = \deg_X P$ (resp. $y = \deg_Y P$) le degré de P en la variable x (resp. y). Si la fonction $\Psi(z) = P(F(z)^{q^s}, e_\Lambda(z))$ a un zéro de multiplicité au moins m en chacun des points de l'ensemble $\{a\omega; a \in A, \deg(a) \leq N\}$, alors on a*

$$mq^N \leq q^s x\big(y(q^2+1) + x(q^{s+\deg_\tau \eta_T})\big).$$

Démonstration. On calcule $\Psi(Tz)$ en utilisant l'équation fonctionnelle de e_Λ:

$$\begin{aligned}\Psi(Tz) &= P\big(F(Tz)^{q^s}, e_\Lambda(Tz)\big)\\ &= P\big((TF(z) + \eta_T \circ e_\Lambda(z))^{q^s}, \Phi_T \circ e_\Lambda(z)\big)\\ &= P\big(T^{q^s} F^{q^s} + (\eta_T \circ e_\Lambda(z))^{q^s}, \Phi_T \circ e_\Lambda(z)\big).\end{aligned}$$

On définit un polynôme $Q \in L[X,Y]$ par

$$Q(X,Y) = P\big(T^{q^s} X + (\eta_T(Y))^{q^s}, \Phi_T(Y)\big).$$

On a alors $\Psi(Tz) = Q(F(z)^{q^s}, e_\Lambda(z))$.

Notons

$$R(X) = \mathrm{Res}_Y(P,Q) \in L[X]$$

le résultant de P et Q par rapport à la variable Y.

a) Si $R = 0$ alors P divise Q car P est irréductible. Donc il existe $S \in L[X,Y]$ tel que $Q(X,Y) = P(X,Y)S(X,Y)$. En particulier on a $P(T^{q^s}X, 0) = P(X,0)S(X,0)$. Or $F(\omega)^{q^s}$ est une racine de $P(X,0)$, donc pour tout $r \in \mathbb{N}$, $T^{rq^s} F^{q^s}(\omega)$ est une racine de $P(X,0)$. Comme $F(\omega)$ n'est pas nul, $P(X,0)$ admet alors une infinité de racines. On en déduit $P(X,0) = 0$, et Y divise P. Or P est irréductible, donc $P = cY$ avec $c \in \overline{k}_\infty$. Mais dans ce cas Ψ n'a qu'un zéro de multiplicité 1 en ω, ce qui est absurde.

b) On peut donc supposer $R \neq 0$. Il existe U, V deux polynômes dans $L[X,Y]$ tels que $R(X) = UP + VQ$. Alors la fonction $R(F(z)^{q^s})$ s'annule avec une multiplicité $\geq m$ en chaque point $aT\omega$, ($a \in A$, $\deg(a) \leq N-1$). Comme $F(aT\omega) = aTF(\omega)$ et $F(\omega) \neq 0$, les nombres $F(aT\omega), \deg(a) \leq N-1$ sont deux à deux distincts. Il existe donc $W \in \overline{k}_\infty[X]$ et des entiers $\alpha_a \geq 0$ tels que

$$R(X) = W(X) \prod_{\deg(a) \leq N-1} (X - F(aT\omega)^{q^s})^{\alpha_a}$$

et W ne s'annule pas en $F(aT\omega)^{q^s}$. On a alors pour tout $a \in A$ avec $\deg(a) \leq N-1$ et pour tout $h \in \overline{k}_\infty$,

$$R(F(aT\omega + h)^{q^s}) = R(F(aT\omega)^{q^s} + F(h)^{q^s})$$

$$= W\big(F(aT\omega + h)^{q^s}\big) \left(\prod_{\substack{\deg(a') \leq N-1 \\ a' \neq 0}} \Big(F\big((a-a')T\omega\big)^{q^s} + F(h)^{q^s}\Big)^{\alpha_{a'}} \right) F(h)^{q^s \alpha_a}.$$

Etant donné que F a un zéro simple à l'origine, il existe une fonction entière G et un élément $\theta \neq 0$ de $\overline{k}_\infty$ tels que $F(h) = -\theta h + h^q G(h)$. Comme tous les coefficients de Taylor de la fonction $R(F(aT\omega + h)^{q^s})$ sont nuls jusqu'à l'ordre m, on a $\alpha_a q^s \geq m$. D'où

$$\deg(R) \geq \sum_{\deg(a) \leq N-1} \alpha_a \geq m q^{N-s}.$$

On a de plus $\deg(R) \leq \deg_X P \deg_Y Q + \deg_X Q \deg_Y P$. Or $\deg_X Q \leq x$ et $\deg_Y Q \leq xq^{s+\deg_T \eta_T} + yq^2$. Le résultat annoncé en résulte.

CQFD

III). Critère de Transcendance

L'anneau $A[X]$ des polynômes en une variable à coefficients dans A est aussi l'anneau $\mathbb{F}_q[T, X]$ des polynômes en deux variables à coefficients dans k; on notera $\text{degtot}(P)$ le degré total (par rapport à X et T) d'un élément de cet anneau.

On définit *la hauteur* d'un élément $P = \sum_{i=0}^{n} p_i X^i \in A[X]$ par

$$H(P) = \max_{0 \leq i \leq n} \deg(p_i).$$

Rappelons que pour tout $P, Q \in A[X]$, on a

$$H(P+Q) \leq \max\{H(P), H(Q)\}$$

et

$$H(PQ) = H(P) + H(Q).$$

Proposition 2. *Soient* Π *un élément de* $\overline{k}_\infty$ *tel que* $\deg(\Pi) \leq 0$, $(R_n)_{n \geq n_0}$ *une suite d'éléments irréductibles de* $A[X]$ *et* $(d_n)_{n \geq n_0}$ *une suite d'entiers strictement positifs. Pour* $n \geq n_0$ *posons* $P_n = {R_n}^{d_n}$. *On suppose que les nombres* $\delta_n = \deg_X(P_n)$ *sont tous différents de 0. On pose*

$$h_n = H(P_n) \qquad \textit{et} \qquad s_n = -\deg P_n(\Pi).$$

Si

$$\min\{s_n, s_{n+1}\} > h_n \delta_{n+1} + h_{n+1} \delta_n \quad \textit{et} \quad \lim_{n \to \infty} \frac{s_n}{\delta_n} = \infty,$$

alors pour tout $n \geq n_0$, *on a* $P_n(\Pi) = 0$.

Démonstration. Pour $n \geq n_0$ définissons $A_n \in A$ par $A_n = \text{Res}(P_n(X), P_{n+1}(X))$. Ecrivons

$$P_n(X) = \sum_{i=0}^{\delta_n} P_{n,i} X^i.$$

On a

$$A_n = \begin{vmatrix} P_{n,\delta_n} & \cdots & P_{n,0} & & X^{\delta_{n+1}-1} P_n(X) \\ & \ddots & & \ddots & \\ & & P_{n,\delta_n} & \cdots & P_n(X) \\ P_{n+1,\delta_{n+1}} & \cdots & P_{n+1,0} & & X^{\delta_n - 1} P_{n+1}(X) \\ & \ddots & & \ddots & \\ & & P_{n+1,\delta_{n+1}} & \cdots & P_{n+1}(X) \end{vmatrix}.$$

Le déterminant ci-dessus est un polynôme en X; en remplaçant X par Π, on trouve

$$\deg A_n \leq \max\{\delta_{n+1}h_n + \delta_n h_{n+1} - s_n, \delta_n h_{n+1} + \delta_{n+1}h_n - s_{n+1}\}$$

et

$$\deg A_n \leq \delta_{n+1}h_n + \delta_n h_{n+1} - \min\{s_n, s_{n+1}\},$$

donc $\deg A_n < 0$, ce qui permet de conclure $A_n = 0$.

On en déduit $P_n = (c_n R)^{d_n}$ avec $R \in A[X]$ indépendant de n et $c_n \in \mathbb{F}_q$. Donc

$$\deg(R(\Pi)) = -\frac{s_n}{d_n} \leq -\frac{s_n}{\delta_n}, \qquad (n \geq n_0),$$

ce qui entraîne $R(\Pi) = 0$.

CQFD

Proposition 3. *Soient* Π *un élément de* $\overline{k}_\infty$ *et* $(P_n)_{n \geq n_0} \in A[X]$ *une suite de polynômes non nuls. On note*

$$\begin{aligned} \delta_n &= \deg_X(P_n), \\ h_n &= H(P_n), \\ s_n &= -\deg P_n(\Pi). \end{aligned}$$

Si, pour tout $n \geq n_0$, *on a*

$$s_n > \max\{h_n\delta_n + h_n\delta_{n+1} + h_{n+1}\delta_n, h_n\delta_n + h_n\delta_{n-1} + h_{n-1}\delta_n\}$$

et

$$\lim\left(\frac{s_n}{\delta_n} - h_n\right) = +\infty,$$

alors pour tout $n \geq n_0$ *on a* $P_n(\Pi) = 0$.

Démonstration. Pour chaque $n \geq n_0$, écrivons $P_n(X) = \prod_{i=1}^{\ell} R_{n,i}^{q_{n,i}}$ avec $q_{n,i} \geq 1$ et les $R_{n,i} \in A[X]$ irréductibles, premiers deux-à-deux.

On ordonne les facteurs de la manière suivante :

$$\deg(R_{n,1}^{q_{n,1}}(\Pi)) \leq \deg(R_{n,2}^{q_{n,2}}(\Pi)) \leq \ldots \leq \deg(R_{n,\ell}^{q_{n,\ell}}(\Pi)).$$

On pose

$$Q_n = R_{n,1}^{q_{n,1}}, \qquad \delta'_n = \deg(Q_n) \leq \delta_n, \qquad h'_n = H(Q_n) \leq H(P_n) = h_n.$$

Montrons que pour tout $n \geq n_0$ on a $\delta'_n \neq 0$. S'il existait $n \geq n_0$ tel que $\delta'_n = 0$, alors Q_n appartiendrait à A ; dans ce cas pour tout i on aurait $\deg(R_{n,i}^{q_{n,i}}(\Pi)) \geq 0$, donc $s_n \leq 0$; mais ceci contredirait l'inégalité $s_n > 0$.

a) Supposons $\deg(\Pi) \leq 0$.

On pose $d_{n,i} = \deg(R_{n,i}^{q_{n,i}})$. Comme pour tout $i \geq 2$ on a $\mathrm{Res}(R_{n,i}^{q_{n,i}}, Q_n) \in A - \{0\}$, on peut écrire

$$\begin{aligned}0 &\le \deg\bigl(\mathrm{Res}(R_{n,i}^{q_{n,i}}, Q_n)\bigr)\\ &\le \max\{\deg(Q_n(\Pi)) + (d_{n,i})H(Q_n) + \delta'_n H(R_{n,i}^{q_{n,i}}),\\ &\qquad \deg(R_{n,i}^{q_{n,i}}(\Pi)) + (\delta'_n)H(R_{n,i}^{q_{n,i}}) + d_{n,i}H(Q_n)\}\\ &\le \deg(R_{n,i}^{q_{n,i}}(\Pi)) + d_{n,i}H(Q_n) + \delta'_n H(R_{n,i}^{q_{n,i}}).\end{aligned}$$

En tenant compte de l'égalité $H(AB) = H(A) + H(B)$, on trouve:

$$\begin{aligned}s'_n = -\deg(Q_n(\Pi)) &= s_n + \sum_{i=2}^{\ell} \deg(R_{n,i}^{q_{n,i}}(\Pi))\\ &\ge s_n - \sum_{i=2}^{\ell}(d_{n,i}H(P_n) + \delta'_n H(R_{n,i}^{q_{n,i}}))\\ &\ge s_n - \bigl(H(P_n)(\delta_n - \delta'_n) + \delta'_n(H(P_n) - H(Q_n))\bigr)\\ &\ge s_n - H(P_n)\delta_n + \delta'_n H(Q_n)\\ &\ge s_n - h_n\delta_n,\end{aligned}$$

donc

$$\frac{s'_n}{\delta'_n} \ge \frac{s_n - \delta_n h_n}{\delta'_n} \ge \frac{s_n}{\delta_n} - h_n.$$

On peut appliquer le Lemme 1 à Q_n et Π pour conclure.

b) Supposons $\deg(\Pi) \ge 0$

On définit $\mathrm{Val}_X(P) \in \mathbb{N}$ comme la plus grande puissance de X divisant P. On pose alors

$$Q_n(X) = X^{\delta_n + \mathrm{Val}_X(P_n)} P_n\left(\frac{1}{X}\right).$$

Comme Q_n est un polynôme, on a $Q_n(\Pi) = 0$ si et seulement si $P_n(\Pi) = 0$. On a trivialement $\deg(Q_n) = \deg(P_n)$ et $H(P_n) = H(Q_n)$. De plus

$$\begin{aligned}-\deg\bigl(Q_n(1/\Pi)\bigr) &= -\deg\bigl(\Pi^{-\delta_n - \mathrm{Val}_X(P_n)} P_n(\Pi)\bigr)\\ &= s_n + \bigl(\delta_n + \mathrm{Val}_X(P_n)\bigr)\deg(\Pi) \ge s_n.\end{aligned}$$

On s'est donc ramené au premier cas.

CQFD

IV). Construction d'une Taille

Soit $L \supset k$ une extension de k de degré de transcendance δ sur k. Considérons $\theta_1, \ldots, \theta_\delta$ une base de transcendance de L sur k. On suppose $[L : k(\theta_1, \ldots, \theta_\delta)]$ fini. Soit L_s la sous-extension séparable maximale. On note $R = A[\theta_1, \ldots, \theta_\delta]$. Il existe alors, par le théorème de l'élément primitif, $\theta_{\delta+1}$ entier sur R tel que

$$L_s = k(\theta_1, \ldots, \theta_\delta)[\theta_{\delta+1}].$$

Désignons par D le degré de $\theta_{\delta+1}$ sur $k(\theta_1, \ldots, \theta_\delta)$. Pour tout $x \in L_s - \{0\}$, on peut écrire

$$x = \frac{\sum_{i=0}^{D-1} P_i(T, \theta_1, \ldots, \theta_\delta)\theta_{\delta+1}^i}{P_D(T, \theta_1, \ldots, \theta_\delta)}$$

avec

$$P_i(T, \theta_1, \ldots, \theta_\delta) \in \mathbb{F}_q[T, \theta_1, \ldots, \theta_\delta],$$
$$\mathrm{PGCD}(P_0, \ldots, P_D) = 1.$$

Cette écriture est unique à multiplication par un élément de $\mathbb{F}_q^*$ près.

Définition. Pour tout $x \in L_s$ non nul, on pose

$$s(x) = \max\{\mathrm{degtot}(P_i) \mid 0 \le i \le D\}.$$

On convient $s(0) = -\infty$.

Propriétés.

1) *Pour tout* $x, y \in R[\theta_{\delta+1}]$,

$$s(x+y) \le \max\{s(x), s(y)\}.$$

2) *Il existe* $C \in \mathbb{R}$ *ne dépendant que de* $\theta_1, \ldots, \theta_{\delta+1}$ *tel que pour tout* $n \in \mathbb{N}$ *et* $x_1, \ldots, x_n \in L_s$

$$s(x_1 x_2 \ldots x_n) \le s(x_1) + s(x_2) + \ldots + s(x_n) + C(n-1).$$

De plus, pour tout $x \in R, y \in R[\theta_{\delta+1}]$

$$s(xy) = \mathrm{degtot}(x) + s(y) = s(x) + s(y).$$

3) *Soit* $x \in k$ *et* Δ *un dénominateur de* x, *c'est-à-dire* $\Delta \in A - \{0\}$ *et* $\Delta x \in A$. *On a alors*

$$s(\Delta x) = \deg(\Delta) + \deg(x).$$

Démonstration.

3) Comme $\Delta x \in A, s(\Delta x) = \deg(\Delta x) = \deg(\Delta) + \deg(x)$.

1) Comme $\theta_{\delta+1}$ est entier sur R, $R[\theta_{\delta+1}]$ est l'ensemble des polynômes en $\theta_{\delta+1}$ de degré inférieur à $D-1$ à coefficients dans R. Le 1) en découle immédiatement.

2) Par récurrence, il suffit de le montrer pour $n = 2$. Soient $x, y \in L_s$

$$x = \frac{\sum_{i=0}^{D-1} P_i \theta_{\delta+1}^i}{P_D}, \quad y = \frac{\sum_{i=0}^{D-1} Q_i \theta_{\delta+1}^i}{Q_D}.$$

D'où

$$s(xy) \le \max\{\mathrm{degtot}(P_D Q_D), \quad s((\sum_{i=0}^{D-1} P_i \theta_{\delta+1}^i)(\sum_{i=0}^{D-1} Q_i \theta_{\delta+1}^i))\}.$$

Or

$$\mathrm{degtot}(P_D Q_D) = \mathrm{degtot}(P_D) + \mathrm{degtot}(Q_D) \le s(x) + s(y).$$

De plus, d'apres le 1)

$$s((\sum_{i=0}^{D-1} P_i\theta_{\delta+1}^i)(\sum_{i=0}^{D-1} Q_i\theta_{\delta+1}^i)) \leq \max_{0\leq i,j\leq D-1}\{s(P_iQ_j\theta_{\delta+1}^{i+j})\}.$$

Donc

$$s(P_iQ_j\theta_{\delta+1}^{i+j}) = \mathrm{degtot}(P_iQ_j) + s(\theta_{\delta+1}^{i+j}).$$

On en déduit immédiatement que $C = \max_{0\leq l\leq 2(D-1)}\{s(\theta_{\delta+1}^l\}$ convient.

CQFD

Notons $K = k(\theta_1, \ldots, \theta_\delta)$. L'extension L_s étant séparable sur K, on a l'opérateur de norme $N_K^{L_s}$. Comme R est intégralement clos, $N_K^{L_s}(R[\theta_{\delta+1}]) \subset R$.

Propriétés. *Il existe* $C \in \mathbb{R}$ *ne dépendant que de* $\theta_1, \ldots, \theta_{\delta+1}$ *tel que pour tout* $x \in R[\theta_{\delta+1}]$

$$s(N_K^{L_s}(x)) \leq C(1 + s(x)),$$

$$\deg(N_K^{L_s}(x)) \leq \deg(x) + C(1 + s(x)).$$

Démonstration. Soient $y_1 = \theta_{\delta+1}$ et $y_2, \ldots, y_D$ les conjugués sur K de $\theta_{\delta+1}$. Posons $x = \sum_{i=0}^{D-1} P_iy_1^i$, on a donc

$$N_K^{L_s}(x) = \prod_{j=1}^{D}(\sum_{i=0}^{D-1} P_iy_j^i) = \sum_{i_1=0}^{D-1} \cdots \sum_{i_D=0}^{D-1}\prod_{j=1}^{D} P_{i_j}y_j^{i_j}.$$

On définit la relation d'équivalence $(i_1, \ldots, i_D) \sim (i'_1, \ldots, i'_D)$ si et seulement si il existe une permutation $\sigma : \{1, \ldots, D\} \longrightarrow \{1, \ldots, D\}$ telle que $(i_{\sigma(1)}, \ldots, i_{\sigma(D)}) = (i'_1, \ldots, i'_D)$. On note E l'ensemble des classes. On remarque que si I est une classe et $(i_1, \ldots, i_D)$ un représentant de cette classe alors le produit $\prod_{j=1}^{D} P_{i_j}$ est indépendant du représentant choisi. On le notera P_I. D'où

$$N_K^{L_s}(x) = \sum_{I\in E}(P_I \sum_{(i_1,\ldots,i_D)\in I}\prod_{j=1}^{D} y_j^{i_j}).$$

Or $\sum_{(i_1,\ldots,i_D)\in I} \prod_{j=1}^{D} y_j^{i_j}$ est invariant par le groupe de Galois de $K[y_1 \ldots y_D]$ sur K, donc appartient à R. On peut donc parler de sa taille. De plus, d'après la propriété 1)

$$s(N_K^{L_s}(x)) \leq \max_{I\in E}\{s(P_I \sum_{(i_1,\ldots,i_D)\in I}\prod_{j=1}^{D} y_j^{i_j})\}.$$

Donc, d'après la propriété 2)

$$s(N_K^{L_s}(x)) \leq Ds(x) + \max_{I \in E}\{s(\sum_{(i_1,\ldots,i_D)\in I} \prod_{j=1}^{D} y_j^{i_j})\}.$$

D'où la première inégalité.

Pour la seconde inégalité, il suffit de montrer que

$$\deg(\sum_{i=0}^{D-1} P_i y_j^i) \leq C'(s(x)+1).$$

Or

$$\begin{aligned}\deg(\sum_{i=0}^{D-1} P_i y_j^i) &\leq \max_{0\leq i\leq D-1}\{\deg(P_i) + \deg(y_j^i)\} \\ &\leq \max_{0\leq i\leq D-1}\{\deg(P_i)\} + C''\end{aligned},$$

avec C'' indépendant de x. Comme $s(P_i) \leq s(x)$, il reste donc à voir que pour $P \in R, \deg(P) \leq C'''s(P)$. Posons $P = \sum_{i_0,\ldots,i_\delta} P_{i_0,\ldots,i_\delta} T^{i_0}\theta_1^{i_1}\ldots\theta_\delta^{i_\delta}$, avec $P_{i_0,\ldots,i_\delta} \in \mathbb{F}_q$ on a alors

$$\begin{aligned}\deg(P) \leq& \max_{P_{i_0,\ldots,i_\delta}\neq 0}\{\deg(T^{i_0}\theta_1^{i_1}\ldots\theta_\delta^{i_\delta})\} \\ \leq& s(x)\max\{1, \deg(\theta_1),\ldots,\deg(\theta_\delta)\}.\end{aligned}$$

CQFD

V). Lemme de "Siegel"

Soit L une extension de k de degré de transcendance δ sur k. On garde les mêmes notations qu'au IV), c'est-à-dire que $\theta_1,\ldots,\theta_\delta$ est une base de transcendance. La sous-extension séparable maximale L_s de L sur $k(\theta_1,\ldots,\theta_\delta)$ s'écrit $L_s = k(\theta_1,\ldots,\theta_\delta)[\theta_{\delta+1}]$ avec $\theta_{\delta+1}$ entier sur $A[\theta_1,\ldots,\theta_\delta]$ de degré D. On construit la taille s comme précédemment.

Proposition 4. (Lemme de "Siegel") *Soient $n, m \in \mathbb{N}$ tels que $n > Dm$ et soient $a_{i,j}$, $1 \leq i \leq n$, $1 \leq j \leq m$ des éléments de $A[\theta_1,\ldots,\theta_\delta,\theta_{\delta+1}]$. Le système linéaire*

$$\sum_{i=1}^{n} a_{i,j}x_i = 0, \textit{ pour tout } 1 \leq j \leq m$$

admet alors une solution non triviale dans $A[\theta_1,\ldots,\theta_\delta]$ avec

$$\max_{1\leq i\leq n}\{s(x_i)\} \leq \frac{(Dm)^{\frac{1}{\delta+1}}}{n^{\frac{1}{\delta+1}} - (Dm)^{\frac{1}{\delta+1}}}(\max_{\substack{1\leq i\leq n\\ 1\leq j\leq m}}\{s(a_{i,j})\}) + 1.$$

Démonstration. Posons $S = \max_{\substack{1\leq i\leq n \\ 1\leq j\leq m}} \{s(a_{i,j})\}$. Soit $s \in \mathbb{N}$. Le cardinal de l'ensemble

$$\{(x_i)_{1\leq i\leq n} \in A[\theta_1, \dots, \theta_\delta] \mid \max_{1\leq i\leq n} s(x_i) \leq s\}$$

vaut $q^{n\binom{\delta+1+s}{\delta+1}}$ car il y a exactement $\binom{\delta+1+s}{\delta+1}$ monômes de degré $\leq s$ en $\delta+1$ variables. De plus

$$\begin{aligned} s(\sum_{i=1}^{n} a_{i,j}x_i) &\leq \max_{1\leq i\leq n} \{s(a_{i,j}x_i)\} \\ &\leq s + S \end{aligned}$$

car $a_{i,j} \in A[\theta_1, \dots, \theta_\delta \theta_{\delta+1}]$ et $x_i \in A[\theta_1, \dots, \theta_\delta]$. Donc le cardinal de l'ensemble

$$\{(\sum_{i=1}^{n} a_{i,j}x_i)_{1\leq j\leq m} \mid x_i \in A[\theta_1, \dots, \theta_\delta] \text{ et } \max_{1\leq i\leq n} s(x_i) \leq s\}$$

est inférieur à $q^{Dm\binom{\delta+1+s+S}{\delta+1}}$. Par le principe des tiroirs de Dirichlet, pour qu'il y ait une solution non triviale avec $\max_{1\leq i\leq n} s(x_i) \leq s$, il suffit

$$n\binom{\delta+1+s}{\delta+1} > Dm\binom{\delta+1+s+S}{\delta+1}.$$

Lemme. *Soient* $a, b, c \in \mathbb{N}$ *tels que* $b \geq a$, *alors*

$$\left(\frac{b}{a}\right)^c \geq \frac{\binom{b+c}{c}}{\binom{a+c}{c}}.$$

Démonstration. Comme $\frac{\binom{b+c}{c}}{\binom{a+c}{c}} = \frac{(b+c)\dots(b+1)}{(a+c)\dots(a+1)}$, il suffit donc de montrer que

$$f(x) = c(\log(x) - \log(a)) - \sum_{i=1}^{c} \log(x+i) + \sum_{i=1}^{c} \log(a+i)$$

est positive pour $x \geq a$. Or $f'(x) = \frac{c}{x} - \sum_{i=1}^{c} \frac{1}{x+i} \geq 0$ si $x > 0$, donc f est croissante. Or $f(a) = 0$, on a donc le résultat.

D'apres ce lemme, il suffit donc que $\frac{n}{Dm} > (\frac{s+S}{s})^{\delta+1}$.

CQFD

VI). Construction d'une Fonction Auxiliaire

Soient $m, N \in \mathbb{N}$. Notons $F(z) = F_\eta(z) - \theta z$ et $\Pi = F(\omega)$. Ainsi pour $a \in A, F(z + a\omega) = F(z) + a\Pi$. On suppose Π non nul et algébrique sur $k(\theta)$. Soit p^s le degré d'inséparabilité de $L = k(\theta)[\Pi]$ sur $k(\theta)$.

Le but de cette partie est de construire un polynôme non nul $P \in A[\theta][X, Y]$ tel que la fonction

$$\Psi(z) = P(e_\Lambda(z), F^{q^s}(z))$$

ait un zéro d'ordre au moins m en chaque point $a\omega, \deg(a) \leq N$.

On introduit la notation suivante: si f et g sont deux fonctions de m telles que $\lim_{m\to\infty} \frac{f}{g} = 0$, on écrira

$$f \ll g.$$

Soit α entier sur $A[\theta]$ tel que $k(\theta)[\alpha] = L_s$, D son degré. On construit la taille s sur L_s avec θ et α. Si $P \in A[\theta][X, Y]$, on pose $s(P) = \max_{i,j\in\mathbb{N}}\{s(P_{i,j})\}$. On a alors la proposition suivante:

Proposition 5. *Soient $N(m) \neq 0, x(m), y(m)$ des fonctions de $m \in \mathbb{N}$ vérifiant*

$$\begin{aligned}\max\{x, yN\} &\ll m\log_q(m)\\ 4Dmq^{N+1} &\leq (1+x)(1+y).\end{aligned}$$

Alors, il existe une constante c_1 telle que, pour tout $m \geq 2$, il existe un polynôme non nul $P \in A[\theta][X, Y]$ tel que

$$\begin{aligned}\deg_X P &\leq x\\ \deg_Y P &\leq y\\ s(P) &\leq c_1 m\log_q(m)\end{aligned}$$

et la fonction $\Psi(z) = P(e_\Lambda(z), F^{q^s}(z))$ ait un zéro d'ordre au moins m en chaque point $a\omega, a \in A$ et $\deg(a) \leq N$.

Démonstration.
1) Calcul des coefficients de Taylor de Ψ à l'ordre $m' \leq m-1$ aux points $a\omega, \deg(a) \leq N$.

Soient

$$\begin{aligned}e_\Lambda(z) &= \sum_{i=0}^{\infty} a_i z^{q^i}\\ F(z) &= -\theta z + \sum_{i=1}^{\infty} b_i z^{q^i}.\end{aligned}$$

Si on convient $q^{-1} = 0$, on a

$$\begin{aligned}F(a\omega + z) &= (a\Pi + F(z))^{q^s}\\ &= a^{q^s}\Pi^{q^s} - \theta^{q^s} z^{q^s} + \sum_{i=1}^{\infty} b_i^{q^s} z^{q^{i+s}}\\ &= \sum_{i=-1}^{\infty} c_i z^{q^i}.\end{aligned}$$

On remarque que les c_i appartiennent à L_s car Π^{q^s} appartient à L_s. Posons $P[X,Y] = \sum_{i,j} P_{i,j}X^iY^j$, d'où $\Psi(a\omega + z) = \sum_{i,j} P_{i,j}(\sum_{k=0}^{\infty} a_k z^{q^k})^i(\sum_{l=-1}^{\infty} c_l z^{q^l})^j$. Donc, pour tout $0 \le m' \le m-1$, le coefficient de Taylor de Ψ d'ordre m' vaut

$$\sum_{i,j}\Big(P_{i,j}\sum_{n=0}^{m'}\big(\sum_{q^{k_1}+\ldots+q^{k_i}=n} a_{k_1}\ldots a_{k_i}\big)\big(\sum_{q^{l_1}+\ldots+q^{l_j}=m'-n} c_{l_1}\ldots c_{l_j}\big)\Big).$$

On obtient donc un système linéaire dont les inconnues sont les $P_{i,j}$. Pour appliquer le lemme de Siegel, il faut que les coefficients du système appartiennent à $A[\theta,\alpha]$. Le but du paragraphe suivant sera donc de calculer un dénominateur commun à ces coefficients.

2) Calcul d'un dénominateur commun aux coefficients.

Soit $\Delta \in A[\theta]$ tel que $\Delta\Pi^{q^s} \in A[\theta,\alpha]$. Comme e_Λ et F_η sont des E_q fonctions, il existe deux suites de dénominateurs $(\alpha_n)_{n\in\mathbb{N}}$ et $(\beta_n)_{n\in\mathbb{N}}$ vérifiant

$$\alpha_n,\ \beta_n \in A - \{0\};$$

$$\alpha_n a_n,\ \beta_n b_n \in A;$$

il existe c_1, pour tout $n \in \mathbb{N}, \max\{\deg(\alpha_n), \deg(\beta_n)\} \le c_1 n q^n$;

pour tout $n_1, \ldots, n_k$ tels que $q^{n_1} + \ldots + q^{n_k} < q^n$ alors

$$\alpha_{n_1}\ldots\alpha_{n_k} \text{ divise } \alpha_n,$$
$$\beta_{n_1}\ldots\beta_{n_k} \text{ divise } \beta_n.$$

(Voir [J].) On en déduit immédiatement que

$$D_m = \alpha_{[\log_q(m-1)]+1}\beta^{q^s}_{[\log_q(m-1)]+1}\Delta^y$$

est un dénominateur commun, dès que $\deg_Y P \le y$.

3) Calcul de la taille des coefficients.

On peut maintenant appliquer le lemme de Siegel, mais, pour avoir une estimation de $s(P)$, il faut majorer la taille des coefficients. Comme les fonctions e_Λ et F sont entières, il existe c_2 tel que, pour tout $n, \max\{\deg(a_n), \deg(c_n)\} \le c_2$. On impose de plus $\deg_X P \le x$. D'où, pour tout i, j, m'

$$s\Big(D_m\sum_{n=0}^{m'}\big(\sum_{q^{k_1}+\ldots+q^{k_i}=n} a_{k_1}\ldots a_{k_i}\big)\big(\sum_{q^{l_1}+\ldots+q^{l_j}=m'-n} c_{l_1}\ldots c_{l_j}\big)\Big)$$

$$\le \max_{\substack{0\le n\le m' \\ q^{k_1}+\ldots+q^{k_i}=n \\ q^{l_1}+\ldots+q^{l_j}=m'-n}} \Big\{s(D_m a_{k_1}\ldots a_{k_i}c_{l_1}\ldots c_{l_j})\Big\}$$

$$\leq \max_{\substack{0 \leq n \leq m' \\ q^{k_1} + \ldots + q^{k_i} = n \\ q^{l_1} + \ldots + q^{l_j} = m' - n}} \left\{ s(\Delta^y a_{k_1} \ldots a_{k_i} \alpha_{[\log_q(m-1)]+1} c_{l_1} \ldots c_{l_j} \beta^{q^s}_{[\log_q(m-1)]+1}) \right\}$$

$$\leq c_2(x+y) + \deg(\alpha_{[\log_q(m-1)]+1}) + q^s \deg(\beta_{[\log_q(m-1)]+1}) + y(s(\Delta) + s(\Pi^{q^s}) + q^s N) + C2y$$

d'après les propriétés de la taille établies au IV. On en déduit que

$$s(\text{ coefficients }) \leq c_3 m \log_q(m).$$

On applique donc directement le lemme de Siegel. La condition $(x+1)(y+1) \geq 4Dmq^{N+1}$ nous assure que $\frac{\sqrt{Dmq^{N+1}}}{\sqrt{(x+1)(y+1)} - \sqrt{Dmq^{N+1}}} \leq 1$. D'où un polynôme vérifiant l'énoncé avec $s(P) \leq (c_3 m \log_q(m)) + 1$.

$CQFD$

VII). Démonstration du Théorème

Proposition 6. *Les nombres θ et Π sont algébriquement indépendants.*

Démonstration. Quitte à changer ω en $b\omega$ et η en $\frac{\eta}{b}$, on peut supposer $\deg(\omega) \geq 0$ et $\deg(\theta) \leq 0$. Supposons Π algébrique sur $k(\theta)$. Définissons x, y, N par

$$q^{2N} \leq m \log_q(m)^2 < q^{2(N+1)}$$

$$x = y = [2\sqrt{Dq} m^{\frac{3}{4}} \log_q(m)^{\frac{1}{2}}].$$

On peut donc appliquer la Proposition 5. Il existe c_1 tel que, pour tout m, on peut trouver $P \in A[\theta][X, Y]$, non nul vérifiant

$$\deg_X P \leq x$$
$$\deg_X P \leq y$$
$$s(P) \leq c_1 m \log_q(m)$$

et $\Psi(z) = P(e_\Lambda(z), F(z)^{q^s})$ a un zéro d'ordre au moins m en chaque point

$$a\omega, \deg(a) \leq N.$$

Le système des 3 inéquations sera noté: (i). Prenons P_m vérifiant les conditions ci-dessus tel que $h(P_m) = s(P_m) + \deg_X P_m + \deg_Y P_m$ soit minimal. Si P_m n'est pas irréductible sur $A[\theta]$, alors il existe $P_1, P_2 \in A[\theta][X, Y]$ vérifiant $P_m = P_1 P_2$ et $h(P_i) < h(P_m), i = 1$ ou 2. On notera Q_m le polynôme P_i tel que la fonction $\Psi_i(z) = P_i(e_\Lambda(z), F(z)^{q^s})$ a au moins $\frac{m}{2} q^{N+1}$ zéros aux points $a\omega, \deg(a) \leq N$. Si P_m est irréductible, on pose $Q_m = P_m$. Dans tous les cas, on a que Q_m vérifie le système d'inéquations (i). Supposons que pour une infinité de valeurs de m, la fonction $\Psi_m(z) = Q_m(e_\Lambda(z), F(z)^{q^s})$ a un zéro d'ordre au moins m en chaque point

$a\omega, \deg(a) \leq N + [\log_q(N)]$. En particulier, Ψ_m a un zéro d'ordre au moins m en chaque point $a\omega, \deg(a) \leq N$. Donc $h(Q_m) \geq h(P_m)$ par minimalité de $h(P_m)$. Par construction même de Q_m, on en déduit que $P_m = Q_m$ et P_m est irréductible sur $A[\theta][X, Y]$, donc sur $k(\theta)[X, Y]$. La Proposition 1 (lemme de zéros) appliquée à P_m, pour une infinité de m, nous donne une contradiction. D'où, si m est assez grand, il existe $m' \leq m - 1$ et $a \in A, \deg(a) \leq N + [\log_q(N)]$ tel que le coefficient de Taylor d'ordre m' de Ψ_m au point $a\omega$ soit non nul. Notons le C_m.

a) Majoration de $\deg(C_m)$.

Posons $r = N + [\log_q(N)] + \deg(\omega)$, $N(\Psi_m, r)$ le nombre de zéros de la fonction Ψ_m dans la boule de centre 0 et de rayon r et $M(f, R)$ le maximum du degré d'une fonction f sur la boule de centre 0 et de rayon R. Comme $r \geq 0$, on a pour tout $R \geq r$

$$\deg(C_m) \leq M(\Psi_m, r) \leq (r - R)N(\Psi_m, r) + M(\Psi_m, R)$$

(voir **[B]**). Les fonctions e_Λ et F_η étant d'ordre de croissance inférieure à 2. Pour tout $\varepsilon > 0$ il existe une constante c_4 telle que

$$M(e_\Lambda, R) \leq c_4 q^{(2+\varepsilon)R}$$
$$M(F, R) \leq c_4 q^{(2+\varepsilon)R}$$

(voir **[J]**). Comme $\deg(\theta) \leq 0$, on a $M(\Psi_m, R) \leq s(Q_m) + (x + y)(c_4 q^{(2+\varepsilon)R})$. Or $s(Q_m) \leq s(P_m) \leq c_1 m \log_q(m)$. D'où

$$M(\Psi_m, R) \leq c_1 m \log_q(m) + (x + y)(c_4 q^{(2+\varepsilon)R}).$$

b) Estimation de $N(\Psi_m, r)$.

On sait, par construction, que Ψ_m a au moins $\frac{m}{2}q^{N+1}$ zéros aux points $a\omega, \deg(a) \leq N$. De plus, F et e_Λ étant $b\lambda, b \in A$ périodiques, la fonction Ψ_m a au moins $\frac{m}{2}q^{N+1}$ zéros aux points $a\omega + b\lambda, \deg(a) \leq N$. Or il existe une constante $c_5 > 0$ telle que le cardinal de l'ensemble $\{b\lambda | \deg(b\lambda) \leq r\}$ soit supérieur à $c_5 N q^N$. Nous aurons donc $N(\Psi_m, r) \geq c_5 N q^N \frac{m}{2} q^{N+1}$ dès que l'on aura montrér que ω et λ sont A linéairement indépendants.

Proposition 7. *Les nombres ω et λ sont A linéairement indépendants.*

Démonstration. Si $a_1\omega + a_2\lambda = 0$ avec $(a_1, a_2) \neq (0, 0)$, alors

$$0 = F_\eta(0) = a_1 F_\eta(\omega) + a_2 F_\eta(\lambda)$$

car F_η restreinte à Λ est A linéaire. a_1 étant non nul car λ l'est, on a

$$\frac{-a_2}{a_1} = \frac{\omega}{\lambda} = \frac{F_\eta(\omega)}{F_\eta(\lambda)}.$$

D'où $\Pi = F_\eta(\omega) - \frac{F_\eta(\lambda)}{\lambda}\omega = 0$, c'est absurde.

En posant $c_6 = \frac{c_5 q}{2}$, on obtient donc

$$N(\Psi_m, r) \geq c_6 N q^{2N} m.$$

Prenons $q^{(2+\varepsilon)R} = \frac{mq^{2N}}{x+y}$, on a alors que $r - R$ est équivalent à $\left(\frac{1}{2} - \frac{5}{u(2+\varepsilon)}\right) \log_q(m)$. En prenant $\varepsilon = \frac{1}{u}$, on déduit qu'il existe une constante $c_7 > 0$ et un entier m_0 tels que si $m \geq m_0$ alors

$$\deg(C_m) \leq -c_7 m^2 \log_q(m)^4 .$$

Soit D_m le dénominateur construit au paragraphe précédent. On a donc que $D_m C_m \in A[\theta, \alpha]$. Comme $\deg(D_m) \leq c_8 m \log_q(m)$, on a

$$\deg(D_m C_m) \leq -c_9 m^2 \log_q(m)^4 .$$

c) Calcul de $s(D_m C_m)$.

$$D_m C_m = \sum_{i,j} Q_{i,j} a_{i,j,a,m'}$$

avec $Q_m = \sum_{i,j} Q_{i,j} X^i Y^j$ et les $a_{i,j,a,m'}$ sont les coefficients du système linéaire obtenu au paragraphe VI. et C_m le coefficient de Taylor d'ordre m' de Ψ_m au point $a\omega$. Comme $y(N + [\log_q(N)]) \ll m \log_q(m)$, on obtient par un calcul identique à celui du VI. $s(a_{i,j,a,m'}) \leq c_{10} m \log_q(m)$. On en déduit alors

$$s(D_m C_m) \leq c_{11} m \log_q(m) .$$

Soit $R_m = N_K^{L_s}(D_m C_m) \in A[\theta] - \{0\}$. On a immédiatement, grâce aux propriétés de la taille,

$$s(R_m) \leq c_{12} m \log_q(m)$$
$$\deg(R_m) \leq -c_{13} m^2 \log_q(m)^4$$

Donc, à partir d'un certain rang m_1, les conditions de la Proposition 3 (critère de transcendance) sont remplies. C'est absurde.

CQFD

Remarque. Le résultat précédent peut s'étendre facilement à des anneaux A autres que $\mathbb{F}_q[T]$, mais qui en sont des extensions finies. Il suffit pour cela de supposer que $\mathbb{F}_q$ est intégralement clos dans A (pour pouvoir construire F_η, voir **[G]**), que le module de Drinfeld Φ est de rang 2 sur A (pour un choix correct des paramètres $m, N, \ldots$) et que Φ_T et η_T sont à coefficients dans k (pour conserver le lemme de zéros).

VIII). Applications

a). Cas Classique. Dans le cas des nombres complexes, Chudnovski a montré, à partir du théorème donné en introduction, que les nombres $\Gamma(\frac{1}{4})$ et $\Gamma(\frac{1}{3})$ étaient transcendants, où Γ est la fonction d'Euler. Je vais maintenant donner une brève explication de ce résultat.

Soient $\wp$ et ζ comme dans l'introduction, ω_1 et ω_2 une base du réseau des périodes de $\wp$. Posons $\eta_i = 2\zeta(\frac{\omega_i}{2})$. On a alors la relation de Legendre:

$$(\eta_1 \omega_2 - \eta_2 \omega_1)^2 = (2i\pi)^2 .$$

Supposons de plus que $\wp$ est à multiplication complexe, on a alors la relation triviale

$$\omega_2 = \gamma\omega_1$$

où γ est quadratique sur $\mathbb{Q}$, ainsi que la relation suivante de Eisenstein

$$\frac{\eta_2 - \overline{\gamma}\eta_1}{\omega_2} \in \overline{\mathbb{Q}}.$$

En plus des trois relations précédentes, le résultat de Chudnovski nous affirme que les deux nombres

$$\eta_1 - \eta_2\frac{\omega_1}{\omega_2} \quad \text{et} \quad \frac{\eta_2}{\omega_2}$$

sont algébriquement indépendants sur $\mathbb{Q}$. Tout ceci implique facilement que π et ω_2 sont aussi algébriquement indépendants sur $\mathbb{Q}$.

Exemple 1. Soit E la courbe elliptique définie par l'équation:

$$y^2 = 4x^3 - 4.$$

Un calcul simple d'intégrale nous donne

$$\omega_2 = \frac{\Gamma(\frac{1}{3})^3}{2^{\frac{4}{3}}\pi}.$$

On en déduit immédiatement que $\Gamma(\frac{1}{3})$ est transcendant.

Exemple 2. Soit E la courbe elliptique définie par l'équation:

$$y^2 = 4x^3 - 4x.$$

Cette fois-ci, nous avons la formule suivante

$$\omega_2 = \frac{\Gamma(\frac{1}{4})^2}{2\sqrt{2\pi}}.$$

On en déduit également que $\Gamma(\frac{1}{4})$ est transcendant.

b). Cas des Modules de Drinfeld. Comme dans le cas complexe, le théorème montré dans ce papier entraîne la transcendance de valeurs particulières de la fonction $\overline{\Gamma}$ de Goss. (Pour une définition de $\overline{\Gamma}$, voir **[Go]**.)

Pour commencer, nous avons besoin de quelques notations supplémentaires. Soient ω_1 et ω_2 une base du réseau Λ associé au module de Drinfeld Φ. Pour toute dérivation η, on notera $\eta_i = F_\eta(\omega_i)$ avec $i = 1, 2$. Nous noterons π un générateur du réseau des périodes du module de Carlitz.

Si on suppose que le coefficient dominant de Φ_T vaut -1 et que $\eta_T = \tau$ le Frobenius, on a la formule suivante, qui a été montrée par Anderson (voir **[G1]**):

$$(\eta_1\omega_2 - \eta_2\omega_1)^{q-1} = \pi^{q-1}.$$

Il s'agit bien là d'un analogue de la formule de Legendre.

Supposons maintenant que Φ est à multiplication complexe. Il existe donc $\gamma \in \overline{k} - k$ tel que $\gamma\Lambda \subset \Lambda$ et il existe $\Phi_\gamma \in \overline{k}\{\tau\}$ tel que pour tout $z \in \overline{k}_\infty$ on ait

$$e_\Lambda(\gamma z) = \Phi_\gamma(e_\Lambda(z)).$$

On a alors la relation triviale suivante

$$\omega_2 = \alpha\omega_1$$

avec α quadratique sur k. Pour avoir une analogie complète avec le cas complexe, il ne manque plus qu'une relation de type Eisenstein.

Quelques notations encore: soient $D(\Phi)$ l'ensemble des dérivations à coefficients dans $\overline{k}$ et $D_i(\Phi)$ l'ensemble des dérivations de la forme

$$\eta_a = am - m \circ \Phi_a, \quad m \in \overline{k}\{\tau\}.$$

Dans ce cas, on a $F_\eta(z) = m'z - m \circ e_\lambda(z)$ avec m' constant car le polynome m est additif.

Proposition 8. *Il existe* $\beta \in \overline{k}$ *tel que pour tout* $\eta \in D(\Phi)$

$$\frac{\eta_2 - \beta\eta_1}{\omega_1} \in \overline{k}.$$

Démonstration. Soit γ^* l'application de $D(\Phi)$ dans $D(\Phi)$ définie par

$$\gamma^*(\eta) = \eta \circ \Phi_\gamma,$$

l'image de $a \in A$ par $\gamma^*(\eta)$ est $\eta_a \circ \Phi_\gamma$. Cette application est bien définie car $\Phi_\gamma \circ \Phi_a = \Phi_a \circ \Phi_\gamma$ pour tout $a \in A$. De plus, elle est $\overline{k}$ linéaire et $\gamma^*(D_i(\Phi)) \subset D_i(\Phi)$. Donc γ^* passe au quotient $\frac{D(\Phi)}{D_i(\Phi)}$. Or ce quotient est de dimension 1 sur $\overline{k}$ (voir Appendice). Donc γ^* agit comme une homothétie sur le quotient: il existe $\lambda \in \overline{k}$ tel que pour tout $\eta \in D(\Phi)$

$$\gamma^*(\eta) - \lambda\eta = d, \quad d \in D_i(\Phi).$$

D'où $F_{\gamma^*(\eta)} = \lambda F_\eta + F_d$. Or

$$F_{\gamma^*(\eta)}(z) = F_\eta(\gamma z)$$

et si $d_a = am - m \circ \Phi_a$ alors

$$F_d(z) = m'z - m \circ e_\Lambda(z).$$

On a alors immédiatement

$$F_\eta(\gamma\omega_1) = \lambda F_\eta(\omega_1) + m'\omega_1.$$

Or $\gamma\omega_1 = a\omega_1 + b\omega_2$ avec $a, b \in A$ et $b \neq 0$. D'où le résultat.

CQFD

Considérons maintenant le module de rang 2 défini par

$$\Phi_T = TX - X^{q^2}.$$

Ce module est bien à multiplication complexe pour γ tel que $\gamma^{q^2} = \gamma$.

Soit η la dérivation telle que $\eta_T = \tau$, le Frobenius. Le théorème principal nous donne, combiné avec les trois relations ci-dessus, l'indépendance algébrique de π et ω_2.

Les relations de Thakur (voir **[T]**)

$$\overline{\Gamma}(1+\frac{1}{1-q^2})^{q^2-1} = \pi^q \omega_2^{-(q-1)} k_1$$
$$\overline{\Gamma}(1+\frac{q}{1-q^2})^{q^2-1} = (\pi\omega_2)^{q-1} k_2$$

avec $k_i \in \overline{k}, i = 1, 2$ impliquent alors le résultat suivant.

Proposition 9. *Les deux nombres*

$$\overline{\Gamma}(1+\frac{1}{1-q^2}) \quad et \quad \overline{\Gamma}(1+\frac{q}{1-q^2})$$

sont transcendants sur $\mathbb{F}_q(T)$.

Appendice

Afin de clarifier un point méconnu, cette partie sera consacrée à la démonstration du résultat de transcendance énoncé dans l'introduction. Sauf contre-indication explicite, les notations sont les mêmes que celles utilisées jusque là.

Notons C un complété de $\overline{k}_\infty$ et $\Phi : A \longrightarrow \overline{k}\{\tau\}$ un module de Drinfeld de rang $d, d \geq 1$. Soient $D(C, \Phi)$ (resp. $D(\overline{k}, \Phi)$) l'ensemble des dérivations η à coefficients dans C (resp. $\overline{k}$) et $D_i(C, \Phi)$ (resp. $D_i(\overline{k}, \Phi)$) l'ensemble des dérivations dites intérieures, c'est-à-dire de la forme $\eta_a = am - m \circ \Phi_a$ avec $m \in C\{\tau\}$ (resp. $m \in \overline{k}\{\tau\}$).

Lemme 1.

$$D_i(\overline{k}, \Phi) = D(\overline{k}, \Phi) \cap D_i(C, \Phi);$$
$$\dim_{\overline{k}} \frac{D(\overline{k}, \Phi)}{D_i(\overline{k}, \Phi)} = d - 1.$$

Démonstration. L'inclusion $D_i(\overline{k}, \Phi) \subset D(\overline{k}, \Phi) \cap D_i(C, \Phi)$ est triviale.

Réciproquement, supposons que pour tout $a \in A, \eta_a \in \overline{k}\{\tau\}$ et qu'il existe $m \in C\{\tau\}$ tel que pour tout $a \in A, \eta_a = am - m \circ \Phi_a$. Comme $\overline{k}\{\tau\}$ est euclidien à droite, en effectuant des divisions par Φ_T, on peut écrire η_T sous la forme:

$$\eta_T = (Tm_1 - m_1 \circ \Phi_T) + P_T$$

avec $m_1, P_T \in \overline{k}\{\tau\}$ et $\deg_\tau P_T < d$. En comparant les degrés, on déduit immédi atement que $m = m_1$ et $P_T = 0$, d'où l'inclusion réciproque.

De plus, l'écriture de η_T sous la forme ci-dessus nous donne le résultat sur la dimension du quotient. (Il faut toutefois remarquer que P_T n'a pas de terme en τ mais commence directement par τ^2 car tel est le cas pour les autres termes de l'égalité.)

CQFD

Lemme 2. $\eta \in D_i(C, \Phi)$ *si et seulement si* F_η *restreinte à* Λ *est une homothétie.*

Démonstration. Il s'agit d'une application directe de l'isomorphisme de de Rham pour les modules de Drinfeld. (Voir **[G]**.)

CQFD

Proposition. *Soient* $\lambda \in \Lambda - \{0\}$ *et* $\eta \in D(\overline{k}, \Phi)$, *alors* $\frac{F_\eta(\lambda)}{\lambda}$ *est algébrique sur* k *si et seulement si* $\eta \in D_i(\overline{k}, \Phi)$.

Démonstration. Si $\eta \in D_i(\overline{k}, \Phi)$ alors

$$F_\eta(z) = m'z - m \circ e_\Lambda(z)$$

où $\eta_a = am - m \circ \Phi_a$ et m' est le polynôme dérivé de m (m' est constant). On a alors immédiatement que $\frac{F_\eta(\lambda)}{\lambda} = m'$ et donc appartient à $\overline{k}$.

Réciproquement, supposons que $\frac{F_\eta(\lambda)}{\lambda}$ est algébrique sur k. Alors la fonction $F(z) = F_\eta(z) - \frac{F_\eta(\lambda)}{\lambda} z$ est une E_q fonction (trivial). Comme F et e_Λ s'annulent simultanément en une infinité de valeurs, à savoir les $a\lambda, a \in A$, le critère de J.Yu nous dit alors que ces deux fonctions sont algébriquement dépendantes (voir **[J]**). Etant de plus additives, il existe $P, Q \in C[X]$ additifs, non tous les deux nuls, tels que

$$P(F(z)) + Q(e_\Lambda(z)) = 0.$$

Donc pour tout $\omega \in \Lambda, F(\omega)$ est une racine de P.

S'il existe ω tel que $F(\omega)$ soit non nul, alors pour tout $a \in A, aF(\omega) = F(a\omega)$ est une racine de P. On a alors une infinité de racines pour le polynôme P. Donc $P = 0$, ce qui entraîne que $Q(e_\Lambda(z)) = 0$, d'où $Q = 0$; absurde.

On en déduit donc que F restreinte à Λ est nulle, c'est-à-dire que F_η restreinte à Λ est une homothétie. Par le Lemme 2, on a $\eta \in D_i(C, \Phi)$. En appliquant le Lemme 1, on obtient immédiatement le résultat annoncé.

CQFD

Remarque. Pour déduire de la proposition précédente le résultat de transcendance donné dans l'introduction, il suffit de vérifier que $\eta \notin D_i(C, \Phi)$ ou, de façon équivalente, que F_η restreinte à Λ n'est pas une homothétie. Cela est trivialement vrai dès que la condition $F_\eta(\omega) - \frac{F_\eta(\lambda)}{\lambda}\omega \neq 0$ est satisfaite.

REFERENCES

[C] G.V. Chudnovsky: *Contributions to the theory of transcendental numbers*, Math. Surveys & Monographs, Amer. Math. Soc. **19**, (1984), 316–320.

[G] E-.U. Gekeler: On the de Rham isomorphism for Drinfeld modules, *J.Reine Angew Math.* **401**, (1989), 188–208.

[G1] E-.U. Gekeler: Quasi periodic functions and Drinfeld modular forms, *Comp. Math.*, **69**, (1989), 277–293.

[J] J. Yu: Transcendence and Drinfeld modules, *Invent. Math.* **83**, (1986), 507–517.

[B] S. Bosch, U. Günzter, R. Remmert: *Non archimedean analysis*, Springer 1984.

[T] D.Thakur: Gamma functions for function fields and Drinfeld modules, *Ann. of Math.* **134**, (1991), 25–64.

[Go] D.Goss: The Gamma function in the arithmetic of function fields, *Duke Math J.* **56**, (1988), 163–191.

U.F.R. de Sciences, Université de Caen, 14032 Caen Cedex, France

Géométrie Diophantienne sur les Modules de Drinfeld

L. Denis

Abstract. The most important theorems recently proven in diophantine geometry on abelian varieties are the proofs of the Lang-Manin-Mumford conjectures, which include that of Mordell (**[6]**). We formulate analogous conjectures on products of Drinfeld modules. The results known before Faltings are then proven: among other things, the analogue of the Manin-Mumford conjectures, under standard Galois assumptions which are known to hold on Carlitz modules.

§1. Position du problème

On désigne par $\mathbb{F}_q[T]$ l'anneau des polynômes en une variable à coefficients dans le corps fini $\mathbb{F}_q$, par $k = \mathbb{F}_q(T)$ son corps des fractions, par $k_\infty = \mathbb{F}_q((1/T))$ le complété de k pour la valuation $(1/T)$-adique v, que l'on prolonge à une clôture algébrique $\overline{k}$ (resp. $\overline{k}_\infty$) de k (resp. k_∞). On notera $|\alpha| = q^{-v(\alpha)}$, la valeur absolue d'un élément de $\overline{k}_\infty$.

On désigne encore par t une indéterminée par θ un élément de $\overline{k}$ transcendant sur $\mathbb{F}_q$. Par t-module de rang d, de dimension m, défini sur $\overline{k}$, on entend la donnée du groupe additif $(\mathbb{G}_a)^m$ et d'un homomorphisme injectif d'anneau Φ de $\mathbb{F}_q[t] = A$ dans l'anneau des endomorphismes du groupe algébrique $(\mathbb{G}_a)^m$ vérifiant:

$$\Phi(t) = a_0 F^0 + \cdots + a_d F^d,$$

où les a_i $(0 \leq i \leq d)$ sont des matrices $m \times m$ à coefficients dans $\overline{k}$, $a_d \neq 0$ et a_0 admet θ pour seule valeur propre. Un sous-t-module est un sous-groupe algébrique de $(\mathbb{G}_a)^m$, stable sous $\Phi(t)$ et connexe.

On désigne par E un t-module de rang $d \geq 1$, de dimension m et d'homomorphisme Φ, défini sur une extension finie K de k. On notera $E = ((\mathbb{G}_a)^m, \Phi)$ et E_{tor} l'ensemble de ses points de torsion. Grâce à l'hypothèse sur le coefficient a_0, on remarque que E_{tor} est défini sur la clôture séparable K^s de K.

On va s'intéresser à des analogues des conjectures de Lang-Manin-Mumford (maintenant connues grâce aux travaux de Faltings **[6]**) concernant les points algébriques sur les sous-variétés de variétés abéliennes définies sur $\overline{\mathbb{Q}}$:

Théorème 1 (Faltings). *Soit V une sous-variété d'une variété abélienne A, Γ un sous-groupe de type fini de $A(\overline{\mathbb{Q}})$. Alors V contient un nombre fini de translatés de sous-variétés abéliennes $\gamma_i + B_i$ $(1 \leq i \leq r)$ tels que:*

$$V(\overline{\mathbb{Q}}) \cap \Gamma = \bigcup_{1 \leq i \leq r} (\gamma_i + B_i \cap \Gamma).$$

Cet énoncé contient en particulier la conjecture de Mordell. Avant qu'il ne soit démontré, M. Hindry (**[8]** et **[9]**) avait prouvé qu'il était équivalent à l'énoncé suivant (conjecture de Lang-Manin-Mumford):

Théorème 2 (Faltings, Hindry). *Soit V une sous-variété d'une variété abélienne A, $\Gamma =$ un sous-groupe de type fini de $A(\overline{\mathbb{Q}})$ et $\overline{\Gamma} = \{x \in A(\overline{\mathbb{Q}}) \mid \exists m > 0$ vérifiant $mx \in \Gamma\}$, alors V contient un nombre fini de translatés de sous variétés abéliennes $\gamma_i + B_i$ tels que:*

$$V(\overline{\mathbb{Q}}) \cap \overline{\Gamma} = \bigcup_{1 \leq i \leq r} (\gamma_i + B_i \cap \overline{\Gamma}).$$

Le premier cas connu de ce théorème était celui des points de torsion (Raynaud **[15]** puis Hindry **[9]**):

Théorème 3 (Raynaud, Hindry). *(cas $\Gamma = 0$ du théorème 2):*

$$V(\overline{\mathbb{Q}}) \cap A_{\text{tor}} = \bigcup_{1 \leq i \leq r} (\gamma_i + (B_i)_{\text{tor}}).$$

Ces conjectures sont fausses en caractéristique finie à cause de l'action du Frobenius et sont également fausses si la variété abélienne est remplacée par $(\mathbb{G}_a)^m$. On se propose de discuter d'analogues sur les t-modules et plus particulièrement sur le produit de modules de Drinfeld.

Proposons comme analogue au théorème 1:

Conjecture 1. *Soit Γ un $\Phi(A)$-module de type fini de $E(\overline{k})$ et V une sous-variété algébrique de E. Il existe un nombre fini de translatés de sous-t_i-modules $\gamma_i + B_i$ ($1 \leq i \leq r$) tels que:*

$$V \cap \Gamma = \bigcup_{1 \leq i \leq r} (\gamma_i + B_i \cap \Gamma).$$

Cette conjecture serait un corollaire immédiat de l'analogue du théorème 2:

Conjecture 2. *Soit $\overline{\Gamma} = \{x \in E(\overline{k}) / \exists m \neq 0 \in A$ vérifiant $\Phi(m)x \in \Gamma\}$ et V une sous-variété algébrique de E. Il existe un nombre fini de translatés de sous t_i-modules $\gamma_i + B_i$ ($1 \leq i \leq r$) tels que:*

$$V \cap \overline{\Gamma} = \bigcup_{1 \leq i \leq r} (\gamma_i + B_i \cap \overline{\Gamma}).$$

(On utilisera l'abréviation $\overline{\Gamma} = \Phi^*(A)\Gamma$.) On a également une version sur les points de torsion:

Conjecture 3. *(cas $\Gamma = 0$ de la conjecture 2):*

$$V \cap E_{\text{tor}} = \bigcup_{i \leq i \leq r} (\gamma_i + (B_i)_{\text{tor}}).$$

Dans toute la suite on désigne par (D, ψ) un module de Drinfeld (t-module de dimension 1) de rang $d \geq 1$ défini sur K. On se placera alors sur le t-module $E = ((\mathbb{G}_a)^m, \Phi)$ produit direct de m copies de D muni de l'action diagonale. Le t-module E est alors de rang d et est défini sur K; quitte à faire une extension finie on supposera également que les ensembles Γ intervenant dans la suite sont également définis sur K.

Suivant une idée de S. Lang (cf. **[10]**) développée par M. Hindry (cf. **[8]** et **[9]**), on va montrer au §2 que sous certaines hypothèses de théorie de Galois-Kummer (dont les analogues sont connus sur les variétés abéliennes), la conjecture 3 est vraie, puis on montrera que la conjecture 1 entraîne la conjecture 2. La démonstration de M. Hindry s'adapte et fournit les résultats précédents mais avec des B_i *sous-groupes algébriques* de E. On montrera ici que c'est bien avec des *sous-t-modules* de E qu'on peut obtenir de telles décompositions. Il serait d'ailleurs intéressant de voir si la démonstration de Raynaud **[15]** s'adapte elle aussi aux t-modules.

Dans les paragraphes qui suivent on discute des hypothèses effectuées en montrant notamment qu'elles sont vérifiées pour le module de Carlitz: au §3 on explicite la théorie de Kummer sur un t-module pour montrer comment on peut venir à bout de l'hypothèse kummerienne modulo certains axiomes de Ribet-Bashmakov; le dernier paragraphe établit des résultats d'indépendance linéaire permettant des prouver un des axiomes précédents. Enfin on a regroupé en appendice deux variations sur les thèmes abordés ici.

L'auteur remercie le referee pour ses utiles suggestions sur la présentation de ce texte.

§2. Points algébriques sur les variétés

a) Points de torsion. Pour tout $a \in A$, on note $E[a] = \{x \in (\mathbb{G}_a)^m \mid \Phi(a)x = 0\}$ l'ensemble des points de $\Phi(a)$ torsion et $G = \mathrm{Gal}(K^s/K)$. On va faire l'hypothèse galoisienne suivante:

GAL 1. *Il existe un entier naturel $c > 0$ tel que pour tout couple (a, b) de polynômes de A premiers entre eux, il existe $\sigma \in G$, tel que $\forall x \in E[a]$: $\Phi(b^c)x = \sigma x$.*

Remarque. On vérifie cette hypothèse quand D est le module de Carlitz (donné par $\psi(t) = TF^0 + F$) grâce aux travaux de Hayes (cf. **[7]**, théorè me 2.3), avec $c = 1$.

L'analogie avec le cas des variétés abéliennes montre qu'en général on ne peut pas prendre $c = 1$ dans l'énoncé précédent (cf. **[9]** lemme 12).

Sous cette condition on démontre le théorème suivant:

Théorème 1. *Sous GAL 1, il existe un nombre fini de points γ_i de E et de sous t-modules B_i de E tels que:*

$$V \cap E_{\text{tor}} = \bigcup_{1 \leq i \leq r} (\gamma_i + (B_i)_{\text{tor}}).$$

On commence par quelques lemmes qui sont les analogues de ceux de M. Hindry **[9]**.

Lemme 1. *Soit E une puissance du module de Drinfeld D défini sur une extension finie K' de k. Alors:*

a) *Il existe une extension finie K de K' telle que tous les sous-t-modules de E soient définis sur K.*
b) *Pour tout $C > 0$, il n'y a qu'un nombre fini de sous-t-modules B tels que* $\deg B \leq C$.
c) *Il existe un réel $C_E > 0$ tel que si B est un sous-t-module de E alors il existe B' sous-t-module de E tel que:*

$$B + B' = E \text{ et } \operatorname{Card}(B \cap B') \leq C_E .$$

Remarque. Si $\operatorname{End}(D) = \mathcal{O}$ est principal, on peut prendre ici $C_E = 1$.

Preuve du lemme 1. a) et b) La preuve se fait par récurrence sur la dimension du sous-t-module. Les propositions a) et b) sont vraies en dimension 0 (rappelons qu'un sous-t-module est connexe par définition). Soit H un sous-t-module de dimension $i > 0$. Il existe un hyperplan de la forme $X_j = 0$ tel que la composante neutre de $H_j = H \cap \{X_j = 0\}$ soit un sous-t-module de dimension $i - 1$.

Soit Σ le morphisme de t-module de D^m dans $D^{m-1} \oplus D$ défini par

$$\Sigma(x_1, \ldots, x_m) = (x_1, \ldots 0, \ldots, x_m) + (0, \ldots, x_j, \ldots, 0).$$

La dernière étape de la récurrence vient d'une part du fait que la restriction de Σ à H est une isogénie entre H et la somme directe $H_j \oplus D$, d'autre part que $\deg H_j \leq \deg H$.

c) La démonstration de b) montre que B est isogène à D^r où $r = \dim B$. Il existe donc un morphisme Ψ de t-module de $(D)^m$ vers E d'image B et Ψ est représenté par un élément de $M_{m\times r}(\mathcal{O})$. Si $\mathcal{O}$ est principal, on peut supposer que les colonnes de Ψ engendrent un sous-$\mathcal{O}$-module primitif de $\mathcal{O}^m$. La théorie des diviseurs élémentaires permet alors de trouver Ψ' dans $M_{m\times(m-r)}(\mathcal{O})$ tel que la matrice complétée (Ψ, Ψ') soit dans $Gl_m(\mathcal{O})$. L'image de Ψ' définit alors B'. Si $\mathcal{O}$ n'est pas principal on ne peut en général plus supposer que les colonnes de Ψ engendrent un réseau primitif, mais le même raisonnement fournit $(\Psi, \Psi') \in M_{m\times m}(\mathcal{O})$ de déterminant non nul et de degré borné en fonction de $\mathcal{O}$.

Sans perte de généralité, on peut supposer que $(\mathbb{G}_a)^m$ est plongé naturellement dans $\mathbb{P}^m$ et les notions de degré utilisées seront relatives à ce plongement. Si V est une variété de E, on note $\deg V$ le degré de sa clôture de Zariski dans $\mathbb{P}^m$, $G_V = \{x \in E \mid x + V = V\}$ son stabilisateur et $(G_V)^0$ la composante neutre de ce dernier.

On dit qu'une variété Y est incomplétement définie dans E par des équations $(P_i)_{1\leq i}$ si Y est union de composantes irréductibles de l'intersection de $(\mathbb{G}_a)^m$ plongé dans $\mathbb{P}^m$ avec les hypersurfaces de $\mathbb{P}^m$ définies par les $(P_i)_{1\leq i}$.

Lemme 2. *Soit C une variété irréductible incomplétement définie dans E par des équations de degré au plus X. Alors le stabilisateur G_C de C est incomplétement défini par des équations de degré au plus X et on a:*

$$\deg(G_C) = (G_C\colon (G_C)^0) \deg(G_C)^0 \leq \deg C \times X^{\dim C - \dim G_C} .$$

Preuve. Pour la commodité du lecteur, on reproduit avec nos notations la preuve de **[9]** (lemme 8):

La variété C est composante de

$$Y = (\mathbb{G}_a)^m \bigcap_{1 \leq i} (Z(P_i)),$$

où $Z(P_i)$ désigne la variété des zéros d'un polynôme P_i de degré $\leq X$. Considérons $H = \bigcap_{y \in C} (Y - y)$.

Comme les translations sont décrites par des formes de degré 1, H est défini par des équations de degré au plus X. Pour conclure la preuve du lemme, il suffit de montrer que G_C est union de composantes de H. Etant donné que G_C est inclus dans H et que H est homogène sous G_C ce dernier point revient à montrer que $(G_C)^0$ est une composante (irréductible) de H. Soit W une composante (irréductible) de H contenant $(G_C)^0$; comme $W \subset H$, $W + y \subset Y$ pour tout $y \in C$. Donc $W + C \subset Y$, et W passant par l'élément neutre de G_C on a aussi $C \subset W + C \subset Y$. Ceci entraîne $C = W + C$ car C est une composante irréductible de Y d'où $W \subset (G_C)^0$ et donc $W = (G_C)^0$.

Lemme 3. *Soit* V *une variété irréductible de* E *de dimension* $\dim V$ *et* a *un polynôme de* A *de degré* $\deg a > 0$. *Alors:*

$$\det \Phi(a)^{-1}(V) = q^{d \operatorname{codim} V \cdot \deg a} \deg(V) \qquad \text{a)}$$

$$\deg \Phi(a)(V) = \frac{q^{d \dim V \cdot \deg a}}{|\mathrm{Ker}\Phi(a) \cap G_V|} \deg(V). \qquad \text{b)}$$

Preuve. Nous allons suivre la démonstration de **[9]** (lemme 6).

Comme dans **[4]** (lemme 1), $\Phi(a)$ a été prolongé en un morphisme de $\mathbb{P}^m$ (noté de la même manière) en posant:

$$\Phi(a)(X_0, \ldots, X_m) = ((X_0)^{q^{d \deg a}}, h\Phi(a)(X_1, \ldots, X_m))$$

avec

$$h_{\Phi(a)}(X_1, \ldots, X_m) = (X_0)^{q^{d \deg a}} \Phi(a)(X_1/X_0, \ldots, X_m/X_0).$$

On a alors le résultat d'équivalence linéaire $\Phi(a)^*(\mathcal{O}_{\mathbb{P}^m}(1)) \sim q^{d \deg a}(\mathcal{O}_{\mathbb{P}^m}(1))$. Comme $\Phi(a)$ est fini et étale $\Phi(a)^*(V)$ est le cycle associé au fermé $\Phi(a)^{-1}(V)$. Désignons par Z la classe d'équivalence rationnelle d'un hyperplan de $\mathbb{P}^m$, on effectue alors le calcul suivant:

$$\begin{aligned}
q^{dm \deg a} \deg(V) &= q^{dm \deg a} \deg(V.Z^{\dim V}) = \deg \Phi(a)^*(V \cdot Z^{\dim V}) \\
&= \deg(\Phi(a)^*(V) \cdot (\Phi(a)^*(Z))^{\dim V}) \\
&= \deg(\Phi(a)^{-1}(V) \cdot (q^{d \deg a} Z)^{\dim V}) \\
&= q^{d \deg a \cdot \dim V} \deg(\Phi(a)^{-1}(V) \cdot (Z)^{\dim V}) \\
&= q^{d \deg a \cdot \dim V} \deg(\Phi(a)^{-1}(V)),
\end{aligned}$$

ce qui démontre a).

On remarque ensuite que $\Phi(a)^{-1}(\Phi(a)(V)) = V + \ker\Phi(a)$. L'invariance du degré par translation et la propriété a) appliquée à $\Phi(a)(V)$ permet alors de prouver b):

$$\deg(\Phi(a)^{-1}(\Phi(a)(V)) = q^{d\,\mathrm{codim}\,V\cdot\deg a}\deg(\Phi(a)(V))$$
$$= \deg(V + \ker\Phi(a)) = \frac{|\mathrm{Ker}\Phi(a)|}{|\mathrm{Ker}\Phi(a)\cap G_V|}\deg(V).$$

Lemme 4. *Si $\Phi(a)(V) = V$ avec V irréductible et $\deg a > 0$ alors V est un translaté d'un t-module.*

Preuve. $\Phi(a)(V) = V$ implique que $(G_V)^0$ est aussi stable sous $\Phi(a)$; c'est donc un a-module. Nous montrerons que c'est un t-module. De là on déduit que si la composante neutre $(G_V)^0$ de G_V est de dimension u, $|\mathrm{Ker}\Phi(a^r)\cap(G_V)^0| = q^{rdu\cdot\deg a}$, et comme le lemme précédent nous dit alors que $|\mathrm{Ker}\Phi(a^r)\cap G_V| = q^{rd\dim V\cdot\deg a}$, on obtient bien $u = \dim V$. Montrons maintenant que $(G_V)^0$ est un t-module. Considérons le module de Drinfeld $D' = (\mathbb{G}_a, \Psi')$ où on a défini $\psi'(t) = \psi(a)$, et désignons par $\mathcal{O}_a$ l'anneau des endomorphismes de D'. Le lemma 1 c), dit que $(G_V)^0$ est l'image de $(\mathbb{G}_a)^m$ par un morphisme de a-module Φ^a représenté par un élément de $M_{m\times m}(\mathcal{O}_a)$. On a $\Phi^a((\mathbb{G}_a)^m) = (G_V)^0$, pour voir que $(G_V)^0$ est un t-module, il suffit donc de montrer que $\psi(t)$ commute aux éléments de $\mathcal{O}_a$. Par définition $\mathcal{O}_a$ est l'ensemble des éléments de $\mathrm{End}^{\mathrm{alg}}(\mathbb{G}_a)$ qui commutent à $\psi(a)$, donc $\psi(t)$ est dans $\mathcal{O}_a$. Enfin la commutativité de $\mathcal{O}_a$ (cf. **[5]**) donne le résultat cherché.

Lemme 5. *Soit G un sous-groupe algébrique de $(\mathbb{G}_a)^m$. Pour tout entier $i \geq 1$, on a:*

$$|\mathrm{Ker}\Phi(a^i)\cap G| \leq q^{di\deg a\cdot\dim G}(\deg G).$$

Preuve. On applique le lemma 3 b) à a^i et à chaque composante de G, en remarquant qu'elles ont toutes le même stabilisateur:

$$\deg\Phi(a^i)G) = \frac{q^{di\dim G\cdot\deg a}}{|\mathrm{Ker}\Phi(a^i)\cap G|}\deg G.$$

La conclusion du lemme traduit alors le fait que $\deg\Phi(a^i)(G) \geq 1$.

On suppose maintenant que V est une variété incomplétement définie par des équations de degré $\leq X$ et est définie sur K (ce qui sera le cas dans notre application, quitte à faire une extension finie du corps de définition de E).

Preuve du Théorème 1. On montre qu'il existe un réel $C_1 > 0$ tel que si B est un sous-t-module de dimension maximale tel qu'il existe x un point de torsion vérifiant $x + B \subset V$, alors $\deg B \leq C_1$.

Soit donc B et un x correspondant. On choisit $a \in A - \mathbb{F}_q$, premier à l'ordre b de x. D'après Gal 1 il existe un élément σ de G tel que $\Phi(a^c)x$ soit conjugué de x par σ. Introduisons les sous-ensembles suivants:

$$V_r = V\cap\Phi(a^c)^*V\cap\ldots\cap\Phi(a^{cr})^*V \quad\text{et}\quad V_\infty = \bigcap_{r=0}^{\infty} V_r.$$

Notons $\dim_x V_r$ la dimension maximale d'une composante irréductible de V_r passant par x. Comme la suite $(V_r)_{1\le r\le\infty}$ est décroissante, la suite $(\dim_x V_r)_{1\le r\le\infty}$ l'est également. Considérons une suite d'entiers consécutifs $s \le r \le s+k$ tels que:

$$\dim_x V_s = \dim_x V_{s+1} = \cdots = \dim_x V_{s+k} = m'.$$

Soit C une composante de dimension m', contenant x, de V_{s+k}. On a, pour $1 \le i \le k$, $\Phi(a^{ci})(C) \subset \Phi(a^{ci})V_{s+i} \subset V_s$. Or $\Phi(a^{ci})(C)$ contient $x_i = \Phi(a^{ci})x = \sigma^i(x)$ donc $\Phi(a^{ci})(C)$ est encore une composante de dimension maximale au point x_i.

Montrons qu'ou bien C est le translaté d'un t-module ou bien que k est borné par une fonction ne dépendant que de s et de X.

Le lemme 3 fournit l'estimation de degrés suivante:

$$\deg \Phi(a^{ck})(C) = \frac{q^{dk \dim C\cdot \deg a^c}}{|\mathrm{Ker}\Phi(a^{ck}) \cap G_C|} \deg(C) \le \deg V_s .$$

Or on a d'après le lemme 5:

$$|\mathrm{Ker}\Phi(a^{ck}) \cap G_C| \le q^{dk \deg a^c \cdot \dim G_C}(\deg G_C),$$

puis grâce au lemme 2:

$$q^{dk \deg a^c \cdot \dim G_C}(\deg G_C) \le q^{dk \deg a^c \cdot \dim G_C} \cdot \deg C((q^{ds \deg a^c} X)^{\dim C - \dim G_C}).$$

On en déduit l'estimation:

$$q^{dk \deg a^c (\dim C - \dim G_C)} \le \deg V_s((q^{ds \deg a^c} X)^{\dim C - \dim G_C}).$$

Comme V_s est incomplétement définie par des équations de degré $\le q^{s \deg a^c} X$, on aboutit à:

$$q^{dk \deg a^c (\dim C - \dim G_C)} \le (q^{ds \deg a^c} X)^{\mathrm{codim}\, V_s}((q^{ds \deg a^c} X)^{\dim C - \dim G_C}).$$

Si $\dim C = \dim G_C$, $C = x + G_C$ et comme tous les $\Phi(a^{ci})(C)$ sont dans V_s ($1 \le i \le k$), il existe $i > j$ tels que $\Phi(a^{ci})(C) = \Phi(a^{cj})(C)$. Donc $\Phi(a^{cj})(C)$ est une variété irréductible stable par $\Phi(a^{c(i-j)})$ et d'après le lemme 1, G_C est un t-module.

Si $\dim C > \dim G_C$, et en supposant:

$$q^{d \deg a^c} > X^m, \tag{h}$$

on déduit de l'inégalité précédente $k \le (\dim C - \dim G_C)^{-1} sm$, majoration qui est bien indépendante de $\deg a$.

Remarquons que si $x + B \subset V_r$ on a $\Phi(a^c)(x+B) = \Phi(a^c)x + B = \sigma x + B = \sigma(x+B) \subset V_r$ (car V et B sont définies sur K). Donc $x + B \subset V_{r+1}$ et par récurrence dans V_∞. Comme $\dim_x V_r$ ne peut prendre que $\dim V + 1$ valeurs, il existe r_0 (indépendant de $\deg a$) tel que pour $r \ge r_0$, $\dim_x V_r = \dim_x V_\infty$.

En particulier on a démontré que l'on a un nombre fini (voir lemme 1 a)) de t-modules B de dimension maximale vérifiant $x + B \subset V$ et on a borné leur degré en fonction de E et de X. Reste à borner l'ordre de ces x modulo ces B.

La variété $Y = \bigcup_{\sigma \in \mathrm{Gal}(K^S/K)} \sigma x + B$ est union de composantes de V_∞. Le lemme de Bézout permet alors d'affirmer que:

$$\deg Y = (\text{Nombre de } \sigma x + B \text{ distincts } \bmod B) \deg B$$
$$\leq (q^{r_0 d \deg a^c} X)^{\operatorname{codim} V_\infty}.$$

Si y est de torsion d'ordre e, on déduit de GAL1 que l'orbite de y sous $\mathrm{Gal}(K^S/K)$ a un cardinal supérieur à $q^{\alpha \deg e}$ (où $\alpha > 0$ ne dépend que de E). On choisit alors B' comme au lemme 1 c, puis x' dans B' congru à x module B. Notons b' l'ordre de x'. Comme B' est de la forme $D^{m'}$ on peut lui appliquer Gal 1 (on a vu qu'il y a un nombre fini de B donc un nombre fini de B'). Pour tout a' premier à b' on obtient alors à l'aide de l'estimation précédente:

$$(q^{r_0 d \deg(a')^c} X)^{\operatorname{codim} V_\infty} \geq (C_E)^{-1} q^{\alpha \deg b'}.$$

Ceci implique que $\deg b'$ est borné, grâce au lemme facile suivant:

Lemme 6. *Pour tout réel $\beta > 0$, dès que $\deg b$ est assez grand, il existe un polynôme a de A irréductible et premier à b vérifiant $\deg a \leq \beta \deg b$.*

On peut alors choisir pour a' un élément irréductible de A premier à tout b'' de degré $\leq \deg b'$ et vérifiant l'hypothèse (h).

On a donc bien majoré le nombre de points et de sous-t-modules intervenant dans le théorème 1 par une fonction effectivement majorable en fonction de E et X.

b) Points de division. Montrons maintenant que sous une hypothèse de théorie de Kummer on a l'implication conjecture 1 $\Rightarrow$ conjecture 2.

Définition. On dira que P est indivisible si pour tout b non constant dans A, $P \notin \Phi(b)E(K)$.

GAL 2. Soient $P \in E(K)$ et a'/a dans A si P est indivisible:

$$\mathrm{Gal}(K(E[a], P/a'): K(E[a])$$

s'identifie à un t-module de $E[a']$ d'indice borné indépendamment de a' (où P/a' désigne un élément Q fixé vérifiant $\Phi(a')Q = P$).

Sous les deux conditions GAL 1 et GAL 2 on peut énoncer:

Théorème 2. *La validité de la conjecture 1 entraîne celle de la conjecture 2.*

Commençons par démontrer un lemme de zéro, analogue à celui de J. Yu **[20]**, mais pour des variétés de dimension quelconque.

Désignons par γ_i ($1 \leq i \leq s$) des éléments de $(\overline{k})^m$, par S un entier naturel $> m$ et posons:

$$\Delta(S) = \{\Phi(f_1)\gamma_1 + \cdots + \Phi(f_s)\gamma_s,\ f_i \in A \ \deg f_i \leq S\},$$

on peut alors démontrer le lemme suivant:

Lemme 7. *Soit V une variété passant par $\Delta(S)$, incomplétement définie dans E par des équations de degré $\leq X$ et ne contenant pas E. Il existe un sous-t-module H de E, différent de E, tel que $H + \Delta(S - m)$ soit inclus dans V et vérifie:*

$$\operatorname{Card}(\Delta(S - m) + H/H) \deg H \leq (q^{dm} X)^{m - \dim H}.$$

Preuve. Considérons $\mathcal{H} = \bigcap_{\gamma \in \Delta(S)} V - \gamma$, cette variété passe par $\Delta(S)$ et reste incomplétement définie par des équations de degré $\leq X$. Son stabilisateur $G_{\mathcal{H}}$ a aussi ces deux propriétés. On considère alors comme dans **[20]** la suite décroissante de sous-groupes de E: $H'_r = G_{\mathcal{H}} \cap \Phi(t)^{-1} G_{\mathcal{H}} \cap \ldots \cap \Phi(t^r)^{-1} G_{\mathcal{H}}$. Le groupe algébrique H'_r passe par $\Delta(S - r)$ et $\Phi(t) H'_{r+1} \subset H'_r$, on choisit alors pour H la composante neutre de H'_r où r est le plus petit indice tel que $\dim H'_{r+1} = \dim H'_r$. D'après le lemme 4, cet H est alors un t-module. Comme H'_r contient $\Delta(S - r) + H$ et est de même dimension, on a la minoration du degré suivante:

$$\operatorname{Card}(\Delta(S - m) + H/H) \deg H \leq \deg H'_r.$$

Enfin, comme H'_r est incomplétement défini par des équations de degré $\leq q^{dm} X$, une estimation de degré semblable à la proposition 3.3 de **[14]**, permet d'obtenir $\deg H'_r \leq (q^{dm} X)^{m - \dim H}$ (estimation qui apparaît aussi dans les lemmes de zéros de **[3]** et de **[20]**).

Preuve du théorème 2. On suppose donc la conjecture 1 vérifiée; montrons d'abord que pour tout Γ de type fini, $V \cap (\Gamma + E_{\text{tor}})$ a une décomposition analogue à celle de $V \cap \Gamma$. En effet si $x \in V \cap (\Gamma + E_{\text{tor}})$, il s'écrit $x = \gamma + z$ avec $\gamma \in \Gamma$ et $z \in E_{\text{tor}}$. Mais alors $V - \gamma$ est encore défini par des équations de degré $\leq X$ et passe par z. La démonstration du théorème 1 appliquée à $V - \gamma$ donne un sous-t-module B_x de dimension maximale, de degré borné et une borne N pour l'ordre de z modulo B_x avec $x + B_x \subset V$, ces bornes ne dépendant que de E et de X.

Pour chaque x, on peut donc choisir un z' dans $E_{[N]}$ tel que $z' + B_x = z + B_x$. On a alors $\gamma + z' + B_x \subset V$. D'après la conjecture 1 appliquée au nombre fini de variétés de la forme $V - z'$, l'élément x appartient donc à un $\gamma_i + z' + B$ et on a bien une décomposition:

$$V \cap (\Gamma + E_{\text{tor}}) = \bigcup_{1 \leq i \leq r} (\gamma_i + B'_i \cap (\Gamma + E_{\text{tor}})). \tag{d}$$

Déduisons en la conjecture 2. Pour tout x appartenant à $V \cap \overline{\Gamma}$, il existe $b \neq 0$ dans A tel que $\Phi(b)x \in \Gamma$. Soit donc a' un élément de A tel que $\Phi(a')x \in \Gamma + E_{\text{tor}}$. On peut écrire $\Phi(a')x = x_0 + t_0$ où $x_0 \in \Gamma$ et $t_0 \in E[a]$, sans perte de généralité on peut supposer que a' divise a et que a' est de degré minimal. Alors il existe $\nu \in \operatorname{End}(E)$, $\nu \neq 0$, ne dépendant que de Γ et de $\operatorname{End}(E)$, tel que $x_0 = \nu(y_0)$, où y_0 est indivisible au sens ci-dessus. Sans perte de généralité, on peut enfin supposer que $\nu = 1$ et que a' est de degré assez grand.

Désignons par $G(a', a)$ l'image de $\operatorname{Gal}(K(E[a], x_0/a'){:}\, K(E[a]))$ dans $E[a']$. Soit $Z = \bigcap_{y \in G(a', a)} V + y$, cette variété est définie sur $K(E[a])$, donc $x + G(a', a) \subset Z \subset V$.

Les méthodes du §3 montrent que $G(a', a)$ est de la forme $E'[b]$, où b est dans l'anneau des multiplications complexes du module de Drinfeld (théorème 5) et E' est un sous-t-module de E. On peut donc appliquer le lemme 5 à $V - x$. Il existe un sous-t-module connexe B_x de E de degré majoré par M et une borne M' pour $((G(a', a)+B_x)/B_x)$.

L'hypothèse GAL 2 appliquée aux t-modules E/B_x (qui sont en nombre fini) montre alors que l'exposant de x dans $\overline{\Gamma}/\Gamma$ modulo B_x est borné par un certain entier N. Prenons pour Γ' le $\Phi(A)$-module de type fini obtenu à partir de Γ en ajoutant ses points de division et ses points de torsion d'ordre a' satisfaisant $\deg a' \leq N$, Γ' est toujours de type fini et est défini sur une extension finie de K. Par construction de Γ', il existe $x' \in \Gamma' + E_{\text{tor}}$ tel que $x = x'$ modulo B_x et $x' + B_x = x + B_x \subset V$.

Enfin, la décomposition (d) montre qu'il n'y a qu'un nombre fini de classes de la forme $y + B$, avec $y \in \Gamma' + E_{\text{tor}}$, $y \in \Gamma' + E_{\text{tor}}$, $y + B \subset V$, on en déduit donc le théorème 2.

Remarque 1. Il est possible de démontrer les théorèmes précédents pour des t-modules plus généraux. Par exemple pour un produit de t-modules simples vérifiant la condition "a_d est inversible".

Remarque 2. Dans le cas du module de Carlitz les groupes de Galois des points de torsion ont été décrits explicitement par Hayes dans **[7]** (théorème 2.3). De même que GAL 1, l'axiome GAL 2 est alors un théorème pour les produits directs du module de Carlitz, comme il résulte de **[7]** joint au théorème 3 du paragraphe suivant.

§3. Théorie de Kummer

On regarde la théorie de Kummer sur $D = (\mathbb{G}_a, \psi)$ un module de Drinfeld (on conserve les notations du §2), défini sur une extension finie k' de k. On désigne par $\mathcal{O}$ l'anneau des multiplications complexes sur $\overline{k}$ du module de Drinfeld, on sait que $\mathcal{O}$ est un A module libre de rang r divisant d (cf. **[5]** et **[21]** Theorem 3.1).

Dans ce paragraphe ℓ est un élément premier non constant de A et a un multiple de ℓ dans A. Désignons par K une extension finie de k' contenant $\mathcal{O}$ et notons abusivement $D(K)$ le groupe des points de $\mathbb{G}_a(K)$.

Posons $G = \mathrm{Gal}(K^S/K)$, $K_a = K(D[a])$, $G_a = \mathrm{Gal}(K_a/K)$ et

$$H_a = \mathrm{Gal}(K^S/K_a).$$

Soient $P_1, \ldots, P_m$ dans K, linéairement indépendant sur $\mathcal{O}$, on pose encore $X = \mathrm{Gal}(K_a(P_1/\ell, \ldots, P_m/\ell)/K)$ et on s'intéresse au groupe de Galois

$$X_\ell = \mathrm{Gal}(K_a(P_1/\ell, \ldots, P_m/\ell)K_a).$$

Comme dans **[16]** on va suivre la méthode de Bashmakov.

Théorème 3. *Supposons les quatre hypothèses suivantes vérifiées:*

B_1) *Pour presque tout* ℓ: $\mathcal{O}/\ell\mathcal{O}$ *est le commutant de* G_ℓ *dans* $\mathrm{End}(D[\ell])$.

B_2) *Pour presque tout* ℓ, $D[\ell]$ *est un* G_ℓ *module semi-simple.*

B_3) *Pour presque tout* ℓ: $H^1(G_a, D[\ell]) = 0$.

B_4) *Pour* $\deg \ell$ *assez grand* $P_1, \dots, P_m$ *sont linéairement indépendants modulo* $\psi(\ell)D(K)$.

Alors pour $\deg \ell \geq \deg \ell_0$*:*

$$X_\ell = \mathrm{Gal}(K_a(P_1/\ell, \dots, P_m/\ell)/K_a) \approx (D[\ell])^m.$$

Preuve. Remarquons une fois pour toutes que comme les polynômes sont séparables, toutes les extensions considérées sont séparables. On a les extensions de corps suivantes:

$$K \subset K_a \subset K_a(P_1/\ell, \dots, P_m/\ell) \subset K^s.$$

Soit $P \in D(K)$ on désigne par P/ℓ un élément Q de $D(K^s)$ vérifiant $\psi(\ell)Q = P$, on définit:

$$\xi_P : H_a \to D[\ell]$$

$$\sigma \to \xi_P(\sigma) = \sigma(P/\ell) - P/\ell.$$

Et on voit qu'on a un homomorphisme injectif:

$$\xi = (\xi_{P_1}, \dots, \xi_{P_m}) : X_\ell \to D[\ell]^m.$$

Regardons la suite exacte:

$$0 \to D[\ell] \to D(K^S) \overset{\times \ell}{\to} D(K^S) \to 0.$$

Elle fournit par inflation-restriction une injection (grâce à B_3):

$$\zeta : D(K)/\ell D(K) \hookrightarrow H^1(H_a, D[\ell])^{G_a} = \mathrm{Hom}(H_a, D[\ell])^{G_a},$$

et l'homomorphisme $\zeta(P)$ n'est autre que ξ_P. Ainsi $\xi_P = 0 \Leftrightarrow P \in \ell D(K)$.

De la $\mathcal{O}$-linéarité de l'application $P \to \xi_P$ et de B_4, il découle que si $P_1, \dots, P_m \in D(K)$ sont linéairement indépendants sur $\Psi(\mathcal{O})$, alors $\xi_{P_1}, \dots, \xi_{P_m}$ sont linéairement indépendants sur $\Psi(\mathcal{O})$ dans $\mathrm{Hom}(H_a, D[\ell])$.

Par ailleurs, $\zeta(P) = \xi_P$ commute à l'action de G_a donc $M = \xi_P(H_a) = \xi_P(X_\ell)$ est un sous-G_ℓ-module de $D[\ell]^m$.

De la semi-simplicité de $D[\ell]$ (axiome B_2) on déduit l' existence de M', sous-G_ℓ-module de $D[\ell]^m$ tel que $D[\ell]^m$ soit somme directe de M et de M'. Alors, il existe un élément $\alpha \in \mathrm{Hom}_{G_\ell}(D[\ell]^m, D[\ell]) = \mathrm{End}_{G_\ell}(D[\ell])^m = (\mathcal{O}/\ell\mathcal{O})^m$ (d'après l'hypothèse B_1), tel que $\alpha(\xi_P) = 0$. De l'hypothèse d'indépendance linéaire de P on tire $\alpha P = 0$ donc $\alpha = 0$. On a donc montré $M = D[\ell]^m$ et prouvé le théorème 3.

Remarque. Les hypothèses B_1, B_2 sont connues pour le module de Carlitz (cf. **[7]**). On déduit également B_3 de **[7]**: on choisit σ dans G_a représentant la multiplication par un élément $\psi(b)$ de A sur le groupe des points de a-torsion du module de Carlitz, en imposant b non congru à 1 modulo a. Le lemme de Sah affirme que $\sigma - 1$ tue $H^1(G_a, D[\ell])$. ; L'hypothèse faite sur b montre que $\sigma - 1$ est un isomorphisme, on a donc bien B_3.

D'après le récent travail de Y. Taguchi, l'hypothèse B_2 est vraie pour tout module de Drinfeld (voir **[17]** et **[18]**). L'hypothèse B_4 sera vérifiée au §4.

Pour tout t-module U, on désigne par $T_\ell U$ la limite projective des $U[\ell^n]$ et par A_ℓ et k_ℓ les complétés ℓ-adique de A et de k.

Démontrons maintenant GAL 2.

Pour obtenir GAL 2 dans le cas où $a' = \ell$ est irréductible de degré assez grand, il suffit d' appliquer le théorème précédent en remplaçant D par $E = D^m$ et en prenant $P = (P_1, \dots, P_m)$. Le résultat pour $a' = \ell^u$ s'obtient en remarquant que $\mathrm{Gal}(K(E[a^\infty], P/\ell^\infty): K(E[a^\infty]))$ est un A_ℓ module fermé de $T_\ell(E)$ (module de Tate de E), se projetant sur $E[\ell]$ d'après le théorème 3, donc égal à $T_\ell(E)$ grâce au lemme de Nakayama. Si a' est produit de facteurs premiers de degré $\geq \deg \ell_0$, la preuve se fait par récurrence sur le nombre de ces facteurs. On vient de traiter le cas d'un seul facteur premier ($a' = \ell^u$), supposons donc que $a' = a_1 a_2$ où a_1 et a_2 sont premiers entre eux, ils ont donc tout deux strictement moins de facteur premier que a'. Le lemme chinois permet de voir que le compositum des extensions $K(E[a], P/a_1): K(E[a])$ et $K(E[a], P/a_2): K(E[a])$ est $K(E[a], P/a'): K(E[a])$. Le groupe $\mathrm{Gal}(K(E[a], P/a'): K(E[a]))$ s'injecte dans $E[a_1 a_2]$ (par la théorie de Kummer). Or son image est stable par multiplication par les éléments de la forme $\Phi(b)$, où b est parcourt les éléments inversibles de $(A/a_1 a_2)$. L'hypothèse de récurrence permet d'affirmer qu'il se projette surjectivement sur $E[a_1]$ et sur $E[a_2]$ (car

$$\mathrm{Gal}(K(E[a], P/a_1): K(E[a]))$$

et $\mathrm{Gal}(K(E[a], P/a_2): K(E[a]))$ sont respectivement isomorphes à $E[a_1]$ et à $E[a_2]$); on en déduit donc que $\mathrm{Gal}(K(E[a], P/a'): K(E[a]))$ est isomorphe à $E[a_1 a_2]$.

On a ainsi montré GAL 2 pour les a' ne contenant que des facteurs premier de degré $\geq \deg \ell_0$. Pour le cas général, on a besoin d'axiomes supplémentaires. Plus précisément on prouve le théorème suivant (cf. **[9]**):

Théorème 4. *On fait les hypothèses du théorème 3 et on suppose de plus que si* $\deg \ell \leq \deg \ell_0$*:*

$B_1)_{\ell^\infty}$: $\mathcal{O} \otimes A_\ell$ *est le commutant de* G_{ℓ^∞} *dans* $\mathrm{End}(T_\ell(E))$.
$B_2)_{\ell^\infty}$: $T_\ell(E) \otimes k_\ell$ *est un* G_{ℓ^∞} *module semi-simple.*
$B_3)_{\ell^\infty}$: $H^1(G_{a^\infty}, T_\ell(E))$ *est fini.*

Alors, $\mathrm{Gal}(K(E[a], P/a'): K(E[a]))$ *s'identifie à un sous-groupe de* $E[a']$ *d'indice borné indépendamment de* a'.

Idée de la preuve. (similaire à celle de **[11]** chap. 5, théorème 5): La finitude de $H^1(G_a, E[\ell^u])$ permet de montrer le résultat pour chaque petit facteur de a, comme ils sont en nombre fini on en déduit bien le théorème (en utilisant encore le fait que deux extensions associées à des points de divisions d'ordre premier entre eux sont linéairement disjointes (cf. **[9]** et **[16]**).

§4. Relations de dépendance linéaire

Dans ce paragraphe on prouve l'hypothèse B_4 du théorème 3. En choisissant une base sur A de l'anneau des multiplications complexes, on remarque qu'il suffit de regarder les relations de dépendance sur $\Phi(A)$.

On a prouvé dans **[4]** l'existence d'une hauteur canonique $\widehat{h}$ sur les t-modules vérifiant la condition "a_d est inversible" (et sur d'autres un peu plus généraux), on en rappellera les propriétés après quelques notations et l'énoncé du théorème 5.

Désignons par $P_1, \ldots, P_N$ des éléments non nuls de $(\mathbb{G}_a)^n(\overline{k})$. On définit $V = \max_{1\leq i\leq N} \widehat{h}(P_i)$, on note k' l'extension algébrique de k engendrée par $P_1, \ldots, P_N$, δ son degré et η le minimum de $\widehat{h}$ sur les points de $(\mathbb{G}_a)^n(k')$ qui ne sont pas de torsion.

Théorème 5. *Si $P_1, \ldots, P_N \in E(k')$ sont linéairement indépendants sur $\Phi(A)$ et si $\mathfrak{L}$ est un élément irréductible de A de degré $> (d\log q)^{-1}N\log(NV/\eta)$; alors $P_1, \ldots, P_N$ restent linéairement indépendants modulo $\Phi(\mathfrak{L})E(k')$.*

Désignons par h la hauteur de Weil sur $(\mathbb{G}_a)^m(\overline{k})$, la hauteur $\widehat{h}$ a alors les propriétés suivantes:

Théorème (cf. [4]). *Il existe une fonction $\widehat{h}$ de $(\mathbb{G}_a)^m(\overline{k})$ dans $\mathbb{R}$, vérifiant les propriétés suivantes:*

1) *pour tout polynôme P de $\mathbb{F}_q[T]$ de degré $r > 0$ et tout élément α de $(\mathbb{G}_a)^m(\overline{k})$:*

$$\widehat{h}(\alpha) = \lim_{n\to\infty} q^{-ndr} \cdot h[\Phi(P^n)(\alpha)];$$

2) *pour tout polynôme P de $\mathbb{F}_q[T]$ de degré $r \geq 0$ et tout α de $(\mathbb{G}_a)^m(\overline{k})$:*

$$\widehat{h}[\Phi(P)(\alpha)] = q^{dr} \cdot \widehat{h}(\alpha);$$

3) *$\widehat{h} - h$ est une fonction bornée sur $(\mathbb{G}_a)^m(\overline{k})$;*
4) *$\widehat{h}$ est l'unique fonction sur $(\mathbb{G}_a)^m(\overline{k})$ vérifiant à la fois la propriété 2) pour un polynôme donné de degré > 0 et la propriété 3);*
5) *pour tout couple de réels $C, C' > 0$, il n'existe qu'un nombre fini d'éléments α de $(\mathbb{G}_a)^m(\overline{k})$ définis sur une extension algébrique de k de degré $\leq C$ tels que $\widehat{h}(\alpha) \leq C'$;*
6) *$\widehat{h}(\alpha) = 0$ équivaut à l'existence d'un polynôme $Q \neq 0$ de $\mathbb{F}_q[T]$ tel que $\Phi(Q)(\alpha) = 0$ (α est un point de torsion);*
7) *si ξ est un point de torsion $\widehat{h}(\alpha + \xi) = \widehat{h}(\alpha)$;*
8) *$\widehat{h}(\alpha + \beta) \leq \widehat{h}(\alpha) + \widehat{h}(\beta)$.*

Dans ce qui suit on notera $\gamma(\Phi) = \sup_{\alpha\in(\mathbb{G}_a)^m(\overline{k})} |\widehat{h}(\alpha) - h(\alpha)|$ la borne supérieure de la différence entre la hauteur de Weil et la hauteur canonique pour le t-module Φ.

Du théorème on tire que si L est une extension finie de k:

a) le nombre de points de torsion de $E(L)$ est fini,

b) il existe une constante η ne dépendant que du degré de $[L:k]$ et de $\gamma(\Phi)$ telle que si P est dans $E(L)$ et n'est pas de torsion $\widehat{h}(P) \geq \eta$.

Dans la preuve du théorème 5, on utilisera le lemme de Dirichlet en caractéristique finie suivant:

Lemme 8. *Soient* $s_1, \ldots, s_N$ *et* u *des éléments de* A, *où* $\deg u \geq 1$. *On suppose* $\deg u \geq \deg s_i$ $(1 \leq i \leq N)$. *Alors, il existe* $(x_0, \ldots, x_N) \neq 0$ *dans* A^{N+1} *tel que pour* $0 \leq i \leq N$ *on ait:*

$$\deg x_i < \deg u,$$
$$\deg(x_0 s_i - u x_i) \leq (1 - 1/N) \deg u.$$

Preuve. Il y a $q^{(N+1)\deg u}$ $N+1$-uplets $(x_0, \ldots, x_N)$ dans A^{N+1} qui vérifient

$$\deg x_i < \deg u.$$

Or

$$\deg(x_0 s_i - u x_i) < \max(\deg u + \deg s_i, 2 \deg u)$$

et comme $\deg u \geq \deg s_i$, on en déduit:

$$\deg(x_0 s_i - u x_i) < 2 \deg u.$$

Les N-uplets $((x_0 s_1 - u x_1), \ldots, (x_0 s_N - u x_N))$ entrent dans $q^{N(2\deg u - \ell)}$ classes de congruence modulo t^ℓ. Par conséquent dès que $(N+1)\deg u > N(2\deg u - \ell)$, deux $N+1$ uplets $(x'_0, \ldots, x'_N)$ et $(x''_0, \ldots, x''_N)$ sont tels que $((x'_0 s_1 - u x'_1), \ldots, (x'_0 s_N - u x'_N))$ et $((x''_0 s_1 - u x''_1), \ldots, (x''_0 s_N - u x''_N))$ soient congrus modulo t^ℓ. On prend pour ℓ le plus petit entier $> (1 - 1/N)\deg u$, on choisit $x_i = x'_i - x''_i$ $(0 \leq i \leq N)$, et on vérifie bien les inégalités du lemme.

Démonstration du théorème 5. On va ici être obligé d'utiliser une astuce due à Cassels (cf. **[16]** Remarque 2.3). En effet l'analogue du théorème de Mordell-Weil faible est faux, car des arguments de degré permettent de montrer que $A/\Phi(t)A$ est un espace vectoriel de dimension infinie sur $\mathbb{F}_q$.

Soit une relation de la forme:

$$\Phi(s_1)P_1 + \cdots + \Phi(s_N)P_N = \Phi(\mathfrak{L})W, \tag{$*$}$$

où W est un point ;de $E(k')$ et les s_i sont dans A non tous nuls. On peut supposer que $s_1, \ldots, s_r$ sont non nuls, et par division euclidienne par $\mathfrak{L}$ qu'ils sont de degré $<$ à celui de $\mathfrak{L}$.

Le lemme de Dirichlet appliqué à $s_1, \ldots, s_r$ et $\mathfrak{L}$ donne alors une relation:

$$\Phi(e_1)P_1 + \cdots + \Phi(e_r)P_r = \Phi(x_0)\Phi(\mathfrak{L})W - \Phi(\mathfrak{L})[\Phi(x_1)P_1 + \cdots + \Phi(x_r)P_r],$$

où les x_i sont des éléments non tous nuls de A de degré $< \deg \mathfrak{L}$ et $e_i = x_0 s_i - \mathfrak{L} x_i$ est de degré $\leq (1 - 1/r)\deg \mathfrak{L}$. Comme $\mathfrak{L}$ est irréductible $e_i = 0 \Rightarrow x_i = 0$ et $x_0 = 0$, donc les e_i sont non tous nuls. On en déduit que le terme de droite n'est pas de torsion grâce à l'indépendance linéaire des P_i.

La comparaison des hauteurs de chaque membre donne:

$$\eta q^{d \deg \mathfrak{L}} \leq r \left[\max_{1 \leq i \leq r} q^{d \deg e_i} \right] V$$
$$\deg \mathfrak{L} \leq (d \log q)^{-1} \log(rV/\eta)^r .$$

Appendice 1

En liaison avec le §2 b), énonçons maintenant un résultat un peu différent, valable sans aucune hypothèse algébrique (sans GAL 1, GAL 2 et sans supposer qu'on est sur une extension finie de k) et qui nous laisse penser que les conjectures 1 et 2 pourraient également être démontrées sans de telles hypothèses.

Théorème 6. *Soit P un polynôme sur $(\mathbb{G}_a)^m$ et x_0 un point qui n'est pas de torsion. Il existe un entier W, effectivement calculable en fonction de m, de d et du degré de P tel que, si P s'annule sur les W premiers itérés de x_0 sous l'action de $\Phi(t)$, alors P s'annule sur un sous-t-module de $(\mathbb{G}_a)^m$ de dimension ≥ 1.*

Preuve. Soit $Z(P)$ la variété des zéros de P. Définissons par récurrence la suite décroissante de variétés $(X_i)_{i \in N}$ par $X_0 = Z(P)$, $X_{i+1} = X_i \cap Z(P \circ \Phi(t^i))$. Ces variétés vérifient $\Phi(t)X_{i+1} \subset X_i$; désignons par N_i le nombre de composantes irréductibles de dimension maximale de X_i.

Soit k_0 un entier tel que $\dim(X_{k_0}) = \dim(X_0) = n - 1$. On va montrer que s'il existe un tel $k_0 \geq N_0 + 1$ le théorème est vrai.

En effet soit Y une composante de X_{k_0}; c'est aussi une composante de X_j $(0 \leq j \leq k_0)$ et on a, pour ces mêmes indices j, $\Phi(t^j)Y \subset X_0$. Comme $\Phi(t^j)Y$ est encore une composante irréductible de X_0, dès que $k_0 \geq N_0$, il existent deux indices $i_1 > i_2$ tels que $\Phi(t^{i_1})Y = \Phi(t^{i_2})Y$. D'où pour tout entier m,

$$\Phi(t^{m(i_1 - i_2)}) \circ \Phi(t^{i_2})Y = \Phi(t^{i_2})Y;$$

et d'après le lemme 4 ceci entraîne que $\Phi(t^{i_2})Y$ contient le translaté d'un sous-t-module et implique donc le théorème.

On peut donc supposer que le plus grand entier k_0 tel que $\dim(X_{k_0}) = \dim(X_0)$ est $\leq N_0$. De proche en proche on voit qu'on peut supposer avoir construit des entiers k_i $(0 \leq i \leq a)$ définis par les deux conditions:

i) k_i est le plus grand entier tel que $\dim(X_{k_i}) = \dim(X_{k_{i-1}+1})$

ii) $\dim(X_{k_a}) = 1$

et tels que $k_i - k_{i-1} \leq N_{k_{i-1}+1}$.

La variété X_{k_a+1} est donc de dimension 0. Soit j tel que $\Phi(t^j)(X_{k_a+1+j}) \subseteq X_{k_a+1}$. Si x_0 est dans X_{k_a+1+j} alors $\Phi(t^j)(x_0)$ est dans X_{k_a+1}. Dès qu'il existe un tel $j \geq N_{k_a+1}$, il existe $j_1 \neq j_2$ tels que $\Phi(t^{j_1})(x_0) = \Phi(t^{j_2})(x_0)$ et donc x_0 est de torsion contrairement à ce qui est supposé.

Finalement on a montré que si P s'annule en les premiers itérés de x_0 jusqu'à un ordre $\geq N_{k_a+1} + \cdots + N_{k_0+1} + N_0$, alors le théorème est vrai.

Il reste à remarquer que ce dernier terme peut être majorée par une fonction ne dépendant que du t-module et du degré de P. C'est en effet le cas car $N_{k_{i-1}+1}$ est inférieur à la fonction de Hilbert de $X_{k_{i-1}+1}$, variété qui est intersection de polynômes de degrés $\leq$ une fonction ne dépendant que du degré de P et de k_{i-1}. Ce dernier argument permet de majorer convenablement k_i grâce à un lemme de Bézout (**[9]**).

Remarque. En dimension 2 on peut prendre $W = 2(q^{dD}D)^2 + D$, où D est le degré total de P. Cette constante peut être remplacée par $2q^d D^2 + 1$ si on suppose de plus P irréductible.

Appendice 2

La méthode employée au §4 conduit également à des estimations des coefficients des relations de dépendance linéaire sur les t-modules. La démonstration qui suit est inspirée de celle donnée dans **[12]** pour traiter le cas des variétés abéliennes. En reprenant les notations du début du paragraphe 4, on montre:

Théorème 7. *Il existe une constante $\beta > 0$, ne dépendant que du t-module E et de δ telle que si $P_1, \dots, P_N$ sont linéairement dépendants sur A, on a une relation:*

$$\Phi(c_1)P_1 + \cdots + \Phi(c_N)P_N = 0,$$

avec des éléments c_i de A, non tous nuls et vérifiant:

$$|c_i|^d \leq \beta(N/\eta)^{N-1}(\max(1, V))^{N-1}.$$

Preuve. Considérons les relations de la forme:

$$\Phi(s_1)P_1 + \cdots + \Phi(s_N)P_N = W, \tag{$*$}$$

où W est un point de torsion et les s_i sont dans A non tous nuls. Choisissons en une telle que $B = \max_{1 \leq i \leq N} q^{d \deg s_i}$ soit minimum. Si $B = 1$, comme le nombre de points de torsion dans k' est effectivement borné en fonction d'un majorant de $\gamma(\Phi)$ et de $[k' \colon k]$, on multiplie cette relation par l'ordre de W et le théorème est prouvé quitte à prendre β plus grand que la puissance d-ième des valeurs absolues de l'ordre d'un point de torsion de k'. On peut donc supposer $B > 1$. Désignons par u un élément de A tel que $\deg u \geq \deg s_i$ et appliquons le lemme 8 à $s_1, \dots, s_N$ et u. Il existe donc $x_0, \dots, x_N$ appartenant à A et non tous nuls, tels que:

$$\deg x_i < \deg u,$$
$$\deg(x_0 s_i - u x_i) \leq (1 - 1/N) \deg u.$$

Posons $e_i = x_0 s_i - u x_i$ $(0 \leq i \leq N)$, alors $(*)$ implique:

$$\Phi(e_1)P_1 + \cdots + \Phi(e_N)P_N = \Phi(x_0)W - \Phi(u)[\Phi(x_1)P_1 + \cdots + \Phi(x_N)P_N].$$

Soit $Z = \Phi(x_1)P_1 + \cdots + \Phi(x_N)P_N$, la hauteur canonique du terme de droite de l'expression précédente est $q^{d \deg u}\widehat{h}(Z)$ (cf. théorème .2) et .6)). Celle du terme de gauche est majorée par $N\left[\max_{1\le i\le N} q^{d\deg e_i}\right] V$ (cf. théorème .2) et .7)). Si Z n'est pas de torsion on en tire les inégalités suivantes:

$$\eta q^{d\deg u} \le N\left[\max_{1\le i\le N} q^{d\deg e_i}\right] V,$$
$$q^{d\deg u} \le (NV/\eta)^N.$$

Ainsi, si u est tel que $q^{d\deg u} > \max[(NV/\eta)^N, B]$, on en déduit que Z est de torsion.

Supposons d'abord $B > (NV/\eta)^N$; alors avec $q^{d\deg u} = B$ on obtient $q^{d\deg e_i} \le B^{(1-1/N)}$. On rappelle que $B = \max_{1\le i\le N} q^{d\deg s_i}$ est minimal et > 1 donc tous les e_i sont nuls, c'est-à-dire $x_0 s_i - u x_i = 0$ $(1 \le i \le N)$. Le même argument montre que les s_i sont premiers entre eux dans leur ensemble, par conséquent il existe $b_1, \ldots, b_N$ dans A tels que $b_1 s_1 + \cdots + b_N s_N = 1$. Le regroupement de ces identités entraîne $x_0 = u(b_1 x_1 + \cdots + b_N x_N)$ ce qui contredit $\deg x_0 < \deg u$.

On est donc ramené au cas $B \le (NV/\eta)^N$. Prenons u tel que $q^{d\deg u} > (NV/\eta)^N$; le point Z est alors un point de torsion de k'. D'après le théorème 1.4), il n'existe qu'un nombre fini J de points $Z_1, \ldots, Z_J$ qui sont de torsion et appartiennent à une extension de k de degré $\le \delta$. Cette même partie du théorème montre que pour tout $1 \le j \le J$, l'ensemble $\{\Phi(t^e)Z_j\}$, où e parcourt l'ensemble des entiers naturels est fini. Ceci assure l'existence d'un entier b, ne dépendant que de E et de δ, tel que $\Phi(t^b)Z_j = 0$ pour tout $1 \le j \le J$. On choisit enfin $c_i = t^b e_i$ et $\beta = q^{db}$ et la conclusion du théorème 7 est satisfaite.

Bibliographie

[1] G. Anderson, t-motives, Duke Math. J. 53, (1986), 457–502.

[2] G. Anderson and D. Thakur, Tensor powers of the Carlitz module and zeta values, Ann. of Math. 132, (1990), 159–191.

[3] L. Denis, Théorème de Baker et modules de Drinfeld, à paraître dans J. Number Theory, (1992).

[4] L. Denis, Hauteurs canoniques et modules de Drinfeld, à paraître dans Math. Ann., (1992).

[5] V. G. Drinfeld, Elliptic modules, Math. USSR-Sb. 23, (1974), 561–592.

[6] G. Faltings, Diophantine approximation on abelian varieties, Ann. Math. 2, 133, (1991).

[7] D. Hayes, Explicit class field theory for rational function fields, Trans. Amer. Math. Soc. 189, (1974), 77–91.

[8] M. Hindry, Géométrie et Hauteurs dans les groupes algébriques, Thèse de l'Unversité Paris 6, (1987) (voir **[9]**).

[9] M. Hindry, Autour d'une conjecture de Serge Lang, Invent.Math. 94, (1988), 575–603.

[10] S. Lang, Fundamentals of Diophantine Geometry, Springer Verlag, 1983.

[11] S. Lang, Elliptic Curves: Diophantine Analysis, Springer Verlag, 1978.

[12] D. Masser, Small value of the quadratic part of the Neron-Tate height, Prog. Math. 12, 1981; Birkhäuser, 1979-1980, 213–222.

[13] D. Masser and G. Wüstholz, Fields of Large Transcendence Degree Generated by Values of Elliptic Functions, Invent. Math. 72, (1983), 407–464.

[14] P. Philippon, Lemmes de zéros sur les groupes algébriques commutatifs, Bull. Soc. Math. France 114, (1986), 355–383. Errata et addenda, Bull. Soc. Math. France 115, (1987), 397–398.

[15] M. Raynaud, Sous-Variétés d'une variété abélienne et points de torsion, Arithmetic and Geometry (dédié à Shafarevic) 1, (1983); Birkhäuser, 327–352.

[16] K. Ribet, Kummer theory on extension of abelian varieties by tori, Duke Math J. 46, (1979), 745–761.

[17] Y. Taguchi, Semi-simplicity of the Galois representations attached to Drinfeld modules over fields of "finite characteristics", Duke Math. J. 62, (1991), 591–599.

[18] Y. Taguchi, Semi-simplicity of the Galois representations attached to Drinfeld modules over fields of "infinite characteristics", J. Number Theory, to appear.

[19] J. Yu, Transcendence and special zeta values in characteristic p, *Ann. Math.* **134**, (1991) 1–23.

[20] J. Yu, Analytic homomorphisms into Drinfeld modules, manuscript, juin 1991.

[21] J. Yu, A six exponentials theorem in finite characteristic, Math. Ann. 272, (1985), 91–98.

Institut Henri Poincaré, 11 rue Pierre et Marie Curie, Paris Cedex 05, 75231 France

Transcendence Properties of Carlitz Zeta-values

G. Damamme

Abstract. The aim of this paper is to give a sketch of proof of the transcendence of zeta-values in the Carlitz case [**6**].

1. Dictionary

Real case	Case of Carlitz module
$\mathbb{Z}$ integers	$Z := A = \mathbb{F}_q[T]$
$\mathbb{Q}$ rationals	$Q := k = \mathbb{F}_q(T)$
$\mathbb{R}$ completion of $\mathbb{Q}$	$R := k_\infty = \mathbb{F}_q((T^{-1}))$
$\mathbb{N}$ positive integers	Z_+ monic polynomials
$\| \ \|$ absolute value	$\| \ \| := \| \ \|_\infty$

Zeta functions :

Riemann	Carlitz [**2**]
$\zeta_{\mathbb{R}}(s) = \sum_{n \in \mathbb{N}} \frac{1}{n^s} \ (s \neq 1)$	$\zeta_c(s) = \sum_{g \in Z_+} \frac{1}{g^s} \ (s \in \mathbb{N})$
Theorem (Euler)	Theorem (Carlitz, Goss) [**3**], [**7**]
$\zeta_{\mathbb{R}}(2m) = \frac{B_{2m}(2\pi)^{2m}}{2(2m)!}$	$\zeta_c((q-1)m) = \frac{B_{(q-1)m}\widetilde{\pi}^{(q-1)m}}{\Pi((q-1)m)}$
Where : $B_{2m} \in \mathbb{Q}$ (Bernoulli numbers) $(2m)! \in \mathbb{Z}$ and π is transcendental over $\mathbb{Q}$ (Lindemann)	Where : $B_{(q-1)m} \in Q$ (Bernoulli – Carlitz numbers) [**3**] $\Pi((q-1)m) \in Z$ and $\widetilde{\pi}$ is transcendental over Q (Wade, [**11**])
Corollary : $\zeta_{\mathbb{R}}(s)$ is transcendental over $\mathbb{Q}$ for $s = 2m$	Corollary : $\zeta_c(s)$ is transcendental over Q for $s = (q-1)m$ (s "even")

<u>Question</u> :

And if s is odd? ?

<u>Answer</u> :	<u>Answer</u> :
$\zeta_{\mathbb{R}}(3) \notin \mathbb{Q}$ (Apery)	$\zeta_c(s)$ is transcendental over Q
... and nothing else	<u>for all</u> $s \in \mathbb{N}$ (Yu, Damamme) **[14]** **[4]**
	and if s is "odd",
	$\frac{\zeta_c(s)}{\tilde{\pi}^s}$ is transcendental (Yu)

Remarks. The proofs of Yu and Damamme are quite different:

- Yu uses his own classical results on transcendency theory for Drinfeld modules and a result of Anderson and Thakur **[1]** that gives directly the transcendence of $\zeta_c(s)$ (for all s) and $\frac{\zeta_c(s)}{\tilde{\pi}^s}$ (for s odd).
- By Wade's method **[12]**, one has to get some intermediate result by D. Thakur **[9]**, Y. Hellegouarch and myself **[5]** from which we obtain the main result.

2. Computing $\zeta_c(s)$

Notation:

$$R \to R$$
$$\sigma : x \to x^\sigma := x^q$$

is the Frobenius endomorphism.

Put

$$[k] = T^{q^k} - T$$
$$L_0 = D_0 = 1 \qquad \text{and} \qquad L_k = [k][k-1]\dots[1]$$
$$D_k = [k][k-1]^\sigma \dots [1]^{\sigma^{k-1}}.$$

We write

$$\zeta_c(s) = \sum_{k=0}^{\infty} \left(\sum_{\substack{\deg G = k \\ G \in Z_+}} \frac{1}{G^s} \right) := \sum_{k=0}^{\infty} S_{k,s}.$$

We can compute $S_{k,s}$ using Carlitz formulas. We obtain $(-1)^{s+k+1} \cdot S_{k,s}$ equals

$$\sum_{\substack{\underline{i}=(i_1,\dots,i_n) \\ \text{s.t:} i_1 q + \dots + i_n q^n < s}} \alpha_{\underline{i}} \frac{[k]^{i_1 q + \dots + i_n q^n} [k-1]^{i_2 q^2 + \dots + i_n q^n} \dots [k-(n-1)]^{i_n q^n}}{L_k^s D_1^{i_1} \dots D_n^{i_n}}$$

with $\alpha_{\underline{i}} \in \mathbb{F}_q$ and $\alpha_{\underline{0}} \neq 0$.

Remark. We cannot easily compute the $\alpha_{\underline{i}}$, but this does not matter. We only need that $\alpha_{\underline{0}}$ is not 0.

Examples.

i) If $s \leq q$

$$\zeta_c(s) = \sum_{k=0}^{\infty} \frac{(-1)^{ks}}{L_k^s}.$$

ii) If $q < s \leq 2q - 1$

$$\zeta_c(s) = 1 + \sum_{k=1}^{\infty} (-1)^{s+k+1} \left(\frac{1}{L_k^s} - \frac{\alpha_k [k]^q}{L_k^s D_1} \right)$$

where

$$\alpha_k \in \mathbb{F}_q \text{ depends on } s.$$

3. Main result

Theorem. *Let* I_s *be the set defined by*

$$I_s = \{ \underline{i} = (i_1, \ldots, i_n) \in \mathbb{N}^n \ |: i_1 q + \cdots + i_n q^n < s \}.$$

For each $\underline{i} \in I_s$, *let* $B_{\underline{i}}$ *be a polynomial of* Z *such that*

$$\deg B_{\underline{0}} \geq \deg B_{\underline{i}} + \sum_{\nu=1}^{n} \nu i_\nu q^\nu \quad \text{for all } \underline{i} \in I_s \tag{1}$$

and let $R_{k,s} \in Z$ *defined by*

$$R_{k,s} = \sum_{\underline{i} \in I_s} B_{\underline{i}} [k]^{i_1 q + \cdots + i_n q^n} \ldots [k-(n-1)]^{i_n q^n}. \tag{2}$$

Then

$$\theta = \sum_{k=0}^{\infty} \frac{R_{k,s}}{L_k^s} \tag{3}$$

is transcendental over Q.

Corollary 1. $\zeta_c(s)$ *is transcendental for all* $s \in \mathbb{N}$.

Proof of corollary. If we put

$$D = D_1^{[(s-1)/q]} \ldots D_n^{[(s-1)/q^n]}$$

where $[x]$ is here the integral part of x and $q^n < s$, then the computing of $\zeta_c(s)$ shows us that $D \cdot \zeta_c(s)$ is of the form (3) with the $B_{\underline{i}}$ satisfying (1). We remark that we need the condition $\alpha_{\underline{0}} \neq 0$ to verify (1). □

4. Outline of the proof of the main result

1°) **First Decomposition.** We will use *Ore's Lemma* ([**8**]):

"θ is transcendental over Q if and only if the $\sigma^n(\theta)$ are linearly independent."

In fact Ore's lemma tells us that all nonzero polynomials P (with coefficients in Z) divide some $\mathbb{F}_q$-linear polynomial

$$T(X) = \sum A_j X^{q^j}.$$

Hence, if we assume θ is algebraic over Q, i.e., there exists a polynomial P such that $P(\theta) = 0$, then there exists an $\mathbb{F}_q$-linear polynomial T of degree m such that

$$T(\theta) = \sum_{j=0}^{m} A_j \theta^{q^j} = 0$$

with $A_j \in Z$.

T being $\mathbb{F}_q$-linear and continuous, we have

$$T(\theta) = \sum_{j=0}^{m} A_j \sum_{k=0}^{\infty} \left(\frac{R_{k,s}}{L_k^s} \right)^{q^j} = 0. \tag{4}$$

To get rid of the denominator of the first terms, we multiply the relation (4) by a big polynomial K_β of Z, where β denotes an arbitrary large positive integer. We choose K_β as follows:

$$K_\beta := L_\beta^{sq^m} \cdot J_{\beta,m}^s$$

where

$$J_{\beta,r} = \prod_{\ell=1}^{r} \prod_{\lambda=(\ell-1)m+1}^{\ell m} [\beta+\lambda]^{q^{m-\ell}}.$$

When

$$\beta \to \infty \qquad \text{and} \qquad |K_\beta| \to \infty,$$

we hope to have a decomposition of $K_\beta T(\theta)$ into an integral part I and a "fractional part" F (i.e., $|F| < 1$). In fact, we have:

Proposition 1. *Put*

$$I_1 := K_\beta \sum_{j=0}^{m} A_j \sum_{k=0}^{\beta+(m-j)m} \left(\frac{R_{k,s}}{L_k^s} \right)^{q^j}$$

and

$$F_1 := K_\beta \sum_{j=0}^{m} A_j \sum_{k=m^2-j+n+1}^{\infty} \left(\frac{R_{\beta+k,s}}{L_{\beta+k}^s} \right)^{q^j}.$$

Then:

i) $I_1 \in Z$;

ii) $I_1 \equiv J^s_{\beta,m} A_m B_0^{q^m} \pmod{[\beta]}$;
iii) $|F_1| < 1$ *for* $m \geq 2$ *and* β *large enough.*

Remarks.
1) The proof of Proposition 1 is in **[6]**.
2) The n in F_1' formula is the biggest natural number such that $q^n < s$.
3) Terms are missing to have $K_\beta T(\theta)$. Accordingly we put

$$S := K_\beta \sum_{j=0}^{m} A_j \sum_{k=(m-j)m+1}^{m^2-j+n} \left(\frac{R_{\beta+k,s}}{L^s_{\beta+k}}\right)^{q^j}$$

to have

$$K_\beta T(\theta) = I_1 + F_1 + S = 0.$$

Now, we want to transform each term of S into a sum of terms which are either integer or terms of absolute value less than 1. For that, we will use a decomposition in "j-simple elements". But first consider a generic term of S; we can write S as

$$S = \sum_{j=0}^{m} \sum_{k=(m-j)m+1}^{m^2-j+n} V_{k,j}$$

with

$$V_{k,j} = A_j L_\beta^{sq^m} J^s_{\beta,m} \left(\frac{B_{\underline{i}}[\beta+k]^{i_1 q+\cdots+i_n q^n} \ldots [\beta+k-(n-1)]^{i_n q^n}}{L^s_{\beta+k}}\right)^{q^j}$$

where: $0 \leq j \leq m$, $(m-j)m+1 \leq k \leq m^2-j+n$, and $i_1 q + \cdots + i_n q^n < s$.

Remark. There are 3 indices of summation

$$j, k \text{ and } \underline{i} = i_1 q + \cdots + i_n q^n$$

because we have decomposed $R^{q^j}_{k,s}$.

Next we write $V_{k,j}$ as

$$V_{k,j} = E_j \cdot H_{k,j}$$

where E_j was chosen to be integral and to get rid of the maximum of the terms in the denominator of $V_{k,j}$. We obtain

$$E_j = A_j B_{\underline{i}}^{q^j} \cdot L_\beta^{sq^m} \cdot J^s_{\beta,m-j} L^{-sq^j}_{\beta+(m-j)m}$$

and

$$H_{k,j} = \prod_{\ell=m-j+1}^{m} \prod_{\lambda=(\ell-1)m+1}^{\ell m} [\beta+\lambda]^{sq^{m-\ell}}$$
$$\left(\frac{[\beta+k]^{i_1 q+\cdots+i_n q^n} \ldots [\beta+k-(n-1)]^{i_n q^n}}{[\beta+(m-j)m+1]^s \ldots [\beta+k]^s}\right)^{q^j}. \qquad (5)$$

Now we deal only with $H_{j,k}$ and we will decompose it in a sum of terms $\mathcal{L}$ as follows:

2°) **Splitting $H_{j,k}$ into "j-simple elements".** We begin by defining a *j-simple element.*

Let β large enough and $\mathcal{L} \in Q$ of the form:

$$\mathcal{L} = \frac{P[\beta + b_1]^{f_1} \dots [\beta + b_u]^{f_u}}{([\beta + \ell_1) \dots [\beta + \ell_v])^{q^j}} \tag{6}$$

where $P \in Z$, independant of β, $f_i > 0$ and

$$b_1 > \cdots > b_u > 0$$
$$\ell_1 \geq \cdots \geq \ell_v > 0.$$

We call $\mathcal{L}$ is *j-simple* if
i) $f_i \leq q - 1$
ii) $\ell_v + j > b_1$

Remarks.
1) The condition i) gives us an order relation between the absolute values of $[\beta + b_i]^{f_i}$ in the numerator of (6).
2) The second condition means that the absolute value of the denominator is always bigger than that of the numerator (except if the denominator is 1 and $\mathcal{L}$ an integer).

It follows immediately that

Property.

$$\deg \mathcal{L} < -q^{\beta + b_u} + \deg P$$

or

$$\deg \mathcal{L} < -q^{\beta + \ell_v + j} + \deg P$$

if the numerator is just P.

(This proprerty helps us to prove that not only $|\mathcal{L}| < 1$ — if $\mathcal{L}$ is not an integer — but also $|E_j \cdot \mathcal{L}| < 1.$)

Now we give a crucial lemma.

Lemma. *Let $\mathcal{H}$ be of the form*

$$\mathcal{H} = \frac{P[\beta + k_1]^{a_1 q^{\alpha_1}} \dots [\beta + k_r]^{a_r q^{\alpha_r}}}{\sigma^j([\beta + \ell_1] \dots [\beta + \ell_t])}$$

where
i) $k_1 > \cdots > k_r \in \mathbb{Z}$;
ii) $\ell_1 \geq \cdots \geq \ell_t \geq 0 \in \mathbb{N}$;
iii) $a_i \not\equiv 0 \pmod q$;
iv) $\min(k_i + \alpha_i) > 0$.

Then $\mathcal{H}$ can be written as a sum of j-simple elements $\mathcal{L}$ of the form (6) where
i) $v \leq t$;

ii) $b_u \geq \min_i(k_i + \alpha_i, \ell_t + j)$;

iii) *If* $\mathcal{L}$ *is integral (then* $v = 0 < t$*)*

$$\deg(\text{res } \mathcal{L}) < \sum_{i=1}^{r} a_i q^{\alpha_i} \sup(1, q^{k_i})$$

(here res *is the residue* mod $[\beta]$ *).*

Remarks.

a) From i) we see that the number of terms of the denominator of (5) is decreasing. And if all the terms of the denominator are vanishing, $\mathcal{L}$ is an integer.

b) Wade has given a proof of this lemma (**[12]**), but only in the case where the k_i are positive. In this case, we can apply the lemma to the terms $H_{k,j}$ of the form (5) only if $n \geq 1$ (i.e., $s \leq q^2$) and we obtain the transcendence of $\zeta(s)$ for $s \leq q^2$ (see **[5]**). If $s > q^2$, then $n \geq 2$ and the k_i are negative but satisfy $\min(k_i + \alpha_i) > 0$.

c) The proof of the lemma is given in **[6]** and examples of this decomposition are in **[4]**.

Now applying lemma to generic term $H_{k,j}$ [1] we obtain

Proposition 2. *$H_{k,j}$ can be written as a sum of j-simple elements $\mathcal{L}$ which verify either*

i) $|E_j \cdot \mathcal{L}| < 1$

or

ii) *$\mathcal{L} \in Z$ and*

$$\deg(\text{res}(\mathrm{E_j} \cdot \mathcal{L})) < \deg(\text{res}\, \mathrm{I_1})$$

where res *A is the residue modulo* $[\beta]$ *of A.*

Remarks.

a) The condition ii) of the lemma and the property of j-simple elements give the i).

b) The denominator of $H_{j,k}$ satisfies the lemma's conditions if we repeat each bracket q^j, s times. (It is possible because the ℓ_μ of the denominator of $\mathcal{H}$ can be equal.)

c) The proof of Proposition 2 is in **[6]**.

3°) **End of the proof.** Now, we can classify the $E_j \cdot \mathcal{L}$ obtained with the elements of I_1 or with the elements of F_1 (see diagram).

We obtain two new sums which verify

$$K_\beta \cdot T(\theta) = I + F$$

with $I \in Z$ and $|F| < 1$.

$$I + F = 0 \quad \text{implies} \quad I = F = 0.$$

Yet

$$I_1 \equiv J^s_{\beta,m} A_m B_0^{q^m} \not\equiv 0 \,(\text{mod}\,[\beta])$$

1 (In some particular cases, we begin to *simplify* numerator and denominator of $H_{k,j}$ before applying the lemma (see **[4]** or **[6]**).)

(Proposition 1. ii)) and

$$\deg(\operatorname{res}(E_j \cdot \mathfrak{L})) < \deg(\operatorname{res} I_1)$$

(Proposition 2. ii)) implies

$$I \equiv I_1 \pmod{[\beta]}.$$

Hence

$$I \neq 0$$

and we have a contradiction. □

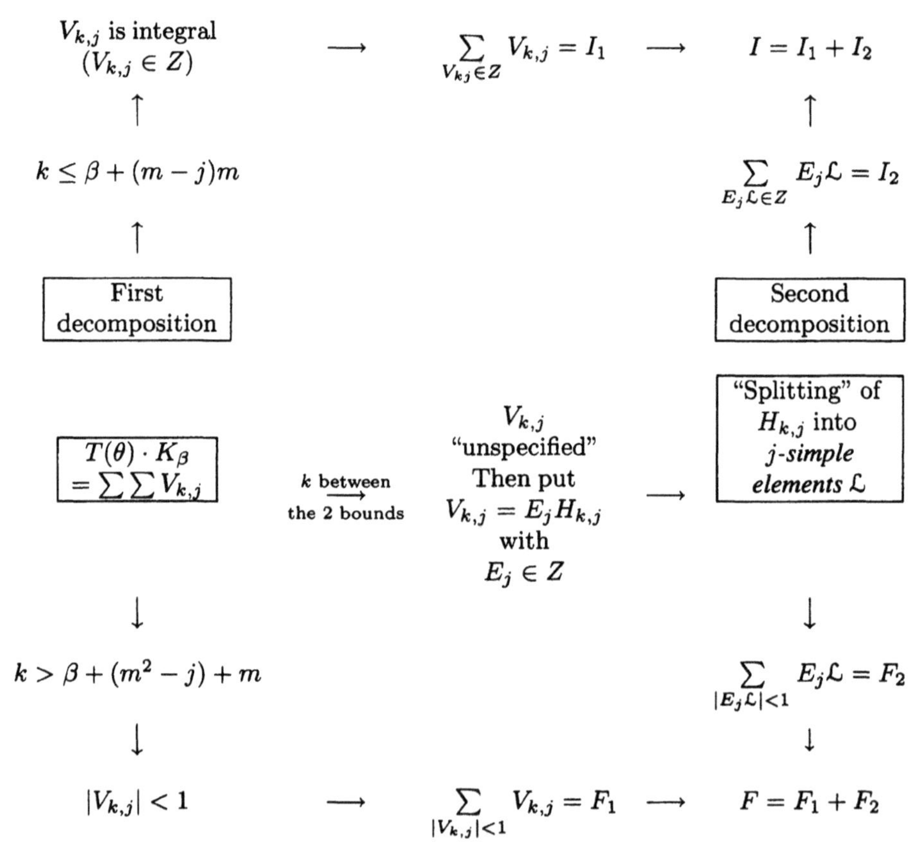

5. Other possible applications of Wade's method

Is it possible to adapt our method to L-values of the Carlitz module or, more generally, to the zeta-values of other Drinfeld modules?

Thakur **[10]** already gave some results for special cases of Drinfeld modules using Yu's results.

References

One can find a very complete list of references on transcendence questions in characteristic p in **[13]**.

[1] G. W. Anderson and D. Thakur, Tensor powers of the Carlitz module and zeta values, Ann. of Math. 132 (1990), 159–191.

[2] L. Carlitz, On certain functions connected with polynomials in a Galois field, Duke Math. J. 1 (1935), 137–168.

[3] L. Carlitz, Analogue of the von Staudt-Clausen theorem, Duke Math. J. 3 (1937), 503–517.

[4] G. Damamme, Transcendence de la fonction zéta de Carlitz par la méthode de Wade, Thèse, Caen (January 1990).

[5] G. Damamme and Y. Hellegouarch, Propriétés de transcendence de la fonction zéta de Carlitz, C. R. Acad. Sci. Paris, (1988) t. 307, Série I, 635–637.

[6] G. Damamme and Y. Hellegouarch, Transcendence of the Values of the Carlitz Zeta Function by Wade's Method, J. Number Theory 39(3), (November 1991), 257–278.

[7] D. Goss, Von Staudt for $\mathbb{F}_q[t]$, Duke Math. J. 45, (1978), 885–910.

[8] O. Ore, On a special class of polynomials, Trans. Amer. Math. Soc. 35, (1933), 559–584.

[9] D. Thakur, Gauss functions and Gauss sums for function fields and periods of Drinfeld Modules, Ph.D. thesis, Harvard (April 1987).

[10] D. Thakur, Drinfeld modules and arithmetic in the function fields, Preprint (1991).

[11] L. I. Wade, Certain quantities transcendental over $GF(p^n, x)$ – I, Duke Math. J. 8 (1941), 701–720.

[12] L. I. Wade, Certain quantities transcendental over $GF(p^n, x)$ – II, Duke Math. J. 10, (1943), 587–594.

[13] M. Waldschmidt, Transcendence problems connected with Drinfeld Modules, Preprint (1990), (to appear in a Turkish revue.)

[14] J. Yu, Transcendence and Special Zeta Values in Characteristic p, *Ann. of Math.* 134, (1991), 1–23.

11, rue Paul Verlaine, 14730 Giberville, France

L-series of t-motives and Drinfeld Modules

David Goss

Abstract

This paper is devoted to a thorough introduction to the L-series of Drinfeld modules and t-modules. The goals are: 1) To set out clearly all the relevant definitions and basic results of the theory in complete generality, and 2) to present the finer theory in the simplest case of the ζ-function of $\mathbf{A} = \mathbb{F}_r[T]$. This particular case contains all sorts of interesting structure that should ultimately apply to *general* L-series.

Table of Contents

Partially supported by NSA Grant MDA 904-91-H-0054. **AMS Classification:** 11G09; **Key Words:** L-series, ζ-function, Γ-functions, Drinfeld Modules, t-motives, Euler-Poincaré characteristic, magic numbers, non-archimedean measures, class-groups.

1. Introduction

Let X be a scheme of finite type over $\mathbb{Z}$. A basic object of study of X is its ζ-function, $\zeta(X, s)$ (see Definition 2.1). This complex valued function possesses an Euler product over Spec($\mathbb{Z}$) of "local" ζ-functions, $\zeta(X_p, s)$, associated to the fibre, X_p, of X at p. In turn, these local factors measure the number of points of X in the various finite dimensional extensions of $\mathbb{Z}/(p)$. When X is smooth and projective, and p is a good prime, one has a completely satisfactory description of the local ζ-function in terms of ℓ-adic cohomology and the action of Frobenius.

The factorization of the local zeta-functions according to cohomology means that its study can be reduced to the study of L-series associated to a *fixed* cohomology group H^i_ℓ. In turn the collection $\{H^i_\ell\}$ ranging over the various primes ℓ gives rise to a "compatible" system of representations of $G = \mathrm{Gal}(\overline{\mathbb{Q}}/\mathbb{Q})$ for a fixed algebraic closure $\overline{\mathbb{Q}}$ of $\mathbb{Q}$. Thus the definition of the L-series (contained in our Section 2) reduces to a generalization of the "usual" definition of Artin L-series associated to representations of G which factor through finite extensions.

The formalism of these L-series is so striking that Grothendieck was lead to conjecture the existence of a "motive" "$X^{(i)}$" associated to H^i. This motive is a geometric realization of H^i with the property (among many such properties) that linear algebra operations on L-series correspond to geometric operations on the motive. For instance, the total cohomology of X — which gives the function $\zeta(X, s)$ — decomposes into the sum of the cohomologies of the motives $X^{(i)}$; thus we obtain again the factorization of the $\zeta(X, s)$. While this example is rather trivial, there are many others like it that are not so prosaic. The motivic formalism allows one to focus on the cohomology of a *fixed* dimension of a variety.

Included in the formalism of zeta-functions of schemes are the usual Dedekind zeta-functions as well as the L-series of abelian varieties. Thus one can well imagine that the *special values* of general L-functions to be a very rich source of theorems and conjectures. Indeed, supplementing the conjectures of Birch and Swinnerton-Dyer, one has the conjectures of Deligne, Beilinson and so on.

Now let $\mathcal{C}$ be a smooth, complete, geometrically connected curve over the finite field $\mathbb{F}_r$, $r = p^m$. Let ∞ be a fixed point of $\mathcal{C}$ and $\mathbf{A}$ the affine algebra of $\mathcal{C} - \infty$. We let $\mathbf{k}$ be the function field of $\mathbf{A}$ and $\mathbf{K}$ the completion of $\mathbf{k}$ at ∞. Thus $\mathbf{A} \subset \mathbf{K}$ *discretely* with compact quotient; exactly as with $\mathbb{Z}$ and $\mathbb{R}$. This discreteness allows one to talk about $\mathbf{A}$-lattices in the algebraic closure $\overline{\mathbf{K}}$ of $\mathbf{K}$ or even in $\overline{\mathbf{K}}^n$. Let M be one such lattice. The quotient, $\overline{\mathbf{K}}^n/M$, can, under very mild conditions, be seen to arise from an algebraic structure called a (uniformizable) "t-module" (or "Drinfeld module" when

$n = 1$). Essentially what happens is that, under these conditions,

$$\overline{\mathbf{K}}^n/M \simeq \overline{\mathbf{K}}^n;$$

thus the natural $\mathbf{A}$-action on the quotient carries over to a *new* $\mathbf{A}$-action on $\overline{\mathbf{K}}^n$ which can be shown to be given by *polynomials* (i.e., sums of linear operators composed with r^j-th power mappings). That is, a t-module or Drinfeld module, is an action of $\mathbf{A}$ on $\mathbb{G}_a^n$ given by polynomials with certain simple coherence properties attached to the action on the Lie algebra. The notion "t-module" comes from the fact that we view the case of general $\mathbf{A}$ as being "complex multiplication" on objects defined with a t-operation. The coherence conditions basically require that the usual action of $\mathbf{A}$ on the Lie algebra should differ from the new action by a *nilpotent* matrix. Morphisms of t-modules are homomorphisms of $\mathbb{G}_a^n \to \mathbb{G}_a^m$ which *commute* with the t-module actions. Since this is an algebraic notion, it makes sense over any $\mathbf{A}$-field $\mathbb{L}$; i.e., there is a morphism from $\mathbf{A} \to \mathbb{L}$. One sees immediately the analogy with abelian varieties. For more along these lines, see **[A1]**, **[A2]**, **[AT1]**,**[Dr1]**, **[G1]**, **[G2]**, **[Ge1]**, **[Y2]**.

Let φ be a t-module over $\mathbb{L}$ and $z \in \overline{\mathbb{L}}^n$ where $\overline{\mathbb{L}}$ is a fixed algebraic closure of $\mathbb{L}$. We shall write $\varphi_a(z)$ for the t-module action of $a \in \mathbf{A}$ on z. An element z is a-torsion if $0 = \varphi_a(z)$ and so on. With this, one is able to define Tate modules $T_v(\varphi)$ associated to primes $v \in \operatorname{Spec}(\mathbf{A})$; these are free $\mathbf{A}_v$-modules of finite rank. A good analog, $H_v^1(\varphi)$, of ℓ-adic cohomology is the *dual* of T_v. In fact, as with elliptic curves and abelian varieties, the other cohomology groups appear as $\wedge^j H_v^1$. There is also a notion of "purity", due to Anderson, **[A1]**, which ensures that the eigenvalues of Frobenius on H_v^1 are pure of a certain weight. The *rank* of the t-module is then the rank of this H_v^1 which does *not* depend on v. All Drinfeld modules are pure (see Section 3).

From the analogy with abelian varieties, one sees that the dual of the lattice of the t-module (which we will now always assume pure *and* uniformizable — i.e., arises from a lattice) forms the first Betti-cohomology group. And, by using universal extensions, a very good analog of the first de Rham group can be defined; see, for example, **[Ge5]**. Thus one has the three basic "cohomological realizations" of classical arithmetic algebraic geometry.

Dual to the notion of "t-module" φ is the notion of "t-motive". The t-motive of φ is the space of $\mathbb{F}_r$-linear morphisms from $\mathbb{G}_a^n$ to $\mathbb{G}_a$ viewed as a *bimodule* via the two actions of $\mathbf{A}$ — the first action is the standard module action, and the second is induced by φ. Since one can tensor such bimodules, one is then able to form the tensor power, exterior product, etc., of t-modules. Anderson has shown that this tensor product preserves purity and uniformizability. So, for example, the tensor product of two Drinfeld modules is both pure and uniformizable. Moreover, the above realizations also carry over to the category of t-motives.

A key observation, due essentially to Drinfeld, is that the highest exterior product of a Drinfeld module *is* a Drinfeld module of rank 1. As an example that will be used later on, let $\mathbf{A} = \mathbb{F}_r[T]$ and let φ be a rank d Drinfeld module with

$$\varphi_T(x) = Tx + a_1 x^r + \cdots + a_d x^{r^d},\ a_d \neq 0.$$

Then $\wedge^d \varphi$ is isomorphic to the rank one module ψ with $\psi_T(x) = Tx + (-1)^{d-1} a_d x^r$.

Since a t-module φ possesses so many of the properties of abelian varieties, it is natural to try to associate an L-series to it. The procedure is completely clear: For simplicity we assume that φ is defined over $\mathbf{k}$. Then a "good prime" $\wp$ for φ is a prime where the reduction has the same rank as φ. Let $\varphi^{\wp}$ denote this reduction, F its Frobenius morphism and v another prime $\neq \wp$. Then the Euler factor at φ is

$$P_{\wp}(u) = \det\left(1 - Fu \mid H^1_v(\varphi^{\wp})\right),$$

which does not depend on v. The "bad" primes are also handled in the classical fashion (Sections 2 and 3). Consequently, $L(\varphi, s)$ *should* be the product over all $\wp$ of $P_{\wp}(\wp^{-s})^{-1}$, *except* that there is no obvious "s". Finding the appropriate spaces of such "powers", and establishing that they have the correct properties, is a major theme of this paper.

One obvious impediment to defining $\wp^s$ is that, in general, $\mathbf{A}$ is not a factorial ring. Thus there will be primes $\wp$ which are not principal. Fortunately this obstacle can be surmounted as follows (see Subsection 3.3):

1) One chooses a "sign" function sgn on $\mathbf{K}$ which gives a notion of "positivity" generalizing "monic polynomial".
2) Let π be a fixed positive uniformizer in $\mathbf{K}$ and let $a \in \mathbf{A}$ be positive. Let $s = (x, y) \in \overline{\mathbf{K}}^* \times \mathbb{Z}_p$. Then we set

$$a^s = x^{\deg(a)}\left(\pi^{\deg(a)}a\right)^y,$$

which makes sense as $\pi^{\deg(a)}a$ is a 1-unit in $\mathbf{K}$. This definition can be easily seen to have all the usual properties of exponentiation.
3) The strict class group of ideals modulo principal and positively generated ideals is finite. *As* $\mathbf{K}$ *is a field of characteristic* p, one can now lift the definition in 2) to *all* ideals. The total space of such powers is called "S_∞".
4) Let $v \in \operatorname{Spec}(\mathbf{A})$ and let a be positive and prime to v. One defines the space of v-adic powers S_v *exactly* as is done classically for $\mathbb{Z}_p$; so one can define a^{s_v} for $s_v \in S_v$. One can now lift this, *in a fashion compatible with* 3), to arbitrary ideals.

The work in all of these constructions is passing from positively generated principal ideals to all ideals. So choices need to be made and so on. Still it is quite remarkable that it can be done. For instance, let t be an integer and I an ideal. Then we have the mapping $I \mapsto I^t$ which agrees with a^t when $I = (a)$ and a is positive. In fact, one sees that this mapping is a *Grössencharakter* of type A_0.

The reader might be well advised to read the paper a first time with the assumption that $\mathbf{A} = \mathbb{F}_r[T]$ in order to avoid the difficulties associated with non-trivial class groups.

Having discussed exponentiation of ideals, the definition of $L(\varphi, s)$, $s \in S_\infty$, is complete. Moreover, with little extra work, we also obtain a function $L_v(\varphi, x_v, s_v)$ with $0 \neq x_v$ in the algebraic closure, $\overline{\mathbf{k}}_v$, of the completion $\mathbf{k}_v$. These functions are easily seen to converge on "half-planes" determined by the valuations of the characteristic p variables (i.e., x or x_v). We say such functions are "entire" if they give rise to an continuous family of convergent power series in x^{-1} or x_v^{-1}. We say that entire functions are "essentially algebraic" if, when the characteristic 0 variable (i.e., y or s_v) is the *negative* of an integer, the resulting function of x^{-1} or x_v^{-1} is a *polynomial.*

We then have the "fond hope" that $L(\varphi, s)$ and $L_v(\varphi, x_v, s_v)$ are essentially algebraic entire functions with the *same* polynomial special-values (up to a small change of variables and a finite number of Euler factors). Via various estimates on certain sums, we will establish this in the case of complex multiplication.

It now also makes sense to search for the meaning of these new L-series. In general the situation remains very mysterious. In Subsection 3.5, we discuss the formalism of Birch and Swinnerton-Dyer in our context. It turns out that convergence properties seem to imply that the *naive* translation of these ideas to characteristic p does *not* hold much promise. However, we also discuss another formalism, based on the local notion of a zeta-function for a Drinfeld module over a "finite" field (due to Gekeler – see Subsection 3.2), which does hold promise. However, much remains to be done (such as finding non-trivial global invariants of t-modules, etc.).

In Subsection 3.7 we introduce a very general type of L-series for families of t-motives. This L-series combines the theory of t-motives with algebraic geometry (classical motives!). For instance, the reduction modulo p of the zeta functions of Section 2 fits into this framework. At this time, one can only speculate at what techniques will be needed to study these functions in general.

If we work with the "trivial" t-motive, H^0, over a finite extension of $\mathbf{k}$, then we obtain an L-series which is really an analog of Dedekind ζ-functions: Let $L/\mathbf{k}$ be a finite separable extension. We associate to the $\mathbf{A}$-integers, $\mathcal{O}_L$, the function

$$\zeta(\mathcal{O}_L, s) = \sum_{I \text{ an ideal}} nI^{-s},$$

where nI is the norm of the ideal down to $\mathbf{A}$. We prove an analog of Siegel's results on ζ-values at positive even integers under the condition that $L/\mathbf{k}$ is abelian. We find, in fact, that the square of the zeta-value — divided by the appropriate power of a period — lies in the Hilbert class field of $\mathbf{k}$, but does *not* seem to lie in $\mathbf{k}$ itself.

In 4.2.5, we establish a fundamental *double* congruence which relates the polynomial special-values of $\zeta(\mathcal{O}_L, s)$ to classical L-series. With this we are able to describe the trivial zeroes for $\zeta(\mathcal{O}_L, s)$. This then allows the construction of a *first* (i.e., "arithmetic") analog of Bernoulli-numbers (4.3.2). These elements describe the non-triviality of p-primary components of the class groups of division fields of rank one Drinfeld modules. We are also able (Definition 4.3.6) to create a faithful $\mathbf{A}$-module which, conjecturally, gives an arithmetic interpretation of the polynomial special-values themselves.

The above mentioned division fields are a good analog of cyclotomic fields. We call them "geometric cyclotomic function fields" as they contain no non-trivial constant field extensions. On the other hand, constant field extensions (i.e., roots of unity) are called "arithmetic cyclotomic extensions". The double congruence, and the use of classical p-adic Tate modules — obtained by constant field extensions — gives a condition for *cyclicity* of class groups. As we have used constant field extensions in the construction of the Tate module, we call this criterion the "arithmetic criterion for cyclicity". It is a general result that can be shown to apply by direct calculation in many cases.

Finally, we restrict ourselves to $\mathbf{A} = \mathbb{F}_r[T]$ and study the simplest possible L-series, $\zeta(\mathbf{A}, s)$. Since $\mathbf{A}$ is now so simple, one is able to do all sorts of calculations. We introduce enough of the theory of factorials and Γ-functions for these calculations. We also introduce *another* (i.e., "geometric") type of Bernoulli-analog, coming from the positive integers. We show the relationship between these elements and class groups of the Dedekind domains of $\mathbf{A}$ *-integers*. We also use them for a *second* cyclicity criterion, called the "geometric criterion", 5.5.19. A conjecture, due to Gekeler, is discussed which would give the analog of Ribet's theorem. Lastly, the relationship between the two types of cyclicity criterion is given.

We are finally ready to enter the fascinating world of "magic numbers". These are integers whose r-adic expansions are particularly simple (like a usual integer whose digits are all 9's except the highest which may be arbitrary — see Definition 5.7.1). They seem to somehow arise in *every* calculation made in the theory!

Every positive integer can be written as a canonical alternating sum of magic numbers. This expansion *commutes* with the construction of factorials. A dual notion, called "magic duals", allows the canonical expansion of p-adic integers as a convergent alternating sum. This notion then commutes with the construction of Γ-functions. The sum of a magic number and its dual is -1, which is exactly the condition for Thakur's functional equation of gamma functions.

We then re-examine some calculations of Gekeler from the viewpoint of magic numbers. We find (after Example 5.7.15) that these calculations look suspiciously like they fit into some unknown functional equation for $\zeta(\mathbf{A}, s)$. Perhaps more calculations will explain this better.

We also show how the magic numbers arise in a "Kummer-Vandiver conjecture" (5.7.12) for $\mathbf{A}$. This conjecture, which can be checked in a number of instances, is based on *both* the arithmetic and geometric criterions for cyclicity.

Finally, we discuss v-adic measures. We develop the formalism that relates them to divided power series (as p-measures are related to formal power series). We then discuss the divided power series arising from $\zeta(\mathbf{A}, s)$. As before, when the smoke clears, *all* that is left are magic numbers (Theorem 5.8.15). This new occurrence is quite remarkable and may, perhaps, lie at the root of the magic numbers phenomena.

We have tried to be as elementary as possible as well. We have also striven to make the paper as easy to follow as possible. For instance, it should be possible to read the material of the last sections for the polynomial case with only a moderate amount of material from the first sections. For any lapses along these lines, I ask the reader's indulgence.

In reading this paper, the reader will notice that classical objects in number theory have *many* analogs in the function field theory. This arises as for function fields we may have non-trivial ideals or cyclotomic extensions *not* arising from roots of unity etc. By abusing language (and science!) the reader can think of the analogy with physics. So, while asking the reader's forgiveness, the above principal can be stated amusingly as: "classical objects break up into their little 'quarks of arithmetic' in the function field theory". If nothing else, this point of view illustrates both how special classical theory is and how general is its function field counterpart!

2. Review of Classical Theory

Let X be a scheme of finite type over $\mathbb{Z}$. We set $t = \dim X$ as a scheme. Let $x \in X$ be a closed point with corresponding residue field $r(x)$. By our assumptions on x, we see that $r(x)$ is finite. We let Nx be the number of elements in $r(x)$.

2.1 Definition. We set

$$\zeta(X,s) = \prod_{\substack{x\in X\\ x \text{ closed}}} (1 - Nx^{-s})^{-1}\,.$$

The function $\zeta(X,s)$ is called the *zeta function of* X. Clearly, it is a massive generalization of the Riemann zeta function $\zeta(s)$. Indeed, it is trivial to see that

$$\zeta(s) = \zeta(\mathbb{Z}, s)\,.$$

In a similar fashion, using the spectrum of the ring of integers, one obtains the zeta function of a number field.

Elementary estimates (see **[S1]**) give

2.2 Theorem. *The product for $\zeta(X,s)$ converges absolutely for $\mathbb{R}e(s) > t = \dim X$.* □

It is conjectured that $\zeta(X,s)$ has a meromorphic continuation to the entire s-plane, etc., (ibid).

Let Y be another scheme and let $f\colon X \to Y$ be a morphism. Clearly

$$\zeta(X,s) = \prod_{\substack{y\in Y\\ y \text{ closed}}} \zeta(X_y, s)\,,$$

where X_y is the fibre of f at y. In particular, when $Y = \operatorname{Spec}(\mathbb{Z})$ (as we may always assume) we obtain

$$\zeta(X,s) = \prod_{p \text{ prime}} \zeta(X_p, s)$$

which is the original definition of Hasse and Weil. The study of $\zeta(X,s)$ is thus reduced to the study of $\zeta(X_p,s)$ for $p \in \operatorname{Spec}(\mathbb{Z})$.

We now make the following assumption: There exists a number field L, with ring of integers $\mathcal{O}_L$, such that X is projective and flat over $\operatorname{Spec}(\mathcal{O}_L)$ and has smooth and geometrically connected general fibre over $\operatorname{Spec}(L)$. Under this assumption, there is a finite set $S \subset \operatorname{Spec}(\mathcal{O}_L)$ of "bad primes" outside of which the map $X \to \operatorname{Spec}(\mathcal{O}_L)$ is smooth. A prime not in S is said to be "good".

Let $\wp$ be a good prime. As a triumph of 20th century mathematics, one has the following remarkable, and quite satisfactory, cohomological description of the geometric points of $X_\wp$: First of all, one makes a very simple charge of variables by setting $u = N\wp^{-s}$, where $N\wp =$ number of elements in $\mathbb{F}_\wp = \mathcal{O}_L/\wp$. Next, let $x \in X_\wp$ be a closed point; one sees easily that $r(x)$ is a finite extension of $\mathbb{F}_\wp$. As such, Nx^{-s} is always an integral multiple of u. We put

$$Z(X_\wp, u) = \zeta(X_\wp, s).$$

Finally, we set $d = \dim(X \otimes \mathrm{Spec}(L)) = \dim X - 1 = \dim(X_\wp)$.

In what follows, we are guided by the exposition in **[Ha]** which also contains a good history of this subject. For $n \geq 1$, let $\mathbb{F}_{\wp^n}$ be the unique (up to isomorphism) extension of $\mathbb{F}_\wp$ of degree n. Let ν_n be the number of points of $X_\wp$ rational over $\mathbb{F}_{\wp^n}$; since $X_\wp$ is projective, there is no problem in showing that ν_n is finite. One then sees directly

2.3 Lemma. $\log Z(X_\wp, u) = \sum_{n=1}^{\infty} \frac{\nu_n}{n} u^n$. □

Now let ℓ be a prime number which is not equal to the characteristic of $\mathbb{F}_\wp$. We let $\overline{\mathbb{F}}_\wp$ be a fixed algebraic closure of $\mathbb{F}_\wp$ which we assume contains $\mathbb{F}_{\wp^n}$ for all n. We put $\overline{X}_\wp = X \otimes \overline{\mathbb{F}}_\wp$. Due to Weil, Grothendieck, M. Artin, and others, one knows the existence of a "good" ℓ-adic cohomology theory $H^i(\overline{X}_\wp, \mathbb{Q}_\ell) = H^i_{\acute{e}t}(\overline{X}_\wp, \mathbb{Q}_\ell)$; in fact such a theory exists for schemes over any algebraically closed field. This theory satisfies the axioms one would expect from singular theory:

a) H^i is finite dimensional over $\mathbb{Q}_\ell$ for all i;
b) $H^i = \{0\}$ for $i > 2d$;
c) H^i is a contravariant functor;
d) Both Poincaré duality (for smooth varieties) and the Lefschetz fixed point formula hold, etc.

Let F be the $N\wp$-th Frobenius morphism on $\overline{X}_\wp$; so if x has homogeneous coordinates $[x_1, \ldots, x_j]$, then

$$F(x) = [x_1^{N\wp}, \ldots, x_j^{N\wp}].$$

The Lefschetz fixed point formula gives

2.4 Theorem.

$$Z(x_\wp, u) = \frac{P_1(u) \cdots P_{2d-1}(u)}{P_0(u) \cdots P_{2d}(u)},$$

where

$$P_i(u) = \det(1 - Fu \mid H^i(\overline{X}_\wp, \mathbb{Q}_\ell)).$$ □

2.5 Theorem (Deligne). *$P_i(u)$ is a polynomial with integer coefficients which is invariant of ℓ. Moreover*

$$P_i(u) = \prod (1 - \alpha_{ij} u)$$

where α_{ij} is an algebraic integer and $|\alpha_{ij}| = N\wp^{i/2}$. □

Finally, put $E = \Sigma(-1)^i B_i$ where $B_i = \dim H^i(\overline{X}_\wp, \mathbb{Q}_\ell)$. Then Poincaré duality gives:

2.6 Theorem.

$$Z\left(X_\wp, \frac{1}{N\wp^d u}\right) = \pm N\wp^{\frac{dE}{2}} u^E Z(X_\wp, u).$$ □

The proof of 2.6 is based on the following elementary lemma. We present the statement as we shall use it again later on.

2.7 Lemma. *Let $\langle v, w\rangle : V \times W \to k$ be a perfect pairing of vector spaces over a field k. Let $\dim V = \dim W = \alpha$. Let $\varphi : V \to V$ and $\psi : W \to W$ be two endomorphisms such that there exists $\lambda \in k$ with*

$$\langle \varphi v, \psi w\rangle = \lambda \langle v, w\rangle ,$$

for all $v \in V$ and $w \in W$. Then

$$\det(1 - \psi u \mid W) = \frac{(-1)^\alpha \lambda^\alpha u^\alpha}{\det(\varphi)} \det\left(1 - \frac{\varphi}{\lambda u} \mid V\right),$$

and

$$\det(\psi) = \frac{\lambda^\alpha}{\det(\varphi)}. \qquad \square$$

The $\mathbb{Q}_\ell$-vector spaces H^i also come equipped with a continuous Galois action. This action arises from the Galois action on coefficients of homogeneous polynomials. In the case of $\overline{X}_\wp$, we let $\sigma \in \mathrm{Gal}(\overline{\mathbb{F}}_\wp/\mathbb{F}_\wp)$ be the $N\wp$-th power mapping on $\overline{\mathbb{F}}_\wp$. One sees that $\sigma \circ F$ acts as the *identity* on H^i; thus σ^{-1} acts as F on $H^i(\overline{X}_\wp, \mathbb{Q}_\ell)$. For this reason, σ^{-1} is called the "geometric Frobenius". The automorphism σ is called the "arithmetic Frobenius".

We remark in passing that the above theorems hold for all projective smooth varieties over $\mathbb{F}_\wp$. It is not necessary that the variety be the reduction of an object over $\mathbb{Q}$.

Return now to X (as opposed to $X_\wp$). Let $\overline{L}$ be a fixed algebraic closure of L; so $\mathrm{Gal}(\overline{L}/L)$ acts on $H^i(\overline{X}, \mathbb{Q}_\ell)$ where $\overline{X} = X \otimes \overline{L}$. A basic result assures us that $H^i(\overline{X}, \mathbb{Q}_\ell) = H^i(\overline{X}_\wp, \mathbb{Q}_\ell)$. (The reader should think of smooth families of compact, complex manifolds over the unit disc and rational singular theory; the cohomology of the fibre over 0 is isomorphic to the general cohomology. Tate modules of abelian varieties and their duals are also good illustrations of this result.) One then has:

a) The action of $\mathrm{Gal}(\overline{L}/L)$ on $H^i(\overline{X}, \mathbb{Q}_\ell)$ is unramified at $\wp$; i.e., the inertia group of any extension of $\wp$ to $\overline{L}$ acts trivially. (N.B.: we are still assuming that ℓ is prime to $N\wp$).
b) Let $\overline{\wp}$ be any extension of $\wp$ to $\overline{L}$ and let $D_{\overline{\wp}} \subseteq \mathrm{Gal}(\overline{L}/L)$ be the decomposition group. Let $\sigma_{\overline{\wp}}$ be the geometric Frobenius in $D_{\overline{\wp}}$. Then the actions of $\sigma_{\overline{\wp}}$ on $H^i(\overline{X}, \mathbb{Q}_\ell)$ and $H^i(\overline{X}_\wp, \mathbb{Q}_\ell)$ are the same. Thus the characteristic polynomial of $\sigma_{\overline{\wp}}$ depends only on $\wp$.

Set

$$\zeta^{\mathrm{good}}(X, s) = \prod_{\wp \text{ good}} \zeta(X_\wp, s).$$

It is now clear that there is a natural factorization of ζ^{good}: For each $i = 0, \ldots, 2d$, define

$$L^{\mathrm{good}}(X^{(i)}, s) = \prod_{\wp \text{ good}} \det(1 - \sigma_\wp N\wp^{-s} \mid H^i(\overline{X}, \mathbb{Q}_\ell))^{-1}$$

where $\sigma_\wp$ is *any* geometric Frobenius at $\wp$. Theorem 2.4 then translates immediately into an expression of ζ^{good} in terms of $\{L^{\text{good}}(X^{(i)}, s)\}$. As with $\zeta(X, s)$, elementary considerations using Theorem 2.5 imply that the above functions converge on half-planes.

Finally, we need to discuss the factors at the bad primes. I am indebted to P. Deligne for illuminating correspondence on this point **[D1]**. Thus, let $\wp$ now be a bad prime; so $X_\wp$ is complete but singular. It remains true that the Lefschetz fixed point formula holds for $H^i(\overline{X}_\wp, \mathbb{Q}_\ell)$; thus $\zeta(X_\wp, s)$ is still factored by cohomology. The question becomes how to find the cohomology of the bad fibre in terms of the cohomology of the general fibre; i.e., the groups $H^i_{\acute{e}t}(\overline{X}, \mathbb{Q}_\ell)$. Let $I_{\overline{\wp}} \subseteq D_{\overline{\wp}}$ be the inertia group. Let ℓ be a prime not dividing $N\wp$. In many cases one has

$$H^i(\overline{X}_\wp, \mathbb{Q}_\ell) = H^i(\overline{X}, \mathbb{Q}_\ell)^{I_{\overline{\wp}}}$$

$=$ fixed space of inertia. Moreover, under our hypotheses, there is always a map

$$H^i(\overline{X}_\wp, \mathbb{Q}_\ell) \to H^i(\overline{X}, \mathbb{Q}_\ell)^{I_{\overline{\wp}}}.$$

(The reader will perhaps find this believable by considering a family of complex manifolds over the pointed disc which degenerates at the origin and using singular cycles.) One "expects" that this map will be surjective when X is a regular scheme (so that any singularities of $X_\wp$ "arise from the map" $X \to \operatorname{Spec}(\mathcal{O}_L)$). Of course, X is generically regular as it is smooth over Spec(L). Thus the regularity of X is a question of what model one uses to extend the generic fibre over the bad primes. Different models may give different bad fibres, etc.

The above discussion leads to the following definition of the L-series at *all* finite primes of L. The reader may view this as a vast generalization of the definition given by E. Artin for finite Galois representations:

$$L(X^{(i)}, s) = \prod_\wp \det(1 - \sigma_\wp N\wp^{-s} \mid H^i(\overline{X}, \mathbb{Q}_\ell)^{I_\wp})^{-1},$$

where ℓ is as before and $\{I_\wp, D_\wp, \sigma_\wp\}$ are *any* choices of decomposition group $D_\wp$, inertia group $I_\wp$ and geometric Frobenius $\sigma_\wp \in D_\wp$. (Elementary linear algebra shows that the determinant is invariant of $\{D_\wp, I_\wp, \sigma_\wp\}$ at the bad primes.) It is expected that the factors at bad primes will have integral coefficients invariant of the choice of ℓ. This can be verified in the case of abelian varieties (see, for instance, the discussion on Néron theory in SGA 7).

The above definition of L-series corresponds to the construction in étale theory which takes the local system given by H^i on $\operatorname{Spec}(\mathcal{O}_L) - S$ and extends it to $\operatorname{Spec}(\mathcal{O}_L)$.

It is expected-hoped that these L-series will have analytic continuations and functional equations. For these, one needs Γ-factors at the infinite primes of L, **[S2]**. The determination of these factors arises from the *complex* cohomology of X.

From the viewpoint of motives, given i, one finds an object whose total cohomology *is* H^i. The reader may consider "$X^{(i)}$" as being this object.

Finally, we mention a "drawback" in the theory discussed in this section: Let X be a curve over a finite field and let x, y be two closed points of X. Suppose, as happens often, that $Nx = Ny$. Then both x and y contribute the *same* Euler factor to $\zeta(X, s)$. The reader will note that this is very different from what happens classically with Dirichlet

L-series over $\mathrm{Spec}(\mathbb{Z})$; indeed, the local factors of such L-series are well-known to be uniquely determined by the L-series. In our next section, we will discuss a theory of L-series in finite characteristic that *does* have unique Euler factors.

3. Generalities on L-series of Drinfeld Modules and t-Motives

3.1. Background. As in the introduction, we let $\mathcal{C}$ be a smooth, projective, geometrically irreducible curve over $\mathbb{F}_r$, $r = p^m$. Let $\infty \in \mathcal{C}$ be a closed point and $\mathbf{A}$ the ring of functions on $\mathcal{C}$ holomorphic outside of ∞. Let $\mathbf{k}$ be the function field of $\mathcal{C}$ and let $\mathbf{K} = \mathbf{k}_\infty$ be its completion at ∞.

Let $v \in \mathrm{Spec}(\mathbf{A})$. We denote by $\mathbf{k}_v$ the completion of $\mathbf{k}$ at v and $\mathbf{A}_v \subseteq \mathbf{k}_v$ the ring of integers. We let $\overline{\mathbf{K}}$ be a fixed algebraic closure of $\mathbf{K}$ which we equip with the canonical extension of the normalized absolute value $|?| = |?|_\infty$ of $\mathbf{K}$. We let $\overline{\mathbf{k}} \subset \overline{\mathbf{K}}$ be the algebraic closure of $\mathbf{k}$ and $\mathbf{k}^s \subset \overline{\mathbf{k}}$ the separable closure.

Let $a \in \mathbf{k}^*$. We set (a) to be the principal divisor generated by a as usual.

Let D be a divisor of $\mathbf{k}$ which is rational over $\mathbb{F}_r$. We let $\deg(D)$ be its degree and set

$$\delta = \deg(\infty).$$

Finally, if $\varrho \subset \mathbf{A}$ is an ideal, then we set $\deg(\varrho)$ to be the degree of the underlying effective divisor.

Let h denote the class number of the *field* $\mathbf{k}$. It is then an easy exercise to check that the class number of $\mathbf{A}$, as a Dedekind domain, is $h\delta$. Moreover, the unit group of $\mathbf{A}$, $\mathbf{A}^*$, is clearly $\mathbb{F}_r^*$.

The following notation will be quite useful: Let R be any ring with finite unit group. Let E be the order of R^*. We say an integer is "R-even" if and only if it is divisible by E; otherwise it is "R-odd". Thus "$\mathbb{Z}$-even" has the same meaning as "even", etc. An integer will be $\mathbf{A}$-even if and only if it is divisible by $r-1$ and so on.

Let $L \subseteq \mathbf{k}^s$ now be a fixed finite separable extension. Let $\mathbb{F} \subseteq L$ be its field of constants and set $t_{\mathbb{F}} = [\mathbb{F}:\mathbb{F}_r]$. Let D be a divisor of L rational over $\mathbb{F}$; we let $\deg_1(D)$ be its degree over $\mathbb{F}$ and $\deg(D) = t_{\mathbb{F}} \deg_1(D)$ its degree over $\mathbb{F}_r$. Finally let $\mathcal{O} = \mathcal{O}_L \subseteq L$ be the ring of $\mathbf{A}$-integers.

Let $\wp \in \mathrm{Spec}(\mathcal{O})$. As in Section 2, we let $N\wp$ be the number of elements in $\mathcal{O}_L/\wp$. We let n be the norm map from $\mathcal{O}_L \to \mathbf{A}$ which we extend to ideals in the usual fashion. Set $\mathcal{B} = \wp \cap \mathbf{A}$. Then

$$n\wp = \mathcal{B}^f$$

where $f = [\mathcal{O}/\wp : \mathbf{A}/\mathcal{B}]$ and $N\wp = (N\mathcal{B})^f$.

We let $\infty_1, \ldots, \infty_\varepsilon$ be the primes of L above ∞ with degrees $\delta_1, \ldots, \delta_\varepsilon$ over $\mathbb{F}$. We use the symbol "w" to denote an arbitrary prime of L and $\wp$ to denote a prime of $\mathcal{O}$. We let L_w, $\mathcal{O}_\wp$ be the completions, etc. We always view L_{∞_i} as being a subfield of $\overline{\mathbf{K}}$ via ∞_i.

The pair $(L, \mathcal{O}_L)$ will play the role taken in Section 2 by the number field L and its ring of integers.

Let φ be a rank d Drinfeld module defined over L that we shall fix for the rest of Section 3. Our goal is to define and study its L-series following Hasse and Weil (see Subsection 3.4). Thus it is clear that we must begin by a discussion of the reduction of φ at the primes of $\mathcal{O}$.

Let $\{\alpha_1, \ldots, \alpha_c\}$ be the generators of $\mathbf{A}$ as an $\mathbb{F}_r$-algebra. Clearly, for almost all $\wp$ in Spec$(\mathcal{O})$, $\{\varphi_{\alpha_1}(x), \ldots, \varphi_{\alpha_c}(x)\}$ have coefficients in $\mathcal{O}_\wp$ with highest coefficient in $\mathcal{O}_\wp^*$; we say that such primes are "good primes" for φ. For those primes not in this set, it may (or may not) happen that a change of variable $\varphi \mapsto \alpha^{-1}\varphi\alpha$, $\alpha \in L^*$, will alter the module so that the above condition again holds. If this happens, we again say that $\wp$ is a "good prime" for φ; otherwise it is a "bad prime", see **[Dr1]**, **[Ta1]**.

Let $\wp$ now be a good prime, and assume that φ has been transformed, as above, so that $\{\varphi_{\alpha_1}(x), \ldots, \varphi_{\alpha_c}(x)\}$ have coefficients in $\mathcal{O}_\wp$ with highest coefficient always in $\mathcal{O}_\wp^*$. By simply reducing the coefficients of these polynomials $(\mathrm{mod}\, \wp)$ we obtain a rank d Drinfeld module over $\mathcal{O}_L/\wp = \mathbb{F}_\wp$. We denote this module "$\varphi^\wp$".

Next we study these "finite" Drinfeld modules.

3.2. The Finite-theory. In this subsection, and this subsection only, we let φ denote a rank d Drinfeld module over the finite field $\mathcal{O}/\wp = \mathbb{F}_\wp$.

Let $v \in$ Spec$(\mathbf{A})$ be prime to $\wp$. We begin by studying the "v-adic cohomology" of φ. The reader will see that this construction is completely elementary in the fashion of the ℓ-adic cohomology of elliptic curves. The first step is the construction of the "v-adic Tate module", $T_v(\varphi)$. This object is defined by

$$T_v(\varphi) = \mathrm{Hom}_{\mathbf{A}}(\mathbf{k}_v/\mathbf{A}_v, \varphi(\overline{\mathbb{F}}_\wp)),$$

where $\overline{\mathbb{F}}_\wp$ is an algebraic closure of the finite field $\mathcal{O}_L/\wp$ and $\varphi(\overline{\mathbb{F}}_\wp) = \overline{\mathbb{F}}_\wp$, but considered as an $\mathbf{A}$-module via φ.

Elementary counting arguments **[Dr1]** show that

$$T_v(\varphi) \simeq \mathbf{A}_v^d$$

and that it is a covariant functor; so it acts as one dimensional "homology" with $\mathbf{A}_v$-coefficients. To get cohomology, as with abelian varieties, we dualize and set

$$\begin{aligned} H_v^1(\varphi, \mathbf{k}_v) &= \mathrm{Hom}_{\mathbf{A}_v}(T_v, \mathbf{k}_v) \\ &\simeq \mathbf{k}_v^d. \end{aligned}$$

Moreover, following the theory of abelian varieties, we put, for $i = 0, \ldots, d$,

$$H_v^i(\varphi, \mathbf{k}_v) = \wedge^i H_v^1(\varphi, \mathbf{k}_v).$$

Clearly, these groups are contravariant functors. Also clearly, the important group is H^1.

Let F be the endomorphism of $\mathbb{G}_a/\mathbb{F}_\wp$ defined by

$$F(x) = x^{r_1},$$

where $r_1 = \#\mathcal{O}_L/\wp = \#\mathbb{F}_\wp$. As the coefficients of $\varphi_a(x)$ are in $\mathbb{F}_\wp$, F commutes with the action of φ. It thus gives rise to an endomorphism of φ called the *Frobenius morphism*.

In what follows next, we are guided by some lovely ideas of Gekeler in **[Ge6]**. Let $\mathbb{F}_{\wp^n}$ be the unique (up to isomorphism) extension of $\mathbb{F}_\wp$ of degree n. Via φ, $\mathbb{F}_{\wp^n}$ obviously becomes a finite torsion $\mathbf{A}$-module.

Now let $\mathbf{M}$ be any finite $\mathbf{A}$-module. We define its *Euler-Poincaré characteristic*, $\chi(\mathbf{M})$, as follows:

1) $\chi(\mathbf{M}) = \mathcal{B}$ if $\mathbf{M} \simeq \mathbf{A}/\mathcal{B}$ for $\mathcal{B}$ prime.
2) Let $0 \to \mathbf{M}_1 \to \mathbf{M} \to \mathbf{M}_2 \to 0$ be exact.

Then

$$\chi(\mathbf{M}) = \chi(\mathbf{M}_1)\chi(\mathbf{M}_2).$$

We let $\chi(\wp^n) = \chi(\mathbb{F}_{\wp^n})$ be the Euler characteristic of $\mathbb{F}_{\wp^n}$ viewed as $\mathbf{A}$-module via φ.

Following classical theory, set

$$P_i(u) = \det\left(1 - Fu \mid H_v^i(\varphi, \mathbf{k}_v)\right),$$

and set $P(u) = P_i(u)$. By our construction of H^i as $\wedge^i H^1$, one sees easily that $P(u)$ determines $P_i(u)$ for all i.

In the next result, we set $\frac{d}{du}\log f(u) =: f'(u)/f(u)$ as usual.

3.2.1. Lemma.

$$\sum_{k\geq 1} \det\left(1 - F^k \mid H_v^1(\varphi, \mathbf{k}_v)\right) u^k = u\frac{d}{du}\log \prod_{0\leq t\leq d} P_t(u)^{(-1)^{t+1}}.$$

Proof. Let ϕ be *any* endomorphism of a d-dimensional vector space V and let $\wedge^i\phi$ act on $\wedge^i V$ in the standard fashion. Let $Q_i(x) = \det(1 - x\wedge^i\phi \mid \wedge^i V)$. Then the statement of the lemma is just an instance of the general equation

$$\sum_{j\geq 1}\det(1-\phi^j)x^j = x\frac{d}{dx}\log\prod_{0\leq i\leq d} Q_i(x)^{(-1)^{i+1}}.$$

On the other hand, the above equation is derived easily from the following well-known (and readily seen) identities:

a) $\det(1 - x\phi) = \sum_{0\leq i\leq d}(-1)^i \mathrm{Tr}(\wedge^i\phi)x^i$ (apply to ϕ^j and set $x = 1$),

b) $-x\frac{d}{dx}\det(1 - x\phi) = \sum_{e\geq 1}\mathrm{Tr}(\phi^e)x^e$, and

c) $\frac{d}{dx}\log(f\cdot g) = \frac{d}{dx}\log(f) + \frac{d}{dx}\log(g)$. □

Note that the classical analog of 3.2.1 (Theorem 2.3) uses the logarithm itself. We are forced to use $\frac{d}{du}\log$ for the simple reason that $\log(u)$ does not exist in characteristic p. Thus, following classical theory, Theorem 2.4, we make the following

3.2.2. Definition (Gekeler). We set

$$Z(\varphi, u) = \prod_{0\leq t\leq d} P_t(u)^{(-1)^{t+1}}.$$

For this definition to have any arithmetic content, we need precisely to interpret the left hand side of 3.2.1 in terms of the structure of φ. This will be done in Corollary 3.2.5.

We begin by noting that F^k is the Frobenius associated to $\mathbb{F}_{\wp^k}$; thus it suffices to set $k=1$ and to discuss $\det\left(1-F \mid H_v^1(\varphi,\mathbf{k}_v)\right)=P(1)$, etc. Let

$$P(u)=\prod(1-\alpha_j u),$$

where $\{\alpha_j\}\subset\overline{\mathbf{k}}_v$. We set

$$\lambda(F^k)=\prod\alpha_j^k=\lambda(F)^k.$$

3.2.3. Theorem.

a) *We have* $(P(1))=\chi(\mathbb{F}_\wp)$.

b) *The polynomial* $P(u)$ *has coefficients in* $\mathbf{A}$ *and is invariant of the choice of* v *(subject to* v *being prime to* $\wp$*).*

c) *The elements* $\{\alpha_j\}\subset\overline{\mathbf{k}}$ *are algebraic integers. Moreover*

$$|\alpha_j|=r_1^{1/d}$$

where $r_1=N\wp$ *is the number of elements in* $\mathbb{F}_\wp$.

d) *We have*

$$(\lambda(F))=n\wp. \qquad \square$$

Theorem 3.2.3 is proved by exploiting the extremely strong analogies between φ and the theory of elliptic curves over finite fields. The result, Theorem 3.2.3.c), is the *Riemann Hypothesis* for φ; it was known to Drinfeld **[Dr2]**. Parts c) and d) obviously put strong restrictions on what polynomials can be candidates for $P(u)$.

Part d) of the above result also shows that $n\wp$ is principal. This fact can also be derived from the existence of a *determinant* Drinfeld module $\det\varphi$ associated to φ. By definition, $\det\varphi$ is a rank one Drinfeld module with the action of F on $H_v^1(\det\varphi,\mathbf{k}_v)$ being given by $\wedge^d F$. But, the coarse moduli of rank one Drinfeld modules is the spectrum of the integers in the Hilbert class field split completely at ∞. Thus the norm must factor through this scheme. Consequently, we see again that the norm is principal. (In characteristic 0, one can use the formalism of the Artin symbol and reverse the argument to show that if the norm is principal, then the field must contain the Hilbert class field. For function fields, the situation is more delicate.)

We now immediately deduce:

3.2.4. Corollary.

a) $P_t(u)$ *has coefficients in* $\mathbf{A}$ *and is invariant of the choice of* v.

b) *Let* $P_t(u)=\prod(1-\alpha_{tj}u)$ *where* $\{\alpha_{tj}\}\subset\overline{\mathbf{k}}$ *are algebraic integers. Then*

$$|\alpha_{tj}|=r_1^{t/d},$$

with r_1 *as in Theorem 3.2.3.* $\square$

3.2.5. Corollary.

$$\chi(\wp^k)=\left(\det\left(1-F^k \mid H_v^1(\varphi,\mathbf{k}_v)\right)\right). \qquad \square$$

Corollary 3.2.5 gives the arithmetic interpretation of $Z(\varphi, u)$. The analogy with Lemma 2.3 will be immediate to the reader.

Let $\lambda = \lambda(F)$. A straightforward computation using Lemma 2.7 gives the following result. (Since we know H^i in terms of H^1, it is easy to determine the signs that appear.)

3.2.6. Proposition. *Let d be the rank of φ as before and put $Z(u) = Z(\varphi, u)$. Then we have:*

a) *Let $d = 1$. Then $Z(u) = \frac{1-\lambda u}{1-u}$ and*

$$Z\left(\frac{1}{\lambda u}\right) = \lambda / Z(u).$$

b) *Let $d > 1$ with d odd. Then*

$$Z\left(\frac{1}{\lambda u}\right) = 1/Z(u).$$

c) *Let d be even. Then*

$$Z\left(\frac{1}{\lambda u}\right) = Z(u).$$ □

3.2.7. Remarks. The following example, pointed out by Dinesh Thakur, illustrates some subtleties in the construction of the Z-function. Let $r = 2$ and put

$$Z_1(u) = \frac{(1 - Tu)}{(1 - u)},$$

and

$$Z_2(u) = \frac{(1 - Tu^2)}{(1 - u)(1 - Tu)}.$$

The functions $Z_i(u)$ are of the form of the Z-function of rank i Drinfeld modules for $i = 1, 2$. However, they both have the *same* logarithmic derivative.

On the other hand, if we *fix* the rank of the Drinfeld module, we have the following pleasing result. I am indebted to Gekeler for his correspondence **[Ge7]** outlining the proof.

3.2.8. Theorem. *The function $Z(\varphi, u)$ is uniquely determined by the rank d of φ, and the numbers*

$$\left\{\det\left(1 - F^k \mid H_v^1(\varphi, \mathbf{k}_v)\right)\right\}_{k=1}^{\infty}.$$

Before giving the proof of 3.2.8, we need to discuss a bit more the formalism involved in the construction of the Z-functions.

3.2.9. Definition. Let $h(x) = \sum a_i x^i$ and $g(x) = \sum b_i x^i$ be two power series in the variable x. We define the *Hadamard product* $(h * g)(x)$ by

$$(h * g)(x) = \sum a_i b_i x^i.$$

Now let $f(x)$ be a polynomial in x of degree d, with $1 = f(0)$ and with coefficients in $\overline{\mathbf{K}}$. The function $Z_f(x)$ is formed in the same fashion as given above for $Z(u)$ (by, e.g., viewing $f(x)$ as the characteristic polynomial of some endomorphism of $\overline{\mathbf{K}}^d$), and we set

$$U_f(x) = x\frac{x}{dx}\log Z_f(x)$$

in the obvious fashion. Note that $U_f(x) \in x\overline{\mathbf{K}}[[x]]$.

3.2.10. Lemma.

$$U_{fg} = U_f * U_g.$$

Proof. Use the identities given in the proof of 3.2.1, and induction. □

3.2.11. Corollary. *Assume that the roots of $f(x)$ have absolute values < 1. Then all coefficients of $U_f(x)$ are nonzero.*

Proof. The lemma allows us to reduce to the linear case where the result is trivial. □

3.2.12. Lemma. *Let f, g be two polynomials of degree d in $\overline{\mathbf{K}}$ with constant term 1 and roots of fixed absolute value $\delta < 1$. If*

$$U_f(x) = U_g(x),$$

then $f(x) = g(x)$.

Proof. The proof follows by induction on d. For $d = 1$ the result is trivial. Thus let $d > 1$. There are then two cases:
Case 1: f and g have some common factor, say, $h(x) = (1 - x/\lambda)$ where $\lambda \in \overline{\mathbf{K}}$ and $|\lambda|_\infty < 1$. Let $f(x) = h(x)f_1(x)$ and $g(x) = h(x)g_1(x)$. From 3.2.10 and 3.2.11 we conclude that

$$U_{f_1}(x) = U_{g_1}(x)$$

whence the result follows by induction. Thus we are reduced to:
Case 2: f and g are relatively prime. Note that a little calculus, together with the equality $U_f = U_g$, shows that

$$\frac{Z_f(x)}{Z_g(x)}$$

must be the p-th power of another power series $\alpha(x)$ (remember, we are working over $\overline{\mathbf{K}}$).

Let

$$\frac{Z_f(x)}{Z_g(x)} = \frac{H(x)}{G(x)},$$

where $H(x)$ and $G(x)$ are relatively prime. From the equality,

$$H(x) = G(X)\alpha(x)^p,$$

and the product rule, we conclude that

$$H'(X) = \alpha(x)^p G'(X),$$

whence

$$\frac{H(x)}{G(x)} = \frac{H'(x)}{G'(x)},$$

or

$$G'(x)H(x) = H'(x)G(X).$$

But, as G and H are relatively prime, this implies $G(x)$ divides $G'(X)$; thus $G'(x) = H'(x) = 0$. We thus conclude, as before, that $G(x)$ and $H(x)$ are p-powers and that $\alpha(x)$ is also a rational function in x.

Because of our restrictions on the absolute values of the roots of $f(x)$ and $g(x)$, it follows directly that

$$\frac{f(x)}{g(x)}$$

is the p-th power of a rational function. Finally, as f and g are relatively prime, we see that $f(x) = f_1(x)^p$ and $g(x) = g_1(x)^p$ and the result follows by induction. □

Proof of 3.2.8.. This result now follows immediately from 3.2.12 and the Riemann Hypothesis for φ. □

3.2.13. Remarks.

a) As happens with abelian varieties over finite fields, the polynomial $P(u)$ determines φ *up to isogeny* **[Ge6]**. Thus, from 3.2.12, we see that the isogeny class of φ is determined by the rank of φ and

$$\left\{\det\left(1 - F^k \mid H_v^1(\varphi, \mathbf{k}_v)\right)\right\}_{k=1}^{\infty}.$$

b) The action of F on $H_v^1(\varphi, \mathbf{k}_v)$ is *semisimple.* Indeed the endomorphisms of φ form a division algebra and $\mathbf{k}(F)$ a commutative subfield. The result then follows by standard lemmas on tensor products of global fields. It is interesting, however, that this action need not be *geometrically* semisimple; i.e., diagonalizable over the algebraic closure $\overline{\mathbf{k}}_v$. In fact, let $r = 2$, $L = \mathbf{k}$ and $\mathbf{A} = \mathbb{F}_r[T]$. Let φ be the rank two module defined by $\varphi(T) = TF^0 + F^2$. Then over $\mathbf{A}/(T)$, we have $F^2 = \varphi(T)$. Thus F satisfies the purely inseparable equation $x^2 - T = 0$. (In general, when there is a power i so that $F^i \in \mathbf{A}$ we are dealing with *supersingular* Drinfeld modules **[Ge6]**).

c) Much of the above goes through in the context of Anderson's t-modules, **[A2]**. That is, for those t-modules that are "pure", one still has the Riemann Hypothesis on the roots of Frobenius. However, no interpretation has yet been given of the associated Z-function in terms of Euler-Poincaré characteristics. If one had such an interpretation, then one should be able to give an interpretation of $P_i(u)$ in terms of the "motive" $\wedge^i H^1 = \varphi^{(i)}$.

One question now naturally arises: to what extent are the numbers

$$\left\{\det\left(1 - F^k \mid H_v^1(\varphi, \mathbf{k}_v)\right)\right\}_{k=1}^{\infty}$$

uniquely determined by the Euler-Poincaré characteristics; i.e., the *ideals* generated by these numbers. This is partially answered by the following result which uses the elementary notion of a "sign function" defined at the very beginning of our next subsection.

3.2.14. Proposition.

a) *Let the rank of* $\varphi = 1$. *Then the Euler-Poincaré characteristics uniquely determine*

$$\left\{\det\left(1 - F^k \mid H_v^1(\varphi, \mathbf{k}_v)\right)\right\}_{k=1}^{\infty}.$$

b) *Let* φ *have arbitrary rank* d. *Let* ψ *be another Drinfeld module of rank* d *with the same Euler-Poincaré characteristics as* φ. *Then* ψ *and* φ *are isogenous over the field* $\mathbb{F}_{\wp^k}$ *with* $k = r^\delta - 1$.

Proof. From 3.2.3.d) we know that $\chi(\wp) = (1 - \lambda)$ where $\lambda \in \mathbf{A}$ and $n\wp = (\lambda)$. Suppose that we have another number, say λ_1, with the same properties and with an equality of ideals

$$(1 - \lambda) = (1 - \lambda_1).$$

Thus there exists $\zeta \in \mathbb{F}_r^*$ with

$$\zeta(1 - \lambda) = 1 - \lambda_1.$$

But $\zeta(1 - \lambda) = 1 - (1 - \zeta) - \lambda$; so

$$\lambda_1 = \lambda + 1 - \zeta.$$

Since (λ_1) also equals $n\wp$, this equation forces $\zeta = 1$. Thus Part a) is established.

To see Part b), we now use a sign function sgn on $\mathbf{K}$. By the Riemann Hypothesis, 3.2.3.c), one sees

$$\operatorname{sgn}\left(\det\left(1 - F^k \mid H_v^1(\varphi, \mathbf{k}_v)\right)\right)$$

is determined by

$$\operatorname{sgn}\left(\lambda(F^k)\right) = \operatorname{sgn}\left(\lambda(F)\right)^k.$$

Thus for $k \equiv 0 \pmod{(r^\delta - 1)}$ we see that

$$\det\left(1 - F^k \mid H_v^1(\varphi, \mathbf{k}_v)\right)$$

is the unique *positive* (i.e., with sign 1) generator of $\chi(\wp^k)$. The exact same argument also holds for ψ and we deduce an equality of the two positive generators of the same (by hypothesis) ideal.

Now base change up to $\mathbb{F}_{\wp^{(r^\delta - 1)}}$. Over this extension, the Frobenius gets replaced by its $r^\delta - 1$-st power, etc. The result now follows by the above argument as in 3.2.13.a). □

As a corollary of the proof of the above result, we see that the isogeny class of φ is determined by its Euler characteristics and $\operatorname{sgn}(\lambda(F))$. It would be interesting to know if 3.2.14.b is the best possible result.

We end this subsection by defining a "mixed" zeta-function that intertwines the theory presented in this subsection with that of Section 2: Let $\wp \in \operatorname{Spec}(\mathbf{A})$ and, as before, put $\mathbb{F}_\wp = \mathbf{A}/\wp$. Let X now be a scheme of finite type over $\mathbb{F}_\wp$. Let d_x, for x a closed point of X, be the degree of the residue class field $r(x)$ over $\mathbb{F}_\wp$.

Now assume that there is a family φ/X of rank d Drinfeld modules over X. Thus, for fixed $x \in X$, there is a rank d Drinfeld module, $\varphi^{(x)}$, defined over $r(x)$ and thus over $\mathbb{F}_\wp$. We can therefore apply the theory just developed. Let F_x be the Frobenius morphism of $\varphi^{(x)}$ over $r(x)$ and let $v \neq \wp \in \operatorname{Spec}(\mathbf{A})$.

3.2.15. Definition. Set

$$L(\varphi/X, u) = \prod_{x \text{ closed}} P_x(u^{d_x})^{-1},$$

where, as before,

$$P_x(u) = \det\left(1 - uF_x \mid H_v^1(\varphi^{(x)}, \mathbf{k}_v)\right).$$

It will be immediate to the reader how one could work with cohomology of all dimensions. Moreover, using the set-up of our next subsection, we can also work with general schemes over $\operatorname{Spec}(\mathbf{A})$.

Classical theory would lead to the expectation that, at the very least, the above functions are rational in u, etc. As trivial evidence in this regard, let $\mathbf{A} = \mathbb{F}_r[T]$ and let X be smooth, geometrically connected and projective over $\mathbb{F}_\wp$. Let φ/X be the "constant" Carlitz-module, C/X, over X. (Recall that C is the rank one module defined over $\mathbf{A} = \mathbb{F}_r[T]$ defined by $C_T(z) = Tz + z^r$.) Finally, let $Z(X, u)$ be the Z-function of Section 2 associated to X which we consider as having coefficients in $\mathbf{A}$ by first reducing modulo p. Then one sees quite trivially that

$$L(C/X, u) = Z(X, fu)$$

where f is the monic generator of $\wp$. Thus, in this most basic case, $L(\varphi^{(x)}, u)$ is indeed rational in u.

The reader should note that the reduction $Z(X, u)$ itself is also contained in the above definition. Indeed it occurs when we use $H^0(\varphi)$.

We will say more along these lines in Subsection 3.7.

3.3. Exponentiation of Ideals. Before we globalize the theory of 3.2 and define the analogs of $L(X^{(i)}, s)$ of Section 2, we need to discuss the domains of these functions. There will be one such space for all primes of $\mathbf{k}$ and we begin, naturally, with ∞.

In what follows, we continually use the following fact: Let G be an abelian pro-p-group and B a field of characteristic p. Let $H \subseteq G$ be a compact open subgroup and let $\psi: G \to B^*$ be a homomorphism. Then ψ is uniquely determined by its restriction to H. The proof is obvious since p^{th} roots in B are unique.

Let $\mathbb{F}_\infty \simeq \mathbb{F}_{r^\delta} \subseteq \mathbf{K}$ be the field of constants.

3.3.1. Definition.

a) A sign function, sgn, on $\mathbf{K}$ is a continuous morphism from $\mathbf{K}^*$ to $\mathbb{F}_\infty^*$ which is the identity on $\mathbb{F}_\infty^*$.
b) Those $x \in \mathbf{K}^*$ with $\operatorname{sgn}(\alpha) = 1$ are called *positive* or *monic*.

Let $U_1 \subseteq \mathbf{K}^*$ be the subgroup of 1-units. As U_1 is topologically a p-group and $\mathbb{F}_\infty^*$ has no p-torsion, we see that $\operatorname{sgn}(U_1) = 1$.

Let us now *fix* sgn as well as π, a positive uniformizing element of $\mathbf{K}$. We let ν_∞ be the additive valuation on $\overline{\mathbf{K}}$ with $\nu_\infty(\pi) = 1$.

3.3.2. Definition. Let $x \in \mathbf{k}^*$. We set $\deg(x)$ to be the degree of the *finite part* of the divisor of x.

One sees easily that $\deg(x) = -\delta\nu_\infty(x)$.

3.3.3. Definition. Let $\alpha \in \mathbf{K}^*$. We set

$$\langle a\rangle = \langle a\rangle_\infty = \operatorname{sgn}(a)^{-1}\pi^{-\nu_\infty(a)}a.$$

It is trivial to see that $\langle ab\rangle = \langle a\rangle\langle b\rangle$. The function $\langle\ \rangle$ is the projection of $\mathbf{K}^*$ onto U_1 determined by π.

The map $\alpha \mapsto \langle\alpha\rangle$ is analogous to the map which takes $z = e^{x+iy} \in \mathbb{C}^*$ to e^{iy}. The next few definitions will allow us to define a mapping somewhat analogous to $z \mapsto e^x$. As the ring $\mathbf{A}$ is completely general, we will have much technical work to do in order to deal with ideals that are not principal and positively generated. Thus the reader may find it profitable to turn first to example 3.3.12 where the simplest case of $\mathbb{F}_r[T]$ is handled.

3.3.4. Definition.

a) Let $\mathfrak{I}$ be the group of $\mathbf{A}$-fractional ideals of $\mathbf{k}$. Let Pr be the group of principal ideals of $\mathbf{A}$.

b) Let $m \in \mathbb{N}$. Set $P(\pi^m)$ = group of principal ideals of $\mathbf{k}$ generated by positive $\alpha \in \mathbf{k}$ with $\langle\alpha\rangle \equiv 1 \bmod (\pi^m)$. Put $P_0 = P(\pi^0)$ = group of principal and *positively generated* ideals.

c) Set $\widehat{\mathfrak{I}} = \varprojlim_m \mathfrak{I}/P(\pi^m)$.

There are obvious exact sequences

$$0 \to \varprojlim P_0/P(\pi^m) \to \widehat{\mathfrak{I}} \to \mathfrak{I}/P_0 \to 0$$

and

$$0 \to \mathbb{F}_\infty^*/\mathbb{F}_r^* \simeq Pr/P_0 \to \mathfrak{I}/P_0 \to \mathfrak{I}/Pr \to 0.$$

Via $\langle\ \ \rangle$, we see that

$$\widehat{P_0} = \varprojlim P_0/P(\pi^m) \simeq U_1.$$

We therefore view $\widehat{\mathfrak{I}}$ as an extension of U_1 by the finite group $\mathfrak{I}/P_0$.

By construction, (the reader can think of $\mathbb{Z}_p^*$)

$$\widehat{\mathfrak{I}} = \mathfrak{I}_1 \times \mathfrak{I}_2$$

where $\mathfrak{I}_1$ is a pro-p-group which contains U_1 and $\mathfrak{I}_2$ is a finite group of order prime to p containing $\mathbb{F}_\infty^*/\mathbb{F}_r^*$. Let $\alpha \in \widehat{\mathfrak{I}}$. Then we write

$$\alpha = \alpha_1\alpha_2$$

with $\alpha_i = \mathfrak{I}_i$, $i = 1, 2$.

Let ϱ be an ideal. Then we let $\langle\varrho\rangle$ be the image of ϱ under the canonical map of $\mathfrak{I}$ to $\widehat{\mathfrak{I}}$. The elements $\langle\varrho\rangle_1$ and $\langle\varrho\rangle_2$ are defined in the obvious fashion. If ϱ is generated by a *positive* element a, then, by definition, $\langle\varrho\rangle_1 = \langle a\rangle$ and $\langle\varrho\rangle_2$ vanishes.

On U_1, we have the continuous endomorphism $u \in U_1 \mapsto u^y$ for $y \in \mathbb{Z}_p$ This follows easily from the binomial theorem. As $\overline{\mathbf{K}}$ has characteristic p, the map $u \mapsto u^y$

lifts uniquely to a $\overline{\mathbf{K}}^*$-valued quasi-character (= continuous morphism) of $\mathfrak{I}_1$. We abuse notation and also denote this map by "α_1^y".

The $\overline{\mathbf{K}}^*$-valued characters of $\mathbb{F}_\infty^*/\mathbb{F}_r^*$ are given by $\chi(x\mathbb{F}_r^*) = x^{j(r-1)}$ for $j \in \mathbb{Z}/\varepsilon$, $\varepsilon = (r^\delta - 1)/(r-1)$. It is elementary that such characters extend to $\mathfrak{I}_2$. We write $\rho(\mathfrak{I}_2)$ for the $\overline{\mathbf{K}}^*$-valued characters of $\mathfrak{I}_2$. If $\chi \in \rho(\mathfrak{I}_2)$ then we set $j(\chi) = j$ where j is as above since χ restricts to a character on $\mathbb{F}_\infty^*/\mathbb{F}_r^*$.

We set $\rho(\widehat{\mathfrak{I}}) = \mathbb{Z}_p \times \rho(\mathfrak{I}_2)$. Let $Y = (y, \chi_Y) = (y, \chi) \in \rho(\widehat{\mathfrak{I}})$ and let $\alpha \in \widehat{\mathfrak{I}}$. Put

$$\alpha^Y = \alpha_1^y \chi(\alpha_2).$$

If ϱ is an ideal, then $\langle\varrho\rangle^Y$ is now defined.

Our next definition gives a space that plays the role of the complex numbers $\mathbb{C}$ in that one is able to *exponentiate* with it.

3.3.5. Definition.

a) We put $S_\infty = \overline{\mathbf{K}}^* \times \rho(\widehat{\mathfrak{I}})$.

b) Let ϱ be an $\mathbf{A}$-fractional ideal of $\mathbf{K}$. Let $s = (x, Y) \in S_\infty$. We set

$$\varrho^s = x^{\deg(\varrho)} \langle\varrho\rangle^Y.$$

The space S_∞ has a natural group structure which we write *additively*. The reader will easily see that

$$\varrho^{s+u} = \varrho^s \varrho^u$$

and

$$(\varrho_0 \varrho_1)^s = \varrho_0^s \varrho_1^s.$$

3.3.6. Lemma. *Let $a \in \mathbf{k}^*$ and $s = (x, Y) \in S_\infty$. We have*

$$\begin{aligned}(a)^s &= x^{\deg(a)} \langle(a)\rangle_1^y \operatorname{sgn}(a)^t \\ &= x^{\deg(a)} \langle a\rangle^y \operatorname{sgn}(a)^t,\end{aligned}$$

where $t = (r-1)j(\chi_Y)$.

Proof. The projection of (a) to $\mathfrak{I}/P_0$ is determined by $\operatorname{sgn}(a)$. Thus we need only prove

$$\langle(a)\rangle_1^1 = \langle a\rangle.$$

(Recall that $\langle(a)\rangle_1$ is an element of $\mathfrak{I}_1$ whereas $\langle(a)\rangle_1^1$ is an element of $\overline{\mathbf{K}}$.) Now $\langle(a)\rangle_1^1$ and $\langle a\rangle$ are 1-units in $\overline{\mathbf{K}}$ and $\langle(a)\rangle_1^\varepsilon = \langle(a^\varepsilon)\rangle_1^1 = \langle a\rangle^\varepsilon$ for $\varepsilon = (r^\delta - 1)/(r-1)$. The result now follows easily. □

Let $\pi^{1/\delta} \in \overline{\mathbf{K}}$ be a fixed root and let $a \in \mathbf{k}^*$ be positive. Let $i \in \mathbb{Z}$ and set $Y = (i, \chi) \in \rho(\widehat{\mathfrak{I}})$ for arbitrary χ. It is trivial to see from 3.13 that

$$a^i = (a)^s \text{ for } s = (\pi^{-i/\delta}, Y).$$

When the context is clear, we shall use $J^{(i,\chi)}$ for J^s, $s = (\pi^{-i/\delta}, i, \chi)$.

More generally, let $a \in \mathbf{k}^*$ and let $i \in \mathbb{Z}$ be $\mathbf{A}$-even. Put $j = i/(r-1)$ and set $Y = (i, \chi)$ where $j(\chi) = j$. Then

$$a^i = (a)^s \quad \text{for} \quad s = (\pi^{-i/\delta}, Y)$$

as again follows from 3.3.6. In the discussion following 3.5.3, we present a "picture" of S_∞.

We turn now to the v-adic version of S_∞ for $v \in \mathrm{Spec}(\mathbf{A})$. The reader will note a closer resemblance between the ∞-adic and v-adic theories than is seen classically between complex and p-adic theories.

3.3.7. Definition.

a) Let $\mu_v \subseteq \mathbf{A}_v^*$ be the group of roots of unity. Let $U_{v,1} \subseteq \mathbf{A}_v^*$ be the group of 1-units and set $U_v = \mathbf{A}_v^*$.

b) Let $u \in U_v$. Write

$$u = \omega_v(u)\langle u\rangle_v$$

where $\omega_v(u) \in \mu_v$ and $\langle u\rangle_v \in U_{v,1}$.

Clearly both ω_v and $\langle\ \rangle_v$ are homomorphisms.

3.3.8. Definition.

a) Let $\mathfrak{I}_v$ be the group of $\mathbf{A}$-fractional ideals of $\mathbf{k}$ which are prime to v. Let $Pr_v \subseteq \mathfrak{I}_v$ be the group of principal ideals.

b) Let $m \in \mathbb{N}$. Let $P(v^m)$ be the group of principal ideals generated by *positive* $\alpha \in \mathbf{k}$ which are prime to v and $\equiv 1 \pmod{v^m}$. We put $P_{v,0} = P(v^0) =$ group of principal and positively generated ideals which are prime to v.

c) Set

$$\widehat{\mathfrak{I}}_v = \lim_{\substack{\longleftarrow \\ m}} \mathfrak{I}_v/P(v^m).$$

As with $\widehat{\mathfrak{I}}$, we have two exact sequences

$$0 \to \lim_{\longleftarrow} P_{v,0}/P(v^m) \to \widehat{\mathfrak{I}}_v \to \mathfrak{I}_v/P_{v,0} \to 0$$

and

$$0 \to \mathbb{F}_\infty^*/\mathbb{F}_r^* \simeq Pr_v/P_{v,0} \to \mathfrak{I}_v/P_{v,0} \to \mathfrak{I}_v/Pr_v \to 0.$$

By choosing representatives prime to v, we see

$$\mathfrak{I}_v/Pr_v \simeq \mathfrak{I}/Pr$$

and, with some care,

$$\mathfrak{I}_v/P_{v,0} \simeq \mathfrak{I}/P_0.$$

Moreover, one readily sees (through the use of the Riemann-Roch theorem), that

$$\lim_{\longleftarrow} P_{v,0}/P(v^m) \simeq U_v;$$

the map takes $\alpha \in P_{v,0}$ to its positive generator, etc.

3.3.9. Definition.

a) Let $c = \deg(v)$. Set

$$\rho_v(U_v) = \varprojlim_j \mathbb{Z}/p^j(r^c - 1) \simeq \mathbb{Z}_p \times \mathbb{Z}/(r^c - 1).$$

b) Let $s_v = (s_{v,0}, s_{v,1}) \in \rho_v(U_v)$ and let $u \in \mathbf{A}_v^*$. Then we set

$$u^{s_v} = \langle u \rangle_v^{s_{v,0}} \omega_v(u)^{s_{v,1}}.$$

The above definition mimics the classical p-adic theory.

The discussion that now follows is a bit convoluted. The reader will benefit by realizing that we are trying to define a v-adic exponentiation of ideals consistent with our previous ∞-adic definitions.

As with $\widehat{\mathfrak{I}}$, we now write

$$\widehat{\mathfrak{I}}_v \simeq \mathfrak{I}_{v,1} \times \mathfrak{I}_{v,2}$$

where $\mathfrak{I}_{v,1}$ is a pro-p-group which contains $U_{v,1}$, and $\mathfrak{I}_{v,2}$ is a finite group of order prime to p. From the exact sequences given above, we know that $\mu_v \subseteq \mathfrak{I}_{v,2}$. Moreover, there is a *surjective* map between $\mathfrak{I}_{v,2}$ and the group of elements of order prime to p in $\mathfrak{I}_v/P_{v,0}$. This group is canonically isomorphic to $\mathfrak{I}_2$. In fact, we have an exact sequence

$$0 \to \mu_v \to \mathfrak{I}_{v,2} \to \mathfrak{I}_2 \to 0.$$

It is a remarkable fact that $\mathfrak{I}_{v,2}$ also contains a subgroup canonically isomorphic to $\mathbb{F}_\infty^*$. Indeed, there is a surjection from $\widehat{\mathfrak{I}}_v$ to $\mathfrak{I}_v/P(v)$. Thus if we can show that $\mathfrak{I}_v/P(v)$ contains a subgroup $\simeq \mathbb{F}_\infty^*$, the result follows by construction. On the other hand, let $P \subseteq \mathfrak{I}_v/P(v)$ be the subgroup generated by (α) with $\alpha \equiv 1 \pmod{v}$. Then sgn gives an isomorphism of P with $\mathbb{F}_\infty^*$ (the reader should note that the element $\equiv 1 \pmod{v}$ is the canonical generator of (α) in this situation).

Let $s_v = (s_{v,0}, s_{v,1}) \in \rho_v(U_v)$. For each such $s_{v,1}$, we *fix* an extension, $\psi_{s_{v,1}}(\alpha)$, of the character of μ_v, $\alpha \mapsto \alpha^{s_{v,1}}$, to a $\overline{\mathbf{k}}_v^*$-valued character of $\mathfrak{I}_{v,2}$.

3.3.10. Definition.

a) Let $\rho_v(\mathfrak{I}_{v,2})$ (resp. $\rho_v(\mathfrak{I}_2)$) be the set of all $\overline{\mathbf{k}}_v^*$-valued characters of $\mathfrak{I}_{v,2}$ (resp. $\mathfrak{I}_2$).

b) Let $S_v = \rho_v(U_v) \times \rho_v(\mathfrak{I}_2)$.

Let $s_v = (s_{v,0}, s_{v,1}, \chi) \in S_v$ and $\alpha \in \widehat{\mathfrak{I}}_v$. From above, we see that $\rho_v(\mathfrak{I}_2)$ injects into $\rho_v(\mathfrak{I}_{v,2})$. Thus if $\alpha \in \widehat{\mathfrak{I}}$, then we can form, as before,

$$\alpha^{s_v}$$

with all the usual formal properties. If ϱ is a fractional ideal of $\mathbf{k}$ prime to v, then we form ϱ^s by first mapping into $\widehat{\mathfrak{I}}_v$, etc.

From the exactness of

$$0 \to \mu_v \to \mathfrak{I}_{v,2} \to \mathfrak{I}_2 \to 0,$$

and a little group theory, we conclude that *all* $\overline{\mathbf{k}}_v^*$-valued characters of $\mathfrak{I}_{v,2}$ are obtained by our construction. Indeed, any character of $\mathfrak{I}_{v,2}$ can be written as $\psi_{s_{v,1}}(\alpha)\chi(\alpha)$ where $\chi \in \rho_v(\mathfrak{I}_2)$.

Let $a \in \mathbf{k}^*$ be positive and prime to v. Let $i \in \mathbb{Z}$ and $s_v = (i, \chi) \in S_v$. Then it is trivial to see that

$$a^i = (a)^{s_v}.$$

This can be extended to *all* $a \in \mathbf{k}^*$ prime to v, and $i \equiv 0 \pmod{(r-1)}$, exactly as we did in the ∞-adic case.

Let J be a fractional ideal of $\mathbf{k}$ which is prime to v and let $i \in \mathbb{Z}$. Let χ be a character of $\mathfrak{I}_2$ (viewed as $\overline{\mathbf{K}}$- or $\overline{k}_v$-valued). We have now given procedures to define $J^{(i,\chi)}$ *both* v-adically and ∞-adically. Moreover, if $J = (a)$ for a positive, then *both* definitions give

$$J^{(i,\chi)} = a^i.$$

Consequently, $J^{(i,\chi)}$ is always algebraic as one can always find t with J^t principal and positively generated.

The reader should note that $J^{(0,\chi)}$ is not necessarily identically 1. Rather, it is a root of unity that equals 1 when J is principal and positively generated.

To go further, *fix* an embedding of $\overline{\mathbf{k}}$ into $\overline{\mathbf{k}}_v$. Via this, we view χ as being either $\overline{\mathbf{K}}^*$-valued *or* $\overline{\mathbf{k}}_v^*$-valued. Finally, after all this work, the compatibility of these two definitions is given by the simple:

3.3.11. Lemma. *Via the embedding of* $\overline{\mathbf{k}}$ *into* $\overline{\mathbf{k}}_v$ *both definitions of* $J^{(i,\chi)}$ *agree.*

Proof. Let $J^{(i,\chi)_1}$ be the definition obtained by first working ∞-adically and then injecting into $\overline{\mathbf{k}}_v^*$. Let $J^{(i,\chi)_2}$ be obtained by first viewing χ as $\overline{\mathbf{k}}_v^*$-valued and then exponentiating v-adically. Set

$$\phi(J) = J^{(i,\chi)_1}/J^{(i,\chi)_2}.$$

By assumption, ϕ is trivial on $P_{v,0}$; thus it gives rise to a morphism $\widetilde{\phi}$ on $\mathfrak{I}_v/P_{v,0}$. But $\widetilde{\phi}$ is also trivial since, by definition,

$$\widetilde{\phi} = \chi/\chi$$

on the prime to p part of $\mathfrak{I}_v/P_{v,0}$ and must be trivial on the p-torsion part. □

In 3.6.4, we shall see that the map $J \mapsto J^{(1,\chi)}$ gives rise to Grössencharakter of type A_0 for $\mathbf{k}$.

3.3.12. Example. Let $\mathbf{A} = \mathbb{F}_r[T]$. In this case there are no longer any problems with non principal or non positively generated principal ideals. We have

$$S_\infty = \overline{\mathbf{K}}^* \times \mathbb{Z}_p,$$

and

$$S_v = \mathbb{Z}_p \times \mathbb{Z}/(r^v - 1).$$

If $a \in \mathbf{k}$ is positive and $s = (x, y) \in S_\infty$, then

$$a^s = x^{\deg(a)} \langle a \rangle^y ,$$

and for $i \in \mathbb{Z}$

$$a^i = a^s , \quad \text{for} \quad s = (\pi^{-i}, i).$$

If a is prime to v and $s_v = (s_{v,0}, s_{v,1}) \in S_v$, then

$$a^{s_v} = \langle a \rangle_v^{s_{v,0}} \omega_v(a)^{s_{v,1}} ,$$

etc.

We are now in a position to discuss "Dirichlet series" based on our exponentiation. We discuss the situation for S_∞; the v-adic theory is developed in a completely similar way.

3.3.13. Definition. Let $\mathbf{K}_1 \subset \overline{\mathbf{K}}$ be a finite extension of $\mathbf{K}$. For each ideal I of $\mathbf{A}$, let $c(I) \in \mathbf{K}_1$. Let $s \in S_\infty$. Then we call a formal series of the form

$$L(s) = \sum_{I \text{ an ideal}} c(I) I^{-s}$$

a *Dirichlet series.*

As in classical theory, it makes sense to ask where $L(s)$ converges. The condition for convergence is precisely that

$$c(I) I^{-s} \to 0$$

as $\deg(I) \to \infty$. If $L(s)$ converges for some s then it converges in a "half-plane" $\mathbb{H} = \{ (x, Y) \mid |x| \gg 0 \}$ of S_∞. Clearly, $\mathbb{H}$ contains all positive integers $\geq N$ for some integer N.

The notion of an "Euler product" is given in the obvious manner.

Let $h_{\mathbf{A}}$ be the class number of $\mathbf{A}$ considered as a Dedekind domain.

3.3.14. Theorem. *The Dirichlet series $L(s)$ is uniquely determined by the values*

$$\{ L(j) \}$$

where j runs over all positive integers $\geq N$ which are divisible by $h_{\mathbf{A}}(r^\delta - 1)$.

Proof. Let j be as in the statement of the proposition. Write

$$j = h_{\mathbf{A}}(r^\delta - 1) j_1 .$$

Then

$$I^j = \left(I^{h_{\mathbf{A}}(r^\delta - 1)} \right)^{j_1}$$

where $I^{h_{\mathbf{A}}(r^\delta - 1)}$ is now principal and positively generated.

By a simple counting argument (see Example 3.6.4) one can show that if I_0, I_1 are two ideals with, using the above notions of exponentiation,

$$I_0^t = I_1^t \in \overline{\mathbf{K}}$$

for some positive integer t, then $I_0 = I_1$. Thus, we are reduced to considering our Dirichlet series as being a sum over positively generated principal ideals.

Clearly it suffices to show that if $L(j) = 0$ for $j \gg 0$ then $L(s)$ is the trivial Dirichlet series. To see this, we will use a version of *Artin's independence of characters*. Thus let

$$L(s) = c(a_1)a_1^{-s} + \cdots + c(a_h)a_h^{-s} + \sum_{|a|>|a_h|} c(a)a^{-s},$$

where all a_h have the same absolute value.

As $a_1 \neq a_2$, for some $j_0 \geq N$ we have $a_1^{j_0} \neq a_2^{j_0}$. Thus, for $j \gg 0$ we find

$$\begin{aligned} 0 &= c(a_1)a_1^{-(j+j_0)} + \cdots + c(a_h)a_h^{-(j+j_0)} + \sum_{|a|>|a_h|} c(a)a^{-(j+j_0)} \\ &= c(a_1)a_1^{-j}a_1^{-j_0} + \cdots + c(a_h)a_h^{-j}a_h^{-j_0} + \sum_{|a|>|a_h|} c(a)a^{-j}a^{-j_0}. \end{aligned}$$

Now divide by $a_2^{-j_0}$ and subtract to find

$$\begin{aligned} 0 = c(a_1)&\left(\left(\frac{a_2}{a_1}\right)^{j_0} - 1\right) a_1^{-j} + 0 + \cdots + c(a_h)\left(\left(\frac{a_2}{a_h}\right)^{j_0} - 1\right) a_h^{-j} \\ &+ \sum_{|a|>|a_h|} c(a)\left(\left(\frac{a_2}{a}\right)^{j_0} - 1\right) a^{-j}. \end{aligned}$$

Upon repeating the process, we arrive finally at an equation of the form, for $j \gg 0$,

$$0 = \bar{c}(a_1)a_1^{-j} + \sum_{|a|>|a_1|} \bar{c}(a)a^{-j},$$

where

$$\bar{c}(a_1) = 0 \Longleftrightarrow c(a_1) = 0.$$

But, for $j \gg 0$,

$$|\bar{c}(a_1)| = \left| \sum_{|a|>|a_1|} \bar{c}(a)\left(\frac{a_1}{a}\right)^j \right|$$

and the right hand side approaches 0 as j tends to ∞. □

3.3.15. Corollary. *The elements* $\{I^{-s}\}$ *are linearly independent.* □

Let $L(s)$ now have an Euler product

$$L(s) = \prod_{\wp \text{ prime of } \mathbf{A}} P_\wp(\wp^{-s})^{-1},$$

where $P_\wp(u)$ is a polynomial in u and $1 = P_\wp(0)$. Then the *same* proof as in the classical case (together with 3.3.15) gives the next result:

3.3.16. Theorem. *The Euler product for* $L(s)$ *is unique.* □

3.4. Global L-series. Return now to L, φ, etc. Before defining the L-series, we need to discuss a bit more the cohomology of φ and its reduction $\varphi^{\wp}$.

Recall that for a good prime $\wp$,

$$H_v^1(\varphi^{\wp}, \mathbf{k}_v) =: \mathrm{Hom}_{A_v}(T_v(\varphi^{\wp}), \mathbf{k}_v)$$

for $(v, \wp) = 1$. This definition also makes sense for φ and we set

$$H_v^1(\varphi, \mathbf{k}_v) =: \mathrm{Hom}_{A_v}(T_v(\varphi), \mathbf{k}_v)$$

where T_v has the obvious meaning and v can now be arbitrary in $\mathrm{Spec}(\mathbf{A})$. We extend this definition to cohomology of arbitrary dimensions by taking exterior products. It is trivial to see that, for $(v, \wp) = 1$, $H_v^i(\varphi, \mathbf{k}_v) \simeq H_v^i(\varphi^{\wp}, \mathbf{k}_v)$. Moreover, both groups come equipped with Galois actions: The group $\mathrm{Gal}(\overline{L}/L)$ acts on $H_v^i(\varphi, \mathbf{k}_v)$ via the dual of its action on the torsion points (i.e., Tate modules) of φ and the same with $\mathrm{Gal}(\overline{\mathbb{F}}_{\wp}/\mathbb{F}_{\wp})$ on $H_v^i(\varphi^{\wp}, \mathbf{k}_v)$. This action is unramified at $\wp$ as $\wp$ is good. In fact, we have the following analog of the criterion of Néron-Ogg-Šafarevič for abelian varieties [T1]:

3.4.1. Theorem (Takahashi). *The prime $\wp$ is good for φ if and only if for $(v, \wp) = 1$, the Galois module $H_v^1(\varphi, \mathbf{k}_v)$ is unramified at $\wp$.* □

Let F be the qth power mapping for $\varphi^{\wp}$ as before where $q = N\wp = \#\mathbb{F}_{\wp}$. By definition, $\mathrm{Gal}(\overline{\mathbb{F}}_{\wp}/\mathbb{F}_{\wp})$ acts on $H_v^1(\varphi^{\wp}, \mathbf{k}_v)$ by

$$\sigma \times f(t) \mapsto f(\sigma^{-1}t).$$

Thus, as we saw in Section 2, the geometric Frobenius acts as F on cohomology, etc.

With the above remarks, the formalism of Section 2 kicks in and we have:

3.4.2. Definition.

a) Let $s \in S_\infty$. Put

$$\zeta^{\text{good}}(\varphi, s) = \prod_{\wp \text{ good}} Z(\varphi^{\wp}, n\wp^{-s}).$$

b) Let $v \in \mathrm{Spec}(\mathbf{A})$. Let $x_v \in \overline{\mathbf{k}}_v^*$, $s_v \in S_v$, and put $t_\wp = \deg n\wp$. Set

$$\zeta_v^{\text{good}}(\varphi, x_v, s_v) = \prod_{\substack{\wp \text{ good} \\ (\wp, v)=1}} Z(\varphi^{\wp}, x_v^{-t_\wp} n\wp^{-s_v}).$$

These definitions make formal sense since $Z(\varphi^{\wp}, u)$ is a rational function with $\mathbf{A}$-coefficients. In Proposition 3.4.6, we will show that these functions converge on non-trivial domains.

As in Section 2, we also make the following

3.4.3. Definition.

a) Let $s \in S_\infty$. Put

$$L^{\text{good}}(\varphi^{(i)}, s) = \prod_{\wp \text{ good}} \det(1 - \sigma_\wp N\wp^{-s} \mid H_v^i(\varphi, \mathbf{k}_v))^{-1}$$

where $\sigma_\wp$ is a geometric Frobenius, etc.

b) Let $\{v, s_v, x_v\}$, etc., be as in 3.4.2.b). Set

$$L_v^{\text{good}}(\varphi^{(i)}, x_v, s_v) = \prod_{\substack{\wp \text{ good} \\ (\wp, v)=1}} \det(1 - \sigma_\wp x_v^{-t_\wp} n\wp^{-s_v} \mid H_v^i(\varphi, \mathbf{k}_v))^{-1}.$$

When we talk about THE L-series, $L(\varphi, s)$, of φ, we shall always mean $L(\varphi^{(1)}, s)$. It will be obvious to the reader how to twist such L-series by abelian characters and so on.

3.4.4. Remarks.

a) The above definitions also work for those motives (e.g., the "pure" ones) where one has good eigenvalues of Frobenius. Such is the case, for instance, for a tensor product of two Drinfeld modules.

b) In order for L to have a non-trivial Drinfeld module defined over it, L must contain H where H is the Hilbert class field of $\mathbf{k}$ totally-split at ∞. (The argument was given after Corollary 3.2.5.) This implies that $n\wp$ must now be principal though not necessarily positively generated if $\deg(\infty) > 1$.

3.4.5. Variant. The trivial motive H^0 makes sense over any field L; by definition H^0 has trivial Galois action. Thus, following the above prescription, one is lead to define

$$\zeta(\mathcal{O}_L, s) = L(H^0, s) = \prod_{\wp}(1 - n\wp^{-s})^{-1} = \sum_{I \text{ ideal}} nI^{-s}$$

and

$$\zeta_v(\mathcal{O}_L, x_v, s_v) = \prod_{(\wp, v)=1} (1 - x_v^{-t_\wp} n\wp^{-s_v})^{-1}.$$

Of course, for arbitrary L, $n\wp$ need not be principal. For more along these lines see Section 4.

Let $|?|_v$ be the normalized absolute value at v.

3.4.6. Proposition.

a) $L^{\text{good}}(\varphi^{(i)}, s)$ *converges for all* $s = (x, Y) \in S_\infty$ *with* $|x| > r^{i/d}$.

b) $L_v^{\text{good}}(\varphi^{(i)}, x_v, s_v)$ *converges for* (x_v, s_v) *with* $|x_v|_v > 1$.

Proof. Part a) follows easily by the Riemann hypothesis 3.2.3.c). Part b) follows since all the coefficients of $P_i(u)$ are in $\mathbf{A}$ and so have v-adic absolute value ≤ 1. □

The subsets given in Proposition 3.4.6 describe "half-planes of convergence" in S_∞ and $\overline{\mathbf{k}}_v^* \times S_v$.

It remains to discuss the bad primes. but the procedure is now clear from the analogy with classical theory: we use invariants of inertia as in Section 2. Thus we have finally the complete definition

$$L(\varphi^{(i)}, s) = \prod_{\wp} \det(1 - \sigma_\wp n\wp^{-s} \mid H^i(\varphi, \mathbf{k}_v)^{I_\wp})^{-1},$$

etc. Unlike the classical situation, outside of the case of complex multiplication, we know nothing about the factors at bad primes. One would expect their coefficients to be in $\mathbf{A}$, etc. But it is unclear what relationship they have to the reduction of φ and so on.

It is clear how to express $\zeta^{\text{good}}(\varphi, s)$, etc., in terms of L^{good} and so on. Thus we can now extend our definitions of the zeta-functions to the *bad* primes also and obtain the functions $\zeta(\varphi, s)$ and $\zeta_v(\varphi, x_v, s_v)$.

3.5. Essentially Algebraic Entire Functions. In classical theory, the goal (with minor exceptions) is to show that L-series can be extended to entire functions on the complex plane which possess functional equations. Moreover, it very often happens that the special-values of these entire functions can be interpolated p-adically to analytic functions on some weight space.

In the function field theory that we are discussing, as we have seen, the L-series are defined on S_∞ and $\overline{\mathbf{k}}_v^* \times S_v$. In this subsection we discuss the appropriate notion of "entire function" for such spaces. The theories on S_∞ and $\overline{\mathbf{k}}_v^* \times S_v$ are then connected via the notion of "essential algebraicity" for such entire functions.

Let $J \subset \overline{\mathbf{K}}$ be a finite extension of $\mathbf{K}$. Let $h(x) = \sum_{n \gg -\infty} a_n x^n$ be a Laurent series with coefficients in J and with a finite number of negative terms. We assume that $h(x)$ converges for all non-zero x. We set

$$\|h(x)\| = \sup\{|a_n|_\infty\}.$$

The function $h(x)$ is called a "finite tailed entire power series".

3.5.1. Definition. Let $F\colon S_\infty \to \overline{\mathbf{K}}$ be a continuous function. We say F is *entire* if and only if

1) $g_{(y,\chi)}(x) = F(x, y, \chi)$ is given by a finite tailed entire power series in x^{-1};
2) The coefficients of all such $g_{(y,\chi)}(x)$ belong to a fixed finite extension of $\mathbf{K}$;
3) Let $\{\alpha_i\}$ be a collection of elements of $\mathbb{Z}_p$ converging to y. We then require that

$$\|g_{(y,\chi)}(x) - g_{(\alpha_i,\chi)}(x)\|$$

tends to zero.

For example, let J be a fractional ideal. Then J^s is entire. The function J^s also satisfies obvious algebraicity properties. These are given by

3.5.2. Definition. Let F be as in 3.5.1. We say F is *essentially algebraic* if for all integers $j \gg 0$, and $x \in \overline{\mathbf{K}}^*$, the function

$$h_F(x, -j, \chi) = F(x\pi^{i/d}, -j, \chi)$$

is given by a polynomial in $\{x, x^{-1}\}$ with algebraic coefficients. We further suppose that all such coefficients generate a finite extension of $\mathbf{k}$.

It is *very important* that the reader understand what is contained in the Definition 3.5.2: The L-series $L(\varphi^{(i)}, s)$ is defined via our notion of exponentiation of ideals. In turn, this exponentiation uses a fixed choice of positive uniformizer π at ∞. The function h_F of the definition is created precisely to *reverse* this process. That is, to obtain h_L

we formally replace $x^{-\deg(\varrho)}\langle\varrho\rangle^{(j,\chi)}$ in the Euler product for $L(\varphi^{(i)},x,-j,\chi)$ with $x^{-\deg(\varrho)}\varrho^{(j,\chi)}$.

Clearly, J^s is also essentially algebraic.

3.5.3. Remarks. It is important to note that essential algebraicity follows *automatically* from entirety in many cases such as the L-series of Drinfeld modules or pure t-modules. Indeed, by definition, the functions $h_F(x,-j,\chi)$ of Definition 3.5.2 are entire power series in x^{-1}. Moreover they have coefficients with the property that, upon being raised to a bounded power, they belong to $\mathbf{A}$. This easily implies that they are polynomials!

The v-adic notions of "entire function" (as a function of (x_v,s_v)) and "essentially algebraic function" are now defined in a straightforward fashion. The only point worth mentioning is that the constant π does not occur in these definitions. The reader should note that if the ∞-adic L-series is essentially algebraic, then so are all the associated v-adic functions.

It is the notion of essentially algebraic that allows us to tie together the v-adic and ∞-adic theories as we explained in 3.3.11 for the values $\{J^j\}$. Indeed, based on the abelian theory (presented in our next subsection) we have the

3.5.4. Fond Hope. The functions $L(\varphi^{(i)},s)$ and $L_v(\varphi^{(i)},x_v,s_v)$, and their one-dimensional twists, are essentially algebraic entire functions. Moreover, for $j\gg 0$, the polynomials obtained at $(-j,\chi)$ (as in Definition 3.5.2) differ only by the factors above v when x is equated with x_v.

Clearly, if the fond hope is true, we deduce the existence of *meromorphic continuations* for the functions $\zeta(\varphi,s)$ and $\zeta_v(\varphi,x_v,s_v)$, etc. Also, for $L(\varphi^{(i)},s)$ (but not necessarily its twists), we may pick $j\geq 0$ as above.

The reader should note that the polynomial special values at $(-j,\chi)$ for $j\gg 0$ *determine* L and L_v by continuity. Moreover, if the "fond hope" is realized, the functions $L(\varphi^{(i)},s)$ will possess a sort of *critical strip* between those values where the Euler series of the function converges and those values of $-j$ where the function has polynomial special values as above. (In fact, like the function J^s for J a fractional ideal, the L-series may also have polynomial special values at $-j$ for j non-positive.)

The following "picture" of S_∞ may help the reader imagine the "critical strip". First of all, we ignore the characters χ. Next, we view $\overline{\mathbf{K}}^*\times\mathbb{Z}_p$ as a plane where the x-axis is given by $\overline{\mathbf{K}}^*$ etc. In this plane the integers are embedded discretely somewhere along the diagonal. With this, one can visualize the half-plane of convergence of the Euler product exactly as one views it for $\mathbb{C}$. On the other hand, choose the largest integer where one has polynomial special values. Draw a vertical line through this integer on S_∞. To the left of this line lies a half-plane and the critical strip is what is between the two half-planes we just drew!

In the v-adic plane, $\overline{\mathbf{k}}_v^*\times S_v$, the integral powers are *dense* in the subgroup

$$\{(1,s_v)\mid s_v\in S_v\}.$$

The analogy with classical theory would lead one to expect that there would be some sort of functional equation for our L-series and that this functional equation would involve "Γ-factors" at infinite primes of L. Moreover, the exact form of these gamma factors should involve "de Rham" cohomology of the motive. In Section 5, we will discuss what is known along these lines as of this writing. We note that, in any case, there is now a good de Rham theory for Drinfeld modules due to Anderson, Deligne, Gekeler and Yu **[Ge5]**. And, there is a growing body of theorems on Γ-functions for **A** **[G3]**, **[Th1]**. The reader should see **[S2]** for the classical theory. Having it in mind may well help when reading Section 5.

The "fond hope" leads to a remarkable formula that globalizes Lemma 3.2.1: We know from 3.4.4.b) that the norms of ideals of $\mathcal{O}_L$ must be principal and we now make the minor assumption that they are also positive. With this assumption the mapping $\wp \mapsto \wp^{(0,\chi)}$ is identically 1. (We can also derive this fact by assuming that $\chi = 1$. If we do not assume it, we can then twist all the L-series by the inverse of the finite character $\wp \mapsto n\wp^0 = \chi(n\wp)$.)

From the fond hope, as mentioned above, it follows that the function $\zeta^{\text{good}}(\varphi, s)$ has a meromorphic continuation. In particular, we can evaluate it at

$$s = (x, 0, \chi).$$

Upon such an evaluation, each L-series in the expression of ζ^{good} gives rise to a polynomial power series in x^{-1} which we denote by $L(\varphi^{(i)}, x)$. Consequently, the function $\zeta(\varphi, x)$ (same notation) is *rational* in x^{-1}.

Let $L(x)$ be one such L-series. Clearly

$$x\frac{d}{dx}\log L(x) = -\frac{1}{x}\frac{d}{dx^{-1}}\log L(x).$$

Thus, the results of Subsection 3.2, imply that the rational function

$$x\frac{d}{dx}\log \zeta^{\text{good}}(x)$$

is the meromorphic continuation of the power series

$$\sum_{\wp \text{ good}} -\deg(n\wp)x^{-\deg n\wp}\sum_{k\geq 1}\det\left(1-(\sigma_\wp^k)^* \mid H_v^1(\varphi, \mathbf{k}_v)\right) x^{-k\deg n\wp}, \tag{3.5.4}$$

where, as before, $\sigma_\wp$ is a geometric Frobenius and v is prime to $\wp$, etc.

One can also look for analogs of the well-known formalism of Birch and Swinnerton-Dyer for elliptic curves **[S-D1]**. It turns out that, in our situation, the ∞-adic analog is *not* formal at all but rather an actual equality! On the other hand, this equality seems rather banal...

We illustrate the above statement in a very simple case: Let $\mathbf{A} = \mathbb{F}_r[T]$ and $\mathbf{k} = \mathbb{F}_r(T)$. Let ϕ be the rank 2 Drinfeld module defined over $\mathbf{A}$ itself by

$$\phi_T(x) = Tx + \alpha x^r - x^{r^2},$$

where $\alpha \in \mathbf{A}$. This module has everywhere good reduction and its determinant module is the Carlitz module given by $C_T(x) = Tx + x^r$; thus we are in the simplest general rank two situation. (See **[G2]** and **[C1]**; the reader should note the change of sign from **[C1]** that is used here.) It is trivial to see that the Galois action on $H_v^1(C, \mathbf{k}_v)$ is given by an $\mathbf{A}_v$-valued quasi-character of $\mathrm{Gal}(\overline{\mathbf{k}}/\mathbf{k})$ where

$$\sigma_g \mapsto g$$

for the geometric Frobenius σ_g at a monic prime $g \neq v$. Now let $f \in \mathbf{A}$ be a monic prime and let $P_f(u)$ be the inverse characteristic polynomial of Frobenius associated to the reduction of ϕ at f as in Theorem 3.2.3. Let

$$P_f(u) = (1-\alpha_1 u)(1-\alpha_2 u).$$

By Theorem 3.2.3 we have

$$\chi(f) = \chi(\mathbb{F}_f) = (1-\alpha_1-\alpha_2+f),$$

as

$$f = \alpha_1\alpha_2$$

from above. From Proposition 3.4.6, we see that the Euler product for the L-series converges at $s=1$. We thus conclude

$$L(\phi,1)^{-1} = \prod_f \frac{\chi(f)}{f}.$$

The conjectures of Birch and Swinnerton-Dyer are built around the analogous *formal* statement for elliptic curves.

One can also examine the v-adic analog of the above equality. In this case, the Euler product for the L-series does not converge at $(1,1)$ (i.e., $(x_v, s_v) = (1,1)$). We thus obtain the formal equation

$$L_v(\phi,1,1)^{-1} = \prod_{f\neq v} \frac{\chi(f)}{f}.$$

It might be of interest to compute the function of positive integers

$$g(n) = \prod_{\substack{f\neq v\\ \deg(f)\le n}} \frac{\chi(f)}{f}$$

in order to get an idea of its v-adic growth. We note that for $f \neq v$

$$\left|\frac{\chi(f)}{f}\right|_v \le 1.$$

In any case, it seems that the study of equation 3.5.4 will prove more fruitful than obvious analogs of Birch and Swinnerton-Dyer.

3.6. The Abelian Case. We will indicate here rather briefly how the "fond hope", 3.5.4, can be shown in the abelian situation. For more information, the reader can consult **[G10]**. We continue with our previous notation.

Let φ now be a Drinfeld module or one of Anderson's motives which we assume uniformizable. The standard notions of complex multiplication apply to φ: This means that there exists a finite extension field $E \subseteq \overline{\mathbf{k}} \subseteq \overline{\mathbf{K}}$ of $\mathbf{k}$ which injects into the rational endomorphisms (i.e., tensored with $\mathbf{k}$) of φ such that $\{\text{lattice }\varphi\}\otimes\overline{\mathbf{k}}$ is 1-dimensional over $E\otimes\overline{\mathbf{k}}$.

Under this assumption, all v-adic representations can be easily seen to be *abelian* and there exists a "Grössencharakter of type A_0", ψ, (or associated "Hecke character", ψ_H) attached to φ in the manner of abelian varieties with similar definitions (see below). The quasi-character ψ is a $\overline{\mathbf{K}}^*$-valued map on the idele class group, J_L/L^*, of L such that at good primes $\wp$, ψ is unramified and $\psi(\wp)\in E^*$ is the Frobenius morphism at $\wp$. (Standard theory tells us that any central endomorphism of $\varphi^{\wp}$ arises from the reduction of a *unique* element of E. As with abelian varieties, the Frobenius morphism of φ is central.)

The L-series of ψ is defined as usual by

$$L(\psi,s)=\prod_{\wp\text{ good}}(1-\psi(\wp)n\wp^{-s})^{-1}.$$

And, as usual, one has

$$L(\varphi,s)=\prod L(\psi^{\tau},s),$$

where $\tau\colon E\to\overline{\mathbf{K}}$ is a $\mathbf{k}$-linear injection and where multiplicities are counted from possible inseparable elements in $E/\mathbf{k}$.

The notion of Grössencharakter of type A_0 is very general and includes finite characters of the idele class group; i.e., those Grössencharakters which arise from finite abelian extensions of L.

3.6.1. Remarks. There is a subtlety concerning local factors that arises from the situation of finite abelian extensions of L as the following example will show: Suppose L_1/L is a finite abelian extension with $[L_1\colon L]=p$ and L_1 is totally ramified at the finite prime $\wp$. Then, as classically,

$$\frac{\zeta(\mathcal{O}_{L_1},s)}{\zeta(\mathcal{O}_L,s)}$$

factors into a product of L-series associated to $\mathbb{Z}/p\simeq G=\mathrm{Gal}(L_1/L)$. But, of course, the only $\overline{\mathbf{K}}^*$-valued character of $\mathbb{Z}/p$ is the trivial character which is *nowhere* ramified. Thus if we *just* use $\overline{\mathbf{K}}$-valued characters, we would have the *wrong* factor at $\wp$! The solution, in this situation, is to use the *complex* characters of G (thought of as taking values in the algebraic closure $\overline{\mathbb{Q}}_p$ of $\mathbb{Q}_p$) and standard theory to determine local factors as usual. Then, if the local factor in the complex theory be $(1-\chi(\wp)u^{\deg(\wp)})$, for $\chi\colon G\to\overline{\mathbb{Q}}_p$, the correct factor in the characteristic-p theory is

$$(1-\overline{\chi}(\wp)\wp^{-s}),$$

where $\overline{\chi}(\wp)$ is the reduction of $\chi(\wp)$ (which is either 0 or a root of unity) viewed as an element of $\overline{\mathbf{K}}^*$. By this procedure, we arrive at an unambiguous definition of $L(\chi,s)$ for $s\in S_\infty$.

Let $\mathrm{Emb}(L/\mathbf{k})$ be the set of distinct embeddings of L into $\overline{\mathbf{k}}$. (Recall that L is separable over $\mathbf{k}$.) By definition, there exists a *conductor* $\mathcal{B}$ for ψ, and primes *not* dividing $\mathcal{B}$ will be *good* for φ. Thus, we can extend the definition of $\psi(\wp)$, $\wp$ good, to fractional ideals J prime to $\mathcal{B}$ by multiplication. We denote this element by $\tilde{\psi}(J)$. The notation is necessary for the following reason: Let $\alpha \in L^*$ be prime to $\mathcal{B}$. Let $(\alpha)_1$ be the fractional ideal generated by α and let $(\alpha)_2$ be the principal idele generated by α. Then, by definition $\psi((\alpha)_2) = 0$ whereas $\tilde{\psi}((\alpha)_1) \neq 0$. From now on whether (α) means principal ideal or idele will be clear from the context of its usage.

We let $\psi_i(\beta)$ be the local component of ψ at the prime ∞_i. Upon choosing a uniformizer $\pi_i \in L^*_{\infty_i}$, we have the decomposition

$$L^*_{\infty_i} = \pi_i^{\mathbf{Z}} \times \mu_i \times U_{1,i},$$

where $\mu_i \subset L^*_{\infty_i}$ is the group of roots of unity and $U_{1,i}$ is the group of 1-units. A continuous morphism from $L^*_{\infty_i} \to \overline{\mathbf{K}}^*$ is determined by its action on the above three factors. Obviously, the image of μ_i is always finite and the image of $\pi_i^{\mathbf{Z}}$ depends *precisely* on the element $\xi \in \overline{\mathbf{K}}^*$ to which it is sent; thus it may or may not have finite image. Finally, as we have seen, *any* morphism on $U_{1,i}$ with finite image must in fact be trivial. We let $\psi_{i,f}(\beta)$ be the *finite part* (i.e., has finite image) of the morphism $\psi_i(\beta)$, and for ideles $a = (a_w)$

$$\psi_{\infty,f}(a) = \prod \psi_{i,f}(a_{\infty_i}).$$

The fact that ψ is of type A_0 means the following: Let $\alpha \in L*$ with $\alpha \equiv 1 \pmod{\mathcal{B}}$. Then there exists $\omega \in \mathbb{Z}[\mathrm{Emb}(L/\mathbf{k})]$ such that

$$\tilde{\psi}((\alpha)) = \psi^{-1}_{\infty,f}(\alpha)\alpha^{\omega}.$$

3.6.2. Remarks. In classical theory, the root of unity $\psi_\infty(\alpha)$ does *not* appear. Instead one assumes that α is *totally positive* and the connectivity of $\mathbf{R}_+$ assures that no non-trivial roots of unity can occur.

We see that $L^*_{\infty_i}$ has finite characters (arising from $\pi_i^{\mathbf{Z}}$) that do *not* factor through *any* choice of sign function. However, as before, we always conclude that such characters (being finite!) are trivial on $U_{1,i}$.

Let $I(\mathcal{B})$ be the group of $\mathcal{O}$-fractional ideals of L prime to $\mathcal{B}$. The group $Pr(\mathcal{B})$ of principal ideals generated by α with $\alpha \equiv 1 \pmod{\mathcal{B}}$ is of finite index in $I(\mathcal{B})$. Via the above discussion, the values of $\tilde{\psi}$ on $Pr(\mathcal{B})$ clearly generate a *finite* extension of L. Thus one concludes readily

3.6.3. Proposition. *The values $\{\tilde{\psi}(J)\}$, J a fractional ideal of L prime to $\mathcal{B}$, generate a finite extension $L(\psi)$ of L.* □

3.6.4. Example. As in the last subsection, given a choice $\chi \in \rho(\widehat{\mathcal{J}})$ we are able to define

$$J^{(1,x)}$$

for J a fractional ideal of $\mathbf{k}$. The assignment

$$\psi: J \mapsto J^{(1,\chi)}$$

gives rise to a Grössencharakter, Ψ_χ, of type A_0 for $\mathbf{k}$ with conductor (1), $\omega = 1$, and infinite factor $\Psi_{\chi,\infty}(x) = \text{sgn}(x)^\beta$, $\beta = 1 - (r-1)j(\chi)$. This follows easily from 3.3.6. Moreover every ideal J of A becomes principal in $\mathbf{k}(\psi)$. Indeed for some positive integer θ, J^θ is principal and positively generated; say

$$J^\theta = (a)$$

for $\text{sgn}(a) = 1$. Thus, by definition

$$J^{(1,\chi)\theta} = a.$$

A simple count now tells us that J and $(J^{(1,\chi)})$ have the same valuations at all finite primes of $\mathbf{k}(\Psi_\chi)$. Thus $J^{(1,\chi)}$ generates J in $\mathbf{k}(\Psi_\chi)$. The reader should note however that $\mathbf{k}(\Psi_\chi)$ may have inseparable elements. Indeed, there are cases where θ is a power of p; so $J^{(1,\chi)}$ satisfies

$$y^{p^j} \in k$$

for some j.

It is a reasonable expectation that *all* Grössencharakters of type A_0 are associated to Anderson's motives. Thus, there should exist a motive M_χ, defined over $\mathbf{k}$, which is associated to $\Psi_{(1,\chi)}$. As we shall see (Remarks 4.1.11) the t-module associated to M_χ should (under reasonable hypotheses) be a $\mathbf{k}$-form of a product of a Hayes normalized rank one Drinfeld module (see Subsection 4.1).

Let $J_L^0 \subset J_L$ be the subgroup of ideles a with $\deg_1(a) = 0$ (under the canonical extension of $\deg_1$ to ideles). It is a standard fact that J_L^0/L^* is *compact*. We conclude that there exists a rational number $g_\psi \in \mathbb{Q}$ (independent of π) so that

$$\psi(a) = \pi^{g_\psi \deg_1(a)} \cdot u_a,$$

where $u_a \in \overline{\mathbf{K}}^*$ is a unit. Using this, the reader will have no difficulty establishing the existence of non-trivial half-plane $\subseteq S_\infty$ where the Euler product for $L(\psi, s)$ converges.

The main result of this subsection is:

3.6.5. Theorem. *$L(\psi, s)$ is an essentially algebraic entire function.*

3.6.6. Corollary. *$L(\varphi, s)$ is an essentially algebraic entire function.*

The remainder of this subsection is dedicated to describing the proof of Theorem 3.6.5 and the obvious v-adic analogs. We begin by presenting some lemmas in multilinear algebra.

3.6.7. Lemma.
a) *Let J, J_1 be two fields over $\mathbb{F}_r$. Let $W \subset J$ be a finite dimensional $\mathbb{F}_r$-vector space of dimension α and let $\{\mathfrak{L}_1, \dots, \mathfrak{L}_t\}$ be $\mathbb{F}_r$-linear maps of J into J_1. Finally let $x \in J$ and let $\{i_1, \dots, i_t\}$ be non-negative integers so that*

$$\sum_{h=1}^t i_h < (r-1)\alpha.$$

Then

$$\sum_{w\in W}\left(\prod_{h=1}^{t}\mathfrak{L}_h(x+w)^{i_h}\right)=0.$$

b) *Assume now that* J_1 *has an additive valuation* ν *with* $\nu(\mathfrak{L}_h(w))>0$ *for all* h *and* w. *Let* $\{i_h\}$ *now be an arbitrary collection of non-negative integers, and for* $j>0$ *put*

$$W_j=\{w\in W\mid \nu(\mathfrak{L}_h(w))\geq j \textit{ for all } h\}.$$

Then

$$\nu\left(\sum_{w\in W}\prod_{h=1}^{t}\mathfrak{L}_h(w)^{i_h}\right)\geq(r-1)Q,$$

for

$$Q=\sum_j \dim_{\mathbb{F}_r}(W_j).$$

Proof. To see Part a), let $\{e_1,\ldots,e_\alpha\}$ be a basis for W over $\mathbb{F}_r$. Then our sum becomes

$$\sum_{c_1,\ldots,c_\alpha\in\mathbb{F}_r}\prod_{h=1}^{t}(\mathfrak{L}_h(x)+c_1\mathfrak{L}_h(e_1)+\cdots+c_\alpha\mathfrak{L}_h(e_\alpha))^{i_h}.$$

By the multinomial theorem we see that $(\mathfrak{L}_h(x)+c_1\mathfrak{L}_h(e_1)\cdots+c_\alpha\mathfrak{L}_h(e_\alpha))^{i_h}$ equals

$$\sum_{j_1^h,\ldots,j_\alpha^h}\frac{i_h!\,\mathfrak{L}_h(x)^{(i_h-j_1^h-\cdots-j_\alpha^h)}}{j_1^h!\cdots j_\alpha^h!(i_h-j_1^h-\cdots-j_\alpha^h)!}\mathfrak{L}_h(e_1)^{j_1^h}\cdots\mathfrak{L}_h(e_\alpha)^{j_\alpha^h}c_1^{j_1^h}\cdots c_\alpha^{j_\alpha^h}.$$

The product then contains elements of the form

$$(\text{junk})c_1^{j_1^1+\cdots+j_1^t}\cdots c_\alpha^{j_\alpha^1+\cdots j_\alpha^t};\tag{$*$}$$

where (junk) belongs to J_1 and the sum of the exponents is $\leq(i_1+\cdots+i_t)$.

On the other hand, the sum over $\{c_1,\ldots,c_\alpha\}$ in $(*)$ will clearly be zero unless each coefficient is divisible by $(r-1)$. Part a) now follows easily.

For the second part, choose j_0 so that $W_{j_0+1}=\{0\}$ but $W_{j_0}\neq\{0\}$. Pick a basis $\{e_1,\ldots,e_\alpha\}$ for W so that $\{e_1,\ldots,e_{\dim_{\mathbb{F}_r}(W_j)}\}$ is a basis for W_j for all j. This choice of basis allows us to write $w\in W$ as w_1+w_0 with w_0 being the projection of w onto W_{j_0}. Now expand out $\mathfrak{L}_h(w)^{i_h}=(\mathfrak{L}_h(w_1)+\mathfrak{L}_h(w_0))^{i_h}$ by the binomial theorem and use Part a) applied to W_{j_0}. The result follows by induction. □

3.6.8. Example. The following self-contained elementary examples illustrate how both parts of Lemma 3.6.7 are used (as well as where they originated!).

Let $\mathbf{A}=\mathbb{F}_r[T]$ and let $M(d)\subset\mathbf{A}$ be the subset of monics of degree d.

3.6.8.a. Claim.

$$\sum_{n\in M(d)}n^i=0 \textit{ for } d>i.$$

Proof. We use induction. The result is trivial for $i = 0$. Thus assume it is true for some $i = i_0$. We want to show that it is also true for $i = i_0 + 1$. Let $d > i_0 + 1$ and write $n \in M(d)$ as

$$n = Th + \alpha,$$

where $\alpha \in \mathbb{F}_r$ and $h \in M(d-1)$. The binomial theorem then gives

$$n^i = \sum_{t=0}^{i} \binom{i}{t} (Th)^t \alpha^{i-t}.$$

Summing over α kills the term $(Th)^i$. The result now follows by induction upon summing over h. □

As a consequence, we see that for i a non-negative integer, the power series

$$h_i(x) = \sum_{d=0}^{\infty} x^{-d} \left(\sum_{n \in M(d)} n^i \right),$$

is actually a *polynomial* in x^{-1}.

Clearly, 3.6.8.a is analogous, though not as sharp as, the first part of Lemma 3.6.7. We turn now to Part b). In order to use our result in the greatest possible generality, we introduce a new variable u. Let $W(d) \subset \mathbb{F}_r[u]$ be the vector space of polynomials $f(u)$ of degree $\leq d$ with $0 = f(0)$. Let j be a non-negative integer and let $\nu = \nu_u$.

3.6.8.b. Claim.

$$\nu_u \left(\sum_{f \in W(d)} f(u)^j \right) \geq \frac{d(d+1)}{2}.$$

Proof. We again use induction on d. The result is trivial for $d = 1$. Let $f(u) \in W(d)$ be written as

$$f(u) = cu^d + h(u),$$

where $c \in \mathbb{F}_r$ and $\deg(h) \leq d - 1$. Then

$$f(u)^j = \sum_{t=0}^{j} \binom{j}{t} (cu^d)^t h(u)^{j-t}.$$

Summing over c kills the term associated to $t = 0$. The result now follows by summing over $h(u)$ and using induction as

$$d + \frac{d(d-1)}{2} = \frac{d(d+1)}{2}.$$

□

Let ν now be the valuation $\nu_{T^{-1}}$ and let $y \in \mathbb{Z}_p$. Set $\pi = \frac{1}{T}$ and, for monic a,

$$\langle a \rangle = \pi^{\deg(a)} a.$$

Define

$$g_y(x) = \sum_{d=0}^{\infty} x^{-d} \left(\sum_{\substack{a \text{ monic } \in \mathbf{A} \\ \deg(a)=d}} \langle a \rangle^y \right) .$$

Then, 3.6.8.b also implies that $g_y(x)$ converges π-adically for all non-zero x. Indeed, write

$$\langle a \rangle = 1 + \{a\},$$

where $\nu(\{a\}) > 0$ (in fact, $\{a\} = f\left(\frac{1}{T}\right)$ for some $f \in W(d)$) and express $\langle a \rangle^y$ in terms of $\{a\}$ by the binomial theorem. Now apply the above claim to *each* power sum appearing in the binomial theorem and notice that the estimate does not depend on the power!

Let $f \in \mathbf{A}$ be a monic prime. Let $S_f = \rho_f(U_f)$ be as in Definition 3.3.10 (note that for $\mathbf{A} = \mathbb{F}_r[T]$, the group $\mathcal{J}_2$ is trivial). Let $s_f \in S_f$. Define

$$g_{s_f}(x) = \sum_{d=0}^{\infty} x^{-d} \left(\sum_{\substack{a \text{ monic } \in \mathbf{A} \\ \deg(a)=d \\ (a,f)=1}} a^{s_v} \right) .$$

Let $\nu = \nu_f$. By working with residue classes $(\text{mod } f)$, 3.6.8.b shows that $g_{s_f}(x)$ is f-adically entire in exactly the same fashion as given above for $g_y(x)$.

We now return to the set-up of Theorem 3.6.5. Let $I(\mathcal{B})$ be the group of $\mathcal{O}_L$-fractional ideals of L which are prime to $\mathcal{B}$ and let $Pr(\mathcal{B})$ be the subgroup generated by (α) with $\alpha \equiv 1 \pmod{\mathcal{B}}$. (The reader should compare with the notation used in 3.3.8; if $L = \mathbf{k}$ and $\mathcal{B} = v$ then $I(\mathcal{B}) = \mathcal{J}_v$ in the earlier notation. The notation of 3.3.8 will be restricted to $\mathbf{k}$ itself. Note that $Pr(\mathcal{B})$ does *not* use any notion of positivity.) The group

$$I(\mathcal{B})/Pr(\mathcal{B})$$

is finite and we let $\{\varrho_1, \ldots, \varrho_e\}$ be fixed representatives. Upon expanding out the Euler product for $L(\psi, s)$ in the usual manner, we can write

$$L(\psi, s) = \sum_{t=1}^{e} \widetilde{\psi}(\varrho_t) n\varrho_t^{-s} \left(\sum_{\substack{\varrho \in Pr(\mathcal{B}) \\ \varrho\varrho_t \text{ integral}}} \widetilde{\psi}(\varrho) n\varrho^{-s} \right) .$$

Let $L_t(\psi, s)$ be the sum in parentheses. If we can establish that $L_t(\psi, s)$ is an essentially algebraic entire function for all $t \in \{1, \ldots, e\}$ then the result follows easily. Thus we fix $t = t_0$ and concentrate our attention on $L_{t_0}(\psi, s)$. It is very important to note that $L_{t_0}(\psi, s)$ is a sum over principal ideals.

Now let $\{\theta_1, \ldots, \theta_\epsilon\}$, $\epsilon = \frac{r^\delta - 1}{r-1}$, be the classes of $\mathbb{F}_\infty^*/\mathbb{F}_r^*$. The sign, $\operatorname{sgn}(J)$, of any fractional ideal J is invariantly given as an element of $\{\theta_1, \ldots, \theta_\epsilon\}$. So we decompose

$$L_{t_0}(\psi, s) = \sum_{u=1}^{\epsilon} \left(\sum_{\substack{\varrho \in Pr(\mathcal{B}) \\ \varrho\varrho_{t_0} \text{ integral} \\ \operatorname{sgn}(\varrho) \in \theta_u}} \widetilde{\psi}(\varrho) n\varrho^{-s} \right)$$
$$= \sum L_{t_0,u}(s).$$

Finally we are reduced, as above, to focusing on a fixed $u = u_0$. Set

$$L(s) = L_{t_0,u_0}(s).$$

The next step is to analyze $L(s)$ with an eye towards using Lemma 3.6.7.b). The idea is the following: Use the Riemann-Roch theorem to construct "enough" functions with prescribed zeros at ∞. Then the power series in x^{-1} for $L(x, Y)$ will converge *exactly* as did the function $g_y(x)$ in Example 3.6.8. Moreover, the reader will note that one constructs such functions in much the same manner as one constructs polynomials of a fixed degree d by adding polynomials of *lower* degree to T^d.

Let $\widehat{\varrho}_{t_0}$ be the inverse fractional ideal of ϱ_{t_0}. Let $\alpha, \beta \in \widehat{\varrho}_{t_0}$. We say

$$\alpha \sim \beta$$

if and only if there exists $u \in \mathcal{O}^*$ with $\alpha = u\beta$. Thus $\alpha\varrho_{t_0} = \beta\varrho_{t_0}$ if and only if $\alpha \sim \beta$. Let $[\widehat{\varrho}_{t_0}]$ be the equivalence classes of non-zero elements of $\widehat{\varrho}_{t_0}$ under $\sim$. Every element in the same equivalence class gives rise to the *same* fractional ideal.

Let $[\alpha] \in [\widehat{\varrho}_{t_0}]$. Clearly the norm of α, $n[\alpha]$, is uniquely determined up to an element of $\mathbb{F}_r^*$. As a consequence, the sign of $n[\alpha]$, $\operatorname{sgn}(n[\alpha])$, is given as an element of $\mathbb{F}_\infty^*/\mathbb{F}_r^*$.

Let $0 \neq \alpha \in \widehat{\varrho}_{t_0}$ with $-\sum_{i=1}^{\varepsilon} c_i^\alpha \infty_i$ being the *infinite part* of the divisor (α) of α. It is easy to see that the divisor of $n\alpha$ is

$$(\sum c_i^\alpha f_i) \cdot \infty,$$

where f_i is the residue class degree of $L_{\infty_i}/\mathbf{K}$. Let $u \in \mathcal{O}^*$. Then $nu \in \mathbf{A}^* = \mathbb{F}_r^*$. Thus

$$\deg n\alpha = \delta(\sum_i c_i^\alpha f_i)$$

depends only on $[\alpha]$. We set

$$D_\infty(\alpha) = \sum_{i=1}^{\varepsilon} c_i^\alpha \infty_i,$$

and put

$$D_\alpha = \sum (c_i^\alpha - 1)\infty_i.$$

(The divisor D_α will play the role of the degree of a polynomial.)

Let $\widehat{\varrho}_{t_0}(D_\alpha)$ be those elements in $\widehat{\varrho}_{t_0}$ which are also in $\mathcal{B}$ and have poles of order $\leq (c_i^\alpha - 1)$ at ∞_i. By standard algebraic geometry, $\widehat{\varrho}_{t_0}(D_\alpha)$ is finite dimensional over $\mathbb{F}$ (and $\mathbb{F}_r$).

3.6.9. Lemma.

a) *Let* $\alpha \in \widehat{\varrho}_{t_0}$ *with* $(\alpha) \in Pr(\mathcal{B})$. *Let* $\beta \in \widehat{\varrho}_{t_0}(D_\alpha)$. *Then* $\alpha + \beta \in \widehat{\varrho}_{t_0}$ *and*

$$(\alpha + \beta) \in Pr(\mathcal{B}).$$

b) $D_\infty(\alpha) = D_\infty(\alpha + \beta)$.

c) *Upon choosing a uniformizer* $\pi_i \in L_{\infty_i}$, *let* $\psi_{i,f}(\alpha)$ *be the root of unity obtained before Remarks 3.6.2. Then*

$$\psi_{i,f}(\alpha) = \psi_{i,f}(\alpha + \beta).$$

d) $\operatorname{sgn}(n\alpha) = \operatorname{sgn}(n(\alpha + \beta))$.

e) *Let* $\kappa \in \widehat{\varrho}_{t_0}$ *with* $(\kappa) \in Pr(\mathcal{B})$ *also. Then there exists* $\xi \in \widehat{\varrho}_{t_0}(D_\kappa)$ *with*

$$[\alpha + \beta] = [\kappa + \xi]$$

if and only if there exists $\beta_1 \in \widehat{\varrho}_{t_0}(D_\alpha)$ *with*

$$[\alpha + \beta_1] = [\kappa].$$

f) *Let* β_1 *and* $\beta_2 \in \widehat{\varrho}_{t_0}(D_\alpha)$ *with* $\beta_1 \neq \beta_2$. *Then*

$$(\alpha + \beta_1)/(\alpha + \beta_2) \notin \mathcal{O}^*.$$

Proof. Parts a) through d) follow easily from the definitions.

To see Part e), suppose

$$[\alpha + \beta] = [\kappa + \xi].$$

Thus there exists $u \in \mathcal{O}^*$ with

$$\alpha + \beta = u\kappa + u\xi.$$

From Part b), we have

$$D_\infty(\alpha + \beta) = D_\infty(\alpha) = D_\infty(u\kappa + u\xi) = D_\infty(u\kappa).$$

Set $\beta_1 = \beta - u\xi$. It is now easy to see that $\beta_1 \in \widehat{\varrho}_{t_0}(D_\alpha)$.

Finally we turn to Part f). Note that Part b) implies that

$$D_\infty(\alpha + \beta_1) = D_\infty(\alpha + \beta_2).$$

Thus if $(\alpha + \beta_1)/(\alpha + \beta_2)$ is a unit, it must actually be *constant*. Solving for α gives a contradiction. □

The above lemma is fundamental and is used in the following fashion: Pick $d_0 \in \mathbb{Z}$ and let $X = \{[\alpha_1], \ldots, [\alpha_i]\}$ be the (possibly empty) collection of those $[\alpha]$ with 1) $\alpha \in \widehat{\varrho}_{t_0}$, 2) $\alpha \equiv 1\ (\mathcal{B})$, 3) $\operatorname{sgn}(n\alpha) \in \theta_{u_0}$, and $\deg(n\alpha) = d_0$. Then, via 3.6.9, we are able to decompose X into *equivalence classes* under translation by the various vector spaces $\widehat{\varrho}_{t_0}(D_\alpha)$. Moreover, by 3.6.9.e) and 3.6.9.f), there is no non-trivial isotropy in each equivalence class.

Let $[\alpha] \in X$ and let $X_\alpha \subseteq X$ be its equivalence class under $\widehat{\varrho}_{t_0}(D_\alpha)$. Let $\kappa \in \widehat{\varrho}_{t_0}(D_\alpha)$ and let α_i, κ_i be the injections of these elements into $L_{\infty_i} \subseteq \overline{\mathbf{K}}$. Let $\pi_i \in L_{\infty_i}$ be a *fixed* uniformizer and set

$$\alpha_i = \pi_i^{n_i} \cdot \{\text{unit}\},$$

and define $m_i(\kappa)$ by

$$m_i(\kappa) = \frac{\kappa_i}{\pi_i^{n_i}}.$$

Clearly, $m_i(\kappa)$ is $\mathbb{F}_r$-linear in κ. Moreover, by construction, $\nu_\infty(m_i(\kappa)) > 0$. Now let $\{M_h(\kappa)\}$ be the finite set of linear maps obtained by composing $\{m_i(\kappa)\}$ with $\mathrm{Gal}(\overline{\mathbf{K}}/\mathbf{K})$. As the action of $\sigma \in \mathrm{Gal}(\overline{\mathbf{K}}/\mathbf{K})$ is continuous, we also have

$$\nu_\infty(M_h(\kappa)) > 0.$$

Let $\beta \in X_\alpha$. The reader can easily see how to express $\langle(n\beta)\rangle$ in terms of $M_h(\kappa)$ and $\langle(n\alpha)\rangle$.

Recall that we put $t_{\mathbb{F}} = [\mathbb{F}:\mathbb{F}_r]$ where $\mathbb{F} \subseteq L$ is the field of constants. We also put $\delta = \deg(\infty)$, and let us now set

$$\delta_i = \deg_1(\infty_i).$$

These numbers are related by the easily seen equation

$$t_{\mathbb{F}}\delta_i = \delta f_i.$$

As

$$\deg(n\alpha) = \delta(\sum c_i^\alpha f_i),$$

we find

3.6.10. Lemma. $\sum c_i^\alpha \delta_i = \deg(n\alpha)/t_{\mathbb{F}} = d_0/t_{\mathbb{F}}$. □

Recall further that

$$\widetilde{\psi}((\alpha)) = \prod \psi_{\infty_i}(\alpha)^{-1},$$

and for ideles, a,

$$\psi(a) = \pi^{g_\psi \deg_1(a)} \cdot \{\text{unit}\}.$$

The next lemma now follows immediately.

3.6.11. Lemma. $\widetilde{\psi}((\alpha)) = \pi^{g_\psi d_0/t_{\mathbb{F}}} \cdot \{\text{unit in } \overline{\mathbf{K}}\}$. □

We are now, finally, able to prove our main theorem.

Proof of 3.6.5. With a little thought, together with Lemmas 3.6.11 and 3.6.9, and the binomial theorem, the reader will see the following: The sum

$$\sum_{[\beta]\in X_\alpha} x^{-\deg(n\beta)}\widetilde{\psi}((\beta))\langle(n\beta)\rangle^{-y}$$

can be expressed as

$$x^{-d_0}\pi^{g_\psi d_0/t_{\mathbf{F}}} \cdot \sum_{\kappa \in \widehat{\varrho_{t_0}}(O_\alpha)} \widehat{L}_y(\kappa),$$

where $\widehat{L}_y(\kappa)$ is the composition of

$$\{M_j(\kappa)\}$$

and a bounded power series in $[L\colon \mathbf{k}]$ variables. For $d_0 \gg 0$, the Riemann-Roch Theorem now constructs enough functions so that one can conclude $L(s)$ is entire as in Example 3.6.8.

To see that $L(s)$ is essentially algebraic, note that for i, a sufficiently large positive integer, the sum

$$\sum_{[\beta]\in X_\alpha} x^{-\deg(n\beta)}\widetilde{\psi}((\beta))(n\beta)^i$$

can be expressed as sum over κ in the form of 3.6.7.a). So for $d_0 \gg 0$, we conclude that these sums *vanish*! This concludes the proof. □

The v-adic theory follows in essentially the same fashion. The L-series are defined as before except that one uses primes *not* lying over v. One then takes representations of $I(\mathcal{B})/Pr(\mathcal{B})$ prime to v and uses 3.6.7.b) *v-adically* to complete the proof.

3.7. General L-functions over an Arbitrary A-scheme. In this subsection we sketch a globalization of the local L-functions of Definition 3.2.15. Let X be a scheme of finite type over $\text{Spec}(\mathbf{A})$ and let φ/X be a Drinfeld module of rank d over X. Let $\wp \in \text{Spec}(\mathbf{A})$ and let $X_\wp$ be the associated fibre. Let $x \in X_\wp$ with residue class degree $d(x)$ over $\mathbb{F}_\wp$. We set

$$nx = \wp^{d(x)}.$$

For each such $\wp$, we let $v \in \text{Spec}(\mathbf{A})$ be some fixed prime $\neq \wp$ and let $s \in S_\infty$. For $x \in X_\wp$ we let $P_x(u)$ be as in Definition 3.2.15 (i.e., the inverse characteristic polynomial of Frobenius acting on the v-adic cohomology of the fibre $\phi^{(x)}$ at x).

3.7.1. Definition. Set

$$L(\varphi/X, s) = \prod_{x \text{ closed}} P_x(nx^{-s})^{-1}.$$

Although I have not checked the details, it is highly likely (using the Riemann Hypothesis for Drinfeld modules over finite fields) that $L(\varphi/X, s)$ converges on a half-plane of S_∞.

The definition of "$L(\varphi^{(i)}/X, s)$" is easily given using the i-th cohomology of φ. Moreover, the reader will have no difficulty giving the appropriate v-adic version of Definition 3.7.1. In this case, the half-plane of convergence of $\overline{\mathbf{k}}_v \times S_v$ will be the same as given in Proposition 3.4.6.b).

The above definition mimics the classical one given in Subsection 2. Perhaps here too there is a "cohomological" interpretation of the local factors (i.e., the product above a fixed prime as in 3.2.15). Clearly this cohomology should be finite dimensional $\mathbf{A}$- (or $\mathbf{A}_v$-) modules etc.

It is not too unreasonable to expect that $L(\varphi/X, s)$ will have a meromorphic continuation to all of S_∞ etc., and that the "fond hope" (3.5.4) also generalizes in the obvious fashion. If all turns out correctly, one could even hope for some sort of functional equation for $L(\varphi/X, s)$. Again classical theory would lead to the expectation that this functional equation would involve Γ-functions (see **[G3]** and, especially, **[Th1]**). Perhaps the description of these factors at ∞ would depend on the reduction of X at the place ∞ of our base curve $\mathcal{C}$.

As an example, let $X = E$ be an elliptic curve over $\mathbf{k}$. We focus on the 0-th cohomology of φ (which does not depend on the choice of φ). Let $s \in S_\infty$. As we saw following 3.2.15,

$$\zeta(E, s) =: L(\varphi^{(0)}/E, s) = \prod_{\wp} Z(E_\wp, \wp^{-s}).$$

The theory of Section 2 thus gives a factorization of $\zeta(E, s)$ in terms of the cohomology groups of E. For each prime $\wp$ of $\mathbf{A}$ let $f_\wp(u)$ be the usual (characteristic-0) factor of the L-series of E. (Thus $f_\wp(u) \in \mathbb{Z}[u]$.) Finally — without introducing new notation to handle things in complete generality — let us set for $s \in S_\infty$

$$L(E, s) = \prod_{\wp} f_\wp(\wp^{-s})^{-1}.$$

(There is no difficulty with this as the coefficients of f are in $\mathbb{Z}$.) Clearly this product converges for all $(x, Y) \in S_\infty$ with $|x| > 1$.

Lastly, let E have split multiplicative reduction at ∞. Then one knows that E is isogenous to a factor of a Jacobian of a Drinfeld modular curve. Thus we obtain a cusp form and a measure by work of Teitelbaum (see **[Te1]** and **[G11]**). By integrating x^{-s} against these measures over the space of *positive elements* of positive degree, we obtain "another" two-variable function, $\tilde{L}(E, s)$, associated to E, **[G11]**. It seems only reasonable to expect that, *somehow*, these two functions are indeed related to each other.

It may also turn out that the simplest case to handle of these very general L-series, will arise from the universal families of Drinfeld modules lying above the modular varieties.

3.8. The Brauer-Nesbitt Theorem. Let φ be Drinfeld module of rank d over $\mathbf{k}$ itself. Let $G = \operatorname{Gal}(\bar{\mathbf{k}}/\mathbf{k})$ and let $M_v^i = H_v^i(\varphi, \mathbf{k}_v)$ for some $i \in \{0, \dots, d\}$. The space M_v^i is a module for the ring $\mathbf{k}_v[G]$.

In Subsection 3.4 we associated to the compatible system $\{M_v^i\}$ the L-series $L(s) = L(\varphi^{(i)}, s)$. By 3.3.16 we know that the Euler factors of $L(s)$ are determined by $L(s)$ (which is the reason that we are working over k). The *Brauer-Nesbitt Theorem* asserts that these Euler factors determine the *semi-simplification* of M_v^i. (N.B: M_v^1 is semi-simple by **[Ta1]**.)

On the other hand, the treatments of the Brauer-Nesbitt theorem in the literature make additional assumptions that are not necessary. It is our purpose here to sketch a proof which uses *no* additional assumptions. Our treatment is due to J.-P. Serre. We are grateful to him for his communication **[S3]** on this subject.

Let k be an arbitrary field and R a k-algebra with unit (e.g., $k = \mathbf{k}_v$ and $R = \mathbf{k}_v[G]$). Let $K(R)$ be the Grothendieck group of R-modules M which are finite dimensional over k. A basis of $K(R)$ is given by a system $\{M_i\}$ of representatives of simple R-modules.

Let k' be an arbitrary field extension of k and put $R' = k' \otimes R$. Extension of scalars then defines a homomorphism $K(R) \to K(R')$.

3.8.1. Lemma. *The above homomorphism $K(R) \to K(R')$ is injective.*

Proof. Let S and T be two non-isomorphic simple R-modules. The result will follow if we can show the following: $S' = k' \otimes S$ and $T' = k' \otimes T$ are disjoint (i.e., have no common Jordan-Hölder factor) R'-modules. To see this, note that by the *Density Theorem* there exists $r \in R$ which acts as 1 on S and 0 on T. This is clearly enough for the result. □

Now let $H \subset R$ be a set of generators of R as a k-vector space. Let M be an R-module as above. For $h \in H$ let

$$P(M, h, u) = \det(1 - u \cdot h_M),$$

where h_M is the action of h on M.

3.8.2. Theorem (Brauer-Nesbitt). *Let M and N be two R-modules which are finite dimensional over k. Assume that $P(M, h, u) = P(N, h, u)$ for all $h \in H$. Then $(M) = (N)$ as elements of $K(R)$; in particular, M and N are isomorphic if they are semi-simple.*

Proof. We may assume that M and N are semi-simple. Moreover, the lemma assures us that we may assume that k is algebraically closed. The proof now follows as in **[C-R1]**. □

A continuity argument allows us to extend the result to the case where $R = \mathbf{k}_v[G]$ and H consists of the geometric Frobenius elements at finite primes $\neq v$. Indeed, this set is *dense* in G.

The reader will note the theorem asserts that the characteristic polynomials of the geometric Frobenius elements determine the semi-simplification of M_v^i over *any* base field L. However, in the general case, one can *not* read these off from L-series defined over L.

4. ζ-functions of $\mathbf{A}$-integers and Complex L-series.

We continue with the notation of Section 3; so L is a finite separable extension of $\mathbf{k} \subset \overline{\mathbf{k}}$. And, as before, we let $\mathcal{O} = \mathcal{O}_L$ be its ring of $\mathbf{A}$-integers. Recall that in 3.4.5 we put

$$\begin{aligned}\zeta(\mathcal{O}, s) &= \prod_{\wp \text{ prime}} (1 - n\wp^{-s})^{-1}, \\ &= \sum_{\mathfrak{I} \text{ an ideal}} n\mathfrak{I}^{-s},\end{aligned}$$

with the obvious v-adic analogs. The analytic properties of these functions are given by the results of Section 3. Moreover, although it is trivial to see, it is important for the reader to note that

$$\zeta(\mathcal{O}, ps) = \zeta(\mathcal{O}, s)^p;$$

etc.

In order to discuss the deeper properties of these functions we need to discuss a construction of rank one Drinfeld modules, and cyclotomic function fields, due to David Hayes. The two papers **[H1]** and **[H2]** are the standard references on this construction.

4.1. Sign Normalized Rank One Drinfeld Modules. Let $\mathbf{A} = \mathbb{F}_r[T]$. As is well known, the Carlitz module C is the simplest rank one Drinfeld module. It plays the role of $\mathbb{G}_m$ for $\mathbf{A}$ in many ways; among them is its use in the explicit construction of class fields for $\mathbf{k} = \mathbb{F}_r(T)$. One simply adjoins to $\mathbf{k}$ *all* g-division points, for $g \in \mathbf{A}$. In this way one obtains the maximal abelian extension contained in $\mathbf{K}_1 = \mathbf{K}(\xi) = \mathbf{K}(\Upsilon)$ where ξ is the period of the Carlitz module and Υ is *any* non-trivial g-division point for $\deg(g) > 1$. (To see that ξ and Υ generate the same extension, we refer the reader to the proof of Lemma 4.1.8.) From this point of view, the field $\mathbf{K}_1$ is analogous to $\mathbb{C}$ (but of course is not algebraically closed!).

On the other hand, in his original paper on the subject **[Dr1]**, Drinfeld showed, *in general*, that the moduli of rank one Drinfeld modules is the spectrum of the maximal abelian extension of $\mathbf{k}$ lying in $\mathbf{K}$; i.e., is totally-split at ∞ or, by abuse of language, "totally-real". Thus there existed a disparity between the polynomial and general cases; in the polynomial case one had a very satisfactory analog of cyclotomic fields while in general one only had the analog of their totally-real subfield. This disparity is precisely what is remedied by Hayes' construction.

Let $J_{\mathbf{k}}$ be the ideles of $\mathbf{k}$. Let $G_0 \subset J_{\mathbf{k}}$ (resp. G_1) be the subgroup of all ideles which are units at finite primes and arbitrary (resp. *positive*) at ∞.

4.1.1. Definition. We let $H_i \subset \overline{\mathbf{k}}$, $i = 0, 1$, be the class field associated to the subgroup $G_i \cdot \mathbf{k}^*$ of $J_{\mathbf{k}}$.

Class field theory tells us that $H = H_0$ is precisely the maximal totally-real Hilbert class field of $\mathbf{k}$. By definition, the field $\mathbb{F}_\infty \subset H$. Moreover, one has, in the notation of Section 3, a canonical isomorphism

$$\mathrm{Gal}(H/\mathbf{k}) \simeq \mathcal{I}/Pr,$$

via Artin symbols. On the other hand, $H_1/\mathbf{k}$ is ramified only at ∞ by definition. We again find a canonical isomorphism

$$\mathrm{Gal}(H_1/\mathbf{k}) \simeq \mathcal{I}/P_0,$$

where P_0 is the group of positively generated principal ideals. Thus

$$\mathrm{Gal}(H_1/H) \simeq \mathbb{F}_\infty^*/\mathbb{F}_r^*.$$

Every infinite prime of H is totally (and tamely) ramified in H_1.

From the formalism of class field theory, one knows that the norm of an ideal of $\mathcal{O}_H$ is principal. Similarly, and importantly, the norm of an ideal of $\mathcal{O}_{H_1}$ is principal and *positively* generated.

Let $\sigma \in \mathrm{Gal}(\mathbb{F}_\infty/\mathbb{F}_r)$ and sgn be our fixed sign function. The mapping

$$\sigma \circ \mathrm{sgn}\colon \mathbf{K}^* \to \mathbb{F}_\infty^*$$

is called a "twist of sgn".

Let ρ be a rank one Drinfeld module and let $a \in \mathbf{A}$. We write

$$\rho_a(x) = \sum_{j=0}^{\deg(a)} \rho_{a,j} x^{r^j},$$

where $a = \rho_{a,0}$ and $0 \neq \rho_{a,\deg(a)}$. The following basic theorem is contained in **[H2]**.

4.1.2. Theorem (Hayes). *Over H_1 there exists a rank 1 Drinfeld module ψ so that:*
a) *The coefficients of ψ are $\mathbf{A}$-integers.*
b) *The mapping $a \in \mathbf{A} \mapsto \psi_{a,\deg(a)}$ is a twist of* sgn. □

In other words, by working over H_1 we are guaranteed analogs of the Carlitz module in complete generality! Rank 1 Drinfeld modules which satisfy 4.1.2.b) are said to be "sgn-normalized" and we now fix one such module for the rest of this subsection. We also denote it by "ψ".

Let $\widehat{\mathrm{sgn}}$ be the twist of sgn which occurs in 4.1.2.b).

Let I be an ideal of $\mathbf{A}$ and let $\sigma_I \in \mathrm{Gal}(H_1/\mathbf{k})$ be its Artin symbol. Let $\sigma_I\psi$ be the twist of ψ given by σ_I; it is also clearly sgn-normalized. The module ψ_I can be explicitly constructed as follows: Let $\psi[I] \subset \bar{\mathbf{k}}$ be the $\mathbf{A}$-module of I-division points of ψ and

$$\psi_I(x) = \prod_{\tau \in \psi[I]} (x - \tau).$$

The function $\psi_I(x)$ gives the isogeny between ψ and the module ψ_1, which is the quotient of ψ by $\psi[I]$.

4.1.3. Proposition.
a) *We have*

$$\sigma_I\psi = \psi_1.$$

b) *Let $I = x\mathbf{A}$ be principal. Then*

$$\sigma_I\psi = \widehat{\mathrm{sgn}}(x)^{-1} \cdot \psi \cdot \widehat{\mathrm{sgn}}(x).$$

Proof. This is Theorem 4.10 and Corollary 4.11 of **[H2]**. □

We follow Hayes and use the notation "$I * \psi$" for $\sigma_I\psi$.

Now let I be an ideal of $\mathbf{A}$. One knows that $\psi[I]$ is $\mathbf{A}$-module isomorphic to $\mathbf{A}/I$. We let $\Upsilon = \Upsilon_I$ be any generator of $\psi[I]$; i.e., a "primitive I-th division point".

4.1.4. Definition. We set $\mathbf{k}(I) = H_1(\Upsilon)$.

Let $v \in \mathrm{Spec}(\mathbf{A})$. A simple calculation, which follows the model of $\mathbb{Q}(\zeta_p)$, shows that $\mathbf{k}(I)/H_1$ is ramified at the primes dividing I and unramified at all other finite primes; if I is a prime power then this ramification is total. Further, there is an isomorphism

$$\mathrm{Gal}(\mathbf{k}(I)/H_1) \simeq \mathbf{A}/I^*.$$

Using the full power of class field theory, Hayes (ibid) shows that $\mathbf{k}(I)$ is actually *abelian* over $\mathbf{k}$. In fact, it corresponds to the group $\mathbf{k}^*G(I) \subset J_{\mathbf{k}}$ where $G(I)$ is the group of ideles which are positive at ∞, unramified at finite primes not dividing I and $\equiv 1 \pmod{I}$. One then sees readily that

$$\mathrm{Gal}(\mathbf{k}(I)/\mathbf{k}) \simeq J_{\mathbf{k}}/\mathbf{k}^*G(I).$$

Let $\mathcal{I}(I)$ be the group of $\mathbf{A}$-fractional ideals prime to I and let $P(I) \subset \mathcal{I}(I)$ be the group of principal ideals generated by *positive* $\alpha \equiv 1 \pmod{I}$. (The notation generalizes that of 3.3.8; so if $I = v^m$, $v \in \mathrm{Spec}(\mathbf{A})$, then $\mathcal{I}(I) = \mathcal{I}_v$, etc.) By standard theory, we then have

$$\mathrm{Gal}(\mathbf{k}(I)/\mathbf{k}) \simeq \mathcal{I}(I)/P(I).$$

The reader should note that, as $\mathbf{k}(I)/\mathbf{k}$ is Galois, it does *not* depend on our original choice of ψ.

Let $P_\infty \subseteq \mathcal{I}(I)/P(I)$ be the group generated by (α) with $\alpha \equiv 1 \pmod{I}$ and let $G_\infty \subseteq \mathrm{Gal}(\mathbf{k}(I)/\mathbf{k})$ be the subgroup generated by the Artin symbols of P_∞. As in the discussion preceding Definition 3.3.10, the reader can see that $P_\infty \simeq G_\infty$ is also isomorphic to $\mathbb{F}_\infty^*$ via the sign function.

Parts b) and c) of the next result are due to Hayes.

4.1.5. Lemma.

a) *Let $\wp$ be an ideal of $\mathcal{O}_{H_1}$ not lying over I and let $\sigma_\wp \in \mathbf{A}/I^*$ be its Artin symbol under the isomorphism* $\mathrm{Gal}(\mathbf{k}(I)/H_1) \simeq \mathbf{A}/I^*$. *Then we have*

$$\sigma_\wp = n\wp + (I).$$

b) *The group G_∞ is both the decomposition and inertia group at ∞. Furthermore, let $\mathbf{K}_1 = \mathbf{K}(\Upsilon)$ and let $\theta\colon \mathbf{K}^* \to G_\infty$ be the norm residue symbol. Then*

$$\theta(x)(\Upsilon) = \widehat{\mathrm{sgn}}(x) \cdot \Upsilon.$$

c) *Let J be an ideal of $\mathbf{A}$ prime to I with Artin symbol*

$$\sigma_J \in \mathrm{Gal}(\mathbf{k}(I)/\mathbf{k}).$$

Then

$$\sigma_J \Upsilon = \psi_J(\Upsilon).$$

□

Next we discuss the lattices associated to ψ and its conjugates $I * \psi$, for I an ideal. Let $\Gamma \subset \overline{\mathbf{K}}$ be the lattice of ψ. General theory tells us that there exists $\xi \neq 0$ and an ideal $\mathcal{B}$ of $\mathbf{A}$ with

$$\Gamma = \xi\mathcal{B}.$$

Of course, by definition, ξ is only determined up to an element of $\mathbb{F}_r^*$. Let

$$D(\psi_I)$$

be the coefficient of x in $\psi_I(x)$. The next result is readily seen.

4.1.6. Lemma. *The lattice associated to $I * \psi$ is $\xi D(\psi_I) I^{-1}\mathcal{B}$.* □

4.1.7. Corollary. *Let $I \subset \mathbf{A}$ be an ideal. Suppose that $\sigma_I \psi = \psi$. Then I is principal and positively generated.*

Proof. From 4.1.6, we see that $\xi\mathcal{B}$ and $\xi D(\psi_I) I^{-1}\mathcal{B}$ must be the same. Thus I must be principal. The result now follows from 4.1.3.b. □

The reader should note that the elements $D(\psi_I)$ lie in $\mathcal{O}_{H_1}$. In fact, as in **[H1]**, we have

$$(D(\psi_I)) = I\mathcal{O}_{H_1}.$$

Thus we have an explicit version of the *principal ideal theorem* over H_1.

Let $w = r^\delta - 1 = \#\mathbb{F}_\infty^*$.

4.1.8. Lemma. *We have*

$$\xi^w \in \mathbf{K}.$$

Proof. Let

$$\mathbf{e}(z) = z \prod_{0 \neq \alpha \in \mathcal{B}} \left(1 - \frac{z}{\xi\alpha}\right)$$

be the exponential of ψ. Then there exists $0 \neq \beta \in I^{-1}\mathcal{B}/\mathcal{B}$ with

$$\Upsilon = \mathbf{e}(\xi\beta).$$

Upon taking logarithmic derivatives, we find

$$\frac{\mathbf{e}'(z)}{\mathbf{e}(z)} = \frac{1}{\mathbf{e}(z)} = \sum_{\alpha \in \mathcal{B}} \frac{1}{z + \xi\alpha}.$$

Thus

$$\Upsilon^{-1} = \sum_{\alpha \in \mathcal{B}} \frac{1}{\xi\beta + \xi\alpha}$$
$$= \xi^{-1} \sum_{\alpha \in \mathcal{B}} \frac{1}{\beta + \alpha}.$$

The second sum is clearly in $\mathbf{K}$. Thus the result follows from 4.1.5.b). □

One can also show the above proposition using an explicit construction of ξ as in **[Th1]**. As a corollary of the proof of 4.1.6, *all* extensions $\mathbf{k}(I)$ are subfields of $\mathbf{K}_1 = \mathbf{K}(\xi)$. In fact, as with the rational case, Hayes' construction now gives the maximal abelian extension of $\mathbf{k}$ contained in $\mathbf{K}_1$.

From the fact that $\mathbf{k}(I)/\mathbf{k}$ is abelian, one deduces that $\mathbf{k}(\Upsilon)/\mathbf{k}$ is also. The question of determining Gal $(\mathbf{k}(\Upsilon)/\mathbf{k})$ is settled by the following very pleasing result. Its elegant proof was shown to me by David Hayes.

4.1.9. Theorem. *We have*

$$\mathbf{k}(I) = \mathbf{k}(\Upsilon).$$

Proof. From elementary Galois Theory, we know that the result follows if the only Galois automorphism that fixes Υ is the identity. By 4.1.5.c), this is translated into showing for M prime to I:

$$\psi_M(\Upsilon) = \Upsilon \Rightarrow M = x\mathbf{A} \text{ with } \operatorname{sgn}(x) = 1 \text{ and } x \equiv 1 \pmod{I}.$$

Let $\psi' = M * \psi$. As $\psi_M(\Upsilon) = \Upsilon$, and M is prime to I, we conclude that ψ and ψ' have the same I-division points; thus

$$\psi_I = \psi'_I.$$

Now let t be chosen so that $I^t = \alpha\mathbf{A}$ for some α. Clearly any α-division point maps to an I-division point upon dividing ψ by $\psi[I^a]$ for some a. If one puts this together with 4.1.3.a), we have

$$\widehat{\operatorname{sgn}}(\alpha)^{-1} \cdot \psi_a = (\psi_I)^{1+\sigma_I+\cdots+\sigma_I^{t-1}}.$$

Thus

$$\psi_\alpha = \psi'_\alpha.$$

Let $\mathbf{e}(z)$ and $\mathbf{log}(z)$ be the exponential and logarithm of ψ. By definition

$$\psi_a(u) = \mathbf{e}\left(a\mathbf{log}(u)\right),$$

for *all* $a \in \mathbf{A}$. On the other hand, a simple computation proves that $\mathbf{e}(z)$ and $\mathbf{log}(z)$ are uniquely determined by the above equation for *any* non-constant $a \in \mathbf{A}$. Thus, ψ and ψ' have the same exponential function and so must be equal. By 4.1.7, we find that M is principal and positively generated; let m be a positive generator of M. Thus,

$$\psi_M = \psi_m$$

and the action on Υ is via the induced $\mathbf{A}/I$ action on division points. The result now follows immediately. □

Hayes' construction also leads to a construction of *"cyclotomic units"* which completely generalizes that of **[G2]**, (see also Subsection 5.5): Let I, Υ be as above.

4.1.10. Theorem.

a) *Let* $\sigma \in \operatorname{Gal}(\mathbf{k}(I)/\mathbf{k})$. *Then*

$$\frac{\sigma(\Upsilon)}{\Upsilon}$$

is a unit in the $\mathbf{A}$*-integers,* $\mathcal{O}_{\mathbf{k}(I)}$, *of* $\mathbf{k}(I)$.

b) *Let* I *be divisible by two distinct primes. Then* Υ *is itself a unit.*

Proof. We show Part b) first. Suppose that Υ is not a unit and let $\wp$ be a prime of $\mathcal{O}_{\mathbf{k}(I)}$ dividing Υ. Let v be a prime of $\mathbf{A}$ lying under $\wp$. Let v_1 be a prime dividing $I \neq v$ and write $I = v_1^j \times I_1$ with I_1 prime to v_1. Let t be chosen so that I_1^t is principal, say generated by α. Then $\psi_\alpha(\Upsilon)$ is a primitive v_1-th division point. Thus it is only divisible by primes above v_1. On the other hand, since Υ divides $\psi_\alpha(\Upsilon)$, we see that it is also divisible by $\wp$. This is a contradiction.

For Part a) we are now clearly reduced to the case where $I = v^m$ for some prime v. Let σ first of all be the Artin symbol of some positive element x which is prime to v. Then we have seen that $\sigma(\Upsilon) = \psi_x(\Upsilon)$; thus $\sigma(\Upsilon)$ is a polynomial in Υ. As such,

$$\frac{\sigma(\Upsilon)}{\Upsilon} \in \mathcal{O}_{\mathbf{k}(I)}\,.$$

On the other hand, we can reverse the roles of Υ and $\sigma(\Upsilon)$; thus the result follows in this case.

The minimal polynomial of Υ over H_1 is

$$\frac{\psi_I(x)}{\psi_{v^{m-1}}(x)}\,.$$

Recall that $D(\psi_I)$ generates the ideal $I\mathcal{O}_{H_1}$. Thus the norm of Υ down to H_1 generates the ideal

$$v\mathcal{O}_{H_1}\,.$$

But the norm differs from Υ^τ, $\tau = r^{m\deg(v)} - r^{(m-1)\deg(v)}$, by a *unit*. We are thus able to compute the valuation of Υ at the primes above v.

Finally, the above computation is invariant of the normalized module used. Thus the valuations of *all* Galois twists of Υ must be the same. The result follows. □

It has been shown recently (by L. Shu and H. Oukhaba) that these units generate groups of maximal rank. In Subsection 5.5 we will discuss this in the case of $\mathbf{A} = \mathbb{F}_r[T]$.

As in classical cyclotomic theory, one can now conclude that the ring of integers in $\mathbf{k}(I)$ is $\mathcal{O}_{H_1}[\Upsilon]$. (It is not clear whether one can replace $\mathcal{O}_{H_1}$ by $\mathbf{A}$ in this result.)

4.1.11. Remarks. In Example 3.6.4, we constructed a Grössencharakter Ψ_χ associated to our exponentiation of ideals. By definition, if I is a principal ideal generated by a *positive* element a then

$$\Psi_\chi(I) = a\,.$$

Moreover, as mentioned, it is reasonable to believe that this Grössencharakter comes from a t-motive M_χ.

Now let L be an extension of $\mathbf{k}$ which contains H_1. Let Ψ_χ^L be the composition of Ψ_χ and the norm map n from $\mathcal{O}_L$ to $\mathbf{A}$. Let J be an ideal of $\mathcal{O}_L$ and let $\alpha(J)$ be the unique *positive* generator of nJ (which exists since L includes H_1). It follows trivially that $\Psi_\chi^L(J) = \alpha(J)$. As a corollary, the reader should note that Ψ_χ^L no longer depends on the choice of χ.

On the other hand, extend scalars and consider the rank one module ψ over L. It is trivial to see that it too is associated to the Grössencharakter Ψ_χ^L. Classical theory again leads to the expectation (*not* yet proven) that M_χ is isogenous to ψ^θ, where θ is the rank (i.e., the dimension of its cohomology) of M_χ.

Finally, we finish this subsection by mentioning a nice observation due to Dinesh Thakur. Let ψ now be an *arbitrary* Drinfeld module of any rank over $\overline{\mathbf{K}}$. Let $\mathrm{e}(z)$ be its exponential and $\mathbf{log}(z)$ its logarithm. It is easy to see that $\mathbf{log}(z)$ has a non-trivial radius of convergence. Recall now the classical result that for $z \in \mathbb{C}$

$$\lim_{n\to\infty}\left(1+\frac{z}{n}\right)^n = e^z .$$

4.1.12 Proposition. *Let* $z \in \overline{\mathbf{K}}$. *Then*

$$\mathbf{e}(z) = \lim_{\deg(a)\to\infty} \psi_a\left(\frac{z}{a}\right).$$

Proof. Choose t sufficiently large so that $\log\left(\frac{z}{a}\right)$ converges for those a with $\deg(a) > t$. By definition, we have

$$\mathbf{e}(az) = \psi_a\left(\mathbf{e}(z)\right);$$

thus

$$\psi_a\left(\frac{z}{a}\right) = \psi_a\left(\mathbf{e}\left(\mathbf{log}\left(\frac{z}{a}\right)\right)\right) = \mathbf{e}\left(a\mathbf{log}\left(\frac{z}{a}\right)\right).$$

The result now follows by the easily established fact that, for a with $\deg(a) > t$,

$$\lim_{\deg(a)\to\infty} a\mathbf{log}\left(\frac{z}{a}\right) = z. \qquad \square$$

4.2. The Connection with Classical L-series and Trivial Zeroes. In this subsection we establish a fundamental connection between the special values of $\zeta(s) =: \zeta(\mathcal{O}, s)$ at negative integers. We are particularly interested in applying this connection in the simplest case where $\mathbf{A} = \mathbb{F}_r[T]$. Thus we shall restrict ourselves to the case where L contains H_1 (as in Example 4.1.11); so all norms are principal and positively generated. In particular, the definition of $\zeta(s)$ does not depend on any choice of χ.

Let i be a positive integer. Following Definition 3.5.2, we set

$$z_L(x, -i) =: h_\zeta(x, -i) = \zeta(\mathcal{O}, x\pi^{i/d}, -i).$$

From Theorem 3.6.5, we know that $z_L(x, -i) \in \mathbf{A}[x^{-1}]$. Similarly, we set

$$z_{L,v}(x_v, -i) = \zeta_v(\mathcal{O}, v_v, -i).$$

Let $\mathcal{B} \in \mathrm{Spec}(\mathbf{A})$. Let Υ a primitive $\mathcal{B}$-th division point and set

$$L(\mathcal{B}) = L(\Upsilon).$$

From the theory of Subsection 4.1, we know that $\mathrm{Gal}(L(\mathcal{B})/L) \subseteq \mathbf{A}/\mathcal{B}^*$.

Now let W be the Witt ring of the finite field $\mathbf{A}/\mathcal{B}$; thus W is a finite unramified extension of $\mathbb{Z}_p$ with $W/pW \simeq \mathbf{A}/\mathcal{B}$.

4.2.1. Definition. We let $\omega = \omega_{\mathcal{B}}$ be the canonical character $\mathbf{A}/\mathcal{B}^* \to W^*$ determined by $\omega(a) + (p) = a$.

We call ω the *Teichmüller character* at $\mathcal{B}$. From the discussion above, we consider it as giving a W-valued character of $\mathrm{Gal}(L(\mathcal{B})/L)$. *Every* W-valued character of this group is a power of ω.

Recall that we let $\{\infty_1, \ldots, \infty_\varepsilon\}$ be infinite primes of L. Since $H_1 \subset L$, 4.1.5.b) implies that

$$\mathrm{Gal}(L_{\infty_j}(\Upsilon)/L_{\infty_j}) \subset \mathbb{F}_r^*,$$

for all $j = 1, \ldots, \varepsilon$.

Because the field L can be so general, it may happen that $L_{\infty_j}(\Upsilon)/L_{\infty_j})$ is not totally ramified. For our choice of i, let

$$\{\infty_{j_1}, \ldots, \infty_{j_k}\}$$

be the infinite primes where ω^i is an *unramified* character of $\mathrm{Gal}(L(\mathcal{B})/L)$. By 4.1.5, this set of primes *does not* depend on the original choice of $\mathcal{B}$. This fact will prove to be crucial in the next result.

4.2.2. Definition. We let $L(\omega^i, u) \in W(u)$ be the classical (in sense of Artin and Weil) L-series in $u = r^{-st_{\mathbb{F}}}$.

4.2.3. Definition. Let i be a positive integer. We set

$$\tilde{z}_L(x, -i) = z_L(x, -i)(1 - \omega^i(\infty_{j_1})x^{-\deg(\infty_{j_1})})^{-1} \cdots (1 - \omega^i(\infty_{j_k})x^{-\deg(\infty_{j_k})})^{-1}.$$

(The use of deg and not $\deg_1$ is deliberate.)

Recall that we have set $\mathbb{F}$ equal to the full field of constants of L and $t_{\mathbb{F}} = [\mathbb{F} : \mathbb{F}_r]$. It follows that $\tilde{z}_L(x, -i)$ is a rational function of $x^{-t_{\mathbb{F}}}$.

4.2.4. Theorem. *Let i be a positive integer. Then $\tilde{z}_L(x, -i) \in \mathbf{A}[x^{-1}]$.*

Proof. Almost all $\mathcal{B} \in \mathrm{Spec}(\mathbf{A})$ are prime to the discriminant of $L/\mathbf{k}$. For such $\mathcal{B}$, simple ramification arguments imply that $L(\mathcal{B})/L$ is totally ramified above $\mathcal{B}$; thus this extension is *geometric* (i.e., does not contain an extension of finite fields). Similarly, we find

$$\mathrm{Gal}(L(\mathcal{B})/L) \simeq \mathbf{A}/\mathcal{B}^*.$$

Thus, for almost all $\mathcal{B}$, the character ω^i is non-trivial and $L(\omega^i, u) \in W[u]$.

The formal power series $\tilde{z}_L(x, -i)$ has coefficients in $\mathbf{A}$. Thus we may reduce its coefficients mod $\mathcal{B}$. On the other hand, we can reduce the coefficients of the *polynomial* $L(\omega^i, u)$ modulo p. We thus obtain two power series over the finite field $\mathbf{A}/\mathcal{B}$.

Recall that $z_L(x, -i)$ is given by an Euler product over all finite primes of $\mathcal{O}_L$. Factors at infinity are given for $\tilde{z}_L(x, -i)$. Recall further that $\mathbb{F} \subset L$ is the full field of constants and $t_{\mathbb{F}} = [\mathbb{F} : \mathbb{F}_r]$. It is then a simple matter to see the following *double congruence*:

$$L(\omega^i, x^{-t_{\mathbb{F}}}) + (p) = \tilde{z}_L(x, -i) + (\mathcal{B}). \tag{4.2.5}$$

We conclude that the rational function (in x^{-1}), $\tilde{z}_L(x, -i)$, is a polynomial modulo an infinite number of primes. The result now follows directly. □

4.2.6. Remarks ("Trivial Zeroes"). Theorem 4.2.4 gives trivial zeroes for $\zeta(\mathcal{O}_L, s)$. Indeed, we have the equality of polynomials

$$\tilde{z}_L(x,-i)(1-\omega^i(\infty_{j_1})x^{-\deg(\infty_{j_1})})\cdots(1-\omega^i(\infty_{j_k})x^{-\deg(\infty_{j_k})}) = z_L(x,-i).$$

Thus, the zeroes of

$$(1-\omega^i(\infty_{j_1})x^{-\deg(\infty_{j_1})})\cdots(1-\omega^i(\infty_{j_k})x^{-\deg(\infty_{j_k})})$$

are *automatically* zeroes of $z_L(x,-i)$ and thus of $\zeta(\mathcal{O}_L, s)$.

The above argument only establishes a *lower bound* for the order of zero at a trivial zero. It is an open question if these bounds are exact. (See Subsection 5.1 where this is established in certain cases.)

The reader will note that the trivial zeroes theory actually follows the model of classical theory. To illustrate this, let $\mathbf{A} = \mathbb{F}_r[T]$ and assume that $L/\mathbf{k}$ is "totally-real" (i.e., totally split at ∞). In this case the extension $L_{\infty_j}(\Upsilon)/L_{\infty_j}$ *is* totally-ramified. Thus in order for ω^i to be unramified at an infinite prime, i must be $\mathbf{A}$-even; in which case the corresponding factor is just $1-x^{-1}$. We conclude that $\zeta(\mathcal{O}_L, s)$ has a zero of order $\geq [L:\mathbf{k}]$ at $-i$ for i as above. If one recalls that $r-1$ is the number of global units of $\mathbf{A}$, as 2 is for $\mathbb{Z}$, the analogy with classical theory will then be obvious.

Our treatment of trivial zeroes does not use any sort of Γ-function. It may, however, happen that such functions are needed in order to bound from above the orders at the trivial zeroes.

4.2.7. Remarks ("v-adic Trivial Zeroes at Positive Integers."). The trivial zeroes are defined at the negative integers. However, v*-adically*, we can also find trivial zeroes at *positive* integers: Recall that the v-adic function, $\zeta_v(\mathcal{O}, x_v, s_v)$ is defined on the space $\overline{\mathbf{k}}_v^* \times S_v$ and is related the ∞-adic version precisely via the polynomials $z_L(x,-i)$ – indeed, in the v-adic version, one obtains $z_{L,v}(x_v,-i)$ by removing from $z_L(x,-i)$ the Euler factors of the primes above v and replacing x with x_v. Note that removing a finite number of Euler factors does *not* remove zeroes. Now let j be a positive integer and let $\{j_u\}$ be a sequence of positive integers converging to $-j$ (correct sign) in S_v (i.e., a sequence which is $\equiv -j \pmod{r-1}$ and which converges p-adically to $-j$). The functions $z_{L,v}(x_v,-j_u)$ uniformly converge to $\zeta_v(\mathcal{O}, x_v, j)$. Moreover, from above, we see that, for all j_u, the polynomials $z_{L,v}(x,-j_u)$ have the same trivial zeroes theory. The result now follows easily.

So that there is no confusion, the reader should realize that the v-adic trivial zeroes at positive integers occur at places where the Euler product does *not* converge.

For applications, it is important that the double congruence, 4.2.5, can be refined as follows: Let $\overline{\mathbb{Q}}_p$ be a fixed algebraic closure of $\mathbb{Q}_p$ and let $\overline{W}$ be its ring of integers. We can factor $L(\omega^i, u)$ in $\overline{W}[u]$ as

$$L(\omega^i, u) = (1-\alpha_1 u)\cdots(1-\alpha_\kappa u).$$

Set

$$L^{\text{unit}}(\omega^i, u) = \prod_{\alpha_j \in \overline{W}^*} (1-\alpha_j u).$$

The function $L^{\text{unit}}(\omega^i, u)$ is called the "unit-root" piece of $L(\omega^i, u)$. Clearly

$$L(\omega^i, u) + (p) = L^{\text{unit}}(\omega^i, u) + (p);$$

thus the double congruence may also be expressed in terms of $L^{\text{unit}}(\omega^i, u)$.

4.3. Applications to Class Groups. In this subsection we give the first applications of the theory to class groups. In the process we will arrive at an analog of classical Bernoulli-numbers. We will be following the exposition in **[G-S1]**.

We continue with the notation of the previous subsection and we continue to assume that $H_1 \subset L$. Also, for the rest of this subsection, we assume that $\mathcal{B} \in \text{Spec}(\mathbf{A})$ is prime to the discriminant of $L/\mathbf{k}$. Let Υ be a primitive $\mathcal{B}$-division point. As before, we put $L(\mathcal{B}) = L(\Upsilon)$ and we now let $\text{Cl}_L^{(p)}(\mathcal{B})$ be the p-primary component of the class group of the *field* $L(\mathcal{B})$. Since $\text{Gal}(L(\mathcal{B})/L) \simeq \mathbf{A}/\mathcal{B}^*$, and $\mathbf{A}/\mathcal{B}^*$ has order prime to p, we may decompose

$$\text{Cl}_L^{(p)}(\mathcal{B}) \otimes_{\mathbf{Z}_p} W$$

into isotypic components for powers of the Teichmüller character:

$$\text{Cl}_L^{(p)}(\mathcal{B}) \otimes_{\mathbf{Z}_p} W = \bigoplus_i \text{Cl}_L^{(p)}(\mathcal{B})(\omega^i).$$

4.3.1. Proposition. *Let* $0 < i < r^{\deg(\mathcal{B})} - 1$. *Then* $\text{ord}_p\left(L(\omega^i, 1)\right)$ *equals the length of* $\text{Cl}_L^{(p)}(\mathcal{B})(\omega^{-i})$ *as* W*-module.*

Proof. It is trivial to check that $\text{ord}_p\left((L(\omega^i, 1))\right) = \text{ord}_p\left(L^{\text{unit}}(\omega^i, 1)\right)$. Let $T_p =: T_p(\mathcal{B})$ be the p-adic Tate module of the curve underlying $L(\mathcal{B})$ and let F be the Frobenius morphism with respect to $\mathbb{F}$ = the full field of constants of L. Recall that we set $t_{\mathbb{F}} = [\mathbb{F} : \mathbb{F}_r]$. There is a well-known relationship between p-adic Tate modules of curves and unit-root L-series, which implies that, upon tensoring with W:

$$T_p(\omega^{-i})/\left((1 - F)T_p(\omega^{-i})\right) \simeq \text{Cl}^{(p)}(\mathcal{B})(\omega^{-i}),$$

and

$$\det\left((1 - F \mid T_p(\omega^{-i})\right) = L^{\text{unit}}(\omega^i, 1).$$

The result now follows. □

4.3.2. Definition. Let $\tilde{z}_L(x, -i)$ be as in Definition 4.2.3. We set $\beta_{\mathcal{O}_L}(0) = 1$ and for $i > 0$

$$\beta_{\mathcal{O}_L}(i) = \tilde{z}_L(1, -i).$$

By Theorem 4.2.4, we know that $\{\beta_{\mathcal{O}_L}(i)\} \subseteq \mathbf{A}$. The reader will also easily check that

$$\beta_{\mathcal{O}_L}(pi) = \beta_{\mathcal{O}_L}(i)^p.$$

Moreover, the v-adic version of Theorem 3.6.5 immediately implies that $\beta_{\mathcal{O}_L}(i)$ interpolates v-adically to a continuous function

$$\beta_{\mathcal{O}_L, v}(s) \colon \mathbb{Z}_p \to \mathbf{A}_v.$$

From 4.3.1, and our double congruence 4.2.5, we find immediately:

4.3.3. Theorem. *Let* $0 < i < r^{\deg(\mathcal{B})} - 1$. *Then*

$$\mathrm{Cl}^{(p)}(\mathcal{B})(\omega^{-i}) \neq \{0\} \Longleftrightarrow \mathcal{B} \mid \beta_{\mathcal{O}_L}(i). \qquad \square$$

In a completely similar fashion the double congruence 4.2.5 and the relationship with p-adic Tate modules gives:

4.3.4. Theorem.

a) *Let i be as in 4.3.3. Then the W-rank of $T_p(\mathcal{B})(\omega^{-i})$ is $\frac{1}{t_{\mathbb{F}}}$ times the degree in x^{-1} of the reduction modulo $\mathcal{B}$ of $\tilde{z}_L(x, -i)$.*

b) *For almost all $\mathcal{B}$, the modules $T_p(\mathcal{B})(\omega^{-i})$ have the same W-rank.* □

Let $J_p(\mathcal{B})$ be the group of all p-division points of the Jacobian of $L(\mathcal{B})$. In the obvious fashion, we can consider $J_p(\mathcal{B}) \otimes_{\mathbf{F}_p} \mathbf{A}/\mathcal{B}$ as a W-module and can decompose into isotypic components, etc.

Let F now be the r-th power mapping; so $F^{t_{\mathbb{F}}}$ is the Frobenius morphism with respect to $\mathbb{F}$. From the relationship between T_p and unit root L-series, we conclude that $F^{t_{\mathbb{F}}}$ acts invertibly on T_p; thus $F^{-t_{\mathbb{F}}}$ makes sense as an endomorphism of T_p. By the double congruence (again!), and the Cayley-Hamilton theorem, we find:

4.3.5. Proposition. *Let i be as above. Then $\tilde{z}_L(F, -i)$ annihilates $J_p(\mathcal{B})(\omega^{-i})$.* □

Note that if i_1 and i_2 are positive integers which are $\equiv \pmod{r^{\deg(\mathcal{B})} - 1}$, then for $a \in \mathbf{A}$

$$a^{i_1} \equiv a^{i_2} \pmod{\mathcal{B}}.$$

Moreover, if $a \notin \mathcal{B}$ then $a^{r^{\deg(\mathcal{B})}-1} \equiv 1 \pmod{\mathcal{B}}$. On the other hand, if $a \in \mathcal{B}$, then obviously, for i *positive*, $a^i \equiv 0 \pmod{\mathcal{B}}$.

Let $0 < i \leq r^{\deg(\mathcal{B})} - 1$ and let i_1 be any other positive integer which is $\equiv i \pmod{r^{\deg(\mathcal{B})} - 1}$. Then we see that

$$\tilde{z}_L(x, -i_1) \equiv \tilde{z}_L(x, -i) \pmod{\mathcal{B}}.$$

On the other hand, if $i = r^{\deg(\mathcal{B})} - 1$, then

$$\tilde{z}_L(x, i) \equiv \tilde{P}(x^{-t_{\mathbb{F}}}) \pmod{\mathcal{B}},$$

where $\tilde{P}(u)$ is the numerator of the zeta function of L with all the primes above $\mathcal{B}$ removed (since $\mathbb{F} \subset L$ it is easy to see that $\tilde{P}(u)$ is a polynomial).

The above discussion allows us to replace the condition on i in the proposition by the simple condition that $i > 0$.

4.3.6. Definition. Let $i > 0$. Set

$$J^i = \prod_{\substack{\mathcal{B} \in \mathrm{Spec}(\mathbf{A}) \\ \mathcal{B} \text{ unramified in } L}} J_p(\mathcal{B})(\omega^{-i}).$$

Note that J^i is a faithful $\mathbf{A}$-module. Moreover, from above, we see how to make it a module over $\mathbf{A}[x^{-t_{\mathbb{F}}}]$. It now follows that $\tilde{z}_L(F, -i)$ annihilates J^i.

4.3.7. Question. Does $\tilde{z}_L(x,-i)$ generate the annihilator ideal of J^i in $\mathbf{A}[x^{-t_{\mathbb{F}}}]$?

A *sufficient* condition for an affirmative answer to 4.3.7, for instance, would be the *simplicity* of the zeroes of $\tilde{z}_L(x,-i)$ viewed as a function of $x^{-t_{\mathbb{F}}}$.

4.4. The Arithmetic Criterion for Cyclicity. Let $p>2$. In classical Iwasawa theory (see **[L1]**), one studies the growth of p-class groups of the tower of fields $\mathbb{Q}(\zeta_{p^j})$. The total Galois group of this extension is $\mathbb{Z}_p^*$. Note that

$$\mathbb{Z}_p^* \simeq \mu_p \times U_{p,1}$$

where $U_{p,1}$ is the group of 1-units and μ_p is the group of roots of unity. By using the norm, one then constructs a canonical module $\mathcal{M}_p$ over the completed group ring $\mathbb{Z}_p[[U_{p,1}]]$. The logarithm assures us that $U_{p,1} \simeq \mathbb{Z}_p$ and an argument of Serre's establishes that

$$\mathbb{Z}_p[[\mathbb{Z}_p]] \simeq \mathbb{Z}_p[[X]].$$

Thus we arrive, finally, at modules over a simple 2-dimensional local ring.

Now let $\mathcal{B} \in \operatorname{Spec}(\mathbf{A})$. In the theory we are discussing, one is lead in a similar fashion to towers with Galois group $\simeq \mathbf{A}_{\mathcal{B}}^* \simeq \mu_{\mathcal{B}} \times U_{\mathcal{B},1}$ and a module $\mathcal{M}_{\mathcal{B}}$ over the completed group ring of the group, $U_{\mathcal{B},1}$; here, as before, $\mu_{\mathcal{B}}$ is the group of roots of unity and $U_{\mathcal{B},1}$ is the group of 1-units. One knows that $U_{\mathcal{B},1}$ is topologically isomorphic to an infinite product of $\mathbb{Z}_p$'s. Thus the completed algebra is quite huge. However, it is *still* a local ring.

Note that both $\mathbb{Z}_p^*$ and $\mathbf{A}_{\mathcal{B}}^*$ split into products of a finite group (of order prime to p) and the 1-units. The completed group algebra of the 1-units thus acts on the isotypic components of the action of the finite group.

A basic tool in studying modules over such completed group algebras is *Nakayama's Lemma*. In particular, we see that if, at the first level (i.e., $\mathbb{Q}(\zeta_p)$ classically and $L(\mathcal{B})$ in the function field theory) an isotypic component is cyclic, then the corresponding module over the completed group is monogenic. This, together with class number formulae, suffices in many cases to give the *exact* description of the module up the tower.

Fortunately, one of the cases where we can deduce the total structure from cyclicity at the first level is the one that we are studying here. Briefly the procedure is the following: The module $\mathcal{M}_{\mathcal{B}}$ is constructed as the inverse limit (via the norm) of the p-class groups of the fields $L(\mathcal{B}^j)$ (i.e, by adjoining a primitive $\mathcal{B}^j$-th division element to L). One then constructs a "Stickelberger element" $\Theta(\mathcal{B}^j)$ (see **[G2]** and **[G8]**) which annihilates the class group of $L(\mathcal{B}^j)$. Moreover, one computes from class number formulae that the colength of the projection of the Stickelberger element into the isotypic component $\mathcal{M}_{\mathcal{B}}(\omega^{-i})$ is the length (as W-module) of $\mathrm{Cl}_L^{(p)}(\mathcal{B}^j)(\omega^{-i})$. Thus, if $\mathrm{Cl}_L^{(p)}(\mathcal{B}^j)(\omega^{-i})$ is cyclic, then it is isomorphic to $W\left[(\mathbf{A}/\mathcal{B}^j)^*\right]/\left(\Theta(\mathcal{B}^j)\right)$. Finally, the Stickelberger elements then fit together to give an element Θ of the full completed algebra and we are finished.

Thus we are reduced to studying cyclicity at the first level. For the rest of this paper we shall be interested only in this situation and will have nothing more to say about the theory up the $\mathcal{B}^j$-tower.

Having finished this short explanation, we can now turn to our criterion for cyclicity at the first level. Actually, we shall eventually have *two* conditions for cyclicity. The first criterion will depend on the use of classical L-series and the double congruence 4.2.5. Since the classical theory is studied via *constant field* extensions, we shall call this first criterion, which applies very generally, the "arithmetic criterion for cyclicity". The second criterion (see Subsection 5.5) is much more limited in the scope of its application; we shall call it "the geometric criterion for cyclicity".

We continue with the previous notation. Let $0 < i < r^{\deg(\mathcal{B})} - 1$ and let $L(\omega^i, u)$ be the classical L-series of ω^i. As before, factor $L(\omega^i, u)$ in $\overline{W}[u]$ as

$$L(\omega^i, u) = (1 - \alpha_1 u) \cdots (1 - \alpha_\kappa u).$$

Let $\overline{M} \subset \overline{W}$ be the maximal ideal.

4.4.1. Proposition. *Suppose that at most one of* $\{\alpha_j\}_{j=1}^{\kappa}$ *is* $\equiv 1 \pmod{\overline{M}}$. *Then* $\mathrm{Cl}_L^{(p)}(\mathcal{B})(\omega^{-i})$ *is a cyclic* W*-module.*

Proof. If there is *no* such element then the result is trivially true. Thus, let us suppose that there is one and only one such element. From the proof of Proposition 4.3.1, we know that, upon putting $T_p = T_p(\mathcal{B})$,

$$\mathrm{Cl}^{(p)}(\mathcal{B})(\omega^{-i}) \simeq T_p(\omega^{-i}) / \left((1 - F) T_p(\omega^{-i})\right).$$

The result now follows from simple considerations of finite rank modules over a principal ideal domain. □

Now let $\tilde{z}_L(x, -i)$ be as above. Let $\overline{\mathbf{k}}_{\mathcal{B}}$ be a fixed algebraic closure of $\mathbf{k}_{\mathcal{B}}$. Let $\overline{\mathbf{A}}_{\mathcal{B}}$ be the integral closure of $\mathbf{A}_{\mathcal{B}}$ in $\overline{\mathbf{k}}_{\mathcal{B}}$ with maximal ideal $\overline{M}_{\mathcal{B}}$. Over $\overline{\mathbf{A}}_{\mathcal{B}}$ we may factor $\tilde{z}_L(x, -i)$ as

$$\tilde{z}_L(x, -i) = (1 - \theta_1 x^{-t_{\mathbb{F}}}) \cdots (1 - \theta_\iota x^{-t_{\mathbb{F}}}).$$

4.4.2. Corollary ("The Arithmetic Criterion for Cyclicity"). *Suppose that at most one of* $\{\theta_j\}_{j=1}^{\iota}$ *is* $\equiv 1 \pmod{\overline{M}_{\mathcal{B}}}$. *Then* $\mathrm{Cl}_L^{(p)}(\mathcal{B})(\omega^{-i})$ *is a cyclic* W*-module.*

Proof. This follows immediately from the proposition and 4.2.5. □

Let D be the operator $\dfrac{d}{dx^{t_{\mathbb{F}}}}$ on functions of $x^{-t_{\mathbb{F}}}$.

4.4.3. Definition. Let $i > 0$. We set

$$\tilde{\beta}_{\mathcal{O}_L}(i) = D\, \tilde{z}_L(x, -i)|_{x=1}.$$

Clearly $\{\tilde{\beta}_{\mathcal{O}_L}(i)\} \subseteq \mathbf{A}$. Elementary algebra then implies

4.4.4. Proposition. *Let* $\mathbf{B}_{\mathcal{O}_L}(i)$ *be the* $\mathbf{A}$*-ideal generated by* $\beta_{\mathcal{O}_L}(i)$ *and* $\tilde{\beta}_{\mathcal{O}_L}(i)$. *Then the arithmetic criterion for cyclicity holds if and only if* $\mathcal{B} \nmid \mathbf{B}_{\mathcal{O}_L}(i)$. □

4.4.5. Remarks. The p-th power mapping readily gives the following equivalence: Let $i > 0$. Then

$$\mathcal{B} \mid \mathbf{B}_{\mathcal{O}_L}(i) \iff \mathcal{B} \mid \mathbf{B}_{\mathcal{O}_L}(pi).$$

Finally, in **[G-S1]** is a refinement of the arithmetic criterion that takes into account the existence of subfields of L which also lie over H_1 and which may work in more circumstances.

4.5. Special Values at Positive Integers. In this subsection we will study special values of $\zeta(s) =: \zeta_{\mathcal{O}_L}(s)$ at the positive integers. As happens in classical algebraic number theory, we will not get results for *all* extensions but rather for those that are "totally-real" (i.e., totally-split at the prime ∞). Moreover, *unlike* classical theory, we shall unfortunately also need to assume that $L/\mathbf{k}$ is an *abelian* extension of the Hilbert class field H.

Let ψ be the basic rank one sign normalized Drinfeld module guaranteed by the theorem of David Hayes (Theorem 4.1.2). Recall that we let $\Gamma \subset \overline{\mathbf{K}}$ be the lattice of ψ which, by general theory, is of the form

$$\Gamma = \xi \mathcal{B},$$

where $\mathcal{B}$ is an $\mathbf{A}$-ideal (not necessarily prime) and ξ is the (unique up to an element of $\mathbb{F}_r^*$) "period" of ψ. Let $\mathbf{e}(z)$ be the exponential of ψ, and choose an algebraic closure $\overline{\mathbb{Q}}_p$ of $\mathbb{Q}_p$.

Let I be an ideal of $\mathbf{A}$ and let $\mathcal{B}$ be as above. Let $\alpha \in I^{-1}\mathcal{B} - \mathcal{B}$. Then we have

4.5.1. Lemma. *Let j be a positive integer. Then we have*

$$\xi^{-j} \sum_{\beta \in \Gamma} (\alpha + \beta)^{-j} \in \mathbf{k}(I).$$

Proof. The element $\mathbf{e}(\alpha\xi)$ is an I-division point of ψ; thus it belongs the field $\mathbf{k}(I)$ by definition. We know that

$$\mathbf{e}(z) = z \prod_{0 \neq \beta \in \mathcal{B}} \left(1 - \frac{z}{\beta\xi}\right).$$

Thus, as in Lemma 4.1.8,

$$\frac{1}{\mathbf{e}(z)} = \sum_{\beta \in \mathcal{B}} \frac{1}{z + \beta\xi},$$

and

$$\frac{1}{\mathbf{e}(z + \alpha\xi)} = \frac{1}{\mathbf{e}(z) + \mathbf{e}(\alpha\xi)} = \sum_{\beta \in \mathcal{B}} \frac{1}{z + \alpha\xi + \beta\xi} = \sum_{\beta \in \mathcal{B}} \frac{1}{z + \xi(\alpha + \beta)}.$$

The coefficients of the exponential function of ψ are in H_1 by virtue of the functional equation which states that, for $a \in \mathbf{A}$,

$$\mathbf{e}(az) = \psi_a(\mathbf{e}(z)).$$

The result now follows directly by computing the Taylor expansion about the origin of $\dfrac{1}{\mathbf{e}(z + \xi\alpha)}$ and using the above equalities. □

The main result in this subsection, Theorem 4.5.2, will be reduced to Lemma 4.5.1 via the use of L-series.

Since we have assumed that L is both totally-real and abelian, it follows that there is some conductor I with

$$L \subset \mathbf{k}(I).$$

Set $G = \mathrm{Gal}(L/\mathbf{k})$ and let $\hat{G}$ be the group of $\overline{\mathbb{Q}}_p^*$-valued characters of G. As was explained in 3.6.1, we have the factorization

$$\zeta_{\mathcal{O}_L}(s) = \prod_{\chi \in \hat{G}} L(\chi, s).$$

Recall that we set $\delta = \deg(\infty)$. Put $t = [L : \mathbf{k}]$. Our main result here is then:

4.5.2. Theorem. *Let j be a positive integer divisible by $r^\delta - 1$. Then*

$$\xi^{-tj}\zeta_{\mathcal{O}_L}(j)$$

is non-zero, contained in the maximal totally-real subfield of $\mathbf{k}(I)$ and

$$\left(\xi^{-tj}\zeta_{\mathcal{O}_L}(j)\right)^2 \in H.$$

Proof. We will use the well-known *Dedekind Determinant Formula* which states the following: Let W be a finite group and let $\hat{W}$ be its $\overline{\mathbb{Q}}_p$-valued character group. Let f be any $\overline{\mathbb{Q}}_p$-valued function on W. Then, upon fixing an ordering O of W,

$$\prod_{\chi \in \hat{W}} \sum_{w \in W} \chi(w) f(w^{-1}) = \det\left(f(w_0^{-1}w_1)\right)_{(w_0, w_1) \in O \times O}.$$

Although the Dedekind Determinant Formula is defined over $\mathbb{Q}_p$ it is clear that, in fact, it works over $\mathbb{Z}/p$ and thus over $\mathbf{k}$ provided that we use $\overline{\mathbb{Q}}_p$-valued characters of W as explained in 3.6.1. In particular, we may use this formula to evaluate L-series.

We note first of all that we may reduce to considering the product $\prod L(\chi, j)_I$ where the subscript I means that we have removed any Euler factors at the primes dividing I. Indeed, we know that this product equals the value of $\zeta(j)$ with the Euler factors above I removed. On the other hand, let $\wp$ be a prime of $\mathcal{O}_L$ dividing I. Then

$$(1 - n\wp^{-j}) \in \mathbf{k}$$

since $n\wp$ is principal and, as $j \equiv 0 \pmod{r^\delta - 1}$, $n\wp^j$ is positive.

Now let $\sigma \in G$ and set

$$f(\sigma) = \sum_{\substack{U \subset \mathbf{A} \\ (U, J) = \mathbf{A} \\ \sigma_U = \sigma}} U^{-j}$$

where σ_U is the Artin symbol at U. Then the determinant formula gives

$$L(\chi, j)_I = \det\left(f(\sigma^{-1}\sigma_1)\right)$$

where (σ, σ_1) are in an ordering of G.

Since $H \subset L$ there is a *surjective* map from G to the ideal classes of $\mathbf{A}$. Let $R = \{\mathcal{B}_1, \ldots, \mathcal{B}_{h\delta}\}$ be a collection of representatives prime to I and where h is the class number of the field $\mathbf{k}$. For each $\sigma \in G$ let $g(\sigma) \in R$ be its representative.

The two determinants

$$\det\left(f(\sigma^{-1}\sigma_1)\right),$$

and

$$\det\left(f(\sigma^{-1}\sigma_1)g(\sigma^{-1})^j g(\sigma_1)^j\right),$$

differ by the factor

$$\prod g(\sigma)^{2j}.$$

But simple arguments in group theory tell us that, in the ideal class group, $\prod g(\sigma)^2$ is the identity; i.e.,

$$\prod g(\sigma)^2 = (a)$$

for some $a \in \mathbf{A}$. Again, since $j \equiv 0 \pmod{r^\delta - 1}$ we see that, by our definition of exponentiation of ideals,

$$\prod g(\sigma)^{2j} = a^j;$$

consequently, we are reduced to computing the second determinant.

Note that α, β in $\mathbf{A}$, we have

$$(\zeta\alpha + \beta)^j = (\alpha + \zeta^{-1}\beta)^j$$

for $\zeta \in \mathbf{A}^*$. Using this elementary observation, and the explicit class field theory of 4.1, the reader can now express $\xi^{-j} f(\sigma^{-1}\sigma_1)g(\sigma^{-1})^j g(\sigma_1)^j$ in terms of the sums of Lemma 4.5.1; in particular, these elements are in $\mathbf{k}(I)$.

Let ι be an element of $\mathrm{Gal}(L/H)$. It is now easy to see that ι permutes the rows of

$$\left(\xi^{-j} f(\sigma^{-1}\sigma_1)g(\sigma^{-1})^j g(\sigma_1)^j\right);$$

thus the determinant gets multiplied by ± 1. The result now follows. □

In classical theory of course, one obtains that the square belongs to $\mathbb{Q}$. In the function field version of Theorem 4.5.2, one would then expect that the square should lie in $\mathbf{k}$ itself. Our result indeed does imply this when the class number of $\mathbf{A}$ is 1 (e.g., $\mathbf{A} = \mathbb{F}_r[T]$). It may indeed be true in the general case though this does appear doubtful. The difficulty lies in Hayes Lemma 4.1.6 since the elements $D(\psi_I)$ themselves do not lie in $\mathbf{k}$; thus the explicit action of the absolute Galois group on the above determinants is quite complicated.

The elements ξ are transcendental by Yu **[Y1]**. Thus we deduce the same for the zeta-values of Theorem 4.5.2.

Let Ω be the algebraic upper half-plane. One knows that there is a good theory of modular forms for Ω and congruence subgroups of $GL(Y)$, Y a projective rank 2 $\mathbf{A}$-module. Moreover, in the classical fashion, one can attach to an *arbitrary* totally-real extension $L/\mathbf{k}$ an *Eisenstein series* on Ω. It seems reasonable to expect that a proof of the analog of 4.5.2 for such L may be found via these forms.

5. The Theory of the ζ-function for $\mathbf{A} = \mathbb{F}_r[T]$

In this last section we will concentrate on the particular case where $\mathbf{A} = \mathbb{F}_r[T]$, $L = \mathbf{k} = \mathbb{F}_r(T)$, $\mathbf{K} = \mathbb{F}_r((1/T))$ etc. Thus we will no longer need to deal with the myriad details arising from non-principal or positively generated ideals. Recall that an integer is $\mathbf{A}$-even if and only if it is divisible by $r-1$; otherwise it is $\mathbf{A}$-odd.

For simplicity, we will make the harmless assumption that our sign function sgn is chosen so that "positive" and "monic" agree on $\mathbf{A}$. Our fixed uniformizer π is then of the form

$$\pi = \frac{1}{T} + \text{higher terms in } \frac{1}{T}.$$

With the above assumption on sgn, the sign normalized module ψ of Subsection 4.1 is none other than the *Carlitz module* C defined by

$$C_T(x) = Tx + x^r.$$

Recall that in Definition 3.3.5, we chose an algebraic closure $\overline{\mathbf{K}}$ of $\mathbf{K}$ and set

$$S_\infty = \overline{\mathbf{K}}^* \times \mathbb{Z}_p,$$

considered as an additive group. For $I = (i)$, i positive, an $\mathbf{A}$-fractional ideal and $s = (x, y) \in S_\infty$, we have

$$I^s =: x^{\deg(i)} \left(\pi^{\deg(i)} i\right)^y = x^{\deg(i)} \langle i \rangle^y,$$

where $\langle i \rangle = \pi^{\deg(i)} i$ is a 1-unit in $\mathbf{K}$. The integer powers lie discretely inside S_∞ as $i = (\pi^{-i}, i)$. For $v \in \operatorname{Spec}(\mathbf{A})$ the space S_v was defined as $\mathbb{Z}_p \times \mathbb{Z}/(r^{\deg(v)} - 1)$. If I is prime to v, and $s_v \in S_v$, then I^{s_v} was defined in 3.3.9 in analogy with classical p-adic theory.

5.1. Review of the Basic Results. Our particular interest will be the zeta function of $\mathbf{A}$ which, we recall, is defined as

$$\begin{aligned} \zeta(s) =: \zeta(\mathbf{A}, s) &= \prod_{\wp \in \operatorname{Spec}(\mathbf{A})} (1 - \wp^{-s})^{-1} \\ &= \sum_{\substack{n \in \mathbf{A} \\ n \text{ monic}}} n^{-s}, \end{aligned}$$

for $s \in S_\infty$. From the considerations of our the previous sections (e.g., Example 3.6.8), we have the following:

a) The Euler product for $\zeta(s)$ converges in the half-plane $\{s = (x, y) \in S_\infty \mid |x|_\infty > 1\}$ where $x \in \overline{\mathbf{K}}^*$.

b) The function $\zeta(s)$ can be analytically continued to an essentially algebraic entire function on S_∞.

c) Let i be a non-negative integer. Then the function

$$z(x,-i) =: z_{\mathbf{A}}(x,-i) = \sum_{j=0}^{\infty} x^{-j} \left(\sum_{\substack{n \in \mathbf{A} \\ n \text{ monic} \\ \deg(n)=j}} n^i \right) = \zeta(x\pi^i, -i).$$

is a polynomial in x^{-1}. Further (as is trivial to see)

$$z(x,0) = \zeta(0) = 1.$$

d) Let i be a positive $\mathbf{A}$-even integer. Then the rational function

$$\tilde{z}(x,-i) = \frac{z(x,-i)}{1-x^{-1}}$$

is also a polynomial in x^{-1}.

e) Let $v \in \operatorname{Spec}(\mathbf{A})$ and let $\overline{\mathbf{k}}_v$ be a fixed algebraic closure of the completion $\mathbf{k}_v$. Let i be a *positive* integer and let $x_v \in \overline{\mathbf{k}}_v^*$. Then the polynomials $z(x_v, -i)$ interpolate to an essentially algebraic function $\zeta_v(x_v, s_v)$ on the space $\overline{\mathbf{k}}_v^* \times S_v$. Moreover, for $|x_v| > 1$, $\zeta_v(x_v, s_v)$ has the obvious Euler product over the primes of $\mathbf{A} \neq v$. There are v-adic trivial zeroes at the positive $\mathbf{A}$-even integers by 4.2.7.

f) Let ξ be the period of the Carlitz module C. Then $\xi^{r-1} \in \mathbf{K}$ and, for i positive and $\mathbf{A}$-even

$$\frac{\zeta(i)}{\xi^i} \in \mathbf{k}.$$

For the rest of Section 5, unless otherwise stated, $\mathbf{e}(z)$ will be the exponential of the Carlitz module, $\mathbf{log}(z)$ its logarithm and ξ its period. We will give a formula for $\mathbf{e}(z)$ in 5.3; however, from its functional equation, it is clear that its Taylor coefficients are contained in $\mathbf{k}$.

Part f) above also has an elementary derivation as follows: We know that $\mathbf{e}(z) = z \prod_{0 \neq \alpha \in \mathbf{A}} \left(1 - \frac{z}{\xi\alpha}\right)$ and that $\mathbf{e}'(z) = 1$. Thus

$$\frac{1}{\mathbf{e}(z)} = \sum_{\alpha \in \mathbf{A}} \frac{1}{z + \alpha\xi}.$$

Note that

$$\frac{1}{z+\alpha} = \frac{1}{\alpha(1+z/\alpha)} = \frac{1}{\alpha}\left(1 - \frac{z}{\alpha} + \left(\frac{z}{\alpha}\right)^2 + \cdots\right)$$

by the geometric series. Note that

$$\sum_{0\neq\alpha\in\mathbf{A}} \alpha^{-j} = \begin{cases} 0 & \text{if } j \text{ is } \mathbf{A}\text{-odd} \\ -\zeta(j) & \text{if } j \text{ is } \mathbf{A}\text{-even} . \end{cases}$$

Thus we have

$$\frac{z}{\mathbf{e}(z)} = 1 + \sum_{j=1}^{\infty} \frac{\zeta(j(r-1))}{\xi^{j(r-1)}} z^{j(r-1)} .$$

The result now follows easily upon equating Taylor coefficients about the origin.

Since $\mathbb{F}_r[T]$ is so simple, one can derive a very useful formula for $z(x,-i)$ as follows: Let $n \in \mathbf{A}$ be monic of degree d and write it as $Th+\alpha$ where h is monic of degree $d-1$ and $\alpha \in \mathbb{F}_r$. Substitute this formula for n into the definition of $z(x,-i)$ and expand out by the binomial theorem. One then finds easily

5.1.1. Lemma. *We have*

$$z(x,-i) = 1 - \sum_{\substack{b=0 \\ (r-1)|(i-b)}}^{i-1} \binom{i}{b} T^b x^{-1} z(x,-b) . \qquad \square$$

By putting $x=1$ we obtain a recursive formula for $\zeta(-i)$. One finds, again, that for i positive and $\mathbf{A}$-even, one has $\zeta(-i)=0$.

Let $D = \frac{d}{dx}$. From Lemma 5.1, we find

$$D\,z(x,-i) = -\left(\sum_{\substack{b=0 \\ (r-1)|(i-b)}}^{i-1} \binom{i}{b} T^b \left(-x^{-2} z(x,-b) + x^{-1} D\,z(x,-b)\right) \right) .$$

If we use the above formula, and Lemma 5.1, we find the following recursive formula for $\beta(i) =: \beta_{\mathbf{A}}(i)$, where we recall that we have set $0 = \beta(0)$.

5.1.2. Proposition. *Let i be a positive integer. Then*

$$\beta(i) = 1 - \sum_{\substack{b=0 \\ (r-1)|(i-b)}}^{i-1} \binom{i}{b} T^b \beta(b) . \qquad \square$$

Note that $\beta(i) \equiv 1 \bmod (T)$ for $i>0$. In consequence we deduce

5.1.3. Corollary. *The function $\zeta(s)$ is non-zero at $-i$ for i positive and $\mathbf{A}$-odd. It has a simple zero at $-i$ for i positive and $\mathbf{A}$-even.* $\square$

We see that the trivial zeroes of $\zeta(s)$ are in line with what one would expect from classical theory. In our next subsection we will discuss what is known for the v-adic trivial zeroes at positive integers.

5.2. The Results of J. Yu. In this subsection, for completeness, we will recall very briefly the results of Jing Yu based on the work of Greg Anderson and Dinesh Thakur **[A-T1]** (see also **[Th4]**).

For i positive $\mathbf{A}$-even, we have seen that $\zeta(i)$ is a "rational" (i.e., via an element of $\mathbf{k}$) multiple of the period ξ^i. It thus follows trivially (as in classical theory) that $\zeta(i)$ is transcendental over $\mathbf{k}$. Yu's results go far beyond this:

5.2.1 Theorem (Yu).

a) *$\zeta(i)$ is transcendental over $\mathbf{k}$ for all $i > 0$.*

b) *Let $i > 0$ be $\mathbf{A}$-even. Then $\frac{\zeta(i)}{\xi^i}$ is also transcendental over $\mathbf{k}$.*

c) *Let $i > 0$ be $\mathbf{A}$-odd and $v \in \operatorname{Spec}(\mathbf{A})$. Then $\zeta_v(1, i)$ is also transcendental over $\mathbf{k}$.* □

These results are established by an analog of the theorem of Hermite-Lindemann which uses "special-points" and the n-th tensor power, $C^{\otimes n}$, of the Carlitz module. In **[Th5]**, some analogous results are established in the case of genus 1.

Dinesh Thakur, in the appendix just below, shows for some v how one can establish the simplicity of the v-adic trivial zeroes at positive integers. For general v, the techniques of **[A-T1]** allow us to express $\beta_v(i)$ as a linear combination of logarithms. Perhaps the techniques of Yu **[Y2]** may be extended to show the non-vanishing of these numbers, and thus the simplicity of the v-adic trivial zeroes for all v.

Appendix. Remarks on v-adic Special Values, by Dinesh Thakur. It is not known for a single (classical) p-adic L-function, whether its values at positive integers are nonzero, when they are supposed to be. It was noted in **[Th4]** that it can be easily seen in our case. In fact, 5.1.1 implies that if v is a prime of degree one and s is an $\mathbf{A}$-odd negative integer, then $\zeta(s) \equiv 1 \bmod v$. So if n is an odd positive integer, then

$$\zeta_v(n) = \lim_{j \to \infty} (1 - v^{-n+(r-1)p^j})\zeta(n - (r-1)p^j) \equiv 1 \bmod v,$$

so that $\zeta_v(n) \neq 0$. Of course, this simple result is completely superceded by 5.2.1.c. But the same reasoning with 5.1.2 in place of 5.1.1 shows that $\beta_v(n) \neq 0$, when n is an $\mathbf{A}$-even positive integer. This implies the simplicity of the zero, when v is of degree one.

To go a little further, let v be a monic prime of degree d and let n be a positive $\mathbf{A}$-odd integer. Let $S_k =: \sum 1/a^n$, where $a \in \mathbf{A}$ runs through monic elements of degree k and relatively prime to v. We have the v-adically convergent expression $\zeta_v(n) = \sum_{k=0}^{\infty} S_k$.

Claim. *We have $S_k \equiv 0 \bmod v$, if $k \geq d$.*

Proof. The range of the summation $\{a\}$ is disjoint union of $\{vm + b\}$, where b runs through nonzero elements of degree less than d and the union is over monic m of degree $k - d$. Then each subsum is congruent to $\sum 1/b^n \equiv 0$ modulo v, as $r^d - 1$ does not divide n. Hence the claim is proved. □

Hence whenever we can show that $\sum_{k=0}^{d-1} S_k \not\equiv 0 \bmod v$ (for example, $r = 3$ and v is of degree 2), we get $\zeta_v(n) \neq 0$. When n is $\mathbf{A}$-even, but not divisible by $r^d - 1$, we can use a similar reasoning to settle simplicity.

This result, or variants with congruences modulo higher powers of v, establish non-vanishing and simplicity for various values, but are probably inadequate to settle the general case.

5.3. A Quick Review of Factorials and Γ-functions. In this subsection, we will give a very brief introduction to *some* aspects of the theory of Γ-functions for $\mathbf{A}$. For more the reader should consult **[G3]** and **[Th1]**.

The first order of business is to give an $\mathbf{A}$-version of "$n!$" which is due to L. Carlitz:

5.3.1. Definition (Carlitz).

a) Let j be a positive integer. We set

$$[j] = T^{r^j} - T.$$

b) Put $D_0 = 1$ and for $j > 0$,

$$D_j = [j][j-1]^r \cdots [1]^{r^{j-1}}.$$

c) Let i be a non-negative integer. Let $i = \sum_{j=0}^{\nu} \alpha_j r^j$ be its r-adic expansion. We set

$$\Pi(i) = \prod_{j=0}^{\nu} D_j^{\alpha_j}.$$

d) Put $L_0 = 1$ and for $j > 0$,

$$L_j = [j][j-1] \cdots [1].$$

It is easy to see that

$$D_j = \prod_{\substack{n \text{ monic} \\ \deg(n)=j}} n,$$

and that L_j is the *least common multiple* of all monics of degree j. Further,

$$D_j = \Pi(r^j),$$

and

$$\Pi(r^t - 1) = (D_0 \cdots D_{t-1})^{r-1} = (-1)^t \prod_{\substack{0 \neq \alpha \in \mathbf{A} \\ \deg(\alpha) < t}} \alpha.$$

Finally, Sinnott has shown the following analog of well-known properties of $n!$:

$$\Pi(j) = \prod_{f \text{ monic prime}} f^{e_f},$$

where

$$e_f = \sum_{t\geq 1} [j/r^{t\deg(f)}]$$

and, here, $[?]$ denotes "greatest integer".

From the functional equation

$$\mathbf{e}(Tz) = T\mathbf{e}(z) + \mathbf{e}(z)^r,$$

and the fact that $\mathbf{e}(z) \equiv z \mod (z^r)$, we find easily that

$$\mathbf{e}(z) = \sum_{t=0}^{\infty} \frac{z^{r^t}}{D_t},$$

and

$$\mathbf{log}(z) = \sum_{t=0}^{\infty} (-1)^t \frac{z^{r^t}}{L_t}.$$

5.3.2. Remarks. The construction of $\Pi(i)$ out of D_j is the most basic example of the "digit expansion procedure". This procedure "fills in the gaps" in natural definitions. For example, the numbers D_t arise from the exponential of the Carlitz module but arise only at r^t. Thus the digit procedure is used to get the full factorial.

As will be seen below, the digit procedure arises in many different situations. Yet it remains rather poorly understood. For instance, a deeper understanding of it might lead to the relationship between Γ-functions, defined via the procedure, and L-series, which are not (at least on the surface) so defined.

There is a lovely formula for ξ due to Carlitz **[C1]**, **[G2]** and **[Th1]**:

5.3.3. Proposition. *We have*

$$\xi = (-T^r)^{\frac{1}{r-1}} \left(1 - T^{1-r}\right)^{\frac{1}{r-1}} \prod_{t=1}^{\infty} \left(1 - \frac{[i]}{[i+1]}\right),$$

where we take the unique 1-unit root of $1 - T^{1-r}$. □

The element $\xi_1 = \xi / (-T^r)^{\frac{1}{r-1}}$ is called "the 1-unit piece of ξ". The reader should note that it has been defined without any use of our uniformizer π.

We can now rapidly define "arithmetic Γ-functions" (so-called since they are ultimately related to constant field extension). The method used here is based on that of Morita for classical p-adic Γ-functions.

5.3.4. Definition. Let $v \in \operatorname{Spec}(\mathbf{A})$. Define

$$D_{v,j} = - \prod_{\substack{n \text{monic} \\ \deg(n)=j \\ n \notin v}} n.$$

The next lemma follows from simple considerations of finite abelian group theory by reducing modulo powers of the maximal ideal in the various completions of $\mathbf{k}$.

5.3.5. Lemma.
a) *We have*

$$\lim_{j\to\infty} \langle D_j\rangle \to 1 \in \mathbf{K}.$$

b) *Let* $v \in \mathrm{Spec}(\mathbf{A})$. *Then*

$$\lim_{j\to\infty} D_{v,j} \to 1 \in \mathbf{k}_v. \qquad \square$$

5.3.6. Definition. Let $y = \sum_{j=0}^{\infty} \alpha_j r^j$, $0 \le \alpha_j < r$ be in $\mathbb{Z}_p$.

a) We set

$$\Pi_\infty(y) = \prod_{j=0}^{\infty} \langle D_j\rangle^{\alpha_j}.$$

b) We set

$$\Pi_v(y) = \prod_{j=0}^{\infty} D_{j,v}^{\alpha_j}.$$

It is clear from Lemma 5.3.5 that $\Pi_\infty(y)$ and $\Pi_v(y)$ are continuous functions on $\mathbb{Z}_p$. To obtain $\Gamma(y)$ and $\Gamma_v(y)$ one uses the change of variables $z \mapsto z-1$.

D. Thakur has established analogs for the functions of 5.3.6 of many classical properties of Γ-functions. There are multiplication formulae, functional equations, etc., see **[Th1]** and **[G3]**. These results follow formally from manipulations of the digits (another indication of the importance of the digit procedure) and the following lemma:

5.3.7. Lemma.
a) $(T\pi)^{-\frac{r}{r-1}}\Gamma(0) = (T\pi)^{-\frac{r}{r-1}}\Pi_\infty(-1)$ *is the 1-unit part of* ξ.
b) *Let* $v \in \mathrm{Spec}(\mathbf{A})$. *Then*

$$\Gamma_v(0) = \Pi_v(-1) = (-1)^{\deg(v)-1}. \qquad \square$$

5.3.8. Example. The functional equations are particularly easy to see. We will discuss the situation here for the function $\Pi_\infty(y)$ – the appropriate statements for $\Pi_v(y)$ and $\Gamma(y)$, etc., are easily derived. The functional equation then states:

$$\Pi_\infty(y)\Pi_\infty(-1-y) = \Pi_\infty(-1).$$

If the reader writes out the r-adic expansions for y, -1, and $1-y$, and uses the definition of $\Pi_\infty(y)$, the result will prove itself! Thus by the lemma, we can compute $\Pi_\infty(-1-y)$ in terms of $\Pi_\infty(y)$ and the period ξ.

Next we turn to the "geometric Γ-function". This function has the same relationship to $\mathbf{e}(x)$ as the classical Γ-function has to $\sin(x)$.

5.3.9. Definition. Let $y \in \mathbb{Z}_p$ be written as in Definition 5.3.6 and let $x \in \overline{\mathbf{K}}$. Then we set

$$g(x,y) = \prod_{j=0}^{\infty} \left(\prod_{\substack{n \text{ monic} \\ \deg(n)=j}} \left(1 + \frac{x}{n}\right) \right)^{\alpha_j}.$$

It is easy to see that $g(x,y)$ is a continuous function on $\overline{\mathbf{K}} \times \mathbb{Z}_p$; in fact, with the obvious analog of Definition 3.5.1, it is "entire". Moreover, $g(x,-y)$ is "essentially algebraic" if we set $\pi = 1$ in the analog of Definition 3.5.2. The reader should also note that the space $g(x,y)$ is *not* defined on the same space (i.e., S_∞) as L-series.

5.3.10. Definition.
a) We put $\Pi_0(x,y) = g(x,y)^{-1}$.
b) We put

$$\Pi(x,y) = \Pi_0(x,y)\Pi_\infty(y).$$

c) We put $\Gamma_0(x,y) = \Pi_0(x,y-1)$.
d) We put

$$\Gamma(x,y) = \frac{\Pi(x,y-1)}{x}.$$

Thakur, in **[Th1]**, gives non-obvious v-adic analogs of the above functions. Moreover, one also finds in **[Th1]** their functional equations (which follow in the same fashion as above), multiplication formulae, etc. These are based on the following elementary and easily seen result:

5.3.11. Proposition.

$$\prod_{\zeta \in \mathbf{F}_r^*} g\left(\zeta x, \frac{1}{1-r}\right) = \frac{\mathrm{e}(\xi x)}{\xi x} = \prod_{0 \neq \alpha \in \mathbf{A}} \left(1 - \frac{x}{\alpha}\right). \qquad \square$$

We will have more to say about $g(x,y)$ in later sections.

5.4 A Lemma of Carlitz. This short subsection is devoted to a very useful lemma of Carlitz's **[C1]**. We state it in some generality: Let R be a finite extension of the field $\mathbf{K}$. Let $N \subset R$ be a discrete $\mathbb{F}_r$-vector space (e.g., an $\mathbf{A}$-lattice) and form the function $f(z) = z \prod_{0 \neq \alpha \in N} \left(1 - \frac{z}{\alpha}\right)$ (so if N is a lattice then $f(z)$ is its exponential function); as N is the union of finite additive groups, $f(z)$ is an $\mathbb{F}_r$-linear function. Let $g(z)$ be the inverse function of f, which is also additive, and

$$h(z) = \frac{z}{f(z)}.$$

We write

$$f(z) = \sum_{i=0}^{\infty} \alpha_i z^{r^i},$$

$$g(z) = \sum_{i=0}^{\infty} \beta_i z^{r^i},$$

and

$$h(z) = \sum_{m=0}^{\infty} \gamma_m z^m.$$

As in the "elementary" proof of Part f) in Subsection 5.1, it follows that $\gamma_m = 0$ unless $(r-1) \mid m$ and that $\gamma_{pm} = \gamma_m^p$. Moreover, we have the evident recursive formula

$$\sum_{r^i \le m} \gamma_{m-r^i}\alpha_i = \begin{cases} 0 & \text{for } m > 1, \\ 1 & \text{for } m = 1. \end{cases} \tag{5.4.1}$$

5.4.2. Lemma (Carlitz). *Let* $m = r^k - r^j$. *Then*

$$\gamma_m = \beta_{k-j}^{r^j}.$$

Proof. Without loss of generality, we may assume that $j = 0$. We now use induction and we note that the case $k = 0$ reduces to $\gamma_0 = \beta_0 = 1$ which follows trivially by 5.4.1. Thus we assume that the result is true up to and including $k-1$. From 5.4.1, we know that

$$0 = \sum_{i \le k} \gamma_{r^k - r^i}\alpha_i,$$

or

$$\begin{aligned} \gamma_{r^k-1} &= -\sum_{i=1}^{k} \alpha_i \gamma_{r^k - r^i} \\ &= -\sum_{i=1}^{k} \alpha_i \gamma_{r^{(k-i)}-1}^{r^i} \\ &= -\sum_{i=1}^{k} \alpha_i \beta_{k-i}^{r^i}, \end{aligned}$$

by the induction hypothesis.

On the other hand, $\{\beta_t\}$ and $\{\alpha_t\}$ satisfy the obvious relations that come from the relationship

$$z = g(f(z)) = f(g(z)).$$

These relationships state that

$$0 = \sum_{i=0}^{k} \alpha_i \beta_{k-i}^{r^i},$$

or

$$\beta_k = -\sum_{i=1}^{k} \alpha_i \beta_{k-i}^{r^i}.$$

The result now follows. □

We will use Lemma 5.4.2 to study ζ-values. However, it also has uses in the theory of modular forms. Indeed, it is precisely Lemma 5.4.2 that allows one to express the modular form Δ (for $\mathbf{A} = \mathbb{F}_r[T]$ and the full modular group) in terms of Eisenstein series.

5.5 Class Groups of A-integers and the Geometric Criterion for Cyclicity. Let $\wp \in \mathrm{Spec}(\mathbf{A})$ be fixed of degree d. In this subsection we shall describe the geometric criterion for cyclicity for the fields $\mathbf{k}(\wp)$. For the reader's convenience, we recall that $\mathbf{k}(\wp)$ is defined as follows: Let Υ be a non-zero $\wp$-division point of C. Then

$$\mathbf{k}(\wp) = \mathbf{k}(\Upsilon).$$

5.5.1. Definition. Let i be a positive $\mathbf{A}$-even integer. We define the i-th *Bernoulli-Carlitz number*, B_i, by

$$B_i = \Pi(i)\zeta(i).$$

In **[G9]** we reported on a *von-Staudt* type result, due to Carlitz, which gives the denominator of B_i (the result of Carlitz, as reported in **[G9]**, is slightly incorrect; a corrected version can be found in **[C2]**). For our purposes, only the following immediate corollary of this result is needed:

5.5.2. Proposition. *Let $0 < i < r^d - 1$ be $\mathbf{A}$-even. Then B_i is $\wp$-integral.* □

Let i be as in 5.5.2. The reader will then note that the proposition, together with obvious integrality properties of $\Pi(i)$, implies that $\zeta(i)$ is also $\wp$-integral. One also obtains the following result:

5.5.3. Proposition. *Let i be an $\mathbf{A}$-even positive integer. Suppose that $pi < r^d - 1$. Then*

$$\wp \mid B_i \Longleftrightarrow \wp \mid B_{pi}.$$ □

5.5.4. Definition. We let $\mathbf{k}(\wp)_+ \subset \mathbf{k}(\wp)$ be the subfield fixed by $\mathbb{F}_r^* \subseteq \mathbf{A}/\wp^* \simeq \mathrm{Gal}(\mathbf{k}(\wp)/\mathbf{k})$.

Clearly $\mathbf{k}(\wp)$ is the maximal *totally-real* (=totally split at ∞) subfield of $\mathbf{k}(\wp)$. Let Υ be any non-trivial $\wp$-division point of C and let g be any monic prime of $\mathbf{A}$ not equal to $\wp$. Let $\sigma_g \in \mathrm{Gal}(\mathbf{k}(\wp)/\mathbf{k})$ be the Artin symbol of (g). From Lemma 4.1.5, we know that

$$\sigma_g(\Upsilon) = C_g(\Upsilon);$$

it follows easily that

$$\mathbf{k}(\wp)_+ = \mathbf{k}\left(\Upsilon^{r-1}\right).$$

As $\mathbf{k}(\wp)$ is totally-real, and ∞ is an $\mathbb{F}_r$-rational prime of $\mathbf{k}$, we see that all the infinite primes of $\mathbf{k}(\wp)_+$ are also rational. Since each prime of $\mathbf{k}(\wp)_+$ ramifies totally in $\mathbf{k}(\wp)$, all infinite primes of $\mathbf{k}(\wp)$ are rational also.

Let $\mathcal{O}(\wp)$ (resp. $\mathcal{O}(\wp)_+$) be the ring of $\mathbf{A}$-integers of $\mathbf{k}(\wp)$ (resp. $\mathbf{k}(\wp)_+$). We also use "$\mathrm{Cl}^{(p)}$" to denote the p-class groups of these Dedekind Domains as well as for the fields themselves. Finally, we will use the symbol "Cl" to denote the *full* class groups etc.

We write $\mathrm{Cl}(\wp)$, etc., for the class group of the field $\mathbf{k}(\wp)$ and $\mathrm{Cl}(\mathcal{O}(\wp))$, etc., for the class group of $\mathcal{O}(\wp)$, and so on.

5.5.5. Definition.

a) Let $R(\wp)$ (resp. $R(\wp)_+$) be the group of divisors of degree 0 supported on the infinite primes of $\mathbf{k}(\wp)$ (resp. $\mathbf{k}(\wp)_+$).

b) Let $U(\wp) \subset R(\wp)$ (resp. $U(\wp)_+$ etc.) be the divisors of the units in $\mathcal{O}(\wp)$ (resp. $\mathcal{O}(\wp)_+$ etc.).

c) Let $r(\wp)$ (resp. $r(\wp)_+$) be $R(\wp)/U(\wp)$ (resp. $R(\wp)_+/U(\wp)_+$).

The groups $r(\wp)$ and $r(\wp)_+$ are called the "regulator groups" of the fields $\mathbf{k}(\wp)$ and $\mathbf{k}(\wp)_+$.

In general, we shall use a superscripted "(p)" to denote p-primary components of the various abelian groups defined in this subsection.

Let P (resp. P_+) be the projection which takes a divisor of degree zero of $\mathbf{k}(\wp)$ (resp. $\mathbf{k}(\wp)_+$) to its part supported on the finite primes (i.e., those primes lying over $\mathrm{Spec}(\mathbf{A})$). We then have two exact sequences:

$$0 \to r(\wp) \to \mathrm{Cl}(\wp) \xrightarrow{P} \mathrm{Cl}(\mathcal{O}(\wp)) \to 0, \tag{5.5.6}$$

and

$$0 \to r(\wp)_+ \to \mathrm{Cl}(\wp)_+ \xrightarrow{P_+} \mathrm{Cl}(\mathcal{O}(\wp))_+ \to 0. \tag{5.5.7}$$

Note that all of the above groups are naturally Galois modules.

The next result is proved in a similar fashion to its counterpart in classical Iwasawa theory.

5.5.8. Lemma. *Let u be a unit in $\mathcal{O}(\wp)$. Then $u \in \mathcal{O}(\wp)_+$.* □

As $r-1$ is prime to p, Lemma 5.5.8 implies

5.5.9. Corollary. *We have*

$$r(\wp)^{(p)} \simeq r(\wp)_+^{(p)}. \qquad \square$$

Let W be the Witt ring of $\mathbf{A}/\wp$ as usual. Upon tensoring with W we can decompose the above groups into components for the Teichmüller character ω (see Definition 4.2.1). Thus we deduce from 5.5.7 and 5.5.9:

5.5.10. Corollary. *Let i be a positive $\mathbf{A}$-odd integer. Then*

$$\mathrm{Cl}^{(p)}(\wp)(\omega^i) \simeq \mathrm{Cl}^{(p)}(\mathcal{O}(\wp))(\omega^i). \qquad \square$$

The above corollary is very important. It states that one can study the $\mathbf{A}$-odd components of the p-class group of the ring of $\mathbf{A}$-integers, by just studying the *much* simpler $\mathbf{A}$-odd components of the p-class group of the field $\mathbf{k}(\wp)$; the non-triviality of a component is then determined by the divisibility of $\beta(j) =: \beta_{\mathbf{A}}(j)$ by 4.3.3. On the other hand, the $\mathbf{A}$-even components are much more subtle. Our next results will begin to describe their structure. Note, however, that we may study $\mathbf{A}$-even components by working with the field $\mathbf{k}(\wp)_+$ (as in classical theory one studies even components through the field $\mathbb{Q}(\zeta_p + \zeta_p^{-1})$).

Recall (Theorem 4.1.10) that $\frac{\sigma(\Upsilon)}{\Upsilon}$ is a unit for any Galois element σ. The group of units that they generate is called "the group of cyclotomic units".

5.5.11. Definition.

a) Let $Q(\wp)_+$ be the group of divisors of degree 0 supported on the infinite primes of $\mathbf{k}(\wp)_+$ modulo the divisors of *cyclotomic units*.

b) Let $H(\wp)_+$ be the group of *all* units of $\mathcal{O}(\wp)_+$ modulo the group of cyclotomic units.

We then have the following obvious exact sequence:

$$0 \to H(\wp)_+ \to Q(\wp)_+ \to r(\wp)_+ \to 0. \tag{5.5.12}$$

5.5.13. Theorem (Rosen and Galovich). *The group $Q(\wp)_+$ has the same order as the class group of $\mathbf{k}(\wp)_+$.* □

The above theorem is proved by *explicitly* computing the divisors of the cyclotomic units and then using *class number formulae* much like in classical theory (see **[G2]**, for example).

Let $I \subset \mathbb{Z}\left[(\mathbf{A}/\wp^*)/\mathbb{F}_r^*\right]$ be the augmentation ideal; i.e., those

$$\iota = \sum c_g g, \quad g \in (\mathbf{A}/\wp^*)/\mathbb{F}_r^*,$$

with $0 = \sum c_g$. It is trivial to see that the Galois module of divisors of degree 0 supported on the infinite primes of $\mathbf{k}(\wp)_+$ is isomorphic to I. Let ω be the Teichmüller character. Upon tensoring with the Witt ring W of $\mathbf{A}/\wp^*$ one then finds immediately that $Q^{(p)}(\wp)_+(\omega^i)$ and $r(\wp)_+(\omega^i)$ are cyclic W-modules. One then has the following refinement of Theorem 5.5.13:

5.5.14. Proposition. *Let i be an $\mathbf{A}$-even integer with $0 < i < r^d - 1$. Then $Q_+^{(p)}(\wp)(\omega^i)$ and $\mathrm{Cl}^{(p)}(\wp)_+(\omega^i)$ have the same length as W-modules.* □

From 4.3.1, and the exact sequences 5.5.7 and 5.5.12, we now conclude:

5.5.15. Theorem.

a) (Rosen and Galovich) *The finite groups $H(\wp)_+$ and $\mathrm{Cl}\left(\mathcal{O}(\wp)\right)_+$ have the same order.*

b) *Let i be an $\mathbf{A}$-even integer between 0 and $r^d - 1$. Then the W-modules $H^{(p)}(\wp)_+(\omega^i)$ and $\mathrm{Cl}^{(p)}\left(\mathcal{O}(\wp)\right)_+(\omega^i)$ have the same length* □

In other words,

$$H^{(p)}(\wp)_+(\omega^i) \neq \{0\} \Longleftrightarrow \mathrm{Cl}^{(p)}\left(\mathcal{O}(\wp)\right)_+(\omega^i) \neq \{0\}.$$

We are thus able, as in classical theory, to study class groups via the much simpler group of units! In fact, our next goal is to define a version of the classical *Kummer homomorphism*, due to S. Okada **[O1]**, in this context.

Recall that $\mathbf{k}(\wp)/\mathbf{k}$ is totally ramified above $\wp$. Let $\mathbf{k}_\wp$ be the completion of $\mathbf{k}$ at $\wp$ and put

$$\mathbf{k}_\wp^{(1)} = \mathbf{k}_\wp \cdot \mathbf{k}(\wp).$$

Let Υ be any non-trivial $\wp$-division point of the Carlitz module C; one then finds (Υ) is the maximal ideal in the ring $\mathcal{O}_\wp^{(1)}$ of integers in $\mathbf{k}_\wp^{(1)}$.

Let α be a unit of $\mathcal{O}_\wp^{(1)}$. From above, we can write

$$\alpha = f(\Upsilon)$$

where $f(u) \in \mathbf{A}[[u]]$. Divisibility properties of the factorials $\Pi(j)$ (*exactly* similar to those of $n!$, see **[G9]**) tell us that

$$\frac{f'(\mathbf{e}(z))}{f(\mathbf{e}(z))} = \sum_{j=0}^{\infty} a_j \frac{z^j}{\Pi(j)},$$

where a_j is $\wp$-integral for all j.

5.5.16. Definition (S. Okada). We set

$$\phi_j(\alpha) = a_{j-1} + \wp,$$

for $j = 1, \ldots, r^d - 2$.

Okada shows that the ϕ_j are well-defined morphisms from the units $\mathcal{O}_\wp^{(1)}$ to $\mathbf{A}/\wp$. Moreover, following the classical approach he shows

5.5.17. Theorem (S. Okada). *Let $0 < i < r^d - 1$ be $\mathbf{A}$-even. Then*

$$H^{(p)}(\wp)_+(\omega^i) \neq \{0\} \Longrightarrow \wp \mid B_i. \qquad \square$$

Thus, by 5.5.15 we have the important

5.5.18. Theorem. *Let $0 < i < r^d - 1$ be $\mathbf{A}$-even. Then*

$$\mathrm{Cl}^{(p)}\left(\mathcal{O}(\wp)\right)_+(\omega^i) \neq \{0\} \Longrightarrow \wp \mid B_i. \qquad \square$$

The reader should recall that the component $\mathrm{Cl}^{(p)}(\wp)(\omega^i)$ of the *field* $\mathbf{k}(\wp)$ is determined by the $\wp$-divisibility of $\beta(r^d - 1 - i)$.

5.5.19. Corollary ("The Geometric Criterion for Cyclicity"). *Let $0 < i < r^d - 1$ be $\mathbf{A}$-even. Suppose that $\wp \nmid B_i$. Then $\mathrm{Cl}^{(p)}(\wp)_+(\omega^i)$ is a cyclic W-module.*

Proof. From the exact sequence 5.5.7, we conclude that $B_i \in \mathbf{A}_\wp^*$ implies

$$\mathrm{Cl}^{(p)}(\wp)_+(\omega^i) \simeq r^{(p)}(\wp)_+(\omega^i).$$

As above, $r^{(p)}(\wp)_+(\omega^i)$ is a cyclic W-module and the result is established. $\square$

In Subsection 5.7 we will give an important application of this criterion. With the arithmetic and geometric criterions in mind, one may refer to the B_i as "geometric Bernoulli numbers" and the $\beta(i)$ as "arithmetic Bernoulli numbers".

We close this subsection by presenting a conjecture due to E.-U. Gekeler **[Ge4]** which gives the implication "$\Longleftarrow$" in Theorem 5.5.18. Gekeler begins with the following observation: The Frobenius automorphism of W permutes the isotypic components of $\mathrm{Cl}^{(p)}(\wp) \otimes W$. Thus these components of an orbit will all live or die together. More formally, let us define the following equivalence relation: Let i and j be two positive numbers. We say

$$i \overset{\wp}{\sim} j \Longleftrightarrow i \equiv p^t j \bmod (r^d - 1),$$

for some t (and where, we recall, $d = \deg(\wp)$). Then

$$i \overset{\wp}{\sim} j \Longrightarrow \mathrm{Cl}^{(p)}(\wp)(\omega^i) \simeq \mathrm{Cl}^{(p)}(\wp)(\omega^j),$$

via the Frobenius. A similar statement obviously holds true for class groups of the rings of $\mathbf{A}$-integers.

Let $a \in \mathbf{A}$ be prime to $\wp$. Let i and j be two positive integers which are congruent modulo $r^d - 1$. Clearly

$$a^i \equiv a^j \bmod \wp;$$

thus, as we have seen,

$$\beta(i) \equiv \beta(j) \bmod \wp.$$

Moreover, we have trivially

$$\beta(pi) = \beta(i)^p.$$

Therefore,

$$i \overset{\wp}{\sim} j \Longrightarrow \wp \mid \beta(i) \Leftrightarrow \wp \mid \beta(j).$$

In other words, the divisibility of the $\beta(j)$ respects the equivalence relation $\overset{\wp}{\sim}$. However, for B_i it is another matter altogether and we have the following conjecture for which there is some numerical evidence:

5.5.20. Conjecture (Gekeler). Let i be an $\mathbf{A}$-even integer $< r^d - 1$. Suppose that for all $\mathbf{A}$-even $j < r^d - 1$, with $j \overset{\wp}{\sim} i$, we have $\wp \mid B_j$. Then

$$\mathrm{Cl}^{(p)}\left(\mathcal{O}(\wp)\right)_+(\omega^i) \neq \{0\}.$$

Should Conjecture 5.5.20 be established, we would then have a weak *functional equation* relating the values of $\zeta(s)$ at positive and negative integers. Indeed, for $\mathbf{A}$-even $i < r^d - 1$, we deduce from 5.5.7 that

$$\wp \mid B_j \text{ for all } j \overset{\wp}{\sim} i \Longrightarrow \wp \mid \beta(r^d - 1 - j) \text{ for all } j \overset{\wp}{\sim} i.$$

5.6. The Relationship between the Arithmetic and Geometric Criterions. In this subsection we will show that the arithmetic criterion for cyclicity implies the geometric one.

We continue with the notation of the previous subsection.

Let us recall the set-up of Corollary 4.4.2 (the arithmetic criterion): Let $\overline{\mathbf{k}}_\wp$ be a fixed algebraic closure of $\mathbf{k}_\wp$. Let $\overline{\mathbf{A}}_\wp \subset \overline{\mathbf{k}}_\wp$ be the ring of integers with maximal ideal $\overline{M}_\wp$.

Let $i < r^d - 1$ be $\mathbf{A}$-even and set $j = r^d - 1 - i$. Over $\overline{\mathbf{A}}_\wp$, we factor $\tilde{z}(x, -j)$ as

$$\tilde{z}(x, -j) = (1 - \theta_1 x^{-1}) \cdots (1 - \theta_\iota x^{-1}).$$

We then have:

5.6.1. Theorem. *The arithmetic criterion implies the geometric criterion. That is:*

$$\wp \nmid B_i \Longrightarrow \textit{at most one } \theta_t \equiv 1 \pmod{\overline{M}_\wp}.$$

Proof. Let $T_p(\wp)_+$ be the p-adic Tate module of the field $\mathbf{k}(\wp)_+$. Let F be its Frobenius endomorphism. Without loss of generality, we may assume that $\mathrm{Cl}^{(p)}(\wp)_+(\omega^i)$ is non-trivial (else the result follows trivially). We shall show that on $T_p(\wp)_+(\omega^i)$ the mapping $F - 1$ has a *unique* non-unit eigenvalue. The result will then follow by the double congruence 4.2.5.

In this proof, for simplicity, we shall drop the symbol "$\wp$" in our notation for various groups. Thus "$\mathrm{Cl}^{(p)}(\mathcal{O})_+$" will be the p-class group of $\mathcal{O}(\wp)_+$ etc.

Let $\overline{\mathbb{F}}_r$ be a fixed algebraic closure of $\mathbb{F}_r$. Let $\overline{\mathrm{Cl}}_+^{(p)}$ be the p-primary class group of the field $\overline{\mathbb{F}}_r \cdot \mathbf{k}(\wp)_+$ and let $\overline{\mathrm{Cl}}^{(p)}(\mathcal{O})_+$ be the p-primary class group of $\overline{\mathbb{F}}_r \cdot \mathcal{O}(\wp)_+$. Recall that the primes at ∞ on $\mathbf{k}(\wp)_+$ are $\mathbb{F}_r$-rational. Thus, as before, we have the exact sequence:

$$0 \to r_+^{(p)}(\omega^i) \to \overline{\mathrm{Cl}}_+^{(p)}(\omega^i) \to \overline{\mathrm{Cl}}^{(p)}(\mathcal{O})_+(\omega^i) \to 0.$$

By taking Tate modules ($= \mathrm{Hom}(\mathbb{Q}_p/\mathbb{Z}_p, ?)$) of the above sequence, we obtain

$$0 \to \mathrm{Hom}\left(\mathbb{Q}_p/Z_p, r_+^{(p)}(\omega^i)\right) \to T_{p,+}(\omega^i) \to \mathrm{Hom}\left(\mathbb{Q}_p/Z_p, \overline{\mathrm{Cl}}^{(p)}(\mathcal{O})_+(\omega^i)\right)$$

$$\to \mathrm{Ext}^1\left(\mathbb{Q}_p/Z_p, r_+^{(p)}(\omega^i)\right) \to \mathrm{Ext}^1\left(\mathbb{Q}_p/Z_p, \overline{\mathrm{Cl}}_+^{(p)}(\omega^i)\right).$$

As $r_+^{(p)}(\omega^i)$ is finite, it is easy to see that the first term vanishes. As $\overline{\mathrm{Cl}}_+^{(p)}(\omega^i)$ is divisible, the last term also vanishes. Let us set $\tilde{T}_{p,+}(\omega^i) = \mathrm{Hom}\left(\mathbb{Q}_p/\mathbb{Z}_p, \overline{\mathrm{Cl}}^{(p)}(\mathcal{O})(\omega^i)\right)$. We therefore have

$$0 \to T_{p,+}(\omega^i) \to \tilde{T}_{p,+}(\omega^i) \to \mathrm{Ext}^1\left(\mathbb{Q}_p/\mathbb{Z}_p, r_+^{(p)}(\omega^i)\right) \to 0.$$

From the exact sequence

$$0 \to \mathbb{Z}_p \to \mathbb{Q}_p \to \mathbb{Q}_p/\mathbb{Z}_p \to 0,$$

we deduce that

$$\mathrm{Ext}^1\left(\mathbb{Q}_p/\mathbb{Z}_p, r_+^{(p)}(\omega^i)\right) \simeq r_+^{(p)}(\omega^i).$$

We then have

$$0 \to T_{p,+}(\omega^i) \to \tilde{T}_{p,+}(\omega^i) \to r_+^{(p)}(\omega^i) \to 0. \tag{5.6.2}$$

Note that $\tilde{T}_{p,+}(\omega^i)$ is *torsion free*; thus

$$T_{p,+}(\omega^i) \otimes \mathbb{Q}_p \simeq \tilde{T}_p(\omega^i) \otimes \mathbb{Q}_p.$$

Consequently, $T_{p,+}(\omega^i)$ is a *sublattice* of $\tilde{T}_{p,+}(\omega^i)$ of maximal rank.

The endomorphism $F-1$ clearly acts on the exact sequence 5.6.2 and has trivial action on $r_+^{(p)}(\omega^i)$. Thus

$$(F-1)\tilde{T}_{p,+}(\omega^i) \subseteq T_{p,+}(\omega^i).$$

We shall show that this is, in fact, an equality.

By the assumption that B_i is a $\wp$-unit, we find

$$T_{p,+}(\omega^i)/\left((F-1)T_{p,+}(\omega^i)\right) = r_+^{(p)}(\omega^i).$$

Thus, $\operatorname{ord}_p(\det(F-1))$ is the W-length of $r_+^{(p)}(\omega^i)$. But then

$$\tilde{T}_{p,+}(\omega^i)/\left((F-1)\tilde{T}_{p,+}(\omega^i)\right)$$

must have the same W-length. As this is also the W-length of $\tilde{T}_{p,+}(\omega^i)/T_{p,+}(\omega^i)$, we conclude

$$(F-1)\tilde{T}_{p,+}(\omega^i) = T_{p,+}(\omega^i).$$

Next, choose a basis for $\tilde{T}_{p,+}(\omega^i)$, say $\{e_1, \ldots, e_h\}$, so that $\{qe_1, \ldots, e_h\}$ is a basis for $T_{p,+}(\omega^i)$, for some $q \in W$. (Recall that $r_+^{(p)}(\omega^i)$ is a cyclic W-module.) Let $A = (a_{i,j})$ be the matrix of $F-1$ with respect to this basis. Let B be the diagonal matrix with $\frac{1}{q}$ in the first row and all other diagonal elements $=1$. One finds that BA is an automorphism of $T_{p,+}(\omega^i)$ and AB is an automorphism of $\tilde{T}_{p,+}(\omega^i)$. Thus one sees that both the first row and column of A must have coefficients divisible by q.

Let A_1 be the minor obtained from A by deleting the first row and column of A. If we now expand the determinant along the first row we find

$$\det(A) \equiv a_{1,1}\det(A_1) \pmod{q^2}.$$

But $\operatorname{ord}_p(q) = \operatorname{ord}_p(\det(A))$; thus $\det(A_1)$ is a W-*unit.*

If we reduce $\det(tI-A)$ modulo (p), we obtain $t\tilde{P}(t)$, where $\tilde{P}(t)$ is the characteristic polynomial of the reduction of A_1. The result now follows. □

5.7. Magic Numbers and Cyclicity. In this subsection we introduce the magic numbers. These numbers play a very mysterious and not well-understood role in the theory. Basically, when one attempts various computations in the theory, the magic numbers appear. Finding out why should have very important consequences.

5.7.1 Definition. A non-negative integer i is said to be a *magic number* if it is 0 or its r-adic expansion has all digits, *except* the highest, $= r-1$.

Let $0 < c \leq r-1$. We write $i = i(c,j) = cr^j + r^j - 1$. We say that i is "even" if i is also $\mathbf{A}$-even ($c = r-1$); otherwise it is "odd" ($c \neq r-1$).

Thus a magic number is like an integer with all decimal digits, except the highest $=9$; e.g., 3999.

Let $\wp$ be a fixed prime of degree d as before. Let a be an element of $\mathbf{A}$ prime to $\wp$ and let t and j be two integers with $t \equiv j \pmod{r^d - 1}$. Then, as we have seen

$$a^t \equiv a^j \pmod{\wp}.$$

It turns out that this sort of congruence on integers also respects magic numbers:

5.7.2. Lemma. *Let* j_1 *and* j_2 *be two non-negative integers which are congruent modulo a positive integer* d. *Then*

$$i(c, j_1) \equiv i(c, j_2) \ (\mathrm{mod}\ r^d - 1). \qquad \square$$

5.7.3. Remarks ("Generating properties of magic numbers"). The magic numbers also have the following curious property: every positive integer can be written as a canonical alternating sum of magic numbers. Indeed, let

$$n = \sum_{\nu=0}^{t} e_\nu r^\nu$$

be a positive integer written r-adically and where we assume that $e_t \neq 0$. Let $i_0 = i(e_t, t)$. Then

$$i_0 - n = \sum_{\nu=0}^{t-1} (r - 1 - e_v) r^\nu ,$$

and we simply repeat the process.

This expansion takes on greater interest upon noticing that it *never* "carries over" from one r-adic digit to the next. Thus the expansion *commutes* with the construction of $\Pi(n)$. That is, one can now write

$$\Pi(n) = \frac{\Pi(i_0)\cdots}{\Pi(i_1)\cdots}.$$

As an example, let $r = 3$ and $n = 13$. We find

$$13 = i(1,2) - i(1,1) + i(1,0) = 17 - 5 + 1.$$

Thus

$$\Pi(13) = \frac{\Pi(17)\cdot\Pi(1)}{\Pi(5)}.$$

There are also an analog of Definition 5.7.1, which uses *negative* numbers:

5.7.4. Definition. We call any integer of the form $-1 - i(c, j)$, for magic $i(c, j)$, a *magic dual number*.

Note that the magic duals are precisely those p-adic integers which have arbitrary *lowest* digit and *all* others $= r - 1$.

5.7.5. Remarks("Generating properties of magic duals"). As in 5.7.3, *every* p-adic integer y has a canonical expansion as a convergent alternating sum of magic duals. And, as above, this immediately gives a convergent *product* for $\Pi_\infty(y)$ by formally exponentiating the sum. Thus, if $\alpha \in \mathbb{Z}_p$ has the expansion $\alpha = \sum (-1)^t i_t$, then

$$\Pi_\infty(\alpha) = \prod \Pi_\infty(i_t)^{(-1)^t}.$$

On the other hand, from the functional equation of $\Pi_\infty(y)$ (Example 5.3.8) one can compute the values of $\Pi_\infty(y)$ at a magic dual in terms of the 1-unit piece of ξ and the value of $\Pi_\infty(y)$ at the original defining magic number. Thus knowledge of the function $\Pi_\infty(y)$ at magic numbers is sufficient, via the functional equation, to express *all* values of $\Pi_\infty(y)$ as in 5.7.5!

The magic numbers originally arose out of a computation of E. Thomas of the degrees of the elements $\beta(j)$. Thomas's formula showed that precisely for the magic numbers one has an equality of degrees between $\beta(j)$ and $\Pi(j)$. For all others, the degree of $\Pi(j)$ was strictly larger.

Let us write for an arbitrary positive t

$$z(x,-t) = \sum_{e=0}^{\infty} S_e(t)x^{-e},$$

where

$$S_e(t) = \sum_{\substack{n \text{ monic} \\ \deg(n)=e}} n^t,$$

and

$$\tilde{z}(x,-t) = \sum_{e=0}^{\infty} \tilde{S}_e(t)x^{-e}.$$

Thus, $S_e(i) = \tilde{S}_e(i) = 0$ for $e \gg 0$. Moreover, if i is $\mathbf{A}$-odd, then

$$S_e(i) = \tilde{S}_e(i).$$

Let $i = i(c,j)$ be a magic number. Upon further examination, and computer calculations, we were lead to the following empirical facts:

a) The degree in x^{-1} of $z(x,-i)$ is j.

b) $\deg\left(\tilde{S}_e(i)\right)$ strictly increases with e for $e \le j$.

c) $\deg\left(\tilde{S}_{j-1}(i)\right) = \deg\left(\tilde{S}_j(i)\right) - cr^j$.

d) $\tilde{S}_j(i) = (-1)^j \Pi(i)$.

These results have been established by E.-U. Gekeler in **[Ge2]**. As it turns out, the results also follow from an old paper of H. L. Lee **[Le1]**. Both Gekeler's paper and Lee's paper (as well as many of Carlitz) contain remarkable formulae for the $S_e(i)$ for arbitrary i. Still, as yet there seems to be no overarching theory which gives a nice closed form for these sums.

We shall return to Gekeler's computations a bit later. We now turn our attention to the arithmetic applications of magic numbers.

5.7.6. Lemma. *Let* $i = i(r-1,j) = r^{j+1} - 1$. *Then*

$$\frac{\zeta(i)}{\xi^i} = (-1)^{j+1}\frac{1}{L_{j+1}}.$$

Proof. This follows immediately from Carlitz's Lemma 5.4.2 and the discussion in Section 5.1. □

5.7.7. Remarks. In **[Ge3]**, one can find more formulae as in 5.7.6. These additional results arose from techniques used to study the de Rham cohomology of Drinfeld modules and are remarkable. However, it is not clear how they fit (*if* they do fit!) into the yoga of magic numbers.

From the simple divisibility properties of $\Pi(j)$ and L_j we conclude:

5.7.8. Corollary. *Let* $j < d-1$ *and* $i = i(r-1, j) = r^{j+1} - 1$. *Then* $\wp \nmid B_i$. □

5.7.9. Corollary. *Let* $n < r^d - 1$ *be an* $\mathbf{A}$*-even integer of the form* $n = p^t \cdot i(r-1, j)$. *Then* $\mathrm{Cl}^{(p)}(\wp)(\omega^n)$ *is a cyclic* W*-module.*

Proof. This follows from the geometric criterion for cyclicity (Corollary 5.5.19) and the obvious p-th power properties of $\zeta(s)$. □

Recall that we defined the element $\tilde{\beta}(t) =: \tilde{\beta}_{\mathbf{A}}(t) \in \mathbf{A}$ in Definition 4.4.3, and, in Proposition 4.4.4, we let $\mathbf{B}(i) =: \mathbf{B}_{\mathbf{A}}(i)$ denote the $\mathbf{A}$-ideal generated by $\beta(i)$ and $\tilde{\beta}(i)$. Thus from Proposition 4.4.4 and Theorem 5.6.1 we conclude the following result:

5.7.10. Corollary. *Let* $j < d-1$ *and* $i = i(r-1, j)$. *Then* $\wp \nmid \mathbf{B}(i)$.

Proof. Set $t = d - j - 2 \geq 0$ and $i_0 = i(r-1, t)$. Note that

$$r^d - 1 - i_0 = r^{d-j-1}(r^{j+1} - 1).$$

From Corollary 5.7.9, Theorem 5.6.1, and the double congruence (4.2.5), we find that $\mathbf{B}\left(r^{d-j-1}(r^{j+1} - 1)\right)$ is not divisible by $\wp$. But then clearly $\mathbf{B}(i)$ is also not divisible by $\wp$. □

In other words, *the arithmetic criterion for cyclicity is true for even magic numbers.*

For odd magic numbers, the arithmetic criterion has been checked on a computer for small cases. In fact, for such numbers the criterion can be given in a more compact form:

5.7.11. Proposition. $\wp \nmid \mathbf{B}(i)$ *for all odd magic* $i < r^{\deg(\wp)} - 1$ *and all* $\wp$ *if and only if* $\mathbf{B}(i) = (1)$ *for all odd magic* i.

Proof. As the odd magic numbers are closed under congruences modulo $r^d - 1$, the result follows immediately. □

The result does not hold for even magic numbers since then the residue used in the proof may be 0.

5.7.12. Remarks ("Kummer-Vandiver for A"). The statement

"$\mathbf{B}(i) = (1)$ for all *odd* magic i"

seems to function as a "Kummer-Vandiver Conjecture" for $\mathbf{A}$. It has been checked for small r and primes $\wp$.

The reader should note that one *can* find examples of *even* magic i and $\mathbf{A}$-odd non-magic i where $\mathbf{B}(i) \neq (1)$.

Let us now return to the paper **[Ge2]** of Gekeler. As we have mentioned, there are various mysterious formulae for the elements

$$S_e(t) = \sum_{\substack{n \text{ monic} \\ \deg(n)=e}} n^t .$$

We will re-interpret one such formula in a way that *may* be a reflection of an underlying functional equation for $\zeta(s)$. We stress the word "may" as the situation is still quite fluid.

Before presenting this calculation, we need some vanishing formulae for the $S_e(t)$. There are many such combinatorial formulae (see **[Th3]**) and we recall just a few. The reader should note that these formulae for $\mathbb{F}_r[T]$ are much sharper than the general vanishing formula of Lemma 3.6.7.

Let $l(i)$ be the sum of the r-adic digits of i.

5.7.13. Lemma.

a) (H. Lee and E.-U. Gekeler) $S_e(i) = 0$ *if* $e > \dfrac{l(i)}{r-1}$.

b) (H. Lee) *Let* $r = p$. *Then* $S_e(i) = 0$ *if and only if* $e > \dfrac{l(i)}{r-1}$. □

In particular, one can see that if $i < r^e - 1$ then $S_e(i) = 0$.

The reader should note that for $r = p$, 5.7.13 *precisely* gives the degree in x^{-1} of $z(x, -i)$.

5.7.14. Theorem (Gekeler). *Let* $i < r^{e+1} - 1$ *have the* r*-adic expansion* $i = \sum_{j=0}^{e} b_j r^j$.

Then

$$S_e(i) = (-1)^t \cdot M \cdot \Pi(i) \cdot \prod_{j \leq e} L_{e-j}^{r^j(b_j-(r-1))},$$

where $t = e + \sum_{j<e} (e-j+1)b_j$, *and* M *is the multinomial coefficient* $\binom{b_e}{b'_0, \ldots, b'_e}$, $b'_j = r - 1 - b_j$, $(j < e)$, *and* $b'_e = l(i) - e(r-1)$. □

It is a simple matter to check that $b_e = \sum b'_j$. The nonvanishing of the formula in the theorem is precisely determined by the nonvanishing of M modulo p.

We have also seen above that if $b'_e < 0$, or $l(i) < e(r-1)$, then $S_e(i) = 0$.

By definition

$$z(x, -i) = \sum S_e(i) x^{-e} .$$

We know that for those e with $i < r^e - 1$ we have $S_e(i) = 0$. Now let e equal the *least* positive integer with

$$i < r^{e+1} - 1 .$$

Then 5.7.14 gives the coefficient of x^{-e} (which may vanish) in $z(x,-i)$ and gives a generalization of Part d) of the computations presented above for magic numbers.

5.7.15. Example. Let $r = p = 3$. Let $i = 71 = 2 + 2 \cdot 3 + 3^2 + 2 \cdot 3^3$. In this case, $e = 3$. By 5.7.14 (or 5.7.13.b)) we know that

$$S_3(71) \neq 0,$$

and that the degree in x^{-1} of $z(x,-71)$ is 3.

On the other hand, let $i = 27 = 3^3$. In this case it is still true that $e = 3$ but clearly $S_3(27) = 0$.

Finally, let i, e etc., be as in Theorem 5.7.14. By using 5.7.8, the fact that $\zeta(0) = 1$, and the r-th power mapping, we have established that

$$S_e(i) = \pm \cdot M \cdot \xi^{-\nu} \cdot \Pi(i) \cdot \prod_{j \leq e} \zeta(r^e - r^j)^{r-1-b_j},$$

where

$$\nu = r^e\,(l(i) - e(r-1)) + r^{e+1} - 1 - i.$$

(If $S_e(i) \neq 0$ then $\nu \geq 0$!)

Perhaps further computations will clarify things and allow one to guess at some underlying formalism that "explains" this result. The reader will note that this computation is in line with both the philosophy of magic numbers and the philosophy of digit expansions (5.3.2).

5.8. v-adic Measures and $\zeta(\mathbf{A}, s)$. Our purpose in this section is to explain how the magic numbers appear in yet another context — that of v-adic measures for $v \in \operatorname{Spec}(\mathbf{A})$. We will confine ourselves to explaining the basics of these measures. We refer the reader to **[G6]** for the beginnings of a theory that *may* explain these calculations (and, hopefully, the other appearances of magic numbers).

We will assume that the reader is familiar with the rudiments of p-adic measures (such as in **[L1]**). The same formalism then applies to v-adic measures (which, by definition, are finitely additive, $\mathbf{A}_v$-valued functions on the compact open subsets of $\mathbf{A}_v$). Thus if μ is a v-adic measure and $f\colon \mathbf{A}_v \to \mathbf{A}_v$ is a *continuous* function, then one can form the "Riemann Integral"

$$\int_{\mathbf{A}_v} f(x)\, d\mu(x).$$

In fact, as $\mathbf{A}_v$ is totally disconnected, one sees that the space of measures is *dual* to the space of such continuous functions.

The simplest measures are the *Dirac measures*, δ_α, associated to elements $\alpha \in \mathbf{A}_v$. By definition,

$$\int_{\mathbf{A}_v} f(x)\, d\delta_\alpha(x) = f(\alpha).$$

5.8.1. Remarks. One is lead to study v-adic measures by the following easy observations in classical number theory: For the moment, let $\zeta(s) = \sum_{n=1}^{\infty} n^{-s}$ be Riemann's ζ-function and let $\Gamma(s)$ be Euler's function. Then, as is well-known, for $\Re(s) > 1$

$$\Gamma(s)\zeta(s) = \int_0^\infty x^s \frac{e^{-x}}{1-e^{-x}} \frac{dx}{x}.$$

Now let G be a compact totally disconnected abelian group (e.g., $\mathbb{Z}_p$, $\mathbf{A}_v$, or a Galois group). One sees easily that the convolution algebra of p-adic measures on G is isomorphic to the completed group algebra $\mathbb{Z}_p[[G]]$; thus, as in Subsection 4.4., a p-adic measure μ on $\mathbb{Z}_p$ corresponds to a power series $f_\mu(X) \in \mathbb{Z}_p[[X]]$. The power series corresponding to the Dirac measure at $x \in \mathbb{Z}_p$ (with the obvious definition) is

$$f_{\delta_x}(X) = (1+X)^x.$$

Put $T = 1 + X$; then, *formally*,

$$\sum_{n=1}^{\infty} \delta_n = \sum_{n=1}^{\infty} T^n = \frac{T}{1-T}.$$

Note that the function $R(T) = \dfrac{T}{1-T}$ is *invariant* of the choice of p used. Moreover, as above,

$$R(e^{-x})$$

occurs in the above complex integral for $\zeta(s)$. Thus, *the nonarchimedean theory gives an "echo" of the complex theory.* For more along these lines, see **[G5]**.

The results of this subsection are presented with 5.8.1 in mind.

One approach to showing $\mathbb{Z}_p[[\mathbb{Z}_p]] \simeq \mathbb{Z}_p[[X]]$ is via an explicit formula for the continuous functions $f(s): \mathbb{Z}_p \to \mathbb{Z}_p$. This is given by *Mahler's Theorem* which states that

$$f(s) = \sum_{j=0}^{\infty} a_j \binom{s}{j},$$

with $a_j \in \mathbb{Z}_p \to 0$. Moreover any such sequence uniquely gives rise to a continuous function; one obtains the a_j via higher differences. A p-adic measure ν, being dual to the space of continuous functions, is then uniquely determined by

$$\left\{ b_j = \int_{\mathbb{Z}_p} \binom{s}{j} \, d\nu(s) \right\} \subset \mathbb{Z}_p.$$

One associates to the measure $\nu \in \mathbb{Z}_p[[\mathbb{Z}_p]]$ the power series

$$f_\nu(X) = \sum b_j X^j,$$

and any such power series gives a measure as is easy to see. Finally, basic properties of the binomial coefficients establish that the convolution ring of measures is isomorphic to $\mathbb{Z}_p[[X]]$.

Therefore the first order of business for $\mathbf{A}$ is to also give an explicit formula for the v-adic measures on $\mathbf{A}_v$. The basic analogs of $\binom{s}{j}$ are also due to Carlitz. The reader should note that, as always, they will be defined via the digit expansion procedure.

5.8.2. Definition (Carlitz). Let j be a non-negative integer. Set $\mathbf{A}(j) = \{\alpha \in \mathbf{A} \mid \deg(\alpha) < j\}$.

a) We set $e_0(x) = x$ and for $j \geq 1$

$$e_j(x) = \prod_{\alpha \in \mathbf{A}(j)} (x - \alpha).$$

b) Let $i = \sum_{t=0}^{\nu} \alpha_t r^t$ be the r-adic expansion of i. We set

$$G_i(x) = \prod_{t=0}^{\nu} e_t(x)^{\alpha_t}.$$

c) Let i be as above. We set

$$g_i(x) = g_{\alpha_t r^t}(x),$$

where,

$$g_{\alpha_t r^t}(x) = \begin{cases} e_t(x)^{\alpha_t} & \text{if } \alpha_t < r-1 \\ e_t(x)^{r-1} - D_t^{r-1} & \text{if } \alpha_t = r-1. \end{cases}$$

Since the roots of the polynomial $e_j(x)$ form an $\mathbb{F}_r$-vector space, standard theory implies that they are $\mathbb{F}_r$-linear; i.e.,

$$e_j(x + \zeta u) = e_j(x) + \zeta e_j(u)$$

for $\zeta \in \mathbb{F}_r$.

The polynomials $G_i(x)$ are called the "Carlitz functions". They have many wonderful properties. We shall mention only a few of these here. For more, see **[G4]**, **[G6]**.

The most important property of the Carlitz functions is the following result. It was established by C. Wagner based on the work of Carlitz; it was also independently found by the present author:

5.8.3. Theorem. *Let $f(x)\colon \mathbf{A}_v \to \mathbf{A}_v$. Then $f(x)$ can be written uniquely as*

$$f(x) = \sum_{t=0}^{\infty} a_t \frac{G_t(x)}{\Pi(t)},$$

with $\{a_t\} \subset \mathbf{A}$ and $a_t \to 0$. Conversely, any such sequence determines a continuous function via the above expansion. □

Implicit in the above result is the lovely fact, due to Carlitz, that

$$\alpha \in \mathbf{A} \Longrightarrow \frac{G_t(\alpha)}{\Pi(t)} \in \mathbf{A} \text{ also.}$$

There is an elegant "integral formalism" for the coefficients a_t given above.

5.8.4. Theorem (Carlitz). *Let $f(x)$, etc., be as in Theorem 5.8.3. Let j be a fixed non-negative integer and m a positive integer with $r^m > j$. Then*

$$(-1)^m a_j = \sum_{\alpha \in \mathbf{A}(m)} f(\alpha) \frac{g_{r^m-1-j}(\alpha)}{\Pi(r^m-1-j)}. \qquad \square$$

We also have the following *orthogonality formula*:

5.8.5. Proposition (Carlitz). *Let $l < r^m$ and j arbitrary. Then*

$$\sum_{\alpha \mathbf{A}(m)} g_l(\alpha) G_j(\alpha) = \begin{cases} 0 & \text{if } l+j \neq r^m - 1, \\ (-1)^m \Pi(r^m-1) & \text{if } l+j = r^m-1. \end{cases}$$

Proof. This follows immediately from Theorem 5.8.4 by expanding the function $f(x) = G_j(x)$. $\square$

The proof of Theorem 5.8.4 is based on Lagrange interpolation and the following result.

5.8.6. Lemma (Carlitz).
a) $G_j(x+u) = \sum_{e+f=j} \binom{j}{e} G_e(x) G_f(x)$.
b) $g_j(x+u) = \sum_{e+f=j} \binom{j}{e} g_e(x) G_f(u)$. $\square$

Proof. Use the additivity of the $e_t(x)$ to express the function of $x+y$ as a sum of products of functions of only x or y. The result is then a (very clever!) consequence of Lucas' formula. $\square$

In other words, we see that the functions $G_j(x)$ satisfy the *binomial theorem*. This fact will be crucial for us a bit later (and also in **[G6]**).

Let $f(x)$ now be a polynomial in x of degree d. The expansion in Theorem 5.8.4 is then trivially seen to be *finite* and, of course, not dependent on the choice of v. From looking at the digit expansions, it is trivial to see that

$$\Pi(r^m-1) = \Pi(j)\Pi(r^m-1-j).$$

Thus, 5.8.4 implies

$$f(x) = \sum_{j=0}^{d} b_j G_j(x),$$

and, for $r^m > j$,

$$(-1)^m \Pi(r^m-1) b_j = \sum_{\alpha \in \mathbf{A}(m)} g_{r^m-1-j}(\alpha) f(\alpha).$$

As an example of the power of Theorem 5.8.4, we have

5.8.7. Theorem. *Let $f(x)$ be a polynomial of degree d which we expand, as above, as*

$$f(x) = \sum_{j=0}^{d} b_j G_j(x).$$

Then $f(x)$ is invariant under translations by elements of $\mathbf{A}(t)$ if and only if $b_j = 0$ for $j \not\equiv 0 \pmod{r^t}$.

Proof. The "if" follows easily from the additivity of the $e_j(x)$. Thus we concentrate on the "only if" part.

Let $j \not\equiv 0 \pmod{r^t}$. Choose m so that $r^m > j$ and $m > t$. Thus $\mathbf{A}(t) \subset \mathbf{A}(m)$. Let

$$\{\beta_1, \dots, \beta_{r^{m-t}}\}$$

be representatives of $\mathbf{A}(m)/\mathbf{A}(t)$. By the invariance of $f(x)$ under $x \mapsto x + \alpha$, $\alpha \in \mathbf{A}(t)$, we have

$$(-1)^m \Pi(r^m - 1) b_j = \sum_{i=0}^{r^{m-t}} f(\beta_i) \sum_{\alpha \in \mathbf{A}(t)} g_{r^m-1-j}(\beta_i + \alpha).$$

From 5.8.6.b), we see that the sum equals

$$\sum_{i=0}^{r^{m-t}} f(\beta_i) \sum_{\alpha \in \mathbf{A}(t)} \sum_{e+f=r^m-1-j} \binom{r^m-1-j}{f} g_e(\beta_i) G_f(\alpha).$$

From the orthogonality formula (5.8.5), we see that the sum over α vanishes *unless* $f = r^t - 1$. But, as $j \not\equiv 0 \pmod{r^t}$,

$$\binom{r^m-1-j}{r^t-1} \equiv 0 \pmod{p}.$$

The result now follows. □

5.8.8. Question. Is it possible to somehow "pass to the limit" and use 5.8.7 to compute q-expansions of modular forms (e.g., Eisenstein series associated to totally-real extensions)?

Before returning to v-adic measures, we present the connection between the Carlitz functions and the reciprocal gamma function $g(x, y)$ of Definition 5.3.9.

5.8.9. Lemma. *Let i be a non-negative integer. Then*

$$\prod_{\substack{n \text{ monic} \\ \deg(n)=i}} \left(1 + \frac{x}{n}\right) = 1 + \frac{e_i(x)}{D_i}.$$

Proof. Let n be monic of degree i. It is then easy to see that

$$D_i = e_i(n).$$

Thus both sides have the same roots and constant term. □

The next result may also be proved from Lucas's formula.

5.8.10. Lemma. *Let $\{z_j\}$ be a collection of elements of $\mathbf{K}$ which tend to 0 as $j \to \infty$. Let $y = \sum \alpha_j r^j$ be a p-adic integer written in its canonical r-adic expansion. Define*

$h(y)\colon \mathbb{Z}_p \to \mathbf{K}^*$ by

$$h(y) = \prod (1+z_j)^{\alpha_j}\,.$$

Then $h(y)$ is continuous and $h(y) = \sum a_t \binom{y}{t}$, where if

$$t = \sum_{e=0}^{\nu} \beta_e r^e, \quad 0 \le \beta_e < r\,,$$

then

$$a_t = \prod_{e=0}^{\nu} z_e^{\beta_e}\,. \qquad \square$$

5.8.11. Theorem. *Let $y \in \mathbb{Z}_p$ be written in its r-adic expansion as in Lemma 5.8.10. Then*

$$\begin{aligned} g(x,y) &= \prod \left(1 + \frac{e_j(x)}{D_j}\right)^{\alpha_j} \\ &= \sum \binom{y}{t} \frac{G_t(x)}{\Pi(t)}\,. \end{aligned}$$

Proof. This follows immediately from the previous two lemmas. $\square$

Now let μ be a v-adic measure. By Theorem 5.8.3, we see that μ is *uniquely* determined by the sequence $\{b_j\} \subset \mathbf{A}_v$ with

$$b_j(\mu) = \int_{\mathbf{A}_v} \frac{G_j(x)}{\Pi(j)}\, d\mu(x)\,.$$

Moreover, any such sequence is easily seen to determine a measure. We shall return to this theme below.

Since $\dfrac{G_j(x)}{\Pi(j)}$ is a polynomial in x, it is also clear that a v-adic measure μ is determined by its *moments*

$$\left\{ m_t(\mu) = \int_{\mathbf{A}_v} x^t\, d\mu(x) \right\}_{t=0}^{\infty}\,.$$

Further, the condition that a sequence of moments actually comes from a measure is precisely that $\displaystyle\int_{\mathbf{A}_v} \frac{G_j(x)}{\Pi(j)}$, *which is read off from the moments,* is contained in $\mathbf{A}_v$ for all j.

Let μ_1 and μ_2 be two v-adic measures.

5.8.12. Theorem-Definition. *There exists a v-adic measure $\mu_1 * \mu_2$ with moments*

$$m_j(\mu_1 * \mu_2) = \begin{cases} m_j(\mu_1) & \text{if } j \text{ is } \mathbf{A}\text{-add} \\ m_j(\mu_2) & \text{if } j \text{ is } \mathbf{A}\text{-even.} \end{cases}$$

Proof. Let $0 \le \alpha < r$ and $e_t(x)^\alpha = \sum c_i x^i$. It follows easily from the multinomial theorem, and the uniqueness of r-adic expansions, that $c_i = 0$ for $i \not\equiv \alpha \pmod{r-1}$.

Thus with a little work one sees that the coefficient of x^j in $G_s(x)$ is non-zero only if $j \equiv s \pmod{r-1}$. As a consequence, the formal integral of $\frac{G_s(x)}{\Pi(s)}$ with respect to $\mu_1 * \mu_2$ is computed by its integral over μ_1 *or* μ_2 depending on whether s is $\mathbf{A}$-even or odd. The result now follows. □

5.8.13. Remarks.

a) Clearly $\mu_1 * \mu_2$ is *not* symmetric in μ_1 and μ_2.

b) The proof of Theorem 5.8.13 also gives

$$b_j(\mu_1 * \mu_2) = \begin{cases} b_j(\mu_1) & \text{if } j \text{ is } \mathbf{A}\text{-add} \\ b_j(\mu_2) & \text{if } j \text{ is } \mathbf{A}\text{-even.} \end{cases}$$

c) The above theorem works in complete generality for *all* $\mathbf{A}$ of Section 2. Indeed, let $v \in \operatorname{Spec}(\mathbf{A})$. Then one can find a prime f_v of $\operatorname{Spec}(\mathbb{F}_r[T])$ such that $\mathbf{A}_v \simeq \mathbb{F}_r[T]_{f_v}$. Thus, as Theorem 5.8.12 is a local statement, it now automatically carries over to $\mathbf{A}_v$.

Recall the definition of the *divided power series* algebra over $\mathbf{A}_v$: This is the collection of all formal sums

$$\sum_{j=0}^{\infty} a_j \frac{X^j}{j!},$$

with the obvious multiplication induced by

$$\frac{X^j}{j!} \cdot \frac{X^t}{t!} = \binom{j+t}{j} \frac{X^{j+t}}{(j+t)!},$$

which has a meaning *precisely* since $\binom{j+t}{j} \in \mathbb{Z}$. We now associate to a v-adic measure μ the divided power series

$$f_\mu(X) = \sum b_j(\mu) \frac{X^j}{j!}.$$

Lemma 5.8.6 then implies that the convolution algebra of v-adic measures is isomorphic to the divided power series algebra over $\mathbf{A}_v$!

5.8.14. Remarks.

a) Let $\alpha, \beta \in \mathbf{A}_v$ and let $f_\alpha(X) =: f_{\delta_\alpha}(X)$. We thus deduce that

$$f_{\alpha+\beta}(X) = f_\alpha(X) f_\beta(X).$$

This gives a new type of exponentiation in characteristic p.

b) Let $\alpha \in \mathbf{k}$. Then

$$f_\alpha(X) = \sum \frac{G_j(\alpha)}{\Pi(j)} \frac{X^j}{j!},$$

which does not depend on v. On the other hand, by Theorem 5.8.11,

$$g(\alpha, y) = \sum \frac{G_j(\alpha)}{\Pi(j)} \binom{y}{j}.$$

Thus we see a remarkable relation between the Dirac measures and $g(x,y)$; to go from one to the other, one simply formally interchanges $\frac{X^j}{j!}$ and $\binom{y}{j}$. This interchanging can be formalized by viewing $f_\alpha(X)$ as an *operator* on $\mathbf{A}_v$-valued power series in U as follows: Let D be $\frac{d}{dU}$. Then

$$\frac{X^j}{j!}\cdot U^t = \binom{t}{j}\cdot U^{t-j} = \frac{D^j}{j!}\cdot U^t.$$

Thus $g(\alpha,y)$ is the constant term of the power series

$$f_\alpha(X)(1+U)^y.$$

For more along these lines, see **[G5]**, **[G6]**.

Let x_v now be an element of $\mathbf{k}_v^*$ with $|x_v|_v > 1$. We form the measure

$$\mu_\zeta(x_v) = \sum_{\substack{n\in\mathbf{A}\\ n \text{ monic}}} x_v^{-deg(n)}\delta_n.$$

It is clear that the moments of this measure are given by

$$m_j(\mu_\zeta) = z(x_v,-j),$$

for all j. We then have the following remarkable result **[Th3]** which can be established by techniques similar to those in this section:

5.8.15. Theorem (Thakur). *Let* $f_{\mu_\zeta}(X) = \sum m(x_v,-i)\frac{X^i}{i!}$ *be the divided power series associated to* mu_ζ. *We then have*

a) $m(x_v,-i)\neq 0$ *if and only if* i *is a magic number.*

b) *Let* $i=i(c,j)>0$ *be a magic number. The we have*

$$m(x_v,-i) = \begin{cases} (-1)^j x_v^{-j} & \text{if } i \text{ is } \mathbf{A}\text{-odd} \\ (-1)^j(1-x_v^{-1})x_v^{-j} & \text{if } i \text{ is } \mathbf{A}\text{-even.}\end{cases}$$

c) $1 = m(x_v,0)$. □

Notice that the computations given in the theorem are invariant of v (upon substituting x for x_v). This is analogous to the case of Riemann's zeta function as in 5.8.1.

Therefore we see that the magic numbers have come up yet again in *another* situation. With Remarks 5.8.1 in mind, it may be that an understanding of this last occurrence will be the key towards an understanding of the others.

Finally, one can now "analytically continue" the measures $\mu_\zeta(x_v)$ to a neighborhood of $x_v=1$. (This can also be seen by an explicit computation of the measures of the basic open subsets $\alpha+(v^n)$.) We obtain a measure, μ_ζ, which has as its moments $\{\zeta(-t)\}$.

Now let $D=\frac{d}{dx_v}$. Upon applying D to $\mu_\zeta(x_v)$, and setting $x_v=1$, we obtain a measure, μ'_ζ whose moments are $\{D\,z(x_v,-t)|_{x_v=1}\}$. Thus the moments of $\mu_\zeta * \mu'_\zeta$ are precisely $\{\beta(t)\}$. This gives another construction of the "beta measure" of **[Th3]**.

References

[A1] G. Anderson: t-motives, *Duke Math. J.* **53**(2), (1986), 457–502.

[A2] G. Anderson: Notes on t-modules, Lectures at IAS – 1987.

[A-T1] G. Anderson and D. Thakur: Tensor Powers of the Carlitz module and zeta values, *Ann. of Math.* **132**, (1990), 159–191.

[C1] L. Carlitz: On Certain Functions connected with polynomials in a Galois field, *Duke Math. J.* **1**, (1935), 137–168.

[C2] L. Carlitz: An analogue of Bernoulli polynomials, *Duke Math. J.* **8**, (1941), 405–412.

[C-R1] C. Curtis and I. Reiner: *Representation theory of finite groups and associative algebras*, New York: Interscience (1962).

[D1] P. Deligne: Letter to the author dated June 5, 1990.

[Dr1] V. G. Drinfeld: Elliptic Modules, (Russian) *Mat. Sb.* **94**, (1974), 594–627; (English transl.) *Math. USSR-Sb.* **23**(4) (1976), 561–592.

[Dr2] V. G. Drinfeld: Elliptic Modules II, *Math. USSR-Sb.* **31**, (1977), 159–170.

[Ge1] E.-U. Gekeler: *Drinfeld Modular Curves*, Lecture Notes in Math.1231, Springer, Berlin-Heidelberg-New York (1986).

[Ge2] E.-U. Gekeler: On Power Sums of Polynomials over Finite Fields, *J. Number Theory* **30**, (1988), 11–26.

[Ge3] E.-U. Gekeler: Some New Identities for Bernoulli-Carlitz Numbers, *J. Number Theory* **33**, (1989), 209–219.

[Ge4] E.-U. Gekeler: On Regularity of Small Primes in Function Fields, *J. Number Theory* **34**, (1990), 114–127.

[Ge5] E.-U. Gekeler: De Rham cohomology for Drinfeld Modules. In: *Séminaire de Théorie des Nombres, Paris 1988-1989*, Birkhäuser, Boston-Basel-Berlin (1990), 57–85.

[Ge6] E.-U. Gekeler: On Finite Drinfeld Modules, *J. Algebra* **141**, (1991), 187–203.

[Ge7] E.-U. Gekeler: Letter to the author dated May 10, 1991.

[G1] D. Goss: The Algebraist's Upper Half-Plane, *Bull. Amer. Math. Soc.* **2**, (1980), 391–415.

[G2] D. Goss: The Arithmetic of Function Fields 2: The 'Cyclotomic' Theory, *J. Algebra* **81**(1), (March 1983), 107–149.

[G3] D. Goss: The Γ-function in the Arithmetic of Function Fields, *Duke Math. J.* **56**(1), (1988), 163–191.

[G4] D. Goss: Fourier Series, Measures and Divided Power Series in the Theory of Function Fields, *K-Theory* **1** (1989), 533–555.

[G5] D. Goss: A Formal Mellin Transform in the Theory of Function Fields, *Trans. Amer. Math. Soc.* **327**, (1991), 567–582.

[G6] D. Goss: Harmonic Analysis and the Flow of a Drinfeld Module, *J.Algebra* **146**(1), (1992), 219–241.

[G7] D. Goss: On a Fermat Equation arising in the Arithmetic Theory of Function Fields, *Math. Ann.* **261**, (1982), 269–286.

[G8] D. Goss: Analogies between Global Fields. In: *Conference Proceedings of the Canadian Mathematical Society*, Vol. 7, (1987), 83–114.

[G9] D. Goss: von Staudt for $\mathbb{F}_q[T]$, *Duke Math. J.* **45**, (1978), 885–910.

[G10] D. Goss: L-series of Grössencharakters of Type A_0 for Function Fields, (to appear in the *Proc. of the Special Sessions on Algebra and Number Theory*, AMS).

[G11] D. Goss: Some Integrals Attached to Modular Forms in the Theory of Function Fields, (this volume).

[G-S1] D. Goss and W. Sinnott: Class-groups of Function Fields, *Duke Math. J.* **52**(2), (1985), 507–516.

[H1] D. Hayes: Explicit Class Field Theory in Global Function Fields. In: G. C. Rota (ed.), *Studies in Algebra and Number Theory*, Academic Press, New York (1979), 173–217.

[H2] D. Hayes: Stickelberger Elements in Function Fields, *Compositio Math.* **55**, (1985), 209–239.

[Ha1] R. Hartshorne: *Algebraic Geometry*, Springer, Berlin-Heidelberg-New York (1977).

[L1] S. Lang: *Cyclotomic Fields*, Springer, Berlin-Heidelberg-New York (1978).

[Le1] H. Lee: Power sums of polynomials in a Galois field, *Duke Math. J.* **10**, (1943), 277–292.

[O1] S. Okada: Kummer's Theory for Function Fields, *J. Number Theory* **38**, (1991), 212–215.

[S1] J.-P. Serre: Zeta and L-functions, In: *Arithmetic Algebraic Geometry, Proc. of a conference held at Purdue Univ.*, Harper and Row, New York(1965), 82–92.

[S2] J.-P. Serre: Facteurs locaux des fonctions zêta des variéteés algébriques (Définitions et conjectures), *Séminaire Delange-Pisot-Poitou* (169/70), no. 19.

[S3] J.-P. Serre: Letter to the author dated April 14, 1990.

[S-D1] H.P.F. Swinnerton-Dyer: An Application of Computing to Class Field Theory. In: J.W.S. Cassels and A. Fröhlich (ed.), *Algebraic Number Theory*, Thompson, Washington D.C. (1967), 280–291.

[Ta1] Y. Taguchi: Semi-simplicity of the Galois representations attached to Drinfeld modules over fields of "infinite characteristics", *Journal of Number Theory*, (to appear).

[T1] T. Takahashi: Good reduction of Elliptic modules, *J. Math. Soc. Japan* **34**(3), (1982), 475–487.

[Te1] J. Teitelbaum: The Poisson Kernel for Drinfeld Modular Curves, *J. Amer. Math. Soc.* **4**, (1991), 491–511.

[Th1] D. Thakur: Gamma functions for function fields and Drinfeld modules, *Ann. of Math.* **134**, (1991), 25–64.

[Th2] D. Thakur: Gauss sums for $\mathbb{F}_q[T]$, *Invent. Math.* **94**, (1988), 105–112.

[Th3] D. Thakur: Zeta Measure Associated to $\mathbb{F}_q[T]$, *J. Number Theory* **35**, (1990), 1–17.

[Th4] D. Thakur: Number fields and function fields (zeta and gamma functions at all primes) In: N. De Grande-De Kimpe and L. Van Hamme (ed.), *Proceedings of a conference on p-adic analysis, Hengelhoef 1986*, Universiteit Brussel, Brussel (1986), 149–157.

[Th5] D. Thakur: Drinfeld Modules and Arithmetic of function fields, (preprint).

[Y1] J. Yu: Transcendence and Drinfeld Modules: *Invent. Math.* **83**, (1986), 507–517.

[Y2] J. Yu: Transcendence and special zeta values in characteristic p, *Ann. of Math.* **134**, (1991), 1–23.

Department of Mathematics, Ohio-State University, Columbus, Ohio 43210
Email: goss@coltrane.mps.ohio-state.edu

Classgroups of Sheaves of Locally Free Modules over Global Function Fields

R. J. Chapman

1. Introduction

In this paper we describe a new kind of classgroup, designed to study integral Galois module structure in global function fields. We define a *function field* K to be a finitely generated field extension, with transcendence degree one, of the finite field $k = \mathbb{F}_q$ where k is algebraically closed in K. The main difference between the number field case and the function field case is that whereas there is a canonically defined ring of integers in a given number field, there is no direct counterpart in a global function field. Given a finite non-empty set S of places of K we may define $\mathcal{O}_{K,S}$, the ring of S*-integers* to be the set of functions in K having no pole outside S. This is a Dedekind domain whose field of fractions is K. Given a Galois extension L of K with Galois group Γ we may study the structure of $\mathcal{O}_{L,S}$ as an $\mathcal{O}_{K,S}\Gamma$-module where $\mathcal{O}_{L,S}$ is the integral closure of $\mathcal{O}_{K,S}$ in L. This viewpoint has been developed by the author in his thesis, **[C1]** and his papers, **[C2]** and **[C3]**.

This method has the obvious drawback that it depends on the (quite arbitrary) choice of S. One way of avoiding this impasse is to let S vary, thereby getting a sheaf of rings. By algebraic geometry K is the function field of a unique, geometrically integral, projective, non-singular curve C over k which we consider as a scheme $(C, \underline{\mathcal{O}}_K)$ over $\mathrm{Spec}(k)$. The underlying space which we shall lazily identify with the curve C, comprises a generic point G and closed points in one-to-one correspondence with the places of K. The structure sheaf $\underline{\mathcal{O}}_K$ is the true analogue of a number field's ring of integers. If S is a finite non-empty set of places of K then $U = C - S$ is open and $\underline{\mathcal{O}}_K(U) = \mathcal{O}_{K,S}$. Given a finite Galois extension L of K then we define a sheaf of rings $\underline{\mathcal{O}}_{L/K}$ on C by letting $\underline{\mathcal{O}}_{L/K}(U)$ be the integral closure of $\underline{\mathcal{O}}_K(U)$ in L. The sheaf $\underline{\mathcal{O}}_{L/K}$ is a $\underline{\mathcal{O}}_K$-algebra; further if $\Gamma = \mathrm{Gal}(L/K)$ then $\underline{\mathcal{O}}_{L/K}$ is a module for the $\underline{\mathcal{O}}_K$-algebra $\underline{\mathcal{O}}_K\Gamma$ which is defined in the obvious way. Note that all sheaves mentioned here are on the same topological space C.

If the order of Γ is coprime to $q = |k|$ then the extension L is at most tamely ramified everywhere and by standard theory the sheaf $\underline{\mathcal{O}}_{L/K}$ is a *locally free* $\underline{\mathcal{O}}_K\Gamma$-module. We develop here a classgroup theory of such modules in terms of a "Hom-description" à la Fröhlich, **[F]**, and describe the relation between this description and the classgroup theory of rings such as $\mathcal{O}_{K,S}$ as developed in **[C1]** and **[C3]**. We find that

$$\mathrm{Cl}(\underline{\mathcal{O}}_K\Gamma) \cong \mathrm{Hom}_\Omega(R_\Gamma, \mathrm{Pic}(K')) \tag{1}$$

where $\Omega = \mathrm{Gal}(\bar{k}/k)$ is the absolute Galois group of k, R_Γ is the additive group of virtual representations of Γ over $\bar{k}$ and $\mathrm{Pic}(K')$ is the Picard group of the curve C considered over $\bar{k}$. We shall give some examples of explicit computations of the class of $\underline{\mathcal{O}}_{L/K}$ in $\mathrm{Cl}(\underline{\mathcal{O}}_K\Gamma)$.

2. Classgroups and Sheaves

In this section we develop a theory of classgroups for locally free sheaves. We start by adopting some conventions and gathering some terminology. Let K be a global function field with constants $k = \mathbb{F}_q$ with q a power of the prime p. Our convention will be that a *place* of K will be an additive valuation with value group $\mathbb{Z}$. To such a field is associated a unique projective non-singular curve C defined over k. We identify C with the associated scheme over $\mathrm{Spec}(k)$ and denote the underlying topological space of this scheme by C, and its structure sheaf by $\underline{\mathcal{O}}_K$. The set $C = \{G\} \cup \{P_v \mid v \text{ a place of } K\}$ and the topology of C is defined by taking the cofinite subsets containing G to be the non-empty open sets. The point G is the generic point of C and the P_v are closed points, which we shall usually identify with the places of K. We define the structure sheaf by setting $\underline{\mathcal{O}}_K(C - S) = \mathcal{O}_{K,S}$, the set of S-integers of K, where S is a finite set of places of K. The stalk at G, $\underline{\mathcal{O}}_{K,G}$, equals K and at a closed point P_v, $\underline{\mathcal{O}}_{K,P_v} = \mathcal{O}_{K,v} = \{x \in K \mid v(x) \geq 0\}$, the (uncompleted) *local ring* at v. Let Γ be a finite group. We define the sheaf of rings $\underline{\mathcal{O}}_K\Gamma$ to be the sheaf with sections $\underline{\mathcal{O}}_K\Gamma(U) = \underline{\mathcal{O}}_K(U)\Gamma$, for all open U, with the obvious restriction maps. We also let a sheaf of (right) $\underline{\mathcal{O}}_K\Gamma$-modules $\mathcal{M}$ be *locally free of rank* r if the stalks $\mathcal{M}_P$ are free $\underline{\mathcal{O}}_{K,P}\Gamma$-modules of rank r. Let $\bar{k}$ be the algebraic closure of k and $\Omega = \mathrm{Gal}(\bar{k}/k)$. We denote by $F : u \mapsto u^q$, the *Frobenius*, which generates Ω topologically. Put $K' = K \otimes_k \bar{k}$ so that K'/K is (infinite) Galois with group Ω. Let $\mathrm{Pic}(K)$ (resp. $\mathrm{Pic}(K')$) denote the Picard group of the field K (resp. K'), that is $\mathrm{Pic}(K) = \mathrm{Div}(K)/\mathrm{Pr}(K)$, where the *divisor group* $\mathrm{Div}(K)$ is the set of $\mathbb{Z}$-linear combinations of places of K and $\mathrm{Pr}(K)$ is the set of principal divisors. We denote by $\mathrm{Div}_S(K)$ the group of divisors of K supported outside S, and similarly for $\mathrm{Pr}_S(K)$ and $\mathrm{Pic}_S(K)$. The group $\mathrm{Pic}(K')$ is naturally an Ω-module. Assume from now on that p does not divide the order of Γ. Let R_Γ be the group of virtual linear representations of Γ over $\bar{k}$. Again this is an Ω-module. If $x \in K'$ denote by $(x) \in \mathrm{Div}(K')$ its divisor, and if S is a finite set of closed points of C denote by $(x)_S$ the divisor obtained from (x) by "forgetting" the part supported above S. Let θ (resp. θ_S) be the map from $\mathrm{Hom}_\Omega(R_\Gamma, K'^*)$ to $\mathrm{Hom}_\Omega(R_\Gamma, \mathrm{Div}(K'))$ (resp. $\mathrm{Hom}_\Omega(R_\Gamma, \mathrm{Div}_S(K'))$) induced by $x \mapsto (x)$ (resp. $x \mapsto (x)_S$).

If S is nonempty then the Hom-description of Fröhlich, **[F]**, applied to the ring $\mathcal{O}_{K,S}\Gamma$ yields

$$\mathrm{Cl}(\mathcal{O}_{K,S}\Gamma) \cong \frac{\mathrm{Hom}_\Omega(R_\Gamma, \mathrm{Div}_S(K'))}{\theta_S(\mathrm{Hom}_\Omega(R_\Gamma, K'^*))} \tag{2}$$

where $\mathrm{Cl}(\mathcal{O}_{K,S}\Gamma)$ is the usual classgroup (Grothendieck group of locally free modules factored out by free modules). Given a locally free $\underline{\mathcal{O}}_K\Gamma$-module $\mathcal{M}$ we shall define its *class* $[\mathcal{M}]$ to be an element of the *classgroup*

$$\mathrm{Cl}(\underline{\mathcal{O}}_K\Gamma) = \frac{\mathrm{Hom}_\Omega(R_\Gamma, \mathrm{Div}(K'))}{\theta(\mathrm{Hom}_\Omega(R_\Gamma, K'^*))} \tag{3}$$

and we shall describe how $[\mathcal{M}]$ determines the class, $[\mathcal{M}(C-S)]$, of the $\mathcal{O}_{K,S}\Gamma$-module $\mathcal{M}(C-S)$, in $\mathrm{Cl}(\mathcal{O}_{K,S}\Gamma)$.

Theorem 1. *The natural surjection* $\mathrm{Div}(K') \to \mathrm{Pic}(K')$ *induces an isomorphism*

$$\mathrm{Cl}(\underline{\mathcal{O}}_K\Gamma) = \frac{\mathrm{Hom}_\Omega(R_\Gamma, \mathrm{Div}(K'))}{\theta(\mathrm{Hom}_\Omega(R_\Gamma, K'^*))} \cong \mathrm{Hom}_\Omega(R_\Gamma, \mathrm{Pic}(K')). \tag{4}$$

Proof. By writing R_Γ in the form $\bigoplus_i A_i$ where each A_i is a free Abelian group generated by an Ω-orbit of irreducible representations, the theorem reduces to the statement that $\mathrm{Pic}(K'^\Delta) \cong \mathrm{Pic}(K')^\Delta$ where Δ is an open subgroup of Ω. For convenience assume that $\Delta = \Omega$, if not then we just need to introduce new notation.

We have the long exact sequence of Abelian groups

$$1 \to \bar{k}^* \to K'^* \to \mathrm{Div}(K') \to \mathrm{Pic}(K') \to 0 \tag{5}$$

with a discrete Ω-action. We split this into two short exact sequences

$$1 \to \bar{k}^* \to K'^* \to A \to 1 \tag{6}$$

and

$$1 \to A \to \mathrm{Div}(K') \to \mathrm{Pic}(K') \to 0 \tag{7}$$

and apply the long exact sequence of cohomology. By Hilbert's theorem 90, both $H^1(\Omega, \bar{k}^*)$ and $H^1(\Omega, K'^*)$ vanish and we get the exact sequences

$$1 \to k^* \to K^* \to A^\Omega \to 1 \tag{8}$$

and

$$1 \to H^1(\Omega, A) \to H^2(\Omega, \bar{k}^*). \tag{9}$$

But $H^2(\Omega, \bar{k}^*)$ is the Brauer group of the finite field k, which vanishes due to Wedderburns's theorem on the non-existence of finite non-commutative division rings. Hence $H^1(\Omega, A)$ vanishes and (7) gives the short exact sequence

$$1 \to A^\Omega \to \mathrm{Div}(K) \to \mathrm{Pic}(K')^\Omega \to 0 \tag{10}$$

and so also the exact sequence

$$1 \to k \to K^* \to \mathrm{Div}(K) \to \mathrm{Pic}(K')^\Omega \to 0. \tag{11}$$

But we already have the exact sequence

$$1 \to k \to K^* \to \mathrm{Div}(K) \to \mathrm{Pic}(K) \to 0 \tag{12}$$

and so $\mathrm{Pic}(K) \cong \mathrm{Pic}(K')^\Omega$, as required. □

Let $\mathcal{M}$ be as above. For convenience we shall describe $[\mathcal{M}]$ in detail only for $r = 1$. By hypothesis the stalks $\mathcal{M}_P$ of $\mathcal{M}$ are free $\mathcal{O}_{K,P}\Gamma$-modules with bases m_P, say. Now the closed points P_v of C are all specializations of the unique generic point G so that for any sheaf $\mathcal{S}$ on C there is a natural map from the stalk $\mathcal{S}_{P_v}$ to the generic stalk $\mathcal{S}_G$. It is easy to see that, applied to $\mathcal{M}$, this makes $\mathcal{M}_{P_v}$ an $\mathcal{O}_{K,P_v}\Gamma$-lattice in the free $K\Gamma$-module $\mathcal{M}_G$. We may thus write $m_{P_v} = m_G \cdot x_v$ for a unique $x_v \in K\Gamma$. Now let $\chi : \Gamma \to \mathrm{GL}_s(\bar{k})$ be a representation of Γ. We can extend χ to a K-algebra homomorphism $\chi : K\Gamma \to \mathrm{M}_s(K')$ and we let $y_{v,\chi} = \mathrm{Det}(\chi(x_v)) \in K'$. Let $v_1, v_2, \ldots, v_t$ be the distinct liftings of the place v to K' and let $n_{v_i,\chi} = v_i(y_{v,\chi})$. Let

$$\mathcal{D}_\chi = \sum_{v'} n_{v',\chi} v' \tag{13}$$

where the summation is over all places v' of K'.

Lemma 1. *We have $\mathcal{D}_\chi \in \mathrm{Div}(K')$, in other words $n_{v',\chi} = 0$ for all but finitely many v'.*

Proof. Let $U = C - S$ be a proper open subset of C and let $\mathcal{O} = \underline{\mathcal{O}}_K(U) = \mathcal{O}_{K,S}$. Now $\mathcal{M}(U)$ is an $\mathcal{O}$-lattice spanning the $K\Gamma$-module $\mathcal{M}_G$. Similarly $m_G \cdot \mathcal{O}\Gamma$ is also such a lattice. Hence at all but finitely many places these coincide so that with finitely many exceptions we have $\mathcal{M}_{P_v} = m \cdot \mathcal{O}_{K,P_v}$. Hence for these v we have that $x_v \in (\mathcal{O}_{K,P_v})^*$ and so $y_{v,\chi}$ is a unit at v and so $n_{v',\chi} = 0$ for all v' above v, as required. □

Lemma 2. *The image $[\mathcal{D}_\chi]$ of $\mathcal{D}_\chi$ in $\mathrm{Pic}(K')$ is independent of the choice of the m_P.*

Proof. First we note that changing one of the m_{P_v} alters x_v by multiplying it by an element of $(\mathcal{O}_{K,P_v}\Gamma)^*$ and so alters $y_{v,\chi}$ by multiplication by an element of K' whose valuations at the v' above v are all trivial. Hence changing the m_{P_v} leaves the divisors $\mathcal{D}_\chi$ invariant. However changing m_G involves multiplying all the x_v by the same element $a \in (K\Gamma)^*$, and so (for fixed χ) multiplying all the $y_{v,\chi}$ by the same element $b_\chi = \mathrm{Det}(\chi(a)) \in K'^*$. This clearly alters $\mathcal{D}_\chi$ by addition of the principal divisor corresponding to b_χ so that the class $[\mathcal{D}_\chi] \in \mathrm{Pic}(K')$ remains invariant. □

Proposition 1. *The map $[\mathcal{M}] : \chi \mapsto [\mathcal{D}_\chi]$ induces an element $[\mathcal{M}] \in \mathrm{Cl}(\underline{\mathcal{O}}_K\Gamma) \cong \mathrm{Hom}_\Omega(R_\Gamma, \mathrm{Pic}(K'))$.*

Proof. It is easy to see that $[\mathcal{D}_\chi]$ only depends on the isomorphism class of the representation χ. Also it is plain (by writing the matrix representation as a diagonal sum) that $\mathcal{D}_{\chi\oplus\psi} = \mathcal{D}_\chi + \mathcal{D}_\psi$ for all χ, ψ. Hence the map $[\mathcal{M}]$ lifts to a group homomorphism from R_Γ to $\mathrm{Pic}(K')$. Let $\omega \in \Omega$. By definition of the action on places of K' we have $v'^\omega(a^\omega) = v'(a)$ for $a \in K'$ and so

$$\begin{aligned} n_{v'^\omega,\chi^\omega} &= v'^\omega(\chi^\omega(x_v)) \\ &= v'^\omega(\chi(x_v)^\omega) \\ &= v'(\chi(x_v)) \\ &= n_{v',\chi} \end{aligned} \tag{14}$$

from which it is plain that $\mathcal{D}_{\chi^\omega} = \mathcal{D}_\chi{}^\omega$ as required. □

We can use a similar technique to determine the class in $\mathrm{Cl}(\mathcal{O}_{K,S}\Gamma)$ of a locally free module over $\mathcal{O}_{K,S}\Gamma$. If M is a locally free module of rank one over $\mathcal{O}_{K,S}\Gamma$ then we can do exactly the same as above to define a map $\chi \mapsto [\mathcal{D}_\chi]$ where now $\mathcal{D}_\chi \in \mathrm{Div}_S(K')$. The map $\chi \mapsto \mathcal{D}_\chi$ induces an element $[M] \in \mathrm{Cl}(\mathcal{O}_{K,S}\Gamma) \cong \mathrm{Hom}_\Omega(R_\Gamma, \mathrm{Pic}_S(K'))/\theta_S(\mathrm{Hom}_\Omega(R_\Gamma, K'^*))$. Clearly the natural map $\mathrm{Div}(K') \to \mathrm{Div}_S(K')$ induces a map $\varphi : \mathrm{Cl}(\underline{\mathcal{O}}_K\Gamma) \to \mathrm{Cl}(\mathcal{O}_{K,S}\Gamma)$ and the proof of the following lemma is trivial.

Lemma 3. *If $\mathcal{M}$ is a locally free $\underline{\mathcal{O}}_K$-module and $U = C - S$ then $\varphi([\mathcal{M}]) = [\mathcal{M}(U)] \in \mathrm{Cl}(\mathcal{O}_{K,S}\Gamma)$.* □

Hence $[\mathcal{M}]$ contains information about Galois module structure over all possible "rings of integers" in K.

Example 1. Let $K = k(T)$ be a rational function field. Then as the principal divisors of K' are precisely those of degree zero, $\mathrm{Pic}(K) \cong \mathbb{Z}$, the isomorphism being given by taking the degree of a divisor and so

$$\mathrm{Cl}(\underline{\mathcal{O}}_K\Gamma) \cong \mathrm{Hom}_\Omega(R_\Gamma, \mathbb{Z}). \tag{15}$$

Hence if $\mathcal{M}$ is a locally free module over $\underline{\mathcal{O}}_K$, $[\mathcal{M}]$ is given by an Ω-equivariant integer-valued function on the irreducible $\bar{k}$-representations of Γ.

Suppose $S = \{v\}$ be a place of degree d of K. Then $R_\Gamma = \bigoplus_i A_i$ where $A_i = \bigoplus_{j=0}^{r_i-1} \mathbb{Z}\chi_i^{F^j}$. Here χ_i ranges over all Ω-orbits of irreducible representations of Γ, and χ_i has stabilizer of index r_i in Ω. Then we see that

$$\mathrm{Hom}_\Omega(R_\Gamma, \mathrm{Div}_S(K')) \cong \bigoplus_i \mathrm{Div}_S(K_{r_i}) \tag{16}$$

where K_s denotes the constant field extension of degree s of K, and

$$\mathrm{Hom}_\Omega(R_\Gamma, K'^*) \cong \bigoplus_i K_{r_i}^*. \tag{17}$$

It follows that

$$\mathrm{Cl}(\mathcal{O}_{K,S}\Gamma) \cong \bigoplus_i \frac{\mathrm{Div}_S(K_{r_i})}{\mathrm{Pr}_S(K_{r_i})} \cong \bigoplus_i \mathbb{Z}/\gcd(d, r_i)\mathbb{Z} \tag{18}$$

by an easy calculation. It is now apparent that if $[\mathcal{M}(C-S)] : \chi_i \mapsto n_i \in \mathbb{Z}/\gcd(d, r_i)\mathbb{Z}$ then $n_i \equiv [\mathcal{M}](\chi_i) \pmod{\gcd(d, r_i)}$.

Example 2. Let K be the function field of an elliptic curve E defined over k, with basepoint O. We have

$$\mathbb{Z} \oplus \mathrm{Pic}^0(K') \cong \mathrm{Pic}(K'), \tag{19}$$

where $\mathrm{Pic}^0(K')$ is the group of degree zero divisors factored out by the principal divisors, by taking $(n, [\mathcal{D}])$, where $\mathcal{D}$ is a degree zero divisor, to the class of the divisor $nO + \mathcal{D}$. The group $\mathrm{Pic}^0(K')$ is naturally isomorphic to $E(K')$, the group of points of E defined over K'.

If we let $S = \{P\}$ where P is a point in $E(K)$ then $\mathcal{O}_{K,S}$ is isomorphic to $\mathcal{O}$, the ring of K-valued functions on E regular outside O, by "translating" by P on the elliptic curve E. If $[\mathcal{M}] : \chi \to (n_\chi, Q_\chi)$, using the isomorphism (19), where $Q_\chi \in E(K')$ then the class $[\mathcal{M}(C - \{O\})]$ is represented by $\chi \mapsto Q_\chi$. Now as $n_\chi O = n_\chi P + n_\chi(O - P)$ then the class $[\mathcal{M}(C - S)]$ is represented by $\chi \mapsto Q_\chi - n_\chi P$. In short, then, the Q_χ determine the structure of $\mathcal{M}(C - \{O\})$, while the n_χ measure the change needed to go from $\mathcal{M}(C - \{O\})$ to $\mathcal{M}(C - S)$.

Let L be a finite extension of K. The field L has associated to it a k-scheme C' and a morphism $\pi : C' \to C$. The direct image of the structure sheaf of C' is a sheaf on C which we shall denote $\underline{\mathcal{O}}_{L/K}$. If U is open then $\underline{\mathcal{O}}_{L/K}(U)$ is the integral closure of $\underline{\mathcal{O}}_K(U)$ in L. It is plain that $\underline{\mathcal{O}}_{L/K}$ is an $\underline{\mathcal{O}}_K\Gamma$-module whenever L/K is Galois with group Γ. In fact we have the following.

Proposition 2. *If the order of* Γ *is coprime to* q*, then* $\underline{\mathcal{O}}_{L/K}$ *is a locally free* $\underline{\mathcal{O}}_K\Gamma$*-module.*

Proof. The hypothesis implies that the extension L/K is at most tamely ramified at all places. Now the stalk $\underline{\mathcal{O}}_{L,G}$ is isomorphic to L which is free over $\underline{\mathcal{O}}_{K,G}\Gamma = K\Gamma$ by the normal basis theorem. Also the stalk $\underline{\mathcal{O}}_{L,P_v} = \mathcal{O}_{L,v}$ the semi-local ring of L at v. (This is naturally the product of the local rings of L at the places of L above v.) By Noether's criterion, **[F**, Theorem 3, §I.3, pp. 26-27**]**, this stalk is free of rank one over $\underline{\mathcal{O}}_{K,P_v}\Gamma = \mathcal{O}_{K,v}\Gamma$ as required. (Note that actually Fröhlich proves that the *completion* $\hat{\mathcal{O}}_{L,v}$ is free over $\hat{\mathcal{O}}_{K,v}\Gamma$ but this is equivalent by Theorem 18.2 in **[R**, pp. 175-176**]**.) □

To facilitate the computation of $[\underline{\mathcal{O}}_{L/K}]$ we introduce the technique of Galois resolvents. For convenience assume that L/K is a *geometric* extension, that is L and K' are linearly disjoint over K. In this case $L' = L \otimes_k \bar{k}$ is a field (in general it is a product of fields). Let $\chi : \Gamma \to \mathrm{GL}_s(\bar{k})$ be a matrix representation and choose free generators a_P for each stalk $\underline{\mathcal{O}}_{L,P}$ as $\underline{\mathcal{O}}_{K,P}\Gamma$-modules. The field L' contains the values of the entries of $\chi(\Gamma)$ and the various a_P. For each $b \in L$ we define the *Galois resolvent* of b with respect to χ to be the value

$$(b \mid \chi) = \det\left(\sum_{\gamma \in \Gamma} \chi(\gamma^{-1}) b^\gamma\right) \in L'. \tag{20}$$

Lemma 4. *If* $L = b \cdot K\Gamma$ *then* $(b \mid \chi) \neq 0$.

Proof. This is the corollary to Proposition 4.1 in **[F**, §I.4, pp. 28-29**]**. □

Now put $b = a_G$, and $a_v = a_{P_v}$ and suppose $a_v = b \cdot x_v$ for $x_v \in K\Gamma$. We then have $(a_v \mid \chi) = y_{v,\chi}(b \mid \chi)$, either by a simple computation or from the corollary to Proposition 4.2 in **[F**, §I.4, pp. 29-30**]**, where as before $y_{v,\chi} = \det(\chi(x_v))$. Hence we have

$$y_{v,\chi} = \frac{(a_v \mid \chi)}{(b \mid \chi)} \in K'^* \tag{21}$$

and so $\mathcal{D}_\chi = \mathcal{D}'_\chi - \mathcal{D}''_\chi$ where

$$\mathcal{D}'_\chi = \sum_{v'} w((a_v \mid \chi))v' \tag{22}$$

and

$$\mathcal{D}''_\chi = \sum_{v'} w((b \mid \chi))v' \tag{23}$$

where the summations are over places v' in K' and where v' lies above the place v of K and w is any prolongation of the place v' to L' (in the sense that w restricted to K' is v' but the value group of w may strictly contain $\mathbb{Z}$). In practice we may not know the a_v and b but we may know $a'_{v'}$ and b' where $\mathcal{O}_{L',v'} = a'_{v'} \cdot \mathcal{O}_{K',v'}\Gamma$ and $L' = b' \cdot K'\Gamma$.

Lemma 5. *We have*

$$\mathcal{D}'_\chi = \sum_{v'} w((a'_{v'} \mid \chi))v' \tag{24}$$

and

$$\mathcal{D}''_\chi = \mathcal{D} + \sum_{v'} w((b' \mid \chi))v' \tag{25}$$

for some principal divisor $\mathcal{D} \in \mathrm{Div}(K')$.

Proof. This essentially follows from the corollary to Proposition 4.2 in **[F**, p. 30**]** and the observation that $\mathcal{O}_{L',v'} = a_v \cdot \mathcal{O}_{K',v'}\Gamma$ and $L' = b \cdot K'\Gamma$. □

We end this section with a useful local result.

Lemma 6. *If the extension* L'/K' *is unramified at* v' *then*

$$v'((a'_{v'} \mid \chi)) = 0 \tag{26}$$

for all χ.

Proof. This is an immediate consequence of Proposition 4.3 in **[F**, §I.4, p. 30**]**. □

3. Some Computations

We shall briefly describe how we may apply the above techniques to compute $[\underline{\mathcal{O}}_{L/K}]$ for the Abelian extensions L/K of function fields dealt with in §5 of **[C3]**.

Let K be the field of functions on an algebraic curve C, defined over k, and let $S_0 = \{P_\infty\}$ where P_∞ is a k-rational point of C. Let $\mathcal{O} = \mathcal{O}_{K,S_0}$. Let h be the class number of $\mathcal{O}$. If $\mathcal{A}$ is an ideal of $\mathcal{O}$ we define $K^{(\mathcal{A})}$ to be the largest Abelian extension of K totally split over P_∞ and over all prime ideals $\wp = \pi\mathcal{O}$ where $\pi \equiv 1 \bmod \mathcal{A}$. We let $H = K^{(\mathcal{O})}$.

Suppose $\wp$ is a prime ideal of $\mathcal{O}$ of degree d. Then $\wp$ corresponds to an Ω-orbit $\{P_1, P_2, \ldots, P_d\}$ of geometric points of C. We may assume that $P_i^F = P_{i+1}$. Let $m = (q^d - 1)/(q-1)$ and suppose that h and m are coprime. Then the extension $K^{(\wp)}$ can be uniquely written as a compositum $L \cdot H$ over K. The degree of L/K is m and by class field theory there is a natural isomorphism $(\mathcal{O}/\wp)^*/k^* \cong \mathrm{Gal}(L/K) = \Gamma$. In **[C2, §5]** the class $[\mathcal{O}_{L,S_0}] \in \mathrm{Cl}(\mathcal{O}_{K,S_0}\Gamma)$ is determined, and we shall describe the class $[\underline{\mathcal{O}}_{L/K}] \in \mathrm{Cl}(\underline{\mathcal{O}}_K\Gamma)$.

The group $\mathrm{Pic}(K')$ is isomorphic to $\mathbb{Z}\oplus\mathrm{Cl}(\mathcal{O}')$, where $\mathcal{O}' = \mathcal{O}_{K',S_0}$, where the first factor corresponds to the divisors of the form nP_∞ and the latter to the group of divisor classes of degree zero. Let $\wp_i$ denote the ideal of $\mathcal{O}'$ corresponding to the divisor P_i of K'. In **[C3]** we single out a generator χ of $\hat{\Gamma}$, characterized by the reciprocity law of class field theory. The class $[\underline{\mathcal{O}}_{L/K}]$ is given by an Ω-equivariant map $\chi^j \to (n_j, [\mathcal{A}_j])$ $(0 \le j < m)$ where $n_j \in \mathbb{Z}$ and the $\mathcal{A}_j$ are ideals in $\mathcal{O}'$. The $\mathcal{A}_j$ are characterized in **[C3, Theorem 4]**, so it only remains to find the n_χ.

Theorem 2. *With the above hypotheses and notation we have*

$$n_j = \sum_{i=0}^{d-1} \left\langle \frac{q^i j}{m} \right\rangle \tag{27}$$

where $\langle x \rangle \in [0,1)$ denotes the fractional part of a real number x.

Proof. Now n_j is the degree of the divisor $\mathcal{D}_{\chi^j}$ of Proposition 1. By Lemma 5 $\mathcal{D}_{\chi^j} = \mathcal{D}'_{\chi^j} - \mathcal{D}''_{\chi^j}$ and the degree of $\mathcal{D}''_{\chi^j}$ vanishes as it is essentially a principal divisor from an extension field of K. Hence

$$n_j = \deg(\mathcal{D}'_{\chi^j}) = \sum_{v'} w((a'_{v'} \mid \chi^j)) \tag{28}$$

in the notation of §2. By Lemma 6, as L/K is unramified except above $\wp$ then

$$n_j = \sum_{i=1}^{d} w_i((a'_{v_i} \mid \chi^j)) \tag{29}$$

where $v_i = v_{\wp_i}$ and w_i is a prolongation of v_i to $L' = K' \cdot L$. The proof of Theorem 3 of **[C3]** gives $w_i((a'_{v_i} \mid \chi^j)) = \langle jr_i/m \rangle$ where r_i is an integer depending on i. But by the choice of the character χ in §5 of **[C3]** we have $r_i = q^{d-i}$. The result is now immediate. □

Example. Let $K = k(T)$ be a rational function field, and let $P \in k[T]$ be an irreducible polynomial. Let L be the maximal real subfield of the cyclotomic function field of conductor P as described in **[GR, §2]**. The extension L/K is an example of the extensions considered in the previous theorem, and we can identify the character χ by means of the explicit class field theory of **[H]**.

Let $f \mapsto [f]$ denote the Carlitz module, and let M denote the cyclotomic function field of conductor P. Then $M = K(\Lambda_P)$ where Λ_P is the set of zeros of $[P]$. The action of $k[T]$ on Λ_P via the Carlitz module yields isomorphisms

$$\mathrm{Gal}(M/K) \cong (k[T]/Pk[T])^* \tag{30}$$

and

$$\mathrm{Gal}(L/K) \cong (k[T]/Pk[T])^*/k^*. \tag{31}$$

In this case it can be shown that the character χ, considered as a function on

$$(k[T]/Pk[T])^*/k^*,$$

takes an element α to $\eta(\alpha)^{q-1}$, where $\eta : k[T]/Pk[T] \to \bar{k}$ is a k-embedding of fields.

Bibliography

[C1] Chapman, R. J.: Galois Module Structure in Global Function Fields, Ph.D. Thesis, Faculty of Technology, University of Manchester, 1988.

[C2] Chapman, R. J.: Carlitz modules and normal integral bases, *J. London Math. Soc.* (to appear).

[C3] Chapman, R. J.: Kummer theory and Galois module structure in global function fields, *Math. Z.* (to appear).

[F] Fröhlich, A.: *Galois Module Structure of Algebraic Integers*, Springer-Verlag, New York, 1983.

[GR] Galovich, S. and Rosen, M.: The class number of cyclotomic function fields, *J. Number Theory* **13** (1981), 363–375.

[H] Hayes, D. R.: Explicit class field theory for rational function fields, *Trans. Amer. Math. Soc.* **189** (1974), 77–91.

[R] Reiner, I.: *Maximal Orders*, Academic Press, London, 1975.

R. J. Chapman, Department of Mathematics, University of Exeter, North Park Road, Exeter, EX4 4QE, U. K.
Email: R.J.Chapman@cen.exeter.ac.uk

Artin-Schreier-Witt Extensions as Limits of Kummer-Lubin-Tate Extensions, and the Explicit Reciprocity Law

E. de Shalit

Introduction

Kazhdan introduced, in the representation theory of reductive groups, the principle saying that the theory over a local field of characteristic $p > 0$ is the "limit" of the theories over local fields in characteristic 0, when their absolute index of ramification tends to infinity.

On the other hand, many years earlier Krasner **[Kr]** had realized that a similar principle holds for representations of the absolute Galois groups of the local field. In **[De]**, Deligne made Krasner's work precise, and in what follows we adopt Deligne's formalism. A review of his results is given in §1.

The formulas giving the effect of the local Artin symbol of Kummer extensions (or, more generally, Kummer-Lubin-Tate extensions) of local fields in characteristic 0, go under the name "explicit reciprocity laws" (ERL). See **[Ar-Ha]** (also discussed in **[Ar-Ta]**), **[Iw]**, **[Co]**, **[Wiles]**, **[Vo]**, **[Sen]**, **[dS]**, a list that is far from being complete. The proofs of these ERL's are rather laborious, and do not shed much light on the meaning of the final result. In the function field case, there are formulas of Schmid and Witt **[Sch]**, **[W]** (see also **[Se1]**, chap. XIV §5) for the effect of the local Artin symbol on Artin-Schreier-Witt extensions of the local field. In contrast to the number field case, they are easy to prove. Serre supplied a global geometric proof in **[Se2]**, Chap. VI, §30.

It is a remarkable fact that in both cases, the main term in the formulas is a pairing of the form $\int f \cdot d\log(g)$ (common in the study of Riemann surfaces too). The purpose of this note is to convince the reader that this is more than an analogy. In fact, we show that the original ERL of Artin and Hasse can be derived, using Deligne's results, from Schmid's formula in characteristic p.

The author learned that similar results were obtained by R. Coleman (in an unpublished letter to Wintenberger). The generalizations of the results proven here to extensions of degree larger than p, and thereby to the ERL's of Iwasawa and Wiles, are currently being investigated by E. Goren.

Throughout this work p is a prime bigger than 2. If K is a local field, its ring of integers and maximal ideal are denoted by $\mathcal{O}_K$ or $\mathcal{O}(K)$, and $\mathfrak{p}_K$ or $\mathfrak{p}(K)$ respectively. If $e \geq 1$, we let $U_e(K) = 1 + \mathfrak{p}_K^e$. The normalized valuation of K is denoted v_K. The letters F, E are reserved for local fields in char p, and their elements are denoted by Roman characters. The letters K, L are reserved for local fields in char 0, and their elements are denoted by Greek characters.

The assumption $p \neq 2$ is not essential, but the prime 2 calls for some modifications in the proof. Since our main purpose was to relate explicit reciprocity laws in different characteristics, rather than to prove a new law, we allowed ourselves to ignore $p = 2$.

§1. Review of Deligne's work

1.1. Let K be a local field in characteristic p or 0, with residual characteristic p. With every integer $e \geq 1$, Deligne attaches to K its eth truncation $Tr_e(K)$, which is the triple $(\mathcal{O}_K/\mathfrak{p}_K^e, \mathfrak{p}_K/\mathfrak{p}_K^{e+1}, \varepsilon_K)$, ε_K being the canonical map of $\mathfrak{p}_K/\mathfrak{p}_K^{e+1}$ to $\mathcal{O}_K/\mathfrak{p}_K^e$. This triple is an object in a category of triples made of a truncated valuation ring, a free rank-one module over it, and a surjection from that module onto the maximal ideal of the ring.

1.2. If L/K is a finite separable extension, we shall say that it is "at most e-ramified" if $\mathrm{Gal}(E/K)^e = \{1\}$ for the normal closure E of L. The index refers to the upper numbering of the ramification groups **[Se1]**. Let $(\mathrm{ext}\ K)^e$ be the category of extensions of this sort.

In **[De]** Deligne constructs, for a triple S of length e in the category of triples described above (such as $Tr_e(K)$), a category $(\mathrm{ext}\ S)^e$ of extensions of S that are "at most e-ramified". He then constructs an equivalence of categories

$$T\colon (\mathrm{ext}\ K)^e \to (\mathrm{ext}\ Tr_e(K))^e$$

(Théorème 2.8). Thus the category of extensions of K that are at most e-ramified depends functorially only on the eth truncation of K. The whole point about the transition between local fields in different characteristics is now that if K is of characteristic 0, but of absolute index of ramification large compared to e, $Tr_eK \approx Tr_eF$ for a local field F in char p. This induces an equivalence of categories between $(\mathrm{ext}\ K)^e$ and $(\mathrm{ext}\ F)^e$.

1.3. The equivalence of categories T induces an isomorphism of Galois groups between $\mathrm{Gal}(L/K)$ and $\mathrm{Gal}(E/F)$ if we are given an isomorphism $Tr_e(K) \approx Tr_e(F)$, and if L and E are at most e-ramified over K and F respectively, and correspond to each other under T. In case these extensions are abelian, Deligne proves that the isomorphism between $\mathrm{Gal}(L/K)$ and $\mathrm{Gal}(E/F)$ is compatible with the local Artin map (Prop. 3.6.1).

1.4. The final result of **[De]** that we shall employ below is the following. Suppose we are given an isomorphism $\phi\colon Tr_e(K) \approx Tr_e(F)$, and L and E are as above, and correspond to each other. Then the isomorphism on the truncations can be continued, in the sense that an $e(1)$ can be computed, for which ϕ induces an isomorphism $Tr_{e(1)}(L) \approx Tr_{e(1)}(E)$ (Construction 3.4.1).

§2. The theory in characteristic p

2.1. Let F be a field of characteristic p, let c be a nonzero element of F, and q a power of p. Then $\ell(X) = cX + X^q$ is an additive polynomial, and we assume that $W = \mathrm{Ker}(\ell)$ is contained in F. This is a one-dimensional vector space over $\mathbb{F}_q$, and we fix a basis w, so that $c = -w^{q-1}$. Let M be the maximal elementary abelian p-extension of F, contained in a fixed separable closure. If $a \in F$, let a' be a root of $\ell(a') = a$, and define a pairing $\mathrm{Gal}(M/F) \times F/\ell(F) \to W$ by $\langle \sigma, a\rangle = \sigma(a') - a'$. Note that $\langle \cdot, \cdot \rangle$ is $\mathbb{F}_q$-linear in the second variable. The theorem of Artin-Schreier, usually phrased for $c = -1$ and $q = p$, says that this is a perfect pairing. In other words, it induces an isomorphism between $\mathrm{Hom}(\mathrm{Gal}(M/F), W)$ and $F/\ell(F)$, and between $\mathrm{Hom}_{\mathbb{F}_q}(F/\ell F, W)$ and $\mathrm{Gal}(M/F)$.

2.2. Assume now that $F = \mathbb{F}_q((t))$. For $b \in F^\times$ denote by σ_b the Artin symbol of b (in $\mathrm{Gal}(M/F)$).

Theorem (Schmid). *Let* Res *denote the residue of a Laurent series (the coefficient of $t^{-1}dt$). Then*

$$\langle \sigma_b, a\rangle = \mathrm{Res}(w^{-q}a \cdot db/b) \cdot w. \tag{2.2.1}$$

This explicit reciprocity law was first proved in **[Sch]**. See also **[Se1]** Chap. XIV §5. Generalizations to abelian extensions of exponent p^m are given in **[W]**, Satz 17. In fact, Witt introduced his famous Witt vectors for this very purpose.

§3. The theory in characteristic 0

3.1. Let k be a finite extension of $\mathbb{Q}_p$, $\mathbb{F}_q$ its residue field, and π a prime element of k. Let $\ell(X) = \pi X + \cdots \equiv X^q \pmod{\pi}$ be a power series in $\mathcal{O}_k[[X]]$, and G_ℓ the corresponding Lubin Tate group. Let $W_n = \mathrm{Ker}(\ell^n)$ and $K_n = k(W_n)$, the nth layer in the Lubin Tate tower of G_ℓ. K_n is totally ramified over k, of degree $(q-1)q^{n-1}$. Fix generators ω_n of W_n as $\mathcal{O}_k$-modules in such a way that $\ell(\omega_n) = \omega_{n-1}$. Write $[+]$ for the addition in G_ℓ, and $[u]$ for the unique endomorphism whose expansion as a power series starts as $uX + \cdots$. Thus, for example, $[\pi] = \ell$. We denote by λ the logarithm of G_ℓ.

Let $\beta \in K_n^\times$, and let $d = v_{K_n}(\beta)$. Then Coleman has associated with β a power series $g(= g_\beta)$ in $t^d\mathcal{O}_k[[t]]^\times$ satisfying

$$g(\omega_i) = N_{n,i}(\beta) \qquad \text{for all } 1 \le \mathrm{i} \le \mathrm{n} \tag{3.1.1}$$

where $N_{n,i}$ is the norm from K_n to K_i. Such a g is unique up to multiplication by a power series congruent to 1 modulo $[\pi^n](t)/t$. The reduction of g modulo π, $\overline{g}$, is therefore well defined mod ${}^\times U_e(F)$ ($F = \mathbb{F}_q((t))$ as in §2) as long as $e < q^n$.

Consider the map $\psi\colon K_n^\times \to F^\times/U_e(F)$, $\psi(\beta) = \overline{g}_\beta \bmod U_e(F)$. It is clearly a multiplicative homomorphism.

3.2. Proposition. *Suppose that* $e \leq (q-1)q^{n-1}$. *Then* ψ *induces an isomorphism (also denoted by* ψ*) of multiplicative groups:* $K_n^\times/U_e(K_n) \approx F^\times/U_e(F)$, *and also an isomorphism on the truncations,*

$$\phi: Tr_e(K_n) \approx Tr_e(F). \tag{3.2.1}$$

In other words, it induces a ring isomorphism

$$\phi_0: \mathcal{O}(K_n)/\mathfrak{p}(K_n)^e \approx \mathcal{O}(F)/\mathfrak{p}(F)^e,$$

and an isomorphism (over ϕ_0*) of free rank-1 modules*

$$\phi_1: \mathfrak{p}(K_n)/\mathfrak{p}(K_n)^{e+1} \approx \mathfrak{p}(F)/\mathfrak{p}(F)^{e+1},$$

compatible with the obvious maps of these modules to the corresponding rings (the maps denoted by ε *in §1).*

From now on we assume that $e \leq (q-1)q^{n-1}$ as above.

3.3. Let α' be a root of $[\pi^n](\alpha') = \alpha$, where $\alpha \in \mathfrak{p}(K_n)$, and write $v_{K_n}(\alpha) = \delta$. The "Kummer-Lubin-Tate" extension $L_n = K_n(\alpha')$ is an abelian p-extension of K_n, and its Galois group is mapped injectively into W_n via $\sigma \mapsto (\sigma\alpha'[-]\alpha')$. For $\beta \in K_n^\times$ we define

$$(\beta, \alpha)_n = \sigma_\beta(\alpha')[-]\alpha' \in W_n \tag{3.3.1}$$

where σ_β is the Artin symbol of β in L_n/K_n. This pairing is linear in α (with respect to $[+]$) and in β (multiplicatively). The explicit reciprocity law gives a formula for (3.3.1). A complete formula was found in **[dS]**. Previous work, however, always assumed either

(a) β is a norm from K_{2n}, or
(b) α is small (i.e., δ is large).

As explained in **[dS]**, the formula for (3.3.1) consists of a main term and a correction term. Under (a) or (b) the correction term disappears, and the ERL takes a particularly nice form, which suggests the analogy with (2.2.1). Now, the two assumptions are not unrelated. If β is a norm from K_{2n}, we may "push" (3.3.1) to level $2n$, replacing α by $[\pi^n](\alpha)$, making it "small", and compute the pairing at that level. Thus we may concentrate on (b), provided our definition of "small" will accommodate for α obtained by the procedure just described. See Lemma 3.4 below. (In the original work of Artin and Hasse, they confined themselves to the first level. As a result, the two cases looked different.)

3.4. Now, if δ is large, L_n/K_n is at most e-ramified in the sense of **[De]**. It therefore corresponds functorially to an extension E/F. Gal(L_n/K_n) and Gal(E/F) are naturally isomorphic, and the isomorphism is compatible with the filtration by the ramification groups, and with the local Artin maps from $K_n^\times$ and $F^\times$. Furthermore, the isomorphism between $Tr_e(K_n)$ and $Tr_e(F)$ induces an isomorphism between $Tr_{e(1)}(L_n)$ and $Tr_{e(1)}(E)$, where $e(1) = \psi(e)$, and ψ is the inverse-Herbrand function for L_n/K_n (or E/F). If we succeed in writing E/F as an Artin-Schreier-Witt extension, we may use Theorem 2.2 (or its generalization due to Witt) to derive the ERL in char 0 from the theory in char p. The next lemma gives two instances in which this program can be carried out.

Lemma.

(a) *(Here $n = 1$, $e = q-1$.) Assume that $\alpha \in \mathfrak{p}(K_1)^2$, and let $L_1 = K_1(\alpha')$ be as above, where $[\pi](\alpha') = \alpha$. Then L_1/K_1 is at most $q-1$ ramified.*

(b) *(Here n is replaced by $2n$, and $e = q^{2n-1} \leq q^{2n-1}(q-1)$.) Let $\alpha \in \mathfrak{p}(K_n)$, $L_n = K_n(\alpha')$ as above, where $[\pi^n](\alpha') = \alpha$, and let $M = L_n K_{2n}$. Then M/K_{2n} is at most q^{2n-1} ramified.*

Proof. We prove (a), the proof of (b) being similar. Write K and L for K_1 and L_1. It is enough to prove (a) when $v_K(\alpha) = 2$, because if $v_K(\alpha) > 2$ and $v_K(\alpha_0) = 2$, then $v_K(\alpha[+]\alpha_0) = 2$, and if L' and L'' are the fields obtained from α_0 and $\alpha_0[+]\alpha$, L is contained in their compositum. However, the compositum of two fields that are at most e-ramified, is also at most e-ramified, and so is any subfield of it.

So assume $v_K(\alpha) = 2$. Then $v_L(\alpha') = 2$, and $\theta = \omega_1^{-1}\alpha'^{\,(q+1)/2}$ is a uniformizer of L. L/K is totally ramified of degree q, and if σ is any nontrivial element of $\Gamma = \mathrm{Gal}(L/K)$, $\sigma\alpha'[-]\alpha' = \omega \in W_1\backslash\{0\}$. Looking at the power series expansion of $\sigma\alpha' = \alpha'[+]\omega$ we see that $v_L(\sigma\alpha' - \alpha') = v_L(\omega) = q$. Write $\sigma\alpha' = \alpha' + \xi$. Then $\sigma\theta = \theta + \omega_1^{-1}(q+1)\alpha'^{\,(q-1)/2}\xi/2 +$ (smaller terms). Thus $v_L(\sigma\theta - \theta) = q-1$. It follows that $\Gamma_e = \Gamma$ for $e \leq q-2$, and Γ_e is trivial for $e > q-2$. The same is therefore true in the upper numbering, for Γ^e. It follows that L/K is at most e-ramified if and only if $e > q-2$. □

3.5. Our purpose is to deduce the following ERL of Artin and Hasse from Schmid's theorem.

Theorem (Artin-Hasse for the multiplicative group over $\mathbb{Q}_p$, Wiles in general). *Let $K = K_1$ and $\alpha \in \mathfrak{p}(K)^2$. Let a' be a root of $[\pi](\alpha') = \alpha$. Let $\beta \in K^\times$, and let g be a Coleman power series of β. Write $\sigma_\beta = (\beta, L/K)$. Then*

$$\sigma_\beta(\alpha')[-]\alpha' = [\pi^{-1}Tr_{K/k}\{\lambda(\alpha)\cdot\lambda'(\omega_1)^{-1}g'(\omega_1)/g(\omega_1)\}]\cdot\omega_1. \qquad (3.5.1)$$

The expression in the square brackets is an integer in K.

See **[Ar-Ha]** and **[Wi]**.

The formula in 3.5.1 is easily seen to be invariant under "change of coordinates" in the formal group (see **[dS1]**, §3). Since G_ℓ is isomorphic to the formal group law obtained from $\ell = \pi X + X^q$, we may assume that ℓ is of this special form.

§4. From Schmid to Artin-Hasse

4.1. From now on we assume that $\ell = \pi X + X^q$, we put ourselves in case (a) of Lemma 3.4, and write K for K_1, L for L_1, and $e = q-1$. Since L/K is at most e-ramified, it corresponds functorially to an extension E/F. Recall the maps ψ and ϕ constructed in §3.1. Clearly $\psi(\omega_1) = t$ and $\psi(\pi) = \phi(-\omega_1^{q-1}) = -t^{q-1}$. Since (3.5.1) is linear in α we may assume that $v_K(\alpha) = 2$, and write f for $\psi(\alpha)$. Thus $f \in t^2\mathbb{F}_q[[t]]^\times$, and it is well defined mod ${}^\times U_{e+2}(F)$.

Lemma. *$E = F(x)$ where x is a solution of $X^q - t^{q-1}X - f = 0$.*

Proof. The problem is that the equations used to define L and E, although they correspond to each other under ϕ, are not Eisenstein, and Deligne's construction employed Eisenstein polynomials. So let $y = t^{-1}x^{(q+1)/2}$. Then $v_E(y) = 1$, and $\mathcal{O}_E = \mathcal{O}_F[y]$. Similarly we had $\mathcal{O}_L = \mathcal{O}_K[\theta]$, $v_L(\theta) = 1$, see the proof of Lemma 3.4. Now $Tr_{qe}(L)$ is the finite flat totally-ramified extension of $Tr_e(K) = (\mathcal{O}_K/\mathfrak{p}_K^e, \mathfrak{p}_K/\mathfrak{p}_K^{e+1}, \varepsilon_K)$ obtained as follows. Let

$$\prod_{\sigma \in \mathrm{Gal}(L/K)} (Y - \omega_1^{-1}\sigma(\alpha')^{(q+1)/2}) = Y^q + \lambda_1 Y^{q-1} + \cdots + \lambda_q$$

be the minimal polynomial of θ over K, so that $\lambda_i \in \mathfrak{p}(K)$, and $\lambda_q = -\omega_1^{-q}\alpha^{(q+1)/2}$. Let $\overline{\lambda}_i = \lambda_i \bmod \mathfrak{p}_K^{e+1}$, and let $A = (\mathcal{O}_K/\mathfrak{p}_k^e)[\widetilde{\theta}]$, where $\widetilde{\theta}$ satisfies

$$Y^q + \varepsilon_K(\overline{\lambda}_1)Y^{q-1} + \cdots + \varepsilon_K(\overline{\lambda}_q) = 0.$$

Let $I = A \cdot \theta'$, and define $\varepsilon_L\colon I \to A$ by $\varepsilon_L(\theta') = \widetilde{\theta}$. Then $(A, I, \varepsilon_L) \approx Tr_{qe}(L)$ canonically.

On the other hand, $Tr_{qe}(E) = (B, J, \varepsilon_E)$, where $B = (\mathcal{O}_F/\mathfrak{p}_F^e)[\widetilde{y}]$, and $\widetilde{y}$ is a solution of

$$Y^q + \varepsilon_F(\overline{h}_1)Y^{q-1} + \cdots + \varepsilon_F(\overline{h}_q) = 0$$

(here $Y^q + h_1Y^{q-1} + \cdots + h_q$ is the minimal polynomial of y over F), $J = B \cdot y'$, and ε_E maps h' and $\widetilde{y}$. To show that the extension $Tr_{qe}(E)/Tr_e(F)$ corresponds, under the correspondence established by Deligne, to $Tr_{qe}(L)/Tr_e(K)$, we have to check that in the isomorphism $\phi\colon Tr_e(K) \approx Tr_e(F)$, ϕ_0 takes $\varepsilon_K(\overline{\lambda}_i)$ to $\varepsilon_F(\overline{h}_i)$. Let $Q_1, \ldots, Q_q$ be the universal polynomials in the elementary symmetric functions such that, if

$$\prod_{1 \le i \le q} (X - x_i) = X^q - \sigma_1 X^{q-1} + \cdots + (-1)^q \sigma_q,$$

then

$$\prod_{1 \le i \le q} (X - z_i^{(q+1)/2}) = X^q - Q_1(\sigma_1, \ldots, \sigma_q)X^{q-1} + \cdots + (-1)^q Q_q(\sigma_1, \ldots, \sigma_q).$$

Then $\lambda_i = \omega_1^{-i}Q_i(0, \ldots, 0, \pi, \alpha)$ and $h_i = t^{-i}Q_i(0, \ldots, 0, -t^{q-1}, f)$. The crucial point in the proof is that $Q_i(0, \ldots, 0, \sigma_{q-1}, \sigma_q)$ is a *monomial* in σ_{q-1} and σ_q. Indeed, if $\sigma_{q-1}^u\sigma_q^u$ appears in it, then weight considerations force $u \cdot (q-1) + v \cdot q = i(q+1)/2$. If there are positive integers u and v solving this equation, they are unique, as can be seen from the congruence mod q that u has to satisfy, and the observation that $i \le q$. Now $\phi_0(\varepsilon_K(\overline{\lambda}_i)) = \phi_0(\omega_1^{-i}\pi^u\alpha^v) = \xi(\omega_1^{-i}\pi^u\alpha^v) \bmod \mathfrak{p}_F^e$. (We may replace the ring homomorphism ϕ_0 by the multiplicative homomorphism ψ, because $\omega_1^{-i}\pi^u\alpha^v$ is an integer, so $\psi(\omega_1^{-i}\pi^u\alpha^v)$, which is well defined mod ${}^\times U_e(F)$, is clearly well defined mod $\mathfrak{p}_F^e$.) But

$$\psi(\omega_1^{-i}\pi^u\alpha^v) = t^{-i}(-t^{q-1})^u f^v = t^{-i}Q_i(0, \ldots, 0, -t^{q-1}, f) = h_i \bmod {}^\times U_e(F),$$

so $\phi_0(\varepsilon_k(\overline{\lambda}_i)) = \varepsilon_F(h_i)$. □

4.2. Let Γ be the Galois group of L/K, and identify it with $\mathrm{Gal}(E/F)$ via

$$\mathrm{Gal}(L/K) = \mathrm{Aut}(Tr_{qe}(L)/Tr_e(K)) = \mathrm{Aut}(Tr_{qe}(E)/Tr_e(F)) = \mathrm{Gal}(E/F).$$

Lemma. *Let $\sigma \in \Gamma$, and let $u, v \in \mathbb{F}_p$ be defined by*

$$\sigma\alpha'[-]\alpha' = [u]\cdot\omega_1, \qquad \sigma x - x = v \cdot t$$

where α' is as in 3.3, and x as in 4.1. Then $u = v$.

Proof. With $e = q-1$, $\psi_{L/K}(e) = 1\cdot(q-2)+q\cdot 1 = 2q-2$, so $\phi\colon Tr_e(K) \approx Tr_e(F)$ induces a unique isomorphism $Tr_{2q-2}(L) \approx Tr_{2q-2}(E)$ as in **[De]** 3.4.1. With v as above, it is enough to check that $\sigma\alpha'[-]\alpha' = [v]\cdot\omega_1 \bmod \mathfrak{p}_L^{2q-2}$, because $v_L(\omega) = q < 2p-2$, for $0 \neq \omega \in W_1$. Let $\sim$ denote reduction mod $\mathfrak{p}_L^{2q-2}$, or mod $\mathfrak{p}_E^{2q-2}$, depending on the circumstances, and identify $Tr_{2q-2}(L)$ with $Tr_{2q-2}(E)$. The proof of the previous lemma shows that $\widetilde{\alpha}' = \widetilde{x}$, and $\widetilde{\omega}_1 = \widetilde{t}$. Thus $\sigma\widetilde{\alpha}' - \widetilde{\alpha}' = v\widetilde{\omega}_1$, and we have to prove $\sigma\widetilde{\alpha}'[-]\widetilde{\alpha}' = [v]\widetilde{\omega}_1$, all the computations performed in $\mathcal{O}_L/\mathfrak{p}_L^{2q-2}$. The validity of the last claim follows from the following observations.

(a) $\widetilde{\omega}_1^2 = 0$, so $[v]\widetilde{\omega}_1 = v\widetilde{\omega}_1$.

(b) The special group law G_ℓ has the property that $X[+]Y = X + Y + \Sigma A_{ij}X^iY^j$, and $A_{ij} = 0$ unless $i + j \geq q$ **[Wi]**. Hence, writing ω for $[v]\omega_1$, $\widetilde{\alpha}'[+]\widetilde{\omega} = \widetilde{\alpha}' + \widetilde{\omega} +$ (terms not larger than $\widetilde{\omega}\widetilde{\alpha}'^{(q-1)}$). However, already $\widetilde{\omega}\widetilde{\alpha}'^{(q-1)} = 0$, so $\widetilde{\alpha}'[+]\widetilde{\omega} = \widetilde{\alpha}' + \widetilde{\omega}$. □

4.3. To derive the ERL of Artin-Hasse from Schmid's formula, it remains to prove the following lemma.

Lemma. *Let f and g be power series as above, $\alpha = f(\omega_1)$, and let $\overline{f}$ and $\overline{g}$ be their reductions modulo π. Then, in $\mathbb{F}_q$, we have the equality*

$$\mathrm{Res}(t^{-q}\overline{f}d\overline{g}/\overline{g}) = [\pi^{-1}Tr_{K/k}\{\lambda(\alpha)\cdot\lambda'(\omega_1)^{-1}g'(\omega_1)/g(\omega_1)\}] \bmod \pi. \quad (4.3.1)$$

Proof. We first need a lemma that allows us, because of the special form of the formal group, to get rid of the logarithm.

4.4. Sublemma. *The following congruence holds:*

$$\pi^{-1}\sum_{\omega\in W_1}\lambda\circ f(\omega)\cdot\lambda'(\omega)^{-1}g'(\omega)/g(\omega) \equiv \pi^{-1}\sum_{\omega\in W_1} f(\omega)\cdot g'(\omega)/g(\omega) \bmod \pi.$$

Proof. Both sides of the equation belong to $\mathcal{O}_k$, so it is enough to check that they are congruent modulo ω_1 in $\mathcal{O}_K$. Now $\lambda(X) = \lim \pi^{-n}[\pi^n](X)$, from where we conclude, since $[\pi](X) = \pi X + X^q$, that $\lambda(X) = X + \pi^{-1}uX^q + \cdots$, u a unit, so $\lambda'(X) \equiv 1 \bmod X^{q-1}$ in $\mathcal{O}_k[[X]]$. We also note that if $v_K(\alpha) = 2$, $\pi^{-1}[\pi](\alpha) \equiv \alpha \bmod \omega_1^{q+1}$, and therefore $\lambda(f(\omega)) \equiv f(\omega) \bmod \omega_1^{q+1}$. Since $v_K(g'(\omega)/g(\omega)) \geq -1$, a simple computation proves the congruence. □

4.5. We now prove Lemma 4.3. Let $h = -t^{1-q}f(t)$, so that $h(\omega) = \pi^{-1}f(\omega)$ for every $\omega \in \widetilde{W}_1 = W_1 \backslash \{0\}$. We have to show that

$$-\text{constant term}\,(\overline{h} \cdot \overline{g}'/\overline{g}) = \sum_{\omega \in \widetilde{W}_1} h(\omega) \cdot g'(\omega)/g(\omega) \bmod \omega_1 .$$

Call $u = hg'/g$, and observe that $u \in t^d \mathcal{O}_k[[t]]^\times$ with $d \geq 2 - q$. But every power series of this form satisfies

$$-\text{constant term}\,(\overline{u}) \equiv \sum_{\omega \in \widetilde{W}_1} u(\omega) \bmod \omega_1 .$$

Indeed, this is trivial if $d \geq 0$, and if $u = t^d$, $2 - q \leq d < 0$, the sum on the right is a circular sum of roots of unity that vanishes. □

References

[Ar-Ha] E. Artin, H. Hasse, Die beiden Ergänzungssätze zum Reziprozitätsgesetz der ℓ^n-ten Einheitzwurzeln, *Abh. Math. Sem. Univ. Hamburg* **6** (1928), 46–162.

[Ar-Ta] E. Artin, J. Tate, *Class field theory*, Benjamin 1967.

[Co] R. Coleman, The dilogarithm and the norm residue symbol, *Bull. Soc. Math. France* **109** (1981), 372–402.

[De] P. Deligne, Les corps locaux de caractéristique p, limites de corps locaux de caractéristique 0, in: Bernstein et al., *Représentations des groupes réductifs sur un corps local*, Hermann, Paris (1984).

[dS] E. de Shalit, The explicit reciprocity law in local class field theory, *Duke Math. J.* **53** (1986), 163–176.

[I] K. Iwasawa, Explicit formulae for the norm residue symbol, *J. Math. Soc. Japan* **20** (1968), 151–164.

[Kr] M. Krasner, Approximation des corps valués complets de caractéristique $p \neq 0$ par ceux de Charactéristique 0, *Coll. d'algèbre supérieure*, Bruxelles, (C.B.R.M.) 1957.

[Sch] H. L. Schmid, Zyklische algebraische Funktionenkörper vom Grad p^n über endlichem Konstantkörper der Charakteristik p, *J. Reine Angew. Math.* **175** (1936).

[Sen] S. Sen, On explicit reciprocity laws, *J. Reine Angew. Math.* I: **313** (1980), 1-26, II: **323** (1981), 68–87.

[Se1] J.-P. Serre, *Local Fields*, Springer-Verlag, New-York (1979).

[Se2] J.-P. Serre, *Groupes Algébriques et Corps de Classe*, Hermann, Paris, (1959).

[V] S. V. Vostokov, Explicit form on the law of reciprocity, *Math. USSR-Izv.* 13 (1979), 557–588.

[Wi] A. Wiles, Higher explicit reciprocity laws, *Ann. of Math.* **107** (1978), 235–254.

[W] E. Witt, Zyklische Körper und Algebren der Charakteristik p vom Grad p^n. *J. Reine Angew. Math.* **176** (1937), 126–140.

Department of Mathematics, Hebrew University, Jerusalem (Givat Ram), 91904 Israel

The Circle Method and the Strict Waring Problem in Function Fields

M. Car

Introduction

Let $\mathbb{F}_q$ be the finite field with q elements. Let p be its characteristic.

Noticing the analogy between the ring $\mathbb{Z}$ and the ring $\mathbb{F}_q[T]$, one is led to study many additive problems in $\mathbb{F}_q[T]$, more particularly, the Waring problem, c.f. **[1]**, **[2]**, **[4]**, **[6]**, **[7]**, **[10]**, **[11]**, or Goldbach problems, c.f. **[4]**, **[5]**.

Under its most general form, the Waring problem may be studied in any ring $\mathbb{A}$. It is the following. Given an integer $k \geq 2$, let $\mathbb{A}^{(k)}$ be the set of elements in $\mathbb{A}$ which are sums of kth powers, and determine $W(k, \mathbb{A})$ which is the least integer s, if it exists, such that for all $a \in \mathbb{A}$ the equation

$$a = a_1^k + \cdots + a_s^k$$

admits a solution $(a_1, \ldots, a_s) \in \mathbb{A}^s$. Like this, the Waring problem is close to what is called easier Waring problem in $\mathbb{Z}$, which has been studied by R.E.A.C. Paley when $\mathbb{A}$ is the ring $\mathbb{F}_q[T]$ and by L. N. Vaserstein when $\mathbb{A}$ is an algebra over the finite field $\mathbb{F}_q$, c.f. **[7]**, **[10]** and **[11]**. This contains the case $\mathbb{A} = O_S$ where O_S is the ring of S-integers of a function field over the finite field $\mathbb{F}_q$ and the bounds provided by **[10]** and **[11]** are valid in these rings.

If we study the Waring problem with the most restrictive degree conditions, we get a problem which is close to what is called the difficult Waring problem in $\mathbb{Z}$. Following G. Effinger and D. Hayes, c.f. **[4]**, we call the Waring problem with the most restrictive degree conditions in $\mathbb{F}_q[T]$, *the strict Waring problem.* In the same manner, if we request the most restrictive valuation conditions, we get a strict Waring problem in the ring O_S. In the sequel we are interested in the strict Waring problem. In the first part we give the known main results concerning the strict Waring problem. In a second part, we give some indications about the proof of these results.

The Main Results

1. The Strict Waring Problem in $\mathbb{F}_q[T]$

Let $A \in \mathbb{F}_q[T]$. If $(A_1, \ldots, A_S) \in \mathbb{F}_q[T]^s$ verifies

$$A = A_1^k + \cdots + A_s^k, \tag{1.1}$$

with

$$\deg A_i \leq n, \tag{1.2}$$

where n is defined by the relation $k(n-1) < \deg A \leq kn$, we say that (1.1) is a *strict representation* of A as a sum of s k^{th}-powers.

We denote by $\mathbf{g}_q(k)$, resp. $\mathbf{G}_q(k)$, the least integer s, if it exists, such that every $A \in \mathbb{F}_q[T]^{(k)}$, resp. every $A \in \mathbb{F}_q[T]^{(k)}$ whose degree is large enough, admits a strict representation as a sum of s k^{th}-powers. Let $\mathbf{R}_{k,s}(A)$ be the number of strict representations of A as a sum of s k^{th}-powers.

Using the circle method, if $\mathbf{k} < \mathbf{p}$, one gets an asymptotic estimate for the number $\mathbf{R}_{k,s}(A)$. From this estimate, one deduces straightforwardly that

$$\mathbf{G}_q(k) \leq k2^{k-1} + 1. \tag{1.3}$$

This bound may be improved. By Hua's method, one gets

$$\mathbf{G}_q(k) \leq 2^k + 1, \tag{1.4}$$

cf. **[6]**, and by Vinogradov's method, one gets,

$$\mathbf{G}_q(k) \leq 2k(k-2)\,\mathrm{Log}(2) + 2k + 3, \tag{1.5}$$

cf. **[1]**. Notice that, in this last case, one has only a lower bound for the number $\mathbf{R}_{k,s}(A)$. A particular study, cf. **[2]**, gives estimates for the numbers $\mathbf{R}_{2,4}(A)$ and $\mathbf{R}_{2,3}(A)$ from which one deduces the bounds

$$\mathbf{G}_q(2) \leq 4 \text{ and } \mathbf{G}_q(2) \leq 3 \text{ for } q \geq 53. \tag{1.6}$$

A better result has been obtained by J. P. Serre who proved that

$$\mathbf{g}_q(2) = 3 \text{ for } q \geq 5. \tag{1.7}$$

A proof of this can be found in **[4]**.

2. The Strict Waring Problem for the Ring of S-integers in a Function Field

Let $\mathbb{L}$ be a finite and separable extension of $\mathbb{F}_q(T)$. The analogy between number fields and algebraic function fields over $\mathbb{F}_q$ allows us to expect that the methods introduced by C. L. Siegel, cf. **[8]**, **[9]**, to deal with Waring's problem in number fields may be fitted to the ring $\mathbb{B}$, the integral closure of $\mathbb{F}_q[T]$ in $\mathbb{L}$.

The main difficulty is to ask the question of Waring's problem in $\mathbb{B}$ in a suitable manner. One approach is to request the sums of kth-powers to be subject to the most restrictive valuation conditions for all valuations extending the ∞-valuation ν_∞ of $\mathbb{F}_q(T)$.

Following Siegel's proof led us to separate P-adic valuations and ∞-valuations of $\mathbb{L}$. The difference between these valuations is artificial. All the results obtained for the ring $\mathbb{B}$ may be generalized to the ring of S-integers of the field $\mathbb{L}$. We ask the question of the strict Waring problem in the ring of S-integers of $\mathbb{L}$ in the following manner.

Let $\mathbb{L}$ be a finite and separable extension of $\mathbb{F}_q(T)$. Let S be a finite and non-empty set of independent valuations of $\mathbb{L}$. Let O_S be the ring of the S-integers of $\mathbb{L}$, i.e., the set of elements $x \in \mathbb{L}$ such that $v(x) \geq 0$ for all valuations $v \notin S$. Let $b \in O_S$. A representation

$$b = b_1^k + \cdots + b_m^k \tag{2.1}$$

with $b_1, \ldots, b_m \in O_S$ subject to the condition

$$w(b_j) \geq \operatorname{Int}\left(\frac{w(b)}{k}\right) \quad \text{for every } w \in S, \tag{2.2}$$

is *a strict representation of* b as a sum of m k^{th}-powers *with respect to* S, or briefly, a strict representation.

We denote by $\mathbf{R}_{k,m}(b) = \mathbf{R}(b)$ the number of these representations, and by $\mathbf{G}(k) = \mathbf{G}_{\mathbb{L},S}(k)$, the least integer m, if it exists, such that $\mathbf{R}_{k,m}(b) > 0$ for all $b \in O_S$ for which the sum

$$h(b) = \sum_{v \in S} f_v v(b) \tag{2.3}$$

is small enough, f_v denoting the residue degree of the place v.

If $\mathbb{L} = \mathbb{F}_q(T)$ and $S = \{\nu_\infty\}$, then $O_S = \mathbb{F}_q[T]$, condition (2.2) is equivalent to condition (1.2), and the function h defined by (2.3) is the opposite of the degree.

If $\mathbb{L}$ is a finite and separable extension of $\mathbb{F}_q(T)$, if S is the set $\{v_1, \ldots, v_r\}$ of valuations extending the valuation ν_∞, O_S is the ring $\mathbb{B}$, and condition (2.2) may be written in the form

$$-v_i(b_j) \leq N_i \text{ if } k(N_i - 1) < -v_i(b) \leq kN_i$$
$$\text{for all } i, j, \ 1 \leq i \leq r \text{ and } 1 \leq j \leq m,$$

and this generalizes (1.2).

We have the following theorem.

Theorem 2.1. *If* $2 \leq k < p$, *then* $\mathbf{G}_q(k) \leq 2^k + 1$.

3. The Waring Problem for Big Exponents

In the above paragraphs, we always have the condition $\mathbf{k} < \mathbf{p}$. This condition seems too restrictive, a better one, being k and p coprime.

The condition $\mathbf{k} < \mathbf{p}$ appears when, bounding some character sums by Weyl's method, we get a factor $k!$ which leads to a trivial estimate when $k \geq p$. Very little is known about the strict Waring problem when $k > p$. In his thesis, cf. **[3]**, J. Cherly proved that every polynomial in $\mathbb{F}_2[T]$, congruent to 0 or 1 modulo $T^2 + T + 1$, of degree large enough, admits a strict representation as a sum of 18 cubes, that is to say $\mathbf{G}_2(3) \leq 18$.

We improve this result and prove the following theorem.

Theorem. *Let $q = 2^m$. Let $\mathcal{M}_q$ be the subset of $\mathbb{F}_q[T]$ defined by*
1) $\mathcal{M}_q = \mathbb{F}_q[T]$ *if* $q \geq 8$
2) $\mathcal{M}_2$ *is the set of polynomials in* $\mathbb{F}_2[T]$ *congruent to* 0 *or* 1 *modulo* $T^2 + T + 1$
3 $\mathcal{M}_4$ *is the set of polynomials in* $\mathbb{F}_4[T]$ *congruent to* 0 *or* 1 *modulo the polynomials of degree* 1, *such that either* 3 *does not divide* $\deg M$ *or* M *is monic.*
Then every polynomial in $\mathcal{M}_q$, *of degree large enough, admits a strict representation as a sum of* 13 *cubes. In other words, we have the bound*

$$\mathbf{G}_{2^m}(3) \leq 13.$$

Sketches of the Proofs

First, we give some details about the circle method in $\mathbb{F}_q((T^{-1}))$ and about the proof of relation (1.3).

4. The Tools of the Circle Method

We put $\mathbb{A} = \mathbb{F}_q[T]$, $\mathbb{K} = \mathbb{F}_q(T)$, $\nu\left(\frac{A}{B}\right) = \deg B - \deg A$, if A and B are non-zero polynomials, ν is the ∞-valuation.

The completion $\mathbb{K}_\infty$ of $\mathbb{K}$ for the valuation ν is identified with the field $\mathbb{F}_q((T^{-1}))$ of the Laurent series in T^{-1}, the valuation ν extending to non-zero elements in $\mathbb{K}_\infty$ by

$$\nu\left(\sum_{s=-\infty}^{+\infty} a_S T^S\right) = -m \quad \text{if } m \text{ is the biggest index such that } a_m \neq 0.$$

We define an absolute value on $\mathbb{K}_\infty$, putting $|y| = q^{-\nu(y)}$, for $y \neq 0$.

Let $\mathcal{P}$ be the valuation ideal of $\mathbb{K}_\infty$. Every $a \in \mathbb{K}_\infty$ admits unique representation as a sum

$$a = [a] + \{a\}, \quad [a] \in \mathbb{A}, \quad \{a\} \in \mathcal{P}.$$

The ideals $\mathcal{P}^j$, $j \in \mathbb{Z}$, are compact. On the locally compact additive group $\mathbb{K}_\infty$, we choose a Haar measure $\mu = dt$ which is 1 on $\mathcal{P}$.

Let Ψ be a non-principal character of the additive group of $\mathbb{F}_q$. We associate to Ψ a non -trivial character E of the additive group of $\mathbb{K}_\infty$ by

$$E(u) = \Psi(\mathrm{Res}_\infty(u)), \text{ i.e., } E\left(\sum_{S=-\infty}^{+\infty} a_S T^S\right) = \Psi(a_{-1}).$$

It is easy to prove that E has the following properties.

$$E(\mathbb{A}) = \{1\}. \tag{4.1}$$

$$\nu(a) \geq 2 \Rightarrow E(a) = 1. \tag{4.2}$$

$$\text{For every } H \in \mathbb{A}, \text{ one has } \int_{\mathcal{P}} E(Ht)dt = \begin{cases} 1 & \text{if } H = 0, \\ 0 & \text{if } H \neq 0. \end{cases} \tag{4.3}$$

For $a \in \mathbb{K}_\infty$ and every positive integer N, one has (4.4)

$$\sum_{\substack{H \in \mathbb{A} \\ \deg H < N}} E(aH) = \begin{cases} q^N & \text{if } \nu(\{a\}) > N, \\ 0 & \text{if not.} \end{cases}$$

For $H \in \mathbb{A}$, $H \neq 0$, for $R \in \mathbb{A}$, one has (4.5)

$$\sum_{K \bmod H} E\left(\frac{KR}{H}\right) = \begin{cases} |H| & \text{if } H \text{ divides } R, \\ 0 & \text{if not.} \end{cases}$$

The circle method is based on equality (4.3). Indeed, let n be a positive integer. For $t \in \mathcal{P}$, let

$$f(t) = f_n(t) = \sum_{\substack{A \in \mathbb{A} \\ \deg A \leq n}} E(tA^k). \tag{4.6}$$

Then the integral

$$I(k, s, n, M) = \int_{\mathcal{P}} f(t)^S E(-Mt)dt \tag{4.7}$$

is equal to the number of solutions $(M_1, \ldots, M_s) \in \mathbb{A}^s$ of the equation

$$M = M_1^k + \cdots + M_s^k,$$

such that $\deg M_i \leq n$, for each index i. If n and M are subject to

$$k(n-1) < \deg M \leq kn, \tag{4.8}$$

then $I(k, s, n, M)$ is equal to the number $\mathbf{R}_{\mathbf{k},\mathbf{s}}(M)$ of strict representations of M as a sum of s k^{th}-powers. In order to estimate the integral $I(k, s, n, M)$, we have to estimate the function f. We use rational approximations of the Laurent series t given by a Farey dissection of $\mathcal{P}$.

5. The Farey Dissection and the Major Arcs

Definition. Let N be a non-negative integer. A *Farey fraction of order* N is a rational function $\frac{G}{H}$ such that $\deg H \leq N$, G and H are coprime, and $\deg G < \deg H$.

Let $\mathcal{F}_N$ be the set of all Farey fractions of order N. If $\frac{G}{H} \in \mathcal{F}_N$, the set

$$\mathcal{U}_{G/H} = \left\{ t \in \mathcal{P} \mid \nu \left(t - \frac{G}{H} \right) > N + \deg H \right\}$$

is the *Farey ball with center* $\frac{G}{H}$.

D. R. Hayes proved the following result, cf. **[5]**.

Theorem. *For every integer* $N \geq 0$, $\mathcal{P}$ *is the union of all Farey balls* $\mathcal{U}_{G/H}$ *with* G/H *running through* $\mathcal{F}_N$. *Moreover, two distinct Farey balls do not overlap.*

Dividing $\mathcal{P}$ in balls $\mathcal{U}_{G/H}$ with G/H running through $\mathcal{F}_N$ is called a *Farey dissection of order* N.

In order to estimate the integral $I(k, s, n, M)$, we divide the ideal $\mathcal{P}$ by a suitable Farey dissection. Following the analogy with the classical Waring problem, we may conjecture that a suitable order N is

$$N = (k-1)n. \tag{5.1}$$

This is true if $k < p$. It seems that it is not true if $k > p$, but very few things are known in that case.

We choose $\mathbf{N = (k-1)n}$ and we divide $\mathcal{P}$ by a Farey dissection of order N. We have a good estimate of $f(t)$ for those t belonging to the Farey balls with $\deg H \leq n$. These balls provide the main term in the integral $I(k, s, n, M)$. We call them the major arcs. The remaining balls will be called the minor arcs. There is no particular difficulty to deal with major arcs whether $k < p$ or not. Hence, I will only give the result. If $s \geq 2k + 1$, the part of $I(k, s, n, M)$ corresponding to the major arcs is of the form

$$C_{q,k,s}(M) q^{n(s-k)} + o(q^{n(s-k)}),$$

where,

- if $k < p$, or if $k > p$ and q is large enough, (i.e., $q \geq q(k)$), $C_{q,k,s}(M) \geq \kappa(q, k, s) > 0$, for all $M \in \mathbb{F}_q[T]$, $\kappa(q, k, s)$ being a constant,
- if $k > p$ and if q is not large enough (i.e., $q < q(k)$), $C_{q,k,s}(M) \geq \kappa(q, k, s) > 0$, for all $M \in \mathbb{F}_q[T]$ verifying some congruence conditions.

In order to prove that $I(k, s, n, M)$ becomes > 0 for n large enough, it suffices to have a non trivial bound for $|f(t)|$ on the minor arcs. Indeed, if $|f(t)| \ll q^{n\delta}$, $0 < \delta < 1$, for t belonging to a minor arc, we shall have the asymptotic estimate

$$I(k, s, n, M) = C_{k,s}(M) q^{n(s-k)} + o(q^{n(s-k)}),$$

as soon as $s > \frac{k}{1-\delta}$.

6. The Farey Dissection and the Minor Arcs

In this section, we show by an example how the condition $k < p$ appears when using Weyl's method. We suppose $k = 3$. We denote by $\mathbb{A}_n$ the set of polynomials $A \in \mathbb{A}$ such that $\deg A \leq n$. Noticing that $\mathbb{A}_n$ is an additive subgroup of $\mathbb{A}$, we have for $t \in \mathcal{P}$,

$$|f(t)|^2 = \sum_{A_1 \in \mathbb{A}_n} \sum_{Y \in \mathbb{A}_n} E(t((A_1 + Y)^3 - Y^3))$$
$$= \sum_{A_1 \in \mathbb{A}_n} E(tA_1^3) \sum_{Y \in \mathbb{A}_n} E(t(3A_1^2 Y + 3A_1 Y^2)),$$

and

$$|f(t)|^2 \leq \sum_{A_1 \in \mathbb{A}_n} \sum_{Y \in \mathbb{A}_n} E(t(3A_1^2 Y + 3A_1 Y^2)). \tag{6.1}$$

(1) If $p = 3$, we get $|f(t)|^2 = q^{n+1} f(t)$, whence $|f(t)| \in \{0, q^{n+1}\}$ and this gives a trivial bound. This is not a good example, for, if $p = 3$, sums of cubes in $\mathbb{A} = \mathbb{F}_q[T]$ are well known, and the circle method is useless in that case.
(2) We suppose $p \neq 3$. Using Schwarz's inequality, we get

$$|f(t)|^4 \leq q^{n+1} \sum_{A_1 \in \mathbb{A}_n} \left| \sum_{Y \in \mathbb{A}_n} E(t(3A_1^2 Y + 3A_1 Y^2)) \right|^2$$
$$\leq q^{n+1} \sum_{A_1 \in \mathbb{A}_n} \sum_{A_2 \in \mathbb{A}_n} \sum_{Y \in \mathbb{A}_n} E(t(3A_1^2(A_2 + Y - Y) + 3A_1((A_2 + Y)^2 - Y^2))).$$

(2a) If $p = 2$, we have

$$|f(t)|^4 \leq q^{2(n+1)} \sum_{A_1 \in \mathbb{A}_n} \sum_{A_2 \in \mathbb{A}_n} E(t(3A_1^2 A_2 + 3A_1 A_2^2)).$$

This gives a bound which is not better than the bound (6.1) and using Schwarz' inequality is useless here.
(2b) If $p > 3$, we get

$$|f(t)|^4 \leq q^{n+1} \sum_{A_1 \in \mathbb{A}_n} \sum_{A_2 \in \mathbb{A}_n} E(t(3A_1^2 A_2 + 3A_1 A_2^2)) \sum_{Y \in \mathbb{A}_n} E(6tA_1 A_2 Y)$$
$$\leq q^{n+1} \sum_{A_1 \in \mathbb{A}_n} \sum_{A_2 \in \mathbb{A}_n} \sum_{Y \in \mathbb{A}_n} E(6tA_1 A_2 Y)$$
$$\leq q^{2(n+1)} \#\{A_1, A_2) \in \mathbb{A}_n^2 \mid \nu(\{tA_1 A_2\}) > n + 1\}.$$

Now, if t belongs to a minor arc, then $t = \frac{G}{H} + u$, with $\nu(u) > 2n + \deg H$ and $\deg H > n$, $f(t) = f\left(\frac{G}{H}\right)$, and, as for the rational Waring's problem, the above relation leads to the bound

$$|f(t)|^4 \ll q^{3n+n\varepsilon}$$

available for any $\varepsilon > 0$.

Whence, if t belongs to a minor arc,

$$|f(t)| \ll q^{n(3/4+\varepsilon)}. \tag{6.2}$$

More generally, for any exponent $k < p$ and t belonging to any minor arc, we obtain the estimate

$$|f(t)| \ll q^{n(1-2^{1-k}+\varepsilon)}, \tag{6.3}$$

and this is sufficient to get an asymptotic estimate for the numbers $I(k,s,n,M) = \mathbf{R}_{k,s}(M)$. From this estimate, we get straightforwardly the bound $\mathbf{G}_q(k) \leq k2^{k-1}+1$. To get the bound $\mathbf{G}_q(k) \leq 2^k+1$, one has to use Hua's method as in the rational case. Using Hua's method also needs the condition $k < p$.

7. The Proof for the Ring of S-Integers

Since the proof of the theorem is too long and too technical, we only give the main steps.
(1) Taking the field $\mathbb{F}_{q_1}$, the algebraic closure of $\mathbb{F}_q$ in $\mathbb{L}$ at the place of $\mathbb{F}_q$ and the field $\mathbb{F}_{q_1}(T)$ at the place of the field $\mathbb{F}_q(T)$, we may suppose that $\mathbb{F}_q$ is algebraically closed in $\mathbb{L}$.
(2) We show that it is sufficient to prove the theorem when $O_S = \mathbb{B}$. For that, we prove that there exists $y \in \mathbb{L}$, such that for every valuation v, $\mathbb{F}_q[y]$ is included in the valuation ring of v if and only if $v \notin S$. This shows that y is transcendental over $\mathbb{F}_q$, that $\mathbb{L}$ is an algebraic extension of $\mathbb{F}_q(y)$, that $\mathbb{L}$ is finite over $\mathbb{F}_q(y)$ and that O_S is the integral closure of the ring $\mathbb{F}_q[y]$.
(3) Now, we prove the theorem when $S = \{v_1, \ldots, v_r\}$, the set of valuations of $\mathbb{L}$ extending the ∞-valuation of $\mathbb{F}_q(T)$ and when $O_S = \mathbb{B}$.

If $\vec{N} = (N_1, \ldots, N_r) \in \mathbb{Z}^r$, $b \in \mathbb{B}$, let $R(k,m,\vec{N},b) = R(\vec{N},b)$ be the number of solutions $(b_1, \ldots, b_m) \in \mathbb{B}^m$ of the equation

$$b = b_1^k + \cdots + b_m^k \tag{7.1}$$

subject to

$$-v_i(b_j) \leq N_i, \quad \text{for all } i,j,\ 1 \leq i \leq r \text{ and } 1 \leq j \leq m. \tag{7.2}$$

If $\vec{N} = (N_1, \ldots N_r) \in \mathbb{Z}^r$, $b \in \mathbb{B}$ are subject to

$$k(N_i - 1) < -v_i(b) \leq N_i, \quad \text{for all } i,j,\ 1 \leq i \leq r \text{ and } 1 \leq j \leq m, \tag{7.3}$$

then, $\mathbf{R}_{k,m}(b) = R(k,m,\vec{N},b)$.

For $\vec{N} = (N_1, \ldots N_r) \in \mathbb{Z}^r$, let

$$S(\vec{N}) = \sum_{i=1}^{r} f_{v_i} N_i .$$

We notice that if $\vec{N} \in \mathbb{Z}^r$ and $b \in \mathbb{B}$ verify (7.3), $S(\vec{N})$ tends to $+\infty$ if and only if $h(b)$, as defined in (2.3), tends to $-\infty$. In order to prove the theorem, we fix an integer $m \geq 2^k+1$. Then, we prove that, for every $b \in \mathbb{B}$ for which $S(\vec{N})$ is large enough, we have $R(k,m,\vec{N},b) > 0$, $\vec{N}$ and b being always subject to (7.3).

We want to get an asymptotic estimate for numbers $R(k, m, \vec{N}, b)$. For that, we have to generalize the work done in Sections 4, 5 and 6 in order to express $\mathbf{R}(k, m, \vec{N}, b)$ with the help of an integral. The main difficulty lies in finding a good definition of major arcs and minor arcs. Our dealing with minor arcs gives us an asymptotic estimate for the numbers $R(k, m, \vec{N}, b)$ corresponding to elements b such that the differences

$$\left| f_{w_i} N_i - \frac{1}{r} S(\vec{N}) \right|$$

remain bounded when $S(\vec{N})$ tends to infinity, $\vec{N}$ and b being always subject to (7.3). For such b, we get an estimate of the form

$$R(k, m, \vec{N}, b) = C_m(b) q^{(m-k)S(\vec{N})} + o(q^{(m-k)S(\vec{N})}), \tag{7.4}$$

with

$$C_m(b \geq \kappa(m, k) > 0, \tag{7.5}$$

$\kappa(m, k)$ being a constant depending only on $\mathbb{L}$, k and m.

Then, using the unit theorem, we prove that there exist constants $a_1, \ldots, a_r$, depending only on $\mathbb{L}$, such that, for every $b \in \mathbb{B}$, there exists a unit U in $\mathbb{B}$ with the following property: if $\vec{y} \in \mathbb{Z}^r$ and bU^k are subject to (7.3), then,

$$\left| f_{v_i} y_i - \frac{1}{r} S(\vec{y}) \right| \leq f_{v_i} a_i .$$

We write (7.4) for bU^k and $\vec{y}$. Then, if b and $\vec{N}$ are subject to (7.3), we have

$$y_i = N_i - v_i(U).$$

By the product formula,

$$S(\vec{y}) = S(\vec{N}).$$

Moreover, for $(b_1, \ldots, b_m) \in \mathbb{B}^m$,

$$b = b_1^k + \cdots + b_m^k \quad \text{if and only if} \quad bU^k = (b_1 U)^k + \cdots + (b_m U)^k,$$

and $R(k, m, \vec{N}, b) = R(k, m, \vec{y}, bU^k)$. So, if $S(\vec{N}) = S(\vec{y})$ is large enough,

$$R(k, m, \vec{N}, b) = R(k, m, \vec{y}, bU^k)$$

is positive.

8. Dealing with Minor Arcs when $k > p$

When $k > p$, Weyl's method fails, and one has to find another method. Very little is known in that case. A first step was done by J. Cherly, who proved in his thesis, that for $q = 2,\ k = 3$,

$$|f(t)| \leq 2^{3 + 19n/20},$$

for t belonging to a minor arc, cf. **[3]**.

It was very easy to generalize the work of Cherly to the ring $\mathbb{A} = \mathbb{F}_q[T]$, when q is a power of 2. Modifying the proof of Cherly a little, one obtains that, for $q = 2^b$, $k = 3$,

$$|f(t)| \ll q^{n(5/6+\varepsilon)},$$

for t belonging to a minor arc.

I give here the main steps of the proof.

We have to assume now that the characteristic is 2.

(1) We divide $\mathcal{P}$ by a Farey dissection of order n. Let $\frac{G}{H} \in \mathcal{F}_n$. Let $t \in \mathcal{U}_{G/H}$. If $\nu\left(t + \frac{G}{H}\right) > 2n + \deg H$, t belongs to a major arc of the Farey dissection of order $N = 2n = (k-1)n$. In that case, we process as for major arcs in the case $k < p$.

(2) We suppose that $\nu\left(t + \frac{G}{H}\right) \leq 2n + \deg H$. We put $u = t + \frac{G}{H}$. We have

$$\begin{aligned} f(t) &= \sum_{R \bmod H} \sum_{\substack{A \in \mathbb{A}_n \\ A \equiv R \bmod H}} E\left(\left(\frac{G}{H} + u\right) A^3\right) \\ &= \sum_{R \bmod H} E\left(\frac{G}{H} R^3\right) \sum_{\substack{A \in \mathbb{A}_n \\ A \equiv R \bmod H}} E(uA^3), \end{aligned}$$

$\mathbb{A}_n$ being as in Section 6. Then, with (4.5), we get

$$|H| f(t) = \sum_{k \bmod H} \sum_{R \bmod H} E\left(\frac{GR^3 + KR}{H}\right) \sum_{A \in \mathbb{A}_n} E\left(uA^3 + \frac{KA}{H}\right),$$

say,

$$|H| f(t) = \sum_{k \bmod H} S(H, G, K) Y(H, u, K). \tag{8.1}$$

(3) We study the sums $S(H, G, K)$.

We prove that $|S(H, G, K)|^2$ is equal to 0 or to the double sum

$$W(H, G) = \sum_{R \bmod H} \sum_{Q \bmod H} E\left(\frac{G(R^2Q + RQ^2)}{H}\right). \tag{8.2}$$

The function $H \mapsto W(H, G)$ being multiplicative, we compute $W(P^m, G)$ for any irreducible P. We find that

$$\begin{cases} W(P^{3m}, G) = |P|^{4m}, \\ W(P^{3m+2}, G) = |P|^{4m+2}, \\ W(P^{3m+1}, G) = y(P)|P|^{4m+1}, \text{ with } y(P) \in \{1, 2, 4\}. \end{cases} \tag{8.3}$$

Further, we prove that, if K is such that $S(H, G, K) \neq 0$, then,

$$S(H, G, K + L) \neq 0 \tag{8.4}$$

if and only if L belongs to a subgroup $\mathcal{L}(H)$ of the group of congruence classes mod H defined as follows.

The set $\mathcal{R}(H)$ of congruence classes R mod H such that $E\left(\frac{G(R^2Q+RQ^2)}{H}\right) = 1$ for every Q mod H, is a subgroup. We say that $L \in \mathcal{L}(H)$ if and only if $E\left(\frac{RL}{H}\right) = 1$

for every $R \in \mathcal{R}(H)$. Moreover, we can prove that

$$\text{modulo } \mathcal{L}(H) \text{ there is one, and only one } K \text{ such that } S(H,G,K) \neq 0, \tag{8.5}$$

and that

$$|H| = \#(\mathcal{R}(H))\#(\mathcal{L}(H)). \tag{8.6}$$

Finally we prove that there exist divisors ϕ and Θ of H such that $\mathcal{R}(H)$ is included in the ideal (ϕ) and such that, for any $\varepsilon > 0$,

$$\#(\mathcal{R}(H)/\mathcal{R}(H) \cap (\Theta)) \ll |H|^{\varepsilon}. \tag{8.7}$$

The functions ϕ and Θ are multiplicative functions of H. They are defined on powers P^j of irreducible polynomials P by

$$\begin{aligned} \Theta(P^{3m+r}) &= P^{m+r}, \qquad 0 \leq r \leq 2, \\ \phi(P^{3m}) &= \phi(P^{3m+1}) = P^m, \\ \phi(P^{3m+2}) &= P^{m+2}. \end{aligned} \tag{8.8}$$

(4) We study the sums $Y(H,u,K)$. We can prove that $|Y(H,u,K)|^2$ is equal to 0 or to another double sum. It is simpler to proceed in another way. We put

$$\nu(u) = 3m + \rho, \qquad \rho \in \{1, 0, -1\}.$$

Then

$$Y(H,u,K) = \sum_{\substack{B \in \mathbb{A}_n \\ T^m | B}} E\left(uB^3 + \frac{KB}{H}\right) \sum_{A \in \mathbb{A}_{m-1}} E\left(u(B^2A + BA^2) + \frac{KA}{H}\right).$$

As the inner sum is nonnegative,

$$\begin{aligned} |Y(H,u,K)| &\leq \sum_{\substack{B \in \mathbb{A}_n \\ T^m | B}} \sum_{A \in \mathbb{A}_{m-1}} E\left(u(B^2A + BA^2) + \frac{KA}{H}\right) \\ &= \sum_{A \in \mathbb{A}_{m-1}} E\left(\frac{KA}{H}\right) \sum_{\substack{B \in \mathbb{A}_n \\ T^m | B}} E(u(B^2A + BA^2)) \\ &= q^{-m} \sum_{A \in \mathbb{A}_{m-1}} E\left(\frac{KA}{H}\right) \sum_{B \in \mathbb{A}_n} E(u(B^2A + BA^2)). \end{aligned}$$

Let $\mathcal{A}(u)$ be the subgroup of polynomials $A \in \mathbb{A}_{m-1}$ such that $E(u(B^2A + BA^2)) = 1$ for each polynomial $B \in \mathbb{A}_n$. Then,

$$|Y(H,u,K)| \leq q^{n+1-m} \sum_{A \in \mathcal{A}(u)} E\left(\frac{KA}{H}\right). \tag{8.9}$$

Let K be a class such that $S(H,G,K) \neq 0$. With (8.1), (8.4) and (8.9), we get

$$|H|\,|f(t)| \leq q^{n+1-m} W(H,G)^{1/2} \sum_{L \in \mathcal{L}(H)} \sum_{A \in \mathcal{A}(u)} E\left(\frac{(K+L)A}{H}\right).$$

Interchanging the order of summation, and recalling the definition of $\mathcal{L}(H)$, we get

$$|H|\,|f(t)| \leq q^{n+1-m} W(H,G)^{1/2} \#(\mathcal{L}(H)) \#\{A \in \mathcal{A}(u) \mid \widehat{A} \in \mathcal{R}(H)\},$$

$\widehat{A}$ denoting the congruence class of A mod H. For every congruence class R mod H, let $\mathcal{A}(u; H, R)$ be the set of $A \in \mathcal{A}(u)$ congruent to R mod H. We prove that the nonempty sets $\mathcal{A}(u; H, R)$ have the same cardinality, equal to $\#(\mathcal{A}(u) \cap (H))$, so,

$$|H|\,|f(t)| \leq q^{n+1-m} W(H,G)^{1/2} \#(\mathcal{L}(H)) \#(\mathcal{A}(u) \cap (H)) \#\{R \in \mathcal{R}(H) \mid \mathcal{A}(u; H, R) \neq \emptyset\}. \tag{8.10}$$

From (8.6), we get in all cases,

$$|f(t)| \leq q^{n+1-m} W(H,G)^{1/2} \#(\mathcal{A}(u) \cap (H)), \tag{8.11}$$

and, if $\deg H \geq m$,

$$|H|\,|f(t)| \leq q^{n+1-m} W(H,G)^{1/2} \#(\mathcal{L}(H)) \#(\mathcal{A}(u) \cap \mathcal{R}(H)).$$

In that case, (8.6) and (8.7) give us

$$\#(\mathcal{R}(H)) |f(t)| \ll |H|^{\varepsilon} q^{n-m} W(H,G)^{1/2} \#(\mathcal{A}(u) \cap (\Theta)). \tag{8.12}$$

If $\deg \Theta \geq m$, (8.12) give us

$$|f(t)| \ll |H|^{\varepsilon+1} q^{n-m} W(H,G)^{-1/2} \ll q^{n-n/3} |H|^{1-1/3+\varepsilon} W(H,G)^{-1/2} \ll q^{n-n/3} |H|^{1/2-1/3+\varepsilon}.$$

So, if $\deg \Theta \geq m$, we have,

$$|f(t)| \ll q^{n(5/6+\varepsilon)}. \tag{8.13}$$

We are able to bound the numbers $\#(\mathcal{A}(u) \cap (D))$, when $\deg D < m$. We get that

$$\begin{cases} \#(\mathcal{A}(u) \cap (D)) \ll (q^m |D|^{-1})^{1/2} & \text{if } \nu(u) \leq 2n+1+\deg D, \\ \#(\mathcal{A}(u) \cap (D)) \ll q^{m/2} |D|^{-1} & \text{if } \nu(u) > 2n+1+\deg D. \end{cases} \tag{8.14}$$

We bound $|f(t)|$ using (8.11) and (8.14) when $\deg H < m$, using (8.12), (8.14) and (8.8) when $\deg H \geq m$ and $\deg \Theta < m$. We prove that the bound (8.13) is true in all cases.

If we use the bound (8.13), we get that $\mathbf{G}_q(3) \leq 19$. The bound $\mathbf{G}_q(3) \leq 13$ follows from (8.12) and from the estimate

$$\int_{\mathcal{P}} |f(t)|^4 dt \ll q^{n(2+\varepsilon)}$$

which is easy to prove.

Remarks. Looking the exponent 3 in this section to be $2+1$, we want to use the same method to bound the sum $f(t)$ for t belonging to a minor arc, for exponent $k = 2^h + 1$, and more generally for exponents $k = p^h + 1$, if the characteristic is p. We obtain the corresponding results about sums $S(H, G, K)$, $W(H, G)$ or sums $Y(H, u, K)$. Unfortunately, these results are not powerful enough, and the above method fails for $k > 3$. Another method is to be found.

References

[1] M. Car, Arithmétique additive dans l'anneau des polynômes à une indéterminée sur un corps fini. Thèse soutenue à l'Université de Provence, (1972).

[2] M. Car, Sommes de carrés dans $\mathbb{F}_q[X]$. Dissertationes Math. 215, (1983).

[3] J. Cherly, Sommes d'exponentielles cubiques dans l'anneau des polynômes en une variable sur le corps à deux éléments et application au problème de Waring. Thèse soutenue à l'Université de Bordeaux I, (1989).

[4] G. W. Effinger & D. R. Hayes, Additive number theory of polynomials over a finite field, Oxford University Press, (1991).

[5] D. R. Hayes, The expression of a polynomial as the sum of three irreducibles, Acta Arith. 11, (1966), 461–488.

[6] R. M. Kubota, Waring's problem for $\mathbb{F}_q[X]$. Dissertationes Math. 117, (1974).

[7] R. E. A. C. Paley, Theorems on polynomials in a Galois field. Quart. J. Math. 4 (1933), 52–63.

[8] C. L. Siegel, Generalization of Waring's problem to algebraic number fields, Amer. J. Math. 66, (1944), 122–136.

[9] C. L. Siegel, Sums of mth powers of algebraic integers, Ann. of Math. 46, (1945), 313–339.

[10] L. N. Vaserstein, Waring's problem for commutative rings, J. Number Theory 26, (1987), 299–307.

[11] L. N. Vaserstein, Waring's problem for algebras over fields, J. Number Theory 26, (1987), 286–298.

Mathématiques, URA 225, Faculté des Sciences de St-Jérôme, Av. Escadrille Normandie Niemen, 13397 Marseille Cedex 13, France

Ramsey's Theorem and Waring's Problem for Algebras over Fields

L. N. Vaserstein

Abstract

We show that for each integer $k \geq 1$ there is an integer $R(k)$ such that every element of every algebra A over every finite field $F = kF$ is a sum or difference of k k-th powers as well as the sum of $3k/2$ k-th powers provided that $\mathrm{card}(F) \geq R(k)$. Moreover, the statement on sums or differences holds also for any infinite field $F = kF$ such that the subgroup of the k-th powers in the multiplicative group F^* is of finite index.

Another result is that in the polynomial ring $F[T]$ in one variable T over any field F, every sum of cubes is the sum of 4 cubes. This result is new in the case when $\mathrm{card}\,(F) = 2$.

1. Statement of results

For any integer $k \geq 1$ and any ring A we denote by $\pm A^k$ the set of all $\pm a^k$ with $a \in A$. We denote by A_k the additive subgroup of A generated by $\pm A^k$, and we denote by $v_k(A)$ the least integer $s \geq 0$ such that every element of A_k is a sum or difference of s k-th powers. Let $w_k(A)$ denote the least integer $s \geq 0$ such that every sum of k-th powers in A is the sum of s k-th powers. Clearly, $v_k(A) \leq 2w_k(A)$. Note that $v_k(A) = w_k(A)$ when -1 is a k-th power in the center C of A, e.g., when k is odd.

Assume now that the center C of A contains a field F. A few upper bounds are given in **[7]** for the numbers $v_k(A)$ and $w_k(A)$). For example, by Theorem 3(d) of **[7]**, $v_k(A) \leq w_k(A) \leq k(k+1)/2$ when $\mathrm{card}(F) \geq k^2 - k + 1$ and $v_k(A) \leq w_k(A) \leq 2k - 1$ when $\mathrm{card}(F) \geq k^4$, provided that F is finite and either $kA \neq 0$ or $A = C$. In this paper we prove that $v_k(A) \leq k$ and $w_k(A) \leq 3k/2$ for sufficiently large $\mathrm{card}(F)$, provided that F is finite and either $kA \neq 0$ or $A = C$. In fact we obtain a more precise result. Namely, we replace the upper bound k by a number $k_{(p)} \leq k$ where p is the characteristic of the field F.

When $p \neq 0$, we set $k_{(p)} = \prod(a_i + 1) - 1$, where a_i are the digits of k in base p (so $k = \sum a_i p^i$). It is clear that $k_{(p)} \leq k$ and that $k_{(p)} = k$ if and only if $1 \leq k \leq p-1$ or $k = p^m - 1$ for some integer $m \geq 1$. We set $k_{(p)} = k$ when $p = 0$.

Theorem 1. *Assume that either $kA \neq 0$ or $C = A$. Assume also that $q = \mathrm{card}(F) < \infty$. Set $r = [F^*: \pm C^k \cap F^*] \leq \min(k, q-1)$. Then:*

(i) $A_k = A$ *when* $q \geq r(k-1) + 1$ *and* $kA \neq 0$*;*

(ii) $v_k(A) \leq 2k - \lceil k/r \rceil \leq 2k - 1$ *when* $q \geq k^4$, *where* $\lceil k/r \rceil$ *means the least integer* $\geq k/r$*;*

(iii) *there is an integer* $R(k)$ *depending only on* k, *such that* $v_k(A) \le k_{(p)} \le k$ *and* $w_k(A) \le 3k_{(p)}/2 \le 3k/2$ *when* $q \ge R(k)$.

We will prove the theorem in the following three sections.

If p divides k we can write $k = k'p^e$ with integers $k', e \ge 1$, and k' not divisible by p. Then $v_k(A) = v_{k'}(A)$ and $w_k(A) = w_{k'}(A)$ when $C = A$. Moreover, then $k_{(p)} = k'_{(p)}$. So it suffices to prove the theorem in the case when p does not divide k.

To prove (iii), we use a result of Ramsey **[6]**. We show that $v_k(A) \le k_{(p)}$ when $q \ge R(k_{(p)} - 1, k, r+1)$, where $R(?,?,?) < \infty$ are the Ramsey numbers **[6**, Theorem B**]**. Since $k_{(p)} \le k$ and, for any finite field F, $r \le k$ we can bound all these numbers $R(k_{(p)} - 1, k, r+1)$ by a function $R(k)$ of k.

Our next goal is to bound $v_k(A)$ for some infinite fields F. Namely, we will assume that the subgroup $\pm C^k \cap F^*$ of F^* is of finite index and we will prove that $v_k(A) \le k_{(p)}$.

Theorem 2. *Assume that either* $kA \ne 0$ *or* $C = A$. *Assume also that* F *is infinite and* $r = [F^*: \pm C^k \cap F^*] < \infty$. *Then every element of* A_k *has the form* $\sum_{m=1}^{k_{(p)}} (-1)^{m-1} y_m^k$ *with* $y_m \in A$. *In particular,* $v_k(A) \le k_{(p)} \le k$.

It follows from **[2]** that under the conditions of Theorem 2 every element of F has the form $y_1^k - y_2^k$. The proof uses Ramsey's theorem, and so does our proof of Theorem 2 in Section 4 below.

Note that when F is algebraic over a finite subfield, $[F^*: \pm F^k \cap F^*] \le k$, so $r \le k$ in this case. When F is algebraically closed, $r = 1$. When F = **R**, the reals, $r \le 2$. More generally, $[F^*: \pm F_k \cap F^*] < \infty$ for any locally compact nondiscrete field F.

Let us now discuss small values of k. When $k = 1$, $v_1(A) = w_1(A) = 1$ for any ring $A \ne 0$. When $k = 2$, $v_2(A) \le 2 = k_{(2)}$ in case $2A = A$ (i.e., $p \ne 2$), and $v_2(A) = w_2(A) = k_{(2)} = 1$ when $2A = 0 \ne A$ and A is commutative. If -1 is the sum of two squares in C (e.g., F is algebraic over a finite subfield,) and $p \ne 2$, then $w_2(A) \le 3$. On the other hand, when $F = \mathbb{R}$, $w_2(A) = \infty$ for some commutative ring A **[3]**. So $w_k(A)$ cannot be bounded under the conditions of Theorem 2.

When $k = 3$, we have $3 = k_{(p)}$ for any $p \ne 3$. By **[8]**, $v_3(A) = w_3(A) \le 3$, without any restrictions on A provided $p \ne 0$, 2. Also, as observed by Serre (see **[8]**), $w_3(A) \le 3$ when $p \ne 3$ (or A is commutative) and there are $\alpha, \beta, \gamma \in \pm C^3 \cap F^*$ with $\alpha + \beta + \gamma = 0$ (see also Theorem 2 and its proof below). Such α, β, γ exist in $\pm F^3$ when F is algebraic over a finite subfield and $\operatorname{card}(F) \ne 2, 4, 7, 13, 16$. They also exist when F is infinite and $[F^*: \pm C^k \cap F^*] < \infty$ by **[2]**. So under the conditions of the theorem, we have $v_3(A) = w_3(A) \le 3_{(p)} \le 3$ when $\operatorname{card}(F) \ne 2, 4, 16$. When $p = 3$, clearly $v_3(A) = w_3(A) = 1$.

In the case when F is a field, which does not contain a subfield of 4 elements, with $\operatorname{char}(F) = 0$ or 2 and there are no $\alpha, \beta, \gamma \in F$ such that $\alpha^3 + \beta^3 + \gamma^3 = 0 \ne \alpha\beta\gamma$, it is unknown whether $w_3(A) = 3$ for the polynomial ring $A = F[T]$ in one variable T. We complete the discussion of the case $k = 3$ by the following result which is new in the case $F = \mathbb{Z}/2\mathbb{Z}$:

Theorem 3. *For the polynomial ring* $A = F[T]$ *in one variable* T *over any field* F, *we have* $w_3(A) \le 4$.

When $k = 4$, by Norrie's formula [**4**, p. 729], every element x of A has the form $a^4 + b^4 - c^4 - d^4$, where each $a, b, c, d \in F[x]$ has degree 1, provided that $p \neq 2$ and $\mathrm{card}(F) \neq 3, 5, 9$. See Theorem 2 and its proof below regarding the case $k = 4 = p+1$. When $p = 2$, clearly $v_4(A) = w_4(A) = 1$.

Since little is known about lower bounds for $v_k(A)$, we can conjecture safely that the upper bounds in Theorems 1(iii) and 2 cannot be improved in general. A more precise conjecture is that $v_k(A) \ge k_{(p)}$ for $A = F[x]$ with an arbitrary field F of characteristic $p \ge 0$ (set $k_{(0)} = k$). Now we check this for very small $k_{(p)}$.

When $k_{(p)} = 1$, k is a power of p, so $v_k(A) = w_k(A) = 1$ for any commutative ring $A \neq 0$. When $k_{(p)} = 2$, we have $k = 2 \neq p$. It is clear that $x \in F[x]$ is not $\pm a^2$ with $A \in F[x]$, so $v_2(A) = 2$ for $A = F[x]$. When $k_{(p)} = 3$, then either $k = 3$ and $p \neq 3$, or $k = p^m + p^s$ with integers $m > s$. Under the conditions of Theorem 1 or 2, we have $v_k(A) = v_{k'}(A)$ with $k' = p^{m-s} + 1$, so we can assume that $s = 0$ in the second case. Since $k \ge 3, x \in F[x] = A$ is not a sum or difference of two k-th powers. So $v_k(F[x]) \ge 3$. By Theorems 1 and 2, $v_k(F[x]) = 3$ when $q = \mathrm{card}(F)$ is sufficiently large. By [**5**], $v_k(F[x]) \le 6$ for all F, and $v_k(A) \le 4$ when $q > (k-1)^2$.

2. Proof of Theorem 1(i)

We take k distinct elements $b_1, \ldots, b_k$ of $\pm C^k \cap F = H \cup \{0\}$, where $H = \pm C^k \cap F^* \subset F^*$ is a subgroup of index r. We claim that there are $c_1, \ldots, c_k \in F$ such that $d = \sum c_i (x + b_i)^k \in F[x]$ has degree 1. Indeed the system of k linear equations

$$\textstyle\sum_{i=1}^k c_i b_i^t = 0 \quad \text{for } t = 0, 1, \ldots, k-2, \quad \sum_{i=1}^k c_i b_i^{k-1} = 1$$

in k unknowns $c_1, \ldots, c_k$ has nonzero Vandermonde determinant

$$\delta = \prod_{i<j} (a_i - a_j).$$

For the solution $c_1, \ldots, c_k$ of the system, $d = \sum c_i (x + b_i)^k$ is a polynomial in x of degree 1 (with leading coefficient $\delta k \neq 0$). When x ranges over A, d also ranges over A. On the other hand, $c_i \in C_k$ because they are polynomials in $b_1, \ldots, b_k$ with integral coefficients and C_k is a subring of A. So d is a sum or difference of k-th powers. Thus, $A_k = A$.

3. Proof of Theorem 1(ii)

Let $b_i, c_i \in F$ be as in the proof of Theorem 1(i). At least $[k/r]$ of k numbers c_i are either 0 or in the same coset of $H = \pm C^k \cap F^* \subset F^*$. Multiplying $d = \sum c_i (x + b_i)^k$ by an element of F^*, we can assume that at least $[k/r]$ coefficients c_i belong to $H \cup 0 \subset \pm C^k$. The polynomial d in x is still of degree 1.

By Weil [**9**, p. 502], every element of F is the difference of two k-th powers when $q \ge k^4$. Thus, d is a sum or difference of $2k - [k/r]$ k-th powers.

4. Proof of Theorems 1 (iii) and 2

By **[6**, Theorem B**]**, given any natural numbers k', k'', and ρ with $k' < k''$ we can find a natural number $R(k', k'', \rho)$ such that if all k'-element subsets of an $R(k', k'', \rho)$-element set are colored into ρ colors, then there is a k''-element subset all of whose k'-element subsets are of the same color.

Since very few Ramsey numbers $R(k', k'', \rho)$ have been computed, we will give three versions of Theorem 1(iii) with different Ramsey numbers. The following theorem also includes Theorem 2.

Theorem 4. *Assume that* $kA \neq 0$ *and* $r = [F^*: \pm C^k \cap F^*] < \infty$. *Set* $r' = [F^*: C^k \cap F^*] \leq 2r < \infty$ *and* $q = \operatorname{card}(F) \leq \infty$.
(i) *If* $q \geq R(k_{(p)} - 1, k, r+1)$, *then every element of* A *is a sum or difference of* $k_{(p)}$ *k-th powers; if in addition* $k^4 \leq q$ *and* F *is algebraic over a finite subfield, then every element of* A *is the sum of* $k_{(p)}$ *k-th powers;*
(ii) *If* $q \geq R(k_{(p)} - 1, k, r' + 1)$, *then every element of* A *has the form* $\sum_{m=1}^{k_{(p)}} (-1)^{m-1} y_m^k$ *with* $y_j \in A$; *if in addition* $k^4 \leq q$ *and* F *is algebraic over a finite subfield, then every element of* A *is the sum of* $k_{(p)}$ *k-th powers;*
(iii) *If* $q \geq R(2, k, r')$ *or* k *is even and* $\operatorname{card}(F) \geq rR(2, k/2, r)$, *then every element of* A *has the form* $\sum_{m=1}^{k} (-1)^{m-1} y_m^k$ *with* $y_m \in A$; *if in addition* $k^4 \leq q$ *and* F *is algebraic over a finite subfield, then every element of* A *is the sum of* $\frac{3k}{2}$ *k-th powers.*

Proof of 4(i). When $k_{(p)} = 1$, k is a power of p, so $v_k(A) = w_k(A) = 1$ when $A \neq$) is commutative. By our assumption, $kF = F$, so $k_{(p)} = 1$ only when $k = 1$. Assume now that $k_{(p)} > 1$. Set $H = \pm C^k \cap F^*$, so $[F^*: H] = r$.

The number of binomial coefficients $t!(k-t)!/k!$ which are not zero in F is $k_{(p)} + 1 = \prod(a_i + 1)$. Let $t(1) = 0$, $t(2), ..., t(k_{(p)}) = k - 1$, $t(k_{(p)} + 1) = k$ be all these coefficients.

For any subset $X = \{b(1), ..., b(k_{(p)} - 1)\}$ of $\operatorname{card}(X) = k_{(p)} - 1$ distinct elements in F, we define its color $k(X) = \det(b(i)^{t(j)} : 1 \leq i, j \leq k_{(p)} - 1)H \in \{0\} \cup F^*/H$. It is well-defined (independent on ordering of X) because $-1 \in H$.

By Ramsey's Theorem B **[6]**, there is a subset Y of F with $\operatorname{card}(Y) = k$ whose subsets X with $\operatorname{card}(X) = k_{(p)} - 1$ all have the same color provided that $\operatorname{card}(F) \geq R(k_{(p)} - 1, k, r+1)$.

Now we can choose $k_{(p)}$ distinct elements $b(1), ..., b(k_{(p)})$ of Y with $\det(b(i)^{t(j)} : 1 \leq i, j \leq k_{(p)}) \neq 0$. Indeed, otherwise we would obtain a nonzero polynomial of degree $t(k_{(p)}) = k - 1$ which vanishes on the set Y of k elements.

Now $d = \sum c(m)(x + b(m))^k$ is a polynomial of degree 1 in x. Moreover all colors $c(i)H$ are the same. If $c(1) = 0$, then all $c(i)$ would be zero, but $d \neq 0$. So $c(1) \neq 0$. Dividing d by $c(1)$, we can assume that $c(i) \in H$ for all i. So d is a sum or difference of $k_{(p)}$ k-th powers. On the other hand, choosing an appropriate x in A, we can make d to be equal to an arbitrary element of A. So every element of A is a sum or difference of at most k k-th powers, hence $v_k(A) \leq k_{(p)}$.

Multiplying, if necessary, d by -1, we can assume that at least half of the coefficients $c(i)$ are k-th powers. When F is algebraic over a finite field and $q \geq k^4$, we can write -1 as the sum of two k-th powers in F **[9]**. So d becomes the sum of at most $3k_{(p)}/2$ k-th powers.

Proof of 4(ii). The case $k_{(p)} = 1$ is trivial, so let $k_{(p)} \geq 2$. Set $H' = C^k \cap F^*$, so $[F^*: H'] = r'$. Let $t(1) = 0, t(2), ..., t(k_{(p)}) = k-1, t(k_{(p)}+1) = k$ be as in the proof of 4(i).

Let $b(1), ..., b(R(k_{(p)} - 1, k, r' + 1))$ be distinct elements of F. For any subset $X = b(s(1)), ..., b(s(k_{(p)} - 1))$ of $\mathrm{card}(X) = k_{(p)} - 1$ distinct elements, we set $C(X) = \det(b(s(i))^{t(j)}: 1 \leq i, j \leq k_{(p)} - 1) \in F$, where $s(i) < s(j)$ for $i < j$, and we define its color $K'(X) = C(X)H' \in \{0\} \cup F^*/H'$. By Ramsey's theorem, there is a subset Y of F with $\mathrm{card}(Y) = k$ whose subsets X with $\mathrm{card}(X) = k_{(p)} - 1$ all have the same color. This color cannot be 0, because otherwise we would obtain a nonzero polynomial of degree $t(k_{(p)}) = k - 1$ which vanishes on the set Y of k elements. Multiplying the set Y' by an element of F^*, we can assume that $C(Z) \in H$ for every $(k_{(p)} - 1)$-element subset Z of Y'.

We take $k_{(p)}$ elements $b(s(1)), ..., b(s(k_{(p)}))$ of Y', where $s(i) < s(j)$ for $i < j$, and we set $c(m) = (-1)^{m-1}C(Y' \backslash b(s(m))) \in (-1)^{m-1}H$ for $m = 1, ..., k_{(p)}$. Now $d = \sum c(m)(x + b(s(m)))^k$ is a polynomial of degree 1 in x, so it ranges over A when x does so. On the other hand d has the form $\sum(-1)^{m-1}y_m^k$.

When F is algebraic over a finite field and $q \geq k^4$, we can write -1 as the sum of two k-th powers in F **[9]**, so d becomes the sum of at most $3k_{(p)}$ k-th powers.

Proof of 4(iii). The case $k = 1$ is trivial, so let $k \geq 2$. We choose a subset Z of F with $R(2, k, r')$ elements and order it linearly. We define an r'-coloring D of the 2-element subsets of Z' by $D\{a, b\} = (b - a)H \in F^*/H$, where $H = C^k \cap F^*$ and $a < b$. There is a k-element subset $Y' = \{b_1, ..., b_k\}$ of Z' with all 2-element subsets of the same color. Let $b_1 < ... < b_k$. Dividing all elements in Y by an element of F^*, we can assume that $b_j - b_i \in H$ for $i < j$. Then $d = \sum_{m=1}^k c_i(x + b_m)^k$ is a degree one polynomial in x, where

$$c_m = (-1)^{m-1} \prod_{\substack{i<j \\ i,j \neq m}} (b_j - b_i).$$

Note that all $(-1)^{m-1}c_m \in H$. So d is of the form $\sum(-1)^{m-1}y_m^k$.

Assume now that k is even. We color 2-element subsets of $H = \pm C^k \cap F^*$ by $D\{a, b\} = (a - b)H$. This is well defined, because $-1 \in H$. Since $\mathrm{card}(H) \geq R(2, k/2, r)$, there is a $k/2$-element subset $Y = \{b_1, ..., b_{k/2}\}$ of H with all 2-element subsets of the same color. Then

$$d = \sum_{m=1}^{k/2} (c_m(x + b_m)^k - c_m(x - b_m)^k)$$

is a degree one polynomial in x, where $b_m c_m = (-1)^{m-1} \prod(b(i)^2 - b(j)^2), m \neq i < j \neq m \leq k/2$.

Note that all c_k are the same modulo H. Dividing d by c_1, we can assume that all $c_k \in H$. So d is of the form $\sum(-1)^{m-1}y_m^k$.

When F is algebraic over a finite field and $q \geq k^4$, we can write -1 as the sum of two k-th powers in F **[9]**, so d is the sum of at most $3k_{(p)}/2$ k-th powers.

Remark 5. If -1 is the sum of s k-th powers in C, then the proof above gives that $w_k(A) \leq (s+1)k_{(p)}/2$ under the conditions of Theorem 1 with large q or Theorem 2.

Remark 6. Theorems 1, 2 and their proofs can be easily generalized, replacing the field F by any commutative local ring with sufficiently large residue field.

5. Proof of Theorem 3

When $\text{card}(F) \neq 2$, this is known **[7]**, **[8]**. So we assume that $F = \mathbb{Z}/2\mathbb{Z}$. Let e be an arbitrary sum of cubes in $A = (\mathbb{Z}/2\mathbb{Z})[T]$. Set $f = T + T^4 \in A$. Note that e is congruent to 0 or 1 modulo T, $T+1$, T^2+T+1, hence $e^3 + e^2 \in fA = T(T+1)(T^2+T+1)A$. Set $h = (e^3+e^2)/f \in A$.

We consider the identity

$$\begin{aligned}(x+d^2x+y)^3+(dx+d^2y+z)^3+(d^2x+d^3x+y+d^2z)^3+(x+d^2x+d^3x+y+d^2z)^3\\ = d^2xy^2+d^5xy^2+y^3+d^6y^3+d^4y^2z+dxz^2+d^4xz^2+d^2yz^2+z^3 =: e_1\end{aligned}$$

in the polynomial ring $(\mathbb{Z}/2\mathbb{Z})[d,x,y,z]$.

We replace d by T^2 and z by $yT+g$ with an indeterminate g in this identity. The left hand side stays a sum of 4 cubes, and the right hand side e_1 becomes

$$\begin{aligned}g^3+g^2T^2x+g^2T^8x+g^2Ty+g^2T^4y+gT^2y^2+gT^8y^2+y^3+T^3y^3+T^6y^3+T^9y^3+T^{12}y^3\\ = xf^2g^2+f^2y^2(g+fy)+fg^2y+g^3+y^3 =: e_2 \in A[g,x,y].\end{aligned}$$

Note that $e_2 \equiv y^3(1+f^3)+gf^2y^2 \,(\text{mod } g^2A[g,x,y])$. We set $y = e+1+vf^2$ and $g = 1+f(1+e+h)+f^2u$ with the above e, h and indeterminates u, v. Then e_2 becomes $e_3 \in A[x,u,v]$ with $e_3 \equiv e\,(\text{mod } f^2A[x,u,v])$ and $e_3 \equiv y^3(1+f^3)+gf^2y^2\,(\text{mod } g^2A[x,u,v])$.

We specialize u to an element u_0 of A so that $g = 1+f(1+e+h)+f^2u \in A[u]$ becomes an irreducible polynomial $g_0 = 1+f(1+e+h)+f^2u_0 \in A$ of odd degree $\deg(g_0) > \max(4, \deg(e))$. This is possible by a theorem on primes in arithmetic progressions (see **[1, A12]**). Then g_0 is coprime with $ef(1+f^3)$ (it is coprime with $1+f^3$ since the irreducible factors of $1+f^3$ have degrees ≤ 4 and g_0 is irreducible with $\deg(g_0) > 4$), and every element of A is a cube modulo g_0 (since g_0 is irreducible of odd degree).

Next we specialize v to an element v_0 of A so that $y = e+1+vf^2$ becomes an element y_0 of A such that $y_0^3(1+f^3)+g_0f^2y_0^2 \equiv 0\,(\text{mod } g_0^2A)$. Namely, modulo g_0 we take y_0 to be the cubic root of $e/(1+f^3)$. Then we can adjust our choice modulo g_0^2 by Hensel's lemma because $3y_0(1+f^3)$ is invertible modulo g_0.

After these specializations of u and v, the polynomial $e_3 \in A[x,u,v]$ becomes an element $e_4 = xf^2g_0^2 + c$ of $A[x]$ (still a sum of 4 cubes) with $c \in A$ such that $c \equiv e$

(mod f^2A) and $c \equiv e$ (mod g_0^2A). Since f and g_0 are coprime, we can now specialize x to an element x_0 of A to obtain that e is the sum of 4 cubes.

References

[1] H. Bass, J. Milnor, and J.-P. Serre: Solution of the congruence subgroup problem for SL_nA $(n \geq 3)$ and $Sp_{2n}A$ $(n \geq 2)$, *Inst. Hautes. Etudes Sci. Publ. Math.* 33, (1967), 59–137.

[2] V. Bergelson and D. B. Shapiro: Multiplicative subgroups of finite index in a ring, preprint 1990.

[3] M. D. Choi, Z. D. Dai, T. Y. Lam, and B. Reznick: The Pythagoras number of some affine algebras and local algebras, *J. Reine Angew. Math.* 336, (1982), 45–82.

[4] L. E . Dickson, History of the theory of numbers, *Carnegie Institution of Washington*, 1923.

[5] R.E.A.C. Paley, Theorems on polynomials in a Galois field: *Quart. J. Math.* 4, (1933), 52–63.

[6] F.P. Ramsey, On a problem of formal logic *Proc. London Math. Soc.* 30, (1930), 269–285.

[7] L. N. Vaserstein: Waring's problem for algebras over fields, *J. Number Theory* 26:3, (July 1987), 286–298.

[8] L. N. Vaserstein: On sums of cubes in polynomial rings, *Math. Comp.* 56, (1991), 349–357.

[9] A. Weil: Number of solutions of equations in finite fields, *Bull. Amer. Math. Soc.* 55, (1949), 427–568.

Department of Mathematics, The Pennsylvania State University, University Park, PA 16802
Email: vstein@math.psu.edu

Sums of Three Cubes

G. Payne and L. Vaserstein

In Section 1, we consider the so-called easier Waring problem for cubes and give tables of numerical data. This leads us to a similar problem for polynomial rings both with integral and rational coefficients. In Section 2 we discuss the Waring problem for polynomials with coefficients in an arbitrary field. In Section 3 we discuss representations of 1 as a sum of 3 rational cubes. We give the parametric representations of Euler, Binet, and others and discuss relations among them. In Section 4 we define a composition of representations. In Section 5 we construct infinite families of representations of 1 as a sum of 3 integral cubes. In Section 6 we show that every cube has infinitely many nontrivial primitive representations as a sum of 3 integral cubes. We pose several open problems (Questions 1-8).

1. Sums of integral cubes

It has been known since 1895 [**1**, p. 719] that every nonnegative integer is the sum of 9 nonnegative integral cubes. On the other hand, 23 is not the sum of 8 such cubes, so the number $g_3 = 9$ is computed. (In fact, all g_k are now known.) It is also known that every sufficiently large integer is the sum of 7 such cubes.

Allowing negative cubes, it is easy to see that every integer is the sum of 5 integral cubes [**9**]. On the other hand, the numbers of the form $9m \pm 4$ are not sums of 3 cubes, so we have the following open problems.

Question 1. *Is every integer the sum of* 4 *integral cubes?*

Question 2. *Is every integer* $\neq 9m \pm 4$ *with integral* m *the sum of* 3 *cubes?*

It was proved in [**6**] and [**2**] that every integer $\neq 18m \pm 4$ with integral m is the sum of four cubes. The affirmative answer to Question 2 would imply the affirmative answer to Question 1 as well as to the following question.

Question 3. *Is there a finite algorithm to recognize whether an integer is the sum of* 3 *integral cubes?*

The monstrous possibility of a negative answer to Question 3 might scare many mathematicians away from working on Questions 1-3. However we are in position to set a computer to work on the problem in hope of finding some patterns. In Table 1 we give representations of nonnegative integers $\neq 9m \pm 4$ as sums of 3 integral cubes.

Table 1. The integral solutions of $a^3 + b^3 + c^3 = d$ with an integral parameter $d \neq 9m \pm 4$ in the interval $0 \leq d \leq 29$ which are lexicographically minimal with respect to $(|a|, |b|, |c|)$.

a	0	0	0	1	-1	0	0	0	1	-2	7	-1
b	0	0	1	1	-1	-1	0	1	1	-2	10	2
c	0	1	1	1	2	2	2	2	2	3	-11	2
d	0	1	2	3	6	7	8	9	10	11	12	15

a	0	1	-1	0	1	-11	2	-1	0	0	0	1
b	2	2	-2	-2	-2	-14	2	-1	-1	0	1	1
c	2	2	3	3	-3	16	2	3	3	3	3	3
d	16	17	18	19	20	21	24	25	26	27	28	29

A computer search showed that there are no integral solutions to $a^3+b^3+c^3 = 30$ with $|a|, |b|, |c| \leq 99,999$ (we thank P. Berman for help with this and other computations). We also found all integral solutions of $a^3 + b^3 + c^3 = d$ with $1 \leq d \leq 12$ and $|a|, |b|, |c| \leq 99,999$. Some of them are given in Tables 2-5.

Table 2. All integral solutions of $a^3 + b^3 + c^3 = 1$ with $38,200 \geq |a| > |b| \geq |c|$ ordered with respect to $|a|$.

a	1	9	-12	-103	144	-150	172	-249	-495	505	577	729	-738	904	1010
b	0	-8	10	94	-133	144	-138	235	438	-426	-486	-720	729	-823	-812
c	0	-6	9	64	-71	73	-135	135	334	-372	-426	-242	244	-566	-791

a	1210	-1544	-1852	-1988	2304	-2316	3097	-4184	3753	-5262	5625	-5640
b	-1207	1537	1738	1897	-2292	2304	-2820	3518	-3230	4528	-5610	5625
c	-236	368	1033	1010	-575	577	-1938	3097	-2676	3753	-1124	1126

a	6081	6756	-8657	8703	-9953	11664	-11682	12884	16849	18649	21609
b	-5984	-6702	8343	-8675	9735	-11646	11664	-11903	-16806	-17328	-21588
c	-2196	-1943	4083	-1851	3987	-1943	1945	-7676	-3318	-10866	3086

a	-21630	24987	29737	-31615	-38599	-36620	36864	-36888	37513	38134
b	21609	-24965	-27630	31180	33412	35442	-36840	36864	-31212	-37887
c	3088	-3453	-17328	10876	27238	16617	-4607	4609	-28182	-10230

Table 3. All primitive integral solutions of $a^3+b^3+c^3 = 8$ with $25,000 \geq |a| > |b| \geq |c|$ ordered with respect to $|a|$.

a	-16	-34	41	-89	-127	-150	-385	-466	-873	995	-1312	-1985
b	15	33	-40	86	106	141	345	459	825	-947	1293	1808
c	9	15	-17	41	95	83	252	165	470	-514	459	1241
a	-1987	-2448	2840	2883	-2908	-2920	-3330	-4073	-4129	-6562	-8786
b	1740	2189	-2831	-2526	2671	2377	2979	4016	3966	5245	7163
c	1371	1611	-601	-1987	1769	2255	2189	1409	2001	5171	6773
a	-10713	11852	-13813	17429	18414	19881	20351	-21184	23145	-23399	23961
b	10289	-10303	13224	-14636	-17217	-18808	-19806	20883	-21076	22268	-20097
c	5196	-8297	6861	-12925	-10447	-10641	-8703	7365	-14481	12095	-17800

Table 4. All integral solutions of $a^3 + b^3 + c^3 = d$ with $99,999 \geq |a| > |b| \geq |c|$ and $d = 3, 6, 7, 9$, lexicographically ordered with respect to $(d, |a|)$ excluding permutations of the solutions given in Table 1.

a	-5	65	236	644	-60355	-105	-169	217	-11329
b	4	-58	-235	-637	60248	104	168	-216	11305
c	4	-43	-55	-205	10529	32	44	-52	2097
d	3	6	6	6	6	7	7	9	9

Table 5. All integral solutions of $a^3 + b^3 + c^3 = d$ with $99,999 \geq |a| > |b| \geq |c|$ and $d = 10, 11, 12$, lexicographically ordered with respect to $(d, |a|)$.

a	2	4	-171	683	3	258	-641	843	6628	-11
b	1	-3	141	-650	-2	-212	619	-695	-5973	10
c	1	-3	130	-353	-2	-197	297	-641	-4274	7
d	10	10	10	10	11	11	11	11	11	12

All integral solutions of $a^3 + b^3 + c^3 = 2$ with $99,999 \geq |a| > |b| \geq |c|$ found by computer can be obtained, up to a permutation of a, b, c, from the polynomial solution $(6t^3 + 1)^3 - (6t^3 - 1)^3 - (6t^2)^3 = 2$ **[4]** by substituting an integer s for $t, 0 \leq s \leq 55$.

This leads to the following question.

Question 4. *Does every integral solution of $a^3 + b^3 + c^3 = 2$, up to a permutation of a, b, c, come from the polynomial solution $(6t^3 + 1)^3 - (6t^3 - 1)^3 - (6t^2)^3 = 2$ by specializing t to an integer?*

Among the integral solutions of $a^3 + b^3 + c^3 = d$ found by computer, the number of those with $d = 3, 7, 9, 10, 11,$ or 12 seems surprisingly small in comparison with the number of those with $d = 1, 2,$ or 8. The question arises whether there exist other solutions. For instance, here is

Question 5. *Are there any integral solutions to* $a^3 + b^3 + c^3 = 3$ *with* $a \neq 1, 4, -5$? *If yes, is the number of such solutions finite?*

To answer affirmatively the analogous question of Mordell for $d = 1$, Mahler **[3]** used the polynomial solution

$$(9t^4)^3 + (3t - 9t^4)^3 + (1 - 9t^3)^3 = 1. \tag{1}$$

This polynomial solution explains several integral solutions in Table 2, but not all of them. This brings us to the following problem.

Question 6. *Is it true that for every integral solution of*

$$a^3 + b^3 + c^3 = 1 \tag{2}$$

there is an integral polynomial solution $a(t), b(t), c(t) \in \mathbb{Z}[t]\backslash\mathbb{Z}$ *(or in* $\mathbb{Q}[t]\backslash\mathbb{Q}$*) such that* $a(0) = a, b(0) = b, c(0) = c$?

We can obtain an affirmative answer to a weaker question if we are allowed to use more than 3 cubes:

$$x_1^3 + x_2^3 + x_3^3 = x_1(s,t)^3 + x_2(s,t)^3 + x_3(s,t)^3 + x_4(s,t)^3 + x_5(s,t)^5,$$

where

$$x_1(s,t) = -4s^2t + x_1 - 8s^3t^3x_1 + 2st^2x_1^2 - 5s^4t^5x_1^2 + s^2t^4x_1^3 - s^5t^7x_1^3 - 8s^3t^3x_2 - 10s^4t^5x_1x_2 + s^2t^4x_1^2x_2 - 3s^5t^7x_1^2x_2 - 2st^2x_2^2 - 5s^4t^5x_2^2 - s^2t^4x_1x_2^2 - 3s^5t^7x_1x_2^2 - s^2t^4x_2^3 - s^5t^7x_2^3 - 4stx_3 - 6s^2t^3x_1x_3 - 2s^3t^5x_1^2x_3 - 6s^2t^3x_2x_3 - 4s^3t^5x_1x_2x_3 - 2s^3t^5x_2^2x_3,$$

$$x_2(s,t) = 4s^2t + 8s^3t^3x_1 - 2st^2x_1^2 + 5s^4t^5x_1^2 - s^2t^4x_1^3 + s^5t^7x_1^3 + x_2 + 8s^3t^3x_2 + 10s^4t^5x_1x_2 - s^2t^4x_1^2x_2 + 3s^5t^7x_1^2x_2 + 2st^2x_2^2 + 5s^4t^5x_2^2 + s^2t^4x_1x_2^2 + 3s^5t^7x_1x_2^2 + s^2t^4x_2^3 + s^5t^7x_2^3 + 4stx_3 + 6s^2t^3x_1x_3 + 2s^3t^5x_1^2x_3 + 6s^2t^3x_2x_3 + 4s^3t^5x_1x_2x_3 + 2s^3t^5x_2^2x_3,$$

$$x_3(s,t) = 2s + s^2t^2x_1 + s^2t^2x_2 + x_3,$$

$$x_4(s,t) = 4s^2t^2x_1 - tx_1^2 + 4s^3t^4x_1^2 - st^3x_1^3 + s^4t^6x_1^3 + 4s^2t^2x_2 + 8s^3t^4x_1x_2 - st^3x_1^2x_2 + 3s^4t^6x_1^2x_2 + tx_2^2 + 4s^3t^4x_2^2 + st^3x_1x_2^2 + 3s^4t^6x_1x_2^2 + st^3x_2^3 + s^4t^6x_2^3 + x_3 + 4st^2x_1x_3 + 2s^2t^4x_1^2x_3 + 4st^2x_2x_3 + 4s^2t^4x_1x_2x_3 + 2s^2t^4x_2^2x_3,$$

$$x_5(s,t) = -2s - 5s^2t^2x_1 + tx_1^2 - 4s^3t^4x_1^2 + st^3x_1^3 - s^4t^6x_1^3 - 5s^2t^2x_2 - 8s^3t^4x_1x_2 + st^3x_1^2x_2 - 3s^4t^6x_1^2x_2 - tx_2^2 - 4s^3t^4x_2^2 - st^3x_1x_2^2 - 3s^4t^6x_1x_2^2 - st^3x_2^3 - s^4t^6x_2^3 - x_3 - 4st^2x_1x_3 - 2s^2t^4x_1^2x_3 - 4st^2x_2x_3 - 4s^2t^4x_1x_2x_3 - 2s^2t^4x_2^2x_3,$$

$$x_1(0,t) = x_1,\ x_2(0,t) = x_2,\ x_3(0,t) = x_3,\ x_4(0,t) = -x_5(0,t).$$

Note that none of $x_k(s,t)$ above is constant. The same is true for $x_k(t)$ in the following similar identity for the sum of 2 cubes :

$$x_1^3 + x_2^3 = x_1(t)^3 + x_2(t)^3 + x_3(t)^3 + x_4(t)^3,$$

where

$$x_1(t) = x_1 + x_1^2t - 2x_1x_2t - 3x_2^2t - 8x_1^3t^2 - 12x_1^2x_2t^2 + 4x_2^3t^2 + x_1^4t^3 + 12x_1^3x_2t^3 + 30x_1^2x_2^2t^3 + 28x_1x_2^3t^3 + 9x_2^4t^3 + 10x_1^5t^4 + 34x_1^4x_2t^4 + 36x_1^3x_2^2t^4 + 4x_1^2x_2^3t^4 - 14x_1x_2^4t^4 - 6x_2^5t^4 - 4x_1^6t^5 - 24x_1^5x_2t^5 - 60x_1^4x_2^2t^5 - 80x_1^3x_2^3t^5 - 60x_1^2x_2^4t^5 - 24x_1x_2^5t^5 - 4x_2^6t^5,$$

$x_2(t) = x_2+3x_1^2t+2x_1x_2t-x_2^2t+4x_1^3t^2-12x_1x_2^2t^2-8x_2^3t^2-9x_1^4t^3-28x_1^3x_2t^3-30x_1^2x_2^2t^3-12x_1x_2^3t^3-x_2^4t^3-6x_1^5t^4-14x_1^4x_2t^4+4x_1^3x_2^2t^4+36x_1^2x_2^3t^4+34x_1x_2^4t^4+10x_2^5t^4+4x_1^6t^5+24x_1^5x_2t^5+60x_1^4x_2^2t^5+80x_1^3x_2^3t^5+60x_1^2x_2^4t^5+24x_1x_2^5t^5+4x_2^6t^5,$

$x_3(t) = -x_1/2+x_2/2-x_1^2t/2+3x_1x_2t+7x_2^2t/2+8x_1^3t^2+12x_1^2x_2t^2-4x_2^3t^2-x_1^4t^3-12x_1^3x_2t^3-30x_1^2x_2^2t^3-28x_1x_2^3t^3-9x_2^4t^3-10x_1^5t^4-34x_1^4x_2t^4-36x_1^3x_2^2t^4-4x_1^2x_2^3t^4+14x_1x_2^4t^4+6x_2^5t^4+4x_1^6t^5+24x_1^5x_2t^5+60x_1^4x_2^2t^5+80x_1^3x_2^3t^5+60x_1^2x_2^4t^5+24x_1x_2^5t^5+4x_2^6t^5,$

$x_4(t) = x_1/2-x_2/2-7x_1^2t/2-3x_1x_2t+x_2^2t/2-4x_1^3t^2+12x_1x_2^2t^2+8x_2^3t^2+9x_1^4t^3+28x_1^3x_2t^3+30x_1^2x_2^2t^3+12x_1x_2^3t^3+x_2^4t^3+6x_1^5t^4+14x_1^4x_2t^4-4x_1^3x_2^2t^4-36x_1^2x_2^3t^4-34x_1x_2^4t^4-10x_2^5t^4-4x_1^6t^5-24x_1^5x_2t^5-60x_1^4x_2^2t^5-80x_1^3x_2^3t^5-60x_1^2x_2^4t^5-24x_1x_2^5t^5-4x_2^6t^5,$

$x_1(0) = x_1, x_2(0) = x_2, x_3(0) = -x_4(0).$

In Section 5 below, using composition of rational solutions, we will construct an infinite exponentially parameterized family of integral solutions of (2). All these solutions are different from those given by (1). Moreover, such a family cannot be covered by a finite number of polynomial solutions. We also give Hickerson's analogous construction of an infinite family of integral polynomial solutions of (2).

2. Sums of polynomial cubes

Often, when we cannot solve a problem for $\mathbb{Z}$ (e.g.,to prove the Riemann hypothesis or Fermat's last theorem) we state a similar problem for polynomial rings and solve it. However in our situation, numerical data for $\mathbb{Z}$ suggest that we look for solutions of the equations $a^3+b^3+c^3=d$ in $\mathbb{Q}[t]$ with the degree of d being 0 or 1. If there is a solution with the degree of d equal to 1, then obviously every polynomial in $\mathbb{Q}[d]$ is the sum of 3 cubes. Our next open problem is

Question 7. *Are there polynomials $a, b, c \in \mathbb{Q}[t]$ satisfying the equation*

$$a^3+b^3+c^3=t? \tag{3}$$

In trying to answer this question, we come to the analogous question for polynomials over local and finite fields instead of polynomials over $\mathbb{Q}$. Serre observed in 1982 that for any field F with $3F \neq 0$ the equation (3) has a solution $a, b, c \in F[t]$ with $\deg(a), \deg(b), \deg(c) \leq 1$ if and only if there are nonzero $\alpha, \beta, \gamma \in F$ such that $\alpha^3+\beta^3+\gamma^3=0$. The 'only if' part is trivial, and the 'if' part follows from the identity

$$(\alpha t+\beta^2)^3+(\beta t-\alpha^2)^3+(\gamma t)^3=\beta^6-\alpha^6-3\alpha\beta\gamma^3t.$$

In particular, the equation (3) has a degree one solution for all finite fields F with $3F \neq 0$ with the exceptions $\mathrm{card}(F) = 2, 4, 7, 13, 16$. Assuming that there are no solutions of degree one, Serre proved that there are no solutions of degree ≤ 2, Hayes (1987) proved that there are no solutions of degree ≤ 3, and Vaserstein (1987) proved that there are no solutions of degree ≤ 4.

Surprisingly, in 1989 solutions of higher degree were found. For example **[8]**,

$(-4t+5t^2+6t^3+t^4)^3+(1+4t+t^2-5t^3-3t^4+5t^5)^3+(-1+5t-5t^2+5t^3+3t^4-5t^5)^3=t \pmod{13}$

and

$(t-3t^3-t^5)^3+(-1-t-t^3+3t^4-t^6)^3+(1-t-t^3-3t^4+t^6)^3=t \pmod 7$.

In fact, all solutions of

$$\alpha a^3+\beta b^3+\gamma c^3=d$$

with $a,b,c\in F[t], \alpha,\beta,\gamma\in F$, $\deg(a)<\deg(b)=\deg(c)\le 4$, $\deg(d)\le 1$, and F an arbitrary field were found by computer **[8]**. All solutions of this equation with $\deg(a)<\deg(b)=\deg(c)\le 6$, $\deg(d)\le 1$, and $\operatorname{card}(F)=5,7,11,13$ were also found **[8]**.

3. Rational solutions for (2) and (3)

The formula

$$t=((t^3-1/27)^3+(t^2+t/3)^3+(-t^3+t/3+1/27)^3)/(t^2+t/3+1/9)^3$$

gives a solution $a,b,c\in\mathbb{Q}(t)$ of (3), and it shows that every element of any field F such that $F=3F$ is the sum of 3 cubes. Several mathematicians **[1**, p.550-552**]**, including Diophantus, Bungus, Vieta, Bachet, Girard, Fermat, Frenniccle, Wallis, Prestet, Euler, Emerson, Grüson, Jones, Young, Poselger, Binet, Bounikowsky, Richaud, Grassmann, Hermite, Several, Korneck, Réalis, Brunel, Schlegel, Martin, Moreau, Werebrusow, Schwening, Kühne, Mirimanoff, Holm, Teilhet, Gérardin, Holden, Jandasek, Petr, Norrie,and Fujiwara have found rational solutions of (2), i.e. integral primitive solutions to

$$a^3+b^3+c^3+d^3=0. \tag{4}$$

Euler was the first to describe all solutions. Up to a rational factor, they are given by

$a=-f^4-6f^2g^2-9g^4+fh^3-3gh^3+3fh^2k+3gh^2k+3fhk^2-9ghk^2+9fk^3+9gk^3$,

$b=f^4+6f^2g^2+9g^4-fh^3-3gh^3+3fh^2k-3gh^2k-3fhk^2-9ghk^2+9fk^3-9gk^3$,

$c=-f^3h+3f^2gh-3fg^2h+9g^3h+h^4-3f^3k-3f^2gk-9fg^2k-9g^3k+6h^2k^2+9k^4$,

$d=f^3h+3f^2gh+3fg^2h+9g^3h-h^4-3f^3k+3f^2gk-9fg^2k+9g^3k-6h^2k^2-9k^4$.

Binet's solution to (4) is

$$(a:b:c:d)=(3\alpha-9\gamma^2:3\beta+9\gamma^2:1-9\alpha\gamma:-1-9\beta\gamma),$$

where α,β are arbitrary and $\gamma=\alpha^2+\beta^2-\alpha\beta$.

Setting $\alpha=0$ and then replacing β by $-t$, we obtain (1). A. Gérardin also found (1).

We describe a general solution to (4) following Dickson **[3**, p. 59**]**. We start with the following remarkable determinant: $\det(M)=16(a^3+b^3+c^3+d^3)$, where

$$M=\begin{pmatrix} a+b+c+d & 3(a-b-c+d) & -3(a-b+c-d)\\ -(a-b-c+d) & a+b+c+d & 3(a+b-c-d)\\ a-b+c-d & -(a+b-c-d) & a+b+c+d\end{pmatrix}.$$

If $a, b, c, d \in F$ is a solution to (4) over a field F, then the determinant vanishes, so there is a nonzero column $(x, y, z)^{\mathrm{T}}$ such that $M(x, y, z)^{\mathrm{T}} = 0$. Solving this for a, b, c, d, we obtain the following solution to (4) (up to a rational factor):

$$\begin{aligned}
a &= -9x^3 + 3y^3 + 3x^2z + 6xyz + yz^2 + z^3 + 27x^3y^3 - +9xy^2z^3,\\
b &= 9x^3 - 9x^2y + 3xy^2 - 3y^3 + 3x^2z + 6xyz + 3y^2z + 3xz^2 - yz^2 + z^3,\\
c &= 9x^3 + 9x^2y + 3xy^2 + 3y^3 - 3x^2z + 6xyz - 3y^2z + 3xz^2 + yz^2 - z^3,\\
d &= -9x^3 - 9x^2y - 3xy^2 - 3y^3 - 3x^2z + 6xyz - 3y^2z - 3xz^2 - yz^2 - z^3.
\end{aligned} \tag{5}$$

To obtain the Euler solution (up to a rational factor) we substitute

$$\begin{aligned}
x &= 6(gh - fk)(f^4 + 6f^2g^2 + 9g^4 + f^2h^2 + 3g^2h^2 + h^4 + 3f^2k^2 + 9g^2k^2 + \\
&\quad + 6h^2k^2 + 9k^4),\\
y &= (f^2 + 3g^2 - h^2 - 3k^2)(f^4 + 6f^2g^2 + 9g^4 + 2f^3h + 6fg^2h + \\
&\quad + 3f^2h^2 + 9g^2h^2 + 2fh^3 + +h^4 + 6f^2gk + 18g^3k + 6gh^2k + \\
&\quad + 9f^2k^2 + 27g^2k^2 + 6fhk^2 + 6h^2k^2 + 18gk^3 + 9k^4),\\
z &= f^6 + 9f^4g^2 + 27f^2g^4 + 27g^6 - 2f^3h^3 - 6fg^2h^3 + h^6 - 6f^2gh^2k - 18g^3h^2k\\
&\quad - 6f^3hk^2 - 18fg^2hk^2 + 9h^4k^2 - 18f^2gk^3 - 54g^3k^3 + 27h^2k^4 + 27k^6.
\end{aligned}$$

To obtain the Binet solution (up to a rational factor) we substitute into (5)

$$\begin{aligned}
x &= -3(\alpha + \beta)(1 + 3\alpha^2 + 9\alpha^4 - 3\alpha\beta - 18\alpha^3\beta + 3\beta^2 + 27\alpha^2\beta^2 - 18\alpha\beta^3 + 9\beta^4),\\
y &= (-1 + 3\alpha^2 - 3\alpha\beta + 3\beta^2)(1 + 3\delta),
\end{aligned}$$

where

$$\begin{aligned}
\delta &= \alpha + 3\alpha^2 + 3\alpha^3 + 3\alpha^4 - \beta - 3\alpha\beta - 6\alpha^2\beta - 6\alpha^3\beta + 3\beta^2 + 6\alpha\beta^2 + \\
&\quad + 9\alpha^2\beta^2 - 3\beta^3 - 6\alpha\beta^3 + 3\beta^4,\\
z &= 1 - 9\alpha^3 + 27\alpha^6 + 18\alpha^2\beta - 81\alpha^5\beta - 18\alpha\beta^2 + 162\alpha^4\beta^2 + 9\beta^3 - 189\alpha^3\beta^3 + \\
&\quad + 162\alpha^2\beta^4 - 81\alpha\beta^5 + 27\beta^6.
\end{aligned}$$

To obtain (1) (up to a rational factor) we substitute into (5)

$$\begin{aligned}
x &= (-1 + 3t^2)(1 + 3t + 9t^2 + 9t^3 + 9t^4),\\
y &= -(1 - 9t^3 + 27t^6),\\
z &= t(1 + 3t^2 + 9t^4).
\end{aligned}$$

The formula (5) gives all rational solutions to (4) (up to a scalar factor). More generally, it gives all solutions (up to a scalar factor) over any field F such that $6F = F$ and F does not contain cubic roots of 1. If $6F = F$ and F contains a cubic root, $\varepsilon \neq 1$ of 1, then for any solution (a, b, c, d) to (4) over F either (a, b, c, d) or $(\varepsilon a, b, c, d)$ can be obtained, up to a scalar factor, from (5) with appropriate x, y, z.

The solution (5) shows that the surface given by (2) is birationally equivalent to the affine plane over any field of characteristic $\neq 2$.

Question 8. *How can all solutions of* (2) *over a field of characteristic* 2 *be described?*

4. Composition of solutions of (4)

For vectors $u = (u_1, u_2, u_3, u_4)$ and $w = (w_1, w_2, w_3, w_4)$ we denote by $u \cdot w$ their dot product, $\sum u_i w_i$. We denote by u^m the vector $(u_1^m, u_2^m, u_3^m, u_4^m)$.

If $u = (u_1, u_2, u_3, u_4)$ and $w = (w_1, w_2, w_3, w_4)$ are solutions of (4), then $u \bullet w$ defined by $u \bullet w = (u^2 \cdot w)w - (u \cdot w^2)u = -w \bullet u$ is also a solution. Note that $u \bullet (cw) = c^2(u \bullet w) = (cu) \bullet w$ for any scalar c.

The set $S(Z)$ of all solutions of (4) over a commutative ring Z includes the following 3 planes of *trivial* solutions:

$$P_1(Z) = \{(a, -a, b, -b) : a, b \in Z\},$$
$$P_2(Z) = \{(a, b, -a, -b) : a, b \in Z\},$$
$$P_3(Z) = \{(a, b, -b, -a) : a, b \in Z\}.$$

The composition of trivial solutions results in trivial solutions:

$$P_1(Z) \bullet P_2(Z) \subset P_3(Z), P_1(Z) \bullet P_3(Z) \subset P_2(Z), \text{and } P_2(Z) \bullet P_3(Z) \subset P_1(Z).$$

In geometric terms, we connect two points of the surface (2) by a straight line and take the third point of the surface on this line as the composition **[5]**. The trivial solutions are 3 straight lines on the surface (2).

Note that $(u \bullet w) \bullet w = du$, where d is a polynomial in the entries of u and w with integral coefficients; in particular, $(u \bullet w) \bullet w = (-u \cdot w^2)u$ when w is a trivial solution to (4).

This operation $\bullet$ is a particular case of the operation which can be defined for any cubic surface of any dimension **[5]**. For curves, this operation on elliptic curves can be modified to a commutative group operation.

Here are some examples of composition which explain some solutions in Tables 2 and 3:

$$(9, -8, -6, -1) \bullet (3, -2, -3, 2) = 9(12, -10, -9, 1),$$
$$(9, -8, -6, -1) \bullet (7, -8, 8, -7) = 168(41, -40, -2, -17),$$
$$(9, -8, -6, -1) \bullet (35, -38, 38, -35) - 1736(89, -86, 2, -41),$$
$$(9, -8, -6, -1) \bullet (35, -35, 2, -2) = 133(94, -103, 64, -1),$$
$$(9, -8, -6, -1) \bullet (35, -35, -16, 16) = 7(904, -823, -566, -1),$$
$$(9, -8, -6, -1) \bullet (19, -19, 1, -1) = 2(1808, -1985, 1241, -2),$$
$$(9, -8, -6, -1) \bullet (151, -151, -69, 69) = 38(3097, -2820, -1938, -1),$$
$$(9, -8, -6, -1) \bullet (8, -8, 1, -1) = 57(15, -16, 9, -2),$$
$$(9, -8, -6, -1) \bullet (19, -19, 1, -1) = 2(1808, -1985, 1241, -2),$$
$$(9, -8, -6, -1) \bullet (-515, 568, -568, 515) = -744(-46481, 45064, 2, 20729),$$
$$(10, -12, 9, -1) \bullet (19, -19, 1, -1) = 42(-172, 138, 135, 1),$$
$$(10, -12, 9, -1) \bullet (20, -20, 1, -1) = 8(-1010, 812, 791, 1),$$
$$(16, -15, -9, 2) \bullet (2, -3, 3, -2) - 72(34, -33, -15, 2),$$

$$(144,-138,-71,-1) \bullet (179869,-116857,116857,-179869)$$
$$= 5756396646(21068,-2,64355,-65099).$$

5. Infinite families of integral solutions for (2)

The equation (2) has 3 infinite families of trivial integral solutions:

$$(a,b,c) = (t,-t,1), \quad (-t,1,t), \quad (1,t,-t).$$

We show how an arbitrary nontrivial integer solution for (2) can be included in an infinite family of nontrivial solutions. The construction depends on composition and the Pell equation.

Let $x = (a,b,c)$ be an integral solution for (2), i.e., let $u = (u_1,u_2,u_3,u_4) = (b,a,c,-1) \in S(\mathbb{Z})$ be an integral solution of (4). We assume that $a \geq b \geq c$ and that (a,b,c) is not trivial, i.e., $a,b,c \neq 1$. Then $abc \neq 0$, both $a+b$ and c^2-c are positive, and $u = (b,a,c,-1)$ is a nontrivial integral solution to (4). We take relatively prime $\alpha, \beta \in \mathbb{Z}$ and consider the trivial solution

$$\begin{aligned} w =&(w_1,w_2,w_3,w_4)\\ =&(\alpha+\beta(a^2-b^2), -\alpha-\beta(a^2-b^2), 2\beta(a+b), -2\beta(a+b)) \in P_1(\mathbb{Z}) \end{aligned}$$

of (4). Taking the composition, we obtain the following solution to (4):

$$v = (w \bullet u)/(a+b) = (v_1,v_2,v_3,v_4),$$

where

$$\begin{aligned} v_1 =\ & 2\alpha\beta - 2\beta^2b^2 + \alpha^2a - 4\beta^2ba + 2\alpha\beta b^2a + \beta^2b^4a - 2\beta^2a^2 - 2\alpha\beta a^3 - 2\beta^2b^2a^3\\ &+ \beta^2a^5 + 4\beta^2b^2c + 4\beta^2bac - 2\alpha\beta c^2 - 2\beta^2b^2c^2 + 2\beta^2a^2c^2,\\ v_2 =\ & -2\alpha\beta + \alpha^2b - 2\beta^2b^2 + 2\alpha\beta b^3 + \beta^2b^5 - 4\beta^2ba - 2\beta^2a^2 - 2\alpha\beta ba^2\\ &- \beta^2b^3a^2 + \beta^2ba^4 + 4\beta^2bac + 4\beta^2a^2c + 2\alpha\beta c^2 + 2\beta^2b^2c^2 - 2\beta^2a^2c^2,\\ v_3 =\ & 4\beta^2b - 2\alpha\beta b^2 - 2\beta^2b^4 + 4\beta^2a + 2\alpha\beta a^2 + 4\beta^2b^2a^2 - 2\beta^2a^4 + \alpha^2c - 4\beta^2bc\\ &+ 2\alpha\beta b^2c + \beta^2b^4c - 4\beta^2ac - 2\alpha\beta a^2c - 2\beta^2b^2a^2c + \beta^2a^4c,\\ v_4 =\ & -\alpha^2 + \beta^2b^4 - 2\beta^2b^2a^2 + \beta^2a^4 - 4\beta^2bc - 4\beta^2ac + 4\beta^2bc^2 + 4\beta^2ac^2. \end{aligned}$$

Note that $-v_4 = \alpha^2 - r\beta^2$, where $r = (b^2-a^2)^2 + 4(b+a)(c^2-c) > 0$.

Let us prove that v is not trivial. If $u \cdot w^2 \neq 0$, then $((a+b)v) \bullet w = (-u \cdot w^2)u$ is not trivial, hence v is not trivial (see Section 4). Assume now that $u \cdot w^2 = 0$. Since $a+b \neq 0$ and $\alpha\mathbb{Z}+\beta\mathbb{Z} = \mathbb{Z}$, we obtain that $\beta \neq 0$. By the definition of the composition, $v = (u^2 \cdot w)w \in P_1(\mathbb{Z})$. Since $a+b \neq 0$, we obtain that $u^2 \cdot w = 0$. For the vector $y = (w_1/u_1, w_2/u_2, w_3/u_3, w_4/u_4)$ we have

$$\begin{aligned} y^3 \cdot u^3 &= w_1^3 + w_2^3 + w_3^3 + w_4^3 = 0,\\ y^2 \cdot u^3 &= w^2 \cdot u = 0,\\ y \cdot u^3 &= u^2 \cdot w = 0,\\ y^0 \cdot u^3 &= a^3 + b^3 + c^3 - 1 = 0, \end{aligned}$$

where $y^0 = (1, 1, 1, 1)$.

Since $u_1u_2u_3u_4 \neq 0$, at least three entries of y are the same. Since w is trivial, this makes u trivial, which contradicts our choice of u.

Since v cannot be trivial and since $v_1^3 + v_2^3 + v_3^3 = 0$ has no integral solutions with $v_1v_2v_3 \neq 0$, by Fermat's theorem for cubes, $-v_4 = \alpha^2 - r\beta^2 \neq 0$, hence r is not a square. Since r is positive, $-v_4 = \alpha^2 - r\beta^2 = 1$ is the Pell equation. It has an infinite sequence $(\alpha_n, \beta_n), n \in \mathbb{Z}$, of integral solutions, where $\alpha_n + \beta_n r^{1/2} = (\alpha_1 + \beta_1 r^{1/2})^n$ (here (α_1, β_1) is the fundamental solution), $\alpha_n > 0$ for all n. Each (α_n, β_n) gives rise to a nontrivial integral solution $v = v(n) = (v_1(n), v_2(n), v_3(n), v_4(n))$ of (4) with $v_4(n) = -1$, i.e., to a nontrivial integral solution $v' = v'(n) = (v_1(n), v_2(n), v_3(n))$ of (2). When $(\alpha, \beta) = (\alpha_0, \beta_0) = (1, 0)$, we obtain the solution $v'(0) = x = (a, b, c)$ as a member of our sequence. Since (α_n, β_n) and (α_m, β_m) are not collinear for $n \neq m$, we see that $v'(n) \neq v'(m)$ for $n \neq m$ with at most one exception (up to switching m and n): $m = 0$ and the vector $w = w(n)$ satisfies the equations $u \cdot w^2 = a + b, w \cdot u^2 = 0$. In particular, all solutions $v'(n)$ with $n \neq 0$ are distinct.

Observe that $u \bullet P_1$ is an affine 3-space containing all solutions $v(n)$ of (4). Thus, we obtain

Theorem 1. *For every nontrivial integral solution* $(a, b, c) = (u_2, u_1, u_3)$ *of* (2) *with* $u_2 \geq u_1 \geq u_3$, *the plane* $(u_2 - 1)(a + b) = (c - 1)(u_1 + u_2)$ *contains infinitely many nontrivial integral solutions* (a, b, c) *of* (2).

The family $v'(n), n \in \mathbb{Z}$, of nontrivial integral solutions to (4) is exponentially parameterized in the following sense. Every entry $v_k(n)$ $(k = 1, 2, 3)$ of $v'(n)$ is a polynomial in $(\alpha_1 + \beta_1 r^{1/2})^{2n}$ and $(\alpha_1 - \beta_1 r^{1/2})^{2n}$. As $n \to \infty, \beta_n/\alpha_n \to r^{-1/2}$, and $v_1(n)/v_2(n)$ converges to an irrational number (otherwise, it will be independent of n, which would make all $v(n)$ the same).

Any polynomial solution $a, b, c \in \mathbb{Q}[t]$ of (2) can cover only finitely many members of such a sequence. So we obtain

Theorem 2. *The set* $S(\mathbb{Z})$ *of integral solutions to* $a^3 + b^3 + c^3 = 1$ *cannot be covered by a finite set of polynomial solutions* $a, b, c \in \mathbb{Q}[t]$.

The above construction works over an arbitrary commutative ring Z rather than $Z = \mathbb{Z}$, provided that the Pell equation $-v_4 = \alpha^2 - r\beta^2 = 1$ has a nontrivial solution (one starts with a solution $x = (a, b, c)$ of (2) over Z rather than an integer solution). The authors are grateful to Dean Hickerson (3/20/90) for pointing out the case $Z = \mathbb{Z}[t], x = (9t^4, 3t - 9t^4, 1 - 9t^3)$. This gives rise to the infinite family $v'(n)$ of polynomial solutions of (2). When $n = 1$, we obtain

$$(1-9t^3+648t^6+3888t^9)^3+(-135t^4+3888t^{10})^3-(-3t+81t^4+1296t^7+3888t^{10})^3 = 1.$$

6. Primitive nontrivial solutions to (4)

A solution (a, b, c, d) of (4) is called *primitive* if $a\mathbb{Z}+b\mathbb{Z}+c\mathbb{Z}+d\mathbb{Z} = \mathbb{Z}$, or, equivalently, $a\mathbb{Z} + b\mathbb{Z} + c\mathbb{Z} = \mathbb{Z}$.

Theorem 3. *For any integer* $d \neq 0$, *the equation* (4) *has infinitely many primitive nontrivial solutions with given* d.

Proof. It suffices to prove this in the case $d > 0$. One primitive nontrivial solution is $u = (a, b, c, d) = (3d^3 + 3d^2 + 2d, -3d^3 - 3d^2 - 2d - 1, 3d^2 + 2d + 1, d)$. To obtain other solutions, we consider the trivial solution $w = (\alpha - \beta - 4\beta d - 6\beta d^2 - 6\beta d^3, -(\alpha - \beta - 4\beta d - 6\beta d^2 - 6\beta d^3), 2d\beta, -2d\beta) \in P_3(\mathbb{Z})$ and the composition $v = u \bullet w = (v_1, v_2, v_3, v_4)$, where

$v_1 = -\alpha^2 + 2\alpha\beta - \beta^2 - 2\alpha^2 d + 14\alpha\beta d - 12\beta^2 d - 3\alpha^2 d^2 + 42\alpha\beta d^2 - 63\beta^2 d^2 - 3\alpha^2 d^3 + 84\alpha\beta d^3 - 213\beta^2 d^3 + 108\alpha\beta d^4 - 504\beta^2 d^4 + 90\alpha\beta d^5 - 846\beta^2 d^5 + 36\alpha\beta d^6 - 1008\beta^2 d^6 - 828\beta^2 d^7 - 432\beta^2 d^8 - 108\beta^2 d^9$,

$v_2 = 2\alpha^2 d - 6\alpha\beta d + 4\beta^2 d + 3\alpha^2 d^2 - 30\alpha\beta d^2 + 39\beta^2 d^2 + 3\alpha^2 d^3 - 72\alpha\beta d^3 + 165\beta^2 d^3 - 108\alpha\beta d^4 + 432\beta^2 d^4 - 90\alpha\beta d^5 + 774\beta^2 d^5 - 36\alpha\beta d^6 + 972\beta^2 d^6 + 828\beta^2 d^7 + 432\beta^2 d^8 + 108\beta^2 d^9$,

$v_3 = \alpha^2 - 2\alpha\beta + \beta^2 + 2\alpha^2 d - 14\alpha\beta d + 12\beta^2 d + 3\alpha^2 d^2 - 42\alpha\beta d^2 + 63\beta^2 d^2 - 72\alpha\beta d^3 + 192\beta^2 d^3 - 72\alpha\beta d^4 + 396\beta^2 d^4 - 36\alpha\beta d^5 + 576\beta^2 d^5 + 576\beta^2 d^6 + 360\beta^2 d^7 + 108\beta^2 d^8$,

$v_4 = \alpha^2 d - \beta^2 d - 12\beta^2 d^2 - 48\beta^2 d^3 - 108\beta^2 d^4 - 144\beta^2 d^5 - 108\beta^2 d^6 - 36\beta^2 d^7 = d(\alpha^2 - r\beta^2)$

with

$$r = 1 + 12d + 48d^2 + 108d^3 + 144d^4 + 108d^5 + 36d^6.$$

Using the integral solutions (α, β) to the Pell equation $\alpha^2 - r\beta^2 = 1$, we obtain infinitely many nontrivial primitive solutions v with $v_4 = d$.

Bibliography

[1] L.E. Dickson, *History of the theory of numbers*. New York, Chelsea Pub. Co., (1966).

[2] V.A.Dem'yanenko, On sums of four cubes, *Izv. Vyssh. Uchebn. Zaved. Mat.* (1966) No.5, 64–69 (in Russian)

[3] L.E. Dickson, *Introduction to the theory of numbers*. [New ed.]. New York, Dover Publications, (1957).

[4] K.Mahler, Note on hypothesis K of Hardy and Littlewood, *J. London Math. Soc.* 11 (1936), 136–138.

[5] Yu. Manin, *Cubic forms, algebra, geometry, arithmetic*. 2nd ed. Amsterdam; New York, North-Holland; Elsevier Science Pub. Co., (1986).

[6] A.Schinzel and W.Sierpinski, Sur les sommes de quatre cubes, *Acta Arith.* 41 (1959).

[7] L. N. Vaserstein, Waring's problem for algebras over fields, *J. Number Theory* 26(3) (1987), 286–298.

[8] L. N. Vaserstein, On sums of cubes in polynomial rings, *Math. Comp.* 56 (1991), 349–357.

[9] E.M.Wright, An easier Waring's problem, *J. London Math. Soc.* 9 (1934), 267–272.

Remarks added in proof. After the paper was submitted, we found the following relevant references:

[M1] L. J. Mordell, On the four integer cubes problem, *J. London Math. Soc.* 11 (1936), 208–218.

[M2] L. J. Mordell, On the integer solutions of the equation $x^2 + y^2 + z^2 + 2xyz = n$, *J. London Math. Soc.* 28 (1953), 500–510.

[K] M. P. Kesava, On the equation $x^3 + y^3 = u^3 + v^3$, *Math. Student* 23 (1955), 101–103.

[MW] J. C. P. Miller and M. F. C. Woollett, Solutions of the Diophantine equation $x^3 + y^3 + z^3 = k$, *J. London Math. Soc.* 30 (1955), 101–110.

[M3] L. J. Mordell, On an infinity of integral solutions of $ax^3 + zy^3 + bz^3 = bc^3$, *J. London Math. Soc.* 30 (1955), 111–113.

[L] D. H. Lehmer, On the Diophantine equation $x^3 + y^3 + z^3 = 1$, *J. London Math. Soc.* 31 (1956), 275–280.

[GLS] V. L. Gardiner, R. B. Lazarus, and R. P. Stein, Solutions of the Diophantine equation $x^3 + y^3 = z^3 - d$, *Math. Comp.* 18 (1964), 408–413.

In **[M1]** (and several other publications of Mordell), our Question 1 was considered.

In **[M2]**, among other things, the first half of our Question 5 is posed on p. 505.

In **[MW]**, all integral solutions of $a^3+b^3+c^3 = d$ with $0 \le d \le 100$, $|a| \le |b| \le |d| \le 3164$ are found. Our Question 1 for special integers and Question 2 are posed explicitly on p. 107.

In **[K]**, rational solution of (2) are parameterized.

Our Theorem 3 is essentially contained in **[M3]**, except that the condition of primitivity is not stated in **[M3]** explicitly.

The infinite family of polynomial solutions of (2) mentioned at the end of Section 5 was found in **[L]**.

In **[GLS]**, all integral solutions of $a^3 + b^3 + c^3 = d$ with $0 \le d \le 999$, d is not a cube, $|a| \le |b| \le |c|$, $|b| \le 65,536$, are found.

Department of Mathematics, The Pennsylvania State University, University Park, PA 16802
Email: pan@math.psu.edu and vstein@math.psu.edu

Heights and Zeta Functions in Function Fields

Daqing Wan

1. Introduction

In this paper, we study the question of counting the number of points with bounded height in a projective space over a function field. We first recall the corresponding results over a number field.

A classical result in number theory states that the probability for two integers to be relatively prime is $6/\pi^2$. More precisely, let $N(B)$ be the number of points in the projective space $\mathbb{P}^1(\mathbb{Q})$ with height at most B. Then (see **[Hu**, p114-115**]**),

$$N(B) = \frac{12}{\pi^2}B^2 + O(B\log B). \tag{1.1}$$

A natural more general question, due to Lang **[La]**, is to estimate the number of points with bounded height in a projective space over a number field.

Let $n > 1$ be a positive integer. Let K be a number field of degree d, and let $H_K(x)$ be the relative multiplicative height function in the projective space $\mathbb{P}^{n-1}(K)$. Let $N_K(n, B)$ be the number of points $x \in \mathbb{P}^{n-1}(K)$ with $H_K(x) \leq B$. Lang's question is then to estimate $N_K(n, B)$ as $B \to \infty$. This was solved by Schanuel **[Sc]**, who showed that

$$N_K(n,B) = \frac{hR/w}{\zeta_K(n)}\gamma(K,n)B^n + \begin{cases} O(B\log B), & \text{if } d = 1 \text{ and } n = 2; \\ O(B^{n-1/d}), & \text{otherwise.} \end{cases}$$

As usual, h is the class number, R is the regulator, w is the number of roots of unity in K, $\zeta_K(s)$ is the Dedekind zeta function, and

$$\gamma(K,n) = \left(\frac{2^{r_1}(2\pi)^{r_2}}{\sqrt{d_K}}\right)^n n^r,$$

where d_K is the absolute value of the discriminant and r_1 (resp. r_2) is the number of real (resp. complex) embeddings of K. In the case $n = 1$, Schanuel's theorem was already proved by Dedekind and Weber in 19th century (see **[He**, p. 139**]** and **[Sc]**). In this special case, a better error term was obtained by Landau **[Lu]**.

Our purpose here is to study a function field analogue of Lang's question. This turns out to be closely related to zeta functions of divisors introduced in **[Wa]**.

Let X be an absolutely irreducible projective variety of dimension r in $\mathbb{P}^m$, non-singular in codimension 1, and defined over a finite field $\mathbb{F}_q$ of q elements. Let $\mathbb{F}_q(X)$ be the function field of X. If P is a point in $\mathbb{P}^n$ rational over $\mathbb{F}_q(X)$ with coordinates $(y_0, y_1, \cdots, y_n)$ in $\mathbb{F}_q(X)$, then the height function (the logarithmic height) is defined by

$$h_X(P) = -\deg \inf_i (y_i),$$

where the inf is taken over all i such that $y_i \neq 0$. When one of the y_i is equal to 1, then

$$h_X(P) = \deg \sup_i (y_i)_\infty .$$

That is, $h_X(P)$ is the projective degree in $\mathbb{P}^m$ of the sup of the polar divisors of the y_j. This height function was first used by Néron in his proof of the theorem of the base, and was taken up again in Lang-Néron **[LN]**. Let $A_d(n)$ be the number of points in $\mathbb{P}^n$ rational over $\mathbb{F}_q(X)$ with height d. A function field analogue of Lang's question is then to study the sequence $A_d(n)$ as d varies or its generating function. In the case that X is a curve, an asymptotic formula for $A_d(n)$ was given without proof by Serre **[Se**, p. 19**]**.

Define the zeta function of the projective space $\mathbb{P}^n$ over the function field $\mathbb{F}_q(X)$ to be the following power series

$$Z(\mathbb{P}^n(\mathbb{F}_q(X)), t) = \sum_{d=0}^{\infty} A_d(n) t^d .$$

We conjecture that under a mild condition, $Z(\mathbb{P}^n(\mathbb{F}_q(X)), t)$ is a p-adic meromorphic function. This is closely related to our earlier conjecture about meromorphic continuation of the zeta function of divisors on X. In fact, $Z(\mathbb{P}^n(\mathbb{F}_q(X)), t)$ can be expressed in terms of certain twists of the zeta function of divisors. Thus, all the results in **[Wa]** carry over to $Z(\mathbb{P}^n(\mathbb{F}_q(X)), t)$. In particular, we conclude that $Z(\mathbb{P}^n(\mathbb{F}_q(X)), t)$ is p-adic meromorphic if X is a nonsingular complete intersection of dimension $r \geq 3$.

In the case when X is a curve, we show that $Z(\mathbb{P}^n(\mathbb{F}_q(X)), t)$ is a rational function. The Riemann-Hypothesis for curves allows us to give a sharp asymptotic formula for $A_d(n)$, which improves Serre's result.

Acknowledgment. I would like to thank Neal Koblitz for several comments and J-P. Serre for many corrections and suggestions.

2. Zeta Functions of Divisors over Finite Fields

The group $\mathrm{Div}(X/\mathbb{F}_q)$ of divisors on $X/\mathbb{F}_q$ is the free abelian group generated by the irreducible $\mathbb{F}_q$-rational subvarieties of X of codimension 1. Each divisor has a unique factorization $D = \sum_i n_i P_i$ into prime divisors. If all $n_i \geq 0$, then D is called effective. If $f \in \mathbb{F}_q(X)$, then the rational function f defines a principal divisor on X, denoted (f). Two divisors are linearly equivalent if their difference is a principal divisor. Let $\mathrm{CL}(X/\mathbb{F}_q)$ be the group of equivalence classes of divisors. It follows from Néron's theorem of the base that $\mathrm{CL}(X/\mathbb{F}_q)$ is a finitely generated abelian group with rank at least 1 (since X is projective).

If D is a divisor on $X/\mathbb{F}_q$, define

$$L(D) = \{f \mid (f) + D \geq 0, f \in \mathbb{F}_q(X)\} \cup \{0\}.$$

$L(D)$ is a finite demensional $\mathbb{F}_q$-vector space, whose dimension is denoted by $l(D)$. The following lemma is well-known, see **[Mo]**.

Lemma 2.1. *The number* $N(D)$ *of all effective divisors linearly equivalent to* D, *is given by*

$$N(D) = \frac{q^{l(D)} - 1}{q - 1}. \qquad \square$$

Let $E(X)$ be the additive monoid of effective divisors on $X/\mathbb{F}_q$. If $D_i \in E(X)$, then $\sup_i(D_i)$ is defined to be the smallest effective divisor D such that all $D - D_i$ are effective. In other words, it is the least common multiple of the D_i. The D_i are called relatively prime if they do not have a common component. We now define the Möbius function and Euler function on $E(X)$. Let $D = \sum_{i=1}^{k} n_i P_i$ be effective. Define the Möbius function by

$$\mu(D) = \begin{cases} 1, & \text{if } D = 0; \\ (-1)^k, & \text{if all } n_i = 1; \\ 0, & \text{if some } n_i > 1. \end{cases}$$

Define the Euler function $\phi(D)$ to be the number of effective divisors which are linearly equivalent to D and which are relatively prime to D.

Lemma 2.2. *Let* D *and* n *be given. Then,*

$$\sum_{\sup(D_1,\cdots,D_n)=D} \prod_{i=1}^{n} \phi(D_i) = \sum_{0 \le D^* \le D} \mu(D - D^*) N(D^*)^n,$$

where the sum on the left is over all effective divisors D_i *'s such that* $\sup(D_1, \cdots, D_n) = D$.

Proof. From the definition of $N(D)$, one checks that

$$\begin{aligned} N(D)^n &= (\sum_{D_1 \le D} \phi(D_1))^n \\ &= \sum_{D_i \le D} \prod_{i=1}^{n} \phi(D_i) \\ &= \sum_{\sup(D_1,\cdots,D_n) \le D} \prod_{i=1}^{n} \phi(D_i). \end{aligned}$$

This and Möbius inversion imply Lemma 2.2. $\square$

Definition 2.3. For a non-negative integer k, we define the k-th zeta function of divisors on $X/\mathbb{F}_q$ to be the following power series

$$Z^{(k)}(X, t) = \sum_{D \in E(X)} N(D)^k t^{\deg D}.$$

Note that

$$Z^{(0)}(X,t) = \sum_{D \in E(X)} t^{\deg D} = \prod_P (1 - t^{\deg P})^{-1}$$

is just the zeta function of divisors on $X/\mathbb{F}_q$ introduced in **[Wa]**, where P runs over all prime divisors on $X/\mathbb{F}_q$. In **[Wa]**, we conjectured the p-adic meromorphic continuation of $Z^{(0)}(X,t)$ under a mild condition. This was proved in many cases. Now, we extend our conjecture to all $Z^{(k)}(X,t)$.

Conjecture 2.4. Assume that the monoid of effective divisor classes in $\mathrm{CL}(X/\mathbb{F}_q)$ is a finitely generated monoid. Then, for all integers $k \geq 0$, $Z^{(k)}(X,t)$ is p-adic meromorphic. Furthermore, the order of pole of $Z^{(k)}(X,t)$ at $t = 1$ equals the rank of $\mathrm{CL}(X/\mathbb{F}_q)$.

All the results in **[Wa]** about $Z^{(0)}(X,t)$ carry over to $Z^{(k)}(X,t)$. The proofs are almost identical. In particular, we have

Theorem 2.5. *Let* $\mathrm{CL}(X/\mathbb{F}_q)$ *have rank* 1. *Then for all integers* $k \geq 0$, $Z^{(k)}(X,t)$ *is* p*-adic meromorphic with a simple pole at* $t = 1$. □

If X is a curve, the condition is automatically satisfied. In this case, all of the $Z^{(k)}(X,t)$ are rational functions. Examples satisfying $\mathrm{CL}(X/\mathbb{F}_q) = \mathbb{Z}$ include: non-singular complete intersections of dimension at least 3 and Grassmannian varieties.

Theorem 2.6. *Let* $\mathrm{CL}(X/\mathbb{F}_q)$ *have rank* ρ *and torsion order* h. *Assume that the monoid of effective divisor classes in* $\mathrm{CL}(X/\mathbb{F}_q)$ *is a finitely generated monoid. Then for all integers* $k \geq 0$, $Z^{(k)}(X,t)$ *is* p*-adic meromorphic on the closed unit disk* $|\, t \,| \leq 1$ *with a pole of order* ρ *at* $t = 1$. *Furthermore, the special value is given by*

$$\lim_{t \to 1^-} (1-t)^{\rho} Z^{(k)}(X,t) = \left(\frac{1}{1-q}\right)^{k+1} hR(X),$$

where $R(X)$ *is a certain regulator.* □

Examples of Theorem 2.6 include product varieties of non-singular complete intersections of dimension at least 3. In this case, all of the $Z^{(k)}(X,t)$ are actually p-adic meromorphic everywhere.

3. Zeta Functions of Projective Spaces in Function Fields

In order to apply the results in §2 to the zeta function $Z(\mathbb{P}^n(\mathbb{F}_q(X)),t)$ in function fields, we now show that $Z(\mathbb{P}^n(\mathbb{F}_q(X)),t)$ can be expressed in terms of the $Z^{(k)}(X,t)$.

Theorem 3.1. *We have*

$$Z(\mathbb{P}^n(\mathbb{F}_q(X)),t) = \frac{1}{Z^{(0)}(X,t)} \sum_{k=0}^{n} \binom{n+1}{k+1} (q-1)^k Z^{(k)}(X,t). \qquad (3.1)$$

Proof. Following Serre's suggestion, let $a_d(n)$ be the number of points in $\mathbb{P}^n(\mathbb{F}_q(X))$ of height d such that all the $n+1$ components are non-zero. Such a point can be written as $(1, y_1, \cdots, y_n)$ with all $y_i \neq 0$. Then, a simple combinatorial argument shows that

$$A_d(n) = \sum_{k=0}^{n} \binom{n+1}{k+1} a_d(k). \tag{3.2}$$

It suffices to prove that

$$\sum_{d=0}^{\infty} a_d(k) t^d = (q-1)^k \frac{Z^{(k)}(X,t)}{Z^{(0)}(X,t)}. \tag{3.3}$$

Let D be an effective divisor. From the definition of the Euler function, one sees that there are exactly $(q-1)\phi(D)$ functions $f \in F_q(X)^*$ such that the polar divisor $(f)_\infty$ is equal to D. This observation and Lemma 2.2 show that (all divisors below are effective)

$$\begin{aligned} a_d(k) &= \sum_{\deg D=d} \sum_{\sup(D_1,\cdots,D_k)=D} (q-1)^k \prod_{i=1}^{k} \phi(D_i) \\ &= (q-1)^k \sum_{\deg D=d} \sum_{0 \le D^* \le D} \mu(D-D^*) N(D^*)^k \\ &= (q-1)^k \sum_{\deg D^* \le d} N(D^*)^k \sum_{\deg D = d - \deg D^*} \mu(D) \\ &= (q-1)^k \sum_{j=0}^{d} \Big(\sum_{\deg D=j} N(D)^k \Big) \Big(\sum_{\deg D=d-j} \mu(D) \Big). \end{aligned} \tag{3.4}$$

Since $\sum_{\deg D=j} \mu(D)$ is the coefficient of t^j in

$$1/Z^{(0)}(X,t) = \prod_{P} (1 - t^{\deg P}),$$

where P runs over all prime divisors on $X/\mathbb{F}_q$, it follows from (3.4) that (3.3) holds. The theorem is proved. □

Applying Theorems 2.5 and 2.6, we have

Theorem 3.2. *Let* $\mathrm{CL}(X/\mathbb{F}_q)$ *have rank* 1. *Then* $Z(\mathbb{P}^n(\mathbb{F}_q(X)), t)$ *is p-adic meromorphic. In particular,* $Z(\mathbb{P}^n(\mathbb{F}_q(X)), t)$ *is rational in the case when* X *is a curve.* □

Theorem 3.3. *Let* $\mathrm{CL}(X/\mathbb{F}_q)$ *have rank* ρ *and torsion order* h. *Assume that the monoid of effective divisor classes in* $\mathrm{CL}(X/\mathbb{F}_q)$ *is a finitely generated monoid. Then*

$$Z(\mathbb{P}^n(\mathbb{F}_q(X)), t)$$

is p-adic analytic on the closed unit disk $|t| \le 1$ *with the special value given by*

$$Z(\mathbb{P}^n(\mathbb{F}_q(X)), 1) = 1.$$

Proof. By Theorem 2.6 and Theorem 3.1, we have

$$\begin{aligned} Z(\mathbb{P}^n(\mathbb{F}_q(X)),1) &= \sum_{k=0}^{n} \binom{n+1}{k+1} (q-1)^k (\frac{1}{1-q})^k \\ &= \sum_{k=0}^{n} \binom{n+1}{k+1} (-1)^k \\ &= 1 - \sum_{k=0}^{n+1} \binom{n+1}{k} (-1)^k \\ &= 1. \end{aligned}$$

Thus the result. □

Remark 3.4. Let Y be a projective variety in $\mathbb{P}^n$, defined over the function field $\mathbb{F}_q(X)$. Let $A_d(Y)$ be the number of $\mathbb{F}_q(X)$-rational points on Y with height d. Define the zeta function of $Y/\mathbb{F}_q(X)$ to be the following power series

$$Z(Y(\mathbb{F}_q(X)),t) = \sum_{d=0}^{\infty} A_d(n) t^d .$$

A general question is then to study p-adic meromorphic continuations of $Z(Y(\mathbb{F}_q(X)),t)$. The present paper corresponds to the case $Y = \mathbb{P}^n$.

4. An Asymptotic Formula

We now prove an asymptotic formula for the number $A_d(n)$ of points in $\mathbb{P}^n(\mathbb{F}_q(X))$ with height d as d varies. We shall assume that X is a non-singular curve. This is the true function field analogue of the number field case. As usual, let $\zeta_X(s) = Z^{(0)}(X, q^{-s})$ be the Dedekind zeta function of the curve $X/\mathbb{F}_q$.

Theorem 4.1. *Let X be a non-singular projective curve of genus g. Then for all $d \geq 1$,*

$$A_d(n) = \frac{hq^{(n+1)(1-g)}}{\zeta_X(n+1)(q-1)} q^{(n+1)d} + \sum_{i=1}^{2g} b_i(d)\alpha_i^d , \tag{4.1}$$

where the $b_i(d)$ are polynomials in d whose coefficients are algebraic numbers and the α_i are algebraic integers satisfying $\mid \alpha_i \mid = \sqrt{q}$.

Proof. Let $D_1, \cdots, D_h$ be the torsion elements of $\mathrm{CL}(X/\mathbb{F}_q)$. Choose a divisor D^* of degree 1 (such a divisor exists, see Monsky [**Mo**]). For $j \geq 0$, let

$$Z(j,t) = \sum_{i=1}^{h} \sum_{d=0}^{\infty} q^{jl(D_i+dD^*)} t^d .$$

The Riemann-Roch theorem for curves implies that for $d > 2g-2$, $l(D_i + dD^*) = d - g + 1$. This shows that $Z(j,t)$ is a rational function of the form

$$Z(j,t) = hq^{j(1-g)}\frac{1}{1-q^j t} + Q_j(t) = \frac{P_j(t)}{(1-q^j t)},$$

where $Q_j(t)$ is an integral polynomial of degree at most $2g-2$ and $P_j(t)$ is an integral polynomial of degree at most $2g-1$. One finds immediately that

$$P_j(q^{-j}) = hq^{j(1-g)}. \tag{4.2}$$

By definition,

$$\begin{aligned} Z^{(n)}(X,t) &= \sum_{i=1}^{h}\sum_{d=0}^{\infty}\Big(\frac{q^{l(D_i+dD^*)}-1}{q-1}\Big)^{n+1}t^d \\ &= \sum_{i=1}^{h}\sum_{d=0}^{\infty}\sum_{j=0}^{n+1}\frac{(-1)^{n+1-j}}{(q-1)^{n+1}}\binom{n+1}{j}q^{jl(D_i+dD^*)}t^d \\ &= \sum_{j=0}^{n+1}\frac{(-1)^{n+1-j}}{(q-1)^{n+1}}\binom{n+1}{j}Z(j,t) \\ &= \sum_{j=0}^{n+1}\frac{(-1)^{n+1-j}}{(q-1)^{n+1}}\binom{n+1}{j}\frac{P_j(t)}{(1-q^j t)}. \end{aligned} \tag{4.3}$$

Since $Z^{(0)}(X,t)$ is the classical zeta function of the curve X over $\mathbb{F}_q$, we have

$$Z^{(0)}(X,t) = \frac{P(t)}{(1-t)(1-qt)},$$

where $P(t) = \prod_{i=1}^{2g}(1-\alpha_i t)$ is an integral polynomial such that $|\alpha_i| = \sqrt{q}$. It follows that for $n \geq 1$,

$$\frac{Z^{(n)}(X,t)}{Z^{(0)}(X,t)} = \sum_{j=0}^{n+1}\frac{(-1)^{n+1-j}}{(q-1)^{n+1}}\binom{n+1}{j}\frac{(1-t)(1-qt)P_j(t)}{(1-q^j t)P(t)}. \tag{4.4}$$

This rational function has at most simple poles at $t = q^{-2}, \cdots, q^{-(n+1)}$. For each term on the right of (4.4), the degree of the numerator is at most the degree of the denominator, i.e., $2g+1$. By partial fraction decomposition and (3.1), we deduce that there are rational numbers a_j $(n+1 \geq j \geq 2)$ and polynomials $b_i(d)$ $(1 \leq i \leq 2g)$ such that for $d \geq 1$,

$$A_d(n) = \sum_{j=2}^{n+1}a_j q^{jd} + \sum_{i=1}^{2g}b_i(d)\alpha_i^d. \tag{4.5}$$

The polynomials $b_i(d)$ are constant if $P(t)$ is squarefree.

We now find explicit values of the coefficients a_j. Let $\zeta_X(s) = Z^{(0)}(X, q^{-s})$ be the Dedekind zeta function of the function field $\mathbb{F}_q(X)$. By (4.2), (4.4), (3.1) and partial fraction decomposition, we deduce that for $j \geq 2$,

$$\begin{aligned}
a_j &= \sum_{k=0}^{n} \binom{n+1}{k+1} (q-1)^k \frac{(-1)^{k+1-j}}{(q-1)^{k+1}} \binom{k+1}{j} \frac{hq^{j(1-g)}}{\zeta_X(j)} \\
&= \frac{hq^{j(1-g)}}{\zeta_X(j)(q-1)} \sum_{k=j}^{n+1} (-1)^{k-j} \binom{n+1}{k} \binom{k}{j} \\
&= \frac{hq^{j(1-g)}}{\zeta_X(j)(q-1)} \sum_{l=0}^{n+1-j} (-1)^l \binom{n+1}{j+l} \binom{j+l}{j} \quad \text{(writing } k = j+l\text{)} \\
&= \frac{hq^{j(1-g)}}{\zeta_X(j)(q-1)} \sum_{l=0}^{n+1-j} (-1)^l \binom{n+1}{j} \binom{n+1-j}{l} \\
&= \frac{hq^{j(1-g)}}{\zeta_X(j)(q-1)} \binom{n+1}{j} \sum_{l=0}^{n+1-j} (-1)^l \binom{n+1-j}{l}.
\end{aligned}$$

This shows that $a_j = 0$ for $j < n+1$ and

$$a_{n+1} = \frac{hq^{(n+1)(1-g)}}{\zeta_X(n+1)(q-1)}. \tag{4.6}$$

The theorem follows from (4.5). □

Corollary 4.2. *For $g = 0$ and $d \geq 1$,*

$$A_d(n) = \frac{hq^{(n+1)(1-g)}}{\zeta_X(n+1)(q-1)} q^{(n+1)d}. \tag{4.7}$$

□

This result is first observed by Serre.

Corollary 4.3. *Let $X/\mathbb{F}_q$ be a non-singular projective curve of genus g. Let n be a positive integer. For any $\epsilon > 0$, we have*

$$A_d(n) = \frac{hq^{(n+1)(1-g)}}{\zeta_X(n+1)(q-1)} q^{(n+1)d} + O(q^{\frac{d}{2}+\epsilon}) \text{ for } d \to \infty. \tag{4.8}$$

If $P(t)$ is squarefree, then we can take $\epsilon = 0$. □

This is a function field analogue of Schanuel's asymptotic formula. Note that (4.8) was already given in **[Se. p.19]** with an error term both weaker and misprinted.

If X is a higher dimensional variety, one can, in certain cases, give a direct estimate of $A_d(n)$ with the help of (3.4). In order to get a sharper result, one needs to estimate the behavior of the arithmetic function

$$f(d) = \sum_{\deg D = d} \mu(D).$$

This last problem can be viewed as a higher dimensional function field analogue of the well-known Merten conjecture which estimates $\sum_{n \leq d} \mu(n)$.

References

[He] E. Hecke, *Lectures on the Theory of Algebraic Numbers*, Springer-Verlag, 1981.

[Hu] L.K. Hua, *Introduction to Number Theory*, Springer-Verlag, 1982.

[La] S. Lang, *Diophantine Geometry*, New York, Interscience Publishers, 1962.

[LN] S. Lang and A. Néron, Rational points of abelian varieties over function fields, *Amer. J. Math.* 81(1959), 95–118.

[Lu] E. Landau, *Einführung in die elementare und analytische Theorie der algebraischen Zahlen und der Ideale*, Leipzig, Teubner, 1927.

[Mo] P. Monsky, *P-adic Analysis and Zeta Functions*, Kinokuniya Book-Store Co., Ltd. 1970.

[Sc] S. Schanuel, Heights in number fields, Bull. Soc. Math. France **107** (1979), 433–449.

[Se] J-P. Serre, *Lectures on the Mordell-Weil Theorem*, Aspects of Math., Vieweg, 1989.

[Wa] D. Wan, Zeta functions of algebraic cycles over finite fields, *Manuscripta Math.*, to appear.

Department of Mathematical Sciences, University of Nevada at Las Vegas, Las Vegas, NV 89154
Email:dwan@nevada.edu

Continued Fraction Characterization and Generic Ideals in Real Quadratic Function Fields

*C. Friesen**

Introduction

In the ordinary number field case, it is a trivial matter to show that those integers m where the continued fraction expansion of $\sqrt{m}$ has a period of 1 are precisely those m of the form x^2+1 for some positive integer x. In this case the continued fraction expansions of period 1 are characterized by the equation x^2+1 in the one integer variable x. If we were to look at $\sqrt{m}$ with continued fraction period of 2 we might come up with the formula $m=(xy/2)^2+y$ where x and y represent two arbitrary positive integers whose product is even. As we increase the period length our characterization equations become more complicated and have more variables. It is possible (see **[Pe**, pp. 96-99**]** or **[Fr1]**) to give a meta-equation that provides the characterization equation for any period.

Such a result is proved here, in Section 1, for the function field case. Our characterization (which is by no means unique) of polynomials with a certain continued fraction period is closely tied in with the partial denominators that arise. Applications of this meta-equation include results on the distribution of real quadratic function fields of given regulator. In Section 2 we will make use of a correspondence between continued fractions and ideals to derive a generic list of principal primitive reduced ideals over quadratic function fields. Section 3 is devoted to a short example of using the characterization theorem and in Section 4 we consider an application of the theory in which the list of generic ideals is used to obtain families of quadratic extensions whose ideal class numbers are divisible by pre-selected integers.

The serious study of quadratic function fields can be said to have started with the doctoral thesis of E. Artin **[Ar]**. Several decades later L. Carlitz **[Ca]** continued the investigation into these fields which have now seen a resurgence of interest. One can, for example, point to recent work of D. R. Hayes **[Ha]**.

Let $\mathbb{F}_q$ be the finite field with q elements and characteristic $p\neq 2$. Let $k=\mathbb{F}_q(t)$ be the field obtained by adjoining a transcendental element, t. Denote by k_∞ the completion of k at ∞. We may write any element of k_∞^* in the form

$$x=\sum_{i=-\infty}^{d} a_i t^i \qquad \text{with } a_i\in\mathbb{F}_q \qquad \text{and } a_d\neq 0.$$

* The author's research was supported by a University of Toronto research grant and an NSERC Canada grant. This paper is a reworking of a chapter of the author's doctoral thesis, Brown University, 1989.

We define

$$\deg(x) \stackrel{\text{def}}{=} d \qquad \text{and} \quad \operatorname{sgn}(x) \stackrel{\text{def}}{=} a_d$$

Definition. We say that x is ***monic*** if $\operatorname{sgn}(x) = 1$.

If x is a monic of even degree d then there exists a unique monic function y, of degree $d/2$, such that $y^2 = x$. We may write $y = \sqrt{x}$. The ***greatest polynomial function*** is defined as follows:

$$\text{If} \quad x = \sum_{i=-\infty}^{d} a_i t^i \qquad \text{then}$$

$$[x] \stackrel{\text{def}}{=} \sum_{i=0}^{d} a_i t^i \qquad \text{if } d \geq 0 \qquad \text{and}$$

$$[x] \stackrel{\text{def}}{=} 0 \qquad \text{if} \quad d < 0.$$

We now define the continued fraction algorithm. Given an element $x \in k_\infty^*$ we can define $F_0(x) \stackrel{\text{def}}{=} x$ and, if $F_i(x) \neq [F_i(x)]$,

$$F_{i+1}(x) \stackrel{\text{def}}{=} (F_i(x) - [F_i(x)])^{-1}.$$

It is easy to see that this sequence of functions terminates if and only if x is a rational function.

Definition. For convenience we shall write $[a_0; a_1, a_2, \ldots, a_{n-1}, a_n]$ for the continued fraction expansion

$$a_0 + \cfrac{1}{a_1 + \cfrac{1}{a_2 + \ddots + \cfrac{1}{a_{n-1} + \cfrac{1}{a_n}}}}$$

where $a_i \in k_\infty^*$ and where $\deg(a_i) > 0$ for $1 \leq i \leq n$.

If, for a_i as above, we have

$$\begin{pmatrix} a_0 & 1 \\ 1 & 0 \end{pmatrix} \begin{pmatrix} a_1 & 1 \\ 1 & 0 \end{pmatrix} \cdots \begin{pmatrix} a_n & 1 \\ 1 & 0 \end{pmatrix} = \begin{pmatrix} A & B \\ C & D \end{pmatrix}$$

then

$$\frac{A}{C} = [a_0, a_1, \ldots, a_n] \qquad \text{and} \tag{0.1}$$

$$\frac{B}{D} = [a_0, a_1, \ldots, a_{n-1}]. \tag{0.2}$$

For a matrix treatment of continued fractions (in the number field case) readers can refer to a paper by A.J. van der Poorten **[Po]**.

Definition. Let us call a finite ordered set of non-zero polynomials $\{a_1, a_2, ..., a_n\}$ ***skew-symmetric*** with skew = $c \in \mathbb{F}_q^*$, if $a_{n-2i} = ca_{2i+1}$ for all $0 \le i < n/2$.

Note that a skew-symmetric set with skew $\neq 1$ must have an even number of elements.

1. Characterization of the Continued Fraction Expansion

Theorem 1.0 (Form Theorem). *Let M be a non-square even-degree monic in $\mathbb{F}_q[t]$. Then the continued fraction expansion (CFE) of $\sqrt{M}$ has the form*

$$\sqrt{M} = [a_0; \overline{a_1, \ldots, a_{m-1}, 2a_0/c, a_{m-1}, \ldots, a_1, 2a_0}]$$

for some positive integer m, some $c \in \mathbb{F}_q^*$ and some set of positive-degree polynomials $\{a_1, a_2, ..., a_{m-1}\}$ that are skew-symmetric with a skew of c and where $a_0 = [\sqrt{M}]$.

Definition. If m is minimal then we call it the ***skew-period*** of the continued fraction expansion of $\sqrt{M}$. If, in addition, we have $c = 1$ then the period of the continued fraction is just m, otherwise the period is $2m$.

This theorem is of interest to us because it restricts the CFEs (continued fraction expansions) of $\sqrt{M}$ to those of a special form. A proof of the Form Theoıem, which can be found in **[Fr2**, Chapter 2**]**, has been omitted since none of the results of the paper depend on this theorem. A similar theorem for the form of the CFE of the square root of a positive integer is well-known (in the number field case, we only have $c = 1$).

Now that we have restricted the CFEs of $\sqrt{M}$ to those of a special form it is natural to ask which combinations of m, c and skew-symmetric $\{a_1, \ldots, a_{m-1}\}$ actually arise. The next theorem shows that all of the possible combinations can occur, and do so infinitely often.

Theorem 1.1 (Characterization Theorem). *Let m be a positive integer, $c \in \mathbb{F}_q^*$ and let $\{a_1, a_2, \ldots, a_{m-1}\}$ be a set of skew-symmetric positive-degree polynomials in $\mathbb{F}_q[t]$ with a skew of c. Then there exist infinitely many monic $M \in \mathbb{F}_q[t]$ with a continued fraction expansion of the form*

$$\sqrt{M} = [a_0; \overline{a_1, \ldots, a_{m-1}, 2a_0/c, a_{m-1}, \ldots, a_1, 2a_0}] \tag{1.2}$$

where $a_0 = [\sqrt{M}]$. Define P, Q, R and S via

$$\begin{pmatrix} a_1 & 1 \\ 1 & 0 \end{pmatrix}\begin{pmatrix} a_2 & 1 \\ 1 & 0 \end{pmatrix}\cdots\begin{pmatrix} a_{m-1} & 1 \\ 1 & 0 \end{pmatrix} = \begin{pmatrix} P & R \\ Q & S \end{pmatrix} \tag{1.3}$$

Then M satisfies (1.2) if and only if

$$M = (P^2)X^2 + (2Q + (-1)^{m-1}PQS)X + (Q^2/4 + (-1)^{m-1}c)S^2$$

for some $X \in \mathbb{F}_q[t]$ satisfying

$$\operatorname{sgn}\left(XP + (-1)^{m-1}QS/2\right) = 1.$$

Remark. It is possible (see **[Pe**, pp. 96-99**]** or **[Fr1]**) to prove a result that is similar to this one for the number field case. But we end up with a weaker theorem as the presence of a modulo 2 congruence prevents some of the combinations of m and $\{a_1, a_2, \ldots, a_{m-1}\}$ from being represented.

Proof. We begin with the comment that, since $\{a_1, a_2, \ldots, a_{m-1}\}$ is a skew-symmetric set with skew = c, it follows that $R = Q/c$ in equation (1.3). The proof, a simple induction left as an exercise, makes use of the fact that, for $m > 3$, the set $\{a_2, a_3, \ldots, a_{m-2}\}$ is skew-symmetric with a skew of $1/c$.

From (1.2) and the definition of the continued fraction we obtain

$$\sqrt{M}/c = [a_0/c; \overline{ca_1, a_2/c, \ldots, a_{m-1}/c, 2a_0, a_{m-1}/c, \ldots, ca_1, 2a_0/c}].$$

We use the skew-symmetry of the $\{a_1, a_2, \ldots, a_{m-1}\}$ together with the fact that skew = 1 if m is even to rewrite the above as

$$\sqrt{M}/c = [a_0/c; \overline{a_{m-1}, a_{m-2}, \ldots, a_1, 2a_0, a_1, \ldots, a_{m-1}, 2a_0/c}].$$

On substituting this into (1.2) we arrive at the following equation, equivalent to (1.2):

$$\sqrt{M} = \left[a_0; a_1, \ldots, a_{m-1}, \frac{a_0 + \sqrt{M}}{c}\right]. \tag{1.4}$$

Multiplying out the associated matrices gives

$$\begin{pmatrix} a_0 & 1 \\ 1 & 0 \end{pmatrix} \begin{pmatrix} P & Q/c \\ Q & S \end{pmatrix} \begin{pmatrix} \frac{a_0+\sqrt{M}}{c} & 1 \\ 1 & 0 \end{pmatrix} = \begin{pmatrix} A & B \\ C & D \end{pmatrix}$$

where

$$\begin{aligned} A &= (a_0^2 P + 2a_0 Q + S + (a_0 P + Q)\sqrt{M})/c, \\ B &= a_0 P + Q, \\ C &= (a_0 P + Q + P\sqrt{M})/c \qquad \text{and} \\ D &= P. \end{aligned}$$

From (0.1) we see that $\sqrt{M} = A/C$ and solving for M leaves us with the equation

$$(M - a_0^2)P - (2a_0)Q = cS.$$

It is clear from the definition of $\begin{pmatrix} P & R \\ Q & S \end{pmatrix}$ that it has determinant equal to $(-1)^{m-1}$. That is, $(S)P - (Q/c)Q = (-1)^{m-1}$, which gives us the general solution to the above equation

$$\begin{aligned} M - a_0^2 &= (-1)^{m-1} cS^2 + 2XQ \\ 2a_0 &= (-1)^{m-1} QS + 2XP \end{aligned} \tag{1.5}$$

where X varies over $\mathbb{F}_q[t]$. (As 2 is invertible in $\mathbb{F}_q$ we use $2X$ here instead of X for cosmetic simplicity of the final answer.)

Solving for M gives us the desired result subject to the condition that $[\sqrt{M}] = a_0$. It is not difficult to verify that this holds whenever a_0 is monic. (We have not attached a meaning to $\sqrt{M}$ for non-monic M and since the square root of a monic is chosen to be monic we must have a_0 monic.) Requiring a_0 to be monic in (1.5) results in the condition $\operatorname{sgn}\left(XP+(-1)^{m-1}QS/2\right)=1$ as noted in the statement of the theorem. It only remains to note that, for X of suitably large degree and of sgn equal to $1/\operatorname{sgn}(P)$, a_0 is monic and we have demonstrated our infinitude of solutions, M, of the form (1.2).

□

It is worth mentioning that equation (1.5) leads us to the conclusion that there is exactly one a_0 with degree less than $\deg P = \sum_{i=1}^{m-1} \deg(a_i)$, and hence only one M, not necessarily monic, with degree less than $2\sum_{i=1}^{m-1} \deg(a_i)$ satisfying equation (1.2). One may further deduce that there are exactly q^n monic polynomials, M, with $\deg(M) = 2n + 2\sum_{i=1}^{m-1} \deg(a_i)$ of the desired form. Also, if M is square-free, and if the CFE of $\sqrt{M}$ has a skew-period of m, then the regulator of the field $\mathbb{F}_q(t, \sqrt{M})$ turns out to be nothing more than $\deg(a_0) + \sum_{i=1}^{m-1} \deg(a_i)$, so our characterization provides information on the distribution of regulators as well.

Definition. Let M be a non-square even-degree monic in $\mathbb{F}_q[t]$ and let

$$\sqrt{M} = [a_0; a_1, a_2, \ldots].$$

We define

$$F_i(\sqrt{M}) \stackrel{\text{def}}{=} [a_i, a_{i+1}, \ldots] \qquad \text{for all} \quad i \geq 0. \tag{1.6}$$

Fact. *Let M be a non-square even-degree monic in $\mathbb{F}_q[t]$. Let $\sqrt{M}$ have a continued fraction period of m. Then*

$$F_{i+m}(\sqrt{M}) = F_i(\sqrt{M}) \qquad \text{for all} \quad i \geq 1 \qquad \text{and}$$

$$F_i(\sqrt{M}) = \frac{p_i + \sqrt{M}}{q_i} \qquad \text{for all} \quad 0 \leq i \leq m \tag{1.7}$$

where p_i and q_i are polynomials in $\mathbb{F}_q[t]$ that depend on M.

A simple induction proves this.

In the following section we will examine the connection that the continued fraction expansion has with ideals.

2. A Generic List of Principal Ideals

We begin with some notation that will be in force throughout this section. Let $M \in \mathbb{F}_q[t] \backslash \mathbb{F}_q$ be a square-free, even-degree monic, let $\mathcal{O}_E$ be the ring of integers in $E = \mathbb{F}_q(t, \sqrt{M})$ and let $\mathcal{O}_k = \mathbb{F}_q[t]$ be the ring of integers in $k = \mathbb{F}_q(t)$. Every ideal, $\mathcal{A}$, in

$\mathcal{O}_E$ can be expressed in the form $\mathcal{A} = \omega_1\mathcal{O}_k + \omega_2\mathcal{O}_k$, with $\omega_1, \omega_2 \in \mathcal{O}_E$, and we define the norm of such an ideal as

$$\mathcal{N}(\mathcal{A}) = \frac{\omega_1\overline{\omega_2} - \omega_2\overline{\omega_1}}{a\sqrt{M}}$$

where $a \in \mathbb{F}_q^*$ is chosen so that the resulting polynomial is monic and where the bars denote conjugation. It is easily checked that this norm function is well-defined and multiplicative. We can define an ideal equivalence as follows. Let $\mathcal{A}$ and $\mathfrak{B}$ be two ideals of $\mathcal{O}_E$. Then

$$\mathcal{A} \sim \mathfrak{B} \qquad \Longleftrightarrow \qquad \mathcal{A} = \gamma\mathfrak{B}$$

for some non-zero element $\gamma \in E$. This equivalence divides the ideals into ideal classes which form a finite group under multiplication. The order of this ideal class group is called the ideal class number, which we shall denote with h.

Definition. An ideal, $\mathcal{A}$, of $\mathcal{O}_E$ without $\mathcal{O}_k$-factors is called ***primitive***. An ideal, $\mathcal{A}$, of $\mathcal{O}_E$ satisfying $\deg(\mathcal{N}(\mathcal{A})) < \deg(\sqrt{M})$ will be called ***reduced***.

Theorem 2.0 (CFE – Ideal Correspondence). *Let $\mathcal{A}$ be a reduced primitive ideal of $\mathcal{O}_E$. Then we may write $\mathcal{A}$ uniquely in the form*

$$\mathcal{A} = (U + \sqrt{M})\mathcal{O}_k + V\mathcal{O}_k$$

where $U, V \in \mathcal{O}_k$, with V monic and $\deg(U - \sqrt{M}) < \deg(V)$. *Let $\mathfrak{B} = \omega_1\mathcal{O}_k + \omega_2\mathcal{O}_k$ be an ideal of $\mathcal{O}_E$ where ω_1 and ω_2 are elements of $\mathcal{O}_E$. Then*

$$\mathcal{A} \sim \mathfrak{B} \qquad \Longleftrightarrow \qquad aF_i\left(\frac{\omega_1}{\omega_2}\right) = \frac{U + \sqrt{M}}{V}$$

for some $F_i(\omega_1/\omega_2)$ (defined in (1.6)*) and some $a \in \mathbb{F}_q^*$.*

Proof. The proof of this correspondence, though straightforward, leans towards the tedious and readers interested in the details are directed to **[Fr2**, Chapter 4**]** for a proof of this result or to **[Go]** for the key ideas involved in proving a similar theorem. □

If we set $\omega_1 = \sqrt{M}$ and $\omega_2 = 1$ in the above theorem, then $\mathfrak{B} = \mathcal{O}_E$ is principal and we immediately obtain the following

Corollary 2.1 (CFE – Principal Ideal Correspondence). *Notation as above. Let $\mathcal{A}$ be a reduced primitive ideal of $\mathcal{O}_E$. Then $\mathcal{A}$ is a principal ideal if and only if*

$$\mathcal{A} = (p_n + \sqrt{M})\mathcal{O}_k + q_n\mathcal{O}_k$$

for some $n \geq 0$ and where p_n and q_n are as defined in (1.7).

That is, the CFE of $\sqrt{M}$ gives all principal reduced primitive ideals of $\mathcal{O}_E$. It is a short exercise to show that $(p_n + \sqrt{M})\mathcal{O}_k + q_n\mathcal{O}_k$ is both reduced and primitive for all $n \geq 0$ and this permits us to conclude that the converse of the corollary holds as well — all ideals that appear in the CFE of $\sqrt{M}$, for square-free M, are principal reduced primitive ideals. Taken together with the Characterization Theorem, it is now possible to list these ideals in generic form, the shape of the expression depending only on the skew-period, m.

3. An Example with a Skew-period of 4

In this section we will determine the continued fraction expansion and a generic list of principal primitive reduced ideals for all square-free, even-degree monics, M, such that the skew-period of the CFE of $\sqrt{M}$ is 4. Applying the Characterization Theorem with $m = 4$ and noting that the skew-symmetry requires $c = 1$ whenever $m - 1$ is odd we find that we can write down the CFE as follows:

$$\sqrt{M} = ([\sqrt{M}]; \overline{a_1, a_2, a_1, 2[\sqrt{M}]}).$$

We perform the matrix product of (1.3) to arrive at the values

$$\begin{aligned} P &= a_1^2 a_2 + 2a_1, \\ Q &= a_1 a_2 + 1 \qquad \text{and} \\ S &= a_2 . \end{aligned}$$

With the aid of the following easily-verified recursion formulas,

$$\begin{aligned} p_{i+1} &= a_i q_i - p_i \qquad \text{and} \\ q_{i+1} &= q_{i-1} + 2a_i p_i - a_i^2 q_i, \end{aligned}$$

and the initial values of $p_0 = 0$, $q_0 = 1$ and $q_1 = M - a_0^2$ we arrive at values for the p_i and the q_i:

$$\begin{aligned} q_0 &= 1, \\ p_0 &= 0, \\ q_1 &= 2(a_1 a_2 + 1)X - a_2^2, \\ p_1 &= a_1(a_1 a_2 + 2)X - a_2(a_1 a_2 + 1)/2, \\ q_2 &= 2a_1^2 X + (1 - a_1 a_2), \\ p_2 &= a_1^2 a_2 X + a_2(1 - a_1 a_2)/2, \\ q_3 &= q_1, \\ p_3 &= p_2, \\ q_4 &= q_0 = 1 \qquad \text{and} \\ p_4 &= p_1 . \end{aligned}$$

There is no need to compute further values of p_i and q_i as they are periodic with period $= 4$. The above results hold for all M where the skew-period of the CFE of $\sqrt{M}$ is equal to 4. But to apply the connection to ideals we need M to be square-free. Under this additional assumption, we can write down all primitive reduced ideals of $\mathcal{O}_E$

as follows:

$$\mathcal{P}_0 = (0 + \sqrt{M})\mathcal{O}_k + 1\mathcal{O}_k = \mathcal{O}_E,$$
$$\mathcal{P}_1 = (p_1 + \sqrt{M})\mathcal{O}_k + q_1\mathcal{O}_k \qquad \text{and}$$
$$\mathcal{P}_2 = (p_2 + \sqrt{M})\mathcal{O}_k + q_2\mathcal{O}_k$$

with the p_i and the q_i as calculated above. The norms of these three ideals (up to multiplication by elements of $\mathbb{F}_q^*$) are 1, q_1 and q_2, respectively.

We can also write down the fundamental unit of the field $\mathbb{F}_q(t, \sqrt{M})$ (here we need M square-free and a skew-period exactly equal to 4 — note that nothing above rules out the possibility that the skew-period could be 1 or 2). The fundamental unit in this case is

$$\epsilon_0 = [a_0, a_1, a_2, a_1] + [a_1, a_2, a_1]\sqrt{M}.$$

Since we have formulas for M and a_0 in terms of a_1, a_2 and X it is now possible to rewrite the expression for the fundamental unit solely in terms of those three parameters. Although such an expression is bound to be quite messy, the regulator of the field (which is nothing more than the degree of the fundamental unit) is simply $\deg(a_0) + \sum_{i=1}^{m-1} \deg(a_i)$, a result that we have used in the previous section.

One possible use of a generic listing of primitive principal reduced ideals is in the following application where we successfully use this information to force class number divisibility by arbitrary integers.

4. An Application to Class Number Divisibility

Let $\mathcal{A} = (p_i + \sqrt{M})\mathcal{O}_k + q_i\mathcal{O}_k$. Recognizing that $\mathcal{N}(\mathcal{A}) = q_i/\mathrm{sgn}(q_i)$ for such an ideal gives us the result that the $\{q_i\}$ in the CFE of $\sqrt{M}$ provide us with the norms of all principal primitive reduced ideals (modulo multiplication with an element of $\mathbb{F}_q^*$). How can we take advantage of this? Well, if $q_i = aF^n$ for some $F \in \mathbb{F}_q[t]\backslash\mathbb{F}_q$, some $a \in \mathbb{F}_q^*$ and some integers $n > 1$ and i then the corresponding primitive reduced ideal, $\mathcal{A} = (p_i + \sqrt{M})\mathcal{O}_k + q_i\mathcal{O}_k$, must have the form $\mathcal{A} = \mathcal{P}^n$ for some primitive ideal $\mathcal{P}$ with norm $= F/\mathrm{sgn}(F)$. Let us further suppose that no other positive power of F occurs as one of the $\{q_j\}$ (again, up to a constant). Since the norms of all smaller powers of $\mathcal{P}$ are smaller than q_i which is smaller than $\sqrt{M}$, it follows that smaller powers of $\mathcal{P}$ are reduced. Since $\mathcal{P}^n$ is primitive, so must smaller powers of $\mathcal{P}$ be primitive. But since they don't appear in the CFE (to do so, their norms would have to be one of the $\{q_j\}$) it follows that none of those powers are principal. On the other hand, we do know that $\mathcal{P}^n$ *is* principal and we conclude that the order of the ideal $\mathcal{P}$ in the ideal class group is n. From this we immediately see that n must divide the order of the ideal class group. That is, n must divide the ideal class number, h.

Computer calculations (performed by FFCalc, a Macintosh program developed by the author for function field computations) led to the discovery of the following sub-family of skew-period-4 polynomials that allow for easy manipulation of the class number (as respects divisibility).

Let $F \in \mathbb{F}_q[t] \backslash \mathbb{F}_q$ be a monic polynomial. Set

$$a_1 = -F/2, \quad a_2 = -2(F^{n+1} - 1)/(F - 1) \quad \text{and} \quad X = 2(F^n - 1)/(F - 1).$$

From the formulas of the previous section we readily verify that

$$q_1 = -4F^n, \quad q_2 = 1 - F \quad \text{and} \quad M = (F^{n+1} + 1)^2 - 4F^n.$$

Here q_1 is a constant times an nth power. Since no other positive power of F appears as one of the q_i (up to a constant) it follows from the argument detailed above that $n|h$ whenever M is square-free.

Putting the above comments together allows us to state

Theorem 4.1. *Let $\mathbb{F}_q$ be a finite field with odd characteristic p. Let $F \in \mathbb{F}_q[t] \backslash \mathbb{F}_q$ be a monic polynomial and let $M = (F^{n+1} + 1)^2 - 4F^n$ where $n \in \mathbb{N}$. If M is square-free then $n|h$ (the ideal class number of $\mathbb{F}_q(t, \sqrt{M})$) and the nth root of the ideal $(F^{n+1} + 1 + \sqrt{M})\mathcal{O}_k + F^n\mathcal{O}_k$ is an ideal with order $= n$ in the ideal class group.*

It is possible to check that such an M is square-free precisely when both

$$\gcd(M, F') = 1 \quad \text{and} \quad n^n(n+2)^{n+2} \not\equiv (n+1)^{2n+2} \pmod{p}$$

where F' denotes the derivative of F with respect to t.

As an example, we may take $F = t$, $n = 3$ and $q = 3$. Since $M = (t^4 + 1)^2 - 4t^3 = (t^4 + 1)^2 + 2t^3$ is square-free over $\mathbb{F}_3$, we expect that $n|h$. A computation revealed that the ideal class number in this case is 3, which is divisible by our choice of n. In this particular example it was actually possible to obtain a class number that was exactly equal to our choice of divisor n. In general this technique will provide class numbers much larger than the choice of n. Readers interested in the results that arise when the skew-period is 1 are directed to **[Fr3]** where specific class number divisibility results are proved without recourse to the Characterization Theorem. Similar results in the number field case (with small period) can be found in **[We]** and **[Lu]**.

Concluding Remarks

The Characterization Theorem shows that there exist polynomials M where the CFE of $\sqrt{M}$ has any desired form (in so far as is allowed by the Form Theorem) and all such M can be generated with the aid of a quadratic expression involving all the partial denominators (up to half of the skew-period), a constant $c \in \mathbb{F}_q^*$ and a polynomial variable X. Furthermore, we have seen that the correspondence between CFEs and ideals is such that the CFE of $\sqrt{M}$ for a square-free M provides us with a list of principal reduced primitive ideals, for any skew-period, that is generic in the sense that it is valid for all $\sqrt{M}$ with that skew-period. It was then possible to use this list to force class number divisibility.

The author would like to thank Alf van der Poorten for suggesting a proof of Theorem 1.1, much more readable than the initial one.

Bibliography

[Ar] E. Artin, Quadratische Körper im Gebiet der höheren Kongruenzen I, II, Math. Z. **19**, (1924), 153–246.

[Ca] L. Carlitz, A class of polynomials, Trans. Amer. Math. Soc. **43**, (1938), 168–182.

[Fr1] C. Friesen, On Continued Fractions of Given Period, Proc. Amer. Math. Soc. **103**, (1988), 9–14.

[Fr2] C. Friesen, Continued Fractions and Real Quadratic Function Fields, doctoral thesis, Brown University, 1989.

[Fr3] C. Friesen, Class Number Divisibility in Real Quadratic Function Fields, to appear in Can. Math. Bull.

[Go] C. D. González, Class Numbers of Quadratic Function Fields and Continued Fractions, to appear in J. Number Theory.

[Ha] D. R. Hayes, Real Quadratic Function Fields, Canadian Mathematical Society Conference Proceedings **7**, (1985), 203–236.

[Lu] Hong Wen Lu, Divisibility of the Class Number of some Real Quadratic Fields, Acta Math. Sinica **28**, (1985), 756–762.

[NZ] I. Niven and H. S. Zuckerman, An Introduction to the Theory of Numbers, edition 4, 1980. John Wiley & Sons, New York.

[Pe] O. Perron, Die Lehre von den Kettenbrüchen, Chelsea Publishing, New York, 1950.

[Po] A.J. van der Poorten, Fractions of the Period of the Continued Fraction Expansion of Quadratic Integers, Bull. Austral. Math. Soc. **44**, (1991), 155–169.

[We] P. J. Weinberger, Real Quadratic Fields with Class Number Divisible by n, J. Number Theory **5**, (1973), 237–241.

Department of Mathematics, University of Toronto, Toronto, Ontario, CANADA

Dictionary

David Goss

Number Fields: $\mathbb{Q}$, the rational numbers.

Function Fields: $\mathbf{k}$ a global function field over $\mathbb{F}_r$, $r = p^m$.

Number Fields: Absolute value $|?|$ on $\mathbb{Q}$.

Function Fields: Any *fixed* place ∞ of $\mathbf{k}$ and the normalized absolute value $|?|_\infty$.

Number Fields: Completion $\mathbb{R} = \mathbb{Q}_\infty$.

Function Fields: Completion $\mathbf{K} = \mathbf{k}_\infty$.

Number Fields: $\mathbb{C}$, the complex numbers. We equip $\mathbb{C}$ with the canonical extension of $|?|_\infty$.

Function Fields: $\overline{\mathbf{K}}$ a *fixed* algebraic closure of $\mathbf{K}$. We equip $\overline{\mathbf{K}}$ with the canonical extension of $|?|_\infty$. N.B., $\overline{\mathbf{K}}$ is *not* locally compact.

Number Fields: $\mathbb{Z} \subset \mathbb{Q}$, the usual integers.

Function Fields: $\mathbf{A} \subset \mathbf{k}$, the functions regular *away* from ∞. The ring $\mathbf{A}$ is a Dedekind domain with finite class group and unit group $= \mathbb{F}_r^*$.

Number Fields: Basic fact: $\mathbb{Z} \subset \mathbb{R}$ discretely and $\mathbb{R}/\mathbb{Z} \simeq S^1$ is compact.

Function Fields: Basic fact: $\mathbf{A} \subset \mathbf{K}$ discretely and $\mathbf{K}/\mathbf{A}$ is compact. (In fact, $\mathbf{K}/\mathbf{A}$ is a piece of a rank 1 Drinfeld module.)

Number Fields: $p \in \operatorname{Spec}(\mathbb{Z})$, with completions $\mathbb{Z}_p$, $\mathbb{Q}_p$.

Function Fields: $v \in \operatorname{Spec}(\mathbf{A})$, with completions $\mathbf{A}_v$, $\mathbf{k}_v$.

Number Fields: Positive integers n and the usual notion of the sign of an integer.

Function Fields: A *sign* function, $\operatorname{sgn}\colon \mathbf{K}^* \to \mathbb{F}^*$, where $\mathbb{F}$ is the field of constants of $\mathbf{K}$. An element $a \in \mathbf{A}$ is said to be "monic" or "positive" if $\operatorname{sgn}(a) = 1$. If $\mathbf{A} = \mathbb{F}_r[T]$, this notion agrees with the usual notion of "monic" for appropriate choice of sgn.

Number Fields: $\mathbb{R}^*/\mathbb{R}_+ \simeq \mathbb{Z}/2$.

Function Fields: A *choice* of $\pi \in \mathbf{K}^*$ which is a *positive* uniformizing element. With this choice, $\mathbf{K}^* \simeq \mathbb{Z} \times \mathbb{F}^* \times U_1$, where U_1 is the group of 1-units. If $a \in \mathbf{A}$ is positive, we have $a = \pi^{v_\infty(a)}\langle a\rangle$, with $\langle a\rangle \in U_1$.

Number Fields: Let $s = x + iy \in \mathbb{C}$ and, as usual,

$$n^s = e^{x\log n}e^{iy\log n}\,.$$

Let $s_p = (s_0, s_1) \in S_p = \mathbb{Z}/(p-1) \times \mathbb{Z}_p$ and let $n \in \mathbb{Z}$ be positive and prime to p. Write $n = \omega_p(n)\langle n\rangle_p$, where $\omega_p(n) \in \mu_{(p-1)}$ and $\langle n\rangle_p$ is a 1-unit. Then we set

$$n^{s_p} = \omega_p(n)^{s_0}\langle n\rangle_p^{s_1}\,.$$

Function Fields: Let $a \in \mathbf{A}$ be positive. Let $s = (x, y) \in S_\infty = \overline{\mathbf{K}}^* \times \mathbb{Z}_p$. We set

$$a^s = x^{\deg(a)} \langle a \rangle^y .$$

This operation can be extended to $\wp^s$ for arbitrary (i.e., *not* principally generated by a positive element) $\wp$. Let $s_v = (s_0, s_1) \in S_v = \mathbb{Z}/(r^{\deg(v)} - 1) \times \mathbb{Z}_p$. Let $a \in \mathbf{A}$ be positive and prime to v. Write $a = \omega_v(a)\langle a \rangle_v$ where $\omega_v(a) \in \mu_{(r^{\deg(v)}-1)}$ and $\langle a \rangle_v$ is a 1-unit. Then we set

$$a^{s_v} = \omega_v(a)^{s_0} \langle a \rangle_v^{s_1} .$$

As before, this can be extended to $\wp^{s_v}$ for general fractional ideals prime to v. For a positive integer $i \in S_\infty$, the assignment $\wp \mapsto \wp^i$ is a Grössencharakter of type A_0.

Number Fields: The base field $\mathbb{Q}$ and the multiplicative group $\mathbb{G}_m$.

Function Fields: Let H be the Hilbert class field of $\mathbf{k}$ which is split totally at ∞, i.e., is "totally-real". Over H there exist $h = \#\mathrm{Pic}(\mathbf{A})$ non-isomorphic rank 1 Drinfeld modules. Each corresponds to an element of $\mathrm{Pic}(\mathbf{A})$ viewed as the isomorphism classes of rank 1 modules. By work of D. Hayes, there is an extension H_1/H, which is abelian down to $\mathbf{k}$ and ramified only over the infinite primes, such that over H_1 there exist *sign normalized* rank one Drinfeld modules. *Each* such module is an analog of $\mathbb{G}_m$. The simplest such module is the *Carlitz module*, C, over $\mathbf{A} = \mathbb{F}_r[T]$ with $C_T(z) = Tz + z^r$.

Number Fields: The cyclotomic fields $\mathbb{Q}(\zeta_m)$ for m a positive integer.

Function Fields: The fields $H_1(\rho[\mathfrak{I}])$, where ρ is sign normalized and $\rho[\mathfrak{I}] \subset \overline{\mathbf{K}}$ is the $\mathbf{A}$-module of $\mathfrak{I}$-division points. Actually, there are two types of "cyclotomic" extensions for function fields; the "geometric" ones just given and the "arithmetic" ones of constant field extensions.

Number Fields: Complex analysis.

Function Fields: The "usual" non-archimedean analysis of power series and entire functions. This is supplemented in two ways: 1) On the space S_∞ one uses continuously varying families of power series. 2) (More subtle) One needs to have a way to piece together locally analytic functions. This cannot be done naively; there are too many locally analytic functions. *Remedy* (basically due to John Tate): Let D^n be the nth polydisc. Subspaces of D^n defined by power series are called "affines". The fundamental result is that with a *finite* cover of an affine by other affines, functions patch locally. *Rigid analytic spaces* are then spaces with a covering by affines and a method (a Grothendieck topology) which forces the covering to be locally finite on affine subspaces.

Number Fields: An elliptic curve E over a number field L.

Function Fields: A Drinfeld (elliptic) module ϕ of rank $d > 1$ over a finite extension $L/\mathbf{k}$.

Number Fields: $\mathbb{C}$ = the universal covering of E.

Function Fields: $\overline{\mathbf{K}}$ = the universal covering of ϕ.

Number Fields: The singular and ℓ-adic cohomologies of E: Let $\mathrm{lat}(E) \subset \mathbb{C}$ be the lattice of E. One knows that $\mathrm{lat}(E) \simeq \mathbb{Z}^2$. We set $H_1(E, \mathbb{Q}) = \mathrm{lat}(E) \otimes \mathbb{Q}$ and $H^1(E, \mathbb{Q}) = \mathrm{Hom}_{\mathbb{Q}}(H_1(E, \mathbb{Q}), \mathbb{Q}) \simeq \mathbb{Q}^2$. Let $\ell \in \mathrm{Spec}(\mathbb{Z})$. Then $H_1^{\acute{e}t}(E, \mathbb{Q}_\ell) = T_\ell \otimes$

$\mathbb{Z}_\ell$ where T_ℓ is the ℓ-adic Tate module. Moreover, $H^1_{ét}(E, \mathbb{Q}_\ell)$ equals $\mathrm{Hom}_{\mathbb{Q}_\ell}(H_1^{ét}(E, \mathbb{Q}_\ell), \mathbb{Q}_\ell)$. These groups come equipped with continuous actions of $\mathrm{Gal}(\overline{L}/L)$. Let $\wp$ be a finite prime of L where E has good reduction. Let $\sigma_\wp \in \mathrm{Gal}(\overline{L}/L)$ be a *geometric Frobenius* at $\wp$. Then, for $(\ell, \wp) = 1$, we have $P(u) = \det\left(1 - \sigma_\wp u \mid H^1_{ét}(E, \mathbb{Q}_\ell)\right)$ having coefficients in $\mathbb{Z}$ which are invariant of ℓ. The roots of $P(u)$ satisfy the *Riemann Hypothesis* as $P(u)$ equals $\det\left(1 - Fu \mid H^1_{ét}(E_\wp, \mathbb{Q}_\ell)\right)$ where F is the $N\wp$-th power morphism of the reduction $E_\wp$ of E at $\wp$. The inertia group $I_\wp$ at $\wp$ acts trivially. (In fact, this last condition is "if and only if".)

Function Fields: The singular and v-adic cohomologies of ϕ: Let $\mathrm{lat}(\phi) \subset \overline{\mathbf{K}}$ be the lattice of ϕ (so $\mathrm{lat}(\phi)$ is discrete and $\mathbf{A}$-projective of rank d for some $d > 0$). We set $H_1(\phi, \mathbf{k}) = \mathrm{lat}(\phi) \otimes \mathbf{k}$ and $H^1(\phi, \mathbf{k}) = \mathrm{Hom}_{\mathbf{k}}(H_1(\phi, \mathbf{k}), \mathbf{k})$. Let $v \in \mathrm{Spec}(\mathbf{A})$. Then we set $H_1^v(\phi, \mathbf{k}_v) = T_v(\phi) \otimes \mathbf{k}_v$, where T_v is the v-adic Tate module. Moreover, we set $H^1_v(\phi, \mathbf{k}_v) = \mathrm{Hom}_{\mathbf{k}_v}(H_1^v(\phi, \mathbf{k}_v), \mathbf{k}_v) = H^1(\phi, \mathbf{k}) \otimes \mathbf{k}_v$. The dimension of these vector spaces is d. The v-adic spaces come equipped with continuous $\mathrm{Gal}(\overline{L}/L)$ actions. Let $\wp$ be a prime of L over $\mathrm{Spec}(\mathbf{A})$ where ϕ has good reduction. Let $\sigma_\wp$ be a *geometric Frobenius* at $\wp$. Then, for $(v, \wp) = 1$ $P(u) = \det(1 - \sigma_\wp u \mid H^1_v(\phi, \mathbf{k}_v))$ has coefficients in $\mathbf{A}$ invariant of the choice of v. The roots of $P(u)$ satisfy the *Riemann Hypothesis* as $P(u)$ is the same as $\det(1 - Fu \mid H^1_v(\phi^\wp, \mathbf{k}_v))$ where F is the $N\wp$-th power morphism of the reduction $\phi^\wp$ of ϕ at $\wp$. The inertia group at $\wp$ acts trivially. (In fact, this is also if and only if.)

Number Fields: The de Rham cohomology $H^1_{DR}(E)$ of E.

Function Fields: The de Rham cohomology $H^1_{DR}(\phi)$ of ϕ. Let $\overline{\mathbf{K}}\{\tau\}$ be the $\mathbf{A}$-bimodule of "twisted" polynomials in $\tau = \{$the r-th power mapping$\}$ considered as the $\mathbb{F}_r$-linear endomorphisms of the additive group $\mathbb{G}_a$. The left action of $\mathbf{A}$ is the usual multiplication and the right action is via ϕ_a. A *derivation* $\mathbf{A} \to \overline{\mathbf{K}}\{\tau\}$ is an $\mathbb{F}_r$-linear map η with

$$\eta_{ab} = a\eta_b + \eta_a\phi_b\,.$$

We say that η is *strictly inner* if there exists $m \in \overline{\mathbf{K}}\{\tau\}\tau$ with $\eta_a = am - m\phi_a$. Then $H^1_{DR}(\phi)$ equals

$$\{\text{All derivations}\}/\{\text{strictly inner derivations}\}.$$

The dimension of this vector space is d. Moreover, there is also a *de Rham isomorphism* in this context.

Number Fields: Factors at bad primes $\wp$ for E. Let $I_\wp$ be the inertia group at $\wp$. Let $\sigma_\wp$ be the geometric Frobenius at $\wp$. Then the factor at $\wp$ is

$$P(u) = \det(1 - \sigma_\wp u \mid H^1_{ét}(E, \mathbb{Q}_\ell)^{I_\wp}).$$

These polynomials have coefficients in $\mathbb{Z}$ which are invariant of ℓ.

Function Fields: Factors at the bad primes $\wp$ for ϕ. Let $I_\wp$ be the inertia group at $\wp$. Let $\sigma_\wp$ be the geometric Frobenius at $\wp$. Then the factor at $\wp$ is

$$P(u) = \det\left(1 - \sigma_\wp u \mid H^1_v(\phi, \mathbf{k}_v)^{I_\wp}\right).$$

Little is known about these polynomials. It is hoped that they will have coefficients in $\mathbf{A}$ which are invariant of v.

Number Fields: The higher groups are defined by $H^i = \wedge^i H^1$.

Function Fields: The higher groups are defined by $H^i = \wedge^i H^1$.

Number Fields: The L-series of E, $L(E,s)$ which is defined as

$$\prod_{\text{all } \wp} \det\left(1 - \sigma_\wp N\wp^{-s} \mid H^1_{\acute{e}t}(E, \mathbb{Q}_\ell)^{I_\wp}\right)^{-1}.$$

If we would have worked with $H^0(E)$ we would have obtained the zeta-function,

$$\zeta_L(s) = \prod (1 - N\wp^{-s})^{-1}.$$

Function Fields: The L-series of ϕ, $L(\phi, s)$ ($s \in S_\infty$) which is defined as

$$\prod_{\text{all } \wp} \det\left(1 - \sigma_\wp n\wp^{-s} \mid H^1_v(\phi, \mathbf{k}_v)^{I_\wp}\right)^{-1}.$$

Here $n\wp$ is the norm map on ideals from the integers of L to $\mathbf{A}$. Similarly, we obtain the zeta-function $\zeta_{\mathcal{O}_L}(s)$ of the integers $\mathcal{O}_L$ of L as

$$\prod (1 - n\wp^{-s})^{-1}.$$

Number Fields: Convergence in a half-plane of $\mathbb{C}$.

Function Fields: Convergence in a half-plane of S_∞; where $|x|_\infty \gg 0$.

Number Fields: Factorials $n! = 1 \cdot 2 \cdots n$.

Function Fields: Factorials for $\mathbf{A}$. We present Carlitz's construction in two steps: 1) Let i be a non-negative integer and $D_i = \prod_{n \in M(i)} n$ where $M(i)$ is the set of all monics of degree i. (One also has $D_i = (T^{r^i} - T)(T^{r^i} - T^r) \cdots (T^{r^i} - T^{r^{i-1}})$.) 2) Let j be a non-negative integer which we write r-adically as $j = \sum c_t r^t$. Then we set

$$\Pi(j) = \prod D_t^{c_t}.$$

This has the usual divisibility properties of $n!$.

Number Fields: The exponential function

$$e^x = \sum \frac{x^j}{j!}$$

with period $2\pi i$. It is well-known that π is transcendental.

Function Fields: The exponential of the Carlitz module

$$e_C(x) = \sum \frac{x^{r^t}}{D_t}$$

with period ξ. By work of Wade (general case by Yu) ξ is transcendental.

Number Fields: Gauss sums for $\mathbb{Z}$: Let $\chi: \mathbb{F}_{p^m} \to \mathbb{C}^*$ be a character and let $\psi: \mathbb{F}_p \to \mathbb{C}^*$ be a non-trivial additive character. Then a Gauss sum is

$$-\sum_{x \in \mathbb{F}_{p^m}} \chi(x)\psi(\operatorname{Tr} x).$$

Function Fields: Gauss sums for $\mathbf{A}$: Let $\mathbb{F}_{v^m}$ be the extension of $\mathbb{F}_v = \mathbf{A}/v$ of degree m. *Choose* an $\mathbb{F}_r$-embedding $\phi: \mathbb{F}_{v^m} \to \overline{\mathbf{K}}$. (Given one such embedding, all others are obtained by applying r^j-power Frobenius elements.) Let $\psi: \mathbb{F}_v \to \rho[v]$ be an $\mathbf{A}$-module isomorphism. Then (Thakur) a "basic" Gauss sum is

$$-\sum_{x \in \mathbb{F}_{v^m}^*} \phi(x^{-1})\psi(\operatorname{Tr} x).$$

(This sum may be computed over $\mathbb{F}_v$ itself.) Other Gauss sums are defined in the fashion of the Carlitz factorial.

Number Fields: Γ-functions for $\mathbb{Z}$: The complex Γ-function is defined originally by

$$\Gamma_{\mathbb{C}}(s) = \int_0^\infty t^{s-1}e^{-t}\,dt.$$

This function then has an analytic continuation with $\Gamma_{\mathbb{C}}(1/2) = \pi^{1/2}$. The p-adic Γ-function, $\Gamma_p(s)$ is a continuous function from $\mathbb{Z}_p \to \mathbb{Z}_p^*$ which interpolates "p-unit" factorials. *All* Γ-functions satisfy functional equations!

Function Fields: Γ-functions for $\mathbf{A}$. There are *two* types of such functions for $\mathbf{A}$ at ∞. The first type is a Morita-style interpolation of the 1-unit parts of D_i; one gets

$$\Gamma_{\text{arith}}(s): \mathbb{Z}_p \to \mathbf{K}^*.$$

By Thakur one has $\Gamma_{\text{arith}}\left(\frac{r-2}{r-1}\right) = \langle\xi\rangle^{\frac{1}{r-1}}$, where $\langle\xi\rangle$ is the "1-unit piece of ξ" (and differs from ξ by an algebraic element). The second Γ-function is $\Gamma_{\text{geom}}(s)$ where $s \in \overline{\mathbf{K}} \times \mathbb{Z}_p$. (Note the difference between this domain and S_∞.) It is related to *exponentials* as $\Gamma_{\mathbb{C}}(s)$ is related to $\sin(s)$. The v-adic Γ-function, $\Gamma_v(s)$, is a continuous function from $\mathbb{Z}_p \to \mathbf{A}_v^*$ which interpolates "v-unit" factorials. *All* Γ-functions satisfy functional equations!

Number Fields: Functional equation for $L(E,s)$: Upon multiplication by appropriate Γ-factors, one *should* have an analytic continuation and invariance (up to ± 1) under $s \mapsto 2-s$.

Function Fields: ???? — However there is some enticing evidence linking certain differential operators arising from ζ-functions and "magic numbers", i.e., certain numbers arising from factorials and an analog of the Kummer-Vandiver conjecture for cyclotomic function fields.

Number Fields: ???? — But perhaps the usual functional equation which takes values at negative integers of Dirichlet series and relates them to values at positive integers where they *determine* the L-series.

Function Fields: The *strong* v-adic interpolation of abelian L-series $L(\psi, s)$. This means that $L(\psi, s)$ has "polynomial" special values (i.e., $\in \overline{\mathbf{k}}[x^{-1}]$) at negative integers that interpolate to an entire function $L_v(\psi, s)$ on $\overline{\mathbf{k}}_v^* \times S_v$. In fact, one can always define an $L_v(\phi, s)$ *independently* of $L(\phi, s)$. They *should* be related in the same fashion as in the abelian case.

Number Fields: The Riemann zeta function

$$\zeta(s) = \sum_{n=1}^{\infty} n^{-s}.$$

Function Fields: Let $\mathbf{A} = \mathbb{F}_r[T]$ (the simplest case). The we set

$$\zeta_{\mathbf{A}}(s) = \sum_{\substack{n \in \mathbf{A} \\ n \text{ monic}}} n^{-s}.$$

Number Fields: Bernoulli numbers B_i. These are essentially $\frac{\zeta(i)}{\pi^i}$ for i even. This can be extended to special values of ζ-functions of *totally-real* extensions.

Function Fields: There are *two* analogs of Bernoulli numbers for $\mathbf{A}$ The first $\{BC_i\}$ ("geometric Bernoulli numbers") are essentially $\frac{\zeta_{\mathbf{A}}(i)}{\xi^i}$ where i is "$\mathbf{A}$-even" — that is, $\equiv 0 \bmod (r-1)$. The second, $\{\beta(i)\}$ for *all* i (the "arithmetic Bernoulli numbers") are essentially $\zeta_{\mathbf{A}}(-i)$. (The phrase "essentially" means that for i an $\mathbf{A}$-even number, $\zeta(s)$ has a trivial zero at $-i$ and we differentiate with respect to x^{-1}.) This can be extended to special values of ζ-functions of the rings of integers of totally-real *abelian* extensions.

Number Fields: Arithmetic interpretation of $\{B_i\}$: The p-divisibility of B_i gives the non-triviality of isotypic components of the p-class group of $\mathbb{Q}(\zeta_p)$. This statement is "if and only if".

Function Fields: Arithmetic interpretation of $\{BC_i, \beta(i)\}$: The numbers $\beta(i)$ give the non-triviality of isotypic components of the p-class group of the *field* $\mathbf{k}(C[f])$ for prime f — this statement is "if and only if". The numbers BC_i give the non-triviality of the p-class groups of the $\mathbf{A}$-integers, $\mathcal{O}$ of $\mathbf{k}(C[f])$. This is not yet "if and only if"; there is a conjecture relating the existence of p-class groups of $\mathcal{O}$ and $\{BC_i\}$ due to Gekeler.

Number Fields: The complex analytic upper half-plane $\mathcal{H} =: \{x + iy \mid y > 0\}$. The manifold $\mathcal{H}$ is simply connected. One obtains modular curves (for elliptic curves with level structure) as $\Gamma \backslash \mathcal{H}$ for $\Gamma \subset SL_2(\mathbb{Z})$.

Function Fields: The rigid analytic "algebraic upper half-plane" $\Omega =: \Omega^2 =: \{x \in \overline{\mathbf{K}} \mid x \notin \mathbf{K}\}$. The space Ω has non-trivial coverings. However, one still obtains the modular curves for rank 2 Drinfeld modules as $\Gamma \backslash \Omega$ for $\Gamma \subseteq GL(Y)$ for Y a projective rank 2 $\mathbf{A}$-module.

Number Fields:

Function Fields: The rigid spaces Ω^d for $d > 2$. These spaces uniformize the moduli spaces of higher rank Drinfeld modules.

Number Fields: The function $q = e^{2\pi i z}$ which is the uniformizer at ∞ for $\Gamma \backslash \mathcal{H}$.

Function Fields: The function $q = e_C(z)^{-1}$ which is the uniformizer at ∞ for $\Gamma\backslash\Omega$.

Number Fields: Modular forms for Γ.

Function Fields: Modular forms for Γ (*same* definition!).

Number Fields: The measure $f(it)\,\frac{dt}{t}$ on $\mathbb{R}_+$ for cusp forms f.

Function Fields: The "measure" (in the sense of Vishik, Amice and Vèlu) $\frac{d_{\mu_f}(t)}{t}$ on the group $\mathcal{U}_1 \subset \mathbf{K}^*$ of 1-units, for $GL_2(\mathbb{F}_r[T])$-cusp forms f. Here μ_f is the measure on $\mathbb{P}^1(\mathbf{K})$ associated to f by Teitelbaum.

Number Fields: The Fourier Transform: Let $f(x)$ be a rapidly decreasing smooth function on $\mathbb{R}$. Set

$$\hat{f}(y) = \int_{\mathbb{R}} e^{2\pi ixy} f(x)\, dx .$$

Function Fields: The Fourier Transform: Let $g(x)\colon \mathbb{Z}_p \to \mathbf{K}$ be continuous. Let $\mathcal{Z}(U)$ be a fixed *additive* formal power series with coefficients in $\mathcal{O}_\infty = \{x \in \mathbf{K} \mid |x|_\infty \leq 1\}$ with $\mathcal{Z}'(U) \in \mathcal{O}_\infty^*$. Set

$$\hat{g}(y) = \int_{\mathbb{Z}_p} g(x)\, d\,(1 + \mathcal{Z}(U))^y\,(x) .$$

This is used, e.g., to compute integrals against $\frac{d_{\mu_f}(t)}{t}$ in terms of q-expansions.

Number Fields: The Fermat equation:

$$x^n + y^n = z^n .$$

Function Fields: The Fermat equation for $\mathbb{F}_r[T]$: Let C be the Carlitz module. Let $a \in \mathbf{A}$ be monic of degree d. Then the inhomogeneous Fermat equation is

$$y^{r^d} C_a\left(\frac{x}{y}\right) = z^p .$$

For $d > 1$ and $p > 2$ this is shown to have no non-trivial solutions by L. Denis. More generally, one can follow the same prescription to define "Fermat families" for general "t-modules" (see next item).

Number Fields: Tensoring H^* of elliptic curves: Let E_1 and E_2 be two elliptic curves. Under the *standard conjectures* one can create a category $\mathcal{MOT}$ (called the category of "motives") such that *each* cohomology group is given by an object of $\mathcal{MOT}$. The tensor product then arises from the *Künneth formula*. The prototypical tensor-category is $\mathcal{REP}_G =$: the category of finite dimensional $\mathbb{Q}$-representations of an affine group scheme G with the standard tensor product. The main theorem of the "Tannakian philosophy" asserts that the automorphisms of $H^*(?, \mathbb{Q})$ on $\mathcal{MOT}$ are represented by an affine group scheme G with an *isomorphism* of $\mathcal{MOT}$ and $\mathcal{REP}_G$. The scheme G is called the "geometric Galois group".

Function Fields: Tensoring H^1 of Drinfeld modules: Let ϕ_1 and ϕ_2 be two Drinfeld modules. Building on the "shtuka" of Drinfeld, Anderson creates a category of "t-modules" containing the category of Drinfeld modules. The dual of a t-module is the t-motive. By definition, one can form the tensor product of the t-motives. The automorphisms of $H^1(?, \Omega)$ then give rise to a group scheme G in a similar fashion to the number field theory. Still there are peculiarities of finite characteristic — for instance, is G reduced?

Number Fields: Classical solitons: Solitons ("solitary waves with particle-like stability") are solutions to certain nonlinear differential equations with large internal symmetry. Special solutions are often constructed by a line bundle on an algebraic curve, or by representations of infinite dimensional Lie algebras.

Function Fields: Function field solitons: Let $X/\mathbb{F}_r$ be a smooth projective geometrically irreducible curve. A *soliton* ϕ is a meromorphic function on $X \times_{\mathbf{F}_r} X$ such that modulo horizontal and vertical components, the divisors of both ϕ and $1 - \phi$ are linear combinations of subvarieties $Z \subset X \times_{\mathbf{F}_r} X$ of the form

$$Z = \text{graph}\ (\sigma \circ \text{Frob}^n \colon X \to X).$$

Here $\sigma \in \text{Aut}(X/\mathbb{F}_r)$, Frob: $X \to X$ is the rth power Frobenius endomorphism, and n is any integer. In many cases explicit formulas may be given for solitons.

Department of Mathematics, Ohio State University, Columbus, Ohio 43210
Email: goss@coltrane.mps.ohio-state.edu

Walter de Gruyter
Berlin · New York

New Series

Ohio State University Mathematical Research Institute Publications

Editors:

Gregory R. Baker, Walter D. Neumann, Karl Rubin Columbus, Ohio, USA

This series is devoted to the publication of research monographs, lecture notes, proceedings, and other mathematical works arising from activities of the International Mathematical Research Institute at Ohio State University. The IMRI was founded in 1989 to support a program of visiting research scholars in mathematics at Ohio State and to run Workshops and Special Emphasis Programs on topics of particular importance and timeliness. The Research Semester on Low Dimensional Topology was the first major program of the Institute. In the meantime other programs have been organized and corresponding publications are in preparation.

Volume 1: Topology '90

Proceedings of the Research Semester in Low Dimensional Topology at Ohio State University

Editors: **Boris Apanasov, Walter D. Neumann, Alan W. Reid, Laurent Siebenmann**

1992. XII, 457 pages. 17 x 24 cm. Cloth. ISBN 3-11-012598-6

This volume consists of contributions from participants in a Research Semester in Low Dimensional Topology which took place under the auspices of the International Mathematical Research Institute at Ohio State University from February through June 1990. The main topics of this volume include: the geometry and topology of 3-manifolds, with particular emphasis on hyperbolic 3-manifolds and their interactions with number theory; the "new" invariants of 3-manifolds related to quantum field theory; plane algebraic curves.

Walter de Gruyter & Co., Genthiner Str. 13, D-1000 Berlin 30, FRG,
Phone (30) 2 60 05-0, Telex 1 84 027, Fax (30) 2 60 05-2 51
Walter de Gruyter, Inc., 200 Saw Mill River Road, Hawthorne, N. Y. 10532, USA,
Phone (914) 747-0110, Telex 64 66 77, Fax (914) 747-1326

Walter de Gruyter
Berlin · New York

Forum Mathematicum

An international journal devoted to pure and applied mathematics as well as mathematical physics

1993. Volume 5. 6 issues.

Editorial Board:

M. Brin (College Park, USA)
F. R. Cohen (Rochester, USA)
V. Enss (Aachen, FRG)
R. Fintushel (East Lansing, USA)
M. Fliess (Gif-sur-Yvette, France)
M. Fukushima (Osaka, Japan)
G. Gallavotti (Rome, Italy)
R. Göbel (Essen, FRG)
K. H. Hofmann (Darmstadt, FRG)
J. Lindenstrauss (Jerusalem, Israel)
D. H. Phong (New York, USA)
D. Ramakrishnan (Pasadena, USA)
A. Ranicki (Edinburgh, GB)
P.-A. Raviart (Palaiseau, France)
P. Sarnak (Stanford, USA)
D. S. Scott (Linz, Austria)
D. Segal (Oxford, GB)
B. Shiffman (Baltimore, USA)
F. Skof (Turin, Italy)
K. Strambach (Erlangen, FRG)
H. J. Sussmann (New Brunswick, USA)
G. Talenti (Florence, Italy)

Walter de Gruyter & Co., Genthiner Str. 13, D-1000 Berlin 30, FRG,
Phone (30) 2 60 05-0, Telex 1 84 027, Fax (30) 2 60 05-2 51
Walter de Gruyter, Inc., 200 Saw Mill River Road, Hawthorne, N. Y. 10532, USA,
Phone (914) 747-0110, Telex 64 66 77, Fax (914) 747-1326

Bei Fragen zur Produktsicherheit wenden Sie sich bitte an:
If you have any questions regarding product safety,
please contact:

Walter de Gruyter GmbH
Genthiner Straße 13
10785 Berlin
productsafety@degruyterbrill.com